Theory of Image Formation

Fully revised and updated, the second edition of this classic text is the definitive guide to the mathematical models underlying imaging from sensed data. Building on fundamental principles derived from the two- and three-dimensional Fourier transform and other key mathematical concepts, it introduces a broad range of imaging modalities within a unified framework, emphasising universal theoretical concepts over specific physical aspects. This expanded edition presents new coverage of optical-coherence microscopy, near-field microscopy, and medical imaging modalities including MRI, CAT, ultrasound, and the imaging of viruses, and introduces additional end-of-chapter problems to support reader understanding. Encapsulating the author's fifty years of experience in the field, this is the ideal introduction for senior undergraduate and graduate students, academic researchers, and professional engineers across engineering and the physical sciences.

Richard E. Blahut is the Emeritus Henry Magnuski Professor in the Department of Electrical and Computer Engineering at the University of Illinois, Urbana-Champaign, having served as the Department Head from 2001 to 2008. He has authored a series of advanced textbooks on the mathematical aspects of theoretical informatics, is a member of the US National Academy of Engineering, and a Fellow of the Institute of Electrical and Electronics Engineers (IEEE).

Theory of Image Formation

RICHARD E. BLAHUT
University of Illinois, Urbana-Champaign

Shaftesbury Road, Cambridge CB2 8EA, United Kingdom

One Liberty Plaza, 20th Floor, New York, NY 10006, USA

477 Williamstown Road, Port Melbourne, VIC 3207, Australia

314–321, 3rd Floor, Plot 3, Splendor Forum, Jasola District Centre,
New Delhi – 110025, India

103 Penang Road, #05–06/07, Visioncrest Commercial, Singapore 238467

Cambridge University Press is part of Cambridge University Press & Assessment,
a department of the University of Cambridge.

We share the University's mission to contribute to society through the pursuit of
education, learning and research at the highest international levels of excellence.

www.cambridge.org
Information on this title: www.cambridge.org/9781009356206

DOI: 10.1017/9781009356190

© Cambridge University Press & Assessment 2025

This publication is in copyright. Subject to statutory exception and to the provisions
of relevant collective licensing agreements, no reproduction of any part may take
place without the written permission of Cambridge University Press & Assessment.

When citing this work, please include a reference to the DOI 10.1017/9781009356190

First published 2025

A catalogue record for this publication is available from the British Library

A Cataloging-in-Publication data record for this book is available from the Library of Congress

ISBN 978-1-009-35620-6 Hardback

Cambridge University Press & Assessment has no responsibility for the persistence
or accuracy of URLs for external or third-party internet websites referred to in this
publication and does not guarantee that any content on such websites is, or will remain,
accurate or appropriate.

This book is dedicated to the memory of Barbara Ann Blahut (1938–2023). A remarkable and loving woman, this book could not have been written without her generous help and support.

"The story is told that young King Solomon was given the choice between wealth and wisdom. When he chose wisdom, God was so pleased that he gave Solomon not only wisdom but wealth also. So it is with science."

Arthur Holly Compton

"Where the telescope ends, the microscope begins."

Victor Hugo

Contents

Preface		*page* xiii
Acknowledgments		xvi

1 Introduction — 1
- 1.1 Image Formation — 2
- 1.2 The History of Image Formation — 4
- 1.3 Baseband and Passband Waveforms — 9
- 1.4 Monodirectional Waves — 11
- 1.5 Wavefront Diffraction — 14
- 1.6 Temporal and Spatial Coherence — 17
- 1.7 Deterministic and Random Models — 19
- 1.8 The Electromagnetic Spectrum — 21
- 1.9 Imaging by Tomography — 23
- 1.10 Radar and Sonar Systems — 25

2 Signals in One Dimension — 30
- 2.1 The One-Dimensional Fourier Transform — 30
- 2.2 Transforms of Some Useful Functions — 35
- 2.3 The Dirichlet Functions — 39
- 2.4 Passband Signals and Passband Filters — 44
- 2.5 Signal Space — 46
- 2.6 Sampling in One Dimension — 53
- 2.7 The Matched Filter — 58
- 2.8 Stationary Random Processes — 63
- 2.9 Resolution and Apodization — 65
- Problems — 67
- Notes — 73

3 Signals in Two Dimensions — 75
- 3.1 The Two-Dimensional Fourier Transform — 75
- 3.2 Transforms of Some Common Functions — 78
- 3.3 Circularly Symmetric Functions — 83
- 3.4 The Projection-Slice Theorem — 88

	3.5	Two Fundamental Fourier Transform Pairs	91
	3.6	Two-Dimensional Pulse Arrays	94
	3.7	Sampling in Two Dimensions	99
	3.8	Two-Dimensional Signals and Filters	107
	3.9	Resolution and Apodization	109
	Problems		111
	Notes		117
4	**Optical Imaging Systems**		119
	4.1	Scalar Diffraction	120
	4.2	The Huygens–Fresnel Principle	122
	4.3	Fresnel and Fraunhofer Approximations	127
	4.4	The Geometrical Optics Approximation	132
	4.5	The Ideal Lens	135
	4.6	Noncoherent Imaging	145
	4.7	Phase-Contrast Imaging	151
	4.8	Wavefront Reconstruction	153
	4.9	Optical Filtering	158
	Problems		160
	Notes		165
5	**Apertures and Radiation Patterns**		167
	5.1	Aperture Illumination and Antenna Pattern	168
	5.2	Antenna Arrays	175
	5.3	Focused Antennas and Arrays	183
	5.4	Nondiffracting Beams	184
	5.5	Interferometry	185
	5.6	Scanning Antenna Patterns	187
	5.7	Wideband Radiation Patterns	188
	5.8	Reciprocity	190
	5.9	Vector Diffraction	194
	Problems		200
	Notes		205
6	**Tomographic Imaging Systems**		207
	6.1	Projection Tomography	208
	6.2	Frequency-Domain Inversion	216
	6.3	Algebraic Inversion	217
	6.4	Image Formation from Magnetic Resonance	221
	6.5	Image Formation from Nuclear Decay Emissions	233
	6.6	Diffraction Tomography	236
	6.7	Diffusion Tomography	241
	6.8	Optical-Coherence Tomography	244
	6.9	Evanescent Wave Tomography	246

	6.10	Merging of Multiple Images	246
	Problems		249
	Notes		252

7 Construction and Reconstruction of Images — 254

- 7.1 The Inverse Problem — 254
- 7.2 Deconvolution and Deblurring — 256
- 7.3 Deconvolution of Nonnegative Images — 263
- 7.4 Diffractive Lenses — 266
- 7.5 Coded Aperture Imaging — 268
- 7.6 Phase Retrieval — 272
- 7.7 Blind Image Deconvolution — 279
- 7.8 Optical Imaging from Point Events — 285
- 7.9 Imaging of a Semitransparent Bulk Material — 289
- 7.10 Near-Field Microscopy — 292
- 7.11 Electron-Beam Microscopy — 293
- 7.12 Spectral Imaging — 294
- Problems — 296
- Notes — 299

8 Likelihood and Information Methods — 301

- 8.1 Likelihood Functions and Decision Rules — 302
- 8.2 The Maximum-Likelihood Principle — 307
- 8.3 Alternating Maximization-Maximization — 309
- 8.4 Other Principles of Inference — 314
- 8.5 Recovery of Nonnegative Signals — 319
- 8.6 Recovery of Missing Phase — 323
- 8.7 Blind Recovery of Nonnegative Signals — 328
- 8.8 Differencing Methods in Photon Imaging — 329
- 8.9 Alternating Expectation-Maximization — 333
- 8.10 Notions of Equivalence — 338
- 8.11 Regularization — 341
- Problems — 344
- Notes — 348

9 Diffraction Imaging Systems — 350

- 9.1 The Three-Dimensional Fourier Transform — 350
- 9.2 Transforms of Some Useful Functions — 353
- 9.3 Sampling in Three Dimensions — 358
- 9.4 Diffraction by a Three-Dimensional Object — 362
- 9.5 Image Formation from Diffraction Data — 366
- 9.6 Lattice Structure of Arrays — 368
- 9.7 Diffraction by Three-Dimensional Arrays — 370
- 9.8 Sampling of an Array — 375

	9.9	Image Formation from Array Diffraction	377
	9.10	Model Construction from Diffraction Data	378
	9.11	Direct Methods from Diffraction Data	380
	9.12	Diffraction from Fiber Arrays	389
	9.13	Diffraction from Excited Arrays	392
	Problems		393
	Notes		398
10	**The Woodward Ambiguity Function**		**399**
	10.1	Theory of the Ambiguity Function	400
	10.2	Ambiguity Functions of Some Simple Pulses	403
	10.3	More Properties of the Ambiguity Function	405
	10.4	Shape and Resolution Parameters	407
	10.5	More Properties of the Ambiguity Function	411
	10.6	Ambiguity Function of a Pulse Train	414
	10.7	Ambiguity Function of a Costas Pulse	417
	10.8	The Cross-Ambiguity Function	422
	10.9	The Sample Cross-Ambiguity Function	425
	10.10	Ambiguity Functions in Signal Space	428
	Problems		430
	Notes		435
11	**Radar Imaging Systems**		**437**
	11.1	The Received Signal	438
	11.2	The Radar Imaging Equation	445
	11.3	Imaging Resolution	449
	11.4	Focusing and Motion Compensation	451
	11.5	Structure of Imaging Systems	454
	11.6	Computing the Cross-Ambiguity Function	459
	11.7	Dual Aperture Imaging	465
	11.8	Radar Imaging of Diffuse Reflectors	466
	11.9	Coherent and Noncoherent Radar Tomography	472
	Problems		475
	Notes		478
12	**Radar Search Systems**		**480**
	12.1	The Radar Range Equation	480
	12.2	Coherent Detection of a Pulse in Noise	482
	12.3	The Neyman–Pearson Theorem	487
	12.4	Rayleigh and Ricean Probability Distributions	491
	12.5	Noncoherent Detection of a Pulse in Noise	494
	12.6	Arrival Time Estimation of a Baseband Pulse	496
	12.7	The Cramer–Rao Theorem	500
	12.8	Noncoherent Estimation of Pulse Parameters	504

		12.9 Clutter	509
		12.10 Detection of Moving Objects	512
		Problems	515
		Notes	519
13	**Passive and Baseband Systems**		521
	13.1	Radio Astronomy	522
	13.2	Magnetic Anomaly Detection	534
	13.3	Passive Location of Radiation Emitters	539
	13.4	Estimation of Differential Parameters	545
	13.5	Estimation of Direction	547
	13.6	Lidar Surveillance	551
		Problems	552
		Notes	555
14	**Data Combination and Tracking**		557
	14.1	Noncoherent Integration	558
	14.2	Sequential Detection	561
	14.3	Multitarget Tracking	562
	14.4	The Assignment Problem	566
	14.5	Lagrangian Relaxation	570
	14.6	Multilateration	571
		Problems	573
		Notes	574
15	**Phase Noise and Phase Distortion**		575
	15.1	Quadratic-Phase Errors	576
	15.2	Phase Noise and Coherence	579
	15.3	Phase Noise and the Fourier Transform	581
	15.4	Phase Noise and the Matched Filter	582
	15.5	Phase Noise and the Ambiguity Function	586
	15.6	Effect of Phase Distortion	587
	15.7	Array Errors	587
		Problems	588
		Notes	589
		Bibliography	590
		Index	609

Preface

As processing technology continues its rapid growth, it occasionally causes us to take a new point of view toward many long-established technological disciplines. It is now possible to record precisely such signals as ultrasonic or X-ray signals or electromagnetic signals in the radio and radar bands and, using advanced digital or optical processors, to process these records to extract information deeply buried within the signal. Such processing requires the development of algorithms of great precision and sophistication. Until recently such algorithms were often incompatible with most processing technologies, and so there was no real impetus to develop a general, unified theory of these algorithms. Consequently, it was scarcely noticed that a general theory might be developed, although some special problems were well studied. Now the time is ripe for a general theory of these algorithms. These are called algorithms for *image formation*. This topic of image formation is a branch of the broad field of *informatics*.

Systems for image formation, or imaging, have developed independently in diverse fields over the years. They are often very much driven by the kind of hardware that is used to sense the raw data or to do the processing. The principles of operation, then, are described in a way that is completely intertwined with a description of the hardware. Recently, there has been interest in abstracting the common signal processing and information-theoretic methods that are used by these sensors to extract the useful information from the raw data. This unification is one way in which to make new advances because then the underlying principles can be more clearly understood, and ideas that have already been developed in one area may prove to be useful in others. A unified formulation is also an efficient way to teach the many topics as one integrated subject.

The theory of image formation, which we regard as comprising the various information-theoretic and computational methods underlying image formation systems, is now emerging as an integrated field of study. This is in marked contrast to the development of the subject of communication theory, in which radio, telephone, television, and telegraphy were seen early on as parts of a general theory of communication. For a long time, communication theory has been treated as an integrated field of study while the theory of image formation has not.

This book is devoted to a unified treatment of the mathematical methods that underlie the various methods of image formation. It is not a book that describes the hardware used to build such systems. Rather, it is a book that describes the mathematics used to design the illumination waveforms and to develop the algorithms that form the images or extract the desired information.

Because my goal is a unified presentation of the mathematical principles underlying the development of image formation, the book is constructed so that the core is the ubiquitous two-dimensional Fourier transform. In addition, the mathematical topics of coherence, correlation functions, the ambiguity function, the Radon transform, and the projection-slice theorem play important roles. The applications, therefore, are introduced as consequences or examples of the mathematical principles, whereas the more traditional treatment of these subjects begins with a description of a sensing apparatus and lets the mathematical principles arise only as needed. While many would prefer that approach – and certainly it is closer to the history of these subjects – I maintain that it is a less transparent approach and does not advance the goal of unification.

Many of the physical aspects of a surveillance system – such as a radar or sonar system or a microwave antenna – are complicated and difficult to model exactly. Some may conclude from this that it is pointless to develop a mathematical theory that is honed to a sharpness beyond the sharpness of the physical situation. I take exactly the opposite view: if a rich mathematical theory can be developed, it should be developed as far as possible. It is in the application of the theory that care must be exercised. For example, the interaction of electromagnetic waves with reflectors or with antennas can be quite complex and incompletely understood. Nevertheless, we can model the interaction in some more simple way and then carry the study of that model as far as it will go. To object to this would be analogous to objecting to the study of linear differential equations under the argument that every physical system has some degree of nonlinearity. Thus, our primary emphasis is on an axiomatic formulation of the mathematical principles of image formation rather than a phenomenological development from the underlying physical laws. The book treats image formation in a formal way as a topic in applied mathematics. The engineering task, then, is to choose judiciously and combine the elements developed here with due regard to the capricious behavior of physical devices.

The two-dimensional Fourier transform (or the three-dimensional Fourier transform) is central to most of the systems developed, and we shall always take pains to bring the role of the two-dimensional Fourier transform to the forefront. Huygens' principle, which is at the core of much of optics and antenna theory, but often is not treated rigorously, will be presented simply as a mathematical consequence of the convolution theorem of two-dimensional Fourier transform theory. In turn, the study of the far-field diffraction pattern of antennas will be largely reformulated as the study of the two-dimensional Fourier transform. Even the near-field diffraction pattern can be studied with the aid of the Fourier transform.

We also describe the behavior of an imaging or search radar system from the vantage point of the two-dimensional Fourier transform. Specifically, we view the output of a radar's front-end signal processing as the two-dimensional convolution of a radar ambiguity function and the reflectivity density function of the illuminated scene. This is the elegant formulation, and it is very powerful in that it submerges all the myriad details of image formation while leaving exposed the underlying limitations on resolution and estimation accuracy, as well as the nature of ambiguities and clutter.

I have regularly taught a one-semester course from the manuscript of this book, starting with Chapters 3 through 12 and ending with portions of Chapter 13. This program would treat Chapters 1 and 2 as introductory reading material, intended only for review and motivation, and would return to topics in Chapter 2, principally Sections 2.3 and 2.5, as they arise. Presumably, the Fourier transform is already known, and the study of the two-dimensional Fourier transform in Chapter 3 implicitly provides a review of the one-dimensional case.

Chapters 3, 4, and 5 are closely connected and can be thought of as a long discourse on the two-dimensional Fourier transform; the theory is in Chapter 3, its role in optics is in Chapter 4, and its role in antenna systems is in Chapter 5. Chapter 9 picks up this thread again, studying X-ray diffraction imaging as an application of the three-dimensional Fourier transform.

The important problems of image construction and reconstruction from fragmentary data are studied in Chapters 6, 7, and 8. Chapter 6 is devoted to the important problem of tomography and to the general task of reconstruction of an image from multiple degraded views. Chapter 7 deals with other topics in image construction and reconstruction including the formation of images from partial Fourier transform data and the formation of images from discrete event data. Chapter 8 introduces the important subject of information-theoretic methods in image formation, a subject which itself can grow to fill a book.

Chapters 10, 11, and 12 form another closely connected sequence centered on the ambiguity function. Chapter 10 develops the theory of the ambiguity function; Chapters 11 and 12 develop the theory of imaging radar and search radar, respectively, from the point of view of the ambiguity function. Chapter 14 deals with data association and tracking as a sequel to Chapter 12; these problems arise in the postprocessing of radar and sonar data. The ambiguity function also plays a role in some parts of Chapter 13, which may be described as a counterpart to Chapter 12 in which the observed data is passive.

Other kinds of image formation system are discussed in Chapter 13. Passive systems, in which the illuminating signal is already in the environment and is not under the control of the system designer, are discussed in Chapter 13. The passive location of emitters may arise as a sonar problem in which the emitters are sources of underwater sound; as a seismic problem in which the emitters may be earthquakes or artificial explosions; or as a radar sensor in which the emitters are noncooperating radars or radios. Passive systems can also be imaging systems; modern radio astronomy is a conspicuous example of a passive imaging system, forming images of distant radio galaxies from passively received radio signals. The book concludes with Chapter 15 where an analysis of phase errors and phase noise appears, and where the phase approximations used elsewhere in the book are studied more closely.

Acknowledgments

It is a pleasure for me to acknowledge the frequent discussions with Professor George C. Papen that helped with the preparation of this book. I am thankful for his generous advice and positive criticism. I am also thankful to Professor Zhi-Pei Liang, Professor Figen S. Oktem, Professor Frank Kschischang, Professor Milan Gembicky, Professor Minh N. Do, and the students of the Grainger College of Engineering at the University of Illinois for their help on this edition, The first edition benefited from comments by many friends and colleagues, in particular Professor Joseph A. O'Sullivan, Professor Donald L. Snyder, Professor Aaron D. Lanterman, Professor P. Scott Carney, Professor Farzad Kamalabadi, Professor David L. Munson, Jr., Professor Timothy J. Schulz, and Professor Negar Kiyavash.

"A picture is worth a thousand words."
Confucius

"A formula is worth a thousand pictures."
Edsger Dijkstra

"An insight is worth a thousand formulas."
Anonymous

1 Introduction

Our immediate environment is a magnificent tapestry of information-bearing signals of many kinds reaching us from many directions and from many sources, both large and small. Some signals arise as a response to some form of human-made illumination probing a scene or object of interest. Other signals occur naturally in the environment. Most signals are not readily compatible with our human senses. Only those signals such as optical signals in the visible band and certain acoustic signals are immediately compatible with our human senses. Other signals – such as electromagnetic signals in the infrared, ultraviolet, and X-ray bands, or waves in the radio and radar bands, or acoustic waves at ultrasonic frequencies – are not compatible with our natural senses, such as they are. To perceive any of these signals, which to our senses are invisible, sophisticated imaging algorithms are used to convert the sensed data into an understandable form such as an image.

A great variety of sensors now exist that collect signals and process these signals to form some kind of image, normally a visual image, of an object or a scene of objects. We refer to these as sensors for forming images, often requiring extensive computation to render the raw data into the form of an image. There are many kinds of sensors collected under this heading, differing in the size of the observed scene, as from classical microscopes to modern radio telescopes; in complexity, as from the simple lens to synthetic-aperture radars; and in the current state of development, as from photography to microscopy, holography, and tomography. Each of these systems collects raw sensor data and processes that sensor data into imagery that is useful to a user. This processing might be done by a digital computer, by an analog computer, or by an optical computer as may consist of a system of lenses. The development and description of a processing algorithm often requires a sophisticated theory and a precise mathematical formulation.

In this book, we bring together a number of signal-processing concepts that will form a background for the study and design of the many kinds of image formation system used for clinical evaluation or for remote surveillance. The signal-processing principles that we study include or adjoin the methods of medical imaging, classical radar and sonar systems, electromagnetic propagation, tomography, and physical optics, as well as estimation and detection theory.

1.1 Image Formation

Mankind has designed a variety of devices that are used to observe the general environment and specific objects of interest in that environment. Images are formed by processing acoustic, pressure, or magnetic variations or by processing electromagnetic radiation in the radio and the microwave frequency bands, the infrared band, and the optical and X-ray bands. The many varieties of such medical imaging systems, and of radar and sonar systems, are examples of such systems. An image formation system may be active, using its own energy to illuminate the environment, or it may be passive, relying on signals already in the environment or signals produced by the scene itself.

The theory of image formation studies the design of signals to probe the environment as well as the design of computational procedures for the extraction of information from received signals within which that information may be deeply buried. As such, this theory comprises that branch of information theory that is explicitly concerned with the design of systems to observe the environment and with their performance of those systems. The theory herein is concerned specifically with the mathematical structure of the image-formation algorithms needed to extract information from the received signals.

An image formation system is any system that collects signals and creates an observable image by processing those signals by computation or otherwise to form that image. Figure 1.1 illustrates a computational image formation system partitioned into the "sensors" and the "algorithms." We will be concerned with the details of the image-formation algorithms and with the performance of those algorithms. We will be concerned with the physics of the sensors only insofar as is necessary to explain the relationship between the object of interest and the observed data or with the development of the algorithms.

The "image," which is the end product of the image formation system, is always some kind of depiction of an "actual" scene, usually a two-dimensional or three-dimensional scene, which we denote as $\rho(x,y)$ or $\rho(x,y,z)$. The scene may emit its own signals that the sensors intercept, or it may be probed with signals generated by the image formation system. Figure 1.2 shows a representative configuration in which the scene $\rho(x,y)$ is probed by a signal generated as a one-dimensional waveform. In this case, the sensors collect one or more reflected one-dimensional waveforms, $s_m(t)$, and from these reflected waveforms, the computational algorithms must form a suitable two-dimensional image of the scene. In this case, the computational task is to estimate the two-dimensional function, $\rho(x,y)$, (or a three-dimensional function, $\rho(x,y,z)$), when given a set of one-dimensional scattered waveforms, $s_m(t)$ for

Figure 1.1 A computational imaging system

1.1 Image Formation

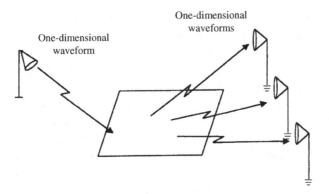

Figure 1.2 Probing a scene with waveforms

$m = 1, \ldots, M$, that depend on $\rho(x, y)$. The task of forming an image of an object or scene from a relevant signal is called an *inverse problem*. Among the most useful mathematical tools that we will develop for this task are the two-dimensional Fourier transform, the projection-slice theorem, and the ambiguity function. Probability theory, especially the notion of the likelihood function, is also an important tool in later chapters.

The many forms of reconstructed imagery, such as medical imagery and radar imagery, may often look very different from the visual imagery or conventional photographs of that same scene or object. This means that the user of that sensor may need training and experience in interpreting the generated image. To the novice, it may seem to be a limitation of that specialized image, but a more sophisticated view is that a new sensor opens a new window in our way of perceiving reality. A bat or a dolphin lives in a world that is perceived in large measure by means of acoustic or sonar data. This kind of sensor has nothing like the high angular resolution of our optical world, yet it does have other attributes, such as a strong doppler shift and the ability to resolve objects instantly by their velocities. Because it uses a different kind of data, the dolphin or the bat undoubtedly perceives the world differently from the way in which we do. Thus, it may be argued that modern image formation does change the way that society sees the world around us.

One way of defining the kind of imaging system to be studied is as a system in which raw signals in the environment that the human cannot sense directly are turned into processed signals that are compatible with one of the human senses. Thus a radar receiver converts an electromagnetic wave into a visual image compatible with the human eye, and a tomographic medical scanner turns an X-ray signal into a visual image of an anatomy. The image need not be a realistic replica of a photograph. Other details may be more important. An X-ray image of the human body does not look like a photograph of a human skeleton, but it may be preferred by the diagnostician because it contains useful information of other kinds.

An important topic also studied herein is the relationship between the illumination at the input of an aperture, such as an antenna or a lens, and the wavefront radiated by that illumination. The reflection of these wavefronts, however, will be modeled in a simple way. Simplified models for both reflection and spontaneous emission will be adequate for most of our purposes. The detailed relationship between the wave incident on a reflecting object or scene and the wave reflected by that object or scene is called the *forward problem*. The forward problem is of interest in this book only insofar as it sets up the inverse problem. This is the problem of forming an image of that object or source from the reflected signal or from other observed data.

Radar and sonar are surveillance systems that are included among the imaging systems that will be studied. Originally, radar and sonar systems used simple waveforms and simple processing techniques that could be implemented with simple electronic circuits such as filters and threshold detectors. But over the years a new level of sophistication began to find its way into many of these systems. By maintaining a precise phase record of long-duration signals, and processing the signals phase-coherently, one can obtain new levels of system performance. Systems that depend on phase coherence over time are called *coherent surveillance* systems. Some early coherent systems in the radar bands were designed to use optical processing. More recently, digital processing of coherent electromagnetic waveforms has become practical.

Imaging algorithms depend on the angles available for viewing the target. Most radar systems view a target from only a single, or limited, viewing angle. The same is usually true for microscopy, photography, and astronomy. Medical imaging, in contrast, is often able to view an object from many angles, and the principles of tomography can be applied. Our goal is to develop a general theory of imaging systems in a common mathematical setting. We will be concerned with a range of processing algorithms, such as those used for forming the images of remote radar reflectors, X-ray or magnetic-resonance tomography, microscopy, and astronomy.

1.2 The History of Image Formation

The subject of image formation consists of the common overlap of a number of well-developed subjects such as physical optics, electromagnetics, and signal processing. From a broad point of view, the historical roots of imaging go back to the roots of these various subjects. We are interested here in a narrower view of this history, especially the history of clinical imaging systems, diffraction imaging systems, surveillance systems, and tomography. Our brief discussion in this section serves only to sketch the historical background of the material in this book.

Many kinds of imaging systems were developed independently, but share common fundamentals of signal processing and a common mathematical framework. These include: optical imaging, holography, medical imaging, radio astronomy, sonar beamforming, microscopy, diffraction crystallography, imaging radars, moving-target detection radars, as well as more recent topics such as seismic processing and passive source location.

Optical image formation systems within the topic of photography are among the earliest, the most developed, and the most familiar to the user. Credit for the invention

1.2 The History of Image Formation

of photography is usually given to Niépce, who produced a photograph in 1826 or 1827 using the three principal components: an aperture, a lens, and a photosensitive medium. Photography passed into common use in 1839 after the work of Daguerre. Photographic imaging systems may be passive, using reflected light and occasionally using radiated light, or they may be active, using a source of light to illuminate the scene. The optics of a basic photographic system is adequately described using geometrical optics, but modern, high-performance multitone photography is based on wave optics. Multitone optical images are usually quite sharp with high resolution and excellent color contrast.

Image formation systems may use passive radiation in the infrared bands. These systems are similar to optical systems, but can form images of temperature variations in a scene because the intensity and wavelength of the radiation emitted by an object varies according to the temperature of that object.

Imaging was introduced into medical diagnostics by Roentgen in 1895 with the invention of X-ray radiography, which exposes a photographic film to X-rays transmitted through a body, then processes that film. Edison, by introducing an X-ray-sensitive fluorescent screen or fluoroscope in 1896, eliminated the delay required to process the film. The development of X-ray tomography in the modern sense of computerized image reconstruction for medical applications began in Great Britain. The key feature, based on the projection-slice theorem and the Radon transform, is the algorithmic reconstruction of images from their X-ray projections, first developed by Cormack in 1963 and reduced to practice by Hounsfield in 1971. The 1979 Nobel prize in physiology and medicine was awarded to Hounsfield and Cormack for the development of computerized tomography. The ideas of tomography are closely related to similar methods used in radio astronomy, especially the formulation of reconstruction algorithms by Bracewell (1956). Other kinds of computerized tomography are now in use for medical diagnostic imaging systems. In addition to the method of projection tomography based on X-ray projections, there are the methods of emission tomography and diffraction tomography. Photon- or positron-emission tomography (PET) based on radioisotope decay was proposed by Kuhl and Edwards (1963).

Magnetic resonance imaging (MRI) is yet another kind of tomographic imaging system based on magnetic excitation of atomic nuclei and the induced magnetization of the hydrogen nuclei distribution. The ground-breaking idea that enables MRI – for which Lauterbur and Mansfield shared the 2003 Nobel prize in medicine – is to use gradients of a magnetic field to encode spatial information into the transient response of a nuclear spin system after excitation by a magnetic pulse. Whereas X-ray tomography gives an image of the electron density, MRI gives an image of the distribution of hydrogen nuclei (isolated protons) in a body, though in principle it can be tuned to observe instead the distribution of other species of nuclei. The physical phenomenon of nuclear magnetic resonance had been observed independently in 1946 by Bloch and Purcell, for which they received the 1952 Nobel prize in physics. It was later realized that magnetic resonance effects varied with the kind of tissue excited but it was not known how to use this effect to make images. Lauterbur conceived and demonstrated his method of using the magnetic resonance phenomenon to form images by spatially encoding the magnetic field, for which Lauterbur shared in the 2003 Nobel prize in medicine. Since then, magnetic resonance imaging has become an important modality in medical diagnosis. By using both static and time-varying magnetic excitation fields, an MRI system causes all nuclei of a given kind – selected by the resonance frequency of those nuclei – to precess (or oscillate), but with an amplitude and frequency modulation that depends on position as determined by the magnetic excitation at that position. The magnetization that is produced by the selected species of nuclei, usually hydrogen nuclei, is measured and sorted by frequency analysis. Because of the spatially-varying magnetic excitation, the frequency distribution of the induced magnetic field corresponds to the spatial distribution of the sources of radiation, which equates to the spatial density distribution of the target nuclei. Repeated scans at different angles allows the methods of tomography to be used to sort the data in other ways. Sophisticated mathematical algorithms based on the methods of tomography have been developed to so extract a high-resolution image of the hydrogen density distribution from the frequency distribution for each of multiple projections of measured magnetic resonance data.

The diffraction of X-rays by crystals was demonstrated in 1912 by Max von Laue, thereby demonstrating the wave properties of X-rays. Sir William Henry Bragg then immediately inverted the point of view to turn this diffraction phenomenon into a way of probing crystals, which has since evolved into a sophisticated imaging technique. The 1914 Nobel prize in physics was awarded to von Laue, and the 1915 Nobel prize in physics was awarded to Bragg and his son, Sir William Lawrence Bragg, who formulated the famous Bragg law of diffraction. This early work was directed toward finding the lattice structure of the crystal as a whole, but was not much concerned with the structure of the individual molecules making up the crystal. Attention soon turned to the finer question of finding the scattering structure within an individual cell of the crystal, and so to form an image of the molecule by processing that signal. A difficulty of this task is that, because of the small wavelength of X-rays, the phase of the diffracted X-ray wavefronts cannot be measured. Only the intensity (or amplitude) can be measured. Herbert Hauptman and Jerome Karle (1953) showed how to bypass this problem of missing phase by using prior knowledge about the molecules that compose the crystal, for which they shared the 1985 Nobel prize in chemistry. Earlier, in 1953, James Watson and Francis Crick – using the X-ray diffraction images produced by Rosalind Franklin – discovered the structure of the DNA molecule, for which they shared the 1962 Nobel prize in medicine.

Closely related to the methods of the Fourier transform and signal processing are many kinds of optical processing, many of them using diffraction phenomena that are describable in terms of the two-dimensional Fourier transform. A method known as the *schlieren method* was proposed by Jean Foucault in 1858 as a way to image density variations of air. Fritz Zernicke in 1935 developed phase-contrast methods to improve microscopy images, for which he was awarded the 1953 Nobel prize in physics. Aaron Klug developed methods for the imaging of viruses using the diffraction of electron microscope images, for which he won the 1982 Nobel prize in chemistry.

1.2 The History of Image Formation

The 1914 Nobel prize in chemistry was awarded to Betzig, Moerner, and Hell for the development of superresolution fluorescence microscopy.

Dennis Gabor, influenced by the techniques used in microscopy and crystallography, proposed the idea of holography in a series of papers in 1948, 1949, and 1951. He originally intended holography as a method of microscopy but later as a replacement for photography. The work earned Gabor the 1971 Nobel prize in physics. Gabor realized that, whereas conventional photography first processes the raw optical wavefront to form an image which is then recorded on film, it is also possible to record the raw optical wavefront on the photographic film directly and place the processing of the optical wavefront in the future with the viewer. He called his method for the photographic recording of the raw optical data a *hologram*. Because the raw optical data contains more information than a final photographic image, in principle, the hologram can be used to create images superior to a photograph. Most striking in this regard is the creation of three-dimensional images from a two-dimensional hologram. Holography is technically much more difficult than photography because recording the raw optical data requires precision on the order of the optical wavelengths. For this reason, the idea of holography did not immediately draw the attention it deserved. Holography became more attractive after the invention of the laser and also after the more practical reformulation of the method by Leith and Upatnieks (1962), which was strongly influenced by Leith's other work on the optical processing used in synthetic-aperture radar.

Early radars used simple electronic circuits for processing the received signal while modern radars may use processing that is quite sophisticated. Sophisticated radar signal processing first appeared in the development of those imaging radars known as synthetic-aperture radars. The principle of such radars has been credited to a suggestion in 1951 by Wiley, although he did not then publish his ideas nor did those ideas then result directly in the construction of such a radar. Wiley observed that, whereas the azimuthal resolution of a conventional radar is limited by the width of the antenna beam, each reflecting element within the antenna beam from a moving radar has a doppler frequency shift that depends on the angle between the velocity vector of the radar and the direction to the reflecting element. Thus he concluded that a precise frequency analysis of the radar reflections would provide finer along-track resolution than the azimuthal resolution defined by the antenna beamwidth. The following year, a group at the University of Illinois arrived at the same idea independently, based upon frequency analysis of experimental radar returns. During the summer of 1953, these ideas were reviewed by the members of a summer study, "Project Wolverine," at the University of Michigan and plans were laid for the development of synthetic-aperture radar. It was recognized that the processing requirements placed extreme demands on the technology of the day. Many kinds of analog processors (filter banks, storage tubes, etc.) were tried. Meanwhile, Emmett Leith, at the University of Michigan, turned to the processing ideas of holography and adapted the optical processing techniques to satisfy the processing requirements for radar. In 1957, by using optical processing, the first synthetic-aperture radar was successfully demonstrated. Later, Green (1962) proposed the use of the range-doppler techniques of synthetic-aperture radar for remote radar imaging of the surface of rotating planetary objects. This method is closely related to synthetic-aperture radar except using the rotation of the object itself to provide relative motion. High-resolution radar images of Venus from the earth gave us our first view of the surface of that planet unobstructed by the cloud cover of that planet.

Optical processors[1] are analog processors using the Fourier transforming property of a lens to compute two-dimensional Fourier transforms. Early on, these have been the processors of choice for imaging radars because of the sheer volume of data that can be handled. However, to form an image with an optical processor requires developing the photographic film twice within the processing, once to create the optical input signal and once to record the output. Optical processors are very sensitive to vibration, and so they are limited by the environment and also by the form of computations that can be included. Hence attention now has turned to other methods for processing. The advent of high-speed, digital array processors has had a large impact on the massive processing needed for synthetic-aperture imagery, and optical processing now plays a diminished or vanishing role.

The development of search radars for the detection of moving targets is spread more broadly, and individual contributions are not as easy to identify. From the first use of radar, it was recognized that the need to detect moving targets could be satisfied by using the doppler shift on the return signal. A moving object causes a doppler-shifted echo. However, the magnitude of the doppler shift is only a very small fraction of the transmitted pulse bandwidth. At that time, the technology did not exist to filter a faint, doppler-shifted signal from a strong background of signals echoed from other stationary emitters. Hence the development of search radars did not depend so much on invention at the conceptual level as it did on the development of technology to support widely understood requirements. By the end of World War II, radars had been developed that used doppler filters to suppress the clutter signal reflected from the stationary background. These early radars used simple delay lines to cancel the stationary return from one pulse with the (nearly identical) return from the previous pulse, thereby rejecting signals with zero doppler shift. In this way, large rapidly moving objects could be detected from stationary radar platforms and the radial velocity of these objects could be estimated.

Later, the requirements for search radars shifted to include moving, airborne radars for observing small, slowly moving target objects at long range. It then became necessary to employ much more delicate techniques for finding a signal return within a large clutter background. These techniques employ coherent processing with the aid of large digital computers.

Like radar, sonar is based on the reflection of a passband waveform from an object or scene, in the case of sonar, it is an acoustic wave. The carrier frequency might typically be between one kilohertz and one megahertz. Although sonar is similar to radar in principle, the speed of propagation is smaller by a factor of approximately 10^6, leading to practical consequences in beamforming. The concept of a synthetic-aperture radar

---

[1] Not to be confused with photonic processors.

leads naturally to the notion of a synthetic-aperture sonar. High-resolution synthetic-aperture sonar using hydrophone arrays for beam steering has been developed for imaging the ocean floor and for other applications.

Meanwhile, astronomers had come to realize that a large amount of astronomical information reaches the earth in the microwave bands. Astronomers are well grounded in optical theory where beamwidths smaller than one arc second are obtained. In the microwave band, a comparable beamwidth requires a reception antenna that is many miles in diameter. Under the impact of wind, ice, and temperature gradients, such an antenna would need to be mechanically rigid. Clearly, such antennas are not practical. Around 1952, Martin Ryle, at the University of Cambridge, began to study methods for artificially creating a kind of an aperture by pairwise combinations of many individual antenna elements, or by allowing the earth's rotation to sweep an array of fixed antenna elements through space. In retrospect, this development of radio astronomy may be viewed as a passive counterpart to the development of active synthetic-aperture radar. The aperture is synthesized by recording the radio signal received at two or more antenna elements and later processing these records coherently pairwise within a digital computer. The first such radio telescope was the Cambridge One-Mile Radio telescope completed in 1964, followed by the Cambridge Five-Kilometer radio telescope in 1971. More recently, other synthetic-aperture radio telescopes have been built and put into operation throughout the world. (The continent-sized Very Large Baseline Array has an angular resolution of 0.0002 arc second.) For the development of synthetic-aperture radio telescopes, Ryle was awarded the 1974 Nobel prize in physics (jointly with Hewish who discovered pulsars with the radio telescope). Much of our knowledge of the extragalactic universe comes from the signal-processing algorithms that form the galactic images from the data gathered by the radio telescope antennas.

1.3 Baseband and Passband Waveforms

We will have frequent occasion to use real or complex baseband signals and also occasions to use passband signals. A real *baseband signal*, $s(t)$, is any real function of time with its spectral energy density concentrated near zero frequency. The baseband signal $s(t)$ may also be called a *baseband waveform* when it is regarded as a complicated signal or a *baseband pulse* when it is regarded as a relatively simple signal of finite energy. The *support* of $s(t)$ is the closure of the set of t for which $s(t)$ is nonzero.

A *complex baseband signal*, $s(t) = s_R(t) + js_I(t)$, where $j = \sqrt{-1}$, is any complex function of time with its spectral energy density concentrated near zero frequency. The *real* (or *in-phase*) component $s_R(t)$ and the imaginary (or *quadrature*) component $s_I(t)$ are both real baseband signals. The complex baseband signal $s(t)$ may also be called a *complex baseband waveform* or a *complex baseband pulse* as may be appropriate.

A *passband signal*, which is denoted by $\widetilde{s}(t)$, with a tilde overbar is a function of the form

$$\widetilde{s}(t) = s_R(t) \cos 2\pi f_0 t + s_I(t) \sin 2\pi f_0 t,$$

where f_0 is a constant known as the *carrier frequency* and $s_R(t)$ and $s_I(t)$ are real functions of time whose Fourier spectra $S_R(f)$ and $S_I(f)$ are zero for $|f| \geq f_0$. The signals $s_R(t)$ and $s_I(t)$ are called the *modulation components* of $\tilde{s}(t)$. For a radar system, the carrier frequency f_0 lies somewhere in the interval from 0.1 to 35 gigahertz and is often in the interval from 1 to 10 gigahertz. For a sonar system, f_0 is usually measured in kilohertz. For an ultrasound system, f_0 may be measured in megahertz.

The passband signal $\tilde{s}(t)$ may also be called a *passband waveform*, usually when it is regarded as a complicated signal; or a *passband pulse*, such as when it is regarded as a relatively simple signal of finite energy.

The *complex baseband signal* $s(t)$ corresponding to the passband signal $\tilde{s}(t)$ is

$$s(t) = s_R(t) + js_I(t).$$

The real passband signal $\tilde{s}(t)$ corresponding to the complex baseband signal $s(t)$ is[2]

$$\tilde{s}(t) = \text{Re}[s(t)e^{-j2\pi f_0 t}].$$

The signals $\tilde{s}(t)$ and $s(t)$ are regarded as essentially the same signal but for the detail of the multiplying complex exponential. To emphasize this, these may be called the *real passband representation* and the *complex baseband representation* of the same signal. It is often convenient to suppress the real part operator and write

$$\tilde{s}(t) = s(t)e^{-j2\pi f_0 t}.$$

In such a case, this is called the *complex passband representation* of the signal.

There are two reasons for replacing the passband signal $\tilde{s}(t)$ with the complex baseband signal $s(t)$. From the notational point of view, the complex baseband signal is preferred because the complex baseband signal is notationally more compact than the passband signal, and mathematical manipulations of complex baseband equations exactly mimic mathematical manipulations of the corresponding passband equations and are much easier. Moreover, within a transmitter or receiver, it is often convenient to translate a real passband signal into the complex baseband representation. Ultimately, the simplest and most rewarding point of view is to think of the complex baseband signal as the more fundamental form which is temporarily represented as a passband signal for purposes of transmission and reception. While we study it and process it, the signal is a complex baseband signal; when we transmit it and receive it, the signal is a passband signal. To convert between the two forms is trivial, and is often the last operation in a transmitter and the first operation in a receiver.

---

[2] The sign in the exponent is arbitrary. It is chosen here so that Fourier transform relationships in optics and antenna theory have the conventional form. This choice leads to the positive sign convention appearing in the passband waveform. However, the opposite sign convention is used in modulation theory.

1.4 Monodirectional Waves

Many imaging modalities are based on the processing of waves that are reflected from objects in a scene. The image formation algorithms are developed based on the structure and behavior of waves. A wavefront propagating in free space may have a complicated structure, both temporally and spatially. To gain an understanding of the general case, one can begin with a study of the simple case of a *plane wave*. More complicated situations can be built up from multiple plane waves. The Huygens–Fresnel principle, which is developed in Chapter 4, describes how any planar surface in free space crossed by a wave can be viewed as the source of that wave.

Physically, a wave may be a time-varying and space-varying electric or magnetic vector field associated with an electromagnetic wave, or it may be the time-varying and space-varying pressure field associated with an acoustic wave. A wave may be a vector function as in the case of the electromagnetic wave, or it may be a scalar function, as in the case of the acoustic wave. Our primary concern is with the mathematical description of the wave. Usually, we are content to deal with scalar-valued waves because of analytical simplicity. Although an electromagnetic wave is a vector-valued wave, this property of the wave does not often affect the properties of propagation that are of interest herein. With some exceptions, the wave can be regarded as a scalar wave for most of our purposes.

The propagation of electromagnetic waves at optical frequencies obeys the same fundamental laws as it does at microwave frequencies. However, the great difference in the wavelengths leads to a difference in the phenomena that we perceive. The wavelength of a microwave is on the order of centimeters, while the wavelength of a light wave is on the order of a micron. A microwave antenna rarely has dimensions of more than a few hundred wavelengths – and usually much less – while an optical lens has dimensions of more than 10^4 wavelengths. Consequently, an everyday optical beam is usually much sharper than a microwave beam and often is described adequately by geometrical optics and ray tracing.

Monochromatic Monodirectional Waves

A *monochromatic* wave is a wave at a single frequency. A *monodirectional wave* is a wave traveling in a single direction. Mathematically, a spatially uniform, monodirectional, monochromatic, scalar plane wave traveling in the z direction is given by

$$\widetilde{s}(t,x,y,z) = A\cos(2\pi f_0(t - z/c) + \theta)$$
$$= A\cos(2\pi f_0 t - kz + \theta),$$

where the constant $k = 2\pi f_0/c = 2\pi/\lambda$ is called the *wave number* and λ is called the *wavelength*. This passband wave is also written as

$$\widetilde{s}(t,x,y,z) = \text{Re}[Ae^{-j\theta}e^{-j(2\pi f_0(t-z/c))}]$$
$$= \text{Re}[Ae^{-j\theta}e^{-j(2\pi f_0(t-kz))}],$$

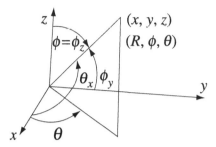

Figure 1.3 Direction cosines and spherical coordinates

where $Ae^{-j\theta}$ is called the *complex amplitude* of the wave at $z = 0$. The *complex baseband representation* of this wave at arbitrary z is

$$s(x,y,z) = Ae^{-j\theta} e^{j2\pi f_0 z/c}$$
$$= Ae^{-j\theta} e^{jkz},$$

with the time dependence now removed. This is the complex baseband representation of a monodirectional, monochromatic wave moving in the z direction. Such a wave is called a *plane wave*. A plane wave has the same value at every point of the wavefront plane.

The most general form of a spatially uniform, monodirectional, monochromatic wave that satisfies the wave equation is given by

$$\tilde{s}(t,x,y,z) = A\cos(2\pi f_0(t - (\alpha x + \beta y + \gamma z)/c) + \theta)$$
$$= A\cos(2\pi f_0 t - \alpha kx - \beta ky - \gamma kz + \theta).$$

The variables α, β, and γ are called *direction cosines*. The direction cosines specify the direction of travel of the plane wave. They are equal, respectively, to the cosines of the angles between the direction of travel of the plane wave and the three coordinate axes:

$$\alpha = \cos\phi_x,$$
$$\beta = \cos\phi_y,$$
$$\gamma = \cos\phi_z.$$

The direction cosines are related to spherical coordinates, as shown in Figure 1.3, by

$$\alpha = \cos\theta \sin\phi = \cos\phi_x,$$
$$\beta = \sin\theta \sin\phi = \cos\phi_y,$$
$$\gamma = \cos\phi = \cos\phi_z,$$

and so they are related by

$$\alpha^2 + \beta^2 + \gamma^2 = 1.$$

1.4 Monodirectional Waves

The angles θ and ϕ are two of the three angles, called eulerian angles, relating two coordinate systems.

Three alternative complex baseband representations of the monodirectional, monochromatic wave with direction cosines α, β, and γ are[3]

$$s(x,y,z) = Ae^{-j\theta}e^{j2\pi f_0(\alpha x+\beta y+\gamma z)/c}$$
$$= Ae^{-j\theta}e^{jk(\alpha x+\beta y+\gamma z)}$$
$$= Ae^{-j\theta}e^{j(k_1 x+k_2 y+k_3 z)}.$$

The quantities k_1, k_2, and k_3 are called the *wave numbers* of the plane wave. The wave numbers are related to the direction cosines by

$$k_1 = (2\pi f_0/c)\alpha = (2\pi/\lambda)\alpha,$$
$$k_2 = (2\pi f_0/c)\beta = (2\pi/\lambda)\beta,$$
$$k_3 = (2\pi f_0/c)\gamma = (2\pi/\lambda)\gamma.$$

The vector $\mathbf{k} = (k_1, k_2, k_3)$ is called the *vector wave number* of the plane wave.

The complex baseband representation using complex exponentials is more convenient to work with than is the passband representation. The real passband representation is recovered by

$$\tilde{s}(x,y,z,t) = \text{Re}\left[s(x,y,z)e^{-j2\pi f_0 t}\right].$$

Time-Varying Monodirectional Waves

When the complex amplitude $Ae^{-j\theta}$ is replaced by a *time-varying* complex amplitude $A(t)e^{-j\theta(t)}$, the waveform is no longer monochromatic. A monodirectional waveform with time-varying amplitude and phase traveling in direction (α, β, γ) has the general form

$$\tilde{s}(x,y,z,t) = A(t - \tau(x,y,z))\cos\left(2\pi f_0(t - \tau(x,y,z)) + \theta(t - \tau(x,y,z))\right),$$

where $\tau(x,y,z) = (\alpha x + \beta y + \gamma z)/c$. This is called the *passband representation* of the time-varying monodirectional wavefront. Using the complex amplitude $A(t)e^{-j\theta(t)}$, the passband representation can be written concisely as

---

[3] A wave of the form

$$\tilde{s}(t,x,y,z) = A(x,y)\cos(2\pi f_0(t - z/c) + \theta),$$

does not satisfy the wave equation

$$\frac{\partial^2 \tilde{s}}{\partial x^2} + \frac{\partial^2 \tilde{s}}{\partial y^2} + \frac{\partial^2 \tilde{s}}{\partial z^2} = \frac{1}{c^2}\frac{\partial^2 \tilde{s}}{\partial t^2}$$

whenever $A(x,y)$ is not the constant A. Consequently, it is not properly among the waves that are studied herein. Sometimes, it may be convenient to write a wave in this form. In such cases, $\tilde{s}(t,x,y,z)$ should be regarded only as an approximation (geometrical optics) of a wave that does satisfy the wave equation. This approximation is studied in Chapter 4.

$$\tilde{s}(x,y,z,t) = \text{Re}\big[A(t-\tau(x,y,z))e^{-j\theta(t-\tau(x,y,z))}e^{-j2\pi f_0(t-\iota(x,y,z))}\big]$$
$$= \text{Re}\big[s(t,x,y,z)e^{-j2\pi f_0 t}\big],$$

where $s(x,y,z,t)$ is the complex baseband representation of the wavefront given by

$$s(x,y,z,t) = A(t-(\alpha x+\beta y+\gamma z)/c)e^{-j\theta(t-(\alpha x+\beta y+\gamma z)/c)}e^{-j2\pi f_0(t-(\alpha x+\beta y+\gamma z)/c)}.$$

At the origin, the real passband waveform is

$$\tilde{s}(0,0,0,t) = \text{Re}\big[A(t)e^{-j\theta(t)}e^{-j2\pi f_0 t}\big].$$

Let $S(f)$ be the Fourier transform of $s(t,0,0,0)$. A narrowband wave is one for which the support of $S(f)$ is narrow compared to f_0 insofar as the needs of the application may require. In most of this book, narrowband waves are approximated as monochromatic waves having a single wavelength λ.

1.5 Wavefront Diffraction

A monochromatic monodirectional wavefront is a wavefront that is traveling in only one direction. A wavefront that is not monodirectional is the superposition of waves traveling in multiple directions. We will see that when the waveform amplitude is spatially varying in every plane, the waveform is no longer monodirectional. It is a superposition of such monodirectional plane waves traveling in multiple directions. The complex amplitude in the x,y plane of each plane wave now depends on the distribution or spectrum of the wavefront directions.

Space-Varying Monochromatic Waves

When multiple monochromatic waves are simultaneously traveling in a finite number of directions, indexed by ℓ, the composite wave at complex baseband is

$$s(x,y,z) = \sum_{\ell=1}^{L} A_\ell e^{-j\theta_\ell} e^{j2\pi f_0(\alpha_\ell x+\beta_\ell y+\gamma_\ell z)/c},$$

where α_ℓ, β_ℓ, and γ_ℓ are the direction cosines specifying the direction of the ℓth plane wave.

When monochromatic waves are simultaneously traveling in all directions with an infinitesimal amplitude in each direction, then the complex baseband representation of the wavefront becomes an integral over the extent of wavefront directions. A monochromatic wavefront that has a continuum of directions has the complex baseband representation

$$s(x,y,z) = \int_{-\infty}^{\infty}\int_{-\infty}^{\infty} a(\alpha,\beta)e^{j2\pi f_0(\alpha x+\beta y+\gamma z)/c}\,d\alpha\,d\beta,$$

where the pair (α,β) of direction cosines specifies a direction and $\gamma = \sqrt{1-\alpha^2-\beta^2}$ is the third direction cosine. The term $a(\alpha,\beta)d\alpha d\beta$ is the infinitesimal complex amplitude of the wave propagation in direction (α,β). Even though the direction cosines range only between −1 and 1, the limits of integration have been written from $-\infty$ to ∞. This allows some important flexibility later. For now, the excess region of integration can be temporarily suppressed by requiring that $a(\alpha, \beta) = 0$ when $\alpha^2 + \beta^2$ is larger than one. This constraint will be dropped later.

When the wave $s_0(x, y) = s(x, y, 0)$ in the plane $z = 0$ is specified, the integral

$$s_0(x, y) = \int_{-\infty}^{\infty} \int_{-\infty}^{\infty} a(\alpha, \beta) e^{j2\pi(\alpha x + \beta y)/\lambda} d\alpha d\beta$$

implicitly defines $a(\alpha, \beta)$ in terms of $s_0(x, y)$. This equation can be interpreted as an instance of the inverse two-dimensional Fourier transform. The function $a(\alpha, \beta)$ is called the *angular spectrum* of the input signal $s_0(x, y) = s(x, y, 0)$ in the plane $z = 0$. The angular spectrum completely describes the propagation of a monochromatic wave. The "input" in the plane at z equal to zero is $s_0(x, y)$, which implicitly determines the angular spectrum $a(\alpha, \beta)$. In turn, in the plane with z equal to d, the complex amplitude $s_d(x, y) = s(x, y, d)$ is given in terms of $a(\alpha, \beta)$ by the expression

$$s_d(x, y) = \int_{-\infty}^{\infty} \int_{-\infty}^{\infty} a(\alpha, \beta) e^{j(2\pi/\lambda)\sqrt{1-\alpha^2-\beta^2}} e^{j2\pi(\alpha x + \beta y)/\lambda} d\alpha d\beta.$$

We follow this line of thought in Chapter 4 to derive the important Huygens–Fresnel principle.

Evanescent Waves

There is also a less-familiar, monochromatic and monodirectional solution of the wave equation called an *evanescent wave*. An evanescent wave is a wave of the form

$$\widetilde{s}(x, y, z, t) = \cos(2\pi f_0 (t - (\alpha x + \beta y)/c)) e^{-2\pi f_0 \gamma z / c},$$

satisfying the wave equation, where now the term involving z is a real decaying exponential. To satisfy the wave equation, (α, β, γ) must satisfy

$$\alpha^2 + \beta^2 - \gamma^2 = 1,$$

where here γ^2 is led by a negative sign.

An evanescent wave has the complex baseband representation

$$s(x, y, z) = e^{j2\pi f_0 (\alpha x + \beta y)/c} e^{-2\pi f_0 \gamma z / c},$$

with γ now defined as

$$\gamma = \begin{cases} \sqrt{1 - \alpha^2 - \beta^2} & \text{for } \alpha^2 + \beta^2 \leq 1, \\ \sqrt{\alpha^2 + \beta^2 - 1} & \text{for } \alpha^2 + \beta^2 \geq 1. \end{cases}$$

The two lines are defined differently so that γ is real in both cases. Thus $\alpha^2 + \beta^2 - \gamma^2 = 1$ for evanescent waves.

The evanescent wave is an exponentially decreasing wave in the z direction. This wave is needed by the mathematics or the physics in order to meet boundary conditions that cannot be met with a propagating wave in the z direction.

Clearly, the amplitude of the evanescent wave becomes infinite as z goes to negative infinity. Therefore an evanescent wave can only exist in a half-space and requires special boundary conditions on the boundary of this half-space. When the half-space is taken to be the half-space for which z is nonnegative, the boundary conditions are on the plane at which $z = 0$. The evanescent wave decays quickly with increasing z, becoming negligible after a few wavelengths. Along the plane at $z = 0$ in the direction specified by α and β runs an evanescent wave with a velocity of $c/\sqrt{\alpha^2 + \beta^2}$, which is smaller than c.

With the introduction of evanescent waves, the equation

$$s(x,y,z) = \int_{-\infty}^{\infty} \int_{-\infty}^{\infty} a(\alpha, \beta) e^{j2\pi f_0 z \sqrt{1-\alpha^2-\beta^2}/c} e^{j2\pi f_0(\alpha x+\beta y)/c} d\alpha d\beta,$$

introduced earlier, can now be interpreted more generally. The infinite limits of integration allow the direction cosines to be larger than one. Physically, this allows evanescent waves to be included within the angular spectrum $a(\alpha, \beta)$.

Transverse Vector Waves

Besides scalar-valued waves, there are also vector-valued waves. A vector-valued wave may be regarded as three scalar-valued waves comprising the three components of the vector in a suitable coordinate system. A monodirectional, monochromatic vector wave at complex baseband has the form

$$s(x,y,z) = [s_x \mathbf{i}_x + s_y \mathbf{i}_y + s_z \mathbf{i}_z] e^{j2\pi f_0(\alpha x+\beta y+\gamma z)/c},$$

where $(\mathbf{i}_x, \mathbf{i}_y, \mathbf{i}_z)$ forms a triad of orthogonal unit vectors along the three axes of the coordinate system. When the three scalar components s_x, s_y, and s_z can be independently specified, then such a wave amounts to nothing more than three independent scalar waves.

There are certain vector waves of widespread physical interest satisfying a special constraint that makes the components dependent. These vector waves, called *transverse vector waves*, are those waves that satisfy an additional constraint. Electromagnetic waves in free space are transverse vector waves.

A transverse vector wave is a vector wave that takes only values perpendicular to its direction of propagation. For the wave to be a transverse wave, the direction of the vector field must be perpendicular to the direction of propagation. The dot product of the field vector and the direction of propagation must be zero. For a plane wave, the direction of propagation $\alpha \mathbf{i}_x + \beta \mathbf{i}_y + \gamma \mathbf{i}_z$ is constant. The direction of the field is $s_x \mathbf{i}_x + s_y \mathbf{i}_y + s_z \mathbf{i}_z$. This means that for a transverse-vector plane wave, the dot product

$$s_x \alpha + s_y \beta + s_z \gamma = 0$$

must be satisfied as a side condition.

1.6 Temporal and Spatial Coherence

The words "coherent" and "noncoherent" continually recur. These words are used to designate the quality of the phase angle of a passband waveform. Every passband signal is a sinusoid that can be expressed in terms of the time-varying amplitude and phase

$$\tilde{s}(t) = A(t)\cos(2\pi f_0 t + \theta(t)),$$

where $A(t)$ is the time-varying amplitude, f_0 is the carrier frequency, and $\theta(t)$ is the time-varying phase. The complex baseband representation then has the form

$$s(t) = A(t)e^{-j\theta(t)}.$$

The phase $\theta(t)$ may be intentional and known, or it may be partially or wholly unintentional and unknown. The phase angle may be random phase noise. When the phase angle $\theta(t)$ of the signal $\tilde{s}(t)$ is known to the extent that knowledge of $\theta(t)$ is critical to the application, the signal $\tilde{s}(t)$ is called a *coherent* signal. Otherwise, $\tilde{s}(t)$ is called a *noncoherent* signal.

The term "coherent" may also arise in connection with the processing of a real passband signal, perhaps in the form of a complex baseband signal. The processing may be the kind known as *coherent processing*, which fully uses both $A(t)$ and $\theta(t)$, or the kind known as *noncoherent processing*, which makes only limited – or no – use of $\theta(t)$.

Coherence not only refers to a deterministic relationship between the phase angles of a waveform at different time instants, but may also refer to a deterministic relationship between the phase angle of two different waveforms, $\tilde{s}_1(t)$ and $\tilde{s}_2(t)$. The former case is then referred to as a *temporally* coherent waveform. The latter case is referred to as a *spatially* coherent waveform when a common wavefront is incident on two antennas or two lenses at different locations, or at two regions of the same antenna or lens. The deterministic relationship between points in a spatially coherent wavefront may be due to the different times at which the wavefront reaches those different points. Two signals, $\tilde{s}_1(t)$ and $\tilde{s}_2(t)$, may be spatially coherent even though they are jointly temporally noncoherent. For example, in photographic systems, the light from a point source incident on a lens may be temporally noncoherent, but across the lens it is spatially coherent. Otherwise, the lens could not focus the light into an image of that point source. Moreover, when there are multiple point sources, the light emitted by the multiple point sources can be mutually spatially noncoherent because the point sources are mutually noncoherent, yet the light reaching the lens from each individual point source can be spatially coherent.

A pulse train is a common example of a passband radar waveform. A pulse train has the form

$$\tilde{p}(t) = \sum_{n=0}^{N-1} s(t - nT_r)\cos(2\pi f_0 t + \theta_0),$$

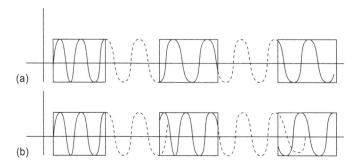

Figure 1.4 A coherent pulse train (a) and a noncoherent pulse train (b)

where $s(t)$ is a single pulse, T_r is a constant called the *pulse repetition interval*, and θ_0 is a constant. The pulse train consists of N uniformly spaced translates of the pulse $s(t)$ modulated onto the carrier $\cos(2\pi f_0 t + \theta_0)$. In complex baseband notation, the pulse train is denoted

$$p(t) = \sum_{n=0}^{N-1} s(t - nT_r)e^{-i\theta_0}.$$

This pulse train might be called coherent to mean that θ_0 remains constant from pulse to pulse even though θ_0 may be unknown. Instead, the waveform might be called coherent to mean that the pulses all have the same known constant phase θ_0. Hence whether or not a given waveform is called coherent might depend on the circumstances of the discussion.

A pulse train in which the phase is not the same from pulse to pulse is given in the passband representation by

$$\tilde{p}(t) = \sum_{n=0}^{N-1} s(t - nT_r)\cos(2\pi f_0 t + \theta_n),$$

and in the complex baseband representation as

$$p(t) = \sum_{n=0}^{N-1} s(t - nT_r)e^{-j\theta_n}.$$

This is called a noncoherent pulse train when the θ_n are random and independent (or weakly correlated). Then the θ_n may form a sequence of independent random variables, perhaps taking values uniformly between 0 and 2π. Figure 1.4 compares the coherent pulse train with a noncoherent pulse train. It is important to the usage here that the phase angles are unknown. When the phase angles are known, even though different, the waveform is a coherent waveform because the known values of the phase angles can be included in the processing of the waveform.

More generally, $\theta(t)$ may separate into two parts: a phase angle that is known, and a phase angle that is unknown. For an arbitrary waveform in the passband representation, this may be written as

$$\tilde{s}(t) = A(t)\cos[2\pi f_0 t + \theta_s(t) + \theta_n(t)].$$

The complex baseband representation is

$$s(t) = A(t)e^{-j[\theta_s(t)+\theta_n(t)]},$$

where $\theta_s(t)$ is the intentional and known part of the phase modulation associated with the signal, and $\theta_n(t)$ is the unintentional and unknown part of the phase modulation. The unknown part is called *phase noise*. When described in this way, a coherent waveform could mean a waveform in which $\theta_n(t)$ is negligible, and a noncoherent waveform could mean a waveform in which $\theta_n(t)$ is not negligible. In some applications, surprisingly large values of unknown phase noise may be acceptable. Even phase errors as large as one radian can sometimes be tolerated, though with a significant loss of system performance. Chapter 15 is devoted to the quantitative analysis of the effect of phase error and phase noise on performance, including the notion of coherence in various situations.

1.7 Deterministic and Random Models

The usual goal of an image formation system is to form the best image of an object based on the available data. However, it is difficult to formulate a precise statement of optimality because a general criterion of optimality can be elusive and prior knowledge may be subjective. This is due partly to the fact that the underlying physical reality is much richer than the desired image, and it is difficult to state the real goal as an abstraction of the physical reality, and also because prior knowledge or prior assumptions about the image must be accommodated. Such considerations require that a model of the problem be developed. Such a model may be either deterministic or random. In the early chapters of this book, deterministic models of the image are usually used. Such models assume that a "true" image does exist, and our task is to estimate that image by processing the observed data. Randomness enters the problem in those chapters only because the measurements can be random or noisy. There is a single underlying image that is to be found.

In later chapters, we turn to a more abstract view of imaging, regarding the task as one of selecting an image from a space of possible images. In that more abstract view, an image is a realization of a random variable characterized by a probability distribution on a predefined space of images. The goal is not to pick the "true" image, but to select that image from the space of images that best explains the observed data. This reformulation of the task of imaging may be seen as nearly the same task as before, but it does suggest alternative approaches.

For these reasons, both to model measurement noise and to model a random image, probability theory inevitably enters the topics of this book. We will need the notions of a random variable and a random process. Here we briefly review some of the fundamentals of probability theory that are used.

The reason for introducing the topic of random variables described in this section and the topic of random processes described in more detail in Section 2.8 of Chapter 2 is to study randomness arising in various situations of imaging. Imaging theory uses quantities such as probability density functions and correlation functions. They must be meaningful and known. Eventually, we even deal with the doubly vague situation in which the probability density function $p(x)$ associated with the random variable X is itself unknown. Such formulations provide structure and lead to useful procedures for image formation.

Random Variables

A *random variable*, X, consists of a set of values that the random variable can take and a probability distribution on this set of values. A random variable, X, may be restricted to a finite, or countable, number of values, in which case it is called a *discrete random variable*. Then it is characterized by the *probability vector* p with a finite, or countable, number of components denoted p_j. Likewise, a pair of discrete random variables, (X, Y), is associated with a joint probability distribution, P, with an array of components denoted P_{jk}. A joint probability distribution is associated with *marginals*, defined by $p_j = \sum_k P_{jk}$ and $q_k = \sum_j P_{jk}$, and *conditionals*, defined by $Q_{k|j} = P_{jk}/p_j$ and $P_{j|k} = P_{jk}/q_k$. This leads to the *Bayes formula*

$$Q_{k|j} = \frac{q_k P_{j|k}}{\sum_k q_k P_{j|k}}$$

as a consequence of the definitions of marginals and conditionals.

A *real random variable* is a random variable that takes values in the set of real numbers. A real random variable may take values in a finite set of real numbers, in which case it is called a *discrete real random variable*, or values in a continuous set of real numbers, in which case it is called a *continuous real random variable*. Whereas a discrete random variable is described by a probability vector p, a continuous random variable is described by a function, $p(x)$, called the *probability density function*, or a conditional function $p(x|y)$, called the *conditional probability density function*. We consider only discrete random variables and continuous random variables. We do not consider mixed random variables.

A discrete or continuous real random variable has a mean, $\bar{x}$, denoted by

$$\bar{x} = \sum_j p_j x_j,$$

or by

$$\bar{x} = \int_{-\infty}^{\infty} x p(x) dx,$$

and a variance, σ^2, denoted by

$$\sigma^2 = \sum_j p_j (x_j - \bar{x})^2,$$

or by

$$\sigma^2 = \int_{-\infty}^{\infty} (x - \bar{x})^2 p(x) dx.$$

Similar definitions can be made for *complex random variables*, which are random variables taking values in the set of complex numbers.

An important random variable is a *gaussian random variable*, which is the only example of a random variable given in this section. Other random variables appear later in the book. The real gaussian random variable is defined by its probability density function

$$p(x) = \frac{1}{\sqrt{2\pi}\sigma} e^{-(x-\bar{x})^2/2\sigma^2}.$$

The gaussian random variable has the mean $\bar{x}$ and variance σ^2. Likewise, the complex gaussian random variable $X = X_R + jX_I$ with circular symmetry has probability density function

$$p(x_R, x_I) = \frac{1}{2\pi\sigma^2} e^{-|x-\bar{x}|^2/2\sigma^2}$$

$$= \frac{1}{2\pi\sigma^2} e^{-x_R^2/2\sigma^2} e^{-x_I^2/2\sigma^2},$$

where $\sigma^2 = \mathrm{E}[X_R^2] = \mathrm{E}[X_I^2] = \mathrm{E}[XX^*]/2$ and $\mathrm{E}[X_R X_I] = 0$ for the circularly symmetric complex gaussian random variable.

A real (or complex) *multivariate random variable*, $X = (X_1, \ldots, X_n)$, also called a *vector random variable*, with zero mean has a probability density function, $p(x_1, \ldots, x_n)$ and a covariance matrix, Σ, whose ij entry is the expectation $\mathrm{E}[X_i X_j]$ (or $\mathrm{E}[X_i X_j^*]$). A covariance matrix is always nonnegative-definite because[4] $a\Sigma a^\dagger = a\mathrm{E}[XX^\dagger]a^\dagger = \mathrm{E}[(aX)^2]$, which is always nonnegative because it is the expectation of a squared term.

Random Processes

A *random process* or a *stochastic process*, $X(t)$, on the variable t consists of a set of functions that the random process can take and a probability distribution on this set of functions. The values that $X(t)$ can take may be continuous or discrete. The independent variable t can be continuous or discrete. Usually, a *discrete random process* refers to a random process for which t is discrete.

1.8 The Electromagnetic Spectrum

Signals throughout the electromagnetic spectrum are everywhere and carry a great deal of information, much of it hidden from our senses. Table 1.1 shows the remarkable twenty orders of magnitude of the electromagnetic spectrum that are of interest. The table shows the spectrum broken into bands annotated with individual names. The

---

[4] The symbol † denotes the transpose of a real-valued matrix or the complex conjugate of the transpose for a complex-valued matrix.

Table 1.1 Common names of electromagnetic spectrum intervals

Radiation type	Frequency range (hertz)
Radio waves	2×10^4 to 1×10^9
Microwaves	1×10^9 to 3×10^{11}
Infrared	3×10^{11} to 4×10^{14}
Near infrared	1×10^{14} to 4×10^{14}
Visible	4×10^{14} to 7.5×10^{14}
Ultraviolet	1×10^{15} to 1×10^{17}
X-rays	1×10^{17} to 1×10^{20}
Gamma rays	1×10^{20} to 1×10^{24}

naming of these frequency bands is not standardized and the usage may vary somewhat. The boundary between the bands is not sharply defined. The various topics in this book range throughout the electromagnetic spectrum shown in Table 1.1.

The electromagnetic spectrum is labeled using one of three measurement units. These three measurement units are the frequency f, the wavelength λ, and the photon energy E, only one of which is mentioned in Table 1.1. These three quantities are related by the expressions

$$\lambda = c/f \qquad E = hf = h\lambda/c,$$

where c is a constant called the *speed of light* and h is a constant called the *Planck constant*. Each of these three measurement units is most convenient to use in a different region of the spectrum according to how the electromagnetic signal presents in that region. Only the frequency designation is given in Table 1.1.

Our senses lack the ability to observe most of the electromagnetic spectrum directly. Our sense of vision allows us to observe only the very narrow range of frequencies known as the visual band. Although our sense of touch is sensitive to frequencies in the near infrared and near ultraviolet, this sensitivity is only a vague awareness of radiation in these frequencies with no awareness of a corresponding image. The remainder of the immense electromagnetic spectrum, though teeming with information of many kinds, is outside of our immediate experience.

The electromagnetic spectrum is indeed full of signals. Both natural signals and man-made signals are present. Many image formation systems are based on sensing these electromagnetic signals throughout the electromagnetic spectrum. Even in the visible spectrum, our natural vision is now augmented by many man-made devices, such as eyeglasses, cameras, microscopes, and telescopes.

Table 1.1 partitions the electromagnetic spectrum according to general terms that are in common usage. A more systematic partition of a part of the spectrum is given in Table 1.2. The terms in this table are also in common use.

Other kinds of image formation systems are based on acoustic (seismic) signals. The acoustic spectrum is also large, extending into the ultrasound frequencies.

Table 1.2 Decades of the electromagnetic spectrum

Decade designation		Frequency range
Very low frequency	(VLF)	3×10^0 to 3×10^1 kHz
Low frequency	(LF)	3×10^1 to 3×10^2 kHz
Medium frequency	(MF)	3×10^2 to 3×10^3 kHz
High frequency	(HF)	3×10^0 to 3×10^1 MHz
Very high frequency	(VHF)	3×10^1 to 3×10^2 MHz
Ultra high frequency	(UHF)	3×10^2 to 3×10^3 MHz
Super high frequency	(SHF)	3×10^0 to 3×10^1 GHz
Extremely high frequency	(EHF)	3×10^1 to 3×10^2 GHz

1.9 Imaging by Tomography

In many situations, such as in medical imaging, it is possible to observe projections of the two-dimensional object $p(x,y)$ (or the three-dimensional object $p(x,y,z)$) of a certain kind, even though direct observations of that object are not possible. Using X-rays, a nuclear beam, or magnetic resonance gradients, the object $p(x,y)$, as perceived by that energy source, can be integrated along lines. The results of these integrations are called *projections*. An image of $p(x,y)$ is computed from the set of its projections. The process of imaging from projections is known as *tomography*.[5] The central theorem of tomography is the *projection-slice theorem*. The set of all projections of $p(x,y)$ is called the *Radon transform* of $p(x,y)$. While $p(x,y)$ is not directly observable, the Radon transform of $p(x,y)$ is observable as a collection of projections.

A familiar example is an elementary X-ray projection of internal body organs. As a single ray passes along a line, say the y axis with x held constant and ignored for now, the intensity is attenuated at each y by an amount described by an attenuation function $p(y)$. That is, the intensity, denoted I', leaving a small interval of width Δy, centered at y_1, is related to the intensity, denoted I, entering that interval by

$$I' = I[1 - p(y_1)\Delta y].$$

This is approximated as

$$I' \approx I e^{-p(y_1)\Delta y}$$

under the condition that the attenuation is small for a sufficiently small interval Δy. Over two consecutive intervals, each of width Δy, the intensity attenuation is described approximately as

$$I'' = I e^{-p(y_1)\Delta y} e^{-p(y_2)\Delta y}.$$

---

[5] The term tomography has been broadened, by some, to include other forms of medical imaging

Consequently, over a sequence of many such intervals, the output intensity I_{out} is related to the input intensity I_{in} by

$$I_{out} = I_{in} e^{-\sum_i \rho(y_i) \Delta y}.$$

In the limit as Δy goes to zero,

$$\log_e \frac{I_{in}}{I_{out}} = \int_{-\infty}^{\infty} \rho(y) dy,$$

where the integration limits can be replaced by the support of the function $\rho(y)$.

By introducing a ray in the y direction at each value of x, passing through the two-dimensional function $\rho(x, y)$, one can define the *projection* $p(x)$ onto the x axis for each value of x:

$$p(x) = \int_{-\infty}^{\infty} \rho(x, y) dy.$$

The function on the left is the projection of $\rho(x, y)$ onto the x axis.

In the general case, the attenuation of an X-ray at angle θ integrates the function $\rho(x, y)$ along each ray in the direction indicated by angle θ. The projection at angle θ

$$p_\theta(t) = \int_{-\infty}^{\infty} \rho(t \cos\theta - r \sin\theta, t \sin\theta + r \cos\theta) dr,$$

at each t, consists of the integration of $s(x, y)$ along each ray in the r direction as a function of t. By varying the viewing angle θ, such projections can be observed from many directions. One wants to process such a set of projections to form an estimate of $\rho(x, y)$, as may show the internal organs of the body. The signal-processing topic of tomography is studied in Chapter 6. The central theorem of signal processing that underlies the methods of tomography is the projection-slice theorem, which is introduced in Chapter 3.

The origins of tomography can be traced back to 1917 when the Austrian mathematician Radon showed that the spatial function $\rho(x, y, z)$ can be reconstructed from the complete set of its projections. Because the reconstruction of images from projections arises in many diverse situations, it is not surprising that this mathematical principle, first discovered by Radon, was independently rediscovered many times and in many fields. It has been used in radio astronomy and in the field of electron microscopy. In the context of medical applications, tomography has led to important advances in the noninvasive imaging techniques available in recent years for clinical practice and medical research. Tomography is used in many other applications, such as geophysical applications, where it can be used for subsurface exploration, or in atmospheric sensing, where it can be used, for example, to form images of pollutant densities in the upper atmosphere.

In Chapter 6, the mathematical principles underlying tomography are studied, especially the projection-slice theorem, which relates the one-dimensional Fourier transform of the projection to the two-dimensional Fourier transform of the object. A number of algorithms for the reconstruction of images are described. The central idea of these algorithms is the method of back projection. Reconstruction of an arbitrary object from projections is exact only when the uncountably infinite set of projections from all angles is known. However, mild prior conditions on the object, such as a spatially bandlimited Fourier transform, can soften this statement. Many good algorithms are known that compute an approximate reconstruction of an image from a finite set of its projections. These images may be satisfactory for practical applications even when the individual projections are weak or noisy.

In addition to projection tomography, many other important forms of tomography go by names such as *emission tomography*, *diffraction tomography*, *diffusion tomography*, and *coherence tomography*. These forms are also studied in Chapter 6. Emission tomography requires that the scene itself emits some form of emission that provides the received signal from which the image is computed. This usually means that the scene must receive an excitation that provides the energy for the emission. There are two methods that are in wide use to excite an object so that it will produce a useful signal. The important method of *magnetic resonance imaging* (MRI) uses a time-varying and spatially varying magnetic field to provide energy to the scene. This time-varying magnetic field causes isolated protons, or perhaps other selected nuclei, to resonate and thus generate signal-dependent magnetic fields. The magnetic field is intercepted, measured, and processed by the methods of tomography to form an image of the density of isolated protons (hydrogen atoms). Another method of excitation, called *positron-emission tomography*, uses a radioactive isotope that is selectively absorbed by a tissue of interest, usually a diseased tissue. The radioactive isotope then decays, thereby releasing radiation energy in the form of positrons. These positrons immediately combine with electrons to produce photons. The photons are captured by an array of photosensors. From the positions and times at which these photons are detected, an image is formed. Through this method, a specific tissue can be selectively imaged as a function of x and y by its tendency to acquire a particular radioactive isotope.

Diffraction tomography and diffusion tomography deal with situations in which the geometrical-optics approximation to propagation is not adequate. It may be necessary to treat wave propagation in a more exact way by considering the effect of diffraction. This is particularly important when observing details that are small compared to the relevant wavelengths. Another difficult instance of tomography is based on the propagation of a wave in a strongly scattering medium. This is the difficult topic of diffusion tomography.

A related form of tomography is *geophysical tomography* in which seismic waves are used to image geophysical features. Then the dispersion of the wave is not caused by diffraction, but rather is caused by scattering anomalies in the propagation medium.

1.10 Radar and Sonar Systems

A radar obtains information about an object or a scene by illuminating the object or the scene with electromagnetic waves, then processing the echo signal that is reflected from that object or scene and intercepted by the radar receiving antenna. A sonar obtains information about an object or a scene by illuminating the object or scene with acoustic waves, then processing the echo signal that is reflected from that object and intercepted by the sonar hydrophones.

By using electromagnetic waves in the microwave bands, a radar is able to penetrate optically opaque media such as clouds, dust, soil, or foliage. In this way, it is possible to form radar images of objects hidden by such obstructions. Similarly, a sonar or ultrasound system can form an image of an object that is optically masked by an opaque medium.

Most instances of radar or sonar are *monostatic*. This means that the transmitter and the receiver are colocated. A monostatic radar may use the same antenna for transmission and reception, as is the usual case. For a *bistatic* radar or sonar, the transmitter and the receiver are at separate locations.

In some cases, the broad beamwidth of a radar antenna or a sonar hydrophone is appropriate as a way of viewing a large region of space. For reasons such as these, radar and sonar have long been popular as imaging systems for surveillance.

While there may be a great deal of difference between the propagation of electromagnetic waves and the propagation of acoustic or pressure waves, there is also a great deal of similarity.[6] This similarity carries over to radar and sonar systems. From our point of view, each is a system that forms a complex baseband pulse, $s(t)$, that is transmitted as the amplitude and phase modulation of the passband pulse $\tilde{s}(t)$, and receives an echo pulse, $\tilde{v}(t)$, that is a composite of delayed and frequency-shifted copies of the passband pulse $\tilde{s}(t)$ and contaminated by noise. The transmitted pulse $\tilde{s}(t)$ propagates at a velocity c over a path of length R_1 from the transmitter to the reflector, and then over a path of length R_2 from the reflector to the receiver. The received pulse $\tilde{v}(t)$ is a superposition of echoes of the transmitted pulse from multiple reflectors. Because the received signal is contaminated by noise and other impairments, it is difficult to recognize individual reflectors. We are interested in methods of processing the received pulse to extract useful information from it. The same basic ideas apply equally to radar pulses and to sonar pulses although the propagation medium is not uniform for sonar. The terminology of the discussion will favor radar systems.

The received signal is distributed both in space across the aperture of an antenna and in time. The distribution in space may be processed by the antenna system to gather all of the received spatially distributed signal into a single, time-dependent signal. Simple linear processing of the signal across the aperture is usually summarized by referring to the shape and width of an antenna "beam." The time variations of the received signal are processed so as to determine the time-varying distance to the reflecting objects. The space distribution may be processed in other ways to determine the direction of arrival of the signal. The processing of the space distribution of the signal across an aperture is studied in Chapter 5. The processing of the time distribution of the signal is studied in Chapters 10, 11, and 12.

---

[6] As a transverse-vector wave, an electromagnetic wave also has the property of polarization which is sometimes useful to a radar system.

1.10 Radar and Sonar Systems

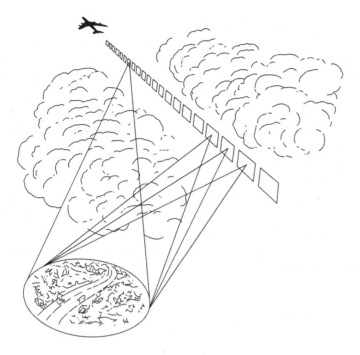

Figure 1.5 A spotlight-mode airborne radar

The typical airborne radar transmits a passband microwave signal consisting of a train of uniform pulses. A moving transmitter illuminates a scene with this waveform. This is illustrated in Figure 1.5, which shows the position of the antenna for each transmitted pulse. The figure shows a spotlight-mode radar in which the antenna or antenna system moves the antenna beam to illuminate a chosen scene. The pulses may be processed individually, in which case a noncoherent pulse train suffices. Such would be a rather simple radar for the detection of fixed or moving objects as described in Chapter 12.

A more advanced radar maintains coherence across the entire pulse train and processes the pulse train echo coherently as a whole. Coherent processing of a received passband signal is the form of processing that employs the carrier phase structure of the waveforms. A coherent system is informally defined as any system for which coherent processing is fundamental to its operation. The extraction of maximum information from a received passband signal requires a waveform that supports coherent processing over long time intervals, and this can lead to the use of sophisticated signal processing.

When the transmitter or receiver is in motion with respect to the object reflecting the signal, coherent processing becomes a potent technique because of the resulting frequency shifts, called *doppler*, in the echo. One system of this kind that is used for imaging is called a *synthetic-aperture radar* because of the heuristic notion of synthesizing a long, fixed antenna by the sequence of positions of a short, moving antenna as suggested by Figure 1.5. A synthetic-aperture imaging system depends on motion.

The motion can be described as a time-varying position. This description is often simplified as a straight line that is specified by an initial position and a constant velocity. Whenever this description suffices, the received electromagnetic signal depends on the scene parameters through the time delay and the doppler frequency shift of that signal.

The received echo signal corresponding to each pulse is converted to a precision optical or digital replica, maintaining both the amplitude modulation and the phase modulation. A history of such received pulses is accumulated, each pulse occurring at a slightly different position along the trajectory of the radar. From this history, an image of the illuminated scene is assembled by coherent processing.

The performance of a waveform for search or imaging is studied with the aid of a two-dimensional function called the *ambiguity function* or the *Woodward function*. The ambiguity function of any pulse $s(t)$ or pulse train $p(t)$ is unique for that pulse or waveform. The ambiguity function of a pulse or waveform is the key to understanding the performance of an imaging or detection radar that uses the pulse or waveform. The ambiguity function is studied in detail in Chapter 10.

Problems

1.1 Show that the gaussian density function

$$p(x) = \frac{1}{\sqrt{2\pi}\sigma} e^{-(x-\bar{x})^2/2\sigma^2}$$

has mean $E[x] = \bar{x}$ and variance $E[(x-\bar{x})^2] = \sigma^2$, and so these constants in the gaussian density function have been aptly named.

1.2 Sketch and label your own diagram of the electromagnetic spectrum on a log scale. Label your diagram with the three systems of units: frequency; wavelength; and frequency. Comment on which unit might be the more appropriate to use in each region of the spectrum.

1.3 Explain how the Bayes formula follows from the definitions of marginals and conditionals.

1.4 Prove that a waveform of the form

$$\widetilde{s}(x, y, z, t) = A(x, y)\cos\left(2\pi f_0(t - z/c) + \theta\right)$$

does not satisfy the wave equation

$$\frac{\partial^2 \widetilde{s}}{\partial x^2} + \frac{\partial^2 \widetilde{s}}{\partial y^2} + \frac{\partial^2 \widetilde{s}}{\partial z^2} = \frac{1}{c^2}\frac{\partial^2 \widetilde{s}}{\partial t^2}$$

unless $A(x,y)$ is a constant, A, and so is independent of x and y. Conclude that a spatially modulated plane wave satisfying the wave equation does not exist.

1.5 A scalar-valued plane wave has the complex baseband representation

$$s(x,y,z) = Ae^{j2\pi f_0(\alpha x + \beta y + \gamma z)/c + \theta},$$

where A is a complex constant and the direction cosines α, β, and γ are constants. Show that, in general, the sum of two scalar-valued plane waves is not a plane wave. When is the sum a plane wave?

1.6 Answer the following:

(a) Is convolution of finite-energy functions commutative? That is, is $f(x) * g(x)$ equal to $g(x) * f(x)$?

(b) Is convolution of finite-energy functions associative? That is, is $\bigl(f(x) * g(x)\bigr) * h(x)$ equal to $f(x) * \bigl(g(x) * h(x)\bigr)$?

1.7 The general form of a bivariate gaussian density function on the vector random variable $x = (x, y)$ is

$$p(x) = \frac{1}{\sqrt{\det(2\pi \Sigma)}} e^{-(x-\bar{x})^\dagger \Sigma^{-1}(x-\bar{x})/2},$$

where

$$\Sigma = \begin{bmatrix} \sigma_x^2 & \rho\sigma_x\sigma_y \\ \rho\sigma_x\sigma_y & \sigma_y^2 \end{bmatrix}$$

and $|\rho| \leq 1$. Find the marginals $p(x)$ and $p(y)$ and find the conditionals $p(x|y)$ and $p(y|x)$.

1.8 The general form of a multivariate gaussian density function on the vector random variable x is

$$p(x) = \frac{1}{\sqrt{\det(2\pi \Sigma)}} e^{-(x-\bar{x})^\dagger \Sigma^{-1}(x-\bar{x})/2},$$

where x is a random vector of length n.

Compute $E[x]$ and $E\bigl[(x - E[x])^2\bigr]$, showing that these are equal to $\bar{x}$ and Σ, respectively. Can we conclude that these two quantities of $p(x)$ are well-named and well-designated as the mean and the covariance matrix?

2 Signals in One Dimension

Waveforms in one dimension are studied in this chapter. Waveforms in two dimensions are studied in Chapter 3. Waveforms in three dimensions are studied in Chapter 9. These waveforms are also called *one-dimensional signals, two-dimensional signals,* and *three-dimensional signals,* respectively.

A waveform that is used by an imaging system to probe an object of interest is usually a function $s(t)$ of the single variable time. The signal may be scattered or reflected in many directions and collected by one or more observation sensors as a set of one or more one-dimensional waveforms. The object being probed is usually a two-dimensional or three-dimensional spatial region of interest on which is defined a two-dimensional signal $s(x,y)$ or a three-dimensional signal $s(x,y,z)$. Thus the usual computational task of image formation consists of estimating an unknown two-dimensional or three-dimensional function when given a measured set of one-dimensional signals.

Every function of finite energy is associated with another function known as its *Fourier transform* which describes a decomposition of the waveform into an infinite continuum of sinusoids called the frequency domain. The Fourier transform constitutes an alternative representation of the function in the frequency domain. The frequency-domain representation of the waveform is often much easier to work with than the original time-domain representation of the waveform. The Fourier transform is an important tool used throughout the book. The Fourier transform in one dimension is studied in this chapter. The Fourier transform in two dimensions is studied in Chapter 3. The Fourier transform in three dimensions is studied in Chapter 9.

2.1 The One-Dimensional Fourier Transform

A real-valued or complex-valued function of time, $s(t)=s_R(t) + js_I(t)$, where $j = \sqrt{-1}$, whose energy

$$E_p = \int_{-\infty}^{\infty} |s(t)|^2 dt$$

is finite is called a *signal* or a *pulse*. The signal $s(t)$ may be called a *waveform* when it is a more elaborate signal. The *support* of $s(t)$ is the closure of the set of all t at which

2.1 The One-Dimensional Fourier Transform

$s(t)$ is nonzero. The pulse $s(t)$ has *bounded support* when the support is contained in a finite interval.

The Fourier transform $S(f)$ of the signal $s(t)$ is defined as

$$S(f) = \int_{-\infty}^{\infty} s(t) e^{-j2\pi ft} dt.$$

The Fourier transform pair will be indicated by the notation

$$s(t) \leftrightarrow S(f).$$

The two-way arrow implies that $s(t)$ determines $S(f)$ and that $S(f)$ determines $s(t)$. The two-way arrow relates a function in the *time domain* to a function in the *frequency domain*.

The alternative functional notation

$$S(f) = \mathcal{F}[s(\cdot)](f)$$

could also be used to denote the Fourier transform. The functional notation

$$s(t) = \mathcal{F}^{-1}[S(\cdot)](t)$$

then denotes the inverse Fourier transform.

The Fourier transform is a linear operation, which means that when $s(t)$ is the linear combination of two pulses given by

$$s(t) = as_1(t) + bs_2(t),$$

then

$$S(f) = aS_1(f) + bS_2(f)$$

for any complex constants, a and b.

A function $S(f)$ that is the Fourier transform of the signal $s(t)$ is not the Fourier transform of any other signal.[1] Consequently, the Fourier transform can be inverted. This is stated in the following theorem, which gives an explicit formula for the inverse Fourier transform.

Theorem 2.1.1 (Inverse Fourier Transform) *The signal $s(t)$ with Fourier transform $S(f)$ satisfies*

$$s(t) = \int_{-\infty}^{\infty} S(f) e^{j2\pi ft} df$$

with equality meaning that the difference between the two sides has zero energy.

---

[1] Technically, there can be violations, known as Gibbs phenomena, on a discrete set of points. Therefore, to make the statement precise, the Fourier transform is defined on the signal space of equivalence classes of square-integrable complex functions. Two functions whose difference has zero energy are regarded as the same element of this complex signal space. In this sense, any two functions with the same Fourier transform are equivalent and herein treated as the same function. One-dimensional signal space of complex functions with finite energy goes by the name $L^2(\mathbb{C})$.

Proof:

Rather than give a rigorous proof, we use the formal symbolism of the impulse function $\delta(t)$ to facilitate an informal derivation. (The impulse function is discussed in Section 2.2.) By definition,

$$S(f) = \int_{-\infty}^{\infty} s(\xi) e^{-j2\pi f \xi} \, d\xi.$$

Therefore

$$\int_{-\infty}^{\infty} S(f) e^{j2\pi ft} df = \int_{-\infty}^{\infty} \left[\int_{-\infty}^{\infty} s(\xi) e^{-j2\pi f \xi} d\xi \right] e^{j2\pi ft} df$$

$$= \int_{-\infty}^{\infty} s(\xi) \left[\int_{-\infty}^{\infty} e^{-j2\pi (\xi-t)f} df \right] d\xi.$$

The inner integral is infinite for $\xi = t$. Otherwise, for $\xi \neq t$, the integral is the integral of a sine wave and integrates to zero. Therefore, the inner integral is the impulse function $\delta(\xi - t)$, and so

$$\int_{-\infty}^{\infty} S(f) e^{j2\pi ft} df = \int_{-\infty}^{\infty} s(\xi) \delta(\xi - t) d\xi$$

$$= s(t).$$

The informal derivation is now complete. ∎

The two properties

$$S(0) = \int_{-\infty}^{\infty} s(t) dt \quad \text{and} \quad s(0) = \int_{-\infty}^{\infty} S(f) df$$

are trivial. Other useful properties are developed in the following theorems.

Theorem 2.1.2 (Scaling Property) *The transform pair*

$$s(t) \leftrightarrow S(f),$$

implies that

$$s(at) \leftrightarrow |a|^{-1} S(f/a).$$

Proof:
The proof consists simply of a change of variables in the defining integral. ∎

Setting $a = -1$ in the theorem leads to the statement that when

$$s(t) \leftrightarrow S(f)$$

is a Fourier transform pair, then

$$s(-t) \leftrightarrow S(-f)$$

is also a Fourier transform pair.

2.1 The One-Dimensional Fourier Transform

The next several theorems develop a general *shift property*, which is known as the *translation property* (or the *delay property*) when used to shift the time origin and as the *modulation property* when used to shift the frequency origin

Theorem 2.1.3 (Translation Property) *When the pulse $s(t)$ has the Fourier transform $S(f)$, the translated pulse $s(t - t_0)$ has the Fourier transform $S(f)e^{-j2\pi t_0 f}$.*

Proof:
Use the change of variable $\tau = t - t_0$ to write

$$\int_{-\infty}^{\infty} s(t - t_0)e^{-j2\pi ft}dt = \int_{-\infty}^{\infty} s(\tau)e^{-j2\pi f(\tau + t_0)}d\tau$$

$$= S(f)e^{-j2\pi t_0 f}$$

as asserted. ∎

Theorem 2.1.4 (Complex Modulation Property) *When the pulse $s(t)$ has the Fourier transform $S(f)$, the pulse $s(t)e^{-j2\pi f_0 t}$ has the Fourier transform $S(f + f_0)$.*

Proof:

$$\int_{-\infty}^{\infty} s(t)e^{-j2\pi f_0 t}e^{-j2\pi ft}dt = \int_{-\infty}^{\infty} s(t)e^{-j2\pi(f + f_0)t}dt$$

$$= S(f + f_0)$$

∎

Corollary 2.1.5 (Modulation Property) *When the pulse $s(t)$ has the Fourier transform $S(f)$, the pulse $s(t)\cos 2\pi f_0 t$ has the Fourier transform $\frac{1}{2}[S(f + f_0) + S(f - f_0)]$ and the pulse $s(t)\sin 2\pi f_0 t$ has the Fourier transform $\frac{1}{2}j[S(f + f_0) - S(f - f_0)]$.*

Proof:
The proof follows immediately from the theorem and the relations

$$\cos 2\pi f_0 t = \tfrac{1}{2}\left(e^{-j2\pi f_0 t} + e^{j2\pi f_0 t}\right),$$
$$\sin 2\pi f_0 t = \tfrac{1}{2}j\left(e^{-j2\pi f_0 t} - e^{j2\pi f_0 t}\right)$$

as asserted. ∎

Theorem 2.1.6 (Energy Relation) *When the pulse $s(t)$ has the Fourier transform $S(f)$, the energy E_p of the pulse $s(t)$ satisfies*

$$E_p = \int_{-\infty}^{\infty} |s(t)|^2 dt = \int_{-\infty}^{\infty} |S(f)|^2 df.$$

Proof:

$$\int_{-\infty}^{\infty} |s(t)|^2 dt = \int_{-\infty}^{\infty} s(t)\left[\int_{-\infty}^{\infty} S(f)e^{j2\pi ft}df\right]^* dt$$

$$= \int_{-\infty}^{\infty} S^*(f)\left[\int_{-\infty}^{\infty} s(t)e^{-j2\pi ft}dt\right] df$$

$$= \int_{-\infty}^{\infty} |S(f)|^2 df$$

as asserted. ∎

The energy theorem is a special case of the following theorem.

Theorem 2.1.7 (Parseval's Formula) *Two pulses $s_1(t)$ and $s_2(t)$ with Fourier transforms $S_1(f)$ and $S_2(f)$, respectively, satisfy the equality*

$$\int_{-\infty}^{\infty} s_1(t) s_2^*(t) dt = \int_{-\infty}^{\infty} S_1(f) S_2^*(f) df.$$

Proof:
The proof is similar to the proof of Theorem 2.1.6. ∎

A linear filter with the impulse response $g(t)$ and the input signal $p(t)$ has the output $s(t)$ given by the *convolution*

$$s(t) = \int_{-\infty}^{\infty} p(\xi) g(t-\xi) d\xi.$$

Because the multiplication of two functions is simpler than the convolution of two functions, it is often easier to treat a filtering problem in the frequency domain. The next theorem is a fundamental theorem for the study of the effect of a linear filter on a signal.

Theorem 2.1.8 (Convolution Theorem) *The functions $p(t)$, $g(t)$, and $s(t)$ with the Fourier transforms $P(f)$, $G(f)$, and $S(f)$, respectively, are related by the convolution*

$$s(t) = \int_{-\infty}^{\infty} p(\xi) g(t-\xi) d\xi$$

if and only if

$$S(f) = P(f) G(f).$$

Proof:

$$s(t) = \int_{-\infty}^{\infty} g(t-\xi) \left[\int_{-\infty}^{\infty} P(f) e^{j2\pi f \xi} df \right] d\xi$$

$$= \int_{-\infty}^{\infty} P(f) e^{j2\pi ft} \left[\int_{-\infty}^{\infty} g(t-\xi) e^{-j2\pi f(t-\xi)} d\xi \right] df.$$

Let $\eta = t - \xi$. Then

$$s(t) = \int_{-\infty}^{\infty} P(f) e^{j2\pi ft} \left[\int_{-\infty}^{\infty} g(\eta) e^{-j2\pi f \eta} d\eta \right] df$$

$$= \int_{-\infty}^{\infty} P(f) G(f) e^{j2\pi ft} df.$$

Consequently, by the invertibility of the Fourier transform,

$$S(f) = P(f) G(f).$$

The theorem is proved in the reverse direction by tracing the argument in the reverse direction. ∎

Closely related to the convolution of two signals is their correlation. The *correlation function* of the complex signal $p(t)$ with the complex signal $g(t)$ is defined by

$$s(t) = \int_{-\infty}^{\infty} p(\xi)g^*(\xi - t)\,d\xi$$

$$= \int_{-\infty}^{\infty} p(\xi + t)g^*(\xi)\,d\xi.$$

The correlation function of $p(t)$ with $g(t)$ is the convolution $p(t) * g^*(-t)$. Because the definition is asymmetric, this is not the same as the correlation of $g(t)$ with $p(t)$. The *correlation* between $p(t)$ and $g(t)$ is the correlation function at $t = 0$. The *autocorrelation function* of $p(t)$ is the correlation function of $p(t)$ with itself.

Corollary 2.1.9 (Correlation Theorem) *The function $s(t)$ is the correlation of $p(t)$ with $g(t)$ if and only if their Fourier transforms $S(f)$, $P(f)$, and $G(f)$ satisfy*

$$S(f) = P(f)G^*(f).$$

Proof:
Replace $G(f)$ by $G^*(f)$ in the theorem, noting that $g^*(-t) \leftrightarrow G^*(f)$. ∎

An immediate consequence of the correlation theorem is that the autocorrelation function $s(t) * s^*(-t)$ is the convolution of $s(t)$ with $s^*(-t)$ and so has the transform $|S(f)|^2$. This is written

$$s(t) * s^*(-t) \leftrightarrow |S(f)|^2.$$

The dual of this Fourier transform pair is

$$|s(t)|^2 \leftrightarrow S(f) * S^*(-f).$$

Theorem 2.1.10 (Differentiation Property) *The derivative ds/dt of the pulse $s(t)$ with Fourier transform $S(f)$ has the Fourier transform $j2\pi f\, S(f)$.*

Proof:
Interchange differentiation and integration. Then

$$\frac{ds(t)}{dt} = \frac{d}{dt}\int_{-\infty}^{\infty} S(f)e^{j2\pi ft}\,df$$

$$= \int_{-\infty}^{\infty} [j2\pi f\, S(f)]e^{j2\pi ft}\,df,$$

as was to be proved. ∎

2.2 Transforms of Some Useful Functions

A list of some useful one-dimensional Fourier transform pairs is given in Table 2.1. Several of these pairs are developed in this section. Others may arise later in the book.

Table 2.1 A table of one-dimensional Fourier transform pairs

$s(t)$	$S(f)$
$\text{rect}(t)$	$\text{sinc}(f)$
$\text{sinc}(t)$	$\text{rect}(f)$
$\delta(t)$	1
1	$\delta(f)$
$\text{comb}(t)$	$\text{comb}(f)$
$\text{comb}_N(t)$	$\text{dirc}_N(f)$
$e^{-\pi t^2}$	$e^{-\pi f^2}$
$e^{j\pi t^2}$	$\frac{1+j}{\sqrt{2}} e^{-j\pi f^2}$
$\cos 2\pi t$	$\frac{1}{2}\delta(f+1) + \frac{1}{2}\delta(f-1)$
$\sin 2\pi t$	$\frac{j}{2}\delta(f+1) - \frac{j}{2}\delta(f-1)$
$e^{j2\pi t}$	$\delta(f-1)$
$e^{-|t|}$	$\frac{2}{1+(2\pi f)^2}$
$J_0(2\pi t)$	$\frac{\text{rect}(f/2)}{\pi(1-f^2)^{\frac{1}{2}}}$
$\text{jinc}(t)$	$\sqrt{1-4f^2}\,\text{rect}(f)$

An elementary pulse is the *rectangle function* or *rectangular pulse*, defined as

$$\text{rect}(t) = \begin{cases} 1 & \text{for} \quad |t| \leq 1/2, \\ 0 & \text{for} \quad |t| > 1/2. \end{cases}$$

The Fourier transform of a rectangular pulse is readily evaluated. Let $s(t) = \text{rect}(t)$. Then

$$S(f) = \int_{-1/2}^{1/2} e^{-j2\pi ft} dt$$
$$= \frac{1}{-j2\pi f}\left[e^{-j\pi f} - e^{j\pi f}\right]$$
$$= \text{sinc}(f),$$

where the *sinc function* or the *sinc pulse* is defined as

$$\text{sinc}(x) = \frac{\sin \pi x}{\pi x}.$$

The magnitude of $\text{sinc}(x)$ satisfies the bound $|\text{sinc}(x)| \leq 1/\pi x$. The function $\text{sinc}(x)$ has all of its zeros at the nonzero integer values of x. Moreover, it follows from

2.2 Transforms of Some Useful Functions

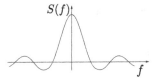

Figure 2.1 Fourier transform of a rectangular pulse

l'Hôpital's rule that sinc(0) = 1. This conclusion also follows from the fact that $S(0)$ is the integral of $s(t)$. The sinc function is shown in Figure 2.1. The central part of the sinc function between two zeros is called the *main lobe*. The peaks corresponding to the secondary maxima are called *sidelobes*.

The Fourier transform of a rectangular pulse of timewidth $T > 0$ is obtained easily by using the scaling property. The Fourier transform pair after scaling is written concisely as

$$\text{rect}\left(\frac{t}{T}\right) \leftrightarrow T\text{sinc}(Tf).$$

For a frequency interval of bandwidth $W > 0$, it follows immediately that

$$W\text{sinc}(Wt) \leftrightarrow \text{rect}(f/W)$$

by the duality property of the Fourier transform.

Impulse Function

The *impulse function* (or the (Dirac) *delta function*) $\delta(t)$ is a useful function[2] defined as the limit of a series of unit-area rectangular functions of shrinking width. Let

$$\text{rect}(t/T) \leftrightarrow T\text{sinc}(fT).$$

The pulse $T^{-1}\text{rect}(t/T)$ has a unit area and becomes infinitely large and infinitely thin as T goes to zero. The impulse function $\delta(t)$ is defined formally as the limit of this sequence of rectangular pulses as T goes to zero,

$$\delta(t) = \lim_{T \to 0} \frac{1}{T}\text{rect}\left(\frac{t}{T}\right).$$

The Fourier transform of $\delta(t)$ does not exist in the sense of the original definition of the Fourier transform, but it can be defined formally as the limit as T goes to zero

---

[2] The impulse function is an example of a *generalized function*. As such, the impulse function is characterized by the "sifting property" $\int_{-\infty}^{\infty} v(t)\delta(t - \tau)dt = v(\tau)$. One of its properties is that $\delta(t/T) = T\delta(t)$ for $T > 0$.

A generalized function is not an ordinary function, but it is a useful and provides immediate insight into many situations. The notion of a generalized function, such as the impulse, does have a formal mathematical theory. This formal theory is not provided herein. Generalized functions should be used with care.

of the sequence of Fourier transforms of increasingly narrow rectangular pulses. The *generalized* Fourier transform of $\delta(t)$ is

$$S(f) = \lim_{T \to 0} \operatorname{sinc} fT$$
$$= 1.$$

This generalized Fourier transform pair and its dual are written concisely as

$$\delta(t) \leftrightarrow 1,$$
$$1 \leftrightarrow \delta(f).$$

These two Fourier transform pairs are quite useful and so are appended to the list of Fourier transform pairs.

The rectangular pulse has finite support. In contrast, the *gaussian pulse*

$$s(t) = e^{-\pi t^2}$$

does not have finite support because it has tails that go on forever. The Fourier transform of the gaussian pulse is

$$S(f) = \int_{-\infty}^{\infty} e^{-\pi t^2} e^{-j2\pi ft} dt.$$

Because $s(t)$ is an even function, this becomes

$$S(f) = \int_{-\infty}^{\infty} e^{-\pi t^2} \cos 2\pi ft \, dt$$
$$= e^{-\pi f^2},$$

where the last line follows from consulting a table of definite integrals. Thus we have the Fourier transform pair

$$e^{-\pi t^2} \leftrightarrow e^{-\pi f^2}.$$

More generally,

$$e^{-at^2} \leftrightarrow \sqrt{\frac{\pi}{a}} e^{-\pi^2 f^2/a},$$

which follows from the scaling property of the Fourier transform.

A pulse $s(t)$ can also have complex values. An important complex-valued pulse is the *chirp pulse*. The chirp pulse is also called the *quadratic-phase pulse*. It is

$$\operatorname{chrp}(t) = e^{j\pi t^2}.$$

The chirp pulse has infinite energy, so the Fourier transform of the chirp pulse is not defined.

Define the *truncated* chirp pulse as

$$s_T(t) = e^{j\pi t^2} \operatorname{rect}\left(\frac{t}{T}\right).$$

The truncated chirp pulse does have finite energy. The Fourier transform

$$S_T(f) = \int_{-T/2}^{T/2} e^{j\pi t^2} e^{-j2\pi ft} dt$$

is defined for every positive value of T. Hence, it is natural to generalize the notion of the Fourier transform by defining the Fourier transform of the chirp pulse simply by letting T go to infinity.

$$S(f) = \int_{-\infty}^{\infty} e^{j\pi t^2} e^{-j2\pi ft} dt.$$

To evaluate the integral, complete the square in the exponent

$$S(f) = e^{-j\pi f^2} \int_{-\infty}^{\infty} e^{j\pi (t-f)^2} dt.$$

By a change of variables in the integral, and using Euler's formula, this becomes

$$S(f) = e^{-j\pi f^2} \left[\int_{-\infty}^{\infty} \cos \pi t^2 dt + j \int_{-\infty}^{\infty} \sin \pi t^2 dt \right].$$

The integrals are now standard tabulated integrals and each is equal to $1/\sqrt{2}$. Therefore the generalized Fourier transform of the chirp pulse is

$$S(f) = \frac{1+j}{\sqrt{2}} e^{-j\pi f^2},$$

which also has infinite energy.

Notice that in both the derivation of the Fourier transform of a chirp pulse and the derivation of the Fourier transform of an impulse, the Fourier transform of a limit of a sequence of functions is defined as the limit of the sequence of Fourier transforms. In this manner, the set of Fourier transform pairs can be enlarged in a natural and satisfying way to include these generalized Fourier transform pairs. We freely enlarge the set of Fourier transform pairs in this way, duly noting that this limiting procedure needs to be justified in each case. This book is not the place to give the formal mathematical setting to make these arguments precise.

2.3 The Dirichlet Functions

One way to construct a complicated waveform is by the repetition of a simple waveform. This kind of construction arises in the study of many engineering subjects, such as antenna arrays or pulse trains, and also in digital signal processing. Common mathematical phenomena, underlying all of these constructions, can be understood in terms of the solutions of some simple Fourier transform constructions.

Let $s(t)$ be any pulse, and let $S(f)$ be its Fourier transform. Consider the pulse doublet

$$s_2(t) = s(t + T/2) + s(t - T/2).$$

2 Signals in One Dimension

The doublet has the Fourier transform

$$S_2(f) = S(f)e^{j\pi fT} + S(f)e^{-j\pi fT}$$
$$= 2S(f)\cos \pi f T.$$

In turn, the pulse quadruplet can be viewed as a doublet of doublets,

$$s_4(t) = s_2(t+T) + s_2(t-T).$$

This has the Fourier transform

$$S_4(f) = 2S_2(f)\cos 2\pi f T$$
$$= 4S(f)\cos \pi f T \cos 2\pi f T.$$

Continuing, the pulse octuplet can be viewed as a doublet of quadruplets,

$$s_8(t) = s_4(t+2T) + s_4(t-2T).$$

This has the Fourier transform

$$S_8(f) = 2S_4(f)\cos 4\pi f T$$
$$= 4S(f)\cos \pi f T \cos 2\pi f T \cos 4\pi f T.$$

To collapse the right side, recall that $\sin 2x = 2\cos x \sin x$, from which it follows that $\sin 8x = \cos 4x \cos 2x \cos x \sin x$. Therefore

$$S_8(f) = S(f)\frac{\sin 8\pi f T}{\sin \pi f T}.$$

This doubling procedure can be repeated multiple times to obtain the Fourier transform whenever the number of pulses in a pulse train is a power of two.

For a uniform pulse train of arbitrary length N centered at the airport, which is given by

$$p(t) = \sum_{\ell=0}^{N-1} s(t - \ell T + \tfrac{1}{2}(N-1)T),$$

a more direct derivation is needed.

Theorem 2.3.1 *A uniform pulse train consisting of N equispaced, identical pulses centered at the origin has the Fourier transform*

$$P(f) = S(f)\frac{\sin N\pi f T}{\sin \pi f T},$$

where T is the pulse spacing and $S(f)$ is the Fourier transform of the individual pulse $s(t)$.

2.3 The Dirichlet Functions

Proof:
Use the translation property on each pulse to write

$$P(f) = \sum_{\ell=0}^{N-1} S(f) e^{-j2\pi f \ell T + j\pi f(N-1)T}$$

$$= S(f) e^{j\pi f(N-1)T} \sum_{\ell=0}^{N-1} e^{-j2\pi f \ell T}.$$

Now use the relationship

$$\sum_{\ell=0}^{N-1} x^\ell = \begin{cases} \frac{1-x^N}{1-x} & x \neq 1 \\ N & x = 1 \end{cases}$$

to obtain the identity

$$\sum_{\ell=0}^{N-1} e^{-j2\pi f \ell T} = \frac{1 - e^{-j2\pi f TN}}{1 - e^{-j2\pi f T}}$$

$$= \left[\frac{e^{j\pi f TN} - e^{-j\pi f TN}}{e^{j\pi f T} - e^{-j\pi f T}} \right] e^{-j\pi f(N-1)T}.$$

Insert this into the expression for $P(f)$. The two phase terms involving $(N-1)$ cancel. Finally, divide the numerator and denominator each by $2j$ to give $\sin(N\pi f T)$ in the numerator and $\sin(\pi f T)$ in the denominator. The result is

$$P(f) = S(f) \frac{\sin N\pi f T}{\sin \pi f T},$$

as was to be proved. ∎

The Fourier transform of a pulse train is the product of two terms. One term $S(f)$ is the Fourier transform of a single pulse. The other term is due to the array. Because the latter term appears frequently, it deserves its own name. For each integer N, define the *dirichlet function* as

$$\text{dirc}_N(x) = \frac{\sin \pi Nx}{\sin \pi x}.$$

The dirichlet functions are illustrated in Figure 2.2.

It is easily shown using l'Hôpital's rule that $\text{dirc}_N(0) = N$. The first zero of $\text{dirc}_N(x)$ occurs at $x = 1/N$. The portion of the dirichlet function between $x = \pm 1/N$ is dubbed the *main lobe* of the dirichlet function. When x is small, the denominator can be approximated by using the small-angle approximation for the sine function. Then

$$\text{dirc}_N(x) \approx \frac{\sin \pi Nx}{\pi x}$$

$$\approx N \text{sinc} Nx.$$

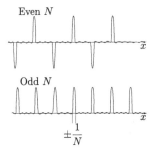

Figure 2.2 Illustrating the dirichlet functions

Thus, near the origin, the dirichlet function looks like a sinc function. (For this reason the dirichlet function has been informally called the *periodic sinc*.) The amplitude at the origin is N, and the first zeros are at $\pm 1/N$.

Next notice that for any integer k,

$$\mathrm{dirc}_N(x+k) = \frac{\sin \pi N(x+k)}{\sin \pi (x+k)}$$
$$= \pm \mathrm{dirc}_N(x),$$

which implies that for any integer k

$$\mathrm{dirc}_N(k) = \pm N.$$

Moreover, when N is an odd integer, the dirichlet function satisfies

$$\mathrm{dirc}_N(x+k) = \mathrm{dirc}_N(x),$$

When N is an even integer

$$\mathrm{dirc}_N(x+k) = (-1)^k \mathrm{dirc}_N(x).$$

In particular, the main lobe of the dirichlet function appears periodically replicated. The periodic copies of the main lobe are called *grating lobes*. For even N, there is a sign reversal on alternate grating lobes.

The statement of Theorem 2.3.1 can now be rewritten as

$$P(f) = S(f)\mathrm{dirc}_N(fT).$$

Thus $P(f)$ displays the grating lobes of the dirichlet function modulated by the complex function $S(f)$. The modulation theorem states that the grating lobes of the dirichlet function can be translated along the frequency axis by phase modulating the pulses. Thus, a uniform, linearly phase-modulated pulse train is a pulse train of the form

$$p(t) = \sum_{\ell=0}^{N-1} s(t - \ell T + (N-1)T/2) e^{j2\pi \ell f_0 T}.$$

The ℓth pulse has phase $\ell f_0 T$.

2.3 The Dirichlet Functions

Corollary 2.3.2 *A uniform, linearly phase-modulated pulse train[3] has the Fourier transform*

$$P(f) = S(f)\text{dirc}_N(f - f_0)T.$$

Proof:
The delay theorem immediately gives

$$P(f) = S(f)e^{j\pi f(N-1)T} \sum_{\ell=0}^{N-1} e^{-j2\pi(f-f_0)\ell T},$$

which reduces to the statement of the theorem. ∎

When $s(t)$ is an impulse $\delta(t)$, the pulse train $p(t)$ is a finite train of N impulses. The finite and the infinite train of impulses are given names as follows.

Definition 2.3.3 *The comb function is given by*

$$\text{comb}(t) = \sum_{\ell=-\infty}^{\infty} \delta(t - \ell).$$

The finite comb function is given by

$$\text{comb}_N(t) = \sum_{n=0}^{N-1} \delta\big(t - \ell + \tfrac{1}{2}(N-1)\big),$$

for any positive integer N. ∎

The Fourier transform pair for the finite comb function

$$\text{comb}_N(t) \leftrightarrow \text{dirc}_N(f)$$

follows from Theorem 2.3.1. Because the grating lobes of $\text{dirc}_N(f)$ become larger and thinner as N goes to infinity, the grating lobes become more like impulses. For an odd N that is large, the grating lobes of the dirichlet function resemble a comb of impulses. This suggests the Fourier transform pair

$$\text{comb}(t) \leftrightarrow \text{comb}(f).$$

To develop this transform pair more directly, define

$$\text{comb}(t) = \lim_{\tau \to 0} \lim_{N \to \infty} \sum_{\ell=-N}^{N} \frac{1}{\tau}\text{rect}\Big(\frac{t-\ell}{\tau}\Big).$$

The Fourier transform of the right side is given by

$$\lim_{\tau \to 0} \lim_{N \to \infty} \text{sinc}(f\tau)\text{dirc}_{2N+1}(f) = \text{comb}(f),$$

---
[3] Not to be confused with a pulse train of linearly phase-modulated pulses.

which gives the Fourier transform pair already mentioned. Of course, comb(*t*) is not an ordinary function. It is a generalized function and is interpreted by its role under an integral sign. Then

$$\int_{-\infty}^{\infty} s(t)\text{comb}(t)\,dt = \sum_{\ell=-\infty}^{\infty} s(\ell),$$

for any $s(t)$ of finite energy.

Applying the scaling property to the Fourier transform pair

$$\text{comb}(t) \leftrightarrow \text{comb}(f)$$

gives the pair

$$\text{comb}\left(\frac{t}{T}\right) \leftrightarrow T\,\text{comb}(Tf)$$

for positive values of T. This is written explicitly as

$$\sum_{\ell=-\infty}^{\infty} \delta\left(\frac{t}{T} - \ell\right) \leftrightarrow T \sum_{\ell=-\infty}^{\infty} \delta(Tf - \ell).$$

By using the scaling properties of the impulse function, this is rewritten in the more convenient form

$$\sum_{\ell=-\infty}^{\infty} \delta(t - \ell T) = \frac{1}{T} \sum_{\ell=-\infty}^{\infty} \delta\left(f - \frac{\ell}{T}\right).$$

2.4 Passband Signals and Passband Filters

A baseband signal $s(t)$ is a signal whose Fourier transform $S(f)$ is concentrated near the frequency origin. The Fourier transform $S(f)$ of a baseband signal can be nonzero only in a finite interval enclosing the frequency origin, as illustrated in Figure 2.3.

A signal $\tilde{s}(t)$ is called a passband signal when its Fourier transform is zero in some interval including the frequency origin and is zero for sufficiently large f, as illustrated in Figure 2.4. From the modulation theorem for the Fourier transform, it is easy to see that one can construct a passband signal $\tilde{s}(t)$ by multiplying the baseband signal $s(t)$ by $\cos 2\pi f_0 t$ or by $\sin 2\pi f_0 t$ for a sufficiently large f_0, provided $S(f)$ has a bounded

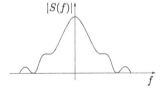

Figure 2.3 Spectrum of a baseband signal

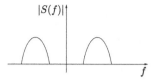

Figure 2.4 Spectrum of a passband signal

support. Moreover, any two baseband signals $s_R(t)$ and $s_I(t)$ whose Fourier transforms $S_R(f)$ and $S_I(f)$ have bounded support can be used to define the passband signal[4]

$$\tilde{s}(t) = s_R(t)\cos 2\pi f_0 t - s_I(t)\sin 2\pi f_0 t,$$

for some sufficiently large f_0. The Fourier transform of $\tilde{s}(t)$ can be formed by using the modulation theorem to write

$$s_R(t)\cos 2\pi f_0 t \leftrightarrow \tfrac{1}{2}[S_R(f+f_0) + S_R(f-f_0)],$$
$$s_I(t)\sin 2\pi f_0 t \leftrightarrow \tfrac{1}{2j}[S_I(f+f_0) - S_I(f-f_0)],$$

so that

$$\tilde{S}(f) = \tfrac{1}{2}[S_R(f-f_0) - jS_I(f-f_0)] + \tfrac{1}{2}[S_R(f+f_0) + jS_I(f+f_0)].$$

This construction of a passband signal from two baseband signals is the most general construction of this kind in the sense that any passband waveform could have been constructed in this way. For any passband waveform $\tilde{s}(t)$, the positive and negative frequency segments of $\tilde{S}(f)$ can be separated and used to solve for $S_R(f)$ and $S_I(f)$, which leads to the desired decomposition. Thus a passband signal $\tilde{s}(t)$ can always be decomposed into $s_R(t)$ and $s_I(t)$ of the stated form. These are called the *in-phase modulation component* and the *quadrature modulation component* of $\tilde{s}(t)$, respectively. Because the modulation components vary much more slowly than does $\tilde{s}(t)$ itself, it may be simpler to work with the modulation components when processing such signals.

The modulation components of the passband waveform $\tilde{s}(t)$ consist of a pair of real waveforms that, taken together, can be used to form a single complex waveform,

$$s(t) = s_R(t) + js_I(t),$$

called the *complex baseband representation* of $\tilde{s}(t)$. The complex baseband representation is equivalent to the passband representation $\tilde{s}(t)$. The expression

$$\tilde{s}(t) = \text{Re}[s(t)e^{j2\pi f_0 t}]$$

recovers the passband waveform.

A *passband filter* is a filter whose impulse response is a passband pulse. A passband waveform has the form

$$\tilde{g}(t) = g_R(t)\cos 2\pi f_0 t - g_I(t)\sin 2\pi f_0 t,$$

---

[4] Now using the sign convention of modulation theory.

which has the complex baseband representation given by

$$g(t) = g_R(t) + jg_I(t).$$

Theorem 2.4.1 *A passband filter, $\widetilde{g}(t)$, with a passband input, $\widetilde{s}(t)$, has a passband output, $\widetilde{r}(t)$, whose corresponding complex baseband representation $r(t)$ is the output of the corresponding complex baseband filter $g(t)$ with the corresponding complex baseband input $s(t)$.*

Proof:
This is a straightforward manipulation in the Fourier transform domain. ∎

Theorem 2.4.2 *When the phase angle of the carrier of a passband signal at the input to a passband filter is changed by θ, the phase angle of the carrier of the signal at the output of the filter is changed by θ as well, and otherwise the output signal is unchanged.*

Proof:
The passband signal with a phase offset in the carrier is

$$\widetilde{s}'(t) = s_R(t)\cos(2\pi f_0 t + \theta) - s_I(t)\sin(2\pi f_0 t + \theta).$$

Theorem 2.4.1 tells us that the passband pulse $\widetilde{s}'(t)$ passed through the passband filter $\widetilde{g}(t)$ can be described by representing the passband signal and the passband filter at complex baseband as $s'(t)$ and $g(t)$. At complex baseband, the phase angle can be factored out to obtain the convenient form

$$s'(t) = [s_R(t) + js_I(t)]e^{j\theta} = s(t)e^{j\theta}.$$

Then

$$s'(t) * g(t) = [s(t) * g(t)]e^{j\theta}.$$

We conclude that the passband signal satisfies

$$\widetilde{s}'(t) * \widetilde{g}(t) = [s(t) * g(t)]_R \cos(2\pi f_0 t + \theta) - [s(t) * g(t)]_I \sin(2\pi f_0 t + \theta),$$

as was to be proved. ∎

The proof of the theorem illustrates the advantage of using the complex baseband representation. The same conclusion can be reached from the passband representation by using trigonometric identities.

2.5 Signal Space

The set of all complex finite energy functions of the single variable t comprises a space of functions called *signal space*. However, this summary definition must be elaborated.

2.5 Signal Space

In defining signal space, it is convenient and conventional to regard any two functions $s(t)$ and $s'(t)$ that satisfy

$$\int_{-\infty}^{\infty} |s(t) - s'(t)|^2 dt = 0$$

to be equivalent. The notion of equivalence partitions the set of finite energy functions into equivalence classes.[5] Two elements in the same equivalence class are declared to be the same element of signal space.

This means that two functions whose difference has zero energy are equivalent and are considered to be the same function. With this notion of equivalence classes, every element $s(t)$ of signal space has a unique Fourier transform, $S(f)$. Signal space has the properties that the sum of two elements of the space is again an element of the space, and a scalar multiple of any element of the space is again an element of the space. A set with these two properties is called a *vector space*. Signal space is a vector space, and thus it has all of the general properties that hold in every vector space. Signal space also has the strong property of orthogonality that is defined in some vector spaces. These are the vector spaces that have an inner product as defined next. Signal space[6] is such a vector space.

The *inner product* $\langle r(t), s(t) \rangle$ of the two signals $r(t)$ and $s(t)$ of complex signal space is

$$\langle r(t), s(t) \rangle = \int_{-\infty}^{\infty} r(t)s^*(t)dt.$$

Two elements $s(t)$ and $r(t)$ of the signal space whose inner product $\langle r(t), s(t) \rangle$ is zero, given by

$$\int_{-\infty}^{\infty} r(t)s^*(t)dt = 0,$$

are called *orthogonal* signals.

Definition 2.5.1 *The norm of $s(t)$ is defined as $\|s(t)\| = \langle s(t), s^*(t) \rangle^{\frac{1}{2}}$. Thus $\|s(t)\|^2 = E_p$, where E_p is the energy in the pulse $s(t)$.*

Theorem 2.5.2 *Given any two signals, $r(t)$ and $s(t)$, the signal $r(t)$ has a unique decomposition of the form*

$$r(t) = \alpha s(t) + s^{\perp}(t),$$

where α is a constant and $s^{\perp}(t)$ is orthogonal to $s(t)$.

---

[5] To validate this definition, one must show that equivalence is a transitive property and that a sum and a scalar multiple are uniquely defined properties of the equivalence classes forming signal space. Problem 2.19 of this chapter asks for a proof of these properties. The remark following Corollary 2.5.5 allows a formal statement.

[6] Signal space over the real numbers $\mathbb{R}$ is denoted $L^2(\mathbb{R})$. Signal space over the complex numbers $\mathbb{C}$ is denoted $L^2(\mathbb{C})$.

Proof:

The theorem is trivial when $s(t) = 0$. For any other $s(t)$, let

$$\alpha = \frac{\langle r(t), s^*(t)\rangle}{\langle s(t), s^*(t)\rangle}$$

and $s^\perp(t) = r(t) - \alpha s(t)$. This means that $r(t) = \alpha s(t) + s^\perp(t)$ as required by the statement of the theorem. It remains to show that $s^\perp(t)$ is orthogonal to $s(t)$ and that the decomposition is unique.

Form the inner product of $s^\perp(t)$ with $s(t)$:

$$\langle s^\perp(t), s^*(t)\rangle = \langle r(t), s^*(t)\rangle - \alpha\langle s(t), s^*(t)\rangle$$
$$= \langle r(t), s^*(t)\rangle - \langle r(t), s^*(t)\rangle = 0,$$

which proves orthogonality.

The uniqueness of the decomposition is shown as follows. For any α_1 and $s_1^\perp(t)$ for which

$$r(t) = \alpha_1 s(t) + s_1^\perp(t),$$

and for which $s_1^\perp(t)$ is orthogonal to $s(t)$, subtract this expression from the expression of the theorem to obtain

$$0 = (\alpha - \alpha_1)s(t) + s^\perp(t) - s_1^\perp(t).$$

Then take the inner product of this expression with $s^*(t)$ to yield

$$0 = (\alpha - \alpha_1)\langle s(t), s^*(t)\rangle + 0,$$

which means that $\alpha_1 = \alpha$. This, in turn, requires that $s_1^\perp(t) = s^\perp(t)$. Therefore the expression of the theorem is unique, as was to be proved. ∎

The following theorem gives the important *Schwarz inequality*[7] in signal space.

Theorem 2.5.3 (Schwarz Inequality) *Let $r(t)$ and $s(t)$ be finite energy pulses, real valued or complex valued. Then*

$$\left|\int_{-\infty}^{\infty} r(t)s^*(t)dt\right|^2 \leq \int_{-\infty}^{\infty} |r(t)|^2 dt \int_{-\infty}^{\infty} |s(t)|^2 dt$$

with equality if and only if $r(t)$ is a constant, real or complex, multiple of $s(t)$.

Proof:

When $s(t) = 0$, the statement is immediate. Otherwise, $\langle s(t), s^*(t)\rangle \neq 0$. Referring to Theorem 2.5.2, let $r(t)$ be decomposed as

$$r(t) = \alpha s(t) + s^\perp(t),$$

---

[7] The Schwarz inequality for two vectors x and y in $\mathbb{R}^n$ is $|x|^2|y|^2 \geq |x \cdot y|^2$. The Schwarz inequality for the expectation operator applied to random variables X and Y is $E[X^2]E[Y^2] \geq (E[XY])^2$.

where the function $s^\perp(t)$ is orthogonal to $s(t)$, and
$$\alpha = \frac{\langle r(t), s^*(t)\rangle}{\langle s(t), s^*(t)\rangle}.$$

Then
$$\begin{aligned}\langle r(t), r^*(t)\rangle &= \langle \alpha s(t) + s^\perp(t), \alpha s(t) + s^\perp(t)\rangle \\ &= \alpha^2 \langle s(t), s^*(t)\rangle + \langle s^\perp(t), s^{*\perp}(t)\rangle \\ &\geq \alpha^2 \langle s(t), s^*(t)\rangle \\ &= \frac{\langle r(t), s^*(t)\rangle^2}{\langle s(t), s^*(t)\rangle}.\end{aligned}$$

Clearing the denominator gives
$$\langle r(t), s^*(t)\rangle^2 \leq \langle r(t), r^*(t)\rangle \langle s(t), s^*(t)\rangle,$$
which is the statement of the theorem. ∎

The correlation between $r(t)$ and $s(t)$ is $\int_{-\infty}^{\infty} r(t) s^*(t) dt$. The *correlation coefficient* γ between the two signals $r(t)$ and $s(t)$ is defined as
$$\gamma = \frac{\int_{-\infty}^{\infty} r(t) s^*(t) dt}{\left[\int_{-\infty}^{\infty} |r(t)|^2 dt \int_{-\infty}^{\infty} |s(t)|^2 dt\right]^{1/2}}.$$

An immediate consequence of the Schwarz inequality is that the magnitude $|\gamma|$ of the correlation coefficient is not larger than one.

Theorem 2.5.4 (Triangle Inequality) *Given any two complex signals, $s(t)$ and $r(t)$,*
$$\|s(t) + r(t)\| \leq \|s(t)\| + \|r(t)\|,$$
$$\|s(t) - r(t)\| \leq \|s(t)\| + \|r(t)\|.$$

Proof:
Using the Schwarz inequality and $\operatorname{Re}[z] \leq |z|$, write
$$\begin{aligned}\|s(t) + r(t)\|^2 &= \langle s(t) + r(t), s^*(t) + r^*(t)\rangle \\ &= \langle s(t), s^*(t)\rangle + \langle r(t), r^*(t)\rangle + 2\operatorname{Re}\left[\langle s(t), r^*(t)\rangle\right] \\ &\leq \|s(t)\|^2 + \|r(t)\|^2 + 2\|s(t)\|\|r(t)\| \\ &= \left(\|s(t)\| + \|r(t)\|\right)^2.\end{aligned}$$

Taking the positive square root of both sides reduces this statement to the first statement of the theorem. The second statement follows by replacing $r(t)$ with $-r(t)$. ∎

Recognizing the terms of the triangle inequality as square roots of energies, and renaming them as such, the statement of the theorem can be expressed in terms of the energies $E_p^{(s+r)}$, $E_p^{(s)}$, and $E_p^{(r)}$ as
$$\sqrt{E_p^{(s+r)}} \leq \sqrt{E_p^{(s)}} + \sqrt{E_p^{(r)}},$$

for the pulses $s(t) + r(t)$, $s(t)$, and $r(t)$.

Corollary 2.5.5 (Triangle Inequality) *Given any three complex pulses $s(t)$, $r(t)$, and $p(t)$,*

$$\|s(t) - r(t)\| \le \|s(t) - p(t)\| + \|r(t) - p(t)\|.$$

Proof:
Replace $s(t)$ by $s(t) - p(t)$ and replace $r(t)$ by $r(t) - p(t)$ in the second line of the theorem. The corollary follows. ■

The corollary informs the definition of signal space, given earlier. Because two functions $s(t)$ and $r(t)$ whose difference has zero energy are considered to be the same function, Corollary 2.5.5 implies that two functions that are equivalent to the same function are equivalent to each other. This means that the partition of the set of all functions of finite energy into equivalence classes by using the notion of equivalence given earlier is well-defined. Signal space is defined as the set of all equivalence classes in the set of all functions of finite energy.

A pulse, $s(t)$, of finite energy must be largely concentrated in some region of the t axis, so every pulse in signal space has the intrinsic notion of width. Often the width is described only in a qualitative way without a precise definition. To make the term precise, one can define the width of the pulse $s(t)$ in various ways. One measure of width is the Gabor timewidth, which we now define. Other measures of width are discussed in Section 2.9.

Definition 2.5.6 *The Gabor timewidth of the pulse $s(t)$ is given by*

$$T_G = \sqrt{\overline{t^2} - \overline{t}^2},$$

where

$$\overline{t^2} = \int_{-\infty}^{\infty} t^2 \frac{|s(t)|^2}{E_p} dt \qquad \overline{t} = \int_{-\infty}^{\infty} t \frac{|s(t)|^2}{E_p} dt.$$

The Gabor timewidth is also called the *root-mean-squared width* of the pulse $s(t)$. The equation for the Gabor timewidth has the same mathematical form as the equation for the standard deviation of the probability density function $p(t)$ given by $p(t) = |s(t)|^2/E_p$.

Similarly, the spectrum $S(f)$ of the pulse $s(t)$ also has the intrinsic notion of width, which can be made precise in a variety of ways. One measure of the width is the Gabor bandwidth.

Definition 2.5.7 *The Gabor bandwidth of the pulse $s(t)$ is defined as*

$$B_G = \sqrt{\overline{f^2} - \overline{f}^2},$$

where

$$\overline{f^2} = \int_{-\infty}^{\infty} f^2 \frac{|S(f)|^2}{E_p} df \qquad \overline{f} = \int_{-\infty}^{\infty} f \frac{|S(f)|^2}{E_p} df.$$

Definition 2.5.8 *The Gabor skew parameter of the pulse $s(t)$ is defined as*

$$\rho = \frac{1}{T_G B_G} \text{Re}\left[\overline{tf} - \overline{t}\,\overline{f}\right],$$

where $\overline{t}$ and $\overline{f}$ are as defined earlier, and

$$\overline{tf} = \frac{j}{2\pi} \int_{-\infty}^{\infty} t \frac{s(t)\dot{s}^*(t)}{E_p} dt = \frac{j}{2\pi} \int_{-\infty}^{\infty} f \frac{S(f)S'^*(f)}{E_p} df,$$

and where $\dot{s}(t) = ds(t)/dt$ and $S'(f) = dS(f)/df$.

The equality of the two forms for $\overline{tf}$ in the definition follows from Parseval's formula. For a purely real pulse $s(t)$ (or for a pulse with constant phase), the skew parameter ρ is zero. This is because, when $s(t)$ is real, $\overline{tf}$ is purely imaginary and $S(f) = S^*(-f)$, so $\overline{f}$ is zero. By duality, when $S(f)$ is purely real (or has any constant phase), then ρ is again zero.

The three parameters B_G, T_G, and ρ are the *Gabor parameters* of the pulse $s(t)$.

Proposition 2.5.9 *For any t_0 and f_0, the pulses $s(t)$ and $s(t-t_0)e^{-j2\pi f_0 t}$ have the same Gabor parameters.*

Proof:
The proposition is proved here only for the Gabor timewidth. The other Gabor parameters are proved similarly. Start with the definition of T_G^2 for the pulse $s(t)$ and manipulate it as follows:

$$T_G^2 = \overline{t^2} - \overline{t}^2 = \overline{(t+t_0)^2} - (\overline{t+t_0})^2$$

$$= \int_{-\infty}^{\infty} (t+t_0)^2 \frac{|s(t)|^2}{E_p} dt - \left[\int_{-\infty}^{\infty} (t+t_0) \frac{|s(t)|^2}{E_p} dt\right]^2$$

$$= \int_{-\infty}^{\infty} t^2 \frac{|s(t-t_0)e^{-j2\pi f_0 t}|^2}{E_p} dt - \left[\int_{-\infty}^{\infty} t \frac{|s(t-t_0)e^{-j2\pi f_0 t}|^2}{E_p} dt\right]^2,$$

which is the Gabor timewidth for the second pulse. ∎

Theorem 2.5.10 (Uncertainty Principle for Pulses[8]) *Let $s(t)$ be a differentiable pulse of finite energy. The Gabor timewidth T_G of $s(t)$ and the Gabor bandwidth B_G of $s(t)$ satisfy*

$$T_G B_G \geq \frac{1}{4\pi}$$

---

[8] Despite the name, there is nothing uncertain about the uncertainty principle for pulses.

2 Signals in One Dimension

with equality if and only if the pulse $s(t)$ is of the form

$$s(t) = Ae^{-a(t-t_0)^2}e^{-j2\pi f_0 t},$$

for any constants A, t_0, f_0, and positive a.

Proof:
The theorem is trivial when either T_G or B_G is infinite, so we may require that both are finite. Without loss of generality, we may require that the time center of the pulse and the frequency center of the pulse have been chosen so that $\bar{t}$ and $\bar{f}$ are zero. We begin with the equation

$$\frac{d}{dt}(t|s(t)|^2) = |s(t)|^2 + ts(t)\frac{ds^*(t)}{dt} + t\frac{ds(t)}{dt}s^*(t).$$

Take the integral of both sides

$$\int_{-\infty}^{\infty}\frac{d}{dt}(t|s(t)|^2)dt = \int_{-\infty}^{\infty}|s(t)|^2 dt + \int_{-\infty}^{\infty}ts(t)\frac{ds^*(t)}{dt}dt + \int_{-\infty}^{\infty}ts^*(t)\frac{ds(t)}{dt}dt.$$

Because the pulse has finite energy, $|s(t)|^2$ must go to zero faster than $1/t$. Thus the term on the left is evaluated as

$$\int_{-\infty}^{\infty} d(t|s(t)|^2) = t|s(t)|^2\Big|_{-\infty}^{\infty} = 0.$$

Therefore, the right side of the expression above becomes

$$E_p = 2\text{Re}\left[-\int_{-\infty}^{\infty} ts(t)\frac{ds^*(t)}{dt}dt\right]$$

$$\leq 2\left|\int_{-\infty}^{\infty} ts(t)\frac{ds^*(t)}{dt}dt\right|,$$

where the inequality holds because, for any complex z, $\text{Re}[z] \leq |z|$. Now use the Schwarz inequality to bound this as

$$E_p^2 \leq 4\left|\int_{-\infty}^{\infty} ts(t)\frac{ds^*(t)}{dt}dt\right|^2$$

$$\leq 4\int_{-\infty}^{\infty}|ts(t)|^2 dt \int_{-\infty}^{\infty}\left|\frac{ds(t)}{dt}\right|^2 dt.$$

The first integral is equal to $E_p T_G^2$ and by Parseval's formula the second integral is equal to $(4\pi)^2 E_p B_G^2$. Therefore

$$\left(\frac{1}{4\pi}\right)^2 \leq T_G^2 B_G^2,$$

which proves the inequality condition of the theorem. Equality occurs in the Schwarz inequality if and only if $\frac{ds(t)}{dt} = cts(t)$. This differential equation is solved by the gaussian pulse which leads to the equality condition of the theorem provided $t_0 = f_0 = 0$. Because the pulse can be translated in time and then in frequency without changing T_G and B_G, the general equality condition follows. ∎

2.6   Sampling in One Dimension

Digitization is the process of converting an analog waveform on a finite time interval into a representation that uses a finite number of bits. Digitization is commonly partitioned into two parts, called *sampling* and *quantization*. Sampling represents a continuous function of time by its values, called *samples*, on a discrete set of time points. Quantization represents a finite block of samples by a finite number of bits. A *scalar quantizer* represents a sample by a finite number of bits with each sample treated independently. A *vector quantizer* represents a block of samples by a finite number of bits, the entire block treated as a whole. Sampling, but not quantization, is studied in this section.

We now examine the conditions under which it is sufficient to sample a waveform $s(t)$ at a discrete set of time values. In particular, the set of samples $s_\ell = s(\ell T)$ for $\ell = 0, \pm 1, \pm 2, \ldots$ completely describes any waveform, $s(t)$, whose spectrum is limited to B hertz, provided $2BT < 1$. Then the waveform $s(t)$ in continuous signal space is represented by the sequence s_ℓ; for $\ell = \ldots -1, 0, +1, \ldots$ in discrete signal space.[9]

Suppose that $s(t)$, a baseband waveform with finite energy, has a spectrum that satisfies $S(f) = 0$ for $f \geq \frac{1}{2}$. The following discussion shows that the set $s_\ell = s(\ell)$ for $\ell = 0, \pm 1, \pm 2, \ldots$ completely describes $s(t)$.

Given $s(t)$ with $S(f) = 0$ for $f \geq \frac{1}{2}$, let

$$s'(t) = s(t)\text{comb}(t) = \sum_{\ell=-\infty}^{\infty} s_\ell \delta(t - \ell).$$

The values s_ℓ are the samples of $s(t)$. The convolution theorem states that the product $s(t)\text{comb}(t)$ transforms into the convolution $S(f) * \text{comb}(f)$. Thus, the spectrum $S'(f)$ of the sampled waveform is

$$S'(f) \doteq S(f) * \text{comb}(f) = \sum_{k=-\infty}^{\infty} S(f - k)$$

in the frequency domain. The spectrum $S'(f)$ consists of an infinite number of translates of $S(f)$ with the kth copy $S(f - k)$ translated in frequency by k. Each of these spectral translates is called a *sampling image*. Sampling images are shown in Figure 2.5. When the condition that $S(f) = 0$ for $f \geq \frac{1}{2}$ holds, these sampling images do not overlap. This means that

$$S'(f)\text{rect}(f) = S(f)$$

so that $S(f)$ can be completely recovered from $S'(f)$. When the images do overlap, the reconstruction of $S(f)$ is not possible without other information. The occurrence of overlapping sampling images is called *aliasing*.

---

[9] This set of functions on discrete time comprises discrete signal space in that it has a countable basis. This discrete signal space is called $\ell^2(\mathbb{R})$ or $l^2(\mathbb{C})$. These spaces can be compared to the set of functions on continuous time which comprises continuous signal space denoted $L^2(\mathbb{R})$ or $L^2(\mathbb{C})$.

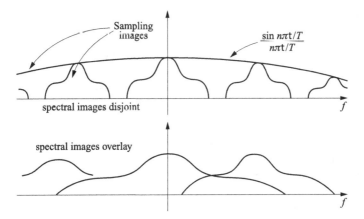

Figure 2.5 Sampling images

As can be seen by invoking the scaling property of the Fourier transform, a similar conclusion holds for any signal confined to a finite bandwidth. For a baseband waveform $s(t)$ whose spectrum satisfies $S(f) = 0$ for $|f| \geq B$ for some constant B, the samples are $s_\ell = s(\ell T)$, where T should be chosen so that $2BT < 1$. Such samples are called *Nyquist samples*. Accordingly, $2B$ is called the *Nyquist rate*. Again, as shown in the top half of Figure 2.5 for Nyquist samples, the sampling images do not overlap so there is no aliasing. Hence an ideal lowpass filter of bandwidth B will reject all sampling images other than $S(f)$ itself. In this way, an ideal lowpass filter will recover $s(t)$.

The following theorem[10] describes how to recover $s(t)$ from the Nyquist samples in the time domain using the sinc function as an interpolating function.

Theorem 2.6.1 (Nyquist–Shannon Theorem) *The signal $s(t)$ whose transform $S(f)$ satisfies $S(f) = 0$ for $|f| > B$ can be recovered from its Nyquist samples at $t = \ell T$ for $\ell = 0, \pm 1, \pm 2, \ldots$ by the Nyquist–Shannon interpolation formula*

$$s(t) = \sum_{\ell=-\infty}^{\infty} s_\ell \operatorname{sinc}\left(\frac{t}{T} - \ell\right),$$

where $2BT < 1$.

Proof:
It is enough to first treat the case with $B = 1/2$ and $T = 1$. The general case then follows from the scaling property. For the case with $B = 1/2$ and $T = 1$, the convolution theorem states that the sampled waveform

---

[10] The theorem presents sampling and interpolation as inverse operations describing an equivalence between a space of discrete functions and a space of continuous functions. These relationships might better be dubbed the Nyquist–Shannon transform and the inverse Nyquist–Shannon transform, respectively, so as to emphasize the bijection between the bandlimited subspace of $L^2(\mathbb{R})$ and the space $\ell^2(\mathbb{R})$. In the study of modulation, the inverse operation presents itself first.

$$s'(t) = s(t)\text{comb}(t) = \sum_{\ell=-\infty}^{\infty} s_\ell \delta(t-\ell)$$

has the Fourier transform

$$S'(f) = S(f) * \text{comb}(f)$$
$$= \sum_{k=-\infty}^{\infty} S(f-k).$$

The condition that $S(f) = 0$ for $|f| > 1/2$ means that these images do not overlap. This means that $S(f)$ can be recovered by writing

$$S(f) = S'(f)\text{rect}(f).$$

The product in the frequency domain becomes a convolution in the time domain

$$s(t) = s'(t) * \text{sinc}(t)$$
$$= \left[\sum_{\ell=-\infty}^{\infty} s_\ell \delta(t-\ell)\right] * \text{sinc}(t)$$
$$= \sum_{\ell=-\infty}^{\infty} s_\ell \text{sinc}(t-\ell).$$

To complete the proof of the theorem for the general case, replace t by t/T and use the scaling property of the Fourier transform. ∎

The set of all infinite sequences of finite energy is a discrete signal space. The Nyquist–Shannon theorem says that this discrete signal space is equivalent to a subspace of signal space consisting of all finite-energy functions whose Fourier transforms satisfy $S(f) = 0$ for $|f| > \frac{1}{2}$. This subspace of signal space $L^2(\mathbb{R})$ or $L^2(\mathbb{C})$ is equivalent to the signal space $\ell^2(\mathbb{R})$ or $\ell^2(\mathbb{C})$ with a countable basis.

Interpolation

Sinc interpolation may be unsatisfactory because the sidelobes of $\text{sinc}(t)$ fall off too slowly. The sidelobe magnitudes of $\text{sinc}(t)$ decay as $1/n$. Recalling that the summation $\sum_n 1/n$ diverges suggests that sinc interpolation can be computationally sensitive. To avoid computational difficulties, a modified interpolation formula can be used. This requires some slack in the sampling rate in order to admit an alternative interpolating function. This slack is described as a *guard interval* on the frequency axis. The guard interval is provided by sampling at a rate higher than the Nyquist rate.

Moreover, to obtain perfect reconstruction, sinc interpolation requires an infinite number of evenly spaced time samples, which implies infinite interpolation delay. When only a finite number of samples are used for a reconstruction, one must be satisfied with less than perfect interpolation. The interpolation error can be managed by oversampling – that is, sampling at a rate higher than the Nyquist rate – and using an interpolating function other than a sinc function as an alternative to the sinc function.

To interpolate oversampled data, it is sufficient to choose any interpolation filter, $h(t)$, such that

$$H(f) = \begin{cases} 1/T & \text{for } |f| \leq B, \\ (\cdot) & \text{for } B \leq |f| \leq 1/2T, \\ 0 & \text{for } |f| \geq 1/2T, \end{cases}$$

where $(\cdot)$ denotes an arbitrary function within the guard interval. For example, $H(f)$ may be a trapezoidal function having slant lines within the guard interval. Then

$$h(t) = \text{sinc}(2B't)\text{sinc}(2Bt)$$

is the interpolating filter, where $2B' = 2B - T^{-1}$.

Passband Sampling

Sometimes prior information about $s(t)$ may be available that can be used to reduce the sampling requirements. One instance of this is the passband waveform $\tilde{s}(t)$. Figure 2.6 shows that a direct application of the sampling theorem requires sampling at a rate of $2(f_0 + B)$ samples per second. On the other hand, $\tilde{s}(t)$ can be converted to its complex baseband representation whose in-phase and quadrature components can each be sampled at $2B$ samples per second. This allows $\tilde{s}(t)$ to be sampled with a total of $4B$ samples per second. The difference is that the first scheme allows one to reconstruct any arbitrary signal whose support is contained in the region $|f| \leq f_0 + B$. The second scheme applies only to functions that are also zero for $|f| \leq f_0 - B$.

Figure 2.7 gives a sampling scheme at complex baseband for passband signals. This scheme requires two baseband channels that must be precisely matched.

Discrete Signal Processing

One reason for sampling is so that the signal can be processed in discrete time instead of continuous time. Many operations defined in continuous signal space $L^2(\mathbb{R})$ or $L^2(\mathbb{C})$ can be executed in the corresponding discrete signal space $\ell^2(\mathbb{R})$ or $\ell^2(\mathbb{C})$ after the signals have been sampled provided the samples are Nyquist samples. This is so because the bandlimited continuous space and the corresponding discrete space are isomorphic as vector spaces. That is, the addition of vectors and multiplication by a scalar are in correspondence in the two signal spaces. It follows that convolution is also in correspondence in the two signal spaces. The correspondence

$$h(t) * s(t) \leftrightarrow \mathbf{h}(k) * \mathbf{s}(k)$$

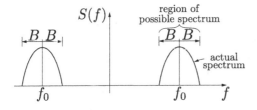

Figure 2.6 A passband spectrum

2.6 Sampling in One Dimension

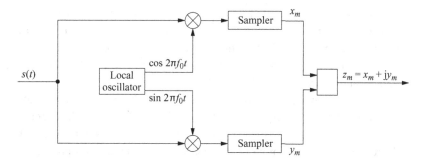

Figure 2.7 Baseband sampling of passband signals

must hold, where $\mathbf{h}(k)$ and $\mathbf{s}(k)$ denote the corresponding discrete sequences obtained by sampling $h(t)$ and $s(t)$, respectively.

To compute the convolution $r(t) = h(t) * s(t)$, one may proceed as follows. Sample both $h(t)$ and $s(t)$ at the larger of their two Nyquist rates. Convolve the discrete sample sequences as convolutions in $\ell^2(\mathbb{R})$, then use the Nyquist–Shannon interpolation formula to recover $r(t)$.

A simple way to demonstrate this equivalence is to first note that $\text{rect}(f)^2 = \text{rect}(f)$, which means that

$$\left(\text{rect}(f)e^{j2\pi k'}\right)\left(\text{rect}(f)e^{j2\pi(k-k')}\right) = \text{rect}(f)e^{j2\pi k}.$$

The convolution theorem now implies that the inverse Fourier transform of the left side is a convolution. Therefore, in the time domain

$$\int_{-\infty}^{\infty} \text{sinc}(t - k')\text{sinc}(t - t' - k + k')dt' = \text{sinc}(t - k).$$

This statement is now used to write a convolution in continuous space as

$$\int_{-\infty}^{\infty} h(t')s(t - t')dt' = \int_{-\infty}^{\infty} \left[\sum_{k'} h_{k'}\text{sinc}(t' - k')\sum_{k} s_{k-k'}\text{sinc}(t - t' - k + k')\right]dt'$$

$$= \sum_{k'} h_{k'} \sum_{k} s_{k-k'} \int_{-\infty}^{\infty} \left[\text{sinc}(t' - k')\text{sinc}(t - t' - k + k')\right]dt'$$

$$= \sum_{k} \left(\sum_{k'} h_{k'} s_{k-k'}\right)\text{sinc}(t - k),$$

as was asserted.

It takes an infinite number of samples to exactly represent a bandlimited signal $s(t)$, but a finite number of samples can represent the signal $s(t)$ to within a negligible error. The set of n samples $\{s_\ell; \ell = 0, \ldots, n - 1\}$ is a discrete approximate representation of $s(t)$ of finite blocklength. This representation can be expressed as a polynomial $\check{s}(z)$ given by

$$\check{s}(z) = \sum_{\ell} s_\ell z^\ell$$

The polynomial $š(z)$ is called the z transform of s. The polynomial $š(z)$ can be studied in terms of its polynomial factors or it can be studied as a function over the complex numbers $\mathbb{C}$. Indeed, evaluating the polynomial $š(z)$ only for those complex numbers of the form $e^{-j2\pi k/n}$ gives the vector $š = š(e^{-j2\pi k/n})$, called the discrete Fourier transform of the vector $s = \{s_\ell; \ell = 0, \ldots, n-1\}$.

As a real polynomial of degree $d \leq n-1$, the polynomial $š(z)$ always factors into at least $\lceil d/2 \rceil$ real polynomials over the real numbers. As a complex polynomial of degree $d \leq n-1$, the polynomial $š(z)$ always factors into d polynomials of degree one over the complex numbers. Similar statements do not exist for bivariate polynomials.

2.7 The Matched Filter

Let $s(t) = s_R(t) + js_I(t)$ be a known complex signal of finite energy E_p. Suppose that the received signal $v(t)$ consists of the signal $s(t)$ in additive complex noise,

$$v(t) = s(t) + n(t).$$

The complex noise

$$n(t) = n_R(t) + jn_I(t)$$

has real and imaginary noise components $n_R(t)$ and $n_I(t)$ that are independent, identically distributed random processes with zero mean. Moreover, suppose that each component of $n(t)$ has the same correlation function, $\phi(\tau)$, and the same power density spectrum, $N(f)$. Often the noise $n(t)$ is gaussian noise, but we do not impose this condition now. Only the second-order properties of $n(t)$ are used in this section.

The noise in the received signal $v(t)$ can be partially removed by passing $v(t)$ through a linear filter, $g(t)$, but this operation also affects the signal $s(t)$. The output of the linear filter is

$$u(t) = g(t) * v(t)$$
$$= g(t) * \big[s(t) + n(t)\big].$$

A compromise between maximizing the signal contribution to $u(t)$ and rejecting the noise contribution to $u(t)$ is needed. Restrict attention to the single sampling instant t_0 and find the filter $g(t)$ that maximizes the signal-to-noise ratio at this time instant. The function $g(t)$ is not restricted to be a causal filter, so the sampling instant can be chosen to be $t_0 = 0$. Any other choice of sampling instant corresponds to a simple translation of the impulse response $g(t)$ by t_0.

Because the real and the imaginary components of the noise entering the filter are uncorrelated and both have the same correlation function, the real and imaginary components of the noise out of the filter are also uncorrelated and each with the same correlation function (see Problem 2.29).

The noise has zero mean and is stationary. The variance of the noise per component at the filter output is called the *noise power* and is denoted N. It is given by

2.7 The Matched Filter

$$N = \tfrac{1}{2}\mathrm{var}[u(t)]$$
$$= \tfrac{1}{2}\mathrm{E}\big[\big|u(t) - \mathrm{E}[u(t)]\big|^2\big].$$

The noise variance N does not depend on time. Then

$$N = \tfrac{1}{2}\mathrm{E}\big[|u(t)|^2\big]$$
$$= \tfrac{1}{2}\mathrm{E}\Big[\int_{-\infty}^{\infty} g(\xi_1)n(-\xi_1)d\xi_1 \int_{-\infty}^{\infty} g^*(\xi_2)n^*(-\xi_2)d\xi_2\Big]$$
$$= \int_{-\infty}^{\infty}\int_{-\infty}^{\infty} g(\xi_1)g^*(\xi_2)\phi(\xi_1 - \xi_2)d\xi_1 d\xi_2,$$

where the last line follows because $\mathrm{E}\big[n(-\xi_1)n^*(-\xi_2)\big] = 2\phi(\xi_1 - \xi_2)$. Next, let $\tau = \xi_1 - \xi_2$ and let $\xi = \xi_1$. Then

$$N = \int_{-\infty}^{\infty} \phi(\tau) \int_{-\infty}^{\infty} g(\xi)g^*(\xi - \tau)d\xi\, d\tau.$$

The Fourier transform of the second integral is $|G(f)|^2$, so by the Parseval formula

$$N = \int_{-\infty}^{\infty} N(f)|G(f)|^2 df.$$

For white noise, $N(f) = N_0/2$ and $\phi(\tau) = (N_0/2)\delta(\tau)$. The energy theorem then gives

$$N = \frac{N_0}{2}\int_{-\infty}^{\infty} |g(\xi)|^2 d\xi$$

as the noise power.

The signal power at the sampling instant, on the other hand, is the square of the magnitude of the expected filter output at $t = 0$ given by

$$S = \big|\mathrm{E}[u(0)]\big|^2$$
$$= \Big|\int_{-\infty}^{\infty} g(-\xi)s(\xi)d\xi\Big|^2.$$

The ratio S/N of signal power to noise power is called the *signal-to-noise ratio*. The filter $g(t)$ is to be chosen to maximize the signal-to-noise ratio at the sample point of the filter output. Theorem 2.5.2 applied to $g^*(-t)$ and $s(t)$ says that any $g(t)$ can be decomposed as

$$g^*(-t) = \alpha s(t) + s^{\perp}(t),$$

where the two terms on the right are orthogonal and

$$\alpha = \frac{\langle g^*(-t), s^*(t)\rangle}{\langle s(t), s^*(t)\rangle}.$$

Then the signal power S at the output of the filter at time zero is

$$S = \Big(\alpha \int_{-\infty}^{\infty} |s(t)|^2 dt + \int_{-\infty}^{\infty} s(t)s^{\perp *}(t)dt\Big)^2$$
$$= \alpha^2 \Big(\int_{-\infty}^{\infty} |s(t)|^2 dt\Big)^2$$

because the second term is zero. Similarly, in white noise, the noise power is

$$N = \frac{N_0}{2} \int_{-\infty}^{\infty} g(\xi) g^*(\xi) d\xi$$

$$= \frac{N_0}{2} \left(\alpha^2 \int_{-\infty}^{\infty} |s(t)|^2 dt + \int_{-\infty}^{\infty} |s^{\perp}(t)|^2 dt \right).$$

With these expressions for the signal power S and the noise power N at the sampling instant, the filter that maximizes the ratio S/N can now be found.

Theorem 2.7.1 *For a signal $s(t)$ in white noise $n(t)$ of power density spectrum N_0 watts per hertz, the filter*

$$g(t) = s^*(-t)$$

achieves the maximum signal-to-noise power ratio.

Proof:
The output signal-to-noise ratio is

$$\frac{S}{N} = \frac{\alpha^2 \left(\int_{-\infty}^{\infty} |s(t)|^2 dt \right)^2}{\left(\alpha^2 \int_{-\infty}^{\infty} |s(t)|^2 dt + \int_{-\infty}^{\infty} |s^{\perp}(t)|^2 dt \right) N_0/2},$$

where $s^{\perp}(t) = g^*(-t) - \alpha s(t)$. Only the second term in the denominator depends on $g(t)$. To maximize the signal-to-noise ratio, that term should be made as small as possible by choice of $g(t)$. This term is made equal to zero by choosing $g^*(-t) = \alpha s(t)$. The scale factor α does not affect the signal-to-noise ratio and can be set equal to one. Therefore $g(t)$ should be equal to $s^*(-t)$ or to a constant multiple of $s^*(-t)$. This completes the proof of the theorem. ∎

The filter $g(t) = s^*(-t)$ is called the *matched filter*. In white noise, the signal-to-noise ratio at the output of the matched filter is

$$\frac{S}{N} = \int_{-\infty}^{\infty} \frac{|S(f)|^2}{N_0/2} df$$

$$= 2E_p/N_0,$$

where E_p is the total energy in the signal pulse.

The conclusion is rather general. It does not depend on the shape of the pulse $s(t)$. The signal pulse contributes to the output signal-to-noise ratio in white noise only through its energy E_p. Because the shape of the signal pulse plays no role in calculating the output signal-to-noise ratio, any pulse shape that is convenient for another reason may be chosen.

An example of the matched-filter response $g(t) = s^*(-t)$ for a real-valued pulse, $s(t)$, is shown in Figure 2.8. This is an example of a matched filter that is not causal. It can be made causal by delaying the sampling instant to time t_0 as shown in Figure 2.9.

2.7 The Matched Filter

Pulse shape   Matched filter response

Figure 2.8 Response of a matched filter

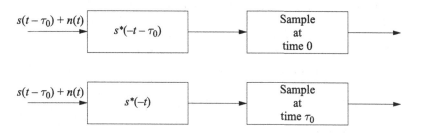

Figure 2.9 Equivalent ways to use a matched filter

To control the output signal or the output noise individually, the matched filter can be written as $Cs^*(-t)$, where C is any complex constant. Then, at the sampling instant, the output signal is

$$S = \left| \int_{-\infty}^{\infty} C|s(t)|^2 dt \right|^2 = |C|^2 E_p^2,$$

and the output noise power is

$$N = \int_{-\infty}^{\infty} N(f) |G(f)|^2 df$$
$$= \frac{N_0}{2} |C|^2 E_p,$$

which does not change the ratio S/N. One attractive choice for the normalizing constant C is $1/\sqrt{E_p}$. With this normalization, the filter output is described by the familiar parameters $S = E_p$ and $N = N_0/2$. When only the ratio of the signal to the noise is of interest, the factor of $1/\sqrt{E_p}$ can be suppressed.

When the noise is not white, to find the maximizing filter, decompose the filter $g(t)$ into two filters, $g(t) = g_1(t) * g_2(t)$, where $g_1(t)$ is chosen to whiten the noise and subsequent filter $g_2(t)$ is the matched filter. This decomposition for an arbitrary noise power density spectrum $N(f)$ leads to the following theorem:

Theorem 2.7.2 *The maximum signal-to-noise power ratio at the output of filter $g(t)$ is*

$$\frac{S}{N} = \int_{-\infty}^{\infty} \frac{|S(f)|^2}{N(f)} df,$$

which is achieved by the filter

$$G(f) = \frac{S^*(f)}{N(f)}.$$

Proof:
The noise is whitened by passing it through the filter $g_1(t)$ such that $G_1(f) = 1/\sqrt{N(f)}$. The signal $g_1(t) * s(t)$ at the output of this filter has Fourier transform $S(f)/\sqrt{N(f)}$. The matched filter for $g_1(t) * s(t)$ is given in the transform domain by

$$G_2(f) = \frac{S^*(f)}{\sqrt{N(f)}}.$$

The cascade of $G_1(f)$ and $G_2(f)$ is then

$$G(f) = \frac{S^*(f)}{N(f)},$$

as was to be proved. ∎

The filter of Theorem 2.7.2 is known as a *whitened matched filter*. It maximizes the signal-to-noise ratio for any covariance-stationary noise of power density spectrum $N(f)$. The noise is not required to be gaussian. For white noise, $N(f) = N_0/2$, so this expression can be reduced to

$$G(f) = S^*(f),$$

which means that $g(t) = s^*(-t)$, as before.

Matched Filter at Passband
The derivation of the matched filter holds for any pulse with finite energy. The real passband pulse

$$\tilde{s}(t) = s_R(t)\cos 2\pi f_0 t - s_I(t)\sin 2\pi f_0 t$$

has finite energy. Therefore,

$$\tilde{g}(t) = \tilde{s}(-t)$$
$$= s_R(-t)\cos 2\pi f_0 t + s_I(-t)\sin 2\pi f_0 t$$

is the matched filter for the real passband pulse $\tilde{s}(t)$.

The corresponding complex baseband pulse has a matched filter that is the complex representation of the matched filter for the passband pulse. Thus, the matched filter for the complex baseband pulse is

$$g(t) = s_R(-t) - js_I(-t)$$
$$= s^*(-t),$$

as before.

Theorem 2.7.3 *Two real filters matched to real orthogonal pulses have uncorrelated noise outputs at time zero. The noise outputs are independent when the noise is also gaussian.*

Proof:
The correlation is

$$E[n_1 n_2^*] = E\left[\frac{1}{E_p}\int_{-\infty}^{\infty} n(\xi)s_1(-\xi)d\xi \int_{-\infty}^{\infty} n^*(\xi')s_2^*(-\xi')d\xi'\right]$$

$$= \frac{1}{E_p}\int_{-\infty}^{\infty}\int_{-\infty}^{\infty} s_1(-\xi)s_2^*(-\xi')\frac{N_0}{2}\delta(\xi - \xi')d\xi\, d\xi'$$

because the noise is white. The sifting property of the delta function gives

$$E[n_1 n_2^*] = \frac{N_0}{2}\frac{1}{E_p}\int_{-\infty}^{\infty} s_1(-\xi)s_2^*(-\xi)d\xi$$

$$= 0,$$

and hence the noise outputs are uncorrelated at $t = 0$. For gaussian noise inputs, the outputs are also gaussian, and therefore independent, because uncorrelated gaussian random variables are independent. ∎

The theorem can be generalized to complex pulses and complex filters.

2.8 Stationary Random Processes

A random process is a concept that is a more elaborate quantity than a random variable. It varies with time. The random process $X(t)$ is characterized by both its amplitude structure and by its time structure. The real random process $X(t)$ is a real random variable for each fixed value of t. As such, for each fixed value of t, the random process $X(t)$ has a probability description on its amplitude. The amplitude structure of $X(t)$ at time t is described by the probability density function, $p_t(x)$. The random process is a *stationary random process* when this probability density function is independent of t.

Correlation Function

The time structure of a stationary random process with zero mean is described, in part, by the covariance function or the correlation function, For a general stationary random process $X(t)$, the *covariance function* is defined as

$$C(\tau) = E[(X(t) - E[X(t)])(X(t+\tau) - E[X(t+\tau)])],$$

and the *correlation function* is defined as

$$\phi(\tau) = E[X(t)X(t+\tau)].$$

The covariance function and the correlation function are the same when the mean is zero, which is usually the case herein.

The *correlation coefficient* of two zero-mean random variables is defined as

$$\rho_{XY} = \frac{E[XY]}{\sqrt{E[X^2]E[Y^2]}}.$$

The *cross-correlation function* of two zero-mean stationary random variables $X(t)$ and $Y(t)$ is defined as $\rho_{X(t)Y(t)} = \mathrm{E}[X(t)Y(t)]/\sqrt{\mathrm{E}[X(t)^2]\mathrm{E}[Y(t)^2]}$.

$$\phi(\tau) = \mathrm{E}[X(t)X(t+\tau)],$$

or by $\mathrm{E}[X(t)X^*(t+\tau)]$ when X is complex.

The Fourier transform of $\phi(\tau)$ is denoted $\Phi(f)$. Thus, by definition,

$$\phi(\tau) \leftrightarrow \Phi(f)$$

is a Fourier transform pair The correlation function $\phi(\tau)$ is an expectation of a product of a stationary random process with its time-delayed self so its Fourier transform has no immediate interpretation.

The Fourier transform $\Phi(f)$ must also have the form of an expectation in the frequency domain, but that expectation is not yet defined and the relationship of that expectation to $\Phi(f)$ is not yet proved.

In anticipation of Theorem 2.8.1, $\Phi(f)$ is called the *power density spectrum*, a name that hints at the nature of $\Phi(f)$. However, simply naming the function does not substitute for a proper definition in the frequency domain and a theorem linking both sides of the Fourier transform. To give $\Phi(f)$ meaning, a precise formulation with a definition and a formal proof of a theorem are necessary. This precise formulation is not trivial.

Power Density Spectrum

The random process $X(t)$ does not have a Fourier transform. This is because it is not a function itself. It is a random ensemble of functions with a probability description. Each element, or realization, of this ensemble has infinite duration and, in general, infinite energy, so the realizations themselves do not have Fourier transforms. However, a cropped realization $x(t)\mathrm{rect}(t/T)$ does have finite energy and a Fourier transform denoted $X_T(f)$, one such Fourier transform for each element of the ensemble. The ensemble of such Fourier transforms is the random process $\mathsf{X}_T(f) = \{X_T(f)\}$ in the frequency domain.

Define the *power density spectrum* $N(f)$ as

$$N(f) = \lim_{t \to \infty} \frac{1}{T}\mathrm{E}[|X_T(f)|^2],$$

where $X_T(t) = X\mathrm{rect}(t/T)$. The statement that this $N(f)$ is equal to the Fourier transform $\Phi(f)$ of the correlation function $\phi(\tau)$ is the statement of the following Wiener–Khintchine theorem.

Theorem 2.8.1 (Wiener–Khintchine) *The power density spectrum $N(f)$ of a well-behaved zero-mean stationary random process is equal to the Fourier transform of the autocorrelation function $\phi(\tau)$.*

Proof

The proof consists of a long sequence of inferences. Some steps of the proof involve the taking of limits. Those limits are straightforward and are accepted in the proof without further discussion:

$$N(f) = \lim_{T\to\infty} \frac{1}{2T} E\left[\left|X_T(f)\right|^2\right]$$

$$= \lim_{T\to\infty} \frac{1}{2T} E\left[\left(\int_{-T}^{T} x(t)e^{-j2\pi ft}\,dt\right)\left(\int_{-T}^{T} x(t')e^{-j2\pi ft'}\,dt'\right)^*\right]$$

$$= \lim_{T\to\infty} \frac{1}{2T} E\left[\int_{-T}^{T}\int_{-T}^{T} x(t)x^*(t')e^{-j2\pi f(t-t')}\,dt\,dt'\right]$$

$$= \lim_{T\to\infty} \frac{1}{2T} \int_{-T}^{T}\int_{-T}^{T} E[x(t)x^*(t')]e^{-j2\pi f(t-t')}\,dt\,dt'$$

$$= \lim_{T\to\infty} \frac{1}{2T} \int_{-T}^{T}\int_{-T}^{T} \phi(t-t')e^{-j2\pi f(t-t')}\,dt\,dt'$$

$$= \lim_{T'\to\infty} \frac{1}{2T'} \int_{-T'}^{T'} \lim_{T\to\infty}\int_{-T}^{T} \phi(t-t')e^{-j2\pi f(t-t')}\,dt\,dt'$$

$$= \lim_{T'\to\infty} \frac{1}{2T'} \int_{-T'}^{T'} \left[\int_{-\infty}^{\infty} \phi(t-t')e^{-j2\pi f(t-t')}\,dt\right]dt'$$

$$= \lim_{T'\to\infty} \frac{1}{2T'} \int_{-T'}^{T'} \left[\int_{-\infty}^{\infty} \phi(t)e^{-j2\pi ft}\,dt\right]dt'$$

$$= \left[\int_{-\infty}^{\infty} \phi(t)e^{-j2\pi ft}\,dt\right]\left[\lim_{T'\to\infty}\frac{1}{2T'}\int_{-T'}^{T'} 1\,dt'\right]$$

$$= \int_{-\infty}^{\infty} \phi(\tau)e^{-j2\pi f\tau}\,d\tau$$

$$= \Phi(f).$$

The last line is the Fourier transform of $\phi(\tau)$, as given in the statement of the theorem. This completes the proof of the theorem. ∎

The theorem states that the power density spectrum $N(f)$ is equal to the Fourier transform $\Phi(f)$ of the correlation function $\phi(\tau)$. When appropriate, the stationary random process is called noise. In particular, when $N(f)$ is a constant, the random process is called a white random process or *white noise*. By convention, the power density spectrum of white noise is written as $N(f) = N_0/2$.

2.9 Resolution and Apodization

The study of resolution was introduced by Lord Rayleigh to quantify how an effect such as a filter impulse response (or, in two dimensions, a point-spread function) prevents one from seeing things as they are. A resolution criterion describes the filter output as it is with no attempt to sharpen the pulse. No single definition of resolution is accepted as the best definition. In different circumstances, different criteria for measuring resolution may be more appropriate. This section discusses resolution in one dimension. Section 3.9 discusses resolution in two dimensions.

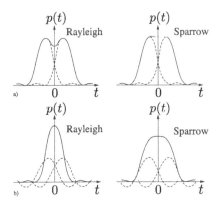

Figure 2.10 A comparison of resolution measures

Let $s(t)$ be a pulse whose maximum value is at the origin. A pair of identical copies of $s(t)$ separated by τ is the function

$$p(t) = s(t - \tau/2) + s(t + \tau/2).$$

A measure of the width of $s(t)$ is wanted so that, whenever this pair of pulses is separated by a value of τ larger than this measure of width, the pair can be readily distinguished from the single pulse $s(t)$. The statement of the problem is too vague to determine a unique definition of resolution. When $s(t)$ is fully known and there is no noise, one can always distinguish between $s(t)$ and $s(t-\tau/2)+s(t+\tau/2)$ for any nonzero ν. However, this distinction requires knowledge of $s(t)$, whereas a discussion of resolution is limited to a less well-defined situation including an incompletely known $s(t)$. A situation may have multiple copies of $s(t)$ attenuated and translated by unspecified amounts and observed in additive noise. Advanced techniques for separating pulses in such an environment are called *superresolution techniques*.

Different criteria for resolution that are in use amount to different definitions of the width of $s(t)$. One criterion for resolution is the *Rayleigh resolution criterion*, which has long been used in optics. This width of pulse $s(t)$ is defined as the difference between the value of t where the maximum of $s(t)$ occurs and the value of t where the zero of $s(t)$ closest to the maximum occurs. This definition has served well in the field of optics, but it fails when $s(t)$ has no zero, as for a gaussian pulse.

Another way to measure resolution is the *Sparrow resolution criterion*. The Sparrow width is the smallest value of ν for which $p(t) = s(t - \tau/2) + s(t + \tau/2)$ has a second derivative at the origin equal to zero. This definition fails for triangular pulses and for square pulses. Figure 2.10 shows a comparison of the Rayleigh and Sparrow resolution criteria for the pulses $s(t) = \text{sinc}(t)$ and $s(t) = \text{sinc}^2(t)$.

A third criterion for resolution is the *half-power width*. For a pulse whose maximum is at $t = 0$, the half-power width is the smallest positive value of t for which $s^2(t) = \frac{1}{2}s^2(0)$. This criterion is well-defined, though the choice of one-half (rather than, say, one-fourth) may be viewed as an arbitrary choice.

A fourth criterion for resolution is the *Woodward resolution criterion*. For a pulse whose maximum is at the origin, the Woodward criterion is

$$\Delta t = \frac{\int_{-\infty}^{\infty} |s(t)|^2 dt}{\max_t |s(t)|^2} = \frac{E_p}{|s(0)|^2}.$$

The Woodward timewidth is the width of a square pulse with the same energy and peak value as $s(t)$. Finally, the Gabor timewidth, which was defined in Section 2.5, can be used as a resolution criterion.

The resolution may also be considered on the frequency axis. The width of the spectrum $S(f)$ can be measured by the Rayleigh resolution criterion, the Sparrow resolution criterion, or the half-power resolution criterion, which are defined in the same way on the frequency axis as on the time axis. Another measure of bandwidth that is sometimes useful for discussing noise is the *noise bandwidth*, defined as

$$B_N = \frac{\int_{-\infty}^{\infty} |S(f)|^2 df}{\max_f |S(f)|^2}.$$

The noise bandwidth is the dual of the Woodward timewidth, and for this reason, might also be called the *Woodward bandwidth*.

Another issue in distinguishing pulses is the structure of the sidelobes of the pulse $s(t)$ because the sidelobes of one copy of $s(t)$ could mask the main lobe of a faint translated copy of $s(t)$. Given the pulse $s(t)$, the optimum filter to discriminate against noise is the matched filter, but the matched filter may have undesirable sidelobes in the response to the signal. Instead of using the matched filter, one may choose a modified matched filter that suppresses the sidelobes of the filter output even though the signal-to-noise ratio may be smaller. Any processing method designed to reduce sidelobes is called *apodization*. It may be possible to satisfy the requirement for apodization using a filter that differs only slightly from the matched filter, so there is only a slight reduction in the output signal-to-noise ratio and a slight reduction in resolution needed to suppress the sidelobes.

Problems

2.1 Prove that when the pulse $s(t)$ has the Fourier transform $S(f)$, the pulse $s(at)$ has the Fourier transform $\frac{1}{|a|} S(f/a)$.

2.2 Prove that when $s(t)$ is a real-valued pulse, its Fourier transform $S(f)$ satisfies $S^*(f) = S(-f)$.

2.3 Let s_i for $i = 0, \ldots, n-1$ be a vector of complex numbers. The *discrete Fourier transform* of the vector s is the vector S given by

$$S_k = \sum_{i=0}^{n-1} e^{-j2\pi ik/n} s_i.$$

a. Prove that the inverse discrete Fourier transform is

$$s_i = \frac{1}{n}\sum_{k=0}^{n-1} e^{j2\pi ik/n} S_k.$$

b. The *cyclic convolution* of two vectors p and q, each of blocklength n, is defined as

$$r_i = \sum_{k=0}^{n-1} p_{((i-k))} q_k$$

and denoted by $r = p * q$, where $((i - k))$ denotes $i - k$ modulo n. Prove the convolution theorem for the discrete Fourier transform. This is the statement that $r = p * q$ if and only if their discrete Fourier transforms satisfy $R_k = P_k Q_k$ for $k = 0, \ldots, n - 1$.

2.4 Prove that the *autocorrelation function* of the pulse $s(t)$ defined as

$$\phi(\tau) = \int_{-\infty}^{\infty} s(t) s^*(t - \tau) dt,$$

has the Fourier transform $|S(f)|^2$.

2.5 Let

$$S(f) = 2\text{sinc} f \cos 4\pi f.$$

Prove that

$$S(f) * S(f) = S(f).$$

2.6 Prove that

$$\text{rect}\left(\frac{t}{T - 2|\tau|}\right) = \text{rect}\left(\frac{\tau}{T - 2|t|}\right).$$

2.7 Evaluate the integral

$$\int_{-\infty}^{\infty} \frac{\sin^2(\pi A x)}{\pi^2 x^2} dx$$

by inspection.

2.8 Pulses exist that are equal to their own Fourier transforms. Verify the following (generalized) Fourier transform pairs:

$$e^{-\pi t^2} \leftrightarrow e^{-\pi f^2}$$

$$\frac{1}{\sqrt{|t|}} \leftrightarrow \frac{1}{\sqrt{|f|}}.$$

Consequently, the integral equation

$$s(y) = \int_{-\infty}^{\infty} s(x) e^{-j2\pi xy} dx$$

has at least two distinct solutions for $s(x)$. Are there others?

2.9 a. Prove that the delta function satisfies the elementary property

$$\delta\left(\frac{t}{T}\right) = T\delta(t)$$

for $T > 0$.

b. Prove that the comb function satisfies the property

$$\sum_{\ell=-\infty}^{\infty} \delta\left(\frac{t}{T} - \ell\right) = T \sum_{\ell=-\infty}^{\infty} \delta(t - \ell T).$$

c. Prove the Fourier transform pair

$$\sum_{\ell=-\infty}^{\infty} \delta(t - \ell T) \leftrightarrow \frac{1}{T} \sum_{\ell=-\infty}^{\infty} \delta\left(f - \frac{\ell}{T}\right)$$

for $T > 0$.

2.10 a. Show that the following

$$s(t) + s(-t) \leftrightarrow 2\text{Re}[S(f)]$$

is a Fourier transform pair. Comment about the case $s(t) = s(-t)$.

b. Show that the following

$$s(t - \tfrac{1}{2}) + s(-t - \tfrac{1}{2}) \leftrightarrow 2\text{Re}[S(f)\cos(\pi f)]$$

is a Fourier transform pair. Comment about the case $s(t) = s(-t)$.

c. Explain the distinction between $s(-t + a)$ and $s(-t - a)$.

d. Find the Fourier transforms of the two expressions $s(t - a) + s(-t - a)$ and $s(t - a) + s(-t + a)$. Which of these is the transform of $2\text{Rm}[S(f)]$?

2.11 The Fourier transform of a train of 2^k uniformly spaced, identical pulses can be obtained in two ways: either by writing down the dirichlet function, or by a process of successive constructions expressing the train of 2^j pulses as a doublet of trains of 2^{j-1} pulses. Show that both of these approaches give the same result.

2.12 A pulse of width 8τ can be thought of as a train of eight pulses, each of width τ and with pulse spacing τ. Starting with the transform of a pulse of width τ, and using the transform of the array, derive the transform of the wide pulse. How does this compare with the transform of the wide pulse computed directly? Now discuss what happens when the spacing between the eight pulses is slightly larger than τ. (The wide pulse is a little wider, but with gaps.)

2.13 Prove that to minimize the Gabor bandwidth of a pulse $s(t) = |s(t)|e^{j\theta(t)}$ over a choice of phase, the phase $\theta(t)$ must be chosen to be of the form $\theta(t) = \theta_0 + \dot{\theta}_0 t$.

2.14 Define the *effective timewidth* of the pulse $s(t)$ as

$$T_E = \frac{\left|\int_{-\infty}^{\infty} s(t)dt\right|}{|s(0)|}.$$

This is the width of a rectangular pulse, $|s(0)|\text{rect}(t/T_E)$, that has the same area as $s(t)$. Define the *effective bandwidth* as

$$B_E = \frac{|\int_{-\infty}^{\infty} S(f) df|}{|S(0)|}.$$

Prove that

$$T_E B_E = 1.$$

2.15 The passband signal,

$$\tilde{s}(t) = s_R(t) \cos 2\pi f_0 t - s_I(t) \sin 2\pi f_0 t,$$

can be filtered by convolving it with the passband filter

$$\tilde{h}(t) = h_R(t) \cos 2\pi f_0 t - h_I(t) \sin 2\pi f_0 t.$$

Alternatively, the passband signal $\tilde{s}(t)$ can be converted to the complex baseband signal $s(t) = s_R(t) + js_I(t)$ and filtered by convolving with the complex baseband filter $h(t) = h_R(t) + jh_I(t)$, then reconverting back to a passband signal. By explicitly writing out the convolutions, show that the result is the same in the two alternatives. Is the exercise a convincing demonstration of the utility of the complex representation?

2.16 Suppose that a passband signal,

$$\tilde{s}(t) = a(t) \cos(2\pi f_c t + \theta(t)),$$

has an amplitude modulation, $a(t)$, a phase modulation, $\theta(t)$, and a carrier frequency, f_c. Find the in-phase and quadrature components that are obtained with respect to the reference frequency f_0.

2.17 Using time-domain arguments in place of frequency-domain arguments, show that the complex representation of a passband signal is unique. That is, show that two different passband signals have different complex representations (and vice versa).

2.18 Show that

$$s_1(t)e^{-j2\pi f_0 t} * s_2(t)e^{-j2\pi f_0 t} = (s_1(t) * s_2(t))e^{-j2\pi f_0 t}.$$

2.19 An equivalence class of finite-energy functions over any domain $\mathbb{A}$ is the set of all functions $s(t)$ over $\mathbb{A}$ such that the difference of any two functions in the same equivalence class has zero energy. Any element of an equivalence class may be used as a representative of that equivalence class.

a. Prove that the addition of equivalence classes by adding representatives does not depend on the choice of representative.
b. Prove that equivalence is a transitive property which means that equivalence forms a partition of the space of functions.
c. Prove that an equivalence class can contain at most one continuous function.

d. Prove that the inner product of two equivalence classes is well-defined.
Hint: The quarter-square multiplier is the identity
$$ab = ((a+b) - (a-b))/4.$$

2.20 The passband filter with impulse response
$$\tilde{h}(t) = te^{-at} \sin 2\pi f_0 t \quad t \geq 0,$$
is excited by the passband pulse,
$$\tilde{s}(t) = \text{rect}(t) \sin 2\pi f_0 t,$$
where f_0 is very large compared to a.

a. Find the output by a direct convolution.
b. Give a complex representation for the filter and the pulse, and find the complex representation for the filter output by complex convolution.
c. Which is easier?

2.21 Use the Schwarz inequality on the integral
$$\int_{-\infty}^{\infty} \left[(t-\bar{t})s(t)\right]\left[\frac{ds(t)}{dt} - 2\pi j \bar{f} s(t)\right] dt$$
to obtain a direct proof of the uncertainty principle when the time center of $s(t)$ and the frequency center of $s(t)$ are arbitrary.

2.22 Prove that the Gabor skew parameter satisfies
$$-1 \leq \rho \leq 1,$$
first for the case where $\bar{t} = \bar{f} = 0$, then for the general case.

2.23 Suppose that $g(t)$ has the maximum frequency B (in hertz). Let $g(t)$ be sampled at a rate of $4B$ samples per second. Define a time-domain interpolation formula based upon the frequency filter with cosine-squared tails, sketched as follows.

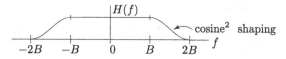

This interpolation filter gives a theoretically perfect reconstruction, as does sinc interpolation. Why might this interpolation rule be preferred to sinc interpolation?

2.24 A system is to be designed to process real-valued signals with spectra confined to the frequency regions shown in the illustration. No other prior information about the signals is known.

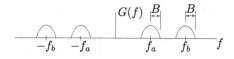

a. What is the minimum number of samples per second that suffice to specify the signal?
b. Sketch at least three different schemes for sampling the signal.

2.25 (**Chirp Filter**) Show that the Fourier transform

$$S(f) = \int_{-\infty}^{\infty} s(t) e^{-j2\pi ft} dt$$

can be written as

$$S(f) = e^{-j\pi f^2} \left(e^{j\pi t^2} * (e^{-j\pi t^2} s(t)) \right).$$

Discuss how this expression can be used to rapidly compute a Fourier transform with three "chirp filters" of the form $h(t) = e^{-j\pi t^2}$ and two multipliers. Sketch a functional block diagram of this circuit. Explain the consequence of using a chirp filter of finite duration, such as

$$h(t) = \begin{cases} e^{j\pi t^2} & |t| \leq T/2, \\ 0 & |t| > T/2. \end{cases}$$

With $s(t) = \operatorname{rect}(t/T)$, sketch the waveform at various points of the circuit.

2.26 (**Chirp Filter**) Show that

$$e^{j\pi t^2} * \left(e^{j2\pi t^2} (e^{j\pi t^2} * s(t)) \right) = e^{j2\pi t^2} s(-t).$$

This expression is closely related to the lens law of optics.

2.27 The inverse Fourier transform of the product $f S(f) H(f)$ can be found by using both the convolution theorem and the differentiation property of the Fourier transform in three different ways as suggested by the three groupings of terms $f(S(f)H(f))$, $(f(S(f))H(f)$, and $S(f)(fH(f))$. Each grouping suggests an order in which the convolution theorem and the differentiation property are invoked. Describe the result of each method. Do these all give the same or equivalent results. Explain.

2.28 An alternative notion of a sample of the function $s(t)$ is the integral of $s(t)$ from k to $k+1$. That is, $s_k = \int_k^{k+1} s(t) dt$.

a. Show that this sampling method is equivalent to sampling the function $\operatorname{rect}(t) * s(t)$.
b. Under what conditions can $s(t)$ be recovered. fully or approximately, from samples of this kind? How?

2.29 The Gabor bandwidth of the rectangular pulse $s(t) = \operatorname{rect}(t)$ is infinite because the sidelobes of $S(f)$ decay as f^{-1}. What is $S(f)$ when $s(t) = \operatorname{rect}(t) * \operatorname{sinc}(2t)$? What is the Gabor bandwidth of $s(t)$?

2.30 (**Pulse Compression**) Show that the filter $g(t) = e^{j\pi\alpha t^2}$ (whose impulse response is a quadratic-phase pulse) can be implemented as

$$u(t) = e^{j\pi\alpha t^2} \int_{-\infty}^{\infty} e^{-j2\pi\alpha\xi t} \left[v(\xi) e^{j2\pi\alpha\xi^2} \right] d\xi.$$

(This expression replaces the filter with two multiplications and a Fourier transform.) What is $r(t) = s(t)e^{-j\pi\alpha t^2}$ when $s(t) = e^{j\pi\alpha(t-\tau)^2}\text{rect}(t/T)$? This operation is called *dechirping*. How does the bandwidth of $r(t)$ compare to the bandwidth of $s(t)$? Explain why one might place a sampling operation after a dechirping operation rather than before.

2.31 Suppose that the complex baseband noise process $n(t) = n_R(t) + jn_I(t)$ is passed through the complex baseband filter $g(t) = g_R(t) + jg_I(t)$. Show that when $n_R(t)$ and $n_I(t)$ have the same correlation function and are uncorrelated, the noise components at the output of the filter have these same properties.

2.32 Derive the whitened matched filter directly as follows. Using Parseval's formula,

$$\int_{-\infty}^{\infty} g(\xi)w^*(\xi)d\xi = \int_{-\infty}^{\infty} G(f)W^*(f)df,$$

show that the signal power out of the linear filter $g(t)$ at time zero is

$$S = \left|\int_{-\infty}^{\infty} G(f)S(f)df\right|^2.$$

Then, with the aid of the Schwarz inequality, show that the maximum signal-to-noise ratio

$$\frac{S}{N} = \int_{-\infty}^{\infty} \frac{|S(f)|^2}{N(f)} df$$

is achieved by the matched filter. Finally, specialize this expression to the case of white noise for which $N(f) = N_0/2$.

2.33 Let $s(t)$, with Fourier transform $S(f)$, be a finite energy pulse with zero mean and finite second moment. Then the energy of $s(t)$ is concentrated near the origin, and

$$s(t-1) \leftrightarrow S(f)e^{j2\pi f}$$

is a Fourier transform pair.

a. What is the Fourier transform of $ts((t-1)/a)$?
b. Because $s((t-1)/a)$ is concentrated near $t=1$ for small values of a, the initial term t in the time-domain pulse of part i) is set equal to one, either because of a mistaken analysis or because of an approximation. What is the Fourier transform of $s((t-1)/a)$?
c. Do parts (a) and (b) agree for small values of a in the sense of the approximation? Describe the error in the frequency domain.

Notes

The uncertainty relationship for signal pulses was first given by Gabor (1946). It is known as the Heisenberg uncertainty relationship in quantum theory. The notation for the rectangular pulse and the sinc pulse was introduced by Woodward (1953). The matched filter was originally introduced in a laboratory report by North (1943). It is sometimes called the *North filter*. The proof of the Wiener–Khintchine theorem given herein is adapted from Bouman (2023).

A form of the sampling theorem was first stated by Whittaker (1915). It reappeared in the engineering literature in papers by Nyquist (1928) and Shannon (1948). It appeared in the Russian literature in a paper by Kotel'nikov (1933). The generalized functions were given a rigorous formulation by L. Schwartz (1952). He called his formal treatment of generalized functions the *theory of distributions.*

3 Signals in Two Dimensions

The Fourier transform of a two-dimensional function – or of an n-dimensional function – is defined in the same way as is the Fourier transform of a one-dimensional function. Because the applications of the two-dimensional Fourier transform discussed herein deal with two-dimensional images, it is common practice, and ours, to refer to the variables of the two-dimensional function as "spatial coordinates" and to refer to the variables of its Fourier transform as "spatial frequencies."

The study of the two-dimensional Fourier transform closely follows the study of the one-dimensional Fourier transform. As the study develops, however, the two-dimensional Fourier transform displays a richness of detail beyond that of the one-dimensional Fourier transform.

3.1 The Two-Dimensional Fourier Transform

The real or complex function $s(x, y)$ of two variables x and y is called a *two-dimensional signal* or a *two-dimensional function*. (More correctly stated, it is a function of a two-dimensional variable.) A common example is an image, such as a monotone photographic image, wherein the variables x and y are the coordinates of the image and $s(x, y)$ is the real or complex amplitude of the image at the point (x, y). In a monotone photographic image, the amplitude is a nonnegative real number. In other examples, such as the complex amplitude of a wavefront in the x, y plane, the function $s(x, y)$ may also take on negative or complex values. Figure 3.1 shows a graphical representation of a two-dimensional complex signal in terms of the real and imaginary parts.

The definition of the two-dimensional Fourier transform is analogous to the definition of the one-dimensional Fourier transform. Given the complex function $s(x, y) = s_R(x, y) + js_I(x, y)$ whose energy

$$E_p = \int_{-\infty}^{\infty} \int_{-\infty}^{\infty} |s(x, y)|^2 dx dy$$

is finite, the *two-dimensional Fourier transform* is defined in terms of the *spatial frequencies*[1] f_x and f_y as

---

[1] The spatial frequencies f_x and f_y in units of cycles per unit distance must not be confused with temporal frequency f in units of cycles per second. Alternative notation ξ_x and ξ_y cycles per unit distance, or k_x and k_y radians per unit distance might be preferred.

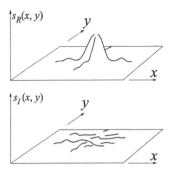

Figure 3.1 Depiction of a two-dimensional complex function

$$S(f_x, f_y) = \int_{-\infty}^{\infty} \int_{-\infty}^{\infty} s(x,y) e^{-j2\pi(f_x x + f_y y)} dx dy.$$

The term "two-dimensional Fourier transform" is used to refer to both the function $S(f_x, f_y)$ itself and the formula that defines the function $S(f_x, f_y)$ in terms of $s(x, y)$. The transform relationship between the function $s(x, y)$ in the *space domain* and the function $S(f_x, f_y)$ in the *frequency domain* is denoted by

$$s(x,y) \Leftrightarrow S(f_x, f_y).$$

Notice the use of the doubly shafted arrow $\Leftrightarrow$ instead of the singly shafted arrow $\leftrightarrow$ used with the one-dimensional Fourier transform.

The inverse two-dimensional Fourier transform is given by

$$s(x,y) = \int_{-\infty}^{\infty} \int_{-\infty}^{\infty} S(f_x, f_y) e^{j2\pi(f_x x + f_y y)} df_x df_y.$$

To verify that this is the inverse Fourier transform, simply view $S(f_x, f_y)$ as formed by a composition[2] of two one-dimensional Fourier transforms. Let

$$\Gamma(f_x, y) = \int_{-\infty}^{\infty} s(x,y) e^{-j2\pi f_x x} dx$$

be the Fourier transform of $s(x, y)$ with respect to x for each value of y. Then

$$S(f_x, f_y) = \int_{-\infty}^{\infty} \Gamma(f_x, y) e^{-j2\pi f_y y} dy$$

and

$$s(x,y) \leftrightarrow \Gamma(f_x, y) \leftrightarrow S(f_x, f_y).$$

The formula for the one-dimensional inverse Fourier transform leads to the two-step procedure

$$\Gamma(f_x, y) = \int_{-\infty}^{\infty} S(f_x, f_y) e^{j2\pi f_y y} df_y$$

$$s(x,y) = \int_{-\infty}^{\infty} \Gamma(f_x, y) e^{j2\pi f_x x} df_x,$$

as was to be proved.

---

[2] The formal conditions under which this statement is valid are stated by *Fubini's theorem*.

Using similar reasoning, it is a simple exercise to prove many basic properties of the two-dimensional Fourier transform. Many of these properties can be verified simply by using the properties of the one-dimensional Fourier transform. The basic properties are as follows:

1. Linearity: For any constants, a and b, possibly complex,
$$as_1(x,y) + bs_2(x,y) \Leftrightarrow aS_1(f_x, f_y) + bS_2(f_x, f_y).$$

2. Sign reversal:
$$s(-x,y) \Leftrightarrow S(-f_x, f_y),$$
$$s(x,-y) \Leftrightarrow S(f_x, -f_y).$$

3. Conjugation:
$$s^*(x,y) \Leftrightarrow S^*(-f_x, -f_y).$$
When $s(x,y)$ is real, $S^*(-f_x, -f_y) = S(f_x, f_y)$.

4. Scaling property[3]: For any real nonzero constants, a and b,
$$s(ax, by) \Leftrightarrow \frac{1}{|ab|} S\left(\frac{f_x}{a}, \frac{f_y}{b}\right).$$

5. Translation: For any real constants, a and b,
$$s(x-a, y-b) \Leftrightarrow S(f_x, f_y) e^{-j2\pi(af_x + bf_y)}.$$

6. Modulation: For any real constants, a and b,
$$s(x,y) e^{j2\pi(ax+by)} \Leftrightarrow S(f_x - a, f_y - b).$$

7. Convolution:
$$g(x,y) ** h(x,y) \Leftrightarrow G(f_x, f_y) H(f_x, f_y),$$
where $**$ denotes the two-dimensional convolution given by
$$g(x,y) ** h(x,y) = \int_{-\infty}^{\infty} \int_{-\infty}^{\infty} g(\xi, \eta) h(x-\xi, y-\eta) d\xi \, d\eta.$$

8. Product:
$$g(x,y) h(x,y) \Leftrightarrow G(f_x, f_y) ** H(f_x, f_y).$$

9. Spatial differentiation:
$$\frac{\partial s(x,y)}{\partial x} \Leftrightarrow j2\pi f_x S(f_x, f_y),$$
$$\frac{\partial s(x,y)}{\partial y} \Leftrightarrow j2\pi f_y S(f_x, f_y).$$

---

[3] A scale change in two dimensions is sometimes called *magnification*; in one dimension, it is sometimes called *dilation*.

10. Frequency differentiation:

$$-j2\pi x s(x,y) \Leftrightarrow \frac{\partial S(f_x, f_y)}{\partial f_x},$$

$$-j2\pi y s(x,y) \Leftrightarrow \frac{\partial S(f_x, f_y)}{\partial f_y}.$$

11. Parseval's formula:

$$\int_{-\infty}^{\infty} \int_{-\infty}^{\infty} g(x,y) h^*(x,y) dx dy = \int_{-\infty}^{\infty} \int_{-\infty}^{\infty} G(f_x, f_y) H^*(f_x, f_y) df_x df_y.$$

12. Energy relation:

$$\int_{-\infty}^{\infty} \int_{-\infty}^{\infty} |s(x,y)|^2 dx dy = \int_{-\infty}^{\infty} \int_{-\infty}^{\infty} |S(f_x, f_y)|^2 df_x df_y.$$

13. Coordinate rotation:

$$s(x \cos \psi - y \sin \psi, x \sin \psi + y \cos \psi) \Leftrightarrow S(f_x \cos \psi - f_y \sin \psi, f_x \sin \psi + f_y \cos \psi).$$

14. Coordinate transformation:

$$s(a_1 x + b_1 y, a_2 x + b_2 y) \Leftrightarrow \frac{1}{|a_1 b_2 - a_2 b_1|} S(A_1 f_x + A_2 f_y, B_1 f_x + B_2 f_y),$$

where

$$\begin{bmatrix} A_1 & B_1 \\ A_2 & B_2 \end{bmatrix} = \begin{bmatrix} a_1 & b_1 \\ a_2 & b_2 \end{bmatrix}^{-1}.$$

The coordinate conversion formula may be rewritten in a more suggestive form as

$$s\left(\begin{bmatrix} a_1 & b_1 \\ a_2 & b_2 \end{bmatrix} \begin{bmatrix} x \\ y \end{bmatrix} \right) \Leftrightarrow \frac{1}{|a_1 b_2 - a_2 b_1|} S\left(\begin{bmatrix} a_1 & a_2 \\ b_1 & b_2 \end{bmatrix}^{-1} \begin{bmatrix} f_x \\ f_y \end{bmatrix} \right),$$

or in a more compact form as

$$s(\boldsymbol{A}\boldsymbol{x}) = \frac{1}{|\det \boldsymbol{A}|} S(\boldsymbol{A}^{-1} \boldsymbol{f}),$$

where $\boldsymbol{x}$ and $\boldsymbol{f}$ are column vectors with components (x,y) and (f_x, f_y).

3.2 Transforms of Some Common Functions

A list of two-dimensional Fourier transform pairs for some common functions is given in Table 3.1. Some of these Fourier transform pairs are developed in this section, while others will be developed in later sections. Whenever $s(x,y)$ factors as

$$s(x,y) = s'(x) s''(y),$$

then the integral defining the two-dimensional Fourier transform separates into a product of two integrals, each of which is a one-dimensional Fourier transform. Hence

$$S(f_x, f_y) = S'(f_x) S''(f_y).$$

3.2 Transforms of Some Common Functions

Table 3.1 A table of two-dimensional Fourier transform pairs

$s(x, y)$	$S(f_x, f_y)$
$\text{rect}(x, y)$	$\text{sinc}(f_x, f_y)$
$\text{sinc}(x, y)$	$\text{rect}(f_x, f_y)$
$\text{circ}(x, y)$	$\text{jinc}(f_x, f_y)$
$\text{jinc}(x, y)$	$\text{circ}(f_x, f_y)$
$\text{lypd}(x, y)$	$\text{sinc}^2(f_x, f_y)$
$\text{chat}(x, y)$	$\text{jinc}^2(f_x, f_y)$
$\delta(x, y)$	1
1	$\delta(f_x, f_y)$
$\delta(x)$	$\delta(f_y)$
$e^{-\pi(x^2+y^2)}$	$e^{-\pi(f_x^2+f_y^2)}$
$e^{j\pi(x^2+y^2)}$	$je^{-j\pi(f_x^2+f_y^2)}$
$\dfrac{1}{\sqrt{x^2+y^2}}$	$\dfrac{1}{\sqrt{f_x^2+f_y^2}}$
$\text{comb}(x, y)$	$\text{comb}(f_x, f_y)$
$\text{comb}(x)$	$\text{comb}(f_x)\delta(f_y)$
$\text{comb}_N(x, y)$	$\text{dirc}_N(f_x, f_y)$
$\text{ring}(x, y)$	$\pi J_0\left(\pi\sqrt{f_x^2+f_y^2}\right)$
$\dfrac{1}{\sqrt{b^2+x^2+y^2}} e^{-2\pi a\sqrt{b^2+x^2+y^2}}$	$\dfrac{1}{\sqrt{a^2+f_x^2+f_y^2}} e^{-2\pi b\sqrt{a^2+f_x^2+f_y^2}}$
$\dfrac{1}{\sqrt{b^2+x^2+y^2}} e^{-j2\pi a\sqrt{b^2+x^2+y^2}}$	$\dfrac{1}{j\sqrt{a^2-f_x^2-f_y^2}} e^{-j2\pi b\sqrt{a^2-f_x^2-f_y^2}}$
$4\pi(1-4x^2-4y^2)\text{circ}(x, y)$	$J_2\left(\pi\sqrt{f_x^2+f_y^2}\right)/(f_x^2+f_y^2)$
$\sqrt{1-4x^2-4y^2}\,\text{circ}(x, y)$	$\dfrac{1}{2\pi^2}\dfrac{\sin\pi\sqrt{f_x^2+f_y^2}-\sqrt{f_x^2+f_y^2}\cos\pi\sqrt{f_x^2+f_y^2}}{\left(\pi\sqrt{f_x^2+f_y^2}\right)^3}$

Figure 3.2 The two-dimensional rectangle function

In this way, many two-dimensional Fourier transforms are easily obtained as products of one-dimensional Fourier transforms.

Figure 3.3 Plan view of the two-dimensional rectangle function

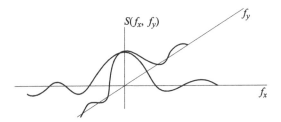

Figure 3.4 The two-dimensional function sinc(f_x, f_y)

An elementary two-dimensional function is the *two-dimensional rectangle function*. The function rect(x, y), as shown in Figure 3.2, is defined as the product of two one-dimensional rectangle functions:

$$\text{rect}(x, y) = \begin{cases} 1 & \text{when } |x| \leq 1/2 \text{ and } |y| \leq 1/2, \\ 0 & \text{otherwise} \end{cases}$$
$$= \text{rect}(x)\text{rect}(y).$$

A simple way to illustrate the two-dimensional rectangle function is by the plan view shown in Figure 3.3.

The *two-dimensional sinc function* is defined as the product of two one-dimensional sinc functions. Thus

$$\text{sinc}(x, y) = \text{sinc}(x)\text{sinc}(y).$$

Because the two-dimensional rectangle function factors as rect(x)rect(y), the appropriate two-dimensional Fourier transform pair involving the sinc function is

$$\text{rect}(x, y) \Leftrightarrow \text{sinc}(f_x, f_y)$$

By the duality property,

$$\text{sinc}(x, y) \Leftrightarrow \text{rect}(f_x, f_y).$$

The two-dimensional sinc function is illustrated in Figure 3.4. It is shown in plan view in Figure 3.5.

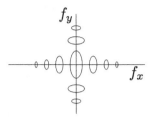

Figure 3.5 Plan view of the function sinc(f_x, f_y)

The *two-dimensional impulse function*[4] is defined by

$$\delta(x,y) = \delta(x)\delta(y).$$

The two-dimensional Fourier transform pairs

$$\delta(x,y) \Leftrightarrow 1$$

and

$$1 \Leftrightarrow \delta(f_x, f_y)$$

follow easily by the separation of variables.

The *two-dimensional gaussian pulse* is defined by

$$e^{-\pi(x^2+y^2)} = e^{-\pi x^2} e^{-\pi y^2}.$$

The two-dimensional Fourier transform pair

$$e^{-\pi(x^2+y^2)} \Leftrightarrow e^{-\pi(f_x^2+f_y^2)}$$

follows easily by the separation of variables.

The *two-dimensional chirp pulse* is

$$e^{-j\pi(x^2+y^2)} = e^{-j\pi x^2} e^{-j\pi y^2}.$$

The two-dimensional Fourier transform pair

$$e^{j\pi(x^2+y^2)} \Leftrightarrow je^{-\pi(f_x^2+f_y^2)}$$

again follows easily by separation of variables.

The *two-dimensional comb function* is defined as

$$\text{comb}(x,y) = \text{comb}(x)\text{comb}(y).$$

---

[4] Mathematically, the two-dimensional impulse function (or delta function) $\delta(x,y)$ is a generalized function defined as the limit of a sequence of smaller and taller two-dimensional rect functions. It has the (two-dimensional) sifting property that, for any continuous (complex-valued) function $f(x,y)$,

$$\int_{\mathcal{A}} \delta(x,y) f(x,y) dx dy = \begin{cases} f(0,0) & \text{for } (0,0) \in \mathcal{A}, \\ 0 & \text{otherwise,} \end{cases}$$

where $\mathcal{A}$ is any region of the plane.

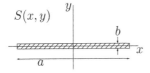

Figure 3.6 A long, thin rectangle function

The separation of variables immediately gives

$$\text{comb}(x,y) \Leftrightarrow \text{comb}(f_x, f_y).$$

Similarly, a *two-dimensional finite comb function* is defined as

$$\text{comb}_{N,M}(x,y) = \text{comb}_N(x)\text{comb}_M(y),$$

and a *two-dimensional dirichlet function* is defined as

$$\text{dirc}_{N,M}(x,y) = \text{dirc}_N(x)\text{dirc}_M(y).$$

Separation of variables gives the two-dimensional Fourier transform pair

$$\text{comb}_{N,M}(x,y) \Leftrightarrow \text{dirc}_{N,M}(f_x, f_y).$$

The properties of the two-dimensional Fourier transform allow us to quickly find the two-dimensional Fourier transforms of many other simple functions. The scaling property is used to find the Fourier transform of a thin strip. The modulation property is used to find the Fourier transforms of a simple array.

An exaggerated example of a rectangle is a long, thin strip. When b is small and a is large, the rectangle function takes the form of a strip along the x axis, as shown in plan view in Figure 3.6. By the scaling property of the Fourier transform,

$$\text{rect}\left(\frac{x}{a}, \frac{y}{b}\right) \Leftrightarrow ab\,\text{sinc}(af_x, bf_y),$$

the Fourier transform then becomes narrow in f_x and wide in f_y. The Fourier transform is approximately a strip along the f_y axis, but exhibiting the sidelobes of the sinc function. The long, thin sinc function is shown in Figure 3.7. The long axis of the Fourier transform is perpendicular to the long axis of the signal in the spatial domain. As the rectangle becomes longer in x and thinner in y, the Fourier transform becomes thinner in f_y and longer in f_x. The sidelobes are long and thin, close to the main lobe, and converging toward an infinitely long, infinitely thin ridge of infinite height as the long, thin rectangle becomes thinner. There is no proper function that serves as the convergent of this process. Instead, the apparent convergent is defined to be the generalized two-dimensional function $\Delta(f_x, f_y) = \delta(f_x) \cdot 1$, taking its properties from the process that defines it.

The two-dimensional function $s(x,y) = \delta(x) \cdot 1$, even though it may be regarded as a function of two variables, is constant in y. The function $\delta(x)$, regarded on two

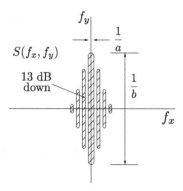

Figure 3.7 Transform of a long, thin rectangle function

dimensions, is then called a *line impulse*.[5] The line impulse is constant in y and should be visualized as an infinitely thin, infinitely high ridge lying along the y axis. The function is an impulse in x. For $x = 0$, it is infinite for all y. For other x, it is zero for all y. Separation of variables allows $\delta(x)$ when written as $\delta(x) \cdot 1$ to be transformed into $1 \cdot \delta(f_y)$. This leads to

$$\delta(x) \Leftrightarrow \delta(f_y)$$

as a (generalized) two-dimensional Fourier transform pair.

Another use of the impulse notation in two dimensions is the *ring impulse* or *ring function*, defined by $\text{ring}(x, y) = \delta\left(\sqrt{x^2 + y^2} - \frac{1}{2}\right)$. The ring impulse should be visualized as an infinitely thin, infinitely high ridge on the circle of unit diameter $\sqrt{x^2 + y^2} = \frac{1}{2}$. The discussion of the two-dimensional (generalized) Fourier transform of $\text{ring}(x, y)$ is deferred to Section 3.3.

3.3 Circularly Symmetric Functions

A function $s(x, y)$ for which

$$s(x, y) = s(x \cos \psi - y \sin \psi, x \sin \psi + y \cos \psi)$$

for all ψ is said to have *circular symmetry*. Consequently, by Property 13 of Section 3.1, the two-dimensional Fourier transform must satisfy

$$S(f_x, f_y) = S(f_x \cos \psi - f_y \sin \psi, f_x \sin \psi + f_y \cos \psi)$$

for any ψ. Therefore $S(f_x, f_y)$ is circularly symmetric whenever $s(x, y)$ is circularly symmetric.

It is convenient and common practice to restate functions with circular symmetry in polar coordinates. This leads to integration in polar coordinates that is expressed

---

[5] The line impulse and the ring impulse are generalized functions

in terms of Bessel functions. Express the x,y plane and the f_x, f_y plane in polar coordinates by

$$x = r\cos\theta,$$
$$y = r\sin\theta,$$
$$f_x = \rho\cos\phi,$$
$$f_y = \rho\sin\phi.$$

Let $s^\circ(r)$ denote the radial dependence of the circularly symmetric function $s(x,y)$:

$$s^\circ(r) = s(r\cos\theta, r\sin\theta).$$

The condition of circular symmetry means that the right side has no θ dependence, so the left side is well-defined as a function of only r. Let

$$S^\circ(\rho) = S(\rho\cos\phi, \rho\sin\phi).$$

The rotation property of the two-dimensional Fourier transform then asserts that the right side has no ϕ dependence. Translating the equation

$$S(f_x, f_y) = \int_{-\infty}^{\infty}\int_{-\infty}^{\infty} s(x,y) e^{-j2\pi(f_x x + f_y y)} dx dy$$

into polar coordinates shows that the functions $s^\circ(r)$ and $S^\circ(\rho)$ are related by

$$S^\circ(\rho) = \int_{-\pi}^{\pi} \left[\int_0^\infty \left[rs^\circ(r) e^{-j2\pi r\rho(\cos\theta\cos\phi + \sin\theta\sin\phi)} dr \right] d\theta \right.$$

$$= \int_0^\infty rs^\circ(r) \left[\int_{-\pi}^{\pi} e^{-j2\pi r\rho\cos(\theta-\phi)} d\theta \right] dr.$$

But the inner integral is over one period of a periodic function. Hence, as expected, that integral does not depend on ϕ, so ϕ can be set equal to any convenient value. Thus, with $\phi = \pi/2$,

$$S^\circ(\rho) = \int_0^\infty rs^\circ(r) \left[\int_{-\pi}^{\pi} e^{-j2\pi r\rho\sin\theta} d\theta \right] dr.$$

The integral on θ cannot be evaluated in closed form; it must be integrated numerically. The integral occurs often and is widely tabulated. It is called the *zero-order Bessel function of the first kind*,[6] and is defined by

$$J_0(t) = \frac{1}{2\pi} \int_{-\pi}^{\pi} e^{-jt\sin\theta} d\theta.$$

---

[6] The integral representation for the nth-order Bessel function of the first kind is

$$J_n(x) = \frac{1}{2\pi} \int_{-\pi}^{\pi} e^{-jx\sin\theta + jn\theta} d\theta.$$

Notice that

$$J_{-n}(x) = J_n(-x) = (-1)^n J_n(x),$$

which implies that $J_0(t)$ has even symmetry and $J_1(t)$ has odd symmetry. In particular, $J_1(0) = 0$.

3.3 Circularly Symmetric Functions

Figure 3.8 The circle function

Thus the two-dimensional Fourier transform of a circularly symmetric function is given in polar coordinates by

$$S^\circ(\rho) = 2\pi \int_0^\infty r s^\circ(r) J_0(2\pi r \rho) dr.$$

The function $S^\circ(\rho)$ is itself a circularly symmetric function on the nonnegative real numbers, as is to be expected.

Example
An important example of a circularly symmetric function is the *circle function*, defined by

$$\text{circ}(x, y) = \begin{cases} 1 & \text{for } \sqrt{x^2 + y^2} \leq \frac{1}{2} \\ 0 & \text{otherwise} \end{cases}$$

$$= \text{rect}\left(\sqrt{x^2 + y^2}\right),$$

and illustrated in Figure 3.8. For the circle function,

$$S^\circ(\rho) = 2\pi \int_0^{\frac{1}{2}} r J_0(2\pi r \rho) dr.$$

This integration can be evaluated by using the standard identity from the theory of Bessel functions:

$$\int_0^x \xi J_0(\xi) d\xi = x J_1(x),$$

where $J_1(x)$ is the first-order Bessel function of the first kind. Then

$$S^\circ(\rho) = \frac{J_1(\pi \rho)}{2\rho}$$

is the radial dependence of the Fourier transform of $s(x, y)$. The term on the right is discussed in detail below.

Inverse Transform
The inverse transform of $S^\circ(\rho)$ can be written in polar coordinates as

$$s^\circ(r) = 2\pi \int_0^\infty \rho S^\circ(\rho) J_0(2\pi r \rho) d\rho$$

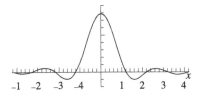

Figure 3.9 The jinc function

because, except for a sign change, there is no difference between the two-dimensional Fourier transform and the inverse two-dimensional Fourier transform. The difference in sign drops out in the transformation to polar coordinates.[7]

The jinc Function
Define the *jinc function* as the pulse

$$\text{jinc}(t) = \frac{J_1(\pi t)}{2t}.$$

The jinc function is illustrated in Figure 3.9. Its general appearance is similar in appearance to the sinc function. However, the zeros of the jinc function are not spaced uniformly, and the magnitudes of the sidelobes of the jinc function fall off more quickly than do the magnitudes of the sidelobes of the sinc function. At $t = 0$, the numerator and the denominator in the definition of jinc(t) are both zero. To evaluate jinc(0), l'Hôpital's rule may be used. A more immediate way to evaluate jinc(0) is to note that jinc(f_x, f_y) is the Fourier transform of circ(x, y), which means that jinc(0) is the area of a circle of unit diameter. Therefore jinc(0) $= \pi/4$. The locations of the zeros and the extrema of the jinc function are given in Table 3.2. A graph is given in Figure 9.1 of Chapter 9 showing a comparison of the sinc function, the jinc function, and a third function called the tinc function.

In polar coordinates, the two-dimensional Fourier transform of circ(x, y) is given succinctly in terms of the jinc function as

$$S^\circ(\rho) = \text{jinc}(\rho).$$

In rectangular coordinates, the Fourier transform of circ(x, y) is

$$S(f_x, f_y) = \text{jinc}\left(\sqrt{f_x^2 + f_y^2}\right).$$

---

[7] The formulas

$$S^\circ(f) = 2\pi \int_0^\infty ts^\circ(t) J_0(2\pi ft) dt$$

and

$$s^\circ(t) = 2\pi \int_0^\infty fS^\circ(f) J_0(2\pi ft) df$$

arise in the study of the two-dimensional Fourier transform of circularly symmetric functions. When the two expressions are themselves regarded as an invertible relationship between the two one-dimensional functions $s^\circ(t)$ and $S^\circ(f)$ on the nonnegative real line, these two integrals are known as the *Hankel transform of order zero* and the *inverse Hankel transform of order zero*.

3.3 Circularly Symmetric Functions

Table 3.2 Locations of zeros and extrema of the jinc function

	x	Normalized amplitude	Decibels
Central maximum ...	0	1.0	
First zero	1.2197	0	
Second peak	1.6347	−0.1323	−17.6
Second zero	2.2331	0.0	
Third peak	2.6793	0.0644	−23.8
Third zero	3.2383	0.0	
Fourth peak	3.6987	−0.0400	−28.0
Fourth zero	4.2411	0.0	
Fifth peak	4.7097	0.0279	−31.1
Fifth zero	5.2428	0.0	

It is sometimes convenient to express the jinc function as a function of two variables. The *two-dimensional jinc function* is defined as

$$\mathrm{jinc}(x, y) = \mathrm{jinc}\left(\sqrt{x^2 + y^2}\right).$$

We now have the two-dimensional Fourier transform pair

$$\mathrm{circ}(x, y) \Leftrightarrow \mathrm{jinc}(f_x, f_y).$$

By duality,

$$\mathrm{jinc}(x, y) \Leftrightarrow \mathrm{circ}(f_x, f_y).$$

The squared magnitude $\mathrm{jinc}^2(x, y)$ is called the *Airy disk* and is denoted $\mathrm{airy}(x, y)$. Thus

$$\mathrm{circ}(x, y) ** \mathrm{circ}(x, y) \Leftrightarrow \mathrm{airy}(f_x, f_y),$$

where $\mathrm{airy}(f_x, f_y) = \mathrm{jinc}^2(f_x, f_y)$

Example

A ring of thickness ϵ and height $1/\epsilon$, centered on a circle of unit diameter, can be written in terms of the difference between a circle function of diameter $1 + \epsilon$ and a circle function of diameter $1 - \epsilon$ as follows:

$$s(x, y) = \frac{1}{\epsilon}\left[\mathrm{circ}\left(\frac{x}{1+\epsilon}, \frac{y}{1+\epsilon}\right) - \mathrm{circ}\left(\frac{x}{1-\epsilon}, \frac{y}{1-\epsilon}\right)\right].$$

When $y = 0$, this is a rectangle of width 2ϵ centered at $x = 1$. This is a circularly symmetric function. In polar coordinates it becomes

$$s^\circ(r) = \frac{1}{\epsilon}\mathrm{rect}\left(\frac{2r - 1}{2\epsilon}\right),$$

where $r = \sqrt{x^2 + y^2}$. The Hankel transform of $s^\circ(r)$ is given by

$$S^\circ(\rho) = 2\pi \int_0^\infty r s^\circ(r) J_0(2\pi r\rho) dr$$

$$= \frac{2\pi}{\epsilon} \int_{\frac{1-\epsilon}{2}}^{\frac{1+\epsilon}{2}} r J_0(2\pi r\rho) dr.$$

In the limit as ϵ goes to zero, this becomes

$$S^\circ(\rho) = \pi J_0\left(\frac{2\pi\rho}{2}\right).$$

To conclude, define the ring impulse as the generalized function

$$\text{ring}(x, y) = \lim_{\epsilon \to 0} \frac{1}{\epsilon} \text{rect}\left[\frac{2\sqrt{x^2 + y^2} - 1}{2\epsilon}\right].$$

Then

$$\text{ring}(x, y) \Leftrightarrow \pi J_0\left(\pi\sqrt{f_x^2 + f_y^2}\right)$$

is a circularly symmetric generalized Fourier transform pair. The ring impulse is infinite on a circle of unit diameter and is zero elsewhere.

3.4 The Projection-Slice Theorem

The *projection-slice theorem* is a theorem relating a two-dimensional Fourier transform to the one-dimensional Fourier transforms of certain one-dimensional functionals called *projections* of the two-dimensional function.

Let $s(x, y)$ be a two-dimensional signal, possibly complex, of finite energy. The projection of $s(x, y)$ onto the x axis is

$$p(x) = \int_{-\infty}^\infty s(x, y) dy.$$

More generally, the projection of $s(x, y)$ onto an axis at angle θ is defined as

$$p_\theta(t) = \int_{-\infty}^\infty s(t\cos\theta - r\sin\theta, t\sin\theta + r\cos\theta) dr.$$

The angle θ specifies a rotation that relates the variables (t, r) to the variables (x, y). Figure 3.10 illustrates the sense in which $s(x, y)$ is "projected" into $p_\theta(t)$ by integrating along lines at an angle θ from the y axis.

The *slice* of the two-dimensional complex signal $S(f_x, f_y)$ of finite energy along the f_x axis is the one-dimensional function $S(f, 0)$. The *slice* of $S(f_x, f_y)$ along an axis at angle θ is $S(f\cos\theta, f\sin\theta)$.

The following theorem says that the one-dimensional Fourier transform of a projection of $s(x, y)$ is a slice of $S(f_x, f_y)$.

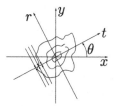

Figure 3.10 Illustrating the projection-slice theorem

Theorem 3.4.1 (Projection-Slice Theorem) *Let $p_\theta(t)$ denote the projection of $s(x,y)$ at angle θ. Then the Fourier transform of $p_\theta(t)$ is given by*

$$P_\theta(f) = S(f\cos\theta, f\sin\theta),$$

where $S(f_x, f_y)$ be the Fourier transform of $s(x,y)$.

Proof:

$$P_\theta(f) = \int_{-\infty}^{\infty} p_\theta(t) e^{-j2\pi ft} dt$$

$$= \int_{-\infty}^{\infty}\int_{-\infty}^{\infty} s(t\cos\theta - r\sin\theta, t\sin\theta + r\cos\theta) e^{-j2\pi ft} dr dt.$$

Now make the change in variables:

$$x = t\cos\theta - r\sin\theta,$$
$$y = t\sin\theta + r\cos\theta.$$

Consequently,

$$t = x\cos\theta + y\sin\theta$$

and

$$drdt = dxdy.$$

Therefore

$$P_\theta(f) = \int_{-\infty}^{\infty}\int_{-\infty}^{\infty} s(x,y) e^{-j2\pi(f\cos\theta x + f\sin\theta y)} dxdy$$
$$= S(f\cos\theta, f\sin\theta),$$

which completes the proof of the theorem. ∎

The projection-slice theorem implies that the Radon transform can be inverted. For each θ, denote the one-dimensional Fourier transform of $p(t,\theta)$ in the variable t by $P(f,\theta)$. Then

$$P(f,\theta) = S(f\cos\theta, f\sin\theta).$$

The two-dimensional Fourier transform of $s(x,y)$ expressed in polar coordinates is $P(f,\theta)$ with f as the radial variable and θ as the polar variable.

3 Signals in Two Dimensions

The projection

$$p_\theta(t) = \int_{-\infty}^{\infty} s(t\cos\theta - r\sin\theta, t\sin\theta + r\cos\theta)dr$$

was introduced with θ considered as a specific angle. A projection is defined in this way for every value of angle θ so $p_\theta(t)$ can be regarded as a function of θ. The projection then becomes a function of two variables, t and θ. The function is called the *Radon transform*[8] of $s(x,y)$.

Definition 3.4.2 *Given the two-dimensional signal $s(x,y)$ of finite energy, the Radon transform of $s(x,y)$ is the bivariate function*

$$p(t,\theta) = \int_{-\infty}^{\infty} s(t\cos\theta - r\sin\theta, t\sin\theta + r\cos\theta)dr.$$

Example
Let

$$s(x,y) = \text{circ}(x,y).$$

The circle function has radius equal to $1/2$ and satisfies

$$\text{circ}(t\cos\theta - r\sin\theta, t\sin\theta + r\cos\theta) = \text{circ}(t,r),$$

so the Radon transform is independent of θ. The projection of the circle function is easily seen to be

$$p(t,\theta) = \begin{cases} \sqrt{1-4t^2} & t \leq 1/2, \\ 0 & t \geq 1/2, \end{cases}$$

which is independent of θ. It follows from the projection-slice theorem that $P(f,\theta) = \text{jinc}(f)$, which is independent of θ because $S(f_x, f_y) = \text{jinc}(\sqrt{f_x^2 + f_y^2})$.

Example
The two-dimensional gaussian pulse

$$s(x,y) = e^{-\pi(x^2+y^2)}$$

has the Radon transform

$$p(t,\theta) = \int_{-\infty}^{\infty} s(t\cos\theta - r\sin\theta, t\sin\theta + r\cos\theta)dr$$

$$= \int_{-\infty}^{\infty} e^{-\pi(t^2+r^2)}dr$$

$$= e^{-\pi t^2},$$

which is independent of θ.

---

[8] The Radon transform can also be defined in multidimensional space. The Radon transform of order (n,k) is defined on the n-dimensional space over the real field or the complex field and maps a function onto its integrals on k-dimensional hyperplanes. A Radon transform of order $(n, n-1)$, also called simply the *Radon transform*, maps an n-dimensional function onto its integral on $(n-1)$ dimensional hyperplanes. The Radon transform of order $(n,1)$, also called the *shadow transform* (or the X-ray transform), maps an n-dimensional function onto its integral on (one-dimensional) lines.

3.5 Two Fundamental Fourier Transform Pairs

Two important two-dimensional Fourier transform pairs of circularly symmetric two-dimensional functions comprise the topic of this section. These Fourier transform pairs are

$$\frac{1}{\sqrt{b^2+x^2+y^2}}e^{-2\pi a\sqrt{b^2+x^2+y^2}} \Leftrightarrow \frac{1}{\sqrt{a^2+f_x^2+f_y^2}}e^{-2\pi b\sqrt{a^2+f_x^2+f_y^2}},$$

$$\frac{1}{\sqrt{b^2+x^2+y^2}}e^{-j2\pi a\sqrt{b^2+x^2+y^2}} \Leftrightarrow \frac{1}{j\sqrt{a^2-f_x^2-f_y^2}}e^{-j2\pi b\sqrt{a^2-f_x^2-f_y^2}}.$$

The second of these Fourier transform pairs is important in the study of propagating waves. The first Fourier transform pair is important in the study of diffusing waves and evanescent waves. The two two-dimensional Fourier transforms given herein are proved separately, but they do have a parallel structure. Replacing the real constant a in the first transform pair by the imaginary constant ja gives the second transform pair, but this formal substitution, without justification, is not a proof.

The two proofs of the two Fourier transform pairs make use of the following two definite integrals, which are developed in the end-of-chapter problems.

The first definite integral,

$$\frac{1}{\sqrt{\pi}}\int_0^\infty e^{-A^2 t^2 - B^2 t^{-2}}\,dt = \frac{1}{2A}e^{-2AB},$$

is developed in Problem 3.19a. To obtain the second definite integral from the first, differentiate both sides of the first expression with respect to B, then interchange the differentiation with the integration. This gives the definite integral

$$\frac{1}{\sqrt{\pi}}\int_0^\infty \frac{1}{t^2}e^{-A^2 t^2 - B^2 t^{-2}}\,dt = \frac{1}{2B}e^{-2AB},$$

where the differentiation has been performed inside the integration.

Each definite integral permits a term on the right side of the expressions in the following propositions involving the terms A and B to be replaced by the integral on the left side involving A^2 and B^2. These definite integrals are useful for studying an expression involving a square-root in the exponent because that square-root becomes squared so that it is more tractable, but this is at the cost of introducing a gratuitous integration.

The idea of the proof of the following theorem is to lift an expression into a larger setting where it can be manipulated, then to drop the result back to the original setting. Use the first definite integral above to replace an expression in $2A = \sqrt{b^2+x^2+y^2}$ by an expression in $4A^2 = b^2+x^2+y^2$ whose two-dimensional Fourier transform is known. Then use the second definite integral to remove the extraneous integration that is introduced during the first step.

3 Signals in Two Dimensions

Proposition 3.5.1 *The following is a two-dimensional Fourier transform pair,*

$$\frac{1}{\sqrt{b^2+x^2+y^2}}e^{-2\pi a\sqrt{b^2+x^2+y^2}} \Leftrightarrow \frac{1}{\sqrt{a^2+f_x^2+f_y^2}}e^{-2\pi b\sqrt{a^2+f_x^2+f_y^2}}.$$

Proof:
Let $2A = \sqrt{b^2+x^2+y^2}$ and $B = 2\pi a$. Use the first definite integral above to obtain

$$s(x,y) = \frac{1}{\sqrt{b^2+x^2+y^2}}e^{-2\pi a\sqrt{b^2+x^2+y^2}}$$

$$= \frac{1}{\sqrt{\pi}}\int_0^\infty e^{-(b^2+x^2+y^2)t^2/4-4\pi^2 a^2 t^{-2}}\,dt$$

$$= \frac{1}{\sqrt{\pi}}\int_0^\infty e^{-\xi(t)}\left[e^{-(x^2+y^2)t^2/4}\right]dt,$$

where $\xi(t) = 4\pi^2 a^2 t^{-2} - b^2 t^2/4$. The third line is written to highlight that the variables x and y appear only in the two-dimensional gaussian pulse contained within the brackets. Take the two-dimensional Fourier transform of $s(x,y)$ and move the integration on x and y inside the integration on t. Then

$$S(f_x,f_y) = \int_{-\infty}^\infty\int_{-\infty}^\infty e^{-j2\pi(f_x x+f_y y)}\left[\frac{1}{\sqrt{\pi}}\int_0^\infty e^{-\xi(t)}e^{-(x^2+y^2)t^2/4}dt\right]dxdy$$

$$= \frac{1}{\sqrt{\pi}}\int_0^\infty e^{-\xi(t)}\int_{-\infty}^\infty\int_{-\infty}^\infty e^{-j2\pi(f_x x+f_y y)}\left[e^{-(x^2+y^2)t^2/4}dxdy\right]dt.$$

Using the two-dimensional Fourier transform pair

$$e^{-C(x^2+y^2)} \Leftrightarrow \frac{\pi}{C}e^{-\pi^2(f_x^2+f_y^2)/C},$$

gives

$$S(f_x,f_y) = \frac{1}{\sqrt{\pi}}\int_{-\infty}^\infty e^{-4\pi^2 a^2 t^{-2}-b^2 t^2/4}\left[\frac{4\pi}{t^2}e^{-4\pi^2(f_x^2+f_y^2)t^{-2}}\right]dt$$

$$= 4\sqrt{\pi}\int_0^\infty \frac{1}{t^2}e^{-(b^2/4)t^2-4\pi^2(a^2+f_x^2+f_y^2)t^{-2}}\,dt.$$

Next use the second definite integral written above, now with $A = b/2$ and $B = 2\pi\sqrt{a^2+f_x^2+f_y^2}$ to evaluate the integral on the right side. This gives

$$S(f_x,f_y) = \frac{1}{\sqrt{a^2+f_x^2+f_y^2}}e^{-2\pi b\sqrt{a^2+f_x^2+f_y^2}},$$

which proves the transform pair given in the statement of the proposition. ∎

The proof of the transform pair of the next proposition is similar to the proof of Proposition 3.5.1. It begins with the definite integral

$$\frac{1-j}{\sqrt{2\pi}}\int_0^\infty e^{-jA^2 t^2 - jB^2 t^{-2}}\,dt = \frac{1}{2A}e^{-j2AB},$$

3.5 Two Fundamental Fourier Transform Pairs

which is developed in end-of-chapter Problem 3.19b. This definite integral is used to expand from the expression for $s(x,y)$. Differentiation of both sides with respect to B gives

$$\frac{1-j}{\sqrt{2\pi}} \int_0^\infty \frac{1}{t^2} e^{-jA^2 t^2 - jB^2 t^{-2}} dt = \frac{1}{2B} e^{-j2AB}.$$

This definite integral will be used to reduce to the expression for $S(f_x, f_y)$.

Proposition 3.5.2 *The following is a two-dimensional Fourier transform pair,*

$$\frac{1}{\sqrt{b^2 + x^2 + y^2}} e^{-j2\pi a \sqrt{b^2 + x^2 + y^2}} \Leftrightarrow \frac{1}{j\sqrt{a^2 - f_x^2 - f_y^2}} e^{-j2\pi b \sqrt{a^2 - f_x^2 - f_y^2}},$$

with the understanding that $j\sqrt{a^2 - \rho^2} = \sqrt{\rho^2 - a^2}$ when ρ is larger than a.

Proof
Let $2A = \sqrt{b^2 + x^2 + y^2}$ and $B = 2\pi a$. Using the first definite integral above to expand $s(x,y)$ gives

$$s(x,y) = \frac{1}{\sqrt{b^2 + x^2 + y^2}} e^{-j2\pi a \sqrt{b^2 + x^2 + y^2}}$$

$$= \frac{1-j}{\sqrt{2\pi}} \int_0^\infty e^{-j(b^2 + x^2 + y^2)(t^2/4) - j4\pi^2 a^2 t^{-2}} dt$$

$$= \frac{1-j}{\sqrt{2\pi}} \int_0^\infty e^{-j(4\pi^2 a^2 t^{-2} + b^2 t^2/4)} \left[e^{-jt^2(x^2+y^2)/4} \right] dt.$$

Take the two-dimensional Fourier transform of $s(x,y)$, interchanging the x and y integrations with the t integration and recalling that

$$e^{-jC(x^2+y^2)} \Leftrightarrow -j\frac{\pi}{C} e^{j\pi^2 (f_x^2+f_y^2)/C}.$$

Therefore

$$S(f_x, f_y) = \frac{1-j}{\sqrt{2\pi}} \int_0^\infty e^{-j(4\pi^2 a^2 t^{-2} + b^2 t^2/4)} \left[\frac{1}{j} \frac{4\pi}{t^2} e^{j4\pi^2 (f_x^2+f_y^2) t^{-2}} \right] dt$$

$$= \frac{4\pi}{j} \frac{(1-j)}{\sqrt{2\pi}} \int_0^\infty \frac{1}{t^2} e^{-j(b^2/4) t^2 - j4\pi^2 (a^2 - f_x^2 - f_y^2) t^{-2}} dt.$$

Finally, provided $a^2 - f_x^2 - f_y^2$ is positive, use the second definite integral written above to reduce the right side, now with $A = b/2$ and $B = 2\pi \sqrt{a^2 - f_x^2 - f_y^2}$. This gives

$$S(f_x, f_y) = \left(\frac{2\pi}{j}\right) \frac{1}{\sqrt{(4\pi^2)(a^2 - f_x^2 - f_y^2)}} e^{-j2\pi b \sqrt{a^2 - f_x^2 - f_y^2}},$$

provided $f_x^2 + f_y^2 \leq a^2$. For the case that $a^2 - f_x^2 - f_y^2$ is negative, the definite integral given in end-of-chapter Problem 3.19c can be used in a similar way to prove that

$$S(f_x, f_y) = \frac{1}{\sqrt{f_x^2 + f_y^2 - a^2}} e^{-2\pi b \sqrt{f_x^2 + f_y^2 - a^2}}.$$

This completes the proof of the proposition. ■

3.6 Two-Dimensional Pulse Arrays

Suppose that we have a doublet consisting of two translated rectangle functions, as shown in plan view in Figure 3.11:

$$s(x, y) = \text{rect}\left(\frac{x - A/2}{a}, \frac{y - B/2}{b}\right) + \text{rect}\left(\frac{x + A/2}{a}, \frac{y + B/2}{b}\right).$$

The translation property of the Fourier transform gives

$$S(f_x, f_y) = ab\,\text{sinc}(af_x, bf_y)e^{-j\pi(Af_x + Bf_y)} + ab\,\text{sinc}(af_x, bf_y)e^{j\pi(Af_x + Bf_y)}$$
$$= 2ab\,\text{sinc}(af_x, bf_y)\cos(\pi(Af_x + Bf_y)).$$

Then the two-dimensional Fourier transform of this pair of rectangles is the transform of a single rectangle function multiplied by a sinusoid in spatial frequency, as is shown symbolically in Figure 3.12. This is analogous to a similar statement for the one-dimensional Fourier transform.

The instance of a one-dimensional pulse train suggests a train of $2N + 1$ two-dimensional rectangles along a common slanted line as given by

$$p(x, y) = \sum_{\ell=-N}^{N} \text{rect}\left(\frac{x - \ell A/2}{a}, \frac{y - \ell B/2}{b}\right)$$

has the Fourier transform

$$P(f_x, f_y) = ab\,\text{sinc}(af_x, bf_y)\text{dirc}_{2N+1}(Af_x + Bf_y).$$

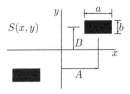

Figure 3.11 A pair of rectangles

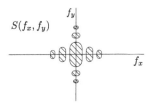

Figure 3.12 Transform of a pair of rectangles

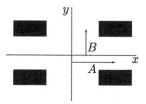

Figure 3.13 An array of rectangles

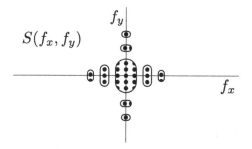

Figure 3.14 Transform of an array of rectangles

This Fourier transform is similar to that shown in Figure 3.12 except that the "cosine ridges" are replaced by the thinner "dirichlet ridges."

Now consider the array of four rectangles shown in Figure 3.13. This can be viewed as a doublet of doublets. The Fourier transform can be computed directly, but it is also instructive to compute the transform in two steps. First, let

$$p'(x,y) = \text{rect}\left(\frac{x-A/2}{a}, \frac{y}{b}\right) + \text{rect}\left(\frac{x+A/2}{a}, \frac{y}{b}\right),$$

which has the Fourier transform

$$P'(f_x, f_y) = 2ab\,\text{sinc}(af_x, bf_y)\cos(\pi A f_x).$$

Then the array of four rectangles is

$$p(x,y) = p'(x, y - B/2) + p'(x, y + B/2),$$

and

$$P(f_x, f_y) = 2P'(f_x, f_y)\cos(\pi B f_y).$$

Consequently,

$$P(f_x, f_y) = 4ab\,\text{sinc}(af_x, bf_y)\cos(\pi A f_x)\cos(\pi B f_y),$$

which is illustrated symbolically in Figure 3.14. Now the two-dimensional sinc function is multiplied by an array of positive and negative bumps created by the cosine-by-cosine function.

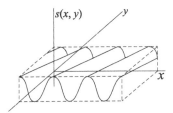

Figure 3.15 A two-dimensional sinusoid

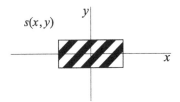

Figure 3.16 Plan view of a two-dimensional sinusoid

More generally, consider a two-dimensional N by N array of copies of the pulse $s(x, y)$ on an A by B grid. The Fourier transform is easily found to be

$$P(f_x, f_y) = S(f_x, f_y)\text{dirc}_N(Af_x)\text{dirc}_N(Bf_y)$$
$$= S(f_x, f_y)\text{dirc}_N(Af_x, Bf_y).$$

The cosine-by-cosine function has been replaced by a two-dimensional dirichlet function.

The next example is a sinusoidally modulated rectangle function. Figure 3.15 shows the function

$$s(x, y) = \text{rect}\left(\frac{x}{a}, \frac{y}{b}\right) \cos(2\pi(Ax + By)).$$

A depiction in plan view is shown in Figure 3.16. The rectangle is modulated by a spatial sinusoid with the frequency $f = \sqrt{A^2 + B^2}$ at an angle of $\tan^{-1}(A/B)$ with respect to the x axis. The modulation property of the Fourier transform then gives

$$S(f_x, f_y) = \tfrac{1}{2}ab\,\text{sinc}(a(f_x - A))\text{sinc}(b(f_y - B))$$
$$+ \tfrac{1}{2}ab\,\text{sinc}(a(f_x + A))\text{sinc}(b(f_y + B)),$$

as shown in Figure 3.17. In the terminology of modulation theory, the "signal" $s(x, y) = \text{rect}(x/a, y/b)$ has been "modulated" by the carrier $\cos 2\pi(Ax + By)$, thereby producing a sinusoidally modulated signal. The Fourier transform consists of two copies of the two-dimensional sinc function centered at the spatial frequencies $\pm(Ax + By)$.

A two-dimensional pulse array is a generalization of a one-dimensional pulse train to two dimensions. Let $s(x, y)$ be a finite-energy pulse. Then an N by M rectangular array of pulses is given by

3.6 Two-Dimensional Pulse Arrays

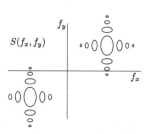

Figure 3.17 Transform of a two-dimensional sinusoid

$$p(x,y) = \sum_{m=0}^{M-1}\sum_{n=0}^{N-1} s(x - nA, y - mB).$$

One way of computing the two-dimensional Fourier transform of $p(x,y)$ is to regard the two-dimensional array as a one-dimensional array of one-dimensional arrays. Thus

$$p(x,y) = \sum_{m=0}^{M-1} p'(x, y - mB),$$

where

$$p'(x,y) = \sum_{n=0}^{N-1} s(x - nA, y).$$

Now use Theorem 2.3.1 twice to write

$$P(f_x, f_y) = P'(f_x, f_y)\text{dirc}_M(Bf_y)e^{-j\pi f_y B(M-1)},$$

where

$$P'(f_x, f_y) = S(f_x, f_y)\text{dirc}_N(Af_x)e^{-j\pi f_x A(N-1)}.$$

When the two-dimensional array is centered at the origin, the phase terms drop out and

$$P(f_x, f_y) = S(f_x, f_y)\text{dirc}_N(Af_x)\text{dirc}_M(Bf_y).$$

For each f_x at which Af_x is an integer, $\text{dirc}_N(Af_x)$ has a grating lobe, and for each f_y at which Bf_y is an integer, $\text{dirc}_M(Bf_y)$ has a grating lobe. The two-dimensional function $P(f_x, f_y)$ has a two-dimensional grating lobe whenever both Af_x and Bf_y are integers.

Example

The two-dimensional nine by nine square array of 81 circle functions,

$$p(x,y) = \sum_{m=-4}^{4}\sum_{n=-4}^{4} \text{circ}(x - n, y - m)$$
$$= \text{circ}(x,y) ** \text{comb}_9(x,y),$$

has the Fourier transform

$$P(f_x, f_y) = \text{jinc}\left(\sqrt{f_x^2 + f_y^2}\right) \text{dirc}_9(f_x)\text{dirc}_9(f_y)$$
$$= \text{jinc}(f_x, f_y) \, \text{dirc}_9(f_x, f_y),$$

where $\text{dirc}_9(f_x, f_y)$ is a two-dimensional dirichlet function. The structure of $P(f_x, f_y)$ is the infinite two-dimensional array of grating lobes on the square grid $\mathbb{Z}^2$ due to $\text{dirc}_9(f_x, f_y)$ with the amplitude of the grating lobes modulated radially by the jinc function and thereby decreasing radially.

Example
Now fill a large circle of radius R with a large number of small circles of radius $\frac{1}{2}$ arranged on a square grid. Place the small circles at each point of the integer lattice $\mathbb{Z}^2$, provided that the small circle lies inside the large circle of radius R centered at the origin. This is approximated by

$$p(x, y) = \left(\sum_{m=-N}^{N} \sum_{n=-N}^{N} \text{circ}(x - n, y - m)\right) \text{circ}\left(\frac{x}{R}, \frac{y}{R}\right).$$

This approximation uses a circle of diameter R as a "cookie cutter" to cut a large circle from the square array of small circles. When N is larger than $R/2$, the pattern of small circles will completely cover the large circle. Some of the small circles may be cut in two by the large circle, keeping only part of those small circles. Because of these partial circles, the expression is only an approximation to the original statement because in the approximation some of the small circles have been cut but the pieces are not discarded.

Under this approximation, the two-dimensional Fourier transform of $p(x, y)$ is

$$P(f_x, f_y) = \big(\text{jinc}(f_x, f_y)\text{dirc}_{2N+1}(f_x, f_y)\big) ** R\,\text{jinc}(Rf_x, Rf_y).$$

Making N larger does not change $p(x, y)$ because the additional small circles outside the large circle are discarded by the cookie cutter. Therefore, making N larger cannot change $P(f_x, f_y)$ either. Even making N infinite does not change $p(x, y)$. Hence

$$P(f_x, f_y) = \big(\text{jinc}(f_x, f_y)\text{comb}(f_x, f_y)\big) ** R\,\text{jinc}(Rf_x, Rf_y).$$

Each of the impulses of the comb function is amplitude weighted by $\text{jinc}(f_x, f_y)$, then convolved with $\text{jinc}(Rf_x, Rf_y)$. We can make the further approximation that

$$P(f_x, f_y) \approx R\,\text{jinc}(f_x, f_y) \sum_{n=-\infty}^{\infty} \sum_{m=-\infty}^{\infty} \text{jinc}(R(f_x - n), R(f_y - m)).$$

This approximation is an infinite square array of thin jinc functions with the amplitude of that array modulated by a wide jinc function. The wide jinc function attenuates the thin jinc functions according to their distance from the origin.

3.7 Sampling in Two Dimensions

Two-dimensional sampling of the complex function $s(x,y)$ is a generalization of one-dimensional sampling of the complex function $s(t)$. It can be described as the multiplication of $s(x,y)$ by the two-dimensional comb function shown in Figure 3.18. Then the function

$$s'(x,y) = \text{comb}(x,y)s(x,y)$$

is a two-dimensional array of amplitude-weighted impulses,

$$s'(x,y) = \sum_i \sum_{i'} s_{ii'} \delta(x-i, y-i').$$

The impulse located at $x = i$ and $y = i'$ has an amplitude given by the sample value $s_{ii'}$. Using the Fourier transform relationship

$$\text{comb}(x,y) \Leftrightarrow \text{comb}(f_x, f_y),$$

together with the convolution theorem, shows that in the frequency domain this becomes

$$S'(f_x, f_y) = \text{comb}(f_x, f_y) ** S(f_x, f_y)$$
$$= \sum_j \sum_{j'} S(f_x - j, f_y - j').$$

This is an array of translated copies, or images, of $S(f_x, f_y)$. For a function $S(f_x, f_y)$ whose support is contained within the unit square, the images do not overlap. There is no aliasing. Therefore, the product

$$S(f_x, f_y) = S'(f_x, f_y)\text{rect}(x,y)$$

recovers $S(f_x, f_y)$.

This motivates the sampling theorem in two dimensions, which states that when the support of $S(f_x, f_y)$ lies inside the unit square, the sampling operation

$$s'(x,y) = \text{comb}(x,y)s(x,y)$$

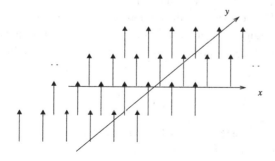

Figure 3.18 A two-dimensional array of impulses

is inverted by the convolution

$$s(x,y) = \text{sinc}(x,y) ** s'(x,y).$$

This is the interpolation operation that recovers the original two-dimensional pulse $s(x,y)$.

Theorem 3.7.1 (Nyquist–Shannon Theorem) *The two-dimensional signal $s(x,y)$ whose transform $S(f_x,f_y)$ satisfies $S(f_x,f_y) = 0$ for $|f_x| > 1/2A$ and $|f_y| > 1/2B$ can be recovered from its two-dimensional Nyquist samples at $x = \ell A$ for $\ell = \ldots, -1, 0, +1, \ldots$ and $y = \ell' B$ for $\ell' = \ldots, -1, 0, +1, \ldots$ by the two-dimensional Nyquist–Shannon interpolation formula*

$$s(x,y) = \sum_{\ell=-\infty}^{\infty} \sum_{\ell'=-\infty}^{\infty} s_{\ell\ell'} \text{sinc}\left(\frac{x}{A} - \ell\right) \text{sinc}\left(\frac{y}{B} - \ell'\right).$$

Proof:
The proof follows the same reasoning as the proof of the one-dimensional Nyquist–Shannon theorem and follows the outline given prior to the statement of the theorem. ∎

To see how sampling works in the general case of two dimensions, suppose that $s(x,y)$ is any signal in two dimensions that has finite energy and whose transform $S(f_x, f_y)$ is zero except within the finite region $\mathcal{R}$ of the f_x, f_y plane. Let A and B denote the length and width of any rectangle centered at the origin of the f_x, f_y plane that encloses $\mathcal{R}$. The signal $s(x,y)$ is represented by the set of its values on a rectangular grid of points spaced in x by $1/A$ and spaced in y by $1/B$. These are the two-dimensional Nyquist samples

$$s_{\ell\ell'} = s\left(\frac{\ell}{A}, \frac{\ell'}{B}\right) \qquad \ell, \ell' = \ldots, -1, 0, +1, \ldots.$$

The convolution with the impulses of the comb function produces images of $S(f_x, f_y)$. These images are translated by integer multiples of A in spatial frequency f_x and by integer multiples of B in spatial frequency f_y. Because of the condition that $S(f_x, f_y)$ is equal to zero outside of the A by B rectangle, the translated copies of $S(f_x, f_y)$ do not overlap, as shown in Figure 3.19. This illustration shows that to recover $S(f_x, f_y)$ as described in Theorem 9.3.1, it is enough to multiply by a rectangle function that covers only the central rectangle of the frequency plane. That is, define the two-dimensional interpolation spectrum as

$$H(f_x, f_y) = \frac{1}{AB} \text{rect}\left(\frac{f_x}{A}, \frac{f_y}{B}\right),$$

corresponding to the two-dimensional interpolation function given by the point-spread function $h(x,y) = \text{sinc}(Ax, By)$. It is clear that

$$S(f_x, f_y) = H(f_x, f_y) S'(f_x, f_y).$$

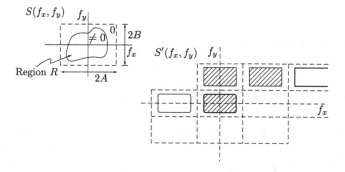

Figure 3.19 Images of a region formed by sampling

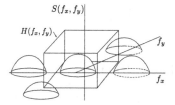

Figure 3.20 Illustrating the derivation of an interpolation formula

This is illustrated in Figure 3.20 showing the sampling images and the rectangle that recovers the original spectrum by cropping the spectrum of the sampled signal. In the spatial domain, this is equivalent to a two-dimensional convolution,

$$s(x,y) = h(x,y) ** s'(x,y),$$

leading to the two-dimensional Nyquist–Shannon interpolation formula

$$s(x,y) = \sum_{\ell=-\infty}^{\infty} \sum_{\ell'=-\infty}^{\infty} s_{\ell\ell'} \operatorname{sinc}(Ax - \ell, By - \ell').$$

as was the case for one-dimensional interpolation.

Jinc Interpolation

Sinc interpolation requires that the support of $S(f_x,f_y)$ is confined to an A by B rectangle as shown in Figure 3.19. When instead the support of $S(f_x,f_y)$ is confined to the interior of a circle, an alternative interpolation formula can be used. When the spectrum $S(f_x,f_y)$ is supported by a unit circle, the spectrum of the sampled data $S'(f_x,f_y)$ is confined to an infinite array of unit circles centered on the points of $\mathbb{Z}^2$. This is an array of translated copies of $S(f_x,f_y)$. Because the support of $S(f_x,f_y)$ is contained within the unit circle, the images do not overlap. There is no aliasing. Therefore, the product

$$S(f_x,f_y) = S'(f_x,f_y)\text{circ}(x,y)$$

recovers $S(f_x,f_y)$. In the x,y plane, this becomes

$$s(x,y) = \text{jinc}(x,y) ** s'(x,y).$$

This alternative interpolation formula is jinc interpolation.

More generally, suppose that $s(x,y)$ is equal to zero everywhere outside a region that can be contained within a circle of radius $A/2$. Because this circle is contained in a square of size A, the signal can be recovered by the Nyquist–Shannon interpolation formula described before. Alternatively, let

$$H(f_x,f_y) = \frac{1}{A^2}\text{circ}\left(\frac{f_x}{A},\frac{f_y}{A}\right),$$

with the transform

$$h(x,y) = \text{jinc}\left(A\sqrt{x^2+y^2}\right).$$

Then the terms in the sum

$$S'(f_x,f_y) = A^2 \sum_{\ell=-\infty}^{\infty} \sum_{\ell'=-\infty}^{\infty} S(f_x - A\ell, f_y - A\ell')$$

have supports that are nonintersecting, so that

$$S(f_x,f_y) = H(f_x,f_y)S'(f_x,f_y).$$

Consequently, in the spatial domain,

$$s(x,y) = h(x,y) **s'(x,y),$$

which results in the alternative Nyquist–Shannon interpolation formula

$$s(x,y) = \sum_{\ell=-\infty}^{\infty}\sum_{\ell'=-\infty}^{\infty} s_{\ell\ell'}\text{jinc}\left[\sqrt{(Ax-\ell)^2+(Ay-\ell')^2}\right],$$

for a two-dimensional signal $s(x,y)$ whose transform $S(f_x,f_y)$ has circular support of radius A.

Although jinc interpolation also has sidelobes, these sidelobes fall off more quickly than those of sinc interpolation. This means that jinc interpolation is less sensitive to error caused by truncation of the summations.

Soft Interpolation

Two-dimensional sinc interpolation may be unsatisfactory because the sidelobes of $\text{sinc}(x,y)$ fall off too slowly. The sidelobe magnitudes of $\text{sinc}(x,y)$ decay as $1/n$. Recalling that the summation $\sum_n 1/n$ diverges suggests that sinc interpolation can be computationally sensitive. To avoid computational difficulties, a modified

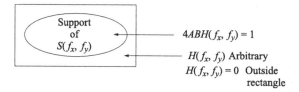

Figure 3.21 Sufficient conditions on $H(f_x, f_y)$

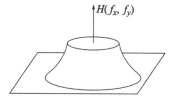

Figure 3.22 A circular interpolating filter

two-dimensional interpolation formula can be used. This requires some slack in the sampling procedure so that an alternative interpolating function can be admitted. This slack is provided by sampling at a rate higher than the Nyquist rate.

It is sufficient to choose any interpolation filter, $h(x, y)$, such that

$$H(f_x, f_y) = \begin{cases} 1/AB & \text{for } S(f_x, f_y) \neq 0, \\ 0 & \text{for } |f_x| \geq A/2 \text{ or } |f_y| \geq B/2, \end{cases}$$

as illustrated in Figure 3.21. In the transition region where $|f_x| < A/2$, $|f_y| < B/2$, and $S(f_x, f_y) = 0$, the function $H(f_x, f_y)$ is arbitrary and can be chosen so as to control the sidelobes of the interpolation formula. One possible choice of $H(f_x, f_y)$, shown in Figure 3.22, is suitable when $S(f_x, f_y)$ is supported on a circle.

Hexagonal Sampling

Figure 3.23 shows how to reduce the unused space between the circles of the interpolating filter $H(f_x, f_y)$. This configuration suggests that by rearranging the spatial samples, the images become placed more closely in the f_x, f_y plane and this would place the spatial-domain samples farther apart. In particular, the spatial samples can be arranged in a hexagonal grid and then the sampling images in the frequency domain will also be hexagonally arranged, as shown in Figure 3.24.

Consider the set of all points of the x, y plane that satisfy

$$\begin{bmatrix} x \\ y \end{bmatrix} = \begin{bmatrix} 2 & 1 \\ 0 & \sqrt{3} \end{bmatrix} \begin{bmatrix} i \\ i' \end{bmatrix},$$

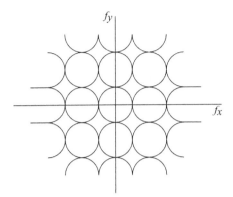

Figure 3.23 Images of a circular region under a rectangular sampling format

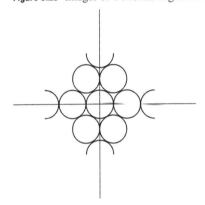

Figure 3.24 Images of a circular region under a hexagonal sampling format

where i and i' are integers. This set of points in the plane is called a *hexagonal lattice*.[9] Figure 3.25 shows the set of points of this lattice; the hexagons are drawn only to show how the lattice derives its name.

Define the two-dimensional array of impulses on the hexagonal lattice

$$\sum_{i=-\infty}^{\infty} \sum_{i'=-\infty}^{\infty} \delta(x - (2i + i'), y - \sqrt{3}i')$$

---

[9] In general, a two-dimensional *lattice*, denoted Λ, is the set of points in the plane that can be written

$$\begin{bmatrix} x \\ y \end{bmatrix} = M \begin{bmatrix} i \\ i' \end{bmatrix},$$

where i and i' are arbitrary integers, and M is any fixed, real-valued, nonsingular, two by two matrix called the *generator matrix* of the lattice. A planar lattice is a regular arrangement of points in the plane. The two lattices formed by a matrix M and its inverse M^{-1} are called *reciprocal lattices*. The *fundamental cell* of the lattice is the parallelepiped defined by four points corresponding to $(i, i') = (0, 0), (1, 0), (0, 1)$, and $(1, 1)$. Other cells of the lattice are defined similarly. A suitable name for the lattice is $M\mathbb{Z}^2$. A lattice in three dimensions is defined similarly as $M\mathbb{Z}^3$, where M is a real full-rank three by three matrix.

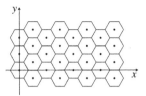

Figure 3.25 The hexagonal lattice

to form the sampling array that is shown in Figure 3.27. For a raster scan of an image, this simply means that alternate horizontal lines are slightly closer and the samples in the alternate horizontal lines are halfway staggered. The sampled function is

$$s'(x,y) = \left[\sum_{i=-\infty}^{\infty}\sum_{i'=-\infty}^{\infty} \delta(x-(2i+i'), y-\sqrt{3}i')\right]s(x,y).$$

To transform this equation into the frequency domain, refer to the Fourier transform pair

$$\text{comb}(x,y) \Leftrightarrow \text{comb}(f_x, f_y),$$

which is written out as

$$\sum_{i=-\infty}^{\infty}\sum_{i'=-\infty}^{\infty} \delta(x-i, y-i') \Leftrightarrow \sum_{j=-\infty}^{\infty}\sum_{j'=-\infty}^{\infty} \delta(x-j, y-j').$$

The coordinate transformation property of the two-dimensional Fourier transform shows that this becomes

$$\sum_{i=-\infty}^{\infty}\sum_{i'=-\infty}^{\infty} \delta(x-(2i+i'), y-\sqrt{3}i') \Leftrightarrow \frac{1}{2\sqrt{3}}\sum_{j=-\infty}^{\infty}\sum_{j'=-\infty}^{\infty} \delta\left(f_x - \frac{j}{2}, f_y + \frac{j-2j'}{2\sqrt{3}}\right)$$

because

$$\begin{bmatrix} 2 & 1 \\ 0 & \sqrt{3} \end{bmatrix}^{-1} = \frac{1}{2\sqrt{3}}\begin{bmatrix} \sqrt{3} & -1 \\ 0 & 2 \end{bmatrix}.$$

A compact statement of this Fourier transform pair is

$$\text{comb}\left(\begin{vmatrix} 2 & 1 \\ 0 & \sqrt{3} \end{vmatrix}\begin{vmatrix} x \\ y \end{vmatrix}\right) \Leftrightarrow \frac{1}{2\sqrt{3}}\text{comb}\left(\frac{1}{2\sqrt{3}}\begin{vmatrix} \sqrt{3} & -1 \\ 0 & 2 \end{vmatrix}\begin{vmatrix} f_x \\ f_y \end{vmatrix}\right).$$

The right side is the hexagonal array of sampling images shown in Figure 3.26. Because the Fourier transform of a hexagonal array of impulses is again a hexagonal array of impulses, the hexagonally sampled signal.

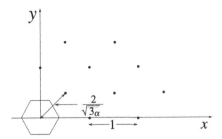

Figure 3.26 Space-domain hexagonal array

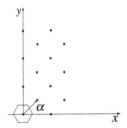

Figure 3.27 Frequency-domain hexagonal array

$$s'(x,y) = \left[\sum_{i=-\infty}^{\infty}\sum_{i'=-\infty}^{\infty} \delta\left(x - (2i+i'), y - \sqrt{3}i'\right)\right] s(x,y)$$

has the Fourier transform

$$S'(f_x, f_y) = \frac{1}{2\sqrt{3}}\left[\sum_{j=-\infty}^{\infty}\sum_{j'=-\infty}^{\infty} \delta\left(f_x - \frac{j}{2}, f_y + \frac{j-2j'}{2\sqrt{3}}\right)\right] ** S(f_x, f_y)$$

$$= \frac{1}{2\sqrt{3}}\sum_{j=-\infty}^{\infty}\sum_{j'=-\infty}^{\infty} S\left(f_x - \frac{j}{2}, f_y + \frac{j-2j'}{2\sqrt{3}}\right).$$

An examination of the hexagons shown in Figure 3.27 suggests that the images are packed as tightly as possible without overlapping. When the x and y axes are scaled by the parameter $A/2$, the spatial-domain samples are $s\left(\frac{4}{A}i + \frac{2}{A}i', \frac{2}{A}\sqrt{3}i'\right)$.

The earlier development of jinc interpolation carries over to hexagonal sampling. This leads to the interpolation formula

$$s(x,y) = \sum_{i=-\infty}^{\infty}\sum_{i'=-\infty}^{\infty} s\left((4i+2i')/A, 2\sqrt{3}i'/A\right) \text{jinc}\left[2\pi\sqrt{(Ax-4i)^2 + (Ay-2i')^2}\right].$$

The number of sampling points needed for hexagonal sampling is less than the number of sampling points needed for rectangular sampling. For rectangular sampling, there is one sampling point for each square of side $1/A$. For hexagonal sampling, there is one sampling point for each hexagon. The area of the square is A^{-2}. The area of a hexagon is $\frac{2}{\sqrt{3}}A^{-2}$. Therefore the number of hexagons needed to cover a large region is 0.866

of the number of squares needed to cover that same region. The number of samples is fewer in this proportion. This is a saving of 13.4 percent of the required number of samples per unit area.

3.8 Two-Dimensional Signals and Filters

A two-dimensional, space-invariant filter is a generalization of a one-dimensional, time-invariant filter. The one-dimensional filter is described by the impulse-response function of the filter. A one-dimensional impulse response is often required to be causal, but causality is not a useful notion in two dimensions. It is common for a two-dimensional spatial filter $h(x, y)$ to be nonzero for all x and y.

The two-dimensional function $h(x, y)$ is usually called a *point-spread function* rather than an impulse-response function. This terminology carries the connotation that $h(x, y)$ can be nonzero for both negative and positive values of both x and y.

The operation of a two-dimensional filter is described by the convolution relationship

$$v(x, y) = \int_{-\infty}^{\infty} \int_{-\infty}^{\infty} h(x - \xi, y - \eta) s(\xi, \eta) d\xi d\eta.$$

This two-dimensional convolution is written compactly as

$$v(x, y) = h(x, y) ** s(x, y).$$

With the change in variables $\xi' = x - \xi$ and $\eta' = y - \eta$, the two-dimensional convolution becomes

$$v(x, y) = \int_{-\infty}^{\infty} \int_{-\infty}^{\infty} s(x - \xi', y - \eta') h(\xi', \eta') d\xi' d\eta'.$$

This shows that two-dimensional convolution is a commutative operation,

$$h(x, y) ** s(x, y) = s(x, y) ** h(x, y).$$

The role of the point-spread function in forming $v(x, y)$ from $s(x, y)$ can be depicted by a block diagram as shown in Figure 3.28. When $s(x, y)$ is the delta function $\delta(x, y)$, the filtering operation

$$v(x, y) = \int_{-\infty}^{\infty} \int_{-\infty}^{\infty} h(x - \xi, y - \eta) \delta(\xi, \eta) d\xi d\eta$$
$$= h(x, y)$$

replaces the impulse $\delta(x, y)$ at its input by the point-spread function $h(x, y)$ at its output. The "point" $\delta(x, y)$ is spread into the function $h(x, y)$. Similarly, the impulse $\delta(x - x_0, y - y_0)$ at the input is replaced by $h(x - x_0, y - y_0)$. This is again the same point-spread function but now located at (x_0, y_0). Consequently, the infinitesimal impulse $s(\xi, \eta) d\xi d\eta \delta(x - \xi, y - \eta)$ of weight $s(\xi, \eta) d\xi d\eta$ located at (ξ, η), when passed through the filter $h(x, y)$, produces the output $s(\xi, \eta) h(x - \xi, y - \eta) d\xi d\eta$. The

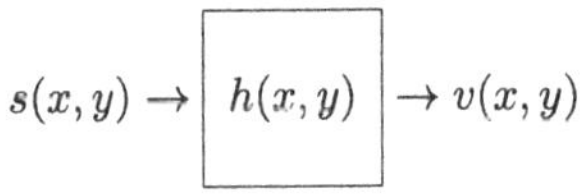

$$s(x,y) \rightarrow \boxed{h(x,y)} \rightarrow v(x,y)$$

Figure 3.28 Two-dimensional, point-spread function as a filter

convolution equation is then the superposition of the response to all such inputs, as expressed by the convolution integral.

The two-dimensional filter is expressed in the frequency domain by using the two-dimensional convolution theorem. In the frequency domain, the output of the two-dimensional filter becomes the multiplication

$$V(f_x, f_y) = H(f_x, f_y)S(f_x, f_y)$$

of the two functions of two-dimensional frequency (f_x, f_y).

Example
The filter with the two-dimensional transfer function

$$H(f_x, f_y) = \text{circ}\left(\frac{f_x}{A}, \frac{f_y}{A}\right)$$

applied to the pulse $s(x,y)$ rejects all two-dimensional frequency components with $f_x^2 + f_y^2 > A^2$ and passes all other frequency components without change. The two-dimensional impulse response of this filter is

$$h(x,y) = A \text{ jinc}\left(A\sqrt{x^2 + y^2}\right).$$

Because the circle function has a sharp cutoff in the frequency domain, points in the space-domain input signal $s(x,y)$ display the sidelobes of the jinc function in the reconstructed output signal $s(x,y)$.

A softer cutoff can be obtained by choosing the two-dimensional filter with the response

$$H_n(f_x, f_y) = \frac{1}{1 + (f_x^2 + f_y^2)^n}.$$

The filter $H_n(f_x, f_y)$ is known as a *two-dimensional Butterworth filter of order n*. Because $H_n(f_x, f_y)$ approaches $\text{circ}(f_x/2, f_y/2)$ as n goes to infinity, the response $h_n(x,y)$ can be described as a softened version of a jinc function with smaller sidelobes. A Butterworth filter might be used to suppress high-frequency noise from a two-dimensional image without introducing significant sidelobes as unwanted artifacts.

Discrete Signals
The output of a sampler of a *continuous* two-dimensional signal is a *discrete* two-dimensional signal. The sampler can be followed by a quantizer, thereby reducing the signal to a digital representation. Filtering of a digital representation that satisfies the

Nyquist criterion can be completely equivalent to the filtering of the original two-dimensional signal on continuous variables.

A two-dimensional sample is closely related to the notion of a pixel. This variation of sampling in two dimensions is also called *pixelization* and the samples are called *pixels*. A pixel is typically defined by reference to a two-dimensional lattice.

The values of the pixelization may be simply the samples of $s(x,y)$ on the points of the lattice. More commonly, however, a pixelization consists of the set of integrations of $s(x,y)$ over the cells of the two-dimensional lattice. The values of the integration values over the cells are the pixel values. For the square lattice $\mathbb{Z}^2$, the cells are the square cells of the lattice. A pixelization of $s(x,y)$ on the square lattice is an array of the values of the integrations over the cells of the square lattice. These values represent $s(x,y)$. This pixelization on the unit square lattice is equivalent to the sampling of the result of the convolution $\text{rect}(x,y) * *s(x,y)$.

The two-dimensional Nyquist–Shannon theorem, given as Theorem 3.7.1, states that either $s(x,y)$ or $\text{rect}(x,y) * *s(x,y)$ can be recovered, depending on the method of pixelization. The latter case is mathematically weaker because the samples of $s(x,y)$ cannot be fully recovered from the samples of $s(x,y) * *\text{rect}(x,y)$.

A signal $s(x,y)$ may be pixelated either for transmission and storage or for discrete processing. The Nyquist–Shannon theorem defines a vector-space equivalence, or isomorphism, between the continuous and discrete versions of a finite-energy bandlimited waveform. Addition of vectors and multiplication by a scalar are preserved across the operations of sampling and interpolation. This immediately implies that linear filtering is preserved. One may filter a continuous function by the operations of first sampling, then discrete filtering, and then interpolation provided that the Nyquist condition on sampling rate is satisfied throughout by both the signal and the filter.

3.9 Resolution and Apodization

The quality of an imaging system is judged, in part, by the resolution of the system. For a linear system, a scene $\rho(x,y)$ is broadened by the point-spread function $h(x,y)$ of the imaging system thereby producing the image $v(x,y) = h(x,y) * * \rho(x,y)$. The resolution of $h(x,y)$ refers to the ability to immediately distinguish the two points $\delta(x,y)$ and $\delta(x-a,y-b)$ by the inspection of the convolution of $\delta(x,y)+\delta(x-a,y-b)$ with the point-spread function $h(x,y)$.

A plan view of the main lobe of a possible point-spread function $h(x,y)$ with a simple shape is shown in Figure 3.29. Figure 3.30 shows a plan view of the output after a pair of impulses is convolved with the point-spread function $h(x,y)$.

The many criteria that are discussed in Section 2.9 for quantifying the resolution of one-dimensional signals can be generalized to quantify the resolution of two-dimensional signals. These criteria include the Rayleigh resolution criterion, the Gabor resolution criterion, the Sparrow resolution criterion, and the Woodward resolution criterion. Of these, the Gabor resolution criterion has the cleanest mathematical

Figure 3.29 Plan view of a point-spread function

Figure 3.30 Convolution of a point-spread function with two impulses

properties, but the Rayleigh resolution criterion is more convenient and is the most popular.

The Rayleigh resolution of the jinc function, $\text{jinc}(\sqrt{x^2 + y^2}/d)$, is defined as the value of the separation between two copies of the function for which the main lobe of the one copy falls on the first zero of the other copy. For the jinc function, this occurs where $2\pi\sqrt{x^2 + y^2}/d = 0.610$. Consequently, the Rayleigh resolution of the jinc function is $0.61d/2\pi$. This is a variation of the Abbe diffraction limit usually expressed in terms of wavelength in the field of optics. The Rayleigh resolution can be defined for any function $h(x,y)$ for which the first zero contour $h(x,y) = 0$ is a circle surrounding the maximum. When the first zero contour is not a circle, then the maximum radial distance of that contour might be used as a measure of resolution. When there is no such closed contour, as for a gaussian point-spread function, the Rayleigh resolution is not meaningful as such.

Another resolution criterion is the half-power resolution criterion. To define the half-power resolution, consider the set of points (x,y) forming the outermost closed contour at which $|h(x,y)|^2$ equals one-half of the maximum value of $|h(x,y)|^2$. This set, when it exists, is the set of (x,y) satisfying

$$|h(x,y)|^2 = \tfrac{1}{2} \max_{x,y} |h(x,y)|^2.$$

The half-power resolution is defined as the maximum distance between this contour and the origin.

Resolution is not the only criterion for judging a point-spread function. A filter selected to maximize the signal-to-noise ratio or optimize a resolution criterion may have unacceptable sidelobes. These sidelobes may result in unwanted artifacts in the image. Then the filter must be modified to suppress or partially suppress these sidelobes. This process is called *apodization*. Apodization of two-dimensional signals is similar to apodization of one-dimensional signals. A point-spread function $h(x,y)$ with unacceptable sidelobes is replaced by a point-spread function with acceptable sidelobes. For example, a jinc function might be replaced by a function with somewhat smaller sidelobes, thereby reducing sidelobes at the cost of a wider and smaller main lobe. Because a jinc function maximally occupies spatial bandwidth, apodization always requires additional spatial bandwidth and a loss of resolution.

Problems

3.1 Find the two-dimensional Fourier transform of the two-dimensional gaussian pulse

$$s(x,y) = e^{-(ax^2+by^2)}.$$

3.2 a. Prove that

$$\text{jinc}(x,y) ** \text{jinc}(x,y) = \text{jinc}(x,y),$$

where

$$\text{jinc}(x,y) = \frac{J_1(\pi\sqrt{x^2+y^2})}{2\sqrt{x^2+y^2}}.$$

b. Prove that

$$\text{jinc}(x,y) ** \text{sinc}(x,y) = \text{jinc}(x,y).$$

c. Prove that

$$\text{jinc}(2x,2y) ** \text{jinc}(x,y) = \text{jinc}(2x,2y).$$

3.3 What is the two-dimensional Fourier transform of the two-dimensional Fourier transform of $s(x,y)$?

3.4 a. Let $s(x,y)$ be a function that is zero outside of the unit square $-\frac{1}{2} \leq x \leq \frac{1}{2}$, $-\frac{1}{2} \leq y \leq \frac{1}{2}$ and inside this unit square takes the shape of a pyramid with a square base (see illustration). Find the two-dimensional Fourier transform of $s(x,y)$.

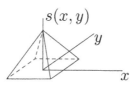

b. Define the lazy pyramid function or pulse, denoted lzpd(x,y), by

$$\text{lzpd}(x,y) = \text{rect}(x,y) ** \text{rect}(x,y).$$

Find lzpd(x,y) and its two-dimensional Fourier transform.

3.5 a. The one-dimensional convolution of two rectangle functions is called the one-dimensional *triangle function* (or *pulse*) and is denoted $\text{trng}(t)$,

$$\text{trng}(t) = \text{rect}(t) * \text{rect}(t).$$

Give an expression for the triangle function $\text{trng}(t)$.

b. Given two overlapping circles of unit radius with centers at distance d, find the area of the lens-shaped intersection.

c. The two-dimensional convolution of two circle functions is called the two-dimensional *circular hat function* and is denoted chat(x, y). That is,

$$\text{chat}(x, y) = \text{circ}(x, y) ** \text{circ}(x, y).$$

Give a closed-form expression for chat(x, y).

d. Give an expression for circ(x, y) $**$ circ(ax, ay). This function may be regarded as a two-dimensional generalization of a one-dimensional trapezoid function. rect(t) $*$ rect(at).

e. For a circular aperture, derive the two Fourier transform pairs

$$\text{circ}(x, y) \Leftrightarrow \text{jinc}(f_x, f_y),$$
$$\text{airy}(x, y) \Leftrightarrow \text{chat}(f_x, f_y),$$

where airy(x, y) = jinc^2(x, y).

f. What is the Fourier transform of chat(x, y)?

3.6 Let $H(f_x, f_y)$ be an ideal lowpass, two-dimensional filter, given by

$$H(f_x, f_y) = \begin{cases} 1 & \text{for } \sqrt{f_x^2 + f_y^2} \leq 1, \\ 0 & \text{otherwise.} \end{cases}$$

Let the filter input be

$$s(x, y) = \text{sinc}(2x)\text{sinc}(2y).$$

Find the filter output.

3.7 Prove that when

$$s(x, y) \Leftrightarrow S(f_x, f_y),$$

then

$$s(x \cos \psi - y \sin \psi, x \sin \psi + y \cos \psi) \Leftrightarrow S(f_x \cos \psi - f_y \sin \psi, f_x \sin \psi + f_y \cos \psi),$$

and more generally,

$$s(a_1 x + b_1 y, a_2 x + b_2 y) \Leftrightarrow \frac{1}{|a_1 b_2 - a_2 b_1|} S(A_1 f_x + A_2 f_y, B_1 f_x + B_2 f_y),$$

where

$$\begin{bmatrix} A_1 & B_1 \\ A_2 & B_2 \end{bmatrix} = \begin{bmatrix} a_1 & b_1 \\ a_2 & b_2 \end{bmatrix}^{-1}.$$

3.8 Find the two-dimensional Fourier transform of $s(x, y) = xye^{-\pi(x^2 + y^2)}$.

3.9 Prove the following two-dimensional Fourier transform relationships:

a. $(a - 2|x|)\text{sinc}(y(a - 2|x|)) \Leftrightarrow (a - 2|f_y|)\text{sinc}(f_x(a - 2|f_y|))$

b. $(a - 2|x|)\text{sinc}((y - bx)(a - 2|x|)) \Leftrightarrow (a - 2|f_y|)\text{sinc}((f_x + bf_y)(a - 2|f_y|)).$

3.10 Some one-dimensional Fourier transform pairs can be derived by starting with a two-dimensional Fourier transform pair. Prove the following one-dimensional Fourier transform pairs:

a. $J_0(2\pi t) \leftrightarrow \dfrac{\text{rect}(f/2)}{\pi(1-f^2)^{1/2}}$,

b. $\text{jinc}(t) \leftrightarrow (1-4f^2)^{1/2}\text{rect}(f)$.

3.11 An "elliptical disc" is given by

$$s(x,y) = \text{circ}(x, ay)$$

for $a \neq 1$. Find the two-dimensional Fourier transform in terms of the jinc function.

3.12 Prove the two-dimensional Fourier transform pair

$$\dfrac{1}{\sqrt{x^2+y^2}} \leftrightarrow \dfrac{1}{\sqrt{f_x^2+f_y^2}}.$$

3.13 a. Show that the following

$$s(x,y) + s(-x,-y) \leftrightarrow 2\text{Re}[S(f_x,f_y)]$$

is a Fourier transform pair.
b. Explain the distinction between $s(-x+a, -y+b)$ and $s(-x+a, -y+b)$.
c. Find the Fourier transforms of the two expressions $s(x-a, y-b) + s(-x-a, -y-b)$ and $s(x-a, y-b) + s(-x+a, -y+b)$. Which of these is the transform of $2\text{Re}[S(f_x,f_y)]$?

3.14 Suppose that $g(x,y)$ is a two-valued function, taking only the values zero and one. The region of the x,y plane where $g(x,y)$ equals one is called the *aperture*. One wants an aperture with small area whose Fourier transform has a narrow (two-dimensional) main lobe. The cross configuration in the illustration has been suggested with the argument that the wide width in x should give a narrow transform in the f_x direction, and the wide width in y should give a narrow transform in the f_y direction. Calculate the Fourier transform and decide whether the suggestion is sound.

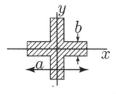

3.15 Using qualitative statements and simple graphical sketches, describe the general nature of the two-dimensional Fourier transform of the thin ring

$$s(x,y) = \text{circ}(x,y) - \text{circ}\left(\dfrac{x}{1-\epsilon}, \dfrac{y}{1-\epsilon}\right),$$

where ϵ is small.

3 Signals in Two Dimensions

3.16 Prove the following Bessel function equality:

$$\int_0^x \xi J_0(\xi) d\xi = x J_1(x),$$

where

$$J_n(x) = \frac{1}{2\pi} \int_{-\pi}^{\pi} e^{-jx \sin\theta + jn\theta} d\theta.$$

3.17 The n-dimensional signal $s(\mathbf{x})$ where $\mathbf{x} = (x_1, \ldots, x_n)$ has an n-dimensional Fourier transform $S(\mathbf{f})$ given by

$$S(\mathbf{f}) = \int_{-\infty}^{\infty} \cdots \int_{-\infty}^{\infty} s(\mathbf{x}) e^{-j2\pi(\mathbf{x}\cdot\mathbf{f})} dx_1 \cdots dx_n,$$

where $\mathbf{f} = (f_1, \ldots, f_n)$.

a. State and prove the inverse multidimensional Fourier transform.
b. State and prove the multidimensional convolution theorem.
c. Prove that when A is an orthogonal matrix (i.e., $A^T = A^{-1}$), the functions $s(\mathbf{x}A)$ and $S(\mathbf{f}A)$ are a multidimensional Fourier transform pair.

3.18 Consider the lattice formed by the matrix

$$M = \begin{bmatrix} 1 & 0 \\ a & b \end{bmatrix}.$$

Find the reciprocal lattice. Sketch the lattice and its reciprocal lattice for $a = b = \sqrt{2}$.

3.19 Generalize the two-dimensional Fourier series to an arbitrary two-dimensional lattice. That is, let

$$S_{ii'} = \iint_{\mathcal{A}} s(x,y) e^{-j2\pi(ix+i'y)} dx dy,$$

where the aperture $\mathcal{A}$ is the fundamental cell of this two-dimensional lattice. Then prove that

$$s(x,y) = \frac{1}{|\mathcal{A}|} \sum_i \sum_{i'} S_{ii'} e^{j2\pi(ix+i'y)}.$$

3.20 a. Prove that

$$\int_0^{\infty} e^{-a^2 t^2 - b^2 t^{-2}} dt = \frac{\sqrt{\pi}}{2a} e^{-2ab},$$

where a and b are both positive. **Hint:** Let $I(b)$ denote the integral on the left, considered as a function of b, and show that $I(b)$ satisfies the first-order differential equation

$$\frac{dI}{db} + 2aI = 0.$$

Find $I(b)$ by solving this differential equation.

b. Prove that
$$\int_0^\infty e^{-ja^2t^2-jb^2t^{-2}}\,dt = \frac{1-j}{\sqrt{2}}\frac{\sqrt{\pi}}{2a}e^{-j2ab}.$$

c. Prove that
$$\int_0^\infty e^{ja^2t^2-jb^2t^{-2}}\,dt = \frac{1-j}{\sqrt{2}}\frac{\sqrt{\pi}}{2a}e^{-2ab}.$$

3.21 a. The two-dimensional Fourier transform of the unit square rect(x,y) is sinc(f_x,f_y). Let $s(t)$ be the triangle pulse
$$s(t) = \begin{cases} 1-|t| & |t| \leq 1, \\ 0 & \text{otherwise.} \end{cases}$$
Use the rotation properties of the two-dimensional Fourier transform applied to rect(x,y) and the projection-slice theorem to derive the Fourier transform of $s(t)$. Compare this to the Fourier transform of $s(t)$ derived directly.

b. Let $s(t)$ be the pulse
$$s(t) = \begin{cases} \sqrt{1-t^2} & |t| \leq 1, \\ 0 & \text{otherwise.} \end{cases}$$
Derive the Fourier transform of $s(t)$ by reference to the two-dimensional Fourier transform of the unit circle and the projection-slice theorem.

3.22 a. Using spherical coordinates, find the three-dimensional Fourier transform of the sphere function. The sphere function is defined as
$$\text{sphr}(x,y,z) = \begin{cases} 1 & \text{when } \sqrt{x^2+y^2+z^2} \leq \tfrac{1}{2}, \\ 0 & \text{otherwise.} \end{cases}$$

Hint: By symmetry, it is enough to compute only $S(0,0,f)$. The indefinite integral $\int x\cos x\,dx = \cos x + x\sin x$ may be helpful. (Whereas the two-dimensional version of this problem requires Bessel functions, the three-dimensional version requires only trigonometric functions. More generally, the n-dimensional version of the problem requires Bessel functions when n is even, but requires only trigonometric functions when n is odd.)

b. Using a three-dimensional generalization of the projection-slice theorem, find the Fourier transform of
$$s(x,y) = \begin{cases} \sqrt{1-x^2-y^2} & x^2+y^2 \leq 1, \\ 0 & x^2+y^2 \geq 1. \end{cases}$$

3.23 State and prove a two-dimensional version of the uncertainty principle.

3.24 Establish the following useful properties of Bessel functions:

a. Prove that
$$J_\nu(a)J_\nu(b) = \frac{1}{2\pi}\int_{-\pi}^{\pi} J_0\left(\sqrt{a^2+b^2+2ab\cos\alpha}\right) e^{-j\nu\alpha}\,d\alpha.$$

b. Prove that

$$J_0\left(\sqrt{a^2 + b^2 - 2ab\cos\alpha}\right) = \sum_{\ell=-\infty}^{\infty} J_\ell(a)J_\ell(b)\cos\ell\alpha.$$

c. Prove that

$$J_n(a+b) = \sum_{\ell=-\infty}^{\infty} J_\ell(a)J_{n-\ell}(b).$$

3.25 Given the infinite array of finite sampling apertures

$$\delta_\epsilon(x,y) = \sum_{\ell''=-\infty}^{\infty}\sum_{\ell'=-\infty}^{\infty} \mathrm{circ}\left(\frac{x-\ell'\Delta}{\epsilon},\frac{y-\ell''\Delta}{\epsilon}\right):$$

a. Describe the sampled signal

$$s_\epsilon(x,y) = \delta_\epsilon(x,y)s(x,y)$$

in the space domain, where $s(x,y)$ is a large, slowly varying image.
b. Describe the function in the spatial frequency domain.
c. What happens as ϵ goes to zero?
d. State a condition under which there is no aliasing.

3.26 A charge-coupled device (CCD) is a light-density to electron-density converter that is widely used as an optical sensor in devices such as camcorders. A CCD is partitioned into nonoverlapping pixels. In each pixel, the total incident light intensity received in a time interval Δt over the area of that pixel is converted to a voltage measurement.

a. Prove that this device can be modeled as

$$s'(x,y) = h(-x,-y) ** s(x,y),$$

where $h(x,y)$ describes the shape of a pixel, followed by a perfect sampler with samples centered on pixel centers. What is $h(x,y)$?
b. When $s(x,y) = \mathrm{circ}\left(\frac{x}{a},\frac{y}{a}\right)$ and the pixels of the rectangular array are also $\mathrm{circ}\left(\frac{x}{a},\frac{y}{a}\right)$, describe how the pixelization of the charge-coupled device degrades the image.

3.27 Suppose that $S(f_x, f_y) = \mathrm{sinc}\sqrt{f_x^2 + f_y^2}$. Is the inverse Fourier transform $s(x,y)$ a cone? What are the projections of $s(x,y)$? What are the projections of a cone? A cone is described by

$$\mathrm{cone}(x,y) = \begin{cases} 1-\sqrt{x^2+y^2} & x^2+y^2 \le 1, \\ 0 & \text{otherwise.} \end{cases}$$

3.28 Suppose that $s(x,y)$ has the property that $S(f_x, f_y) = s(-f_x, f_y)$. (Except for a sign change, $s(x,y)$ is its own two-dimensional Fourier transform.) Use the projection-slice theorem to derive a constraint on the shape of $s(x,y)$. Specifically, in order for $s(x,y)$ to have a narrow projection at angle θ, it must be wide at a slice $90°$ from θ.

3.29 (Centrosymmetric Property) Let $s(x,y)$ be a real nonnegative two-dimensional function with finite energy. An infinite array of such functions is given by

$$s_a(x,y) = s(x,y) **\text{comb}(x,y).$$

a. How should $s(x,y)$ be constrained so that the replicas of $s(x,y)$ do not overlap? Then the *unit cell* with vertices at $(\pm\frac{1}{2}, \pm\frac{1}{2})$ contains $s(x,y)$. Otherwise, the replicas of $s(x,y)$ in $s_a(x,y)$ are overlapping.

b. The region $\mathcal{R} \subset \mathbb{R}^2$ has the *centrosymmetric property* when there is a point designated $(0,0)$ of $\mathcal{R}$, used as the origin and called the symmetric center, such that whenever $(x,y) \in \mathcal{R}$, then $(-x,-y) \in \mathcal{R}$ (with all points defined with respect to $(0,0)$). Does the unit cell of part a) satisfy the centrosymmetric property?

c. The Fourier transform of $s_a(x,y)$ is given by

$$S_a(f_x, f_y) = S(f_x, f_y)\text{comb}(f_x, f_y).$$

Show that the nonzero values of $S_a(f_x, f_y)$ are impulses whose weight is real and can be written

$$S_a(\ell_x, \ell_y) = \int_{-\frac{1}{2}}^{+\frac{1}{2}} \int_{-\frac{1}{2}}^{+\frac{1}{2}} s(x,y) \cos(2\pi(\ell_x x + \ell_y y)) dx dy.$$

d. Given a tessellation of $\mathbb{R}^2$ such that a unit cell $\mathcal{R}$ has a symmetric center and so satisfies the centrosymmetric property. Does the expression of part c) apply with the double integration now over $\mathcal{R}$.

$$S_a(\ell_x, \ell_y) = \int\int_\mathcal{R} s(x,y) \cos(2\pi(\ell_x x + \ell_y y)) dx dy.$$

e. A real nonnegative pulse $s(x,y,z)$ in three variables forms the infinite three-dimensional array using

$$s_a(x,y,z) = s(x,y,z) ***\text{comb}(x,y,z).$$

Repeat the above discussion for $s_a(x,y,z)$.

Notes

The two-dimensional Fourier transform and many of its properties form an obvious generalization of the theory of the one-dimensional Fourier transform. This topic plays an important role in the subjects of computational image formation, of Fourier optics, and of Fourier antenna theory.

The square of the two-dimensional Fourier transform of the circle function is widely used in connection with problems in optical astronomy and was studied by Airy in 1835. Rayleigh, in 1879, defined his resolution limit in terms of the Airy disk to show that optical resolution by conventional methods using monochromatic light is on the order of the wavelength of the light.

The Radon transform was introduced by Johann Radon in a 1917 paper that went unnoticed for many years. The transform and its properties were rediscovered through the years by many people in many fields and in varying degrees of explicitness. The Radon transform and its relationship with the Fourier transform were explicitly stated and formalized in papers by Bracewell (1956, 1958a, 1958b) and by Lighthill (1958). The projection-slice theorem was first stated in the context of Fourier transform theory by Bracewell (1956), although it was implicit earlier in many applications. The shadow transform was discussed by Solomon (1976).

The sampling theorem does not have a satisfactory statement in polar coordinates. This topic was studied by Stark (1979), with comments by Fan and Sanz (1985). Two-dimensional sampling on a nonrectangular array was discussed by Petersen and Middleton (1962). Two-dimensional sampling on a hexagonal grid was further studied by Mersereau (1979). The sampling theorem can be generalized to any lattice in n-dimensional space.

4 Optical Imaging Systems

The earliest imaging systems were optical imaging systems, and optical systems are still the most common imaging systems. Optical imaging systems that employ the simple lens are widespread. They are found both in biological organisms and in man-made devices. Much of optics, including the properties of the ideal lens, can be understood using the language of signal processing in terms of point-spread functions, convolutions, and Fourier transforms. To this purpose, the propagation and diffraction of waves is described in terms of a two-dimensional point-spread function. In this setting, the Huygens–Fresnel principle of optics is presented simply as a consequence of the convolution theorem of the two-dimensional Fourier transform.

In principle, the diffraction of electromagnetic waves could be explained directly from the set of Maxwell's equations, which gives a complete classical description of electromagnetic fields. However, when starting from first principles there may be mathematical difficulties because of concerns about how to model a given problem or how to specify a consistent and accurate set of boundary conditions. It may be difficult to formulate the boundary conditions at a level of detail needed to apply Maxwell's equations, while the weaker conditions needed for diffraction theory may be readily apparent. This is why the theories of diffraction is formulated here as distinct from, but subservient to, electromagnetic theory.

Diffraction is developed herein as an immediate geometric property of wave propagation in a homogeneous medium. The Huygens–Fresnel principle then becomes a systems-theoretic description of a signal when that signal has the form of a propagating wave of constant velocity in an ideal medium. As such, the Huygens–Fresnel principle applies within many subjects involving wave propagation in an ideal medium, including optics and infrared optics, X-rays, antenna theory (both for radar and for communications), sonar hydrophones, acoustics (including speaker design), ultrasonics, and seismics. The mathematical study of diffraction given herein applies whenever the wave equation applies. It need not be restricted to any particular instance of wave propagation. The terminology of this chapter, however, does favor the terminology of optics because applications within optics are emphasized.

4.1 Scalar Diffraction

Diffraction is the study of the relationship between the complex amplitude distribution of a propagating wave in one spatial plane through which the wavefront passes and the complex amplitude distribution of that same wavefront at a subsequent spatial plane. The terms *diffraction* and *interference* refer to the same basic wave phenomenon. The term interference is usually chosen for the interaction of a finite number of wavefronts such as when the source consists of a finite number of point sources or line sources. The term diffraction is preferred for the general situation such as when there is a continuum of wavefronts.

Scalar diffraction theory is a mathematical formulation that provides a satisfactory description of the physical phenomenon for scalar waves such as acoustic waves in an ideal medium. The scalar theory of diffraction also gives satisfactory results for many vector waves such as electromagnetic waves, especially when the significant dimensions within the problem are large in comparison to a wavelength. There are cases of vector waves, however, where scalar diffraction theory is inadequate and fails to describe the physical phenomenon. Then a richer theory, known as the vector diffraction theory, must be used. Vector diffraction theory is described in Section 5.9.

Scalar diffraction is precisely described by the *Huygens–Fresnel principle*, which describes a monochromatic wavefront, $s(x,y,z)$, on the x,y plane at $z=d$ in terms of that monochromatic wavefront on a previous x,y plane at $z=0$. Thus, the complex amplitude of that monochromatic waveform at any point (x,y,z) is $s(x,y,z)$.

Setting $s_0(x,y) = s(x,y,0)$ and $s_d(x,y) = s(x,y,d)$, the Huygens–Fresnel principle is the two-dimensional convolution

$$s_d(x,y) = h_d(x,y) ** s_0(x,y),$$

where $h(x,y)$ is a function known as the *Huygens–Fresnel point-spread function*[1] or the *point-spread function of free space*, and is given by

$$h_d(x,y) = \left[\frac{-jd/\lambda}{(d^2+x^2+y^2)} + \frac{d/2\pi}{(d^2+x^2+y^2)^{3/2}} \right] e^{j2\pi\sqrt{d^2+x^2+y^2}/\lambda}.$$

Mention of the distance d on the left side may be omitted as implicit and this point-spread function may then be written $h(x,y)$.

The Huygens–Fresnel point-spread function is derived in Section 4.2 as a direct consequence of a confluence of plane waves. The study of the point-spread function $h(x,y)$ and of the many simplifying approximations to $h(x,y)$ and their applications will occupy much of this chapter.

The Huygens–Fresnel point-spread function describes the relationship between a wavefront in free space in two parallel x,y planes at two different values of z. There need not be any special significance to the two planes at $z=0$ and $z=d$. The

---

[1] The signal-theoretic term *point-spread function* may be called a *Green's function* in the study of wave propagation, though the term Green's function usually refers to the kernel of a three-dimensional convolution rather than that of a two-dimensional convolution.

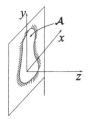

Figure 4.1 An aperture

wavefront need not be generated or sensed at either of these planes. Indeed, since the coordinate origin is arbitrary, the Huygens–Fresnel principle describes the relationship between the complex amplitude of the wavefront between *any* two parallel planes and so it describes the evolution of the wavefront as a function of d. The boundary conditions need not be mentioned. The propagation of a wave does not depend on the realizability of the physical boundary conditions necessary to generate that particular wavefront.

For the purposes of this chapter, any desired wavefront with complex amplitude $s_0(x, y)$ leaving the x, y plane at $z = 0$ can be generated by passing a monodirectional plane wave through an ideal "transparency" in the plane at $z = 0$. Actually, it may be physically very difficult (or even impossible) to generate a wave with a specified complex amplitude in the plane $z = 0$. For example, amplitude and phase details smaller than the wavelength of light may not be easy, or even possible, to generate by passing light through a transparency. Nevertheless, the propagation of all such waves is considered. Mathematically, there is no reason to exclude them. Physically, could such waves be generated, they would propagate as described herein and our only goal is to describe the waves during propagation.

Consider an opaque screen in the x, y plane at $z = 0$ with an aperture, $\mathcal{A}$, cut out of it, as shown in Figure 4.1. The aperture can be described as a closed set of points $\mathcal{A}$ such that the screen at (x, y) is removed for all $(x, y) \in \mathcal{A}$. Passing a plane wave through the aperture can be described as multiplying a two-dimensional signal by a two-dimensional function, called a *transmittance function*, $t(x, y)$, defined in this case by[2]

$$t(x, y) = \begin{cases} 1 & (x, y) \in \mathcal{A}, \\ 0 & (x, y) \notin \mathcal{A}. \end{cases}$$

For an incoming spatial signal $c(x, y)$, called the *illumination function* incident on the aperture, the outgoing spatial signal $s_0(x, y)$ (at the output of the aperture) is

$$s_0(x, y) = t(x, y)c(x, y).$$

---

[2] The branch of optics that postulates the existence of complex transmittance functions and studies their many forms of interaction with the Huygens–Fresnel principle, as studied in this chapter, is called *Fourier optics*.

4 Optical Imaging Systems

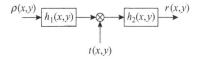

Figure 4.2 Functional model of an optical system

In the general case, $t(x,y)$ is regarded as any complex function of x and y. This transmittance $t(x,y)$ is visualized physically as a transparency whose opacity and thickness vary with x and y. The opacity of the transparency determines the magnitude of $t(x,y)$, and the thickness determines the phase of $t(x,y)$. This way of visualizing $t(x,y)$ is physically reasonable as long as the detail in $t(x,y)$ is large compared to a wavelength.

The signal $s_0(x,y)$ leaves the planar aperture and propagates beyond it. The signal arriving at a plane located at distance d beyond the aperture is described by the Huygens–Fresnel principle

$$s_d(x,y) = h(x,y) \ast\ast s_0(x,y)$$
$$= h(x,y) \ast\ast t(x,y)c(x,y),$$

where $h(x,y)$ is the Huygens–Fresnel point-spread function at distance d. When $c(x,y)$ represents a plane wave moving along the z axis with amplitude one, then $c(x,y) = 1$ and

$$s_d(x,y) = h(x,y) \ast\ast t(x,y)$$

is the spatial signal in the plane at $z = d$.

A more complicated situation is shown in Figure 4.2. The transparency $t(x,y)$ is illuminated by an incident wavefront that has the complex amplitude $p(x,y)$ at distance d_1 prior to the transparency. (This wavefront might have been formed by illuminating a previous transparency with transmittance $p(x,y)$ with the plane wave $c(x,y) = 1$ to produce the signal $p(x,y)$ at the output of that previous transparency.) The signal at the distance d_1 is $h_{d_1}(x,y) \ast\ast p(x,y)$. This is where the transparency $t(x,y)$ is located. Then, at distance d_2 from transparency $t(x,y)$, the signal in the x,y plane is

$$r(x,y) = h_{d_2}(x,y) \ast\ast \left[t(x,y)\left(h_{d_1}(x,y) \ast\ast p(x,y)\right)\right],$$

where $h_{d_1}(x,y)$ and $h_{d_2}(x,y)$ denote the Huygens–Fresnel point-spread functions at distances d_1 and d_2, respectively. In this way, the optical situation is described using the language of two-dimensional signal processing.

4.2 The Huygens–Fresnel Principle

A monochromatic plane wave satisfying the wave equation and moving entirely in the z direction in free space cannot have a complex amplitude that varies in the x, y plane, as was discussed in Section 1.4. The complex amplitude of that wavefront must be constant in the x,y plane. Similarly, a monochromatic plane wave moving in an arbitrary direction described by direction cosines α and β must have a complex amplitude of the form $s(x,y,0) = Ae^{-j\theta}e^{j2\pi(\alpha x + \beta y)}$ in the x,y plane at $z = 0$ for some α and β. No other complex amplitude is possible for a plane wave moving in that direction. Any monochromatic plane wave passing through the x,y plane at $z = 0$ with a spatially varying amplitude across that plane must be a composite of plane waves moving in different directions. This simple geometric fact underlies diffraction and the Huygens–Fresnel principle. These consequences will be developed in detail.

Angular Spectrum

A monochromatic, monodirectional plane wave of amplitude A and phase θ propagating in a single direction, as discussed in Section 1.4, has the complex baseband representation

$$s(x,y,z) = Ae^{-j\theta}e^{j2\pi f_0(\alpha x + \beta y + \gamma z)/c},$$

where α, β, and γ are the direction cosines[3] describing the direction of propagation of the plane wave. Because $\alpha^2 + \beta^2 + \gamma^2 = 1$, where $|\alpha| \leq 1$ and $|\beta| \leq 1$, the direction of the wave is fully described by α and β. Then $\gamma = \sqrt{1 - \alpha^2 - \beta^2}$.

As described here, a plane wave has constant amplitude and phase across a plane orthogonal to the direction of propagation. This property can be taken to simply be the definition of a plane wave. Instead, one can state that a plane wave must satisfy the wave equation,[4] which then leads to the constant amplitude and phase.

As was discussed in Section 1.5, a general monochromatic wave is an angular continuum of superimposed plane waves propagating in all directions. The infinitesimal monochromatic wave with infinitesimal complex amplitude $a(\alpha, \beta)d\alpha d\beta$ propagating in the direction designated by (α, β) is the infinitesimal $ds(x,y,z)$. This wave at location (x,y,z) may be composed of a linear superposition of a continuum of such monochromatic monodirectional plane waves:

$$s(x,y,z) = \int_{-1}^{1} \int_{-1}^{1} a(\alpha, \beta) e^{j2\pi f_0 \sqrt{1-\alpha^2-\beta^2}\, z/c} e^{j2\pi f_0(\alpha x + \beta y)/c} d\alpha d\beta,$$

where $a(\alpha, \beta)$ is known as the *angular spectrum* of the wave and $\gamma = \sqrt{1 - \alpha^2 - \beta^2}$. In this form, the z dependence of the wave is a function of the transverse direction cosines α and β.

---

[3] The direction cosines are $\alpha = \cos \phi_x$, $\beta = \cos \phi_y$, and $\gamma = \cos \phi_z$, where ϕ_x, ϕ_y, and ϕ_z are the angles between the direction of propagation and the x, y, and z axes, respectively. The direction cosines satisfy $\alpha^2 + \beta^2 + \gamma^2 = 1$, where $|\alpha| \leq 1$, $|\beta| \leq 1$, and $|\gamma| \leq 1$.

[4] The wave equation is the partial differential equation

$$\left(\nabla^2 - \frac{1}{c^2} \frac{\partial^2}{\partial^2 t} \right) \Psi(x,y,z,t) = 0,$$

where

$$\nabla^2 = \frac{\partial^2}{\partial^2 x} + \frac{\partial^2}{\partial^2 y} + \frac{\partial^2}{\partial^2 z}.$$

Evanescent Waves

At this point, it is convenient to replace the limits of integration by $\pm\infty$. Then

$$s(x,y,z) = \int_{-\infty}^{\infty}\int_{-\infty}^{\infty} \left(a(\alpha,\beta)e^{j2\pi f_0\sqrt{1-\alpha^2-\beta^2}z/c}\right)e^{j2\pi f_0(\alpha x+\beta y)/c}d\alpha d\beta.$$

This puts the expression in the familiar form of an inverse Fourier transform. By requiring that $a(\alpha,\beta) = 0$ for $\alpha^2 + \beta^2 \geq 1$, nothing is changed while introducing this standard form of the inverse Fourier transform. However, there is actually a good reason for allowing the direction cosines to range from $-\infty$ to ∞ without constraining $a(\alpha,\beta)$. Nonpropagating waves known as *evanescent waves* do exist with $a(\alpha,\beta)$ nonzero for $\alpha^2 + \beta^2 \geq 1$. The illumination and the aperture may produce such an $a(\alpha,\beta)$ thereby launching evanescent waves, but these evanescent waves do not propagate.

Complex Wave Amplitude

Let $s_0(x,y) = s(x,y,0)$ denote the wave $s(x,y,z)$ at the x,y plane with z set equal to zero. Then, with $\lambda = c/f_0$,

$$s_0(x,y) = \int_{-\infty}^{\infty}\int_{-\infty}^{\infty} a(\alpha,\beta)e^{j2\pi\left(\frac{\alpha}{\lambda}x+\frac{\beta}{\lambda}y\right)}d\alpha d\beta.$$

But $s_0(x,y)$ can also be written as an inverse two-dimensional Fourier transform

$$s_0(x,y) = \int_{-\infty}^{\infty}\int_{-\infty}^{\infty} S_0(f_x,f_y)e^{j2\pi(f_xx+f_yy)}df_xdf_y,$$

where $S_0(f_x,f_y)$ is the Fourier transform of $s_0(x,y)$. Comparing the two expressions shows that with the substitutions $f_x = \alpha/\lambda$, and $f_y = \beta/\lambda$, the angular spectrum is equal to the Fourier transform of the complex wave amplitude as it appears in the x,y plane with $z = 0$. Therefore, the angular spectrum can be interpreted as

$$a(\alpha,\beta)d\alpha d\beta = S_0\left(\frac{\alpha}{\lambda},\frac{\beta}{\lambda}\right)d\frac{\alpha}{\lambda}d\frac{\beta}{\lambda},$$

which means that $S_0(f_x,f_y) = \lambda^2 a(\lambda f_x,\lambda f_y)$.

Diffraction

Now consider the wavefront as it appears in the x,y plane at $z = d$. Let $s_d(x,y) = s(x,y,d)$ in the equation for $s(x,y,z)$. Use the variables f_x and f_y to replace α/λ and β/λ. Using similar reasoning with $S_0(f_x,f_y)$ now multiplied by $e^{j2\pi\frac{d}{\lambda}\sqrt{1-\lambda^2 f_x^2-\lambda^2 f_y^2}}$ then gives

$$s_d(x,y) = \int_{-\infty}^{\infty}\int_{-\infty}^{\infty} \left(S_0(f_x,f_y)e^{j2\pi\frac{d}{\lambda}\sqrt{1-\lambda^2 f_x^2-\lambda^2 f_y^2}}\right)e^{j2\pi(f_xx+f_yy)}df_xdf_y,$$

with the complex exponential separated into the product of two terms. This expression is now seen to be the inverse Fourier transform of the product

$$S_d(f_x,f_y) = S_0(f_x,f_y)e^{j2\pi d\sqrt{\lambda^{-2}-f_x^2-f_y^2}}.$$

The term $e^{j2\pi d\sqrt{\lambda^{-2}-f_x^2-f_y^2}}$ multiplies $S_0(f_x,f_y)$ in the frequency domain, so it is appropriate to refer to the term $e^{j2\pi d\sqrt{\lambda^{-2}-f_x^2-f_y^2}}$ as the *transfer function of free space*. The convolution theorem then asserts that the product relationship in the frequency domain transforms into a convolution relationship in the space domain. Thus

$$s_d(x,y) = h(x,y) ** s_0(x,y),$$

where $h(x,y)$ is the point-spread function defined by the two-dimensional Fourier transform relationship

$$h(x,y) \Leftrightarrow e^{j2\pi d\sqrt{\lambda^{-2}-f_x^2-f_y^2}}.$$

This convolution equation expresses the *diffraction* of $s_0(x,y)$ in terms of the point-spread function $h(x,y)$ to result in $s_d(x,y)$.

The remaining task of this section is to derive an expression for $h(x,y)$ by finding an inverse Fourier transform.

Proposition 4.2.1 *The following is a two-dimensional Fourier transform pair*

$$\left[\frac{-j2\pi ab}{(b^2+x^2+y^2)} + \frac{b}{(b^2+x^2+y^2)^{3/2}}\right]e^{j2\pi u\sqrt{b^2+x^2+y^2}} \Leftrightarrow 2\pi e^{j2\pi b\sqrt{a^2-f_x^2-f_y^2}}.$$

Proof
First conjugate the Fourier transform pair given in Proposition 3.5.2. This gives the two-dimensional Fourier transform pair

$$\frac{-1}{\sqrt{b^2+x^2+y^2}}e^{j2\pi a\sqrt{b^2+x^2+y^2}} \Leftrightarrow \frac{1}{j\sqrt{a^2-f_x^2-f_y^2}}e^{j2\pi b\sqrt{a^2-f_x^2-f_y^2}}.$$

Take the derivative of both sides with respect to b to obtain the statement of the proposition. ∎

Setting $b = d$ and $a = \lambda^{-1}$, the following definition names the point-spread function on the left side of the Fourier transform pair given in the theorem, denoting it $h(x,y)$. Then the significance of this point-spread function is summarized in a theorem.

Definition 4.2.2 *The Huygens–Fresnel point-spread function of free space at distance d is*

$$h(x,y) = \left[\frac{-jd/\lambda}{(d^2+x^2+y^2)} + \frac{d/2\pi}{(d^2+x^2+y^2)^{3/2}}\right]e^{j2\pi\sqrt{d^2+x^2+y^2}/\lambda}.$$

Theorem 4.2.3 (Huygens–Fresnel Principle) *A monochromatic wave, $s(x,y,z)$, in free space that has complex amplitude $s_0(x,y)$ in the x,y plane at $z = 0$ has complex amplitude*

$$s_d(x,y) = h(x,y) ** s_0(x,y)$$

in the x,y plane at $z = d$, where $h(x,y)$ is the Huygens–Fresnel point-spread function of free space at distance d.

Proof:
The theorem is proved by referring to the discussion appearing at the beginning of the section. ∎

The mathematically complete form of $h(x,y)$ given in Definition 4.2.1 has two terms. One term has the inverse square of $\sqrt{d^2+x^2+y^2}$ in the denominator; the other term has the inverse cube in the denominator. The second term, sometimes called the *reactive term*, can be important in situations where d is only a few wavelengths. In most common situations of optics, d is larger than a few wavelengths and the reactive term can be neglected. For such cases, the Huygens–Fresnel point-spread function is commonly restated in the approximate form as

$$h(x,y) = \frac{-jd/\lambda}{d^2+x^2+y^2} e^{j2\pi\sqrt{d^2+x^2+y^2}/\lambda}.$$

This approximate form, now known more simply as the *Huygens point-spread function*, is adequate for most applications. It may not be adequate when d is comparable to λ.

The Huygens point-spread function can be motivated by a satisfying intuitive interpretation. First, consider a monochromatic, complex passband, spherical wave radiating omnidirectionally from the origin. It is given by

$$\tilde{c}(t,x,y,z) = \frac{e^{-j2\pi f_0(t-\sqrt{x^2+y^2+z^2}/c)}}{\sqrt{x^2+y^2+z^2}}.$$

The attenuation term in the denominator assures the conservation of energy. It implies that a sphere of any radius about the origin has the same energy passing through it per unit time.

The complex baseband form of this wave crossing the x,y plane at $z = d$ is

$$c(x,y,d) = \frac{1}{\sqrt{x^2+y^2+d^2}} e^{j2\pi\sqrt{x^2+y^2+d^2}/\lambda} \frac{d}{\sqrt{x^2+y^2+d^2}},$$

where the rightmost term $d/\sqrt{x^2+y^2+d^2}$ is an "obliquity factor." This factor is the cosine of the angle between the spherical wavefront at (x,y,d) and the x,y plane passing through the point (x,y,d). One can think of the obliquity factor as due to the projection of the spherical wavefront onto that x,y plane.

The Huygens point-spread function $h(x,y)$ has now been described in a qualitative way which states that each point of an advancing wavefront can be regarded as the source of an infinitesimal secondary spherical wavelet; the subsequent advancing wavefront is then the cumulative effect of the secondary wavelets.

The term $c(x,y,d)$ should be compared to the Huygens point-spread function $h(x,y)$. We arrange $h(x,y)$ as follows:

$$h(x,y,d) = -j\frac{1}{\lambda}\left[\frac{1}{\sqrt{x^2+y^2+d^2}} e^{j2\pi\sqrt{x^2+y^2+d^2}/\lambda}\right] \frac{d}{\sqrt{x^2+y^2+d^2}}$$

so as to mimic the structure of $c(x, y, d)$. Thus the radiated Huygens wavelet is (1) a spherical wavelet, (2) phase-shifted by 90° because of the term $-j$, (3) attenuated by the wavelength described by the term $1/\lambda$.

The intuitive interpretation of $c(x, y, d)$ is often put forward as the basis of a physical "derivation" of Huygens' principle. Each point of an advancing wavefront becomes the source of a secondary Huygens wavelet that radiates spherically. The wave at a subsequent surface is then the integral of these secondary wavelets. This argument is a useful derivation, but it is flawed because it is "intuitively obvious" only in retrospect and to a willing listener. It does not readily explain the term j/λ, nor does it explain why the secondary Huygens wavelets do not propagate in the backwards direction. Most importantly, it gives the wrong answer. The complete form of the Huygens–Fresnel point-spread function $h(x, y)$, given in Definition 4.2.1, is not obtained.

Recapitulation
The central development of this section can be concisely summarized as follows:

- The propagation of the angular spectrum to distance d is

$$a(\alpha, \beta) \longrightarrow a(\alpha, \beta) e^{j2\pi f_0 \sqrt{1-\alpha^2-\beta^2} d/c}.$$

- Setting $f_x = \alpha/\lambda$ and $f_y = \beta/\lambda$, the geometric term $e^{j2\pi f_0 \sqrt{1-\alpha^2-\beta^2} d/c}$ is renamed as the *transfer function of free space*. With this alternative notation, the angular spectrum becomes

$$S(f_x, f_y) \longrightarrow S(f_x, f_y) e^{j2\pi \frac{d}{\lambda}\sqrt{1-\lambda^2 f_x^2 - \lambda^2 f_y^2}}.$$

This frequency-domain expression propagates the Fourier transform $S(f_x, f_y)$ of the complex amplitude distribution $s(x, y)$ from $z = 0$ to $z = d$ where it becomes $S_d(f_x, f_y)$.

- A product in the frequency domain becomes a convolution in the space domain. Taking the inverse Fourier transform $S_d(f_x, f_y)$ yields the propagation of the complex amplitude to distance d as

$$s_0(x, y) \longrightarrow s_d(x, y) = h(x, y) ** s_0(x, y),$$

where

$$h(x, y) = \left[\frac{-jd/\lambda}{(d^2 + x^2 + y^2)} + \frac{d/2\pi}{(d^2 + x^2 + y^2)^{3/2}} \right] e^{j2\pi\sqrt{d^2+x^2+y^2}/\lambda}$$

is the Huygens–Fresnel point-spread function.

4.3 Fresnel and Fraunhofer Approximations

The general theory of scalar diffraction is based on the Huygens–Fresnel principle. In its exact form, the Huygens–Fresnel principle may be needlessly complicated for most practical applications. Even the approximated Huygens principle is usually too complicated. Various approximations are in common use, primarily *Fresnel diffraction* or *near-field diffraction*, and *Fraunhofer diffraction* or *far-field diffraction*. Both approximations are developed in this section.

As we have seen, Huygens' principle

$$s_d(x, y) = h(x, y) ** s_0(x, y),$$

with $h(x, y)$ now representing the simplified Huygens point-spread function with the second term neglected. has the form of the convolution integral

$$s_d(x, y) = \int_{-\infty}^{\infty}\int_{-\infty}^{\infty} s_0(\xi, \eta) \frac{d/j\lambda}{(x-\xi)^2 + (y-\eta)^2 + d^2} e^{j(2\pi/\lambda)\sqrt{(x-\xi)^2+(y-\eta)^2+d^2}} d\xi\, d\eta$$

$$= \int_{-\infty}^{\infty}\int_{-\infty}^{\infty} s_0(x-\xi, y-\eta) \frac{d/j\lambda}{\xi^2 + \eta^2 + d^2} e^{j(2\pi/\lambda)\sqrt{\xi^2+\eta^2+d^2}} d\xi\, d\eta.$$

This integral expression is rather inscrutable as it stands, especially because of the square root in the exponent. In common practice, one or more of several standard simplifying approximations make the expression much more useful. First, assume that the range of x, y, d, ξ, and η are such that the denominator of the point-spread function $h(x, y)$ is adequately[5] approximated by replacing $x^2 + y^2 + d^2$ by the constant d^2. The justification is implied by the statement that d is large compared to x, y, ξ, and η. Then the simplified Huygens point-spread function

$$h(x, y) = \frac{1}{j\lambda d} e^{j(2\pi/\lambda)\sqrt{x^2+y^2+d^2}}$$

can be used as an approximation. Now the amplitude term of $1/j\lambda d$ is a constant, which is not of immediate interest. By redefining $s_0(\xi, \eta)$, the amplitude term $1/j\lambda d$ can be absorbed into a scaling of $s_0(x, y)$. For brevity, this factor is not explicitly written. It can be resurrected when necessary. Then the Huygens principle is written in the approximated form

$$s_d(x, y) = \int_{-\infty}^{\infty}\int_{-\infty}^{\infty} s_0(\xi, \eta) e^{j(2\pi/\lambda)\sqrt{(x-\xi)^2+(y-\eta)^2+d^2}} d\xi\, d\eta,$$

now corresponding to the simplified point-spread function

$$h(x, y) = e^{j(2\pi/\lambda)\sqrt{x^2+y^2+d^2}}.$$

Next, we turn our attention to the square root in the exponent. The square root makes the expression unwieldy and almost useless for an elementary analysis of many optical systems. The two standard approximations to the exponent lead to the *Fresnel approximation* and the *Fraunhofer approximation* as described herein. Each is used in appropriate situations.

In Chapter 15, various criteria are described for judging the quality of phase approximations. For the present, an approximation is made – without further comment – that it is sufficiently accurate when the exponent is approximated by an expression that

---

[5] Notice that this same approximation is not justified in the exponential term because $2\pi/\lambda$ can be large, which means that small errors in the square-root term can lead to large phase errors.

is in error by a phase angle not larger than some maximum phase error θ_{max}, where θ_{max} is given as appropriate to whatever accuracy may be required. Even rather large values of phase error such as $\pi/4$ might be acceptable in many situations, though not in others.

Fresnel Approximation

To obtain the Fresnel approximation, the exponent in the Huygens point-spread function is expanded as the series

$$\sqrt{x^2+y^2+d^2} = d\left[1 + \frac{1}{2}\left(\frac{x^2+y^2}{d^2}\right) - \frac{1}{8}\left(\frac{x^2+y^2}{d^2}\right)^2 + \cdots\right].$$

For large d, the right side may be approximated by the terms in x and y up to the quadratic terms. This leads to the Fresnel approximation given by the point-spread function

$$h(x,y) = e^{j2\pi d/\lambda} e^{j\pi(x^2+y^2)/d\lambda}.$$

The Fresnel approximation may appear to be a rather severe approximation to the exact expression for $h(x,y)$ given in Definition 4.2.2, but it is accurate enough to give much of classical diffraction theory.

Under the Fresnel approximation, $s_d(x,y)$ has the form of a two-dimensional convolution of $s_0(x,y)$ with a two-dimensional quadratic-phase pulse,

$$s_d(x,y) = e^{jkd}\left[s_0(x,y) ** e^{j\pi(x^2+y^2)/d\lambda}\right].$$

where $k = 2\pi/\lambda$. The leading term e^{jkd} describes a simple phase shift between a plane wave at $z = 0$ and that same plane wave at $z = d$. It is often convenient to suppress this term as uninteresting, though it is understood to be there. The point-spread function

$$h(x,y) = e^{j\pi(x^2+y^2)/\lambda d},$$

or, with the amplitude term explicit,

$$h(x,y) = \frac{1}{j\lambda d} e^{j\pi(x^2+y^2)/\lambda d},$$

is called the *point-spread function of free space in the Fresnel approximation* or the *Fresnel point-spread function*.

To determine a sufficient condition on the region in which the Fresnel approximation is adequate, the largest neglected term of the series expansion is compared to $\pi/4$:

$$\frac{2\pi}{\lambda} d \frac{1}{8}\left(\frac{(x-\xi)^2+(y-\eta)^2}{d^2}\right)^2 \leq \frac{\pi}{4}.$$

From this, we conclude that the Fresnel approximation is satisfactory up to phase error $\pi/4$ when

$$[(x-\xi)^2 + (y-\eta)^2]^2 \leq \lambda d^3,$$

for all $x, \xi, y,$ and η. Such a statement is only a guide. The Fresnel approximation may sometimes be satisfactory under a weaker condition, and sometimes unsatisfactory even under a stronger condition. The Fresnel approximation is entirely satisfactory in most instances of classical optical diffraction.

Alternative Paraxial Derivation

An alternative way to derive the Fresnel approximation is to reason in the Fourier domain. Recall that the point-spread function of free space and the transfer function of free space are related by the two-dimensional Fourier transform as

$$h(x, y) \Leftrightarrow e^{j2\pi d \sqrt{\lambda^{-2} - f_x^2 - f_y^2}}.$$

The *paraxial approximation*[6] to the transfer function is based on the first two terms of the series expansion $\sqrt{1-t} = 1 - t/2 - t^2/8 + \cdots$. This is

$$e^{j2\pi d \sqrt{\lambda^{-2} - f_x^2 - f_y^2}} = e^{j2\pi d / \lambda \sqrt{1-(\lambda f_x)^2 - (\lambda f_y)^2}}$$
$$\approx e^{j2\pi d/\lambda} e^{-j\pi \left((\lambda f_x)^2 + (\lambda f_y)^2\right)},$$

neglecting a phase-only term. The two-dimensional Fourier transform pair

$$e^{j\pi(x^2+y^2)} \Leftrightarrow je^{-j\pi(f_x^2+f_y^2)},$$

after scaling as

$$e^{j\pi((x/\lambda)^2+(y/\lambda)^2)} \Leftrightarrow j\lambda^2 e^{-j\pi((\lambda f_x)^2+(\lambda f_y)^2)},$$

now immediately gives the Fresnel approximation.

Fraunhofer Approximation

The Fraunhofer approximation is an alternative approximation to the Huygens convolution equation. Within this approximation, the term "Fraunhofer diffraction pattern" can be translated into the term "two-dimensional Fourier transform," thereby replacing the terminology of optics by that of signal processing, but without otherwise changing the facts.

Unlike the Fresnel approximation, the Fraunhofer approximation is not an approximation to the point-spread function $h(x, y)$ alone. Rather, it is an approximation to the

---

[6] The term "paraxial" means "along the axis." The term applies because the approximation amounts to the statement that α and β are small.

4.3 Fresnel and Fraunhofer Approximations

entire convolution integral in which $h(x,y)$ appears. The approximation replaces the exponent within the convolution equation

$$s_d(x,y) = \int_{-\infty}^{\infty}\int_{-\infty}^{\infty} s_0(\xi,\eta) e^{j(2\pi/\lambda)\sqrt{(x-\xi)^2+(y-\eta)^2+d^2}} d\xi d\eta$$

by a term that is linear in x and y. Let $R^2 = x^2 + y^2 + z^2$, where z is large compared to ξ and η. Then

$$\sqrt{(x-\xi)^2+(y-\eta)^2+z^2} = R\sqrt{1 - \frac{2x\xi}{R^2} - \frac{2y\eta}{R^2} + \frac{\xi^2+\eta^2}{R^2}}$$

$$\approx R\sqrt{1 - \frac{2x\xi}{R^2} - \frac{2y\eta}{R^2}},$$

where the approximation is that R^2 is large compared to $\xi^2 + \eta^2$. Now use the first term in the series expansion $\sqrt{1-t} = 1 - t/2 - t^2/8 + \cdots$ as an approximation, leading to

$$\sqrt{(x-\xi)^2+(y-\eta)^2+z^2} \approx R - \frac{x}{R}\xi - \frac{y}{R}\eta.$$

All quadratic terms inside the integral have been neglected with R taken to be sufficiently large. Therefore, in the Fraunhofer approximation,

$$s(x,y) = e^{j2\pi R/\lambda} \int_{-\infty}^{\infty}\int_{-\infty}^{\infty} s_0(\xi,\eta) e^{-j(2\pi/\lambda)(\alpha\xi+\beta\eta)} d\xi d\eta$$

$$= e^{j2\pi R/\lambda} S_0\left(\frac{\alpha}{\lambda}, \frac{\beta}{\lambda}\right),$$

where $\alpha = x/R$ and $\beta = y/R$ are the direction cosines. Thus, in the Fraunhofer approximation, the Huygens–Fresnel convolution integral is approximated by a Fourier transform. This can be an excellent approximation in the far field.

The region in which the Fraunhofer approximation is adequate is the region where the quadratic term in the preceding series approximation can be neglected. Suppose that the quadratic term in the series expansion can be neglected when it is not larger than $\pi/4$. Examining those quadratic terms in ξ and η that are neglected gives

$$\left|\frac{2\pi R}{\lambda}\left[\frac{1}{2}\left(\frac{\xi^2+\eta^2}{R^2}\right) - \frac{1}{2}\left(\frac{x\xi+y\eta}{R^2}\right)^2\right]\right| \leq \frac{\pi}{4},$$

which can be rewritten as

$$\left|\xi^2 + \eta^2 - \left(\frac{x}{R}\xi + \frac{y}{R}\eta\right)^2\right| \leq \frac{\lambda R}{4}.$$

For x and y small compared to R, this becomes

$$\xi^2 + \eta^2 \leq \frac{\lambda R}{4}.$$

To simplify this conclusion, let r be the radius of a circle enclosing the input aperture. The Fraunhofer approximation may be satisfactory when the range R satisfies

$$r^2 \leq \frac{\lambda R}{4}.$$

In a typical optics application, we may have $r = 0.03$ meters and $\lambda = 0.4 \times 10^{-6}$ meters. The condition above requires that R be greater than 9000 meters. This shows that the Fraunhofer approximation is not appropriate in a typical optics application.

In a typical radar application, we may have $r = 1$ meter and $\lambda = 0.1$ meters. The condition above requires that R be greater than 40 meters. This suggests that the Fraunhofer approximation is appropriate in many radar applications. It would fail this condition, however, with $r = 100$ meters, $\lambda = 0.01$ meters, and $R = 10^6$ meters.

4.4 The Geometrical Optics Approximation

The Fresnel approximation and the Fraunhofer approximation to the Huygens–Fresnel principle are each valid in appropriate situations depending on the relevant physical dimensions. An approximation of a much different kind, called the *geometrical optics approximation*, is obtained by taking λ to be vanishingly small. When the geometry is large compared to λ, the local vector normal to the local plane wave defines the notion of a *ray*. Geometrical optics replaces the study of wave propagation with the study of such rays. Because optical wavelengths are smaller than 10^{-6} meters, this approximation is usually well justified in most everyday situations. Indeed, the geometrical optics approximation, leading to the subject of geometrical optics and ray tracing, is the justification for much of practical everyday optics.

The geometrical optics approximation says that, for λ sufficiently small in comparison to the dimensions of interest, the Huygens–Fresnel point-spread function can be approximated as

$$h(x, y) \approx \delta(x, y) e^{j2\pi d/\lambda},$$

which means that diffraction is neglected. The impulse function $\delta(x, y)$ in the approximation is to be understood, as usual, in terms of its behavior as an operator under an integral sign. In the geometrical optics approximation, propagation of a wave to a distance d reduces to multiplication of the wavefront by $e^{j2\pi d/\lambda}$.

The geometrical optics approximation follows from the Huygens–Fresnel principle by using the *principle of stationary phase*. This principle makes the observation that, in a time-domain integral of the form

$$I = \int_{-\infty}^{\infty} a(t) e^{jb(t)} dt$$
$$= \int_{-\infty}^{\infty} a(t) \big(\cos b(t) + j \sin b(t)\big) dt,$$

there will be little contribution to the integral in regions of the time axis where $a(t)$ is changing slowly and $b(t)$ is large. For a smooth function $b(t)$ with a minimum that

4.4 The Geometrical Optics Approximation

grows large compared to 2π on either side of its minimum, the primary contributions to the integral comes from an interval surrounding the minimum of $b(t)$ provided that $a(t)$ is varying sufficiently slowly with t. For many integrals of this form, it is possible to obtain reasonably good approximations by using the principle of stationary phase.[7]

Suppose that $b(t)$ takes its minimum at $t = t_0$. Then $b'(t_0) = 0$. Expand $b(t)$ in a series about this minimum as

$$b(t) = b(t_0) + \tfrac{1}{2}b''(t_0)(t-t_0)^2 + \tfrac{1}{6}b'''(t_0)(t-t_0)^3 + \cdots.$$

Then

$$\int_{-\infty}^{\infty} a(t)e^{jb(t)}\,dt = e^{jb(t_0)} \int_{-\infty}^{\infty} a(t)e^{j\left[\tfrac{1}{2}b''(t_0)(t-t_0)^2 + \cdots\right]}\,dt.$$

The principle of stationary phase makes the approximation

$$\int_{-\infty}^{\infty} a(t)e^{jb(t)}\,dt \approx e^{jb(t_0)} a(t_0) \int_{-\infty}^{\infty} e^{j\tfrac{1}{2}b''(t_0)(t-t_0)^2}\,dt$$

for slowly varying $a(t)$,

The principle of stationary phase is used to develop the geometrical optics approximation starting with the Huygens–Fresnel principle in the form

$$s_d(x,y) = \int_{-\infty}^{\infty}\int_{-\infty}^{\infty} s_0(x-\xi, y-\eta) a(\xi,\eta) e^{j(2\pi/\lambda)\sqrt{\xi^2+\eta^2+d^2}}\,d\xi\,d\eta,$$

where

$$a(x,y) = \frac{-jd/\lambda}{(x^2+y^2+d^2)} + \frac{d/2\pi}{(x^2+y^2+d^2)^{3/2}}.$$

Expand the term in the exponent using a series in the exponent as

$$e^{j(2\pi/\lambda)\sqrt{\xi^2+\eta^2+d^2}} = e^{j(2\pi/\lambda)(d+\xi^2/2d+\eta^2/2d+\cdots)}$$

$$\approx e^{j2\pi d/\lambda} e^{j\pi(\xi^2+\eta^2)/d\lambda}.$$

The exponent takes its minimum at $\xi = \eta = 0$. By the principle of stationary phase,

$$s_d(x,y) \approx e^{j2\pi d/\lambda} s_0(x,y) a(0,0) \int_{-\infty}^{\infty}\int_{-\infty}^{\infty} e^{j(\pi/\lambda d)(\xi^2+\eta^2)}\,d\xi\,d\eta$$

$$\approx e^{j2\pi d/\lambda} s_0(x,y) \left[\frac{-jd/\lambda}{d^2} + \frac{d/2\pi}{d^3}\right] \int_{-\infty}^{\infty} e^{j(\pi/\lambda d)\xi^2}\,d\xi \int_{-\infty}^{\infty} e^{j(\pi/\lambda d)\eta^2}\,d\eta.$$

Next, use the standard definite integrals

$$\int_{-\infty}^{\infty} \cos x^2\,dx = \int_{-\infty}^{\infty} \sin x^2\,dx = \sqrt{\frac{\pi}{2}}$$

---

[7] A requirement of this discussion is that $s(x,y)$ must be slowly varying in space as compared to the wavelength λ. This condition permits the notion of rays and the methods of geometrical optics.

to write
$$\int_{-\infty}^{\infty} e^{j(\pi/\lambda d)\xi^2}\,d\xi = \int_{-\infty}^{\infty} e^{j(\pi/\lambda d)\eta^2}\,d\eta = \sqrt{\frac{\lambda d}{2}}(1+j).$$

Then
$$s_d(x,y) \approx e^{j2\pi d/\lambda} s_0(x,y) \left[\frac{-j}{\lambda d} + \frac{1}{2\pi d^2}\right]\left(\sqrt{\frac{\lambda d}{2}}(1+j)\right)^2$$
$$\approx e^{j2\pi d/\lambda} s_0(x,y)\left[1 + \frac{j\lambda}{2\pi d}\right].$$

When λ is negligibly small compared to d, this becomes
$$s_d(x,y) \approx e^{j2\pi d/\lambda} s_0(x,y).$$

Then the point-spread function, in the geometrical optics approximation, is
$$h(x,y) = e^{j2\pi d/\lambda}\delta(x,y),$$

and the complex passband wave is described as
$$s(t,x,y,z) = s_0(x,y)e^{j2\pi f_0(t-z/c)}.$$

In the geometrical optics approximation, the wave $s_0(x,y)$ propagates without diffraction and with a phase shift due only to the propagation distance. Ergo, the ray theory applies as a simplified surrogate for the wave theory.

Specular and Diffuse Reflection

A plane wave incident on an optically smooth planar surface at any angle other than normal is partially reflected. In the geometrical optics approximation, a bundle of parallel rays incident on an optically smooth planar surface in a direction described by direction cosines (α_0, β_0) is reflected as a bundle of parallel rays at an angle described by direction cosines $(-\alpha_0, -\beta_0)$. A formal expression for this might involve the delta function $\delta(\alpha - \alpha_0, \beta - \beta_0)$ to describe the incident ray and the delta function $\delta(\alpha + \alpha_0, \beta + \beta_0)$ to describe the reflected wave. This is called *specular reflection*. Specular reflection of a plane wave is described in the same way as specular reflection of a bundle of parallel rays.

On the other hand, a bundle of parallel rays incident on an optically rough surface in a direction described by direction cosines (α_0, β_0) is scattered as a collection of rays at many angles or a continuum of angles. This is called *diffuse reflection*. Diffuse reflection of a plane wave is described in the same way as diffuse reflection of a ray bundle. The scattered diffuse wave is an instance of a spatially noncoherent lightwave.

The reflection from a planar surface may consist of a combination of specular and diffuse reflections. That reflection may have properties between the two cases.

An optically smooth curved surface will reflect light in a continuum of directions. Each infinitesimal element of the curved surface will reflect light at its own appropriate direction, so the light is reflected into a continuum of directions. An optically rough surface scatters the light as noncoherent light. Realizations of the scattered diffuse light into different angles are only weakly correlated.

4.5 The Ideal Lens

The design of a high-quality lens is a sophisticated subject involving both the properties of the physical material and the laws of wave propagation. Within the topic of Fourier optics, only the mathematical description of the input/output relationship of the lens is of interest, and only for an ideal lens. For this purpose, the important processing properties of the simple lens are adequately explained by using a simple mathematical lens model called the *ideal lens*.

An ideal lens, depicted symbolically in Figure 4.3, is a mathematical function that provides an idealized model of certain physical lenses. An ideal lens is a transmittance function, $t(x,y)$, that is nonzero only within a finite aperture, $\mathcal{A}$. Within the aperture $\mathcal{A}$, the transmittance function $t(x,y)$ has constant amplitude and introduces a phase factor across the aperture that varies quadratically in the radial direction of the lens. Specifically, on the aperture $\mathcal{A}$, described by the aperture function

$$A(x,y) = \begin{cases} 1 & (x,y) \in \mathcal{A}, \\ 0 & (x,y) \notin \mathcal{A}, \end{cases}$$

the transmittance is given by

$$t(x,y) = e^{-j(2\pi/\lambda)(x^2+y^2)/2f_\ell} A(x,y),$$

where f_ℓ is a constant known as the *focal length*[8] of the ideal lens. For positive f_ℓ, the lens is called a *positive lens*. For negative f_ℓ, the lens is called a *negative lens*. A plane wave traveling in the z direction, when multiplied by the transmittance $t(x,y)$, has a phase change at coordinate (x,y) that is proportional to $x^2 + y^2$.

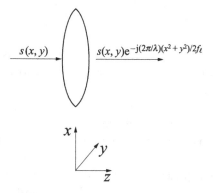

Figure 4.3 The ideal lens

---

[8] The focal length f_ℓ should not be confused with frequency, which is also denoted by a subscripted f. The use of f for focal length has a precedence that cannot be ignored. This leads to the bizarre and perhaps confusing notational coincidence that the spatial coordinates in the focal plane, which might be denoted as (x_f, y_f), represent the spatial frequencies (f_x, f_y) because of the Fourier transforming property of the lens.

4 Optical Imaging Systems

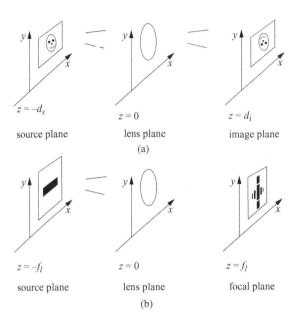

Figure 4.4 Two functions of an ideal lens

For a circular aperture $\mathcal{A}$ of radius r, the ideal lens is

$$t(x,y) = e^{-j(2\pi/\lambda)(x^2+y^2)/2f_\ell} \operatorname{circ}\left(\frac{x}{r}, \frac{y}{r}\right).$$

Notice that the phase of the ideal lens at position (x,y) depends on the wavelength λ and the focal length f_ℓ only through the product λf_ℓ.

The ideal lens can be used to process a waveform in certain ways. It is an example of a special kind of processor – an optical processor. In spite of its apparent simplicity, the processing power of an ideal lens can be enormous when the processing task is exactly the right one. A lens can perform either of two processing tasks, as shown in Figure 4.4. It can form an image, or it can form a two-dimensional Fourier transform.

The two functions of a lens are described in x, y planes along the z axis. The source plane and image plane are any two planes related by the lens law. The wavefronts in that arbitrary pair of planes will be a pair of like images (insofar as the approximations permit). As the source plane moves closer to its focal plane, the image plane moves away from its focal plane. As either the source plane or the image plane approaches its focal plane, the other approaches infinity. Furthermore, the wavefronts in the two focal planes are related by the two-dimensional Fourier transform (insofar as the approximations hold).

A lens can be cascaded with other optical elements to create a processor of enormous throughput, but of quite limited flexibility. At the end of this section, the relationship between $t(x,y)$ and the geometrical shape of a practical lens is discussed. But first, as of primary interest, the properties of the ideal lens as a circuit element are studied.

Lens as an Imager

The ideal lens can be used to form images either with coherent light, which is studied in this section, or with noncoherent light, which is studied in the next section. A *geometrical image* of the function $\rho(x,y)$, in general, is any function of the form $\sigma(x,y) = B e^{j\theta(x,y)} \rho(\pm x/M, \pm y/M)$. This means that an image is a magnified and scaled version of $\rho(x,y)$ with a position-dependent phase shift $\theta(x,y)$. The *magnification* M may be larger than one or smaller than one. The image is inverted whenever there is a sign reversal on both x and y. The real quantity $i(x,y) = |\sigma(x,y)|^2$ is called the *intensity* of the image of the function $\rho(x,y)$. The intensity of the image is proportional to $|\rho(\pm x/M, \pm y/M)|^2$.

The formation of an image by a lens is described by the lens law. First, the lens law is given for the simplified case in which the aperture is infinitely large.[9] Then the analysis is revised to incorporate a finite aperture.

Theorem 4.5.1 (Lens Law) *Within the Fresnel approximation, for d_s and d_i larger than f_ℓ and satisfying*

$$\frac{1}{d_s} + \frac{1}{d_i} = \frac{1}{f_\ell},$$

the complex wavefront $\rho(x,y)$ originating at the distance d_s on one side of an ideal lens with an infinitely large aperture will produce the geometrical image

$$\sigma(x,y) = -\frac{1}{M} e^{ja(x^2+y^2)} \rho\left(-\frac{x}{M}, -\frac{y}{M}\right),$$

at distance d_i on the other side where the magnification is $M = d_i/d_s$ and $a = \pi(d_i + d_s)/\lambda d_i^2$.

Proof:
A coherent monochromatic wavefront with complex amplitude $\rho(x,y)$ at $z = -d_s$ propagates toward an ideal lens placed at $z=0$. The signal that is incident on the lens in the Fresnel approximation is

$$r(x,y) = e^{j(\pi/\lambda d_s)(x^2+y^2)} ** \rho(x,y).$$

The transmittance of the ideal lens with infinite aperture is

$$t(x,y) = e^{-j(2\pi/\lambda)(x^2+y^2)/2f_\ell}.$$

Multiplying $r(x,y)$ by $t(x,y)$ gives $r(x,y)e^{-j(\pi/\lambda f_\ell)(x^2+y^2)}$ as the output of the lens leaving the plane $z = 0$. Using the Fresnel approximation again, now to propagate the signal from the plane at $z = 0$ to the plane $z = d_i$, gives

---

[9] Of course, the Fresnel approximation is not valid for an infinitely large aperture. The lens law for an infinite aperture in the Fresnel approximation is a mathematical statement, not a physical statement. The assumption of an infinite aperture will be removed later, so this point is moot.

$$\sigma(x,y) = \left[r(x,y)e^{-j(\pi/\lambda f_\ell)(x^2+y^2)}\right] ** e^{j(\pi/\lambda d_i)(x^2+y^2)}$$

$$= \left[\left(\rho(x,y) ** e^{j(\pi/\lambda d_s)(x^2+y^2)}\right)e^{-j(\pi/\lambda f_\ell)(x^2+y^2)}\right] ** e^{j(\pi/\lambda d_i)(x^2+y^2)}.$$

To find the image $r(x,y)$, expand the two convolutions in this equation. For simplicity of exposition, expand instead the analogous equation in one variable,

$$\sigma(x) = \left[\left(\rho(x) * e^{j(\pi/\lambda d_s)x^2}\right)e^{-j(\pi/\lambda f_\ell)x^2}\right] * e^{j(\pi/\lambda d_i)x^2}.$$

The original two-dimensional equation expands in the same way.

In the image plane, z is equal to d_i. Writing out the two convolutions gives

$$\sigma(x) = \int_{-\infty}^{\infty} e^{j(\pi/\lambda d_i)(x-\xi)^2}\left[e^{-j(\pi/\lambda f_\ell)\xi^2}\int_{-\infty}^{\infty}\rho(\eta)e^{j(\pi/\lambda d_s)(\eta-\xi)^2}d\eta\right]d\xi.$$

Open the squares in the first and third exponentials and note that the condition of the theorem is that $1/d_s + 1/d_i = 1/f_\ell$. After canceling the relevant terms, the expression reduces to

$$\sigma(x) = e^{j(\pi/\lambda d_i)x^2}\int_{-\infty}^{\infty} e^{-j2\pi x\xi/\lambda d_i}\left[\int_{-\infty}^{\infty}\left(\rho(\eta)e^{j(\pi/\lambda d_s)\eta^2}\right)e^{-j2\pi\eta\xi/\lambda d_s}d\eta\right]d\xi.$$

The inner integral on η has the form of a Fourier transform of $\rho(\eta)e^{j(\pi/\lambda d_s)\eta^2}$. The outer integral on ξ also has the form of a Fourier transform. So that this form is explicit, make the change in variables that $\eta = -\frac{d_s}{d_i}\tau$ and $\xi = -\lambda d_i f_x$. Then

$$\sigma(x) = \lambda d_s e^{j(\pi/\lambda d_i)x^2}\int_{-\infty}^{\infty} e^{j2\pi f_x x}\int_{-\infty}^{\infty}\left[\rho\left(-\frac{d_s}{d_i}\tau\right)e^{j(\pi d_s/d_i^2\lambda)\tau^2}\right]e^{-j2\pi f_x\tau}d\tau df_x.$$

The expression now has the form of an inverse Fourier transform of a Fourier transform. These two operations cancel, leaving x in place of τ. The conclusion is that

$$\sigma(x) = \lambda d_s e^{j(\pi/\lambda d_i)x^2}\left[\rho\left(-\frac{d_s}{d_i}x\right)e^{j(\pi d_s/\lambda d_i^2)x^2}\right].$$

This means that the image $r(x)$ is a complex multiple of $\rho(x)$ under magnification and sign reversal.

The equation for the two-dimensional signal $r(x,y)$ expands in the same way as does the equation for the one-dimensional signal $r(x)$. The proof of the theorem is now complete. ∎

Clearly, to satisfy the condition of the theorem, both d_s and d_i must be larger than the focal length f_ℓ. Then the object $\rho(x,y)$ in the object plane $z = d_s$ produces an image, $\sigma(x,y)$, in the image plane $z = d_i$ that is a magnified, inverted, and phase-shifted copy of $\rho(x,y)$. Moreover, the image intensity is

$$i(x,y) = |\sigma(x,y)|^2$$
$$= \lambda^2 d_s^2|\rho(-x/M,-y/M)|^2,$$

which is an inverted, magnified, and scaled copy of $|\rho(x,y)|^2$.

4.5 The Ideal Lens

Theorem 4.5.1 does not include the diffraction caused by a finite aperture. Every lens has a finite aperture, so diffraction by the aperture must always occur. To include the effect of the finite aperture, write the transmittance of the aperture as

$$t(x,y) = e^{-j(\pi/\lambda f_\ell)(x^2+y^2)} A(x,y),$$

where the aperture indicator function[10] is

$$A(x,y) = \begin{cases} 1 & (x,y) \in \mathcal{A}, \\ 0 & (x,y) \notin \mathcal{A}. \end{cases}$$

Let $s(x,y)$ denote the image in the presence of the finite aperture $\mathcal{A}$. To determine the diffracted image $s(x,y)$, the expression for the undiffracted image $\sigma(x,y)$ that is given in Theorem 4.5.1 must be modified.

To this purpose, examining the proof of Theorem 4.5.1 reveals that, in the equation for the one-dimensional $\sigma(x)$, the term $e^{-j(\pi/\lambda f_\ell)\xi^2}$ must be replaced by the term $A(\xi) e^{-j(\pi/\lambda f_\ell)\xi^2}$. As before, in the image plane, the equality $1/d_s + 1/d_i = 1/f_\ell$ holds. With the term for the aperture $A(\xi)$ now inserted in its proper place, the expression for $s(x)$ becomes

$$s(x) = e^{j(\pi/\lambda d_i)x^2} \int_{-\infty}^{\infty} e^{-j2\pi x \xi/\lambda d_i} \left[A(\xi) \int_{-\infty}^{\infty} \rho(\eta) e^{j(\pi/\lambda d_s)\eta^2} e^{-j2\pi \eta \xi/\lambda d_s} d\eta \right] d\xi.$$

Again make the change of variables $\eta = -\frac{d_s}{d_i}\tau$ and $\xi = -\lambda d_i f_x$,

$$s(x) = \lambda d_s e^{j(\pi/\lambda d_i)x^2} \int_{-\infty}^{\infty} e^{j2\pi f_x x} \left[A(-\lambda d_i f_x) \int_{-\infty}^{\infty} \left[\rho\left(-\frac{d_s}{d_i}\tau\right) e^{j(\pi d_s/\lambda d_i^2)\tau^2} \right] \right.$$
$$\left. e^{-j2\pi f_x \tau} d\tau \right] df_x.$$

To recognize the role that the aperture function plays in this equation, notice that the outer integral is an inverse Fourier transform. Thus, moving the outer phase term to the left side of the equation and taking the Fourier transform gives

$$\int_{-\infty}^{\infty} \left[s(x) e^{-j(\pi/\lambda d_i)x^2} \right] e^{-j2\pi f_x x} dx = A(-\lambda d_i f_x) \int_{-\infty}^{\infty} \left[\lambda d_s \rho\left(-\frac{d_s}{d_i}\tau\right) e^{j(\pi d_s/\lambda d_i^2)\tau^2} \right]$$
$$e^{-j2\pi f_x \tau} d\tau$$
$$= A(-\lambda d_i f_x) \int_{-\infty}^{\infty} \left[\sigma(\tau) e^{-j(\pi/\lambda d_i)\tau^2} \right] e^{-j2\pi f_x \tau} d\tau.$$

The second line follows because

$$\sigma(x) = \lambda d_s e^{j(\pi/\lambda d_i)x^2} \left[\rho\left(-\frac{d_s}{d_i}x\right) e^{j(\pi d_s/\lambda d_i^2)x^2} \right]$$

is the undiffracted image.

---

[10] The aperture indicator function $A(x,y)$ is actually a function of space coordinates but enters the mathematical analysis representing a function in frequency coordinates $A(f_x, f_y)$ whose inverse Fourier transform is the coherent point-spread function $g_c(x,y)$.

Extending this conclusion to two dimensions shows that the aperture, when expressed in the form $A(-\lambda d_i f_x, -\lambda d_i f_y)$, multiplies the Fourier transform of the geometrical (undiffracted) image $\sigma(x,y)e^{-j(\pi/\lambda d_i)(x^2+y^2)}$. Therefore this image itself is convolved with the inverse Fourier transform of the aperture. Thus, because of diffraction, the Fourier transform of the aperture, appropriately scaled, is seen as a point-spread function[11] blurring the image.

Coherent Point-Spread Function

The role of the aperture under coherent illumination can be described as a point-spread function. Express the aperture function $A(f_x,f_y)$ using frequency domain variables f_x and f_y as before. Then let $a(x,y)$ be the inverse Fourier transform of $A(f_x,f_y)$. Define the *coherent optical point-spread function*

$$g_c(x,y) = a\left(-\frac{x}{\lambda d_i}, -\frac{y}{\lambda d_i}\right).$$

The Fourier transform $G_c(f_x,f_y)$ of $g_c(x,y)$ is given by

$$G_c(f_x,f_y) = (\lambda d_i)^2 A(-\lambda d_i f_x, -\lambda d_i f_y),$$

and is known as the *coherent optical transfer function*. It is an ideal lowpass spatial filter.

Example
The most common aperture is a circle:

$$A(x,y) = \text{circ}(x/R, y/R).$$

Then the jinc function

$$g_c(x,y) = R \operatorname{jinc}\left(\frac{R}{\lambda d_i}\sqrt{x^2+y^2}\right)$$

is the coherent optical point-spread function of that circle function. ∎

The coherent optical point-spread function blurs the image as follows.

Proposition 4.5.2 *Within the Fresnel approximation, a finite aperture $A(x,y)$ produces the diffracted image $s(x,y)$ given by*

$$s(x,y) = e^{j(\pi/\lambda d_i)(x^2+y^2)}\left[g_c(x,y) \ast\ast \sigma(x,y)e^{-j(\pi/\lambda d_i)(x^2+y^2)}\right],$$

where the undiffracted geometrical image is

$$\sigma(x,y) = -\frac{1}{M}e^{j(\pi/\lambda d_i^2)(d_s+d_i)(x^2+y^2)}\rho\left(-\frac{x}{M}, -\frac{y}{M}\right),$$

and $g_c(x,y)$ is the coherent optical point-spread function and $M = d_i/d_s$.

---

[11] More generally, to soften the diffraction sidelobes, an aperture can include some amplitude attenuation at the edges, such as $A(x,y) = f(\sqrt{x^2+y^2})$, where $f(\cdot)$ is a two-dimensional rectangle function with rounded edges.

Proof:
The geometrical image $\sigma(x,y)$ was given in Theorem 4.5.1. The modification needed to describe the diffracted image follows from the discussion after the theorem. The proof of the proposition consists of a reference to that discussion. ∎

Were $g_c(x,y)$ an impulse, one phase term in Proposition 4.5.2 could be pulled through the convolution to cancel the other phase term. Cancelling the phase terms may give a convenient approximation either because $g_c(x,y)$ is sufficiently like an impulse or because $e^{-j(\pi/\lambda d_i)(x^2+y^2)}$ is varying sufficiently slowly. Then

$$s(x,y) \approx g_c(x,y) ** \sigma(x,y),$$

when such an approximation is appropriate.

In the language of signal theory, an ideal lens used to form an image rescales by a complex function and inverts the signal $\rho(x,y)$ to form $\sigma(x,y)$, then passes $\sigma(x,y)$ through the two-dimensional filter $g_c(x,y)$. The output of this filtering operation is the diffracted image $s(x,y)$. Consequently, coherent imaging can be seen simplistically as the convolution of the true image with the inverse Fourier transform of the system aperture $A(x,y)$. Diffraction by the aperture of finite size limits the resolution of an image. The larger the aperture, the smaller is the diffraction provided the Fresnel approximation remains valid.[12]

The Lens as a Fourier Transformer

The ideal lens can also be used to perform another useful function. The ideal lens can form a two-dimensional Fourier transform. Figure 4.3 shows the ideal lens in its two major roles, imaging and transforming. In both cases, the ideal lens consists of a perfect (infinitely thin) transparency with a quadratic phase across an aperture, and the source is modeled as a transparency illuminated by a plane wave. The lens lies in the x,y plane at $z = 0$. Theorem 4.5.1 states that for a source that lies in an x,y plane at $z = -d_s$, an image satisfying the lens law is formed in the x,y plane at the value $z = d_i$. But this is not the only noteworthy property.

There are two other x,y planes, one on each side of the lens at $z = \pm f_\ell$, called the *focal planes*, as shown in Figure 4.3. The next proposition states that the focal plane to the right of the lens (at $z = f_\ell$) contains the two-dimensional Fourier transform $P(f_x, f_y)$ of the wave amplitude $\rho(x,y)$ that lies in the focal plane to the left of the lens at $z = -f_\ell$.

---

[12] Two parameters are in use to describe the resolution power of a lens. These are the f-*number* and the *numerical aperture*. The f-*number* is defined as f_ℓ/r, where f_ℓ is the focal length and r is the radius of the lens. (The reciprocal r/f_ℓ is called the *relative aperture*.) The numerical aperture of an optical system is a dimensionless number characterizing the range of angles over which the system can accept or provide light in the geometrical optics approximation. For a thin lens, the numerical aperture is usually defined as $NA = n\sin\theta = n\sin(\arctan\frac{r}{2f_\ell})$, where n is the index of refraction.

4 Optical Imaging Systems

Proposition 4.5.3 *The complex spatial signals in the two x, y planes at distance $\pm f_\ell$ on the two sides of an ideal lens are Fourier transforms of each other in the Fresnel approximation.*

Proof:
With source $\rho(x,y)$ placed at the focal distance f_ℓ from the ideal lens, the signal $\sigma(x,y)$ incident on the lens is $\rho(x,y) ** h_{f_\ell}(x,y)$. The signal in the other focal plane is

$$\sigma(x,y) = \left[[\rho(x,y) ** e^{-j\pi(x^2+y^2)/\lambda f_\ell}] e^{j\pi(x^2+y^2)/\lambda f_\ell}\right] ** e^{-j\pi(x^2+y^2)/\lambda f_\ell}.$$

The exponential terms in the expression are all the same.

All that remains is to expand the two convolutions and cancel like terms. To make the derivation more instructive, the case in which the source plane is at $z = 0$ is treated first. This means that the source plane is coincident with the plane of the ideal lens.[13] Later, the source plane is repositioned to $z = -f_\ell$.

The output of the lens with an aperture function $A(x,y)$ when the source is at $z = 0$ is $\rho(x,y)A(x,y)e^{-j(\pi/\lambda f_\ell)(x^2+y^2)}$ for all x and y. For an aperture $A(x,y)$ that is large enough to include the support of $\rho(x,y)$, one can write $\rho(x,y)A(x,y) = \rho(x,y)$ and ignore the aperture. Under the Fresnel approximation, the signal $\sigma(x,y)$ at distance $z = d$ is

$$\sigma(x,y) = \left[\rho(x,y)e^{-j(\pi/\lambda f_\ell)(x^2+y^2)}\right] ** e^{j(\pi/\lambda d)(x^2+y^2)}.$$

Expanding the convolution with d equal to f_ℓ gives

$$\sigma(x,y) = \int_{-\infty}^{\infty}\int_{-\infty}^{\infty} \rho(\xi,\eta) e^{-j(\pi/\lambda f_\ell)(\xi^2+\eta^2)} e^{j(\pi/\lambda f_\ell)[(x-\xi)^2+(y-\eta)^2]} d\xi\, d\eta.$$

This reduces to

$$\sigma(x,y) = e^{j(\pi/\lambda f_\ell)(x^2+y^2)} \int_{-\infty}^{\infty}\int_{-\infty}^{\infty} \rho(\xi,\eta) e^{-j(2\pi/\lambda f_\ell)(x\xi+y\eta)} d\xi\, d\eta$$

$$= e^{j(\pi/\lambda f_\ell)(x^2+y^2)} P\left(\frac{x}{\lambda f_\ell}, \frac{y}{\lambda f_\ell}\right),$$

where $P(f_x, f_y)$ is the Fourier transform of $\rho(x,y)$.

Thus, the spatial signal in the x, y plane at $z = 0$ is the Fourier transform $P(f_x, f_y)$ of $\rho(x,y)$. After scaling and multiplication by the extraneous phase term $e^{j(\pi/\lambda f_\ell)(x^2+y^2)}$, this is $\sigma(x,y)$.

Now reposition the source away from the x, y plane at $z = 0$, moving it to the x, y plane at $z = -f_\ell$. This means that $\rho(x,y)$ is convolved with the point-spread function

$$h(x,y) = e^{j(\pi/\lambda f_\ell)(x^2+y^2)}$$

---

[13] This requires the lens to be infinitely thin, as is the case with an ideal lens. This physically unreasonable condition is introduced temporarily only to allow the explanation to be given in two steps. It will disappear when the source is moved away from the lens.

(in the Fresnel approximation). The signal that is incident on the lens is $\rho'(x,y) = h(x,y) ** \rho(x,y)$. The Fourier transform pair

$$e^{j\pi(x^2+y^2)} \Leftrightarrow je^{-j\pi(f_x^2+f_y^2)}$$

shows that the point-spread function $h(x,y)$ has the Fourier transform

$$H(f_x,f_y) = \lambda f_\ell j e^{-j(\pi\lambda f_\ell)(f_x^2+f_y^2)}.$$

Thus, for a source at $z = -f_\ell$, the Fourier transform of the signal that is incident on the lens is

$$P'(f_x,f_y) = \lambda f_\ell j e^{-j(\pi\lambda f_\ell)(f_x^2+f_y^2)} P(f_x,f_y).$$

Therefore $P'(f_x,f_y)$ must be used in place of $P(f_x,f_y)$ in the equation for $\sigma(x,y)$, which gives

$$\sigma(x,y) = e^{j(\pi/\lambda f_\ell)(x^2+y^2)} P'\left(\frac{x}{\lambda f_\ell}, \frac{y}{\lambda f_\ell}\right)$$

$$= \lambda f_\ell j P\left(\frac{x}{\lambda f_\ell}, \frac{y}{\lambda f_\ell}\right).$$

Thus placing the source at $z = -f_\ell$ creates a phase term that cancels the extraneous spatially-varying phase term that would be present were the source at $z = 0$. Only the constant $\lambda f_\ell j$ remains multiplying the Fourier transform. ∎

In summary,[14] suppose that a partially transparent screen at $z = -f_\ell$ has the complex transmittance $\rho(x,y)$. A plane wave passing through the screen along the z axis takes on the complex amplitude $\rho(x,y)$ as a function of x and y. When this plane wave is incident on the lens, a wave at the second focal plane has a complex amplitude proportional to

$$P(f_x,f_y) = \int_{-\infty}^{\infty}\int_{-\infty}^{\infty} \rho(x,y) e^{-j\frac{2\pi}{\lambda f_\ell}(xf_x+yf_y)} dxdy.$$

with f_x and f_y represented by the terms $x/\lambda f_\ell$ and $y/\lambda f_\ell$. By placing an appropriate intensity sensor, such as photographic film, in that focal plane, the intensity $|P(f_x,f_y)|^2$ of the Fourier transform can be recorded.

There is one detail that has been overlooked in this discussion. Moving the source plane from $z=0$ to $z=-f_\ell$ means that a finite aperture, $A(x,y)$, at $z=0$ will fail to capture some of the signal because $\rho(x,y) ** h(x,y)$ has infinite support even though $\rho(x,y)$ has finite support. This means that there will always be some error, usually negligible, in the computed Fourier transform because of the truncation of the propagating signal at the lens due to the finite aperture. Making the aperture larger reduces this error, but increases other errors including the error arising from the Fresnel approximation.

---

[14] The Fraunhofer diffraction pattern is itself a Fourier transform in the far field. The lens used in the Fresnel diffraction pattern is not a Fourier transform but becomes a Fourier transform when combined with the quadratic phase term of an ideal lens.

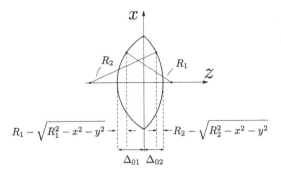

Figure 4.5 A lens

Lens Construction

The ideal lens can be closely approximated by the *thin lens*. The thin lens is a kind of physical lens as described in this section. It is a special shape of a transparent medium such as a piece of glass. The design of lenses is a highly developed branch of the subject of optics, which is touched on only briefly to describe the thin lens. The thin lens will be described within the geometrical optics approximation. This approximation is based on the notion of rays.

Figure 4.5 shows the cross section of a lens for which the lens material has a propagation velocity that is different than the propagation velocity of the material outside of the lens. The lens completely fills an aperture $A(x, y)$. A ray entering the lens will move with a smaller velocity within the lens and may change direction. The propagation velocity within the lens is c/n where the constant n is the *refractive index* of the lens material and c is the speed of light in free space. The lens is to be shaped so that its transmittance approximates the transmittance of the ideal lens.

A simple geometry for a lens has spherically shaped faces. The lens in Figure 4.4 has spherically shaped faces with radii R_1 and R_2, respectively. The thickness $\Delta(x, y)$ is the sum of two terms, $\Delta_1(x, y)$ and $\Delta_2(x, y)$. These terms are given by

$$\Delta_1(x, y) = \Delta_{01} - \left(R_1 - \sqrt{R_1^2 - x^2 - y^2} \right),$$

$$\Delta_2(x, y) = \Delta_{02} - \left(R_2 - \sqrt{R_2^2 - x^2 - y^2} \right),$$

where the usual convention is that R_2 is negative because the spherical surface is in the $-z$ direction with respect to the focus. Consequently,

$$\Delta(x, y) = \Delta_0 - R_1 \left(1 - \sqrt{1 - (x^2 + y^2)/R_1^2} \right) - R_2 \left(1 - \sqrt{1 - (x^2 + y^2)/R_2^2} \right),$$

where $\Delta_0 = \Delta_{01} + \Delta_{02}$.

Two conditions are now imposed so that this expression can be approximated. These conditions are known as the thin-lens approximation and the paraxial approximation.

The *thin-lens approximation* results from the requirement that the thickness of the lens is small compared to both R_1 and R_2. This condition is equivalent to the condition that the radius of the lens aperture is small compared to R_1 and R_2.

The *paraxial approximation* states that almost all rays of significance are *paraxial rays*, meaning rays that are approximately parallel to the z axis. For such rays, the phase delay depends only on the thickness. Every ray entering approximately along the z axis at one face at coordinates (x, y) emerges from the lens at (approximately) the same coordinates and (approximately) the same direction at the other face. There is a negligible displacement of the ray within a thin lens.

Within these approximations, one can use the series expansion $\sqrt{1-t} = 1 - \frac{1}{2}t + \cdots$ to write

$$1 - \sqrt{1 - (x^2 + y^2)/R_1^2} \approx \frac{x^2 + y^2}{2R_1^2},$$

$$1 - \sqrt{1 - (x^2 + y^2)/R_2^2} \approx \frac{x^2 + y^2}{2R_2^2},$$

retaining only the quadratic terms. Within the paraxial approximation, the thickness is

$$\Delta(x, y) = \Delta_0 - \frac{x^2 + y^2}{2}\left(\frac{1}{R_1} + \frac{1}{R_2}\right),$$

and the phase delay is $2\pi f_0(n-1)\Delta(x,y)/c$. In the absence of the lens, the phase delay in a distance Δ_0 would be $2\pi f_0 \Delta_0/c$. Consequently, because $k = 2\pi f_0/c$, the transmittance is

$$t(x, y) = e^{jkn\Delta_0} e^{-jk(n-1)\frac{x^2+y^2}{2}\left(\frac{1}{R_1}+\frac{1}{R_2}\right)},$$

and the focal length of this expression is[15]

$$\frac{1}{f_\ell} = (n-1)\left(\frac{1}{R_1} + \frac{1}{R_2}\right).$$

Then

$$t(x, y) = e^{jkn\Delta_0} e^{-j\frac{k}{2f_\ell}(x^2+y^2)},$$

where $k = 2\pi/\lambda$. Within the limitations of the thin-lens approximation and the paraxial approximation, the thin lens with spherical surfaces has the transmittance of the ideal lens.

4.6 Noncoherent Imaging

We have seen that the ideal lens will process a spatially coherent wave originating at the object $\rho(x, y)$ to produce an image of that object. However, in everyday circumstances, light from a source is both *spatially* noncoherent and *temporally* noncoherent. The first term refers to random phase that is independent at two points of space at the

---

[15] This expression is sometimes called the *lens maker's equation* for a thin lens.

same instant of time. The second term refers to random phase that is independent at two instants of time at the same point of space.[16]

This section shows that for spatially noncoherent illumination the ideal lens still forms images, but only in intensity, not in amplitude. It will be seen that noncoherent imaging is described by replacing the actual squared aperture $|A(x,y)|^2$ by the virtual aperture $A(x,y) \ast\ast A(x,y)$.

Noncoherence

In contrast to the ideal wave structure studied so far, the complex amplitude of a wave may have incidental random fluctuations in amplitude and phase that are not considered to be part of the signal-bearing complex amplitude of that wave. These amplitude and phase fluctuations can be caused by amplitude and phase fluctuations already in the illumination, or can be caused by unresolved roughness in the object being imaged. These fluctuations are the cause of noncoherence in an imaging procedure. The noncoherence can be spatial noncoherence or temporal noncoherence.

A *spatially noncoherent wave* originating from the source $\rho(x,y)$ in the plane at $z = 0$ as a complex baseband wavefront of the form

$$v(x,y,t) = \rho(x,y)u(x,y,t),$$

where $u(x,y,t)$ is modeled as a spatially white, zero-mean random process defined by the correlation function

$$E[u(x,y,t)u^*(x',y',t)] = \delta(x-x', y-y').$$

Consequently,

$$E[v(x,y,t)v^*(x',y',t)] = |\rho(x,y)|^2 \delta(x-x', y-y').$$

Because the time variable t has the same value in both copies of $v(x,y,t)$ in the expectation, the definition of spatial noncoherence is silent with respect to randomness on the time axis.

A *temporally noncoherent* wave is a time-stationary process defined by the correlation function

$$E[u(x,y,t)u^*(x,y,t')] = \delta(t-t').$$

Consequently,

$$E[v(x,y,t)v^*(x,y,t')] = |\rho(x,y)|^2 \delta(t-t').$$

Because the space variables x and y have the same values in both copies of $v(x,y,t)$ in the expectation, the definition of temporal noncoherence is silent with regard to randomness in the space coordinates.

A process can satisfy both definitions of noncoherence and so can be both temporally noncoherent and spatially noncoherent.

---

[16] A deeper study of noncoherence would invoke the van Cittert–Zernike theorem. This theorem is given as a formal mathematical statement in Section 13.1.5 of Chapter 13.

4.6 Noncoherent Imaging

Partially (spatially) coherent imaging is intermediate between the cases of fully (spatially) coherent imaging and fully (spatially) noncoherent imaging. A spatially partially coherent waveform satisfies

$$E[v(x,y,t)v^*(x',y',t)] = |\rho(x,y)|^2 E[u(x,y,t)u^*(x',y',t)]$$
$$= |\rho(x,y)|^2 \phi(x-x', y-y'),$$

where, for partial coherence, the two-dimensional correlation function $\phi(x,y)$ is not an impulse. The difficult case of imaging with partially coherent waveforms is not considered herein.

Noncoherent Imaging

The proposition to be given next says that the expected value of a noncoherent image is a convolution of intensities[17] rather than a convolution of amplitudes. The diffraction point-spread function is no longer a scaled version of the Fourier transform of the aperture. Instead, the diffraction point-spread function is a scaled version of the Fourier transform of the autocorrelation function of the aperture.

Proposition 4.6.1 *For noncoherent imaging in the Fresnel approximation, the intensity*

$$i(x,y) = E|\sigma(x,y,t)|^2$$

is given by the intensity convolution equation

$$i(x,y) = |g_c(x,y)|^2 ** \left|\rho\left(-\frac{x}{M}, -\frac{y}{M}\right)\right|^2,$$

where $g_c(x,y)$ is the coherent optical point-spread function.

Proof:
The coherent imaging equation

$$s(x,y)e^{-j(\pi/\lambda d_i)(x^2+y^2)} = g_c(x,y) ** \sigma(x,y)e^{-j(\pi/\lambda d_i)(x^2+y^2)}$$

was given in Proposition 4.5.2, where

$$\sigma(x,y) = -\frac{1}{M}e^{j(\pi/\lambda d_i^2)(d_s+d_i)(x^2+y^2)}\rho\left(-\frac{x}{M}, -\frac{y}{M}\right),$$

---

[17] The intensity recorded in a stationary random system is the sample time average

$$i(x,y) = \frac{1}{T}\int_0^T |r(x,y,t)|^2 dt,$$

whereas we are dealing with the ensemble average

$$i(x,y) = E[|r(x,y,t)|^2].$$

When the illumination is temporally coherent, as with a laser source, the time average and the ensemble average need not be equal or even approximately equal. This causes *speckle* for temporally coherent light. Speckle is a significant visual difference between imaging with temporally coherent light and temporally noncoherent light.

and $g_c(x, y)$ has the Fourier transform

$$G_c(f_x, f_y) = (\lambda d_i)^2 A(-\lambda d_i f_x, -\lambda d_i f_y).$$

This is now modified for spatially noncoherent illumination as

$$s(x,y) e^{-j(\pi/\lambda d_i)(x^2+y^2)} = g_c(x,y) ** v'(x,y,t) e^{-j(\pi/\lambda d_i)(x^2+y^2)},$$

where

$$v'(x,y,t) = -\frac{1}{M} e^{j(\pi/\lambda d_i^2)(d_s+d_i)(x^2+y^2)} v\left(-\frac{x}{M}, -\frac{y}{M}, t\right),$$

and $v(x, y, t) = \rho(x, y) u(x, y, t)$. The image intensity is given by the expected value of the squared magnitude of this expression:

$$i(x,y) = E[|s(x,y)|^2]$$
$$= E[s(x,y) ** s^*(-x,-y)].$$

Writing the modified imaging equation into this expression twice, then moving the expectation inside the integrations leads to

$$i(x,y) = \int_{-\infty}^{\infty}\int_{-\infty}^{\infty}\int_{-\infty}^{\infty}\int_{-\infty}^{\infty} g_c(x-\xi,y-\eta) g_c^*(x-\xi',y-\eta') \phi(\xi,\eta,\xi',\eta') d\xi\, d\eta\, d\xi'\, d\eta',$$

where the correlation function is

$$\phi(\xi,\eta,\xi',\eta') = E\left[v'(\xi,\eta,t) e^{-j(\pi/\lambda d_i)(\xi^2+\eta^2)} v'^*(\xi',\eta',t) e^{j(\pi/\lambda d_i)(\xi'^2+\eta'^2)}\right].$$

To evaluate this expectation, recall that

$$E[v(x,y,t) v^*(x',y',t)] = |\rho(x,y)|^2 \delta(x-x', y-y'),$$

so

$$\phi(\xi,\eta,\xi',\eta') = E[v'(x,y,t) v'^*(x',y',t)]$$
$$= \left|\rho\left(-\frac{\xi}{M}, -\frac{\eta}{M}\right)\right|^2 \delta(\xi-\xi', \eta-\eta').$$

Then

$$i(x,y) = \frac{1}{M^2} \int_{-\infty}^{\infty}\int_{-\infty}^{\infty} |g_c(x-\xi, y-\eta)|^2 |\rho(-\xi/M, -\eta/M)|^2 d\xi\, d\eta,$$

which completes the proof of the proposition. ∎

Noncoherent Point-Spread Function
To describe the role of the aperture under noncoherent illumination, let $a(x, y)$ be the inverse Fourier transform of the aperture function $A(f_x, f_y)$ expressed in the frequency domain. The *noncoherent optical point-spread function* is defined as the square of the coherent optical point-spread function. That is,

$$g_n(x,y) = |g_c(x,y)|^2$$
$$= \left|a\left(-\frac{x}{\lambda d_i}, -\frac{y}{\lambda d_i}\right)\right|^2.$$

4.6 Noncoherent Imaging

This may be divided by the term $\int_{-\infty}^{\infty}\int_{-\infty}^{\infty}|g_c(x,y)|^2 dxdy$ to give an alternative normalized form of $g_n(x,y)$ that integrates to one.

The *noncoherent optical transfer function* $G_n(f_x,f_y)$ is defined as the Fourier transform of $g_n(x,y)$. By definition

$$g_n(x,y) \leftrightarrow G_n(f_x,f_y)$$

is a two-dimensional Fourier transform pair.

Example

The most common aperture is a circle. For a circular aperture of radius R, the aperture function $A(x,y)$ and its transform are given by

$$A(x,y) = \text{circ}(x/R, y/R),$$
$$a(f_x,f_y) = R^2 \text{jinc}(Rf_x, Rf_y).$$

Then the coherent optical point-spread function is a jinc function given by

$$g_c(x,y) = R^2 \text{jinc}\left(\frac{R}{\lambda d_i}x, \frac{R}{\lambda d_i}y\right).$$

The noncoherent optical point-spread function is an airy function given by

$$\text{airy}(x,y) = \text{jinc}(x,y)^2$$

and has Fourier transform

$$\text{airy}(x,y) \leftrightarrow \text{chat}(f_x,f_y),$$

where the right side is the circular hat function. The functions on the two sides have radial symmetry and the noncoherent optical point-spread function can be abbreviated as

$$g_n(x,y) = R^2 \text{airy}\left(\frac{R}{\lambda d_i}x, \frac{R}{\lambda d_i}y\right)$$

and the noncoherent optical transfer function is

$$G_n(f_x,f_y) = (\lambda d_i)^2 \text{chat}\left(\frac{\lambda d_i}{R}f_x, \frac{\lambda d_i}{R}f_y\right).$$

This noncoherent optical point-spread function corresponds to a coherent optical point-spread function for the hypothetical aperture,

$$A(x,y) = \text{chat}\left(\frac{x}{R}, \frac{y}{R}\right).$$

In spatially noncoherent light, the circular aperture presents itself as a circular hat function. The support of this virtual aperture is a circle with radius *twice* that of the actual aperture, a conclusion that should be highlighted.

For temporally noncoherent light, the ensemble expectation can be replaced by a time average because a photodetection is not instantaneous. For temporally coherent light, time averaging is not appropriate and the image takes on an appearance called speckle as discussed below.

Diffraction-Limited Imaging

The earliest and most common form of optical imaging is noncoherent imaging. This is well-described using the thin-lens approximation and the Fresnel approximation. In this case, the objective is observed through a point-spread function determined by the nature of the aperture containing the lens. This is called *diffraction-limited imaging* because the point-spread function is due to the diffraction of the wave that is spatially limited by the aperture.

For a clear circular aperture, the point-spread function of diffraction-limited noncoherent imaging is the airy function. The airy function, informally, is also called the airy disk. The term "airy disk" suggests that the main lobe of the airy function can be regarded simply as a disk and the volume of the airy function beyond the first null can be ignored.

Whether two separated point objects[18] that are blurred by a point-spread function can be resolved is an ill-posed question because, absent other objects in the scene or noise and impairments in the signal, these point objects can always be resolved, in principle, when the point-spread function is known. However, when the two point objects are close, small impairments, such as noise, may make resolution extremely difficult. The separation at which resolution becomes difficult can be quantified only by a subjective criterion. The Rayleigh resolution criterion states that two point objects blurred by a point-spread function can be resolved provided they are separated by at least the distance between the center of the point-spread function and its first zero along the line of separation.

The Rayleigh resolution criterion can be applied to the common optical imaging system for which an airy function is the point-spread function. In this special case, the Rayleigh resolution criterion is called the *Abbe resolution criterion* or the *Abbe diffraction limit*. The airy function has its first zero at $\sqrt{x^2 + y^2} = 1.2197$. Inspection of the far-field point-spread function then leads to

$$\sin\theta = 1.22 \frac{\lambda}{d}$$

as the Abbe resolution criterion for conventional optical imaging, where the wavelength λ and the aperture diameter d are expressed using the same units of distance.

Speckle

An image of a spatially noncoherent object that is formed with temporally coherent illumination, such as laser illumination or coherent ultrasound illumination, can have a granular appearance called *speckle*. Speckle is a statistical temporally constant spatial fluctuation of the image intensity caused by unresolved roughness in the surface of the imaged object, or caused by irregularities in the propagation medium. Each reflecting

---

[18] Such as two stars.

element of an object returns an echo of the incident passband waveform altered both in amplitude and phase. When many small unresolved complex reflecting elements within one resolution cell of an object are random and independent, the central limit theorem of probability theory implies that the in-phase and quadrature components of the received signal from that resolution cell tend towards complex gaussian random variables and are treated as such. The variance in each cell depends on the reflectors in that cell. Consequently, the amplitude in each cell tends toward a Rayleigh random variable and is treated as such.

When the illumination is temporally noncoherent, the amplitude in each cell changes with time and is time-averaged under normal viewing circumstances. That cell would display the time-average of the Rayleigh-distributed amplitudes. When the illumination is partially temporally coherent, the speckle will fluctuate with the time constant of the partial coherence.

When the illumination is temporally coherent, the amplitude in each cell does not change with time and is not averaged. That cell displays a single-instance of the Rayleigh-distributed amplitude. The visual effect of this variance is called speckle.

Speckle is usually considered to be an impairment in the quality of an image, and the image may be further processed to smooth the speckle. On the other hand, the statistics of the speckle may be observed as a useful way to characterize the smoothness and structure of a reflector at a level finer than the resolution of the image.

4.7 Phase-Contrast Imaging

Transparent objects (such as some microscopic biological organisms) affect light, not by absorption, but by a spatially dependent delay that appears in the light signal as a spatially dependent phase shift $\phi(x,y)$. Because an optical sensor, such as a recording film or the eye, detects only the intensity of light but not its phase, objects that are completely transparent are impossible to see directly. However, such objects can be seen by placing an obstruction in an appropriate optical path that will convert phase modulation into amplitude modulation. A simple technique is the method known as the *central dark ground method* in which a small spot in the frequency domain is used to modify the components of the Fourier transform near the origin of the f_x, f_y plane.

A transparent object with a complex transmittance of the form

$$\rho(x,y) = e^{j\phi(x,y)}$$

consists of only a spatially varying phase. When a transparent object is coherently illuminated in an image forming system, the intensity image of $\rho(x,y)$ produced by a conventional optical imaging system with negligible diffraction would have the intensity

$$|\rho(x,y)|^2 = 1.$$

Because the intensity $|\rho(x,y)|^2$ does not depend on the phase $\phi(x,y)$, the signal $\rho(x,y)$ is not visible. To obtain the desired image, $\rho(x,y)$ is first passed through a filter, called

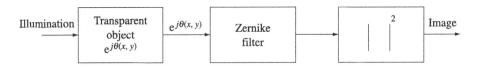

Figure 4.6 Imaging a transparent object

a *Zernicke filter*, prior to computing the magnitude. The Zernicke filter is defined in the frequency domain as

$$H(f_x,f_y) = \begin{cases} ja & \text{for } \sqrt{f_x^2+f_y^2} \le \epsilon \\ 1 & \text{otherwise,} \end{cases}$$

where ϵ is a small number and a is a real constant. Imaging a transparent object with the Zernicke filter is shown in Figure 4.6.

The Zernicke filter can be implemented optically using two lenses, one to compute the Fourier transform, and one to compute the inverse Fourier transform. This requires a total optical path of length $4f_\ell$. A glass slide is placed in the Fourier transform plane between the two lenses. The light from the object is Fourier transformed by the first lens where it is obstructed by a small phase-shifting dot placed at the origin $(f_x,f_y) = (0,0)$ of the glass slide. For a Zernicke filter, the optical phase shift of the dot is $\pi/2$ radians,

The output of the filter is $h(x,y) ** \rho(x,y)$. For a case in which the phase shift $\phi(x,y)$ is sufficiently smaller than one radian, $e^{j\phi(x,y)}$ can be approximated as

$$\rho(x,y) \approx 1 + j\phi(x,y).$$

In the transform domain, this is

$$R(f_x,f_y) \approx \delta(f_x,f_y) + j\Phi(f_x,f_y).$$

Because $\Phi(f_x,f_y)$ is negligible compared to $\delta(f_x,f_y)$ at the origin, the filter output in the transform domain is approximately

$$H(f_x,f_y)R(f_x,f_y) \approx ja\delta(f_x,f_y) + j\Phi(f_x,f_y).$$

Consequently,

$$h(x,y) ** \rho(x,y) \approx j(a + \phi(x,y)),$$

which has the approximate intensity

$$|a + \phi(x,y)|^2 \approx a^2 + 2\phi(x,y).$$

Whereas the intensity of the unfiltered signal is independent of the object $\phi(x,y)$, the intensity of the filtered signal depends on the object seen against the constant background signal denoted by a^2.

Phase-contrast imaging can also be used in applications such as wind tunnel photographs in which the air stream has a nonuniform index of refraction, such as may be caused by pressure differences due to the details of the air flow. The nonuniform index of refraction introduces phase modulation into the optical field leading to the complex transmittance

$$p(x,y) = e^{j\phi(x,y)} \approx 1 + j\phi(x,y).$$

The phase modulation $\phi(x,y)$ can be changed into amplitude modulation by the use of a *knife-edge filter*.[19] This knife-edge filter is defined in the frequency domain as

$$H(f_x, f_y) = \tfrac{1}{2}(1 + \text{sgn}(f_x)).$$

This is a two-dimensional version of the *Hilbert filter*. The knife-edge filter can be constructed optically by placing a knife edge in the Fourier transform plane between two lenses. The output of the knife-edge filter with $1 + \phi(x,y)$ at the input is

$$I(x,y) = \frac{1}{4}\left[1 - \frac{2}{\pi}\int_{-\infty}^{\infty} \frac{\phi(x',y')}{x-x'}dx'\right],$$

which is the Hilbert transform of the phase variation translated into a real intensity variation. In this way, the phase modulation is changed to amplitude modulation and becomes visible.

4.8 Wavefront Reconstruction

Let $p(x,y)$ be the source of a waveform in the x,y plane at $z=0$. After free-space propagation through the Huygens–Fresnel point-spread function of free space $h(x,y)$, the waveform appears as the spatial signal $s(x,y) = h(x,y) ** p(x,y)$ in the reference plane at $z=d$ as shown in Figure 4.7. The complex spatial signal $s(x,y)$ represents the spatially baseband, temporally passband optical signal $\tilde{s}(x,y,t) = s(x,y)e^{j2\pi f_0 t}$.

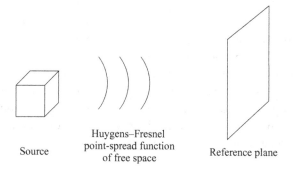

Figure 4.7 Free-space propagation

---

[19] The use of a filter in this way for converting phase into amplitude is known as the *schlieren method* from a German word meaning streak or striation.

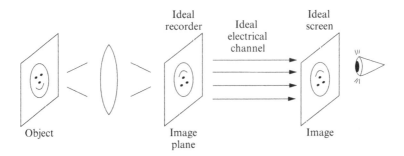

Figure 4.8 Transferring a recorded image

The complex baseband signal $s(x, y)$ as it appears in the reference plane is

$$s(x, y) = |s(x, y)| e^{j\theta(x,y)}.$$

To confine the discussion to a finite region of the x, y plane, the signal $s(x, y)$ is passed undisturbed through a large aperture. The signal can form an image at a subsequent plane by inserting a suitable lens into the reference plane. When the aperture is large, the resulting diffraction has little effect on the signal $s(x, y)$ in the reference plane. Diffraction is ignored in this section.

A camera can record the intensity $|s(x, y)|^2$ of the signal $s(x, y)$ by forming an image on a recording medium in the image plane of the lens. Figure 4.8 shows the situation. A suitable recording medium is a photographic film or an array of optoelectronic photosensors. The recorded image can then be scanned and stored, or the recorded image can be sent to another place through a communication channel. However, even when the intensity in the image plane is recorded perfectly, the displayed image would fail to have all the properties of the original visual field. The visual field in any x, y plane displays depth. The recorded image, in contrast, is flat.

Remove the lens. The scene can now be observed directly through the aperture. The viewer can change the viewing angle through the aperture to change the view of a three-dimensional scene behind the aperture.[20] In contrast, the recorded intensity image lacks this depth. The intensity image is only a two-dimensional image. What happened to the missing information about the image?

The explanation for this difference is that the conventional sensors record only the intensity of the signal in the image plane. The x, y plane at any distance z has a signal with a meaningful phase angle at each point and this phase angle is used by the propagating waveform to reconstruct itself in different ways as it travels in different directions. The three-dimensional effect arises through the phase.

---

[20] The visual field in the reference plane contains a real three-dimensional image as viewed through that aperture. In contrast, an artificial three-dimensional image does appear to be three-dimensional to a viewer using special eye glasses designed for this purpose, but the three-dimensional image lacks parallax or perspective. Although the image appears to have depth, the image does not change with the viewing angle as the viewer moves to a different location.

4.8 Wavefront Reconstruction

A more ambitious goal is to record both the magnitude and the phase of the advancing wavefront $s(x,y)$ because there is useful three-dimensional information in the wavefront phase as a function of x and y. However, a medium that records optical phase directly is difficult to develop, in large part because the wavelength of visible light is so small.

One way of recording phase is to turn the phase information into the amplitude modulation of another wavefront and then record the intensity of that alternative wavefront. This method is called *holography*. In the simplest approach, the signal of interest and a coherent reference signal are added. The intensity of the sum is recorded, as described below. Usually, the recording medium is photographic film, which is developed as a transparency called a *hologram*. The hologram does not contain an image; the image is formed later by first illuminating the hologram with a coherent reference thereby reconstructing the original optical field. A lens is then used to form an image or the optical field is viewed directly.

The source $\rho(x,y)$ becomes the wavefront $s(x,y) = h(x,y) ** \rho(x,y)$ at the recording plane. A reference signal

$$a(x,y) = |a(x,y)|e^{j\psi(x,y)}$$

must also appear in the recording plane. A duplicate or a suitable variation of this reference signal must be regenerated at a later time when the hologram is observed. The total signal in the recording plane is

$$v(x,y) = s(x,y) + a(x,y).$$

The intensity $|v(x,y)|^2$ is recorded in the form of the transmittance of a transparency. The intensity consists of four terms:

$$|v(x,y)|^2 = |s(x,y) + a(x,y)|^2$$
$$= |s(x,y)|^2 + a^*(x,y)s(x,y) + a(x,y)s^*(x,y) + |a(x,y)|^2.$$

The complex wavefront $s(x,y)$ can be recovered from either the second term or the third term. In either case, the other three terms create spatial interference in the regenerated spatial signal.

To view the hologram transparency, illuminate it with a waveform, $b(x,y)$, to produce the signal

$$r(x,y) = b(x,y)|v(x,y)|^2$$
$$= b(x,y)|s(x,y)|^2 + b(x,y)a^*(x,y)s(x,y)$$
$$+ b(x,y)a(x,y)s^*(x,y) + b(x,y)|a(x,y)|^2,$$

leaving the transparency. Now specify that $b(x,y) = a(x,y)$ and choose $a(x,y)$ to be a monochromatic plane wave satisfying $|a(x,y)| = 1$. Then

$$r(x,y) = a(x,y)|s(x,y)|^2 + s(x,y) + (a(x,y))^2 s^*(x,y) + a(x,y).$$

The second term is the desired reconstructed signal containing amplitude and phase in full. An observer of $r(x,y)$ would see the signal $s(x,y)$ within $r(x,y)$ just as though the original source of $s(x,y)$ were still in place behind the hologram. Thus we say that the waveform $r(x,y)$ contains a copy of $s(x,y)$ apparently diverging from a *virtual image* of $\rho(x,y)$ that would appear quite real to the viewer.

The other three terms of $r(x,y)$ are interference terms appearing as artifacts in the visual field and the viewer would want to suppress them. This can be accomplished by a form of spatial heterodyning[21] whereby a tilted reference wave is used to spatially separate the interfering terms as will be discussed below.

Alternatively, again specify that $|a(x,y)| = 1$, but now choose $b(x,y) = a^*(x,y)$. Then

$$r(x,y) = a^*(x,y)|s(x,y)|^2 + (a^*(x,y))^2 s(x,y) + s^*(x,y) + a^*(x,y).$$

In this case, the third term is the conjugate of signal $s(x,y)$. Therefore this gives a method for reversing the sign of the phase, which leads to some interesting properties. As before, the three remaining terms are to be regarded as interference in the visual field that the viewer would want to suppress.

Two variations of holography are now described. These are the *Gabor hologram*, which was the first to be proposed, and the *Leith–Upatnieks hologram*, which has better properties and is more practical. The difference between the two is the way in which the reference is combined with the signal. The Gabor hologram relies on the existence of an ambient coherent illumination that leaks past the object $\rho(x,y)$. The Leith–Upatnieks hologram explicitly provides the reference signal by a system of mirrors.

Let the coherent, monochromatic, monodirectional wave given by $a(x,y) = 1$ be used to illuminate the object $\rho(x,y)$ and to also form the reference. The signal in the reference plane is[22]

$$v(x,y) = s(x,y) + 1,$$

where $s(x,y)$ is the signal originating as $\rho(x,y)$ as it appears in the recording plane. That is, $s(x,y) = h(x,y) ** \rho(x,y)$. The signal $s(x,y)$ may be complex and $|s(x,y)|$ may be much smaller than one. The signal intensity recorded at the photographic film is

$$|v(x,y)|^2 = |s(x,y) + 1|^2$$
$$= 1 + s(x,y) + s^*(x,y) + |s(x,y)|^2.$$

Consequently, the recorded intensity consists of a strong, uniform plane wave; two signal terms; and a small term, $|s(x,y)|^2$, which is neglected under the assumption

---

[21] The multiplication of a passband time-varying waveform at carrier frequency f_0 by a sinusoidal waveform at frequency $f_0 - f_1$ to change the carrier frequency f_0 to frequency f_1 is called *heterodyning*.
[22] This notion and the approximation are similar to those used to develop the Zernicke filter.

4.8 Wavefront Reconstruction

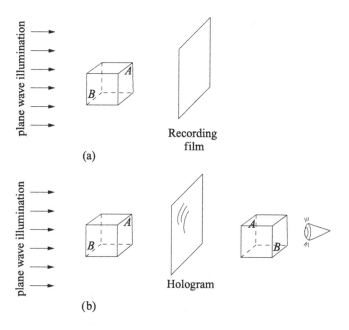

Figure 4.9 The Gabor transmission hologram

that $|s(x,y)|$ is small. When the latter term cannot be neglected, the Gabor hologram may be unsatisfactory. The two signal terms correspond to "twin images": one real image and one virtual image. Either image can be focused, but the other image, which is far out of focus, is superimposed on the focused image. The out-of-focus image interfering with the in-focus image is a disadvantage of the Gabor hologram.

A transparency $|v(x,y)|^2$ is made as the recorded signal. This is the hologram. The hologram is viewed later by passing a uniform plane wave of unit amplitude through the transparency, as shown in Figure 4.9. Then the viewed signal is

$$|v(x,y)|^2 = (s(x,y) + 1)(s^*(x,y) + 1)$$
$$\approx 1 + s(x,y) + s^*(x,y) + |s(x,y)|^2.$$

The small quadratic term is uninteresting and can be ignored. Within the plane of the hologram, the signal $s(x,y)$ is just as it appeared at the hologram when it was made. The signal reaching the viewer due to $s(x,y)$ is just as it would appear were the original subject still behind the hologram. A viewer focusing on this distance sees a virtual image of $\rho(x,y)$ at distance d behind the hologram.

The second term $s^*(x,y)$ has the sign of its phase angle reversed. Thus the term appears as it would appear for an object located at distance d in front of the hologram. The illumination causes a signal that converges to a *real image* to the right of the hologram. A blank screen placed in this plane would display an image of $\rho(x,y)$. A viewer further to the right of this point can focus on this point to see the real image located at distance d in front of the hologram.

4 Optical Imaging Systems

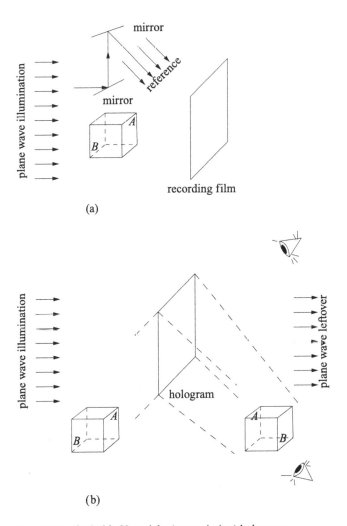

Figure 4.10 The Leith–Upatnieks (transmission) hologram

The Gabor hologram is usually unsatisfactory because of the four waves reaching the viewer, corresponding to the four complex amplitudes 1, $s(x,y)$, $s^*(x,y)$, and $|s(x,y)|^2$. Of these, only the wave $s(x,y)$ is of interest. It can be isolated by focusing. The other three terms are unfocused and appear as visual interference.

This limitation of the Gabor hologram can be circumvented by introducing the reference waveform from another direction. (This is a spatial analog of heterodyning.) This gives the Leith–Upatnieks hologram, shown in Figure 4.10. Now the constant term in the output, which appears as a bright light, appears to come from a different direction and is less intrusive.

4.9 Optical Filtering

The mathematical operation of two-dimensional filtering may be performed optically. A two-dimensional filter

4.9 Optical Filtering

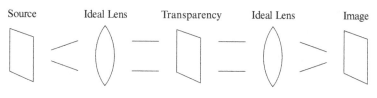

Figure 4.11 An optical processor for spatial filtering

$$r(x,y) = g(x,y) ** s(x,y)$$

can be implemented optically by using the ideal lens as an optical processing component, as will be described. A two-dimensional filtering operation performed by a pair of lenses uses the convolution theorem to multiply $S(f_x,f_y)$ by $G(f_x,f_y)$ in the Fourier transform domain. Figure 4.11 shows a four focal-length arrangement that is used to compute a two-dimensional Fourier transform of the wavefront $s(x,y)$ in the source plane. The first lens is used to compute a Fourier transform. It requires one focal length on each side of that lens. The second lens is used to compute an inverse Fourier transform. It also requires one focal length on each side of its lens. Between the two lenses is a transparency with the transmittance $G(f_x,f_y)$ of the filter. Within the approximations of this chapter, the resulting relationship between the image plane and the source plane is

$$r(x,y) = g(x,y) ** s(x,y)$$

for a coherent system, and

$$|r(x,y)|^2 = |g(x,y)|^2 ** |s(x,y)|^2$$

for a noncoherent system. For the noncoherent case, only spatial filters that have the form of a squared magnitude of a function can be realized.

Spatial filtering of a blurred image can be used to suppress the effects of an unwanted blurring point-spread function and so can sharpen the image. Details in the image can be enhanced by de-emphasizing the low spatial frequencies. Spatial filtering can also suppress the effects of noise, usually by de-emphasizing the high spatial frequencies. For example, the artifacts inherent in halftone imagery can be suppressed by using a spatial lowpass filter.

When the filtering operation is used to remove the effect of an unwanted point-spread function, such as a blur on the image, it is called *spatial equalization* or *deconvolution*. The general topic of deconvolution is studied in Chapter 7.

Many egregious impairments in photographs can be modeled as blurs arising in the system that produced the photographs. Blurring can be caused by poor focusing or by camera movement. A two-dimensional signal of interest, $s(x,y)$, received as a blurred signal is

$$v(x,y) = h(x,y) ** s(x,y),$$

where $h(x,y)$ is a point-spread function that blurs the signal $s(x,y)$.

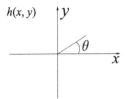

Figure 4.12 Point-spread function for camera movement

To improve the blurred image $v(x,y)$ when the blurring point-spread function is known, one may pass it through a compensating filter, $g(x,y)$, and re-record it. Then

$$u(x,y) = g(x,y) ** v(x,y)$$
$$= \bigl(g(x,y) ** h(x,y)\bigr) ** s(x,y).$$

The choice of a suitable compensating point-spread function $g(x,y)$ is discussed in Section 7.2.

Figure 4.12 shows a point-spread function, $h(x,y)$, as a narrow line that may be used to model streaking due to camera motion. The motion may cause a point object to be reproduced by a narrow streak at the angle θ. The streak can be represented by a two-dimensional rectangle function $h(x,y)$ of width b and length a. With a coordinate system chosen so that the streak is along the x axis, the point-spread function in the frequency domain is

$$H(f_x,f_y) = ab\,\mathrm{sinc}(af_x)\mathrm{sinc}(bf_y),$$

which is zero whenever af_x is a nonzero integer.

For another example, suppose that the signal $s(x,y)$ is transmitted or recorded by some process that inserts false lines into the image or inserts multiple streaks parallel to the x axis. This is an additive noise process. It is different from the point-spread function due to camera movement because the horizontal lines are not caused by details of the image. One instance is an image of a large section of the earth that is formed by mosaicing a number of small photographs. When this is done, there may be visible lines in the mosaic where the small pictures abut. These lines may be modeled as additive noise and removed by a suitable filter.

Problems

4.1 Prove that the Huygens–Fresnel point-spread function is consistent in the following sense. The signal $s_0(x,y) = s(x,y,0)$ in a plane at $z = 0$ propagates into $s_{d_1}(x,y) = s(x,y,d_1)$ in a plane at $z = d_1$. The signal $s_{d_1}(x,y)$ in a plane at $z = d_1$ propagates into $s_{d_2}(x,y) = s(x,y,d_2)$ in the plane at $z = d_2 > d_1$. Prove that $s_{d_2}(x,y)$ is the same as the signal obtained by propagating $s_0(x,y)$ to the plane at $z = d_2$ directly. In particular, the signal in any plane at any intermediate distance between a source and an observation can be considered as an alternative source of that signal.

4.2 (**Conjugation**) Suppose that $s_0(x, y) = s(x, y, 0)$ is a signal at $z = 0$. Let $s(x, y, d_1)$ be the signal in a plane at $z = d_1$ obtained by propagating $s_0(x, y)$. What signal is received in a plane at $z = 2d_1$ when the conjugate signal $s^*(x, y, z)$ is propagated from the plane at $z = d_1$? (A hypothetical device in the plane at $z = d_1$ conjugates the complex baseband optical signal.)

What signal will be observed in the x, y plane at $z = 3d_1$?

4.3 (**Very Near Field**) a. Show that for large λ the Huygens–Fresnel point-spread function reduces to

$$h(x, y) = \frac{d/2\pi}{(x^2 + y^2 + d^2)^{3/2}}.$$

Show that this approximation holds when λ is much larger than $2\pi\sqrt{x^2 + y^2 + d^2}$.

b. Show that for large λ the Fourier transform of the Huygens–Fresnel point-spread function reduces to

$$H(f_x, f_y) = -e^{-2\pi d\sqrt{f_x^2 + f_y^2}}.$$

c. Conclude that the secondary term of the Huygens–Fresnel point-spread function is primarily associated with evanescent waves. Can this correspondence be made exact?

4.4 A transparency consists of opaque squares and transparent squares in an eight by eight checkerboard pattern. What is the far-field (Fraunhofer) diffraction pattern?

4.5 Prove the lens law in the Fresnel approximation by means of an analysis in the transform domain.

4.6 An ideal lens in the zero plane $z = 0$ is used to form an image of the object $s_0(x, y)$ in the plane $z = -d_0$. Prove that, for any x, y, the line defined by $(x, y, -d_0)$ and $(0, 0, 0)$, and the line defined by $(x, y, 0)$ and $(0, 0, f_\ell)$, cross in the image plane and moreover cross at the image of $s_0(x, y)$. Give a sketch of a simple ray-tracing method of locating the image of an object. Give a "derivation" of the lens law from the ray-tracing sketch. Give an explanation of the magnification. Describe how to decompose the analysis of an imaging system into the computation of a diffraction-free image by ray tracing and the computation of a point-spread function by Fourier optics.

4.7 The f-*number* of a simple lens is the ratio of the focal length to the diameter of the aperture. A *fast lens* is a simple lens whose f-number is smaller than one. Will the Fresnel approximation be adequate for analyzing a fast lens?

4.8 A *diffraction grating* consists of an aperture consisting of narrow slits ruled across it. Suppose that a diffraction grating is 1 centimeters wide and has 10,000 evenly spaced slits ruled across it. What is the wavelength of the light whose first-order grating lobe is at 45° when the diffraction grating is illuminated from far behind? What is the wavelength of light whose second-order and third-order grating lobes are at 45°? Are any of these wavelengths visible light? If so, what is the color?

4.9 Derive the Fraunhofer approximation by rewriting the Fresnel point-spread function in terms of chirp filters. Under what condition can the quadratic-phase term inside the integration be neglected?

4.10 (**Focusing Error**) Instead of the lens law,
$$\frac{1}{d_s} + \frac{1}{d_i} - \frac{1}{f_\ell} = 0,$$
suppose that a noncoherent optical-imaging system is out-of-focus, satisfying the expression
$$\frac{1}{d_s} + \frac{1}{d_i} - \frac{1}{f_\ell} = \Delta.$$
Show that the effect of this can be described by cascading the ideal lens with the two-dimensional function
$$G(f_x, f_y) = e^{j\pi(\Delta/\lambda)(f_x^2 + f_y^2)}.$$

4.11 (**Focusing Error**) A one-dimensional ideal lens,
$$t(x) = e^{-j(\pi/\lambda f_\ell)x^2} \text{rect}\left(\frac{x}{a}\right),$$
is used to form a simple image. Suppose that the image is out of focus with
$$\frac{1}{d_s} + \frac{1}{d_i} - \frac{1}{f_\ell} = \Delta.$$

a. Find the noncoherent optical transfer function by deriving and evaluating the following expression,
$$G_n\left(\frac{x}{-\lambda d_i}\right) = \int_{-\infty}^{\infty} A\left(\xi + \frac{x}{2}\right) A^*\left(\xi - \frac{x}{2}\right) d\xi,$$
for an appropriate choice of the "aperture function" $A(x)$.

b. Find the two-dimensional noncoherent optical transfer function for an out-of-focus image formed by an ideal lens with a square aperture.

c. Show that when
$$\frac{1}{8}\left(\frac{\Delta}{\lambda}\right) \ll \left(\frac{\lambda}{a}\right)^2,$$
the noncoherent optical point-spread function reduces to the geometrical optics approximation obtained by ray tracing.

d. What is the noncoherent, optical point-spread function in the geometrical optics approximation of an out-of-focus image formed by an ideal lens with a circular aperture?

4.12 Based on the Rayleigh resolution criterion, what is the angular resolution at wavelength λ of a simple single-lens telescope with a uniformly illuminated circular aperture of diameter a?

4.13 A pinhole can be used to produce images of low intensity. Using geometrical optics and ray tracing, explain how the set of rays originating at the source $p(x,y)$ at $z = -d$, and passing through the point $(0, 0, 0)$ will produce an image at $z = d'$. Why are pinhole cameras not in common use?

4.14 A *Fresnel zone plate* is a two-dimensional weighting function,

$$t(x,y) = \tfrac{1}{2}[1 + \text{sgn}(\cos\alpha(x^2 + y^2))]\text{circ}(x/a, y/a).$$

By using the Fourier series expansion,

$$1 + \text{sgn}(\cos 2\pi t) = \sum_{n=-\infty}^{\infty} \frac{\sin(\pi n/2)}{\pi n} e^{j2\pi nt},$$

show that $t(x,y)$ acts like a composite of ideal lenses with different focal lengths. What are the focal lengths? Describe the appearance of the Fresnel zone plate. Give an expression for the widths of the rings of the Fresnel zone plate. What is the area of each ring?

Describe the output of the Fresnel zone plate when the input is the plane wave with amplitude $s(x,y)$.

4.15 A *Fresnel phase plate* is a two-dimensional weighting function,

$$t(x,y) = e^{j(\pi/2)[1+\text{sgn}(\cos\alpha(x^2+y^2))]}\text{circ}(x/a, y/a).$$

Compare the Fresnel phase plate with the Fresnel zone plate. In particular, compare the intensity of the corresponding images.

4.16 A thin lens, here called a thin *refractive lens*, has the relative phase shift $(2\pi/\lambda)(x^2 + y^2)/2f_\ell$ at the point (x,y). A *diffractive lens* has the relative phase shift $(2\pi/\lambda)(x^2 + y^2)/2f_\ell \pmod{2\pi}$ at the point (x,y). Sketch a thin refractive lens and a thin diffractive lens in cross section. Do a thin refractive lens and a thin diffractive lens both provide the same approximation to an ideal lens? Do the two lenses behave the same under a perturbation of the wavelength?

4.17 (Babinet's Principle) One version of *Babinet's principle* says that for two functions $t_1(x,y)$ and $t_2(x,y)$ summing as

$$t_1(x,y) + t_2(x,y) = \text{circ}(x,y)$$

to a circle function, their diffraction patterns sum to the diffraction pattern of a circle function.

a. Prove Babinet's principle and state its form in the Fraunhofer region.
b. Use Babinet's principle to state the behavior in the Fresnel region of the weighting function

$$t(x,y) = \tfrac{1}{2}[1 - \text{sgn}(\cos\alpha(x^2 + y^2))]\text{circ}(x/a, y/a).$$

4.18 (Huygens Point-Spread Function) Derive the form of the Huygens point-spread function

$$\tilde{c}(t,x,y,z) = \frac{e^{-j2\pi f_0(t-\sqrt{x^2+y^2+z^2}/c)}}{\sqrt{x^2+y^2+z^2}}$$

directly from the transfer function of free space

$$S(f_x,f_y) = e^{j2\pi \frac{d}{\lambda}\sqrt{1-\lambda^2 f_x^2 - \lambda^2 f_y^2}}.$$

4.19 (Gaussian Beams) A spatially coherent, monochromatic wave propagating nominally in the z direction has a gaussian amplitude distribution in the x,y plane at $z=0$,

$$s(x,y,0) = e^{-(x^2+y^2)/w^2}.$$

The parameter w is called the "waist" of the beam.[23] What is the intensity distribution $|s(x,y,z)|^2$ in the plane $z=d$? Can the intensity at $z=d$ be characterized by a waist? Is the paraxial approximation needed to rationalize this question?

4.20 Is the following statement correct? A gaussian beam remains gaussian as it propagates, and a nongaussian beam tends to become more gaussian as it propagates.

4.21 The *Talbot effect* is the following self-imaging property of a periodic transparency. A transparency at $z=0$, periodic in the x direction with period Δ and illuminated by a normally incident plane wave of wavelength λ, reproduces itself periodically in planes at $z = \ell z_T$, where ℓ is an integer and $z_T = 2\Delta^2/\lambda$ is a constant called the *Talbot distance*.

Derive the Talbot self-imaging effect from the Fresnel approximation for an infinitely large aperture. Is any further approximation necessary?

Describe how the effect changes for a finite aperture.

4.22 Given the transmittance $t(x,y)$ in the plane at $z=0$, let $\rho(x,y)$ be the complex amplitude of a monochromatic wave at $z = -d_s$. This wave passes through the transmittance and is observed in the plane at $z = d_i$. Show that, in the Fresnel approximation, the complex amplitude $r(x,y)$ in this plane is described as a scaled and magnified copy of $\rho(x,y)$ passed through a point-spread function given by the Fourier transform of $t(x,y)e^{j2\pi(x^2+y^2)/2\lambda f_\ell}$, where f_ℓ is the constant satisfying

$$\frac{1}{f_\ell} = \frac{1}{d_s} + \frac{1}{d_i}.$$

4.23 a. A transmittance function,

$$t(x,y) = e^{-j(2\pi/\lambda)(x^2+y^2)/2f_\ell} e^{-j2\pi(a+b\cos 2\pi x/\Delta)} \text{circ}(x/r, y/r),$$

---

[23] A *beam* of light is a narrowly concentrated wave moving nominally in the same direction. A beam of light is a term midway between a wave of light and a ray of light.

is placed at the origin. An object, $\rho(x,y)$, is placed at $z = -d_s$. Under the Fresnel approximation, describe the noncoherent image observed in the plane at $z = d_i$, where d_i satisfies

$$\frac{1}{d_i} = \frac{1}{f_\ell} - \frac{1}{d_s}.$$

b. What is the point-spread function through which the magnified image is observed?

4.24 a. A transmittance function,

$$t(x,y) = e^{-j(2\pi/\lambda)(x^2+y^2)/2f_\ell} \operatorname{sinc}(ax, ay),$$

is placed at the origin. An object, $\rho(x,y)$, is placed at $z = -d_s$. Under the Fresnel approximation, describe the noncoherent image observed in the plane at $z = d_i$, where d_i satisfies

$$\frac{1}{d_i} = \frac{1}{f_\ell} - \frac{1}{d_s}.$$

b. What is the point-spread function through which the magnified image is observed?

4.25 Explain how one can magnify an image by the choice of the optical frequencies with which to form and to view a holograph. Can a microscope be formed in this way?

4.26 Explain the nature of a "pseudoscopic image" that may occur in certain forms of holographic aberration correction.

4.27 (**Halftone Imaging**) Halftone imaging is a technique, commonly used in newspaper publication, to simulate a gray scale in printed images using only black dots on a white background. The dots may vary in size or may vary in density to simulate gray.

a. Black dots of varying size are placed on a square grid, thereby sampling the gray-scale image. How fine must the grid be? Does the Nyquist–Shannon sampling theorem apply in any sense? Describe the two-dimensional Fourier transform. Describe an optical filter that recovers the gray-scale image. Describe an optical filter not centered at zero spatial frequency that recovers the gray-scale image.

b. Black dots of the same size are placed on a finer square grid, thereby sampling the gray-scale image. Gray scale is represented by the spatially varying density of dots. Repeat the above questions for this case.

Notes

In 1678, Christian Huygens first put forth the view that each point on an advancing wavefront can be regarded as the source of a secondary wavefront. Huygens stated his principle only in qualitative terms. Much later, in 1818, Augustin Fresnel stated this principle more formally and explained diffraction as an interference phenomenon using precise mathematics, although his formulation was incomplete. Because of continuing concern about the proper statement of consistent boundary conditions, there have been many subsequent derivations of the Huygens–Fresnel principle by Kirchhoff, by Rayleigh, and by Sommerfeld. Herein, instead of the usual treatment of the Huygens–Fresnel principle, we have provided an original alternative approach that sidesteps questions of boundary conditions by stating the Huygens–Fresnel principle as a relationship between a wave in two planes, then deriving it as a simple consequence of a certain two-dimensional Fourier transform pair. This gives a precise mathematical relationship between a wavefront passing through one plane and that same wavefront passing through a subsequent parallel plane. The only condition of the analysis is that the wave must satisfy the wave equation.

The use of the lens for magnification, as in microscopes and telescopes, has been widespread for some time as has been the use of the lens to readjust focal length of biological lenses, such as in eyeglasses.

The first study of the lens in the spirit of this chapter was the Abbe–Porter experiment begun by Abbe (1873) and continued by Porter (1906). They experimented with placing simple obstructions in the Fourier transform plane of a lens to modify the image. Their experiments became the basis for many later developments including Abbe's notion (1873) of a diffraction-limited system. The phase-contrast microscope was proposed by Zernicke (1935). From a mathematical point of view, the Zernicke filter is the two-dimensional analog of the FM radio demodulator proposed by Armstrong (1936). The schlieren method had been proposed much earlier by Foucault (1858).

The intersection of the topic of optics with the topics of information theory and signal processing comes from the role of the Fourier transform in optics. This connection was recognized early on by Duffieux (1946). The formal development began with Elias (1953), who was motivated by his earlier paper with Grey and Robinson (Elias et al., 1952). This was developed considerably by O'Neill and others (1956, 1962). Goodman consolidated the topic of Fourier optics in his highly popular 1968 textbook.

Vander Lugt (1964) proposed the use of an optical processor to calculate a two-dimensional convolution as a product in the Fourier transform domain, thereby providing a two-dimensional filter. Lohmann (1977) described a method to bypass the limitations of noncoherent optics used for filtering. At the same time that this restatement of the theory of optics was underway, applications emerged for the use of optics to perform certain two-dimensional Fourier transforms for imaging with microwave signals, as discussed by Cutrona, Leith, Palermo, and Porcello (1960). The use of the Fourier transform to derive the Fresnel diffraction formula was discussed by Banerjee (1985). The use of the ambiguity function to design cubic-phase apertures for large depth-of-field imaging was proposed by Dowski and Cathey (1995).

Systems for reconstructing images using signals in the Fourier transform domain greatly influenced Gabor (1948), who, in turn, worked out the ideas of holography as a lensless imaging process for electron microscopy. Sometime later, holography was seen to be an important method for optical signals. Various versions of holography have been formulated since Gabor's work. An important advance was by Leith and Upatnieks (1962, 1964).

5 Apertures and Radiation Patterns

The interface between a guided wave within a transmitter or a receiver and an unguided propagating wave in free space or a homogeneous medium is described using the notion of an aperture. The aperture may be realized in concrete form as an antenna, as a hydraphone, or as an acoustic speaker. An antenna is a linear device that forms the interface between guided propagation of electromagnetic signals and unguided propagation of electromagnetic waves. Similarly, a hydrophone or an acoustic transmitter forms the interface between guided electrical signals and unguided acoustic waves in water or air. The terminology appropriate to electromagnetic waves is favored in this chapter, though the theory applies broadly.

An antenna can be used either to transmit an electromagnetic wave or to receive an electromagnetic wave. During transmission, the function of the antenna is to concentrate the guided electromagnetic wave into an unguided beam that nominally points in the desired spatial direction. During reception, the function of the antenna is to collect the incident signal on that antenna and deliver it to the receiver. An important principle of antenna theory, known as the *reciprocity principle* states an equivalence between the antenna pattern for transmission and the antenna pattern for reception. The reciprocity principle allows us to describe the antenna either as a transmitting device or as a receiving device depending on which description is more convenient for a particular discussion.

The primary aspect of antennas studied in this chapter is the propagation and diffraction of waves traveling between the antenna aperture and points in the Fraunhofer far field. During transmission, a signal source presents a time-varying and spatially distributed signal across the antenna aperture to form a wave that then propagates as an unguided wave through free space. The spatial distribution of the signal across the antenna aperture is called the *illumination function*. The distribution in the far field of the waveform amplitude and phase over the spherical coordinate angles is called the *antenna radiation pattern* or the *antenna pattern* or the *radiation pattern*. The relationship between the antenna pattern and the aperture illumination function can be described with the aid of the two-dimensional Fourier transform, as is appropriate to Fraunhofer diffraction.

The formal relationship between the aperture illumination function $s_0(x, y)$ and the antenna radiation pattern is studied herein. The many physical effects involved in generating the illumination function for real antennas and arrays – and the design of those antennas – will not be considered.

5 Apertures and Radiation Patterns

Large antennas can be formed by an array of simpler antenna elements by sharing the signals across the multiple elements of that array. This is analogous to forming a pulse train by combining many copies of a simple pulse. Similar techniques are needed to form the antenna pattern of an antenna array. Again, the concern is only with the mathematical relationship between the illumination function and the antenna radiation pattern.

The wavefront launched by an illumination function on an aperture is completely described by the Huygens–Fresnel principle, which was discussed in detail in Chapter 4. This chapter gives an alternative and looser development of diffraction describing diffraction simply as a two-dimensional Fourier transform,[1] as is suitable in the Fraunhofer far-field diffraction region. The heuristic approach to diffraction in the far field leads to a simple and intuitive method of deriving radiation patterns. It complements and reinforces the formal approach to diffraction given in Chapter 4.

5.1 Aperture Illumination and Antenna Pattern

A physical antenna is the interface between the guided propagation of signals and the propagation of waves in free space. As far as its effect on the signal is concerned, an antenna is described by the antenna radiation pattern. The antenna radiation pattern is a function, $E(\phi, \psi)$, of the spherical coordinates (ϕ, ψ) defining direction. The function $E(\phi, \psi)$ is a complex function that gives the magnitude and phase of the antenna's effect on the signal radiated in direction ϕ, ψ. The squared magnitude of $E(\phi, \psi)$, denoted $P(\phi, \psi) = |E(\phi, \psi)|^2$, is also called the antenna radiation pattern. To distinguish $E(\phi, \psi)$ from $P(\phi, \psi)$, we may speak of the antenna signal radiation pattern or the antenna power radiation pattern. An *omnidirectional antenna*, were such to exist, is an antenna for which $|E(\phi, \psi)|^2$ is a constant for all ϕ and ψ.

An antenna can be described as a device for setting up an illumination function, possibly complex, on a region of the x, y plane near the antenna called the *aperture*. The aperture is a designated planar region at or near the physical antenna where the illumination function is defined. The signal radiated into each spatial direction by this illumination function then constitutes the antenna radiation pattern. Only planar apertures and their radiation patterns are studied here.

The province of the antenna designer is to configure an arrangement of conducting objects and their electrical feeds, as shown in Figure 5.1, to generate the illumination function $s_0(x, y)$ within the aperture.

At each point (x, y) of an aperture at $z = 0$, the illumination function launches an infinitesimal spherical Huygens passband wavelet, $s_0(x, y) e^{-j2\pi f_0 t} dxdy$. The magnitude and phase of the complex baseband representation $s_0(x, y) dxdy$ describe the magnitude and phase of the infinitesimal wave launched from this infinitesimal region at the point (x, y). One may visualize the illumination function as formed by a

---

[1] The study of the relationship between the antenna illumination function and the antenna pattern might be aptly dubbed *Fourier antenna theory*.

5.1 Aperture Illumination and Antenna Pattern

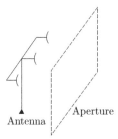

Figure 5.1 Portraying an antenna by an aperture

hypothetical screen with a hole in it that is illuminated from behind by a plane wave of frequency f_0. Visualizing that aperture filled with a semitransparent screen with the attenuation and phase described by $s_0(x, y)$ completes the abstract model of the aperture. This description may help to visualize the role of the illumination function $s_0(x, y)$.

One-Dimensional Aperture

Before discussing the antenna radiation pattern formed by a two-dimensional aperture, the radiation pattern formed by a hypothetical one-dimensional aperture to a point in the hypothetical one-dimensional Fraunhofer far-field is analyzed. A one-dimensional radiation pattern in a given direction formed by a one-dimensional illumination function, $s_0(x)$, is formed by the integration of all signals propagating in the given direction. A signal arriving in the far field from an infinitesimal element of the aperture at the point x has a weight given by the infinitesimal illumination function $s_0(x)dx$ at that x location and is delayed by an amount depending on the path length to a point far from the aperture. At that far point, the straight lines from points of the aperture are essentially parallel lines, and the path from the point x is shorter than the path from the origin by the amount $x \sin \phi$ as shown in Figure 5.2. Let τ denote the time that it takes for a signal from the origin to reach a point in the far field. The differential time delay of the infinitesimal signal coming from an infinitesimal element at x with respect to the delay from the infinitesimal element at the origin is

$$\Delta \tau = -\frac{x}{c} \sin \phi$$

because that infinitesimal signal travels a path that is longer by the distance $x \sin \phi$. Then

$$v(t, \phi) e^{-j 2\pi f_0 (t-\tau)} = \int_{-\infty}^{\infty} s_0(x) e^{-j 2\pi f_0 \left(t - \tau + \frac{x}{c} \sin \phi\right)} dx$$

is the complex passband signal received at this distant point from all points of the aperture.

More generally, a time-varying modulation signal, $m(t)$, may modulate the carrier. The waveform $m(t)$ at the carrier frequency f_0 gives rise to a spatially distributed signal within the aperture given by $m(t) e^{-j 2\pi f_0 t} s_0(x)$. This leads to the composite signal radiated at angle ϕ given by

5 Apertures and Radiation Patterns

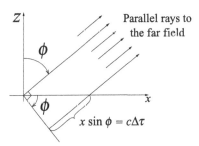

Figure 5.2 Incident wave in two dimensions

$$v(t,\phi)e^{-j2\pi f_0(t-\tau)} = \int_{-\infty}^{\infty} s_0(x) m\left(t - \tau + \frac{x}{c}\sin\phi\right) e^{-j2\pi f_0\left(t-\tau+\frac{x}{c}\sin\phi\right)} dx.$$

This section requires that the bandwidth of $m(t)$ is small compared to f_0, which permits an approximation called the *narrowband approximation*.[2] This is the approximation that $m(t)$ changes slowly enough so that the change during the time $(x\sin\phi)/c$ is negligible. Then $m(t + (x/c)\sin\phi)$ can be set equal to $m(t)$ with negligible error, which leads to

$$v(t,\phi) = m(t-\tau) \int_{-\infty}^{\infty} s_0(x) e^{-j2\pi \frac{x}{\lambda}\sin\phi} dx$$

$$= m(t-\tau) S_0\left(\frac{\sin\phi}{\lambda}\right),$$

where $S_0(f_x)$ is the Fourier transform of $s_0(x)$. The narrowband approximation allows the time-varying function to be separated from the space-varying function. The function $E(\phi)$, defined as

$$E(\phi) = \frac{1}{\lambda} S_0\left(\frac{\sin\phi}{\lambda}\right),$$

is the *one-dimensional antenna pattern* of the one-dimensional antenna as a function of ϕ. Then $v(t,\phi) = m(t-\tau)\lambda E(\phi)$. The multiplying term $1/\lambda$ is introduced into the definition of $E(\phi)$ so that the antenna pattern depends only on the ratio of the spatial dimensions of the antenna to the wavelength λ.

Example
A one-dimensional, uniformly illuminated antenna of length L is

$$s_0(x) = \begin{cases} 1 & |x| \leq L/2, \\ 0 & \text{otherwise.} \end{cases}$$

The transform of a rectangle pulse is a sinc pulse. Therefore

$$E(\phi) = \frac{1}{\lambda} S_0\left(\frac{\sin\phi}{\lambda}\right)$$

---

[2] The narrowband approximation does not always hold in such analyses, especially in sonar applications, and possibly in high-resolution radar imaging.

5.1 Aperture Illumination and Antenna Pattern

$$= \frac{L}{\lambda}\operatorname{sinc}\left(\frac{L}{\lambda}\sin\phi\right).$$

Notice that $E(\phi)$ depends only on the ratio L/λ, not on L or λ individually. This is why $1/\lambda$ is included in the definition of $E(\phi)$.

The antenna pattern of the illumination function $\operatorname{rect}(x/L)$ has the main lobe and the sidelobes of a sinc function. The main lobe may be called the *antenna main beam* or simply the *main beam* of the antenna. The zeros of the antenna pattern are uniformly spaced in $\sin\phi$, not in ϕ. A zero occurs when $(L/\lambda)\sin\phi$ is a nonzero integer. For $L < \lambda$, the antenna pattern has no zero. When L is only a little larger than λ, there is only one zero, or perhaps several. Finally, for large $\frac{L}{\lambda}$, there are many zeros and the antenna pattern is concentrated near $\sin\phi = 0$ with surrounding sidelobes. In the case of large $\frac{L}{\lambda}$, the approximation

$$E(\phi) = \frac{L}{\lambda}\operatorname{sinc}\left(\frac{L}{\lambda}\phi\right)$$

can be made by approximating $\sin\phi$ as ϕ.

Two-Dimensional Aperture

A two-dimensional radiation pattern is determined by a two-dimensional illumination function in the same way that a one-dimensional radiation pattern is determined by a one-dimensional illumination function. The differential delay in the signal transmitted in the direction coordinates ϕ and ψ from the point (x,y) relative to the signal transmitted from the origin is $(x\sin\phi\cos\psi + y\sin\phi\sin\psi)/c$. To see this more clearly, temporarily rotate the coordinate system, shown in Figure 5.3, by the angle ψ about the z axis into a new (x',y',z) coordinate system such that the far-field point is in the x',z plane. The differential delay then is $(x'\sin\phi)/c$. Replacing x' by $x\cos\psi + y\sin\psi$ gives the stated delay. Consequently, as before,

$$v(t,\phi,\psi) = m(t-\tau)\int_{-\infty}^{\infty}\int_{-\infty}^{\infty} s_0(x,y)e^{-j2\pi(\sin\phi\cos\psi\frac{x}{\lambda}+\sin\phi\sin\psi\frac{y}{\lambda})}dxdy$$

$$= m(t-\tau)S_0\left(\frac{\sin\phi\cos\psi}{\lambda},\frac{\sin\phi\sin\psi}{\lambda}\right),$$

where $S_0(f_x,f_y)$ is the two-dimensional Fourier transform of $s_0(x,y)$.

Definition 5.1.1 *The two-dimensional antenna pattern corresponding to the illumination function $s_0(x,y)$ is a function of the spherical angles defined as*

$$E(\phi,\psi) = \frac{1}{\lambda^2}S_0\left(\frac{\sin\phi\cos\psi}{\lambda},\frac{\sin\phi\sin\psi}{\lambda}\right),$$

where $S_0(f_x,f_y)$ is the Fourier transform of $s_0(x,y)$. ∎

Example

Figure 5.4 shows the main beam and sidelobes of a two-dimensional antenna pattern formed by an illumination function that is circularly symmetric. For an illumination function that is a uniformly illuminated circle of diameter $2r$,

$$s_0(x,y) = \operatorname{circ}\left(\frac{x}{2r},\frac{y}{2r}\right),$$

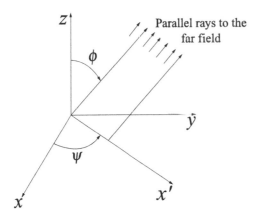

Figure 5.3 Incident wave in three dimensions

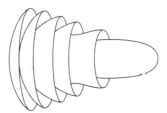

Figure 5.4 A circularly symmetric antenna pattern

then the antenna pattern is

$$E(\phi, \psi) = \frac{4r^2}{\lambda^2}\text{jinc}\left(\frac{2r}{\lambda}\sin\phi\right).$$

In this case, the sidelobes shown in Figure 5.4 are the sidelobes of the jinc function expressed as a function of $\sin\phi$. Because $\sin\phi$ has magnitude at most one, there are only a finite number of sidelobes. The number of sidelobes depends on the ratio of r/λ. The first zero of the jinc function occurs when $(2r/\lambda)\sin\phi = 1.22$, or $\phi = \sin^{-1}(0.61\lambda/r)$. The first zero of the jinc function will not be seen for any value of ϕ when the inequality

$$r \leq 0.61\lambda$$

is satisfied. Note in the example that $E(\phi, \psi)$ does not use all of $S_0(f_x, f_y)$. It displays $S_0(f_x, f_y)$ only over the range of frequencies such that $f_x^2 \leq \lambda^{-2}$ and $f_y^2 \leq \lambda^{-2}$ because otherwise the direction cosines would be larger than one. ∎

Direction Cosines

Recall that the trigonometric quantities $\sin\phi\cos\psi$ and $\sin\phi\sin\psi$ are two of the three direction cosines. The direction cosines are the cosines, respectively, of the angles between the direction of propagation and the x and y axes. The equations

$$\cos\phi_x = \sin\phi\cos\psi,$$
$$\cos\phi_y = \sin\phi\sin\psi,$$
$$\cos\phi_z = \cos\phi$$

express the direction cosines in terms of the angles ϕ and ψ.[3] The two other angles ϕ_x and ϕ_y are defined as the angles between the x and y axes and the direction of propagation, respectively. In terms of the compliments $\overline{\phi}_x$ and $\overline{\phi}_y$ of the angles ϕ_x and ϕ_y between the direction of propagation and the x and y axes, these are

$$\sin\overline{\phi}_x = \sin\phi\cos\psi,$$
$$\sin\overline{\phi}_y = \sin\phi\sin\psi.$$

Then

$$E'(\overline{\phi}_x, \overline{\phi}_y) = \frac{1}{\lambda^2} S_0 \left(\frac{\sin\overline{\phi}_x}{\lambda}, \frac{\sin\overline{\phi}_y}{\lambda} \right).$$

Sometimes $E'(\overline{\phi}_x, \overline{\phi}_y)$ is also called the antenna pattern, though it is not the same mathematical function as $E(\phi, \psi)$, but is nearly the same for small angles. For directions close to the z axis, the small-angle approximation to the sine gives

$$\overline{\phi}_x \approx \phi\cos\psi,$$
$$\overline{\phi}_y \approx \phi\sin\psi.$$

With these approximations, a highly directive antenna pattern can be expressed as

$$E'(\overline{\phi}_x, \overline{\phi}_y) = \frac{1}{\lambda^2} S_0 \left(\frac{\phi\cos\psi}{\lambda}, \frac{\phi\sin\psi}{\lambda} \right),$$

which is a small-angle approximation to the spherical trigonometry.

Gain Pattern

The antenna radiation pattern $E(\phi, \psi)$ is a complex function of spherical angular coordinates. Often the normalized squared-magnitude, given in the following definition, may be preferred to the full complex function.

Definition 5.1.2 *The directivity (or gain pattern) of an antenna in the direction with spherical coordinates ϕ, ψ is defined as*

$$G(\phi, \psi) = \frac{|E(\phi, \psi)|^2}{\frac{1}{4\pi} \int \int_A |s_0(x, y)|^2 dx dy}.$$

The peak-of-beam gain (or gain) is defined as $G = \max_{\phi, \psi} G(\phi, \psi)$. ∎

Theorem 5.1.3 *The peak-of-beam gain G of an aperture of area A satisfies*

$$G \le 4\pi \frac{A}{\lambda^2}$$

---

[3] The angles ϕ and ψ are two of the three angles in one form of the three *eulerian angles*

with equality if and only if the aperture illumination function is of the form

$$s_0(x,y) = e^{j2\pi(x\sin\phi_0\cos\psi_0 + y\sin\phi_0\sin\psi_0)/\lambda}.$$

Proof:
The proof uses the Schwarz inequality. The function $s_0(x,y)$ is nonzero only for those (x,y) in the region $\mathcal{A}$. Hence

$$|\lambda E(\phi,\psi)|^2 = \left|\int\!\!\int_{\mathcal{A}} s_0(x,y) e^{j2\pi(x\sin\phi\cos\psi + y\sin\phi\sin\psi)/\lambda} dxdy\right|^2$$

$$\leq \int\!\!\int_{\mathcal{A}} |s_0(x,y)|^2 dxdy \int\!\!\int_{\mathcal{A}} dxdy,$$

from which the inequality of the theorem follows. The Schwarz inequality is satisfied with equality if and only if $s_0(x,y)$ has the form

$$s_0(x,y) = e^{j2\pi(x\sin\phi_0\cos\psi_0 + y\sin\phi_0\sin\psi_0)/\lambda},$$

and $(\phi,\psi) = (\phi_0,\psi_0)$. ∎

Definition 5.1.4 *The effective area of an antenna aperture is the value A_e satisfying the expression*

$$G = 4\pi \frac{A_e}{\lambda^2}.$$

The effective area of an antenna is the area of a uniformly illuminated aperture that has the same gain as the antenna. Theorem 5.1.3 says that the effective area of an antenna is never larger than the actual area.

The *main beam* of $|E(\phi,\psi)|^2$ is the prominent region surrounding the maximum. The *resolution* of an antenna is determined by the width of the main beam, which varies inversely with the peak-of-beam gain of the antenna. The directivity, or gain, of an antenna is defined in a normalized form by dividing out the total energy in the illumination function $s_0(x,y)$. The normalization is included so that the antenna gain is not increased by scaling the amplitude of the illumination function.

The directivity of an antenna is sometimes displayed in a spherical coordinate system, as shown in Figure 5.5. Each contour line on the sphere denotes the locus of points for which $G(\phi,\psi)$ is equal to the same constant.

Quality Factor
Definition 5.1.1 expresses the antenna radiation pattern $E(\phi,\psi)$ as the Fourier transform of the illumination function. However, the direction cosines $\sin\phi\cos\psi$ and $\sin\phi\sin\psi$ cannot be larger than one in magnitude. Moreover, the sum of the squares of these two terms cannot be larger than one.

The antenna pattern is defined only for $(\sin\phi\cos\psi)^2 + (\sin\phi\sin\psi)^2 \leq 1$. Consequently, the *visible region* of the frequency domain for this definition consists of those spatial frequencies satisfying $|f_x|^2 + |f_y|^2 \leq \lambda^{-1}$. In particular, energy that is incident

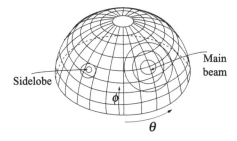

Figure 5.5 A gain pattern in spherical coordinates

on the aperture but is not in the visible region does not radiate.[4] This limitation in the representation of the antenna pattern as a Fourier transform is related to the Fraunhofer approximation. It is not significant for an aperture large compared to a wavelength. Within the Fraunhofer approximation, this consideration is accommodated by the heuristic aphorism that the invisible energy is "stored" in or near the antenna and is not transmitted. To quantify this effect, the *antenna quality factor* is defined as

$$Q = \frac{\int_{-\infty}^{\infty}\int_{-\infty}^{\infty} |S_0(f_x,f_y)|^2 \mathrm{d}f_x \mathrm{d}f_y}{\int_{-\lambda^{-1}}^{\lambda^{-1}}\int_{-\lambda^{-1}}^{\lambda^{-1}} |S_0(f_x,f_y)|^2 \mathrm{d}f_x \mathrm{d}f_y}.$$

For an antenna aperture that is large in comparison with the wavelength λ, the antenna quality factor is nearly equal to 1.

The heuristic notion of the quality factor corresponds to the formal notion of the evanescent wave defined in Chapter 4.

5.2 Antenna Arrays

An antenna pattern can be changed only by changing the illumination function. For an elementary antenna the illumination function is fixed. Therefore the antenna pattern is fixed for such an antenna. The illumination function cannot be changed to steer the beam of the antenna pattern. To steer the beam, the antenna itself may be rotated. An alternative to rotating the antenna is to pixelate the aperture into an array of multiple smaller antenna elements, illuminating each element separately and variably to steer the beam. This is an antenna array.

A large antenna can be constructed by a suitable arrangement of small antennas. In this context, each small antenna is called an *antenna element*, and the large antenna is called an *antenna array*. The relationship between the element and the array is the same as the relationship between a pulse and a pulse train. The larger array has a

---

[4] This chapter accepts the Fraunhofer approximation at the outset. This approach precludes any discussion of the evanescent waves that were introduced in Chapter 4. The antenna quality factor is the term that accounts for this limitation of the Fraunhofer approximation.

5 Apertures and Radiation Patterns

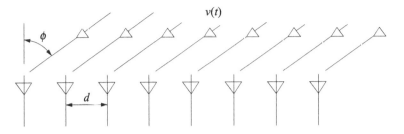

Figure 5.6 Antenna array

narrower Fourier transform and hence a narrower main beam, but it also can be shaped or steered. For this reason, the signal to or from each antenna element is controlled in phase and amplitude in order to create a desired beam. This is called *beamforming*. For an ideal array, each element of the array radiates with the same pattern as it does when it is physically isolated; the total radiation pattern is the sum of the individual patterns as adjusted for their individual placement.[5] The amplitude and phase of the illumination function at each element can be individually controlled so as to form the antenna pattern of the array.

One-dimensional arrays are studied first. The antenna pattern for a two-dimensional array is then a straightforward generalization. Figure 5.6 shows a one-dimensional antenna array. Denote the radiation pattern of the element by $E_e(\phi)$. It is a complex function, in general. The radiation pattern of an ideal array, denoted $E(\phi)$, is the superposition of the complex radiation patterns of all elements of the array.

The proof of the following theorem considers the array antenna pattern under the *reception* of a signal. The proof would be nearly the same when considering the array antenna pattern under the *transmission* of a signal. This is an early instance of the antenna reciprocity principle given in Proposition 5.8.1.

Theorem 5.2.1 *The linear antenna pattern of an array of N equispaced antenna elements, centered at the origin and directly combined, is given by*

$$E(\phi) = E_e(\phi)\mathrm{dirc}_N\left(\frac{d}{\lambda}\sin\phi\right),$$

where d is the element spacing.

Proof:
Suppose that the plane wave $v(t)$ at frequency f_0 and wavelength $\lambda = c/f_0$ is incident on the array at angle ϕ. The output of the nth antenna element is the signal $v(t)$ at this frequency but at a phase angle $2\pi f_0 \tau$ based on the path delay $\tau = nd\sin\phi/c$. Hence the output of the nth antenna element is

---

[5] Direct interaction between real antenna elements is ignored. Only the ideal case in which the antennas do not affect each other and superposition applies is studied.

$$g_n(t) = v(t)E_e(\phi)e^{j(2\pi(d/\lambda)\sin\phi)n}.$$

These N complex signals are combined to produce the composite signal

$$g(t) = \sum_{n=0}^{N-1} g_n(t)$$

$$= v(t)E_e(\phi) \sum_{n=0}^{N-1} e^{j(2\pi(d/\lambda)\sin\phi)n}.$$

The sum has been encountered previously in the definition of the dirichlet function. Thus, the array radiation pattern is

$$E(\phi) = E_e(\phi) \sum_{n=0}^{N-1} e^{j(2\pi(d/\lambda)\sin\phi)n}$$

$$= E_e(\phi)\text{dirc}_N\left(\frac{d}{\lambda}\sin\phi\right)e^{j(N-1)\pi(d/\lambda)\sin\phi}.$$

Centering the array at the origin eliminates the phase term. The proof of the theorem is now complete. ∎

The squared magnitude of the amplitude radiation pattern is the power radiation pattern

$$P(\phi) = |E(\phi)|^2 \text{dirc}_N^2\left(\frac{d}{\lambda}\sin\phi\right).$$

The square of the dirichlet function determines the main beam of the array. This beam is broadside to the array and has its first null at the angle ϕ satisfying $N(d/\lambda)\sin\phi = 1$. This angle is

$$\phi = \sin^{-1}(\lambda/(Nd)),$$

from which is seen that the width of the main beam depends inversely on the length of the array Nd, not on N or d individually.

The dirichlet function has its first grating lobe at the angle ϕ satisfying $(d/\lambda)\sin\phi = 1$, or

$$\phi = \sin^{-1}\left(\frac{\lambda}{d}\right).$$

One would like to choose the element illumination function so that the element pattern $E_e(\phi)$ is small at the grating lobes. This would require the element size to be equal to the spacing d, which would require the antenna elements to touch.

Figure 5.7 shows a two-dimensional K by N array of antennas placed on a rectangular grid. Regard this grid as a one-dimensional array of antennas along the y axis, each of which is an array of antenna elements along the x axis. Consequently, with the origin at the center of the array so that the main beam is real, the antenna pattern for the two-dimensional array is

$$E(\phi,\psi) = E_e(\phi,\psi)\text{dirc}_N\left(\frac{d_1}{\lambda}\sin\phi\cos\psi\right)\text{dirc}_K\left(\frac{d_2}{\lambda}\sin\phi\sin\psi\right).$$

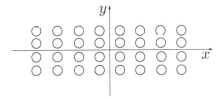

Figure 5.7 A rectangular array of antennas

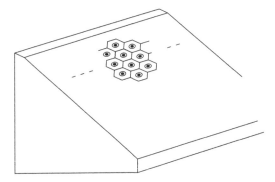

Figure 5.8 Hexagonal antenna array

This pattern has a two-dimensional grating lobe whenever both dirichlet functions have a one-dimensional grating lobe. This occurs whenever both dirichlet functions have an argument that is an integer. Thus, grating lobes occur at the angular coordinates (ϕ, ψ) whenever

$$\sin\phi\cos\psi = \ell_1 \frac{\lambda}{d_1},$$

$$\sin\phi\sin\psi = \ell_2 \frac{\lambda}{d_2},$$

where ℓ_1 and ℓ_2 are arbitrary integers.[6] The main lobe occurs where $\ell_1 = \ell_2 = 0$.

Instead of a rectangular array of antenna elements, a two-dimensional array of antenna elements can be based on the hexagonal lattice, as shown in Figure 5.8. Because the calculation of an antenna pattern from an illumination function is essentially the calculation of a two-dimensional Fourier transform, the calculation of the pattern of a hexagonal array is straightforward and proceeds in a way that is similar to the analysis of hexagonal sampling in Section 3.7. A hexagonal antenna array has the advantage that, for a specified size of the entire aperture occupied by the array, which determines the width of the main lobe, fewer antenna elements are needed to satisfy a required angular separation of grating lobes.

---

[6] These conditions are a form of the equations known as the *Bragg–Laue equations* within the subject of crystallography.

Sequentially Formed Array

It is instructive to contrast the usual way of using an array for a two-way observation with an alternative, less efficient way of using the array for that purpose. The usual way is to use the same array as described by Theorem 5.2.1 twice, once for transmission and once for reception, The signal passes through the same antenna radiation pattern twice. Thus, as a consequence of the reciprocity principle given in Proposition 5.8.1, the received signal (not the received power) is proportional to the square of the antenna radiation pattern. The signal is affected by the square of the expression in Theorem 5.2.1. Thus, for the standard way of using a square array centered at the origin, this squared radiation pattern, which is

$$E^2(\phi) = E_e^2(\phi)\mathrm{dirc}_N^2\left(\frac{d}{\lambda}\sin\phi\right),$$

describes the effect of the antenna array on the signal. The received power is proportional to the fourth power of the antenna radiation pattern.

Rather than using all N antenna elements in this way to simultaneously transmit a single pulse, one pulse at a time is both transmitted and received using a single antenna element. The nth copy of the pulse is transmitted from the nth antenna element and is received only at the nth antenna element. All transmitted pulses share a common phase reference. Each return is recorded as it is received as a complex signal. After all pulses are transmitted, one by one, and the returns are received, all recorded returns are coherently added together. This is a *sequentially formed array*.[7] The nth sequential return is proportional to the square of $E_e(\phi)e^{j[2\pi(d/\lambda)\sin\phi]n}$. Thus the sum of the received echoes after each is adjusted for phase delay gives

$$E_e^2(\phi) = e^{-j\theta}\sum_{n=0}^{N-1}e^{2j[2\pi(d/\lambda)\sin\phi]n}$$

$$= E_e^2(\phi)\mathrm{dirc}_N\left(\frac{2d}{\lambda}\sin\phi\right),$$

where $\theta = (N-1)(d/\lambda)\sin\phi$. This expression differs from the earlier expression in that d is replaced by $2d$, and the dirichlet function is not squared. Because d is replaced by $2d$, the main beam is half as wide so that the resolution is twice as good.[8] The resolution of a sequentially formed array is better than the resolution of a simultaneously formed array. At the same time, the grating lobes become more closely spaced. Because the dirichlet function is not squared, the peak value of the beam is N rather than N^2. This reduction can be attributed to the fact that each reflected signal is incident on all N antennas but is recorded on only one, so only one Nth of the available energy is used.

---

[7] This is implicit in applications in which a single antenna is moved from pulse to pulse in order to synthesize an array.

[8] This improvement in resolution could be important in applications, such as ultrasound transducers, in which the array aperture is constrained in size and power is a secondary concern. The signal strength, however, is greatly reduced.

Another option is to transmit on one antenna at a time, and for each transmission to store the returns seen at all antenna elements. These stored returns are later coherently added. This case is intermediate between the two previous cases.

Phased Arrays

Thus far, only the simple summing of the outputs of the antenna elements to produce the array pattern has been described. This places the main beam of the array pattern along the z axis and perpendicular to the array. The main beam of an antenna array can also be steered by means of an appropriate phase shift on both the received and transmitted signal at each array element. In this case, the antenna array is called a *phased array*.

To steer the main beam of a one-dimensional phased array to the angle ϕ_0, the signal at the nth array element is phase-shifted by the phase angle $2\pi n(d/\lambda)\sin\phi_0$. For the receiving array, the beamsteering is described by the equation

$$G(t) = \sum_{n=0}^{N-1} g_n(t) e^{-j(2\pi n(d/\lambda)\sin\phi_0)},$$

where $g_n(t)$ is the signal received at the nth array element and $G(t)$ is the composite received signal. The received signal $g_n(t)$ at the nth antenna element is equal to the transmitted signal[9] $g(t)$ multiplied by the phase shift unique to the position of that antenna element. This leads to the antenna radiation pattern for the steered array, denoted $E_{\phi_0}(\phi)$, and given by

$$E_{\phi_0}(\phi) = E_e(\phi) \sum_{n=0}^{N-1} e^{j2\pi(d/\lambda)(\sin\phi - \sin\phi_0)n}$$

$$= E_e(\phi) e^{j(N-1)\pi(d/\lambda)(\sin\phi - \sin\phi_0)} \mathrm{dirc}_N\left(\frac{d}{\lambda}(\sin\phi - \sin\phi_0)\right),$$

where $E_e(\phi)$ is the radiation pattern of an array element. The same expression applies to the transmission of the signal. The two-way effect is

$$|E_{\phi_0}(\phi)|^2 = |E_e(\phi)|^2 \mathrm{dirc}_N^2\left(\frac{d}{\lambda}(\sin\phi - \sin\phi_0)\right).$$

The main lobe of the dirichlet function now occurs at ϕ equal to ϕ_0. The first null occurs at

$$\phi = \sin^{-1}\left(\frac{\lambda}{Nd} + \sin\phi_0\right).$$

Because of the shape of the sine function, the angle from the main beam to the first null is greater for the steered array when ϕ_0 is nonzero than when ϕ_0 is zero. As the main beam is steered farther off the boresight where ($\phi_0 = 0$), the width of the main beam increases.

---

[9] Attenuation of the reflected signal energy is ignored in this discussion of antenna patterns.

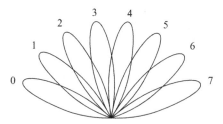

Figure 5.9 Simultaneous beams of an array antenna

Simultaneous Beams

A useful property of a phased array is that a number of steered beams can be formed simultaneously from the same aperture array, both during transmission and during reception. To form several beams during *reception*, the signals for the several beams at the elements of the array are simply processed simultaneously through several sets of beamforming phase angles, thereby forming several beams. To form several beams during *transmission*, use superposition at each element of the array to combine the several phased signals needed at that element by the several beams.

In this way, multiple beams can be formed as suggested by Figure 5.9. Let the spatial angles $\phi_0, \phi_1, \phi_2, \ldots, \phi_{N'-1}$ denote a set of spatial angles for which beams are to be formed. The set of antenna radiation patterns

$$|E_{\phi_\ell}(\phi)|^2 = |E_e(\phi)|^2 \text{dirc}_N^2 \left(\frac{d}{\lambda}(\sin\phi - \sin\phi_\ell) \right) \qquad \ell = 0, \ldots, N'-1$$

describes N' beams, which can be formed by the simultaneous set of beamforming equations

$$G_\ell(t) = \sum_{n=0}^{N-1} g_n(t) e^{-j[2\pi(d/\lambda)\sin\phi_\ell]n} \qquad \ell = 0, \ldots, N'-1.$$

This set of equations has the form of a Fourier transform. Choose N' equal to N and choose ϕ_ℓ to satisfy

$$(d/\lambda)\sin\phi_\ell = \frac{\ell}{N} \qquad \ell = 0, \ldots, N-1.$$

Then N beams at the angles $\phi_{n'} = \sin^{-1}(\lambda n'/dN)$ for $n' = 0, \ldots, N-1$ are formed simultaneously by the N equations

$$G_{n'} = \sum_{n=0}^{N-1} g_n e^{-j2\pi nn'/N}, \qquad n' = 0, \ldots, N-1.$$

By spacing the steering angles appropriately, the form of a *discrete* Fourier transform is obtained.

The Butler Array

The discrete Fourier transform is a mathematical transformation that often appears in the processing of signals. As the expression for $G_{n'}$ is written, it appears that nearly

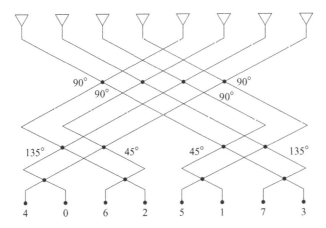

Figure 5.10 An eight-element Butler matrix

N^2 complex multiplications by phase terms are necessary to compute the set of beams because there are nearly N complex multiplications for each value of n'. However, by clever factoring of the equation, it is possible to use far fewer phase multiplications in the calculations.

An eight-element *Butler array*, shown in Figure 5.10, is an example of a *decimation algorithm*.[10] The diagram in the figure denotes the computation graphically. Lines that require a phase shift are labeled with the value of that phase shift. Additions of terms are indicated by a dot where lines cross.

An index $n = 0, \ldots, N - 1$ for which N factors as $N = KI$ can be broken into a coarse index, i, and a vernier index, k, in the same way that a long time interval is broken into hours and minutes. Count with the index k from 0 to $K - 1$. When k overflows, increase the index i by one and return k to zero. The original index n is written $n = k + Ki$, where $i = 0, \ldots, I - 1$, $k = 0, \ldots, K - 1$, and $n = 0, \ldots, IK - 1$. The same components are counted with the double indices k and i as are counted with the single index n.

Express the output index in a similar way by $n' = i' + Ik'$, where $i' = 0, \ldots, I - 1$, $k' = 0, \ldots, K - 1$, and $n' = 0, \ldots, IK - 1$. Then

$$G_{n'} = \sum_{n=0}^{N-1} e^{-j2\pi nn'/N} g_n \qquad n' = 0, \ldots, N - 1$$

becomes

$$G_{i'+Ik'} = \sum_{k=0}^{K-1}\sum_{i=0}^{I-1} e^{-j2\pi (IK)^{-1}(k+Ki)(i'+Ik')} g_{k+Ki}$$

$$= \sum_{k=0}^{K-1} e^{-j2\pi K^{-1}kk'} \left[e^{-j2\pi I^{-1}K^{-1}ki'} \sum_{i=0}^{I-1} e^{-j2\pi I^{-1}ii'} g_{k+Ki} \right],$$

[10] In the study of signal processing, this is known as the *fast Fourier transform*.

for $i' = 0, \ldots, I-1$ and $k' = 0, \ldots, K-1$. The term $e^{-j2\pi nik'}$ equals one for all i and k' and has been dropped.

The inner sum is an I-point discrete Fourier transform for each value of k and the outer sum is a K-point discrete Fourier transform for each value of i'. Although the indexing scheme is now more complicated, the elementary phase-shifting operations are fewer. The original formulation is traded for one that is more complicated but is more efficient computationally.

When either K or I can be factored, the reduction can be repeated to reduce the sum on i or the sum on k. In particular, for $N = 2^n$, choose $K = 2$ and $I = 2^{n-1}$, then repeat for I. By this recursion, the expressions for the Butler array shown in Figure 5.10 are obtained.

5.3 Focused Antennas and Arrays

The antenna radiation pattern of a single aperture is defined for the far-field region and is a consequence of the Fraunhofer approximation. The antenna pattern of an array of apertures is also defined in the Fraunhofer far-field region because the array can be regarded as one large aperture. The phase angles at the individual elements of a phased array aim the antenna beam as seen in the far field. Such a beam can be described as a beam in focus at infinity or as an unfocused beam.

Instead, the beam can be focused at a finite range by choosing the phase angles at the array elements corresponding to the point at this finite range. This technique is referred to as beamforming in the near-field region. Whereas, for a fixed illumination function of magnitude $|s_0(x, y)|$, the beam is sharpest at infinite range when the phase is a linear function of x and y across the aperture. At a fixed finite range, the beam is sharpest when the phase is a quadratic function of x and y across the aperture. It is suggestive to refer to the computation that provides the focusing at each element of a phased array as a "processing lens." A different quadratic function is required to maximally focus for each range and for each angle. A phased array with variable phase compensation within the processing can be used to focus at a variable range and a variable angle. Indeed, by processing multiple copies of the received signal with different quadratic-phase compensations, a receiving array can be used to focus beams simultaneously on many ranges and angles.

Suppose that the magnitude $|s_0(\xi, \eta)|$ of the illumination function is specified. Let

$$s_0(\xi, \eta) = |s_0(\xi, \eta)| e^{j\theta(\xi, \eta)},$$

where $\theta(\xi, \eta)$ is to be specified so as to achieve a desired focus. A Huygens wavelet, originating at the point ξ, η, reaches the point x, y, z with the infinitesimal value

$$dc(x, y, z) = s_0(\xi, \eta) e^{-j2\pi (f_0/c)\sqrt{(x-\xi)^2 + (y-\eta)^2 + z^2}} d\xi\, d\eta.$$

The total signal at x, y, z is

$$c(x, y, z) = \int_{-\infty}^{\infty} \int_{-\infty}^{\infty} s_0(\xi, \eta) e^{-j2\pi (f_0/c)\sqrt{(x-\xi)^2 + (y-\eta)^2 + z^2}} d\xi\, d\eta.$$

In the Fresnel approximation,

$$c(x,y,z) = e^{jkz} \int_{-\infty}^{\infty}\int_{-\infty}^{\infty} |s_0(\xi,\eta)|\, e^{j\theta(\xi,\eta)} e^{-j\frac{k}{2z}[(x-\xi)^2+(y-\eta)^2]}\, d\xi\, d\eta.$$

The integral will be largest when all contributions add in phase. This means that

$$\theta(\xi,\eta) = \frac{k}{2z}[(x-\xi)^2 + (y-\eta)^2]$$

should be chosen as the phase function across the aperture to focus the wavefront at the point (x,y,z). Because the phase function depends on (x,y,z), the phase term cannot be expressed in spherical coordinates as a simple function of angle that is independent of range. Each value of range R requires a different quadratic-phase function for focusing.

5.4 Nondiffracting Beams

A nondiffracting beam keeps the same (x,y) cross section as it propagates. No finite aperture can launch a nondiffracting beam because the far-field pattern is the Fourier transform of the aperture, and a signal and its Fourier transform cannot both be confined. However, an infinite aperture can launch a nondiffracting beam with an illumination function that has most of its signal concentrated near the origin. This suggests that it may be possible to approximate this infinite-aperture illumination function to form a beam that nearly approximates a nondiffracting beam well into the near field. In some applications, a beam holding its shape well into the near field can be desirable.

The wave equation is a partial differential equation whose solutions are plane waves and linear combinations of plane waves. Any wave that satisfies the wave equation also satisfies the Huygens–Fresnel point-spread function. For the infinite planar aperture, it is convenient to work with the wave equation in cylindrical coordinates. The wave equation in cylindrical coordinates is

$$\left[\frac{1}{r}\frac{\partial}{\partial r}\left(r\frac{\partial}{\partial r}\right) + \frac{1}{r^2}\frac{\partial^2}{\partial\phi^2} + \frac{\partial^2}{\partial z^2} - \frac{1}{c^2}\frac{\partial^2}{\partial t^2}\right]\tilde{s}(x,y,z,t) = 0.$$

The zero-order Bessel function solves the differential equation

$$\frac{1}{r}\frac{d}{dr}\left(r\frac{dJ_0(r)}{dr}\right) + J_0(r) = 0.$$

By a simple change of variables, this becomes

$$\frac{1}{r}\frac{d}{dr}\left(r\frac{dJ_0(2\pi\alpha r/\lambda)}{dr}\right) + (2\pi\alpha/\lambda)^2 J_0(2\pi\alpha r/\lambda) = 0.$$

Therefore the function

$$\tilde{s}(x,y,z,t) = J_0(2\pi\alpha r/\lambda)e^{-j2\pi f_0(t-\gamma z/c)}$$

satisfies the wave equation, provided that

$$\alpha^2 + \gamma^2 = 1.$$

Because the (x,y) dependence of this wavefront is independent of z, the illumination function $J_0(2\pi\alpha r/\lambda)$ will launch the nondiffracting beam,

$$s(x,y,z) = J_0(2\pi\alpha r/\lambda)e^{j2\pi\gamma z/\lambda},$$

provided that α is smaller than one, where $r = \sqrt{x^2 + y^2}$.

The function $J_0(2\pi\alpha r/\lambda)$ is supported over the entire x,y plane. It is launched into the z direction and maintains its shape forever. Of course, the illumination function $J_0(2\pi\alpha r/\lambda)$ cannot be realized because it requires an infinite aperture. However, the cropped illumination function

$$s_0(x,y) = J_0(2\pi\alpha r/\lambda)\text{circ}\left(\frac{r}{R}\right)$$

with R large will be a good approximation in the vicinity of the origin and will produce a nearly nondiffracting beam in the near field. Eventually the beam will spread, as it must.

There are other examples of nondiffracting beams, all of which must have an infinite aperture because a signal and its Fourier transform cannot both be confined. Because we are interested primarily in circularly symmetric beams, it is appropriate to express the wave equation in cylindrical coordinates, which implies that all such beams are best described in terms of Bessel functions or sums of Bessel functions. The description of these beams is the topic of Problem 5.15.

5.5 Interferometry

An antenna aperture can be used as a linear component that collects all of the signal that is incident on it. The response of the antenna is described as the antenna radiation pattern, or as the *beam* of the antenna, as studied in Sections 5.1 and Section 5.2. Beamforming is a straightforward way to process an incoming signal across the aperture when the phase structure can be controlled. An antenna or antenna array, as such, does not measure the spatial distribution of phase across the aperture. That information is lost. However, an array can so measure the phase distribution.

The distribution of the phase of the signal across the aperture can be used to estimate the direction of arrival. Many devices are in use that depend on the relationship between the angle of arrival of a wavefront at two points and the relative phase angle of the signals at those points. These devices are used to measure the direction of arrival of a received signal. Such devices use nonlinear methods of processing and are referred to under the general term *interferometer*. This is a broad term and includes most nonlinear methods of processing a phase distribution across an aperture. An interferometer for a microwave waveform is any measurement device based on comparing the phase of the radiation from a point source that arrives at two different points with a phase difference dependent on the angle of arrival and the wavelength.

Definition 5.5.1 *An interferometer is a device for the measurement of direction of arrival or waveform parameters based on the relative phase of an incident wavefront at different positions.*

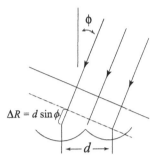

Figure 5.11 An elementary interferometer

The usual interferometer breaks a larger aperture into two or more subapertures, receives a signal in each subaperture, and compares the phase of those two signals. The phase comparison is a nonlinear operation. Consequently, an interferometer is distinctly different from a beamforming system.

The simplest radar-frequency interferometer uses two identical subapertures separated by the distance d. Regard the two subapertures as ideal antennas. The incoming signal from the far field is received by the two antennas as shown in Figure 5.11. For a radiation source that is far from the aperture in comparison to the separation d, the difference in path length is $d \sin \phi$. Hence the phase difference (in radians) in the signal received at the two antennas is

$$\Delta\theta = 2\pi \frac{d}{\lambda} \sin \phi.$$

This relationship connects ϕ to the phase difference $\Delta\theta$.

Interferometry Equations
The equation for $\Delta\theta$ can be inverted to give

$$\phi = \sin^{-1}\left(\frac{\lambda}{d} \frac{\Delta\theta}{2\pi}\right),$$

from which the direction of arrival ϕ can be computed from the phase difference $\Delta\theta$. The phase difference $\Delta\theta$ is measured modulo 2π, so whenever d/λ is larger than 1, more than one value of the angle ϕ will correspond to the given value of $\Delta\theta$ resulting in an ambiguity. There are ambiguities whenever d/λ is larger than 1. Ambiguities in an interferometric, angle-of-arrival measurement are quite routine and are regularly resolved by using a second interferometer (possibly using only one additional antenna) with a different value of the spacing d. Now there are two calculations of phase:

$$\phi = \sin^{-1}\left(\frac{\lambda}{d_1} \frac{\Delta\theta_1 + 2\pi\ell_1}{2\pi}\right),$$

$$\phi = \sin^{-1}\left(\frac{\lambda}{d_2} \frac{\Delta\theta_2 + 2\pi\ell_2}{2\pi}\right),$$

where ℓ_1 and ℓ_2 are unknown integers to be selected so that the two equations yield the same value of ϕ. The two phase differences $\Delta\theta_1$ and $\Delta\theta_2$ are measured by the two interferometers. The spacings d_1 and d_2 of the two interferometers are chosen so that the pair of equations has a unique solution.

Phase Center

The interferometry equations include the term d, which is the distance between the two antennas. For this purpose, it is necessary to understand the meaning of the distance between the two antennas, so the specification of this quantity requires more care. When the two antennas are identical, it may be enough to choose two identical points on the two antennas and define d as the distance between these identical points. When the two antennas are different, a more definite definition is needed. A proper statement is that the distance d is the distance between the *phase centers* of the two antennas, defined as follows.

Definition 5.5.2 *The phase center of an antenna at a given angle is the point of the aperture from which the phase of the signal from a point in the far field at that angle is equal to the phase of the antenna pattern at that angle.*

Under reception, the antenna reduces the wavefront to a single complex number by a weighted integral across the aperture. The phase center of the antenna is the point at which the phase of the incident wavefront is equal to this complex number.

In general, the phase center may vary with the angle of arrival and with the wavelength. Often, it does little harm to use the geometrical center of the aperture as an approximation to the phase center.

Interfering Signals

When multiple simultaneous signals arrive from a multitude of sources, interferometry will break down due to interference unless the signals can first be sorted from one another. The simplest way of separating the signals is by frequency sorting when the signals do not overlap significantly on the frequency axis; by time sorting when the signals normally arrive at different times; or by antenna directivity when interfering signals arrive at different angles and can be sorted by the steerable beam of the two antenna subapertures. Section 11.7 discusses a more elaborate instance in which time delay and doppler shift are used to sort signals prior to an interferometric direction-of-arrival measurement. Chapter 13 will show that in a passive system with decentralized receivers, time difference of arrival, and frequency difference of arrival can also be used to sort signals prior to an interferometric direction-of-arrival measurement.

5.6 Scanning Antenna Patterns

The radar antenna pattern, as studied in Section 5.1, or the sonar hydrophone pattern has no dependence on time. It is a straightforward generalization to introduce time dependence into the antenna pattern. An aperture may be moving, either rotating in the x, y plane or translating in the x, y plane. Then the antenna radiation pattern will rotate or translate in lockstep with the aperture. Such a situation occurs when an antenna is mechanically rotated or carried on a moving vehicle. An antenna aperture that is rotating is called a *scanning aperture*. In contrast, an antenna aperture that is not rotating is called a *staring aperture*.

A radar antenna pattern or sonar hydrophone pattern can also be scanned by changing the phase distribution of the illumination function. There are various ways of doing this. One method is by moving the feed element that is illuminating the aperture to change the illumination function. For a phased array, one may appropriately phase shift the signal fed to each element to steer the beam, as discussed in Section 5.2. By varying the phase shifts with time, the direction of the beam can change with time. By using phase shifts indexed by n of the form

$$\theta_n(t) = 2\pi n \left(\frac{d}{\lambda}\right) \sin \phi_0(t),$$

for the elements of the array, it is straightforward to form an antenna pattern of the form

$$E_{\phi_0(t)}(\phi) = E_e(\phi) \text{dirc}_N \left(\frac{d}{\lambda}(\sin \phi - \sin \phi_0(t))\right).$$

The main beam of this pattern is at the time-varying angle $\phi_0(t)$. This is called an electronically scanned beam or a computationally scanned beam. Referring to the principle of superposition, one can even use a single array to simultaneously form multiple scanning beams with an arbitrary time-varying trajectory for each beam. Each time-varying beam can carry its own time-varying signal.

5.7 Wideband Radiation Patterns

Antenna patterns have been studied in this chapter under the narrowband approximation. The narrowband approximation requires that the modulation $m(t)$ has a spectrum $M(f)$ that is sufficiently narrow so that time delay can be treated simply as a phase shift at the carrier frequency. The delay in the modulation can be neglected. In the narrowband approximation, the spectrum of the transmitted or received modulation at the spherical angle (ϕ, ψ) is

$$M(f, \phi, \psi) = M(f) S_0 \left(\frac{\sin \phi \cos \psi}{\lambda_0}, \frac{\sin \phi \sin \psi}{\lambda_0}\right),$$

where $\lambda_0 = c/f_0$.

In this section, the approximation that $c(t)$ is a narrowband transmitted signal is dropped and now called $m(t)$. The Fourier transform $M(f)$ of the modulation is now nonzero in a wider range of frequencies. The time delay results in a different phase shift at each frequency. The delay in the modulation is now significant.

Time-Domain Analysis

For brevity, the time-domain analysis is given for the one-dimensional case. The time-domain analysis for the two-dimensional case is the same.

5.7 Wideband Radiation Patterns

For a one-dimensional antenna, when the complex signal is sent, the infinitesimal complex passband signal in direction ϕ is $\tilde{m}(t,\phi)dx = m(t,\phi)e^{j2\pi f_0 t}dx$, where $m(t,\phi)$ is the complex baseband signal transmitted at angle ϕ, which is evaluated as follows:

$$m(t,\phi)e^{j2\pi f_0 t} = \int_{-\infty}^{\infty} s(x) m\left(t - \frac{x}{c}\sin\phi\right) e^{j2\pi f_0(t - \frac{x}{c}\sin\phi)} dx$$

$$= \int_{-\infty}^{\infty} s(x) \left[\int_{-\infty}^{\infty} e^{-j2\pi f \frac{x}{c}\sin\phi} M(f) e^{j2\pi f t} df\right] e^{j2\pi f_0(t - \frac{x}{c}\sin\phi)} dx$$

$$= \int_{-\infty}^{\infty} M(f) e^{j2\pi(f_0 + f)t} \left[\int_{-\infty}^{\infty} s(x) e^{-j2\pi(f_0 + f)\frac{x}{c}\sin\phi} dx\right] df$$

$$= e^{j2\pi f_0 t} \int_{-\infty}^{\infty} M(f) S\left(\frac{f_0 + f}{c}\sin\phi\right) e^{j2\pi f t} df.$$

This means that

$$M(f,\phi) = M(f) S\left(\frac{f_0 + f}{c}\sin\phi\right)$$

is the Fourier transform of the signal in the beam at angle ϕ when $M(f)$ is the Fourier transform of the signal. A similar analysis of the two-dimensional case goes through in the same way leading to

$$M(f,\phi,\psi) = M(f) S\left(\frac{f + f_0}{c}\sin\phi\cos\psi, \frac{f + f_0}{c}\sin\phi\sin\psi\right)$$

in two dimensions.

Frequency-Domain Analysis

The two-dimensional antenna pattern at frequency f_0 given in Definition 5.1.1 of Section 5.1 is

$$E(\phi,\psi) = \left(\frac{f_0}{c}\right)^2 S_0\left(\frac{f_0}{c}\sin\phi\cos\psi, \frac{f_0}{c}\sin\phi\sin\psi\right),$$

where $S_0(f_x, f_y)$ is the spatial Fourier transform of the aperture illumination function $s_0(x, y)$. The modulation at the spherical angle (ϕ, ψ) is

$$M(f,\phi,\psi) = M(f) S_0\left(\frac{f_0}{c}\sin\phi\cos\psi, \frac{f_0}{c}\sin\phi\sin\psi\right).$$

Likewise, the two-dimensional antenna pattern at frequency $f + f_0$ is

$$M(f,\phi,\psi) = \left(\frac{f + f_0}{c}\right)^2 S\left(\frac{f + f_0}{c}\sin\phi\cos\psi, \frac{f + f_0}{c}\sin\phi\sin\psi\right).$$

The contribution to the signal $m(t,\phi,\psi)$ in the far field at angular coordinates (ϕ, ψ) is given in the frequency domain by

$$M(f,\phi,\psi) = M(f) S\left(\frac{f + f_0}{c}\sin\phi\cos\psi, \frac{f + f_0}{c}\sin\phi\sin\psi\right).$$

Thus, for each pair of angular coordinates (ϕ, ψ), the inverse Fourier transform of $M(f, \phi, \psi)$,

$$m(t, \phi, \psi) = \int_{-\infty}^{\infty} M(f, \phi, \psi) e^{j2\pi ft} df,$$

gives the signal $m(t, \phi, \psi)$ that is radiated into direction specified by (ϕ, ψ).

Example
A uniformly illuminated one-dimensional aperture of length L is given by

$$s(x) = \frac{1}{L} \text{rect}\left(\frac{x}{L}\right),$$

where $1/L$ is a convenient normalizing constant. Then

$$S\left(\frac{f_0 + f}{c} \sin\phi\right) = \text{sinc}\left(L\frac{f_0 + f}{c} \sin\phi\right),$$

and

$$M(f, \phi) = M(f)\text{sinc}\left(L\frac{f_0 + f}{c} \sin\phi\right).$$

At every value of ϕ, the spectrum of $M(f)$ will be changed differently to form $M(f, \phi)$. This can be put in the form of an angle-dependent linear filter. By the convolution theorem

$$m(t, \phi) = m(t) * h(t),$$

where

$$h(t) \leftrightarrow \text{sinc}\left(L\frac{f_0 + f}{c} \sin\phi\right).$$

Therefore

$$h(t) = \frac{c}{L \sin\phi} \text{rect}\left(\frac{ct}{L \sin\phi}\right) e^{-j2\phi f_0 t}.$$

As $(L \sin\phi)/c$ goes to zero, this goes to $\delta(t) e^{-j2\phi f_0 t}$.

5.8 Reciprocity

An antenna can be used both to transmit to the far field and to receive from the far field. The antenna pattern is the same in both situations. This statement is known as the *antenna reciprocity principle*. The reciprocity principle is not immediately obvious, although some superficial treatments might claim that it is obvious.

This section studies the reciprocity between the transmission antenna pattern and the reception antenna pattern, showing that this reciprocity does hold provided reciprocity holds between the antenna illumination function and the antenna electronics. This latter condition does generally hold but it is not addressed herein.

5.8 Reciprocity

The following two paragraphs compare transmission and reception so as to demonstrate that the principle of reciprocity does require proof.

The antenna is used to *transmit* a waveform to the far field. That waveform is simultaneously transmitted to *every* point of the far field. The illumination function $s(x,y)$ launches a propagating wave reaching all points of the far field. The phase distribution across the aperture is defined by $s(x,y)$. The *transmit antenna pattern* as a function of solid angle describes the angular dependence of the complex amplitude of the waveform transmitted to the far field as a function of solid angle.

The antenna is used to *receive* a signal from the far field. The waveform originates at a *single* point of the far field. A waveform from the far field at solid angle (ϕ, ψ) that is intercepted by the antenna aperture is weighted by the illumination function $s(x,y)$. The phase distribution across the aperture varies with the angle of arrival. The weighted signal is then integrated across the aperture to become the signal seen by the receiver. The *receive antenna pattern* as a function of solid angle is defined to describe the ability of the antenna to turn the intercepted waveform from a single direction into a signal to the receiver.

The transmission antenna pattern describes the signal sent to the far field by the antenna in response to a signal from the transmitter electronics. The reception antenna pattern describes the signal sent to the receiver electronics by the antenna in response to a signal from the far field.

The two definitions of the antenna pattern, one for transmission and one for reception, are quite different. However, the important *reciprocity principle* of antenna theory states that these two notions of antenna pattern are equivalent. The two notions define the same antenna pattern and need not be distinguished. The antenna reciprocity principle, loosely stated, says that the properties of an antenna in the far field are the same whether the antenna is used for transmission or for reception.[11] Specifically, the antenna pattern $G(\phi, \psi)$ of an antenna used to transmit is equal to the antenna pattern of that same antenna used to receive. The relationship between the antenna pattern and the illumination function within the aperture is the same for reception as for transmission even though the wavefront in the two cases is distinctly different. This equivalence is the level at which reciprocity is treated herein.

At a deeper level, and really beyond our scope, is the reciprocity between the illumination function and the transmitter/receiver electronics. This reciprocity is the premise upon which the reciprocity principle for antennas is based.

The reciprocity principle for antennas requires that the relationship between the antenna transmitter/receiver host circuitry and the antenna illumination function is that of a linear reciprocal transducer. A reciprocal transducer, shown symbolically in Figure 5.12, is one with two ports such that when the signal s is the input at the first

---

[11] Sometimes laconically stated "If I can see you, then you can see me," although this statement is not really an analogy.

5 Apertures and Radiation Patterns

Figure 5.12 A linear reciprocal transducer

port, then Gs is the output at the second port. When the signal r is the input at the second port, Gr is the output at the first port.

To clarify the discussion it may be helpful to recall that the Fraunhofer far field has the form of a Fourier transform that decomposes the illumination function into a continuum of angular directions. Thus, the antenna pattern at the specific direction (ϕ, ψ) is

$$\lambda^2 E(\phi, \psi) = \int_{-\infty}^{\infty} \int_{-\infty}^{\infty} s(x,y) e^{-j2\pi(x \sin\phi \cos\psi + y \sin\phi \sin\psi)/\lambda} dxdy.$$

On the other hand, the signal that is seen by the receiver from a waveform from the far field at an infinitesimal region of the aperture is $s(x,y)e^{-j2\pi(xf_x + yf_y)}dxdy$. Then the total signal seen by the receiver from the entire aperture is the integral over the aperture, which gives

$$\lambda^2 E(\phi, \psi) = \int_{-\infty}^{\infty} \int_{-\infty}^{\infty} s(x,y) e^{-j2\pi(x \sin\phi \cos\psi + y \sin\phi \sin\psi)/\lambda} dxdy,$$

where the left side is labeled $\lambda^2 E(\phi, \psi)$ to agree with the earlier expression. The expressions are the same, but are understood differently in their purpose. In the first instance, ϕ and ψ vary over all spherical angles and radiation occurs over all such directions. In the second instance, there is understood to be only a single direction corresponding to a single pair of ϕ and ψ. Nevertheless, the pair of identical equation may be regarded as a satisfactory proof of the reciprocity principle. Nevertheless, the principle is examined more closely.

A linear reciprocal relationship between a host signal and an illumination function is the relationship in which the signal $u(t)$ feeding the antenna generates the illumination function $u(t)s_0(x,y)$, and the input illumination function $v(t,x,y)$ incident on the aperture generates the signal

$$r(t) = \int_{-\infty}^{\infty} \int_{-\infty}^{\infty} v(x,y,t) s_0(x,y) dxdy$$

captured by the antenna. More specifically, as shown in Figure 5.13, for the input $u(t)$ at the signal port, the output at antenna position (x,y) is $u(t)s_0(x,y)dxdy$, while for the received signal $v(x,y,t)$ at antenna position (x,y), its contribution to the signal port is $v(x,y,t)s_0(x,y)dxdy$.

The following proposition says that reciprocity in the host transmitter/receiver structure and the antenna structure implies a reciprocity in the antenna pattern.

5.8 Reciprocity

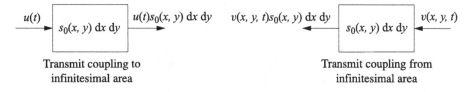

Figure 5.13 Antenna as a reciprocal transducer

Proposition 5.8.1 (Antenna Reciprocity Principle) *The transmit antenna pattern and the receive antenna pattern are equal.*

Proof:
The proof consists of comparing the transmit antenna pattern and the receive antenna pattern.

The transmit antenna pattern was given in Chapter 4 in connection with the Fraunhofer far field. The signal reaching a point in the far field with spherical coordinates ϕ, ψ is

$$v(t,\phi,\psi)e^{j2\pi f_0 t} = c(t-\tau)e^{-j2\pi f_0 t} S_0\left(\frac{\sin\phi\cos\psi}{\lambda}, \frac{\sin\phi\sin\psi}{\lambda}\right),$$

when transmitting the signal $s(t)$.

The receive antenna pattern was derived in Section 5.1. For reception, the signal $c(t)e^{-j2\pi f_0 t}$ originates from a point in the far field with the spherical coordinates ϕ, ψ and is seen at the point (x,y) of the aperture with a relative delay, $\tau = -(x\sin\phi\cos\psi + y\sin\phi\sin\psi)/c$, as referenced to the origin. Consequently, the signal incident at the point (x,y) of the aperture is

$$v(x,y,t)e^{-j2\pi f_0 t} = c(t-\tau)e^{-j2\pi f_0(t-\tau)}$$
$$= c(t-\tau)e^{-j2\pi f_0 t}e^{-j2\pi(x\sin\phi\cos\psi + y\sin\phi\sin\psi)/\lambda},$$

and the contribution to the received signal is $v(x,y,t)s_0(x,y)dxdy$. But the reciprocal relationship between the antenna illumination function and the transmitter circuitry gives

$$r(t) = \int_{-\infty}^{\infty}\int_{-\infty}^{\infty} v(x,y,t)s_0(x,y)dxdy.$$

From this it follows that

$$r(t)e^{-j2\pi f_0 t} = c(t-\tau)e^{-j2\pi f_0 t}\int_{-\infty}^{\infty}\int_{-\infty}^{\infty} s_0(x,y)e^{-j2\pi(x\sin\phi\cos\psi + y\sin\phi\sin\psi)/\lambda}dxdy$$
$$= c(t-\tau)e^{-j2\pi f_0 t}S_0\left(\frac{\sin\phi\cos\psi}{\lambda}, \frac{\sin\phi\sin\psi}{\lambda}\right),$$

as is stated in the proposition. ∎

5.9 Vector Diffraction

Some propagating waves are scalar-valued. Some propagating waves are vector-valued. A *transverse vector-valued plane wave* in free space is a plane wave taking vector values orthogonal to the direction of propagation. Vector-valued waves diffract during propagation. The diffraction of a *transverse vector-valued wave* in free space is studied in this section. Scalar-valued diffraction and vector-valued diffraction both describe how a wave concentrated as a narrow beam spreads as it propagates. The difficult topic of vector diffraction is introduced in this section.

The diffraction of a *scalar-valued wave* in free space is studied in Sections 4.1 and 4.2 of Chapter 4. The diffraction of a scalar-valued wave is completely described by the Huygens–Fresnel principle as a convolution of the wavefront with the Huygens–Fresnel point-spread function of free space. The diffraction of a transverse-vector-valued wave is also described by a form of the Huygens–Fresnel principle but presents differently because vectors add differently than do scalars.

A scalar-valued waveform in the form of a circularly symmetric beam spreads as it propagates in free space but maintains its circular symmetry. In contrast, a linearly-polarized transverse vector-valued waveform in the form of a circularly symmetric beam spreads as it propagates in free space but, as shown in this section, it loses its circular symmetry, spreading less quickly in the direction of polarization and so becoming elliptical in shape. The narrower the beam waist, the stronger is the beam spreading. The wider the waist, the weaker the beam spreading, and the vector diffraction comes to resemble scalar diffraction.

The conditions imposed in Section 4.2 to derive the Huygens–Fresnel principle are that the scalar-valued wave can be decomposed into a continuum of plane waves that each satisfy the scalar wave equation and have a complex scalar amplitude that is constant in the plane orthogonal to the direction of propagation of that plane wave. The plane waves combine by scalar addition. These conditions follow from the properties of the plane-wave solutions of the scalar wave equation.

The premise underlying the derivation of the scalar Huygens–Fresnel principle is that diffraction is a general mathematical property of scalar-valued waves. It is a mathematical consequence of the properties of any scalar wave satisfying the scalar wave equation. There is no need to refer to the underlying physical nature of that scalar-valued wave.

As is the case for scalar diffraction, vector diffraction is also a mathematical property of any transverse vector-valued wave satisfying the wave equation. There is no need to inquire more deeply into the physical nature of the wave. It is enough to posit only that it is a transverse vector-valued wave satisfying the wave equation. The vector diffraction formula then must follow. Moreover, for both the scalar case and the vector case, the analysis of free-space diffraction is treated as a topic separate from the discussion of the boundary conditions on an aperture needed to set up that wave.

A transverse vector-valued wave cannot be usefully regarded as simply three scalar-valued waves comprising the three components of the vector in a suitable coordinate system. A vector wave whose three scalar components could be independently specified amounts to nothing more than three independent scalar waves. No new theory is needed. Componentwise scalar diffraction would then be adequate. For a transverse vector-valued wave, however, the three scalar components are not independent. A more general form of the Huygens–Fresnel principle is necessary.

Premise
The premises of the section are two-fold. Transverse vector-valued plane waves satisfy the wave equation and superposition holds, so any transverse vector-valued wave can be decomposed into transverse vector-valued *plane* waves.

Polarization
A *transverse vector-valued wave* in free space is a vector-valued wave that at each point takes a vector value that is perpendicular to its direction of propagation. This transverse direction of the wave amplitude is called the *polarization* of the wave. The simplest case is a monochromatic, monodirectional transverse vector-valued *plane* wave. A transverse vector-valued plane wave in direction (α, β, γ) has a complex-baseband vector structure

$$s(x, y, z) = [s_x \mathbf{i}_x + s_y \mathbf{i}_y + s_z \mathbf{i}_z] e^{j2\pi f_0(\alpha x + \beta y + \gamma z)/c},$$

where $(\mathbf{i}_x, \mathbf{i}_y, \mathbf{i}_z)$ is a triple of orthogonal unit vectors along the three axes of the coordinate system and (s_x, s_y, s_z) is a triple of scalar constants specifying the vector-valued amplitude of the wave.

For a plane wave to be a transverse vector-valued wave in free space, the direction of the vector must be perpendicular to the direction of propagation. This means that the inner product of the polarization vector $s_x \mathbf{i}_x + s_y \mathbf{i}_y + s_z \mathbf{i}_z$ and the direction vector $\alpha \mathbf{i}_x + \beta \mathbf{i}_y + \gamma \mathbf{i}_z$ must equal zero. Thus, for a monodirectional transverse plane wave traveling in the direction $\alpha \mathbf{i}_x + \beta \mathbf{i}_y + \gamma \mathbf{i}_z$, the side condition

$$s_x \alpha + s_y \beta + s_z \gamma = 0$$

must be satisfied. This side condition creates a linkage in the three vector components. This linkage is the reason that vector diffraction is different from scalar diffraction.

Two polarization vectors that are orthogonal to direction (α, β, γ) are the polarization vector $(\gamma, 0, -\alpha)$ and the polarization vector $(0, \gamma, -\beta)$. These two vectors are not themselves orthogonal, in general, but each is orthogonal to the direction of propagation. Any polarization of a vector traveling in direction (α, β, γ) can be described as a linear combination of these two linearly dependent vectors. Thus $(\gamma, 0, -\alpha)$ and $(0, \gamma, -\beta)$ form a *nonorthogonal* polarization basis. In contrast, $(\gamma, 0, -\alpha)$ and $(0, 1, 0)$ form an *orthogonal* polarization basis.

Two Scalar Plane Waves
To anticipate the distinction between scalar diffraction and vector diffraction, first consider the two scalar plane waves of the complex baseband form $e^{-j2\pi f_0(\pm \alpha x + \gamma z)}$ traveling in the two symmetric directions $(+\alpha, 0, \gamma)$ and $(-\alpha, 0, \gamma)$, respectfully. The two scalar plane waves have the opposite signs in the x direction and have the

same sign in the z direction. The superposition of the two complex-baseband scalar plane waves as a sum is

$$s(t) = e^{-j2\pi f_0(\alpha x+\gamma z)/c} + e^{-j2\pi f_0(-\alpha x+\gamma z)/c}$$
$$= \left(e^{-j2\pi f_0 \alpha x/c} + e^{+j2\pi f_0 \alpha x/c}\right) e^{-j2\pi f_0 \gamma z/c}$$
$$= 2\cos(2\pi\alpha x/\lambda) e^{-j2\pi \gamma z/\lambda}.$$

Similarly, the difference of the two plane waves is $2j\sin(2\pi\alpha x/\lambda)e^{-j2\pi\gamma z/\lambda}$.

The sum of the two complex baseband plane waves results in a standing wave $2\cos(2\pi\alpha x/\lambda)$ in the x direction and a traveling wave $e^{-j2\pi\gamma z/\lambda}$ in the z direction. The complex passband waveform $\widetilde{s}(t) \doteq \widetilde{s}(x,y,z,t)$ and the real passband waveforms $s(t)$ are

$$\widetilde{s}(t) = 2\cos(2\pi\alpha x/\lambda) e^{-j2\pi f_0(t-\gamma z/c)},$$
$$s(t) = 2\cos(2\pi\alpha x/\lambda)\cos\left(2\pi f_0(t - \gamma z/c)\right).$$

The wavelength projected into the z direction λ/γ is increased by $1/\gamma$. The wavelength of the standing wave is λ/α. The composite passband wave remains, of course, at frequency f_0.

It is convenient to replace the cosine by an exponential and to write the space term of this scalar wave in the form of a complex baseband representation as

$$s(x,z) = 2e^{-2\pi\alpha x/\lambda} e^{-j2\pi\sqrt{1-\alpha^2}z/\lambda}.$$

Two Transverse-Vector Plane Waves

Now consider two transverse vector-valued plane waves at frequency f_0 that are traveling in the two directions $(+\alpha, 0, \gamma)$ and $(-\alpha, 0, \gamma)$ as for the scalar case. The two transverse vector-valued waves are polarized. The vector values of each plane wave must be perpendicular to the direction of wave travel which is $(\pm\alpha, 0, \gamma)$. Two orthogonal polarization directions, either $(\gamma, 0, +\alpha)$ and $(0, 1, 0)$ or $(\gamma, 0, -\alpha)$ and $(0, 1, 0)$, respectively, span the set of possible polarization directions for one of the two waves. These two polarization vectors form a basis spanning the possible polarizations. Each of two plane waves can have either or both of the two basis polarization vectors, thereby making four cases in all.

For the first case of polarization, the polarization vectors with components $(\gamma, 0, +\alpha)$ and $(\gamma, 0, -\alpha)$ are not the same. These do not add as scalars add; they add as vectors add. This vector addition is

$$(\gamma,\ 0,\ \alpha) + (\gamma,\ 0,\ -\alpha) = (2\gamma,\ 0,\ 0).$$

The sum wave remains polarized entirely in the x direction. The wavelength projection into the z direction λ/γ is increased by $1/\gamma$. This is the same as for the scalar case. The polarized complex passband waveform $\widetilde{s}(t)$ is

$$\widetilde{s}(t) = \left[2\gamma i_x + 0i_y + 0i_z\right]\cos(2\pi\alpha x/\lambda)e^{-j2\pi\gamma f_0 z/c}e^{-j2\pi f_0 t}.$$

For two transverse vector plane waves at frequency f_0 that are traveling in directions $(0, +\beta, \gamma)$ and $(0, -\beta, \gamma)$ in the y, z plane, the polarization vectors sum to a wavefront

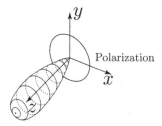

Figure 5.14 Vector Diffraction

polarized in the x direction. When compared to the scalar case, this vector case is reduced in amplitude by the factor γ. This factor is the reason that vector diffraction differs from scalar diffraction.

For the second case of polarization, both plane waves have polarization vector $(0, 1, 0)$. This polarization is in the y direction. These polarization vectors are the same and the two wave amplitudes add as

$$(0, 1, 0) + (0, 1, 0) = (0, 2, 0).$$

This vector addition

$$\tilde{s}(t) = \left[0\boldsymbol{i}_x + 2\boldsymbol{i}_y + 0\boldsymbol{i}_z\right] \cos(2\pi \alpha x/\lambda) e^{-j2\pi \gamma z/\lambda} e^{-j2\pi f_0 t}$$

results in a polarized complex passband waveform with the same magnitude as in the scalar case.

For the third and fourth cases of polarization, the polarizations of the two plane waves are orthogonal. There is no addition of vector components.

Continuum of Plane Waves

The next step is to go from a single pair of transverse vector-valued plane waves to a continuum of transverse vector-valued plane waves. To simplify the discussion, the wavefront $s_0(x, y)$ in the plane at $z = 0$ is a circularly symmetric function such as $\text{circ}(x, y)$ or $e^{-(x^2+y^2)}$. For a waveform that is not circularly symmetric, the relationship between the direction of polarization and the asymmetry of a noncircular beam would be more complicated.

Figure 5.14 depicts the main lobe of a diffracting waveform for a circularly symmetric function at $z = 0$ with the polarization in the x direction and a point-spread function $s_0(x, y)$. Even though the waveform may be circularly symmetric, the existence of polarization on the x axis spoils the overall symmetry of the diffraction pattern. One may expect that the diffracting waveform is described by a vector-valued point-spread function that is not circularly symmetric. This is indeed the case.

Angular Spectrum

A lightwave beam polarized in the x direction as it leaves the x, y plane at $z = 0$ has the form

$$s_0(x, y) = \left[1\boldsymbol{i}_x + 0\boldsymbol{i}_y + 0\boldsymbol{i}_z\right] s_0(x, y)$$

in the x, y plane at $z = 0$.

The vector-valued waveform at $z = 0$ is

$$s(x, y, 0) = s_0(x, y)$$
$$= [1i_x + 0i_y + 0i_z]s_0(x, y),$$

where $s(x, y, 0) = s_0(x, y)$ is a circularly symmetric function.

The propagating wave with complex amplitude $s_0(x, y) = s(x, y, 0)$ can be decomposed into an angular spectrum of plane waves that superimpose in the $z = 0$ plane to form the complex transverse-vector wave amplitude $s_0(x, y)$ with polarization in the x direction. Because $s_0(x, y)$ has no polarization component in the y direction, no plane wave of the angular spectrum forming $s_0(x, y)$ has a polarization with a component in the y direction. This means that a wave in direction (α, β, γ) must have a polarization vector proportional to $(\gamma, 0, -\alpha)$, so that the inner product of the direction vector and the polarization vector $\gamma\alpha - \alpha\gamma$ is equal to zero.

Section 4.2 defined the angular spectrum $A(\alpha, \beta)$ of the wave amplitude $s_0(x, y)$ at $z = 0$ in terms of the Fourier decomposition $S_0(f_x, f_y)$ evaluated at $f_x = \alpha/\lambda$ and $f_y = \beta/\lambda$ and scaled by λ^{-2}. This gives the *angular spectrum* consisting of the plane waves $A(\alpha, \beta) = \lambda^{-2} S_0(\alpha/\lambda, \beta/\lambda)$.

For vector waves, the angular spectrum $A(\alpha, \beta)$ is similar. For a circularly symmetric complex baseband wave amplitude $s_0(x, y)i_x$ at $z = 0$ polarized in the x direction, each plane wave of the angular spectrum has a polarization in the x direction. The angular spectrum is the vector

$$A(\alpha, \beta) = [s_x i_x + 0 i_y + 0 i_z] \lambda^{-2} S_0(\alpha/\lambda, \beta/\lambda)$$
$$= A_x(\alpha, \beta) i_x + A_z(\alpha, \beta) i_z,$$

which describes the output as a continuum of polarized vector plane waves as a function of the direction cosines α and β.

Scalar Point-Spread Function

The scalar point-spread function given in Definition 4.2.2 of Chapter 4 is

$$h(x, y, z) = \left[\frac{-jz/\lambda}{(x^2 + y^2 + z^2)} + \frac{z/2\pi}{(x^2 + y^2 + z^2)^{3/2}}\right] e^{j2\pi\sqrt{x^2+y^2+z^2}/\lambda},$$

here written with the variable distance z replacing the fixed distance d. The vector point-spread function for diffraction is a modification of the scalar point-spread function for diffraction.

Vector Point-Spread Function

The vector point-spread function of free space is defined first in Definition 5.9.1 and validated in Proposition 5.9.2. In the proof of that proposition, the diffraction of the transverse vector-valued wave is analyzed and stated so as to suit the definition of the vector point-spread function.

5.9 Vector Diffraction

Definition 5.9.1 *The Huygens–Fresnel vector point-spread function of free space between the two x,y planes at $z = 0$ and $z = d$ for a transverse vector-valued wave linearly polarized in the x direction is*

$$\mathbf{h}_d(x,y) = h_{d,x}(x,y)\mathbf{i}_x + h_{d,z}(x,y)\mathbf{i}_z$$

with

$$h_{d,x}(x,y) = \frac{\lambda}{j2\pi}\frac{\partial}{\partial z}h(x,y,z)\bigg|_{z=d} \qquad \lambda h_{d,z}(x,y) = \frac{\lambda}{j2\pi}\frac{\partial}{\partial x}h(x,y,z)\bigg|_{z=d},$$

where $h(x,y,z)$ is the scalar Huygens–Fresnel point-spread function $h_z(x,y)$.

The vector point-spread function describes the diffraction of a transverse vector-valued wave in the x,y plane at $z = 0$ as the expression

$$\mathbf{s}_d(x,y) = \left[h_{d,x}(x,y)\mathbf{i}_x + h_{d,z}(x,y)\mathbf{i}_z\right] ** s_0(x,y)$$

for a source waveform $s_0(x,y)$ in the x,y plane at $z = 0$ that is polarized in the x direction. This is the assertion of the forthcoming Proposition 5.9.2.

Reduction to Scalar Diffraction

Vector diffraction reduces to scalar diffraction when $s_0(x,y)$ is large compared to the wavelength. This is best seen by using the frequency-domain counterpart of Definition 5.9.1 written as follows:

$$H_{d,x}(f_x,f_y) = \gamma e^{j2\pi\sqrt{\lambda^{-2}-f_x^2-f_y^2}\,z} \qquad H_{d,z}(f_x,f_y) = \alpha e^{j2\pi\sqrt{\lambda^{-2}-f_x^2-f_y^2}\,z},$$

where $\alpha = \lambda f_x$ and $\beta = \lambda f_y$, and $\gamma = \sqrt{1-\alpha^2-\beta^2}$. When the diffraction is small, the leading terms multiplying the exponentials of these expressions can be approximated as $\gamma \approx 1$ and $\alpha \approx 0$. (see Problem 2.31 of Chapter 2.) With these approximations, the terms are

$$H_{d,x}(f_x,f_y) \approx e^{j2\pi\sqrt{\lambda^{-2}-f_x^2-f_y^2}\,z} \qquad H_{d,z}(f_x,f_y) \approx 0,$$

which corresponds to the scalar Huygens–Fresnel point-spread function, as expected.

Proposition 5.9.2 *The diffracted wave in the x,y plane at distance d is*

$$\mathbf{s}_d(x,y) = \left[h_{d,x}(x,y)\mathbf{i}_x + h_{d,z}(x,y)\mathbf{i}_z\right] ** s_0(x,y).$$

where the transverse vector-valued wave is $\mathbf{s}_0(x,y) = s_0(x,y)\mathbf{i}_x$ in the x,y plane at $z = 0$ which is polarized in the x direction.

Proof:
The specified transverse vector-valued wave traveling in direction (α, β, γ) has the polarization vector[12]

---

[12] To be clear about notation, $s_z(x,y|z)$ denotes the z component of the vector field at coordinate (x,y) of the x,y plane at distance z from the origin, and $s_x(x,y|z)$ denotes the x component of the field at that same point. The bold symbols are unit vectors in the x and z directions. The subscripts on s denote the direction of the vector. The arguments of s denote a point in space in the x,y plane at distance z from the origin.

$$s(x,y|z) = s_x(x,y|z)\mathbf{i}_x + s_z(x,y|z)\mathbf{i}_z$$
$$= \int_{-\infty}^{\infty}\int_{-\infty}^{\infty} [\gamma \mathbf{i}_x - \alpha \mathbf{i}_z] A_x(\alpha,\beta) e^{j2\pi f_0(\alpha x+\beta y+\gamma z)/c} d\alpha d\beta.$$

The x component $s_x(x,y|z)$ and the z component $s_z(x,y|z)$ are evaluated separately as follows.

The z component of $s(x,y|z)$ is the inverse Fourier transform of a product, given by

$$s_z(x,y|z) = \int_{-\infty}^{\infty}\int_{-\infty}^{\infty} \left(A_x(\alpha,\beta)\right)\left(\alpha e^{j2\pi f_0\sqrt{1-\alpha^2-\beta^2}z/c}\right)e^{j2\pi(\alpha x+\beta y)/\lambda} d\alpha d\beta$$
$$= \int_{-\infty}^{\infty}\int_{-\infty}^{\infty} \left(\lambda^2 A_x(\lambda f_x,\lambda f_y)\right)\left(\lambda f_x e^{j2\pi\sqrt{\lambda^{-2}-f_x^2-f_y^2}z}\right)e^{j2\pi(f_x x+f_y y)/\lambda} df_x df_y$$
$$= a_x(x,y) ** \frac{\lambda}{j2\pi}\frac{d}{dx} h(x,y,z),$$

where the third line follows from the convolution property and the differentiation property of the two-dimensional Fourier transform.

The x component of $s(x,y|z)$ with γ replaced by $\sqrt{1-\alpha^2-\beta^2}$ is the inverse Fourier transform of the product of two terms given by

$$s_x(x,y|z) = \int_{-\infty}^{\infty}\int_{-\infty}^{\infty} (A_x(\alpha,\beta))\left(\sqrt{1-\alpha^2-\beta^2}e^{j2\pi f_0\sqrt{1-\alpha^2-\beta^2}z/c}\right)$$
$$e^{j2\pi f_0(\alpha x+\beta y)/c} d\alpha d\beta$$
$$= \int_{-\infty}^{\infty}\int_{-\infty}^{\infty} (A_x(\alpha,\beta))\left(\frac{\lambda}{j2\pi}\frac{d}{dz}e^{j2\pi f_0\sqrt{1-\alpha^2-\beta^2}z/c}\right)e^{j2\pi(\alpha x+\beta y)/\lambda} d\alpha d\beta$$
$$= \frac{\lambda}{j2\pi}\frac{d}{dz}\int_{-\infty}^{\infty}\int_{-\infty}^{\infty}\left(\lambda^2 A_x(\lambda f_x,\lambda f_y)\right)\left(e^{j2\pi\sqrt{\lambda^{-2}-f_x^2-f_y^2}z}\right)e^{j2\pi(f_x x+f_y y)/\lambda} df_x df_y$$
$$= \frac{\lambda}{j2\pi}\frac{d}{dz}\left[a_x(x,y) ** h(x,y,z)\right]$$
$$= a_x(x,y) ** \frac{\lambda}{j2\pi}\frac{d}{dz}h(x,y,z)\bigg|_{z=0}$$

as follows from the properties of the Fourier transform, where $a_x(x,y)$ is the inverse Fourier transform of $A_x(x,y)$.

Both the x component $s_x(x,y|z)$ and the z component $s_z(x,y|z)$ have now been derived. The proof of the proposition is complete. ∎

Problems

5.1 a. The illumination function of one element of a given antenna array is a circle function of radius one. That is,

$$s_0(x,y) = \begin{cases} 1 & \text{for } \sqrt{x^2+y^2} \leq 1, \\ 0 & \text{otherwise.} \end{cases}$$

What is the element pattern as a function of ϕ and ψ?

b. Consider the four-element array of element $s_0(x,y)$ with the combined illumination function

$$p_0(x,y) = s_0(x - \Delta/2, y - \Delta/2) + s_0(x - \Delta/2, y + \Delta/2)$$
$$+ s_0(x + \Delta/2, y - \Delta/2) + s_0(x + \Delta/2, y + \Delta/2).$$

What is the antenna pattern of the four-element array?

c. Consider the four-element "phased" array of element $s_0(x,y)$ with the combined illumination function

$$p_0(x,y) = e^{j\theta_{00}} s_0(x - \Delta/2, y - \Delta/2) + e^{j\theta_{01}} s_0(x - \Delta/2, y + \Delta/2)$$
$$+ e^{j\theta_{10}} s_0(x + \Delta/2, y - \Delta/2) + e^{j\theta_{11}} s_0(x + \Delta/2, y + \Delta/2).$$

How should the phases $\theta_{00}, \theta_{01}, \theta_{10}, \theta_{11}$ be chosen to steer the peak of the beam into the desired spatial direction defined by ϕ_0 and ψ_0?

5.2 An antenna pattern is formed by placing sixteen antenna elements in the pattern shown below.

Find the array antenna pattern by considering the array as a two by two array of two by two arrays.

5.3 Because of the requirements imposed by a mechanical support or because of the position of waveguides or feeds, a radar antenna may have part of its aperture blocked. To analyze the effect of blockage, one may simply subtract the "antenna pattern" of the blockage from the antenna pattern of the unblocked aperture.

Find the antenna pattern of a uniformly illuminated rectangular aperture of dimension A by B that is blocked at the center by an obstruction of dimension a by b.

5.4 A uniformly illuminated circular aperture of diameter D forms a jinc-shaped beam. In order to enlarge the aperture, it is replaced by a D by D square aperture. Does this improve the beam, degrade the beam, or is the answer more complicated?

5.5 A point of a one-dimensional illumination function on a bounded aperture is called the *phase center* of the antenna when it is that point nearest the center of the antenna for which the antenna pattern along the boresight is real when that point is chosen as the origin. Show that every one-dimensional illumination function has a phase center. Does this generalize to two dimensions? The boresight is the line perpendicular to the illumination function at the origin.

5.6 a. An L by L square aperture is illuminated with a phase-only illumination $e^{-j2\pi(ax+by)}$. Describe the resulting antenna pattern.

b. An L by L square aperture is pixelated into n by n contiguous square subapertures, each of size L/n by L/n that can be illuminated individually. Describe the resulting antenna patterns.

c. Discuss in detail the merits and limitations of each approach.

5.7 a. Determine the antenna pattern of the "cross antenna" shown in the illustration with constant illumination over the aperture.

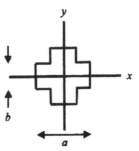

b. Instead of the procedure in part a, consider the vertical strip and the horizontal strip as two separate apertures:

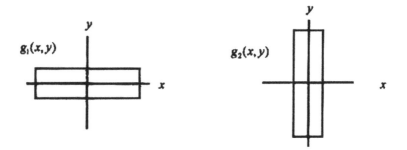

and define

$$M(f_x, f_y) = G_1(f_x, f_y) G_2(f_x, f_y).$$

An antenna array in this shape that uses this kind of multiplicative processing is known as a *Mills cross*. Describe the main lobe of $M(f_x, f_y)$ and describe the sidelobes. Describe the output of the Mills cross when a single signal arrives from a direction with the spherical coordinates (ϕ, ψ). Describe the output when two signals arrive simultaneously from directions (ϕ, ψ) and (ϕ', ψ').

5.8 An interferometer is used to measure the direction of arrival of a signal by the computation

$$\phi = \sin^{-1}\left[\frac{\lambda}{d} \frac{\Delta\theta}{2\pi}\right].$$

Suppose that the received signal is actually the sum of two signals:

$$v_1(t) = A_1 \sin 2\pi \left(f_0 t + \frac{d}{2\lambda} \sin \phi_1 \right) + A_2 \sin 2\pi \left(f_0 t + \frac{d}{2\lambda} \sin \phi_2 \right),$$

$$v_2(t) = A_1 \sin 2\pi \left(f_0 t - \frac{d}{2\lambda} \sin \phi_1 \right) + A_2 \sin 2\pi \left(f_0 t - \frac{d}{2\lambda} \sin \phi_2 \right).$$

Let $A_2/A_1 = 0.1$ and $d/\lambda = 10$. By using the dominant terms in a linearized analysis, find an approximate expression for the error in an estimate of ϕ_1 because of the interfering signal.

5.9 The RMS beamwidth of the antenna pattern $E(\phi)$ is defined as

$$B_\phi^2 = \frac{\int_{-\infty}^{\infty} (\phi - \overline{\phi})^2 |E(\phi)|^2 d\phi}{\int_{-\infty}^{\infty} |E(\phi)|^2 d\phi},$$

where

$$\overline{\phi} = \frac{\int_{-\infty}^{\infty} \phi |E(\phi)|^2 d\phi}{\int_{-\infty}^{\infty} |E(\phi)|^2 d\phi},$$

and where $E(\phi) = P(\sin \phi)$ and $p(u) = s_0(\lambda u)$. Using the Schwarz inequality, prove that within the approximation $\sin \phi = \phi$. the RMS beamwidth of the beam created by the illumination function $s_0(u) = |s_0(u)| e^{j\theta(u)}$ is minimized over a choice of phase by a linear phase function. A linear phase function has the form $\theta(u) = \theta_0 + \theta_0' u$.

5.10 a. Use the properties of the Fourier series to prove the relationship

$$e^{-ja \sin \psi} = \sum_{n=-\infty}^{\infty} J_n(a) e^{-jn\psi}.$$

b. A circular antenna array consists of N identical antenna elements equispaced on the circumference of a circle of radius a (and aligned in the same direction). Find the antenna pattern expressed as an infinite sum of Bessel functions.

c. Discuss which terms of the sum are significant for moderate to large N.

5.11 An antenna illumination function is given by

$$s(x, y) = \text{circ}\left(\frac{x}{D_1}, \frac{y}{D_1}\right) ** \text{circ}\left(\frac{x}{D_2}, \frac{y}{D_2}\right).$$

a. What is the antenna pattern in the far field? What is the area of the active aperture of the illumination function? What is the effective area?

b. Assuming λ is small, what is the asymptotic decay of the sidelobe magnitudes?

c. When the active area is held fixed, how should D_1 and D_2 be chosen to maximize antenna gain? What happens to the sidelobes?

5.12 A one-dimensional array of N uniformly spaced identical antenna elements can be used as follows. A pulse is transmitted from only one antenna element at a time and the return from that transmission is received at all N antenna elements and coherently recorded individually. This is repeated for every antenna element. How should the N^2

recorded returns be combined subsequently? Describe the resulting pattern. How does this differ from a conventional phased array?

5.13 a. Graph the antenna quality factor as a function of D/λ for a uniformly illuminated, one-dimensional aperture of width D.

b. Graph the antenna quality factor as a function of D/λ for a uniformly illuminated, two-dimensional circular aperture of diameter D.

5.14 What is the far-field (Fraunhofer) diffraction pattern for the illumination function

$$s_0(x,y) = J_0(2\pi ar/\lambda)\text{circ}\left(\frac{r}{R}\right)?$$

5.15 Calculate the energy of the two-dimensional function

$$s(x,y) = J_0\left(2\pi\alpha\sqrt{x^2+y^2}/\lambda\right).$$

5.16 Describe a condition under which a waveform $c(t)$ transmitted by a one-dimensional illumination function $s(x)$ can be considered to be a narrowband waveform. Repeat for a two-dimensional illumination function $s(x,y)$.

5.17 Show that the illumination function

$$s_0(x,y) = J_n\left(2\pi\alpha\sqrt{x^2+y^2}/\lambda\right)e^{jn\phi}$$

produces a nondiffracting beam, as does any linear combination of such illumination functions.

5.18 A dual array antenna system uses two arrays, one for transmission and one for reception. Explain how the element spacing of the arrays should be chosen to suppress the effect of grating lobes. What is the two-way $G_t G_r$ product antenna pattern?

5.19 Describe how to produce a desired phase distribution illuminating a planar surface by exciting a shaped surface placed prior to that planar aperture. One instance of this configuration is called a Cassegrain antenna.

5.20 Can autofocusing of a radar antenna array be used to estimate the range to a target? For an antenna array of diameter d and range R, determine the accuracy of such an estimate as a function of the signal-to-noise ratio. Use a carrier frequency f_0 of 10 gigahertz.

5.21 A quadratic-phase illumination function is used to focus a radar beam at (x,y,z). Describe the radar beam at $(x,y,2z)$.

5.22 Two transverse vector plane waves of the complex baseband form $e^{-j2\pi f_0(\pm\alpha x+\gamma z)}$ are traveling in directions $(+\alpha, 0, \gamma)$ and $(-\alpha, 0, \gamma)$. Both waves have polarization vector $0i_x + 1i_y + 0i_z$, respectively. The superposition of the two waves is

$$s(t) = 2\cos(2\pi\alpha x/\lambda)e^{-j2\pi\gamma z/\lambda}$$

with sum polarization vector $0i_x + 2i_y + 0i_z$.

Now suppose that the two vector waves have polarization vectors $0i_x + 1i_y + 0i_z$ and $0i_x + (-1)i_y + 0i_z$. Does this mean that the sum polarization is $0i_x + 0i_y + 0i_z$? Explain.

Notes

The first antenna was designed around 1887 by the German physicist Heinrich Hertz as part of his work demonstrating the validity of the electromagnetic theory that had been recently proposed by the British physicist James Clerk Maxwell. Heuristic methods of antenna design were developed by Guglielmo Marconi. At that time, the choice of carrier frequency was subservient to the consideration of antenna design. Now the antenna considerations are usually subservient to other system considerations when choosing a carrier frequency.

An antenna array is a straightforward extension of an antenna and the idea came easily to many and was developed early. It would be difficult to assign credit for the original ideas. The extant literature is vast. Methods for both beamforming and nullforming are widely described. Sonar beamforming is a critical technology in submarine surveillance. Because the narrowband approximation is often not appropriate for sonar, the literature of wideband beamformers in sonar applications is extensive.

The Bulter array was described in a paper by Butler and Lowe (1961) and has the same decimation structure as the Cooley–Tukey (1965) fast Fourier transform for signal processing, but preceded that work by several years. The Cooley–Tukey paper, however, is credited with first annunciating the decimation structure as a general mathematical identity, unencumbered by addressing one specific application. It seems, however, that the idea had been in use implicitly for many years, such as by radio astronomers, and also as far back as Gauss.

Radiation patterns of ultrasound transducers, sonar hydrophones, and arrays of sonar hydrophones can be treated with the same methods used to study antennas for a propagation medium that is homogeneous and isotropic. Murino and Trucco (2000) discussed imaging using acoustic waves. For sonar applications in inhomogeneous media, propagation models are at best only first approximations. A better performance can be obtained by incorporating a model of the propagation medium into the processing equations, a technique known as *matched-field processing*, as discussed by Baggeroer, Kuperman, and Mikhalevsky (1993).

The uncertainty principle for pulses also pertains to the relationship between the width of an aperture and the width of its antenna pattern. Rhodes (1974) discussed the role in antenna theory of prolate spheroidal wave functions, first studied by Slepian and Pollak (1961). McEwan and Goldsmith (1989) described the theory of illuminating small reflectors with gaussian beams to obtain very high efficiency. In ultrasound applications, gaussian beams are formed in the near field to sharpen resolution, but with limited depth of focus. To counter this limitation, Durnin (1987) introduced a nondiffracting solution to the scalar wave equation given an infinite aperture, which is attractive because it is in the same focus at all depths.

The reciprocity principle is deep and has an immense literature. It was first stated by Hendrik Lorentz (1896) and has been discussed by many authors since then. The treatment of vector diffraction based only on propagation, as given in this chapter, is apparently original.

6   Tomographic Imaging Systems

Observations of a two-dimensional or a three-dimensional object or scene may be in the form of projections onto a lower-dimensional space, or may be in some other form of reduced data. One wants to reconstruct a two-dimensional or a three-dimensional image of that object by combining many such observations, even though the detail of each of these observations is limited in some way. For example, when the only observations of an object are the shadows cast by that object, one may get an idea of the object by viewing many shadows cast at different angles. By using suitable signal-processing techniques, many such shadows of the same object can be combined to construct a single enhanced image.

Techniques for combining multiple one-dimensional projections into a single two-dimensional image are known collectively as *tomography* (Greek *tomo*: a cut + *graphy*). The term may also be used to describe techniques for combining several images of poor quality into a single improved image. This is not the same as the enhancement of a single image by signal-processing techniques, although the two tasks are closely related.

Many kinds of observations, such as absorption, emission, scattering, or diffusion, may be treated by the methods of tomography. *Absorption* refers to the attenuation of an incident signal such as an X-ray while passing through an object. The negative logarithm of the attenuation at an infinitesimal region at (x, y, z) is denoted $\rho(x, y, z)\mathrm{d}x\mathrm{d}y\mathrm{d}z$. *Emission* refers to the generation of a signal such as a wave or a stream of radioactive particles from within an object; the emission from an infinitesimal region at (x, y, z) is denoted $\rho(x, y, z)\mathrm{d}x\mathrm{d}y\mathrm{d}z$. *Scattering* refers to the dispersal of a signal from the interior of an object into various directions; the scattering from an infinitesimal region at location (x, y, z) might again be denoted $\rho(x, y, z)\mathrm{d}x\mathrm{d}y\mathrm{d}z$, although angular dependence must also be included in a fuller description. *Diffusion* refers to the multiple and continuous scattering from a continuum of infinitesimal scattering centers $\rho(x, y, z)\mathrm{d}x\mathrm{d}y\mathrm{d}z$. In each of these situations, we want to estimate $\rho(x, y, z)$ or $\rho(x, y, d)$ at $z = d$ from the observed data but the observed data is not the same from case to case and $\rho(x, y, z)$ in each case is a different physical quantity.

In many applications of tomography, the task is to form a two-dimensional cross section of $\rho(x, y, z)$ at the single value of $z = d$, sometimes called an *axial image* or a *slice*. A three-dimensional function may be replaced by a collection of two-dimensional cross sections by slicing, or sampling, $\rho(x, y, z)$ into discrete planes

$\rho(x, y, z_k)$ in the z direction where $z_k = k\Delta$. The kth two-dimensional cross section, or slice, is defined as the section $s_k(x, y) = s(x, y, k\Delta z)$. Each cross section $s_k(x, y)$ can be treated individually as a two-dimensional function $\rho(x, y)$ and imaged as such. For this reason, we consider herein only the problem of forming a two-dimensional image is considered herein. Methods to form a three-dimensional image directly can be developed in a similar way.

The most widespread and best developed form of tomography, *projection tomography*, reconstructs an image from its projections. Projection tomography based on absorption has a simple mathematical structure. The most familiar instance of projection tomography uses X-rays as the source of illumination and the spatially varying absorption of X-rays by an object as the observed physical phenomenon. The way that the X-ray illumination is used in projection tomography is quite different from the way it is used in diffraction imaging as discussed in Chapter 9. In X-ray projection tomography, the observation is based on the attenuation of the illuminating rays in the geometrical optics approximation.

In all instances of tomography, the function $\rho(x, y, z)$, or $\rho(x, y)$, refers to a specific property of the object being imaged. The estimated image $\hat{\rho}(x, y)$ is a representation of $\rho(x, y)$ as inferred from the observations. The function $\rho(x, y)$ could represent the absorption density, or the scattering density, or the emission density of the targeted object. The image may be different for each of these forms, even for the same object. Tomography itself as described herein is not concerned with the interpretation of $\rho(x, y)$ for some higher purpose of inference. That is a task for the application.

6.1 Projection Tomography

The reconstruction of images from projections is called *projection tomography* (or *transmission tomography*). Projection tomography is familiar to us through important applications in the field of medical imaging.[1] Within a designated region of the human body, an unknown function, $\rho(x, y, z)$, such as absorption density, is of interest. The distribution of the absorption density reveals details of the organs and the anatomy. The three-dimensional absorption density $\rho(x, y, z)$ or the slice $\rho(x, y)$ is to be estimated, in part, based on observations of the projections under a geometrical optics model of the X-ray illumination.

Mathematically, the two-dimensional form of the tomography problem is as follows. Given the projections $p_\theta(t)$ of $\rho(x, y)$ as defined earlier in Section 3.4 for some values of the parameter θ, find the function $\rho(x, y)$. The projections are described in more detail as follows. The projection at angle zero is

---

[1] The most common and most familiar form of illumination for medical tomography is an X-ray beam but other forms of illumination, such as a neutron beam, can also be used. An X-ray beam and a neutron beam are attenuated by different physical mechanisms and so provide images emphasizing different details.

6.1 Projection Tomography

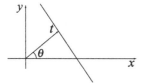

Figure 6.1 A ray at angle θ

Figure 6.2 More rays at angle θ

$$p_0(x) = \int_{-\infty}^{\infty} \rho(x,y) dy.$$

This is the integral in the y direction along each of the parallel lines parameterized by the values of x.

The projection at angle θ is obtained by rotating the coordinate system by θ and proceeding as before. The coordinate rotation is

$$x = t\cos\theta - r\sin\theta,$$
$$y = t\sin\theta + r\cos\theta.$$

A line at angle θ parameterized by t, as shown in Figure 6.1, is

$$t = x\cos\theta + y\sin\theta.$$

With θ fixed, each value of t defines a set of lines, or rays, all of which are parallel. The projection of $\rho(x,y)$ at angle θ on a line with parameter t is the integral of $\rho(x,y)$ along the line. The projection is given by

$$p_\theta(t) = \int_{-\infty}^{\infty} \rho(t\cos\theta - r\sin\theta, t\sin\theta + r\cos\theta) dr.$$

Different values of t give different lines in the x,y plane parallel to a line with $t = 0$. Integration along each of the parallel lines, as shown in Figure 6.2, results in the projection at angle θ for a fixed value of t. The projection $p_\theta(t)$ at angle θ is a function of the variable t.

Finally the value of angle θ is varied, as shown in Figure 6.3, to vary the direction of the bundle of parallel rays. Computation of the function $\rho(x,y)$ from the set of projections $p_\theta(t)$ at all values of θ is known as *parallel-beam* projection tomography.

The set of all projections $p_\theta(t)$ for all θ from 0 to π constitutes the *Radon transform* of $\rho(x,y)$. The Radon transform is the set of functions

$$p_\theta(t) = \int_{-\infty}^{\infty} \rho(t\cos\theta - r\sin\theta, t\sin\theta + r\cos\theta) dr,$$

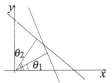

Figure 6.3 A change in angle

for all $\theta \in [0, 2\pi)$. Notice that for all such instances of projection tomography, $p_{\theta+\pi}(t) = p_\theta(-t)$, which means that knowing $p_\theta(t)$ for all $\theta \in [0, \pi)$ is equivalent to knowing $p_\theta(t)$ for all θ.

The task of reconstructing the image $\rho(x, y)$ from its projections is the task of inverting the Radon transform. When the projections are known for all θ, the inversion problem has an exact mathematical inverse.

Recall that the projection-slice theorem, given as Theorem 3.4.1 of Chapter 3, says that $P_\theta(f)$, the Fourier transform of projection $p_\theta(t)$, is given by

$$P_\theta(f) = R(f\cos\theta, f\sin\theta),$$

where $R(f_x, f_y)$ is the two-dimensional Fourier transform of $\rho(x, y)$. The task of inverting the Radon transform $p_\theta(t)$ for all values of θ to find $\rho(x, y)$ is equivalent to the task of inverting $P_\theta(f)$ for all values of θ to find $R(f_x, f_y)$.

The projection-slice theorem gives a straightforward approach, in principle, to computing $\rho(x, y)$ from $p_\theta(t)$. First compute $P_\theta(f)$ from $p_\theta(t)$ for all θ. Then, recalling that $f_x = f\cos\theta$ and $f_y = f\sin\theta$, form the function $R(f\cos\theta, f\sin\theta)$ from $P_\theta(f)$. Finally, compute the inverse two-dimensional Fourier transform of $R(f_x, f_y)$ to find $\rho(x, y)$. Other ways to find $\rho(x, y)$ from its projections are known and may be preferred.

Parallel-Beam Observations

A parallel-beam tomography system collects, at each angle θ, a set of projections along each of a large set of parallel rays as described. Each ray uses one X-ray source and one X-ray detector, so this approach uses a large number of such sources and detectors. It is enough to provide a single array of source/detector pairs at only one angle θ. Then measurements are taken sequentially at many angles by rotating the configuration of arrays with respect to the target, or rotating the target with respect to the configuration of arrays. This may be simplified even further at each angle by holding at that angle and mechanically stepping a single source/detector pair transversely, thereby measuring one ray at a time at each angle. Then the data collection time would be much greater. It may be undesirable to extend the collection time in this way by measuring, one by one, each ray at each angle, especially when the target might move.

Fan-Beam Observations

An alternative to parallel-beam tomography is fan-beam tomography. Fan-beam tomography uses a single wide-beam X-ray source at each angle θ, as shown in Figure 6.4. At each angle θ, a fan beam of X-ray illumination is generated at the

6.1 Projection Tomography

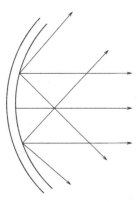

Figure 6.4 Finding parallel beams in fan beams

radiation source. The multiple rays from one source are detected by multiple X-ray sensors distributed in angle.

In this case, the projection data in each observation consists of a groups of rays in the form of a fan, rather than a group of parallel rays. However, inspection of Figure 6.4 makes it clear that, provided the sampling intervals are properly planned, the set of all data in all rays of all fans can be regrouped into groups of parallel rays by choosing one ray from each fan. In this way, data that is collected as fan-beam data can be later rebinned into the form of parallel-beam data, one ray from each fan beam. The parallel beam data is processed by any suitable algorithm for parallel-beam data such as those described below. Moreover, as the discussion suggests, instead of restructuring the data, it may be possible to restructure the algorithm to accept the fan-beam data directly.

Space-Domain Inversion

The Radon transform of $\rho(x,y)$, which consists of the set of all projections $p_\theta(t)$ of $\rho(x,y)$ can be inverted as is asserted by the projection-slice theorem given as Theorem 3.4 of Chapter 3. The following theorem gives one form of the inverse Radon transform.

Theorem 6.1.1 (Back-Projection Theorem) *The space-domain function $\rho(x,y)$ is related directly to the set of its projections by the expression*

$$\rho(x,y) = \int_0^\pi \int_{-\infty}^\infty P_\theta(f) e^{j2\pi f(x\cos\theta + y\sin\theta)} |f| \mathrm{d}f \mathrm{d}\theta.$$

Proof:
The inverse Fourier transform of $R(f_x, f_y)$ is

$$\rho(x,y) = \int_{-\infty}^\infty \int_{-\infty}^\infty R(f_x, f_y) e^{j2\pi(f_x x + f_y y)} \mathrm{d}f_x \mathrm{d}f_y.$$

Change the integration to polar coordinates by setting

$$f_x = f \cos\theta,$$
$$f_y = f \sin\theta,$$

and $df_x df_y = |f| df d\theta$. Then

$$\rho(x,y) = \int_0^\pi \int_{-\infty}^{\infty} R(f\cos\theta, f\sin\theta) e^{j2\pi f(x\cos\theta + y\sin\theta)} |f| df d\theta.$$

But $R(f\cos\theta, f\sin\theta)$ is equal to $P_\theta(f)$. Making this substitution completes the proof of the theorem. ∎

By setting $x = r\cos\phi$ and $y = r\sin\phi$, the back-projection theorem can be expressed in polar coordinates. This is the following corollary.

Corollary 6.1.2

$$\rho(r\cos\phi, r\sin\phi) = \int_0^\pi \left[\int_{-\infty}^{\infty} |f| P_\theta(f) e^{j2\pi fr\cos(\theta - \phi)} df \right] d\theta.$$

Proof:
The proof follows immediately from the change of variables to polar coordinates $x = r\cos\phi$ and $y = r\sin\phi$. ∎

Any reconstruction method based on the back-projection theorem is referred to as *back projection*. The structure of the back-projection double integral may be easier to see when the expression is decomposed into two separate integrals as follows:

$$g_\theta(t) = \int_{-\infty}^{\infty} |f| P_\theta(f) e^{j2\pi ft} df,$$

$$\rho(x,y) = \int_0^\pi g_\theta(x\cos\theta + y\sin\theta) d\theta.$$

This is illustrated in the flow chart of Figure 6.5. The term $g_\theta(t)$ is the inverse Fourier transform of $|f| P_\theta(f)$ for each θ. Therefore one might conclude that $g_\theta(t) = h(t) * p_\theta(t)$, where $h(t)$ is the inverse Fourier transform of $|f|$. This suggestive conclusion, however, is flawed because $|f|$ has infinite energy and does not have an inverse Fourier transform.

Although the inverse Fourier transform of $|f|$ does not exist as a proper function, the inverse Fourier transform of $|f|$ does exist as a generalized function and does behave under formal manipulations. The second derivative of $|f|$ is $2\delta(f)$ where $\delta(f)$ is the impulse function. Because $2\delta(t) \leftrightarrow 2$ is a Fourier transform pair and the derivative property says that twice-differentiation multiplies the Fourier transform by $(j2\pi)^2$, it can be concluded that the function $|f|$ has the *generalized* inverse Fourier transform $-2(2\pi t)^{-2}$. This gives the two dual *generalized* one-dimensional Fourier transform pairs

$$-\frac{2}{(2\pi t)^2} \leftrightarrow |f| \qquad \text{and} \qquad |t| \leftrightarrow -\frac{2}{(2\pi f)^2}.$$

6.1 Projection Tomography

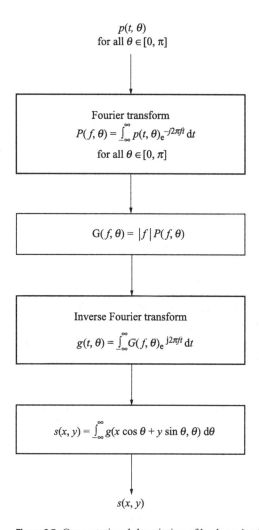

Figure 6.5 Computational description of back-projection theorem

Thus, with $h(t) = -\frac{1}{2}(\pi t)^{-2}$, the back-projection operation can be written formally as

$$g_\theta(t) = h(t) * p_\theta(t),$$

to be understood as a generalized expression. The name "back-projection" refers to the notion that $p_\theta(t)$ is spread along the line by the function $h(t)$.

As described, this formulation of back projection as a generalized expression has the weakness that $h(t)$ is a generalized function and has a singularity. The use of a generalized function can be avoided whenever $R(f_x, f_y)$ has finite support. Let r be the radius of a circle that encloses the support of the spectrum $R(f_x, f_y)$. Then it is enough to replace $|f|$ by any function of finite energy for which $H(f) = |f|$ for $|f| \leq R$ and is otherwise arbitrary. The function $H(f) = |f|\text{rect}(f/r)$ is such a function. This function, however, has an abrupt edge. A smoother function is

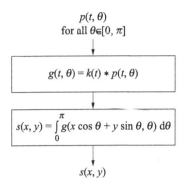

Figure 6.6 Filtered back projection

$$H(f) = \begin{cases} |f| & \text{for } |f| \leq r, \\ \cos\pi \frac{|f|-r}{\Delta} & \text{for } r \leq |f| \leq r+\Delta, \\ 0 & \text{for } |f| \geq r+\Delta, \end{cases}$$

for any positive Δ. Such an $H(f)$ will ensure that the two-dimensional Fourier transform falls off with frequency sufficiently quickly to have a proper inverse Fourier transform $h(t)$. Then we may replace the $|f|$ in the back-projection formula by this smoother $H(f)$, noting that

$$h(t) * p_\theta(t) \leftrightarrow H(f)P_\theta(f),$$

where $H(f)$ is any complex function satisfying $H(f)P_\theta(f) = |f|P_\theta(f)$.

With the aid of Corollary 6.1.2, the expression

$$\rho(r\cos\theta, r\sin\theta) = \int_0^\pi \left[\int_{-\infty}^\infty H(f)P_\theta(f)e^{j2\pi fr\cos(\theta-\phi)}df\right]d\theta$$

can now be used as an alternative form of back-projection. This alternative form, shown in Figure 6.6, is known as *filtered back-projection*.

Rather than filter the projection $p_\theta(t)$ as $h(t) * p_\theta(t)$, one may choose to filter the signal $\rho(x,y)$ as $h(x,y) * * \rho(x,y)$. The following proposition gives a condition under which this is equivalent. Let $h(t)$ be a symmetric one-dimensional filter with Fourier transform $H(f)$ and let $h(x,y)$ be the two-dimensional circularly symmetric filter whose circularly symmetric Fourier transform $H(f_x, f_y) = H\left(\sqrt{f_x^2 + f_y^2}\right)$ is defined in terms of $H(f)$. The one-dimensional Fourier transform pair $h(t) \leftrightarrow H(f)$ implies the two-dimensional Fourier transform pair $h(x,y) \leftrightarrow H\left(\sqrt{f_x^2 + f_y^2}\right)$.

Proposition 6.1.3 *The projection $p_\theta(t)$ of $\rho(x,y)$ at angle θ filtered as $h(t) * p_\theta(t)$ is the projection of $h(x,y) * * \rho(x,y)$.*

Proof:
The projection of $g(x,y)$ at angle θ has the Fourier transform

$$G(f\cos\theta, f\sin\theta) = H(f\cos\theta, f\sin\theta)R(f\cos\theta, f\sin\theta)$$
$$= H(f)P_\theta(f),$$

because $H(f\cos\theta, f\sin\theta) = H(f)$ and $R(f\cos\theta, f\sin\theta) = P_\theta(f)$. This completes the proof of the theorem. ∎

To summarize, the filter $H(f)$ is chosen to equal one at all frequencies of interest contained in the Fourier transform of $\rho(x,y)$. That filter has no effect on $\rho(x,y)$. Then $g_\theta(t) = h(t) * p_\theta(t)$ where

$$h(t) = \int_{-\infty}^{\infty} H(f)e^{j2\pi ft}dt.$$

Finally, the integral

$$\rho(x,y) = \int_0^\pi g_\theta(x\cos\theta + y\sin\theta)d\theta$$

completes the reconstruction of $\rho(x,y)$.

The back-projection computation is now broken into a two-step process given by these two equations. First every projection is passed through a filter with the impulse response $h(t)$. Then, for each (x,y) pair, an integral is evaluated to form $\rho(x,y)$.

Limited Data

The back-projection algorithm provides an exact mathematical expression for projection tomography. As described, the algorithm requires full data. Practical considerations that arise may limit the data set in some ways. For example, actual projection data is always sampled data. The number of angles is finite. Data is collected only at a finite number of θ, and at each θ, data is sampled at only a finite number of t. This means that the data is collected on a two-dimensional sampling grid in polar coordinates. The computations of back projection can be described as the digital inversion of discrete data sampled in polar coordinates. The sampling theorem does not have a completely satisfactory version in polar coordinates. Moreover, the back-projection procedure is formally exact only when $p_\theta(t)$ is known for all θ.

Possibly the data has some angular gaps. When $p_\theta(t)$ is known only for partial data, then one needs to compensate for the missing data or suffer processing artifacts in the computed image. The back-projection theorem and Fourier reconstruction each require data with full angular coverage. When data is not available for some angles, then an algorithm may be modified to account for the missing data.

Another consideration is noise in the projections. In the simplest case of additive noise, the measured data at angle θ is

$$v_\theta(t) = p_\theta(t) + n_\theta(t),$$

where the additive term $n_\theta(t)$ is a noise process whose mean and variance are known. Filtering may be included in the computations to suppress the effect of noise. It is straightforward to incorporate a circularly symmetric noise filter with transform $H\left(\sqrt{f_x^2 + f_y^2}\right)$ into filtered back-projection as discussed earlier. In other models of noise, the mean and the variance of $n_\theta(t)$ may depend on $p_\theta(t)$, in which case the processing is more complicated.

In some situations, prior data may be available. One may have a partial prior knowledge of $\rho(x,y)$. Perhaps a region that contains the support of $\rho(x,y)$ is known or it is known that $\rho(x,y)$ is nonnegative. One may then try to exploit this prior knowledge to improve the processing. Prior knowledge about $\rho(x,y)$ may allow one to compensate for the lack of projection data at some values of θ.

To illustrate with a trivial example that prior knowledge may be beneficial, suppose it is known that $\rho(x,y)$ factors as

$$\rho(x,y) = \rho'(x)\rho''(y).$$

We may readily expect that a projection along the x axis and a projection along the y axis should suffice to define $\rho(x,y)$.

6.2 Frequency-Domain Inversion

Filtered back-projection has complexity proportional to n^3 and so can be computationally unattractive in its elementary form. Moreover, images may need to be formed from partial data or from noisy data. Alternative computational methods of image formation may be preferred.

Back-projection image formation views the processing of tomography data in terms of the polar coordinates of the Radon transform. An alternative to back projection is Fourier reconstruction. Fourier reconstruction views the problem in terms of rectangular coordinates. The polar data is interpolated onto a rectangular grid followed by an inverse two-dimensional Fourier transform.

The projection-slice theorem creates slices of the two-dimensional Fourier transform $R(f_x, f_y)$ from the projections $p_\theta(t)$. These slices are in polar coordinates. It may be computationally awkward to Fourier transform these samples to obtain the image $\rho(x,y)$ because the usual fast Fourier transform algorithms are formulated for data that is sampled on a rectangular grid. To use these fast algorithms, the data samples must be converted from a polar grid to a rectangular grid. Both grids are shown in Figure 6.7. This requires a two-dimensional interpolation of the polar samples into the samples on the rectangular grid. A simple two-dimensional interpolation scheme computes each sample on the rectangular grid as the weighted average of the four nearest samples on the polar grid. The interpolation error directly affects the quality of the image and may be the dominant source of computational error in an image. Furthermore, because the polar samples are further apart at high frequencies, interpolation errors will be more severe at high frequencies unless the interpolation procedure varies across the f_x, f_y plane.

An alternative to the conventional discrete polar grid is the *consecutive-squares grid*, shown in Figure 6.8, which uses samples on each slice that are spaced uniformly in $\tan\theta$ (or in $\cot\theta$) rather than in θ. In addition, the radial sample spacing varies with the angle in such a way that the sample points line up horizontally and vertically.

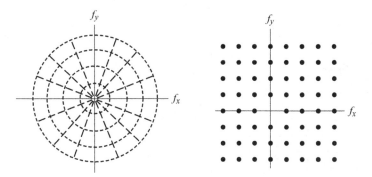

Figure 6.7 A polar grid and a rectangular grid

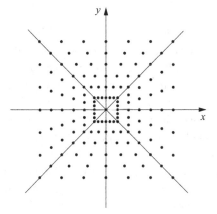

Figure 6.8 A concentric squares grid

The reason for using the consecutive-squares grid is so that a one-dimensional interpolation can be used on each coordinate to compute the samples even though these samples fall on a two-dimensional grid.

6.3 Algebraic Inversion

Fourier reconstruction works best when the Fourier sampling grid satisfies the Nyquist criterion and the entire Fourier sampling grid is filled with measured data. When some Fourier data samples are not measured, as is often the case, they might be replaced by zeros, but unacceptable artifacts may then occur in the image.

An alternative approach to back projection and Fourier reconstruction is *algebraic reconstruction*. Algebraic reconstruction is a computationally intensive approach that can be used for sparse data sets, or to incorporate unusual prior constraints on the support of the image. The approach is to set up a large system of linear equations in the unknown pixels, denoted ρ_{ij}, and the known rays, then to solve for the pixel values

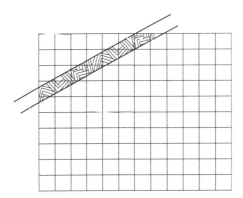

Figure 6.9 Rays overlapping pixels

using the methods of linear algebra. Consider the image to be an n by n grid of pixels with value ρ_{ij} in the ij pixel. Regard a ray to have a width comparable to the pixel size, but oriented at angle θ. A ray overlaps each pixel by a different amount and does not overlap most pixels at all. Let $\gamma_{ij\ell\theta}$ denote the overlap of the ℓth ray at angle θ with the ijth pixel as shown in Figure 6.9. Then projection $p_{\ell\theta}$ is related to the pixel values by $p_{\ell\theta} = \sum_{ij} \gamma_{ij\ell\theta} \rho_{ij}$. There is one such linear equation for each ray. The number of linear equations is equal to the number of rays, which must be comparable to the number of unknown pixels and may be very large. For a 512 by 512 image, there are 262,144 pixels, so the number of linear equations must be a similar number. Indexing the I pixels by a single index i, say by a raster scan, the image can be treated as a single one-dimensional vector, in this case a vector of approximate length 262,144. After reindexing the projections by a single index ℓ, the system of linear equations then becomes

$$p_\ell = \sum_i \gamma_{i\ell} \rho_i.$$

Now the task of image formation is seen as the task of solving the matrix-vector equation of the form $p = \Gamma \rho$ where Γ is a large matrix, but it is a sparse matrix. In this formulation, the projection data set is simply the point p in the large space, and the image is the point ρ, where $\rho = \Gamma^{-1} p$. Of course, when Γ has dimension 262, 144, or even a much smaller dimension, conventional techniques for solving a linear system of equations cannot be used. However, attractive iterative projection methods, suitable for sparse matrices, are available. To explain such methods, consider first only two equations in the two unknowns ρ_1 and ρ_2 as follows:

$$\gamma_{11}\rho_1 + \gamma_{12}\rho_2 = p_1,$$
$$\gamma_{21}\rho_1 + \gamma_{22}\rho_2 = p_2.$$

Replacing (ρ_1, ρ_2) by (x, y), each of these two equations describes a line in the x, y plane as shown in Figure 6.10. One way to find the intersection of the two lines is to start at any point and project that point onto either of the lines, where the projection of

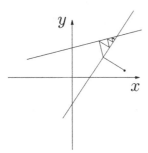

Figure 6.10 Projection operations in the plane

point (x, y) is defined as the point on the line at which the perpendicular intersects the point (x, y). Then project the new point onto the other line. By repeating this process of successive projections, as shown in Figure 6.10, the point gradually moves to the point of intersection.

To derive this recursion, a simple problem is posed in the familiar x, y plane. Which point on the line $x \cos \theta + y \sin \theta = c$ is closest to a given point (x_0, y_0)? It is simple to compute that the point given by

$$x = c \cos \theta + (x_0 \sin \theta - y_0 \cos \theta) \sin \theta,$$
$$y = c \sin \theta - (x_0 \sin \theta - y_0 \cos \theta) \cos \theta$$

is the answer to this question. Similarly, the closest point on the line $Ax + By = c$ to the point x_0, y_0 is given by

$$x = \frac{Ac}{A^2 + B^2} + (x_0 B - y_0 A) \frac{B}{A^2 + B^2},$$
$$y = \frac{Bc}{A^2 + B^2} - (x_0 B - y_0 A) \frac{A}{A^2 + B^2}.$$

The alternating projection method for two linear equations in the plane is now evident. Simply apply these projection equations alternately to these two lines, first to one line, then to the other, and repeat.

The same iterative process can be used on large systems of linear equations. Any system of m linear equations

$$\gamma_{11} p_1 + \cdots + \gamma_{1n} p_n = p_1$$
$$\vdots$$
$$\gamma_{m1} p_1 + \cdots + \gamma_{mn} p_n = p_m$$

in n unknowns defines a set of hyperplanes in $\mathbb{R}^n$. Choose any point of $\mathbb{R}^n$ and project that point onto the first hyperplane, then project the new point onto the second hyperplane. After projecting onto the last hyperplane, start over with the first hyperplane, and so on. The projected point will gradually move towards the point of common intersection provided the system of equations is invertible. To derive the iteration, one uses the following proposition in n-dimensional euclidian space $\mathbb{R}^n$.

Proposition 6.3.1 *The point of the $(n-1)$-dimensional hyperplane $\sum_i a_i x_i = c$ closest to a given point $X° = (x_1°, x_2°, \ldots, x_n°)$ is given by*

$$x_i = \lambda a_i + x_i°,$$

where

$$\lambda = \frac{c - \sum_i a_i x_i°}{\sum_i a_i^2}.$$

Proof:
The expression $\sum_i (x_i - x_i°)^2$ is to be minimized over choice of $x_1, x_2, \ldots, x_n$ subject to the constraint $\sum_i a_i x_i = c$. Introduce the lagrange multiplier 2λ and write the partial derivatives of the augmented objective function as

$$\frac{\partial}{\partial x_i} \left[\sum_i (x_i - x_i°)^2 - 2\lambda \left(\sum_i a_i x_i - c \right) \right] = 0.$$

This reduces to the set of equations

$$x_i - x_i° = \lambda a_i,$$

from which the first expression of the proposition follows immediately. To evaluate λ, first square and sum both sides of this expression over i. Also multiply the same expression by $x_i - x_i°$ and sum. This gives the two equations

$$\sum_i (x_i - x_i°)^2 = \lambda^2 \sum_i a_i^2,$$

$$\sum_i (x_i - x_i°)^2 = \lambda \sum_i a_i (x_i - x_i°) = \lambda c - \lambda \sum_i a_i x_i°.$$

These combine to give

$$c - \sum_i a_i x_i° = \lambda \sum_i a_i^2,$$

from which the second expression of the proposition follows. This completes the proof of the proposition. ∎

A variation of this method is to group together r constraint equations corresponding to $(n - r)$-dimensional hyperplanes. By using this group of r constraint equations at once, each step is a projection onto an $(n-r)$-dimensional hyperspace. In this variation, there are fewer projection steps, but each projection step is more complicated.

The method of successive projections can be applied to the problem of image formation because this problem can be regarded as the task of inverting the system of linear equations given by $p = \Gamma \rho$. The image $\rho(x, y)$ is then obtained from ρ by the usual interpolation formula

$$\rho(x, y) = \sum_i \sum_j \rho_{ij} h(x - i\Delta x, y - j\Delta y),$$

where $h(x, y)$ is an appropriate interpolation function.

Model-Based Imaging

Although algebraic reconstruction is computationally intensive, it can easily tolerate missing data and can accommodate prior information about the support of the image simply by setting certain pixel values ρ_{ij} to zero. Indeed, the method is more general and is not limited to representations in terms of pixels. The method still can be applied even when the image is represented in terms of any basis of functions $b_i(x, y)$ as

$$\rho(x, y) = \sum_i a_i b_i(x, y),$$

where the basis $\{b_i(x, y)\}$ is a set of functions spanning the space of possible images. This is called *model-based imaging*. Prior constraints restricting this space are easily accommodated. For example, when $\rho(x, y)$ is known to be an image of a section of a human brain, then the basis $\{b_i(x, y)\}$ must be adequate to represent such images as a linear combination of basis functions, but other kinds of images need not be representable in terms of that basis. Such a basis for a given class of images might be found by a training process involving a large number of instances of such images. Then a set of projection data is expressed in terms of that learned basis. This again, in principle, can be stated as the inversion of a large system of linear equations and solved by successive projections.

6.4 Image Formation from Magnetic Resonance

The important imaging modality known as *magnetic resonance imaging* is discussed in this section. Magnetic resonance images are formed using the methods of tomography and its variations based on data obtained from magnetic fields. This imaging modality uses measurement data that is stimulated by externally generated, spatially varying, and temporally varying magnetic fields. The external magnetic field induces the target object to generate its own observable magnetic field that depends on a space-varying property of interest within that object. This section provides an introduction to the topic of magnetic resonance imaging while staying within the scope of this book.

An image to be formed using magnetic resonance depends on the density $\rho(x, y, z)$ of the target nuclei[2] within the subject of interest. The required *macroscopic* magnetic-resonance data signals of the target are obtained by manipulating and aligning the inherent *microscopic* magnetic elements within the target object through the use of three *external* magnetic fields: a strong static magnetic field B_0, here called the *bias field*; a time-varying magnetic field B_1, called the *excitation field*; and a space-varying magnetic field B_g, called the *gradient field*. The bias field B_0 creates spin polarization at the microscopic level leading to bulk magnetism at the macroscopic level; the excitation field B_1 creates magnetic resonance that enables detection of the induced bulk magnetism; the gradient field B_g encodes spatial information into the magnetic

---

[2] As may be modulated by the various relaxation times of the induced magnetic fields.

field to obtain the data for image formation from the induced magnetic response. The three external magnetic fields are vector fields that together establish a forward transform relating the desired image to the measured magnetic signals through the Radon transform or the Fourier transform. The image is then formed using the usual signal processing techniques of tomography as discussed earlier in this chapter or using modifications of these techniques.

An image of a three-dimensional target object may be obtained by imaging two-dimensional slices taken along the z axis, one slice at a time. The slice $\rho_z(x, y) = \rho(x, y, z)$ is isolated at, or near, a selected value of z by spatial modulation of the gradient field, as will be described.

Magnetic resonance is a data-rich imaging modality. It can also be used to obtain other information about the structure, about the function, and about the metabolism of a biological system using special data acquisition schemes to capture biological processes through magnetic-field relaxation transients known componentwise as T_1 relaxation and T_2 relaxation, as well as other effects known as the chemical shift effect, the diffusion effect, and so forth. A detailed discussion of these topics of supplementary data acquisition is outside of the image formation topic of this book.

The basic concepts underlying image formation that use magnetic resonance to generate the electrical signals in the sensing coils are introduced in the following subsections – first dealing with the underlying physics, then dealing with the development of the signals needed for image formation. The section concludes with the study of the projections needed to form an image of the target.

Nuclear Magnetism

Fundamental particles like protons and neutrons (nucleons) possess an intrinsic vector quantity called the *angular momentum* $\boldsymbol{J}$. The angular momentum of a fundamental particle displays quantum properties that do not have a clear classical analog. The intrinsic angular momenta of two nucleons in the atomic nucleus may cancel each other thereby reducing to a net zero angular momentum. Only atomic nuclei with an odd number of nucleons have a definite net angular momentum which is conceptualized in terms of the notion of nuclear spin.[3] Based on quantum theory, spin takes integer or half-integer values as given by the *spin quantum number* I. For nuclei with an odd number of nucleons, the spin quantum number takes half-integer values as given by the set $I \in \{1/2, 3/2, 5/2, \ldots\}$. The magnitude $|\boldsymbol{J}|$ of the angular momentum $\boldsymbol{J}$ of a nucleus is determined by the spin quantum number I and is given as

$$|\boldsymbol{J}| = \hbar\sqrt{I(I+1)},$$

where $\hbar = h/2\pi$ and h is Planck's constant (6.6×10^{-34} joule-seconds).

---

[3] Nuclear spin is a semiclassical notion that conceptualizes a core concept of quantum theory using a semiclassical model that can be described and treated using the simpler concepts of classical or semiclassical physics.

The Hydrogen Nuclei

The nucleus of the hydrogen atom, consisting of a single proton with $I = 1/2$ is the simplest instance of an atomic nucleus. The hydrogen nucleus is widely used for magnetic resonance because hydrogen is the most common element found in the human body and a nucleus with a single proton responds well to external magnetic fields. Because $I = 1/2$, the magnitude of the angular momentum of a hydrogen nucleus can take only the value $|J| = \hbar\sqrt{3}/2$.

The density function $\rho(x,y,z)$ of hydrogen atoms in an organism is the density distribution of the single protons of the hydrogen nuclei in that organism. The soft tissue dominates in the image because soft tissue is dominated by water molecules. A water molecule contains two hydrogen atoms so a water molecule contains the two isolated protons of the two hydrogen nuclei. An image of $\rho(x,y,z)$ in an organism is largely an image of the water density distribution in that organism.[4]

Gyromagnetic Ratio

A nucleus with a nonzero nuclear spin has a *magnetic moment* μ. The magnetic moment μ is a vector related to the angular momentum J of the nucleus by $\mu = \gamma J$, where γ is a physical constant called the *gyromagnetic ratio*. This means that the magnitude $\mu = |\mu|$ of the magnetic moment is related to the spin by $\mu = \gamma \hbar \sqrt{I(I+1)}$.

The gyromagnetic ratio depends on the atomic species. The gyromagnetic ratio of a hydrogen nucleus is 42.58 megahertz per tesla. The gyromagnetic ratio of a carbon nucleus is 10.71 megahertz per tesla.

Larmor Frequency

Magnetic resonance imaging uses the fact that the spin axis of an atomic nucleus with an intrinsic spin will precess about the direction of an external magnetic field B_0 of magnitude $B_0 = |B_0|$. The frequency of precession for each atomic species is known as the *Larmor frequency* for that species in that magnetic field. The Larmor frequency of an atomic species depends linearly on the strength of the magnetic field. The Larmor frequency for any nucleus in a field of magnitude $B_0 = |B_0|$ is given by the *Larmor equation*[5] $f_0 = \gamma B_0/2\pi = \chi B_0$, where the constant γ is the gyromagnetic ratio and $\chi = \gamma/2\pi$. Because the Larmor frequencies are different, the radio frequency magnetic signals coming from carbon atoms or from other atomic species are not confused with radio frequency magnetic signals coming from hydrogen atoms.

The performance of magnetic resonance imaging depends on the Larmor frequency. To obtain a high Larmor frequency, a strong bias field B_0 must be used, perhaps with

---

[4] The number of hydrogen nuclei even in a very small volume is immense. Avogadro's number is 6×10^{23}. Representing every hydrogen nucleus in one mole of water by a grain of common table salt would require a cube approximately one-hundred kilometers on a side to hold all the table salt.

[5] The Larmor frequency is seen by the electrical sensing system in the role of the *carrier frequency* or *center frequency* of a received passband waveform, as to be described

strength 1.5 or 3 tesla, corresponding to a Larmor frequency of 63.87 or 127.74 megahertz for hydrogen nuclei. These Larmor frequencies are in the UHF band. Some systems now use magnetic fields of as high as 7 tesla, which corresponds to a Larmor frequency of 298.06 megahertz for hydrogen nuclei. Stronger magnetic fields within a central cavity that is large enough to hold a subject for clinical diagnostics are in use but are expensive and difficult to construct.

Spin Polarization

For hydrogen nuclei, the spin magnetic moment satisfies $|\boldsymbol{\mu}| = \gamma\hbar\sqrt{3}/2$. In the absence of an external magnetic field, each hydrogen nucleus has a random orientation of the magnetic moment vector $\boldsymbol{\mu}$ and so these random orientations of the many such individual magnetic moments $\boldsymbol{\mu}$ of the nuclei average to zero. As a result, when there is no external field and the system is in thermal equilibrium, an object of interest has (almost) no macroscopic magnetism. To generate signals from that object, the first step is to create the nonzero macroscopic magnetism density from the many microscopic magnetic nuclei. This is accomplished by placing the subject inside the bore of a powerful magnet[6] that generates the bias field $\boldsymbol{B}_0$ so that the nuclear spins of the hydrogen nuclei become aligned.

In the presence of the external bias field $\boldsymbol{B}_0 = B_0 \boldsymbol{i}_z$ in the z direction, quantum theory states that the direction of the magnetic moment $\boldsymbol{\mu}$ of a nucleus becomes quantized with regard to the direction of $\boldsymbol{B}_0$ and takes one of $2I + 1$ possible orientations such that the z component μ_z of $\boldsymbol{\mu}$ is $\mu_z = m\gamma\hbar$ for $m \in \{-I, -I+1, \ldots, I-1, I\}$.

For a hydrogen nucleus, the spin value I is equal to $1/2$. Then $m \in \{-1/2, +1/2\}$ and the z component of $\boldsymbol{\mu}$ for each hydrogen nucleus takes one of two possible values, $\mu_z = -\gamma\hbar/2$ or $\mu_z = +\gamma\hbar/2$. These correspond to the two spin states of the hydrogen nucleus. The two spin states have different energies.

Spins with $\mu_z = -\gamma\hbar/2$ are in a high-energy state. Spins with $\mu_z = +\gamma\hbar/2$ are in a low-energy state. Were $\boldsymbol{\mu}$ a classical magnetic moment vector, those spins in a low-energy state would point along with the $\boldsymbol{B}_0$ field while those in a high-energy state would point against the $\boldsymbol{B}_0$ field. This is the effect called *spin polarization*.

Treating $\boldsymbol{\mu}$ more correctly by the rules of quantum theory, the quantization of μ_z requires the magnetic moments to align themselves at specific angles with respect to the direction of the ambient magnetic field. For hydrogen nuclei, because $|\boldsymbol{\mu}| = \gamma\hbar\sqrt{3}/2$ and $\mu_z = \pm\gamma\hbar/2$, the allowed angles between $\boldsymbol{\mu}$ and the bias field $\boldsymbol{B}_0$ are given by $\theta = \cos^{-1}(\mu_z/\mu) = \cos^{-1}(1/\sqrt{3}) = \pm 54.73$ degrees. Thus, the spin axis is 54.73 degrees from the positive or the negative direction of the ambient magnetic field,[7] with a slight preference for the positive direction. Referring to the z component of the spin, the low-energy state is called the *parallel* or the *pointing-up* state. The high-energy state is called the *antiparallel* or the *pointing-down* state.

---

[6] Earth's magnetic field is about 0.5 gauss; a refrigerator motor magnetic field is about 50 gauss; the bias field for common clinical magnetic resonance scanners is from 0.5 tesla to 3.0 tesla. One tesla is equal to 10,000 gauss.

[7] At angle 54.73 degrees, the transverse component of the spin is slightly larger than the z component.

Although the z component of μ is constrained by the bias field, the transverse component of μ remains unconstrained and points in a random direction of the x, y plane. Accordingly, the transverse components of the many nuclei average to zero until the excitation magnetic field is turned on to bring their rotations into synchronization.

Energy Distribution

The statistical distribution of any large population among the available energy states is governed by the Boltzmann probability distribution. This is a general statement of statistical mechanics that applies broadly. It applies here for a population of two states having $N_s = N_\uparrow + N_\downarrow$ nuclei, with $N_\uparrow$ being the number of spins pointing up (the low-energy state) and $N_\downarrow$ being the number of spins pointing down (the high-energy state). Specifically, for a system with spin $1/2$, the Boltzmann distribution requires that

$$\frac{N_\uparrow}{N_\downarrow} = e^{\Delta E/k_B T_s},$$

with the energy difference between the two spin states given by $\Delta E = E_\downarrow - E_\uparrow = \gamma \hbar B_0$, and where k_B is the Boltzmann constant (1.38×10^{-23} joules/kelvin) and T_s is the absolute temperature of the spin system. The exponent on the right side is small, so the exponential can be replaced by a first-order approximation to give

$$N_\uparrow \approx N_\downarrow \left(1 + \frac{\Delta E}{k_B T_s}\right) = N_\downarrow \left(1 + \frac{\gamma \hbar B_0}{k_B T_s}\right)$$

$$\approx N_\downarrow + N_s \frac{\gamma \hbar B_0}{2 k_B T_s},$$

where $N_\downarrow$ in the second term has been approximated by $N_s/2$.

Magnetization Density

A soft magnetic material can be magnetized by an external magnetic field but does not retain that magnetization in the absence of an external field. The vector sum of the magnetic moments μ of all nuclei in the target object is a macroscopic vector field called the *magnetization* M. Then $M = \sum_i \mu_i$ where the sum is over all nuclei in that object. The magnetization M is itself a magnetic field produced by the material itself in response to the external magnetic field B.

When there is no magnetic field, the magnetic moments of the individual nuclei point in random directions. In this case, the vector sum of the magnetic moments has an expected value of zero and a negligibly small variance. Then $M = 0$.

When the bias field is present and is the only external magnetic field, the magnetic moments become aligned so as to have a nonzero sum in the direction of the bias field. The transverse magnetic components of the many magnetic moments remain in random directions and average to zero with a variance that is negligible. The net macroscopic magnetization vector M of the nuclei is given by

$$M = (N_\uparrow - N_\downarrow)\mu_z i_z,$$

where i_z is the unit vector in the z direction. Because $\mu_z = \gamma \hbar / 2$, the z component is

$$M_z = N_s \frac{\gamma^2 \hbar^2 B_0}{4 k_B T_s}.$$

Only the z component is nonzero.

The macroscopic magnetization vector M is nonzero and behaves like a classical magnetization vector. It is aligned with the B_0 field in the direction of i_z. The magnitude of M is directly proportional to the magnitude $B_0 = |B_0|$ of the bias vector B_0 and to the total number of spins N_s. The total number of spins N_s is characteristic of the object being imaged and cannot normally be changed. The only controllable parameters are B_0 and T_s. Therefore, for a given object, one can increase the magnitude of M by increasing the magnitude of B_0 or by decreasing the average temperature T_s. Because magnetic resonance imaging usually takes place at room temperature, the bulk magnetization is increased only by increasing the magnitude of the bias field B_0.

The number N_s of hydrogen nuclei is huge. Even in a microscopic volume of the form $\Delta x \Delta y \Delta z$ centered at (x, y, z), the number $N_s(x, y, z) \Delta x \Delta y \Delta z$ of hydrogen nuclei remains huge and represented as $\rho(x, y, z) \Delta x \Delta y \Delta z$. Thus, replacing N_s by $\rho(x, y, z) \Delta x \Delta y \Delta z$ and letting the increments become arbitrarily small, the above expression is written in terms of the density functions as

$$M_z(x, y, z) = \frac{\gamma^2 \hbar^2 B_0}{4 k_B T_s} \rho(x, y, z),$$

where $\rho(x, y, z)$ is the space-varying density of hydrogen nuclei and $M(x, y, z)$ is the space-varying density of the magnetization density vector. Thus, an image of the z component $M_z(x, y, z)$ of the vector $M(x, y, z)$ is an image of $\rho(x, y, z)$, where $\rho(x, y, z)$ is the density of the hydrogen nuclei.

Signal Generation

Creation of a nonzero macroscopic magnetization vector M – or, more to the point, a macroscopic magnetization density function $M(x, y, z)$ – through the effect of the bias field B_0 is only the first step in the generation of the resonance signal. The bias field alone creates a macroscopic magnetization density vector $M(x, y, z)$ only along the z direction because the spins are not synchronized and the microscopic transverse magnetic field components of the many nuclear spins average to zero. The next step is to modulate the magnetic field.

Magnetic Resonance

Although the precessing magnetic vector of each individual hydrogen nuclei has a transverse component, there are a huge number of such protons and, under the bias field alone, their precessions are not naturally synchronized, which means that the net transverse magnetic field is essentially zero. The next step is to tip the macroscopic magnetization density $M(x, y, z)$ away from the B_0 field by bringing the precession of the microscopic nuclei into synchronization. This makes the macroscopic magnetization density $M(x, y, z)$ precess about the bias field B_0 as shown symbolically in

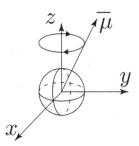

Figure 6.11 Proton precession

Figure 6.11. To bring these many precessing nuclei into synchronization, the transverse excitation field B_1 is rotated in the x,y plane at the Larmor frequency.[8] The *macroscopic* vector is tipped by synchronizing the *microscopic* precessing spin vectors that produce the microscopic magnetizations of the individual nuclei. The resulting macroscopic magnetic field is precessing so it is a time-varying magnetic field. This phenomenon is called *magnetic resonance*. The precessing protons in resonance then produce a net magnetization vector field whose x and y components are time-varying at the Larmor frequency, which is normally a radio frequency in the UHF band. The precessing time-varying field will induce electrical signals in the receiver coils surrounding the object

Excitation Field

Each individual nucleus has a spin that is tipped away from the z axis by the angle $\theta = \pm 54.73$ degrees. As long as the excitation field is turned off, the directions of the transverse components in the x,y plane are random and average to zero. When turned on, the time-varying excitation field B_1 brings the transverse components of the many nuclei into resonance. The direction of the B_1 excitation field rotates in the x,y plane at a rotation rate equal to the Larmor frequency of the bias field. This excitation pulse, rotating at the Larmor frequency, is applied for a brief period of time – on the order of microseconds or milliseconds – which means that the magnetic pulse contains a range of frequencies centered on the Larmor frequency.[9]

The excitation field B_1 usually consists of two orthogonal components in the x,y plane. These transverse components are $B_{1,x}(t) = B_1^e(t)\cos 2\pi f_0 t$ and $B_{1,y}(t) = B_1^e(t)\sin 2\pi f_0 t$ for $0 \leq t \leq \tau_p$, where $f_0 = \gamma B_0$ is the Larmor frequency of the

---

[8] As modeled, the material is transparent to the magnetic field and time variations occur everywhere instantaneously.

[9] The usual Larmor wavelengths are longer than a meter and the sensing coils are much closer than a meter to the source of the magnetic fields. Because the dimensions of a magnetic resonance imaging system are smaller than the wavelength of the radio frequencies in use, propagation delay is insignificant and need not be recognized as such. When the magnetic field changes, it changes everywhere instantly. Therefore, the radio frequency pulse is regarded simply as a pulse of a time-varying magnetization. The time-varying magnetic fields from the precessing nuclei are coupled directly into the sensing coils of the system electronics.

bias field, τ_p is the pulse duration, and $B_1^e(t)$ is the pulse envelope.[10] An idealized possibility for the pulse envelope $B_1^e(t)$ is the rect function $\text{rect}(t/\tau_p)$. The rotating $\boldsymbol{B}_1$ field brings the transverse components of the precessing nuclei of the target object into resonance. The rotation frequency f_0 of the excitation field is set equal the Larmor frequency of the bias field, but the short pulse duration means that the Fourier spectrum of the pulse is wide.

The excitation field is small compared to the bias field and is orthogonal to it. The magnitude of the sum of the two fields is $\sqrt{B_0^2 + B_1^2} \approx B_0 + \frac{1}{2}B_1^2/B_0$, which is approximately the same as the magnitude B_0 of the bias field, but a small adjustment in the value of the Larmor frequency f_0 may be considered.

In the absence of the gradient field, all protons would have the Larmor frequency f_0 of the bias field $\boldsymbol{B}_0$. Rotating the transverse components of the magnetic field at the frequency f_0 causes all protons to synchronize and so to produce a macroscopic magnetic field rotating in the x,y plane at the carrier frequency f_0. As a function of time, the excitation magnetic field takes the form of a periodic pulse train. Each pulse of the excitation pulse train is to be followed by a pulse of the gradient field. Each pulse of the gradient field is individually modulated so as to result in a different projection.

Slice Selection

The spin density $\rho(x,y,z)$ is three-dimensional but the usual method of projection tomography produces two-dimensional images. To this purpose, the three-dimensional spin density is rendered into a set of L two-dimensional slices $\rho_{z_\ell}(x,y) = \rho(x,y,z_\ell)$ in image space for $\ell = 1, \ldots, L$. An image is then formed for the two-dimensional slice at $z = z_\ell$ for each ℓ. Slice selection is described herein. The discussion that follows after the method of slice selection is limited to the imaging of a two-dimensional slice.

To provide selective excitation for a slice along the z direction (i.e., generation of signals only from a slice of an object, say a slice at $z = z_\ell \pm \delta z$), one simply offsets the Larmor frequency for only that slice and uses the offset Larmor frequency as the rotation rate of the excitation field. To this purpose, the z component of the gradient field is nonzero and equal to $B_{g,z}(z-z_\ell)$ in the vicinity of $z = z_\ell$. The Larmor frequency in this slice is $f_0 = \gamma B_0 + \gamma B_{g,z}(z - z_\ell)$. The purpose of the z dependence of the term $B_{g,z}(z - z_\ell)$ is to confine the term to a slice of width δz such as could be done ideally by choosing $B_{g,z}(z) = B\text{rect}(z/\delta z)$. Then the two-dimensional slice is

$$\rho_{z_\ell}(x,y) = \int_{|z-z_\ell|<\delta z/2} \rho(x,y,z)\,dz.$$

The passband carrier frequency f_0 is redefined as $f_0 = \gamma \left(B_0 + B_{g,z}(z-z_\ell)\right)$ and this is the passband rate of the excitation field. This means that the electrical signal received from this slice and only this slice will be at the redefined Larmor frequency f_0. Protons not in this slice have a different Larmor frequency and do not respond well to an excitation

---

[10] In vector notation, $\boldsymbol{B}_1(t) = B_{1,x}(t)\boldsymbol{i}_x + B_{1,y}(t)\boldsymbol{i}_y + B_{1,z}(t)\boldsymbol{i}_z$, where $B_{1,z}$ is typically zero in many systems. The transverse field component $\boldsymbol{B}_{1,xy}(t) = B_{1,x}(t)\boldsymbol{i}_x + B_{1,y}(t)\boldsymbol{i}_y$ may be represented in complex notation as $\boldsymbol{B}_1(t) = B_{1,x}(t) + jB_{1,y}(t)$.

field rotating at a frequency f_0. To the extent that they do respond, the response is not at the carrier frequency f_0.

With the slice now selected, the z integration becomes implicit and the imaging process is two-dimensional. The magnetization density is now two-dimensional given by

$$M_z(x,y) = \frac{\gamma^2 \hbar^2 B_0}{4 k_B T_s} \rho(x,y),$$

where the integration over z is implicit and is not mentioned again.

Gradient Field

The total external magnetic field is written in vector form as $\boldsymbol{B}(x,y,z) = \boldsymbol{B}_0 + \boldsymbol{B}_1(x,y,t) + \boldsymbol{B}_g(x,y,t)$, where the bias field $\boldsymbol{B}_0 = B_0 \boldsymbol{i}_z$ is a constant field lying in the z direction and the gradient field is a spatially varying and temporally varying magnetic field much smaller than the bias field. For the basic case discussed here,[11] the expression for the gradient field vector is separated as $\boldsymbol{B}_g(x,y,z,t) = B_{g,z}(x,y) B^e_{g,z}(t) \boldsymbol{i}_z$. Because the gradient field and the excitation field are both small, the magnitude $|\boldsymbol{B}(x,y)|$ of the total magnetic field is closely approximated by the z component of the total magnetic field. Considering only the effect of the z component of the gradient field on the bias field, the total external magnetic field is $(B_0 + B_{g,z}(x,y,t)) \boldsymbol{i}_z$, which means that the space-varying Larmor frequency in this approximation can be taken as

$$f(x,y,z) = \gamma B_0 + \gamma B_{g,z}(x,y,t).$$

The leading term $f_0 = \gamma B_0$ plays the role of a carrier frequency for the received electrical signal. The term $\gamma B_{g,z}(x,y,t)$ provides the data-dependent modulation as a function of spatial coordinates. For each gradient pulse, the term $B_{g,z}(x,y)$ in the z direction is spatially varying as permitted by the design of the secondary magnets.

Image Formation

Spatial information is embedded into a magnetic resonance signal by the spatially varying gradient field $\boldsymbol{B}_g$ in any of several ways. Each of these ways uses a series of excitation pulses with each excitation pulse followed by a gradient pulse. The gradient pulse spatially modulates the magnetic field across the x,y plane of the selected slice so that the Larmor frequency varies across the x,y plane. When stimulated by the excitation pulse, the induced transverse magnetic field at each (x,y) point rotates at the local Larmor frequency of that (x,y) point. The distribution of frequencies is measured

---

[11] A fully general form of the gradient field would be

$$\boldsymbol{B}_g(x,y,z,t) = B_{g,x}(x,y,z,t) \boldsymbol{i}_x + B_{g,y}(x,y,z,t) \boldsymbol{i}_y + B_{g,z}(x,y,z,t) \boldsymbol{i}_z.$$

The three spatial variables x, y, and z appear twice in this expression, once to specify vector components and once to specify a spatial point. The exemplar instance considered here does not require this generality.

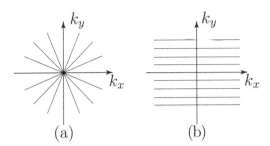

Figure 6.12 Two formats for filling frequency space: (a) polar sampling; (b) rectangular sampling

by the electrical sensing coils. The measured data for each pulse of the gradient field is then the temporal frequency distribution of the sensed electrical signal.

The frequency distribution of a detected pulse is a one-dimensional signal. That one-dimensional signal is an observation of a two-dimensional image. This is why multiple gradient pulses must be used with each pulse making a different one-dimensional observation resulting in the values of $R(f_x, f_y)$ projected along a different line in frequency space.

The detected response to each gradient pulse is used in turn, pulse by pulse, to gradually fill components of $R(f_x, f_y)$ in frequency space or k space. Each gradient pulse consists of a gradient field that is spatially varied in a different way so that each pulse results in a different set of values of $R(f_x, f_y)$ as is suggested by Figure 6.12. Each gradient pulse results in values of $R(f_x, f_y)$ along one of the lines in frequency space.

Figure 6.12 shows two basic formats in frequency space. Figure 6.1(a) shows a polar sampling format. This format has the same form as does projection tomography. The polar sampling format is the exemplar sampling format and is the sampling format described herein. Figure 6.1(b) shows a rectangular sampling format, which is a more elaborate implementation that is often preferred.

In both the polar format and the rectangular format, the k-space samples take the form of samples on a discrete set of lines described as one-dimensional slices in two-dimensional frequency space. A different one-dimensional slice is measured by each gradient pulse. Each slice of $R(f_x, f_y)$ is the Fourier transform of a projection of $\rho(x, y)$ denoted $p_\theta(t)$. Each direction of the projection of $\rho(x, y)$ is obtained by the integration of $\rho(x, y)$ along a line perpendicular to the direction of the slice. There is one such integration for each sampled point t_ℓ along the t axis of each projection $p_\theta(t)$.

Polar Sampling Format

The polar sampling format of magnetic resonance imaging takes the form of projection tomography based on the Radon transform. The usual procedure for projection tomography is to gather projection data at angle θ for all lines crossing the target $\rho(x, y)$ at angle θ. The projection $p_\theta(t)$ for each value of t is given by the integral

$$p_\theta(t) = \int_{-\infty}^{\infty} \rho(t\cos\theta - r\sin\theta, t\sin\theta + r\cos\theta)\,dr,$$

where $t = x\cos\theta + y\sin\theta$ and $r = -x\sin\theta + y\cos\theta$. The polar format of magnetic resonance imaging follows this procedure.

The Fourier transform of the projection $p_\theta(t)$ is the slice $P_\theta(f) = R(f\cos\theta, f\sin\theta)$. The measurement of a projection is repeated, in turn, for every θ in a discrete set of θ. The polar angle format of magnetic resonance imaging has this structure of projection tomography modulating the Larmor frequency distribution at angle θ to obtain the projection at angle θ.

The z component $B_{g,z}(x,y)$ of the gradient field adds to the bias field and changes the Larmor frequency as a function of x and y. To this purpose, the z component of the gradient magnetic field $\mathbf{B}_g(x,y,z,t)$ has the separated form $B_{g,z}(x,y,z,t) = B_{g,z}(x,y)B^e_{g,z}(t)$, where $B^e_{g,z}(t)$ is the time-varying pulse shape that describes each pulse. In the polar format, the spatial part of the gradient field in the z direction for each projection uses linear spatial gradients in polar coordinates of the form

$$B_{g,z}(x,y)\mathbf{i}_z = (G_{z,x}x + G_{y,z},y)\mathbf{i}_z$$
$$= G(x\cos\theta + y\sin\theta)\mathbf{i}_z$$
$$= Gt\mathbf{i}_z,$$

where G_x and G_y are the spatial rates of change of the field in the x and y directions, respectively, and $G = \sqrt{G^2_{z,x} + G^2_{z,y}}$. Each scan uses a different value of θ. Thus, for $\theta = 0$, this reduces to $B_{g,z}(x,y) = Gx$.

The gradient field $B_{g,z}(x,y)$ is the same for all points on the line defined by a constant value of $x\cos\theta + y\sin\theta$, but is different for each such constant value of $x\cos\theta + y\sin\theta$. Accordingly, for the gradient pulse at angle θ, the Larmor frequency is different for every line at angle θ. The Larmor frequency at the point (x,y) is

$$f_0 + f(x,y) = \gamma B_0 + \gamma B_{g,z}(x,y)$$
$$= f_0 + \gamma G(x\cos\theta + y\sin\theta),$$

where $f(x,y)$ is the frequency offset at the point (x,y) from the center frequency f_0 due to the gradient field. For a gradient field of the form $B_{g,z}(x,y) = Gx$, all points of the x,y plane with the same x have the same Larmor frequency.

The infinitesimal area $dxdy$ at x,y contributes

$$M_z(x,y)dxdy = \frac{\gamma^2\hbar^2 B_0}{4k_B T_s}\rho(x,y)dxdy$$

to the macroscopic field. This means that the temporal frequency spectrum $S(f)$ of the sensed magnetic field when $\theta = 0$ satisfies

$$S(f) = S(\gamma G(x\cos\theta - y\sin\theta)) = S(\gamma G(x))$$
$$\sim \int M_z(x,y)dx$$
$$\sim \int \frac{\gamma^2\hbar^2 B_0}{4k_B T_s}\rho(x,y)dx.$$

This is the projection $p_\theta(y)$ at angle $\theta = 0$. A Fourier transform $P_\theta(f)$ of $p_\theta(y)$ then gives the values of the Fourier transform $R(f_x, f_y)$ of $\rho(x, y)$ along the vertical line in k space as shown in Figure 6.12.

A simple rotation of the coordinate system shows that a similar process with the gradient at angle θ gives the values of $R(f_x, f_y)$ along the line at angle θ.

Thus, when collecting data at angle θ for each parallel line at angle θ, the gradient field induces all target nuclei on that line at angle θ to resonate at a frequency that is unique to that line. By making the magnitude of the magnetic field unique along each line at angle θ, the Larmor frequency is unique to each such line. The magnitude of the magnetic field on each line is made unique by means of the gradient magnetic field that is linearly varying in space in the direction $90°$ from θ.

The measured signal strength at each unique Larmor frequency is proportional to the integral of $\rho(x, y)$ along the corresponding line at angle θ. Each such integral along a line at angle θ has the form of a projection. The set of all projections at all angles θ is then converted to an image of $\rho(x, y)$ by one of the usual algorithms of projection tomography. Any algorithm supporting the projection-slice theorem can be used to form an image from the resulting data.

The Bloch Equation

A more extensive discussion of magnetic resonance imaging would require a full discussion of the Bloch equation. A detailed discussion of the Bloch equation is not given herein. Only a brief mention of these topics is given to close the section.

The excitation field causes a pulse rotation of the external field that synchronizes the radio-frequency precessions of the nuclei spin axes. After the excitation pulse ends, the transverse components of the magnetization vector $\mathbf{M}$ will desynchronize and relax to the thermal equilibrium state. This relaxation is characterized by a precession of the spin vectors about the $\mathbf{B}_0$ field called *free precession* consisting of a recovery of the longitudinal magnetization M_z called *longitudinal relaxation* and the decay of the transverse bulk magnetization M_{xy} called *transverse relaxation*.

The free precession and relaxation processes of $\mathbf{M}$ are described by the *Bloch equation*, which is

$$\frac{d\mathbf{M}}{dt} = \gamma \mathbf{M} \times \mathbf{B}_0 - \frac{M_x \mathbf{i}_x + M_y \mathbf{i}_y}{T_2} - \frac{(M_z - M_z^0)\mathbf{i}_z}{T_1},$$

where T_1 and T_2 are parameters that depend on the local properties of the material and where M_z^0 is the thermal equilibrium value of M. Expressing the transverse magnetization in complex notation as $M_{xy} = M_x + jM_y$ and the post-pulse condition as $M_{xy} = M_{xy}(0)$, $M_z = M_z(0)$, and $\mathbf{B} = B_0 \mathbf{i}_z$, the solution to the Bloch equation is given by the pair of equations

$$M_{xy}(t) = M_{xy}(0)e^{-t/T_2}e^{-j2\pi f_0 t},$$
$$M_z(t) = M_z^0 \left(1 - e^{-t/T_1}\right) + M_z(0)e^{-t/T_1}.$$

With this model, the longitudinal component grows exponentially with the time constant T_1, while the transverse component precesses at the Larmo frequency about the

B_0 field and decays exponentially with time constant T_2. The precessing components induce electrical signals in the receiver coils surrounding the object being imaged.

6.5 Image Formation from Nuclear Decay Emissions

Emission tomography depends on radiation that is emitted by an object of interest. The previous section describes a medical imaging modality in which that radiation is induced by magnetic fields. This section describes a medical imaging modality using gamma-ray radiation induced by nuclear decay events. Depending on the kind of decay, the source emits either gamma photons or positrons. In the second case, each positron is immediately converted to a pair of gamma photons. In either case, the source of the observed radiation is a two-dimensional poisson process whose two-dimensional emission rate $\lambda(x,y)$ is proportional to the concentration of the radionuclide $\rho(x,y)$.

Emission tomography from decay events is based on the detection of a large number of point events resulting from the arrival of particles produced by nuclear decay.[12] To produce these decay particles, the target object must be infused with a suitable radionuclide as a source of such radiation. One may use either a radionuclide that emits a single photon or a radionuclide that emits a positron. The first case is called *photon-emission tomography*; the second case is called *positron-emission tomography*.

Positron-emission tomography is an alternative to X-ray tomography as an imaging modality for medical applications. Projection tomography using X-ray absorption to form images of absorption density gives an image primarily of structure. In contrast, projection tomography using positron emission can be used to form images of biological activity such as growth by imaging any tissue or substance that attracts the chosen tracer to which that radionuclide is attached. To this point, a malignant tissue attracts nutrients because it is growing, so it is highlighted in positron-emission tomography with a suitable tracer. The nutrient that the tissue or malignancy is known to attract is artificially tagged with a chosen radionuclide that emits photons or positrons.

The computational task of both photon-emission tomography and positron-emission tomography is to form an estimate of the emission density $\rho(x,y)$ when given a large file of detected photons or photon pairs, each is described by the relevant detection data. Both photon-emission events and positron-emission events are described by a poisson emission rate $\lambda(x,y)$. The emission rate of each is proportional to $\rho(x,y)$ but the data that is detected is different. For single photon emission, the data consists of the direction of arrival of each arriving photon. For positron emission, each positron immediately seeks an electron and converts to a pair of photons. The data for such a pair consists of both the joint line of travel and the differential time of arrival provided events are sparse enough that photon pairs can be recognized as such. Positron-emission data is much richer than single-photon data.

---

[12] In Section 7.8, we study imaging in weak light as a different instance of imaging from a large number of point events.

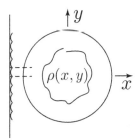

Figure 6.13 Single photon detection

Single Photon Emission

Single photon emission is the less informative kind of emission because for a single photon, only the direction of arrival at the sensor is the available data. The usual single-photon sources emit radiation in the gamma-ray region. The advantage of single-photon tomography is that a larger variety of radioactive sources are available to use as tracers for single photon-emission tomography. For photon-emission events, the detected data has the general form of projections, with little spread in the rays. This is because the viewing angle of each gamma-ray photocell is confined to a narrow angle because the sensor beams at gamma-ray frequencies easily satisfy the geometrical optics approximation.

The detector may be a single one-dimensional array of photosensors that collects projection data from $\rho(x, y)$ for a single viewing angle at a time. Figure 6.13 shows the array of detection gamma-ray photocells arranged on a straight line in the y direction with each photocell staring in the x direction along a line at the angle θ equal to zero. Each photocell detects the gamma-ray photons within its beam. The rays of adjacent sensor beams may overlap slightly. Only photons traveling in the negative x direction are captured. All other photons are lost. The sensor array or the target is then rotated to capture data at the next value of θ. This is repeated to obtain projection data for all θ.

For columnated data detected by nonoverlapping sensor beams, the detected data consists of projections and can be processed by the algorithms of projection tomography. Otherwise, for sensors that are not well-columnated, the sensor beams overlap and the projections are not well-defined. Let (x, y) denote the location of a gamma detector. Let $p(x, y|u, v)$ be the probability that a gamma photon is detected at the location (x, y) given that a photon is emitted at the point (u, v). At the photodetector, the density of arrival events at the point (x, y) is

$$\mu(x, y) = \int_{-\infty}^{\infty} \int_{-\infty}^{\infty} p(x, y \mid u, v) \lambda(u, v) du dv + n(x, y),$$

where $n(x, y)$ is an independent particle noise process at sensor (x, y). The maximum-likelihood image is then

$$\widehat{\lambda}(x, y) = \underset{\lambda(x,y)}{\operatorname{argmax}} \left[\int_{-\infty}^{\infty} \int_{-\infty}^{\infty} p(x, y|u, v) \lambda(u, v) du dv + n(x, y) \right].$$

6.5 Image Formation from Nuclear Decay Emissions

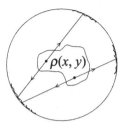

Figure 6.14 Positron-emission tomography

This expression apparently does not have an analytic solution. The argmax can be determined numerically by using an iterated algorithm of the kind described in Chapter 8. Because it is desirable to minimize the amount of radionuclide infusing the target object, the emitted signal is weak. This means that a long observation interval may be required, and the target may not remain stationary during the observation time. Motion of the target will degrade the image. It may be necessary to include motion compensation in the imaging algorithm such as is described in Section 8.8.

Positron Emission

Positron emission is the more informative kind of emission because each positron immediately interacts with an electron to create a pair of photons. The fact that these secondary photons occur in pairs traveling in equal but opposite directions provides considerably more information from each decay event.

Positron-emission tomography images the distribution of positron sources indirectly by imaging the density of photon-pair productions created by positron decay.[13] This equivalence is valid because a photon pair is created quite close to the emission of the positron. Positron density $\rho(x, y, z)$ means that $\rho(x, y, z)dxdydz$ is proportional to the probability density that the infinitesimal volume $dxdydz$ centered at the point (x, y, z) emits a positron in a unit time interval. This probability is proportional to the density of radionuclide molecules. Each positron is almost immediately annihilated by an electron, thereby producing two photons that are required by the laws of conservation of energy and momentum to travel in opposite directions. The two photons are detected by an array of photodetectors surrounding the targeted object as shown for two photon pairs in Figure 6.14.

The figure shows two independent positrons emitted from separate points within the object $\rho(x, y)$. Every such positron is immediately annihilated by combining with an electron thereby producing two photons traveling in opposite directions. Every pair of photons is parameterized by the line of travel, which is the line connecting the two detectors sensing the two photons, and by the differential time of arrival, which locates the emission event on that line. In this way, each decay event is located in the

---

[13] Under nonrelativistic velocities for the colliding particles, the laws of conservation of energy and momentum require the photon wavelength to be 1.21324 picometers, which is a wavelength in the gamma-ray band.

x, y plane. The rate of positron emission must be low enough so that photon pairs can be recognized as such.

Target motion at the time of data collection blurs the image. Particle differencing, as described in Section 8.8, is a method to remove the effect of target motion by recognizing that particles detected consecutively are effected by nearly the same offset.

6.6 Diffraction Tomography

The methods of projection tomography discussed earlier are based on the premise that geometrical optics is an adequate model of wave propagation. This premise does hold in many common situations in which the propagation medium affects the illuminating signal only by the integrated attenuation along the rays. There are other situations in which this premise does not hold. Then a more complete description of the interaction must be used and other forms of tomography, such as *diffraction tomography* or *diffusion tomography*, may be appropriate. Diffraction tomography is studied in this section. Diffusion tomography is studied in the next section.

Diffraction tomography is an alternative to projection tomography that is needed for wavelengths at which the geometrical optics approximation is not valid. In such a case, diffraction cannot be ignored. Applications[14] for which diffraction could be significant include seismic waves, ultrasound, and microwaves. These are applications in which observed details are comparable to or smaller than the wavelength. In such a case, diffraction must be considered. The distinction between diffraction tomography and projection tomography is that a deconvolution of the diffraction point-spread function must somehow be embedded into the back-projection algorithm. The theory of back-projection is modified to include an appropriate reformulation of the projection-slice theorem.

In projection tomography, the projection-slice theorem equates each projection to a slice of the Fourier transform. This equivalence is appropriate under the geometrical optics approximation. When the projection is perturbed by a small amount of diffraction, then it becomes evident that the slice must also be perturbed in some way. First recall the analysis of projection in the absence of diffraction and with only attenuation, as was given in Section 6.1. That analysis is reviewed here, and is then revised to include diffraction. The wave propagation is in the z direction, which means that the wave amplitude can be written as a function of both x and y at each z. The attenuation of a line traversing a small interval of length Δz centered at z_1 is $s(z_1)\Delta z$, where $s(z)$ is the attenuation density at z. The relationship between the input intensity I and the output intensity I' of the line passing through this small interval is

$$I' = I[1 - s(z_1)\Delta z].$$

---

[14] Only an ideal propagation medium is considered here, which is unrealistic for many such applications.

6.6 Diffraction Tomography

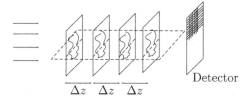

Figure 6.15 Multiple attenuating screens

More generally, the attenuation depends on x and y. Suppose that a wave is passing through a thin attenuating screen in the x, y plane at z_1 with intensity attenuation $\rho(x, y, z_1)\Delta z_1$. Then

$$I_{out}(x,y) = I_{in}(x,y)[1 - \rho(x,y,z_1)\Delta z]$$
$$\approx I_{in}(x,y)e^{-\rho(x,y,z_i)\Delta z}.$$

where the second line approximates the bracketed term of the first line by an exponential. Then, after traversing many such attenuating screens, the screens separated by intervals of length Δz, the full attenuation is the product of the exponential terms leading to

$$I_{out}(x,y) \approx I_{in}(x,y)e^{-\sum_i \rho(x,y,z_i)\Delta z}.$$

In the limit as Δz goes to zero, the projection in the z direction is

$$p(x,y) = \log \frac{I_{in}(x,y)}{I_{out}(x,y)}$$
$$= \int_{-\infty}^{\infty} \rho(x,y,z)\mathrm{d}z.$$

This equation is the basis of projection tomography.

This analysis based on geometrical optics must be modified to describe the case of diffraction with the Born approximation retained. The first change is that the input illumination is now the amplitude signal $a_{in}(x,y)$, rather than the intensity. The medium weakly attenuates the amplitude signal as described by an attenuation function $\rho(x,y,z)$. The input illumination signal is here required to be a constant a_{in} independent of x and y.

We regard the medium $\rho(x, y, z)$ to be approximated by a sequence of thin screens as shown in Figure 6.15. The screen at z_i has amplitude attenuation $\rho(x, y, z_i)\Delta z$, where Δz is the separation between screens. First consider only a single screen at position z_1. The wave reaching this screen is phase-delayed by $2\pi z_1/\lambda$. After the wave passes through this thin attenuating screen at position z_1, it is diffracted. The signal reaching the x, y plane at position d is

$$a_{out}(x,y) = a_{in}e^{j2\pi z_1/\lambda}[1 - \rho(x,y,z_1)\Delta z] ** h_{d-z_1}(x,y),$$

where $h_d(x, y)$ is the Huygens–Fresnel point-spread function of free space for distance d. The complex exponential on the left of the right side accounts for the propagation phase delay prior to the screen. Because a uniform plane wave does not diffract, the diffraction does not change the first term other than by a phase shift by $(d - z_1)/\lambda$. Next, by distributing the term $h_{d-z_1}(x,y)$, this expression can be written as

$$a_{out}(x,y) = a_{in} e^{j2\pi z_1/\lambda}\left[e^{j2\pi(d-z_1)/\lambda} - \rho(x,y,z_1)\Delta z **h_{d-z_1}(x,y)\right]$$
$$= a_{in} e^{j2\pi d/\lambda}\left[1 - e^{-j2\pi(d-z_1)/\lambda}\rho(x,y,z_1)\Delta z **h_{d-z_1}(x,y)\right].$$

With two phase screens, one screen at z_1 and a second screen at z_2, the illumination launches a diffracting wave from each screen. There is one diffracting term from each phase screen. Because the Born approximation has been asserted, the same illuminating plane wave is incident on each screen and the responses of the two screens interact only in second-order terms that are neglected. Thus,

$$a_{out}(x,y) \approx a_{in} e^{j2\pi d/\lambda}\Big[1 - e^{-j2\pi(d-z_1)/\lambda}\rho(x,y,z_1)\Delta z **h_{d-z_1}(x,y)$$
$$-e^{-j2\pi(d-z_2)/\lambda}\rho(x,y,z_2)\Delta z **h_{d-z_2}(x,y)\Big].$$

Generalizing to a sequence of screens at positions z_i, indexed by i, the approximation is written

$$a_{out}(x,y) \approx a_{in} e^{j2\pi d/\lambda}\left[1 - \sum_i \left(\rho(x,y,z_i)\Delta z **e^{-j2\pi(d-z_i)/\lambda}h_{d-z_i}(x,y)\right)\right].$$

Under the Born approximation, the cumulative loss of signal continues to be small as the wave passes through the sequence of screens. In the limit as Δz goes to zero and under the condition of a weakly attenuating medium that admits the Born approximation, the diffracted signal in the plane $z = d$ is

$$\frac{a_{out}(x,y)}{a_{in}} = e^{j2\pi d/\lambda}\left[1 - \int_{-\infty}^{\infty}\rho(x,y,z) **e^{-j2\pi(d-z)/\lambda}h(x,y,d-z)dz\right]$$

with $h(x,y,z) = h_z(x,y)$. The first term is of no interest and is ignored. The double asterisk in the second term refers to a two-dimensional convolution in the x,y plane and the integration is a convolution in the z direction. Therefore, the diffracted projection in the x,y plane at any z is defined as the attenuated and diffracted signal

$$p(x,y,z) = e^{j2\pi z/\lambda}\left[\rho(x,y,z) ***e^{-j2\pi z/\lambda}h(x,y,z)\right],$$

where $h(x,y,z) = h_z(x,y)$ is the Huygens–Fresnel point-spread function of free space. The expression now has the form of a three-dimensional convolution. A simple consequence of the three-dimensional convolution theorem is that the frequency-domain representation of a projection is

$$P(f_x,f_y,f_z + \lambda^{-1}) = R(f_x,f_y,f_z)H(f_x,f_y,f_z + \lambda^{-1})$$

or

$$P(f_x,f_y,f_z) = R(f_x,f_y,f_z - \lambda^{-1})H(f_x,f_y,f_z)$$

in the frequency domain.

The apparent symmetry of this three-dimensional convolution formula is misleading because that symmetry may make it seem that the direction of the wave propagation effects the equation only by the offset by λ^{-1}. However, the diffraction point-spread function $h(x,y,z)$ has a much different dependence on z than on the other two variables. Moreover, the formula is not exact; approximations were made in the derivation.

To verify that this equation reduces to the right form in the absence of diffraction, recall that the point-spread function in the absence of diffraction is given by the geometrical optics approximation

$$h(x,y,z) = e^{j2\pi z/\lambda}\delta(z).$$

Therefore

$$p(x,y) = \int_{-\infty}^{\infty} \rho(x,y,z)dz.$$

Thus, in the absence of diffraction, the statement reduces to the amplitude form of projection tomography in three dimensions, as it must.

Next, to describe the formulation of diffraction tomography in a form parallel to that of projection tomography in two dimensions, the y coordinate is suppressed from the situation and from the notation. This amounts to the condition that $\rho(x,y,z)$ is independent of y. Then the diffracted projection $p(x)$ at distance d is the convolution

$$p(x) = e^{j2\pi d/\lambda}\int_{-\infty}^{\infty}\int_{-\infty}^{\infty}\rho(\xi,\eta)e^{-j2\pi(d-\eta)}h(x-\xi,d-\eta)d\xi d\eta.$$

when the angle θ is equal to zero.

The discussion now ready to give a version of the projection-slice theorem for a diffracted projection. The following theorem gives the formula for the case where θ equals zero. The general case is obtained afterward by a coordinate rotation and is stated as a corollary.

Theorem 6.6.1 (Diffracting Projection-Slice Theorem) *The Fourier transform of a diffracted projection along the x axis is*

$$P(f) = R\left(f,\sqrt{\lambda^{-2}-f^2}+\lambda^{-1}\right)e^{j2\pi d\sqrt{\lambda^{-2}-f_x^2-f_y^2}}.$$

Proof:
The two-dimensional Huygens–Fresnel point-spread function $h(x,y,d)$ has two-dimensional Fourier transform

$$H_d(f_x,f_y) = e^{j2\pi d\sqrt{\lambda^{-2}-f_x^2-f_y^2}}.$$

This holds with d replaced by an arbitrary z. Define the three-dimensional function $h(x,y,z)$ as $h_z(x,y)$. This function has the three-dimensional Fourier transform

$$H(f_x,f_y,f_z) = \int_{-\infty}^{\infty} e^{j2\pi z\sqrt{\lambda^{-2}-f_x^2-f_y^2}}e^{-j2\pi f_z z}dz$$

$$= \delta\left(f_z - \sqrt{\lambda^{-2}-f_x^2-f_y^2}\right).$$

Therefore

$$P(f_x,f_y,f_z) = R(f_x,f_y,f_z - \lambda^{-1})\delta\left(f_z - \sqrt{\lambda^{-2} - f_x^2 - f_y^2}\right).$$

The two-dimensional projection $p(x,y)$ in the plane at $z = d$ has the two-dimensional Fourier transform $P(f_x,f_y)$ given by

$$\begin{aligned}P(f_x,f_y) &= P(f_x,f_y,d) \\ &= \left.\int_{-\infty}^{\infty} P(f_x,f_y,f_z)e^{j2\pi f_z z}df_z\right|_{z=d} \\ &= \int_{-\infty}^{\infty} S(f_x,f_y,f_z - \lambda^{-1})\delta\left(f_z - \sqrt{\lambda^{-2} - f_x^2 - f_y^2}\right)e^{j2\pi f_z z}df_z \\ &= S\left(f_x,f_y,\sqrt{\lambda^{-2} - f_x^2 - f_y^2} - \lambda^{-1}\right)e^{j2\pi d\sqrt{\lambda^{-2}-f_x^2-f_y^2}}.\end{aligned}$$

To restrict the treatment to a projection in two dimensions under the condition that $\rho(x,y,z)$ is independent of y, set $f_x = f$ and drop f_y from the notation. Then

$$P(f) = R\left(f,\sqrt{\lambda^{-2} - f^2} - \lambda^{-1}\right)e^{j2\pi d\sqrt{\lambda^{-2}-f^2}}.$$

The projection $p(x)$ has a Fourier transform $P(f)$ equal to $R\left(f,\sqrt{\lambda^{-2} - f^2} - \lambda^{-1}\right)$ multiplied by a phase term as in the statement of the theorem. ∎

The Fourier transform $P(f)$ is a slice of $R(f_x,f_z)$ along the curve $f_z = \sqrt{\lambda^{-2} - f_x^2} - \lambda^{-1}$. Rewritten in the standard form of a circle, this circle is $(\lambda f_z + 1)^2 + \lambda^2 f_x^2 = 1$. Notice that for small λ, the expression $\lambda^{-1}\sqrt{1 - \lambda^2 f^2} - \lambda^{-1}$ is approximately λf^2. By neglecting this term, $|P(f)|$ becomes $|R(f,0)|$. Thus, within this approximation, the diffracting projection-slice theorem reduces to the situation of attenuation without diffraction.

Corollary 6.6.2 *The Fourier transform of a diffracted projection $p_\theta(t)$ at angle θ is*

$$P_\theta(f) = R(f\cos\theta - F\sin\theta, f\sin\theta + F\cos\theta)e^{j\psi(\cdot)},$$

where

$$F = \sqrt{\lambda^{-2} - f^2} - \lambda^{-1}$$

for a function $\psi(\cdot)$ with an argument similar to the argument of $R(\cdot)$.

Proof:
This follows from the theorem as an immediate consequence of the rotation property of the Fourier transform. Simply rotate the coordinate system by angle θ. ∎

The theorem states that because of diffraction, the "slice" of $R(f_x,f_z)$ now takes place on the circle

$$\lambda^2 f_x^2 + (\lambda f_z - 1)^2 = 1 \qquad f_z \geq 0.$$

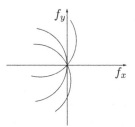

Figure 6.16 Diffracted slices

This is the two-dimensional version of an Ewald sphere, defined and studied in three-dimensions in Section 9.5 of Chapter 9. Because f_x must be positive, this is the equation of a half circle in the f_x, f_z plane, as shown in Figure 6.16. The half-circle extends to a point $\sqrt{2}/\lambda$ from the origin. By varying the angle of the projection, as described in the corollary, this semicircular slice can be reoriented. In this way, values of the two-dimensional Fourier transform that can be measured are those for which the point (f_x, f_z) is interior to the circle

$$f_x^2 + f_z^2 = 2\lambda^{-2}.$$

When $R(f_x, f_y)$ has support inside this circle, the image $\rho(x, y)$ can be recovered, in principle, as the inverse two-dimensional Fourier transform of $R(f_x, f_y)$, which can itself be recovered, diffracted slice by diffracted slice.

When $R(f_x, f_y)$ has support that extends outside of the observable circle in the frequency plane, the function $R(f_x, f_y)\text{circ}(\lambda f_x/\sqrt{2}, \lambda f_x/\sqrt{2})$ can be measured only for those (f_x, f_y) inside the circle. When the interior of the circle is fully observed, the image $\rho(x, y)$ is estimated as the inverse Fourier transform of the observed portion of $R(f_x, f_y)$ given by $R(f_x, f_y)\text{circ}(\lambda f_x/\sqrt{2}, \lambda f_x/\sqrt{2})$. Then what is actually computed is the function $\rho(x, y)$ blurred by a jinc function.

6.7 Diffusion Tomography

A wave can be scattered, rescattered, then rescattered again. The extreme situation in which the wave passes through a continuum of infinitesimal scattering events is called *diffusion*. For example, near infrared electromagnetic radiation at a wavelength of about 850 nanometers interacts with human tissue by diffusion. A bright beam at this wavelength entered into one side of a thickness of such a diffusing material[15] will emerge as a diffuse glow at the other side that is visible in a dark room. A diffused wavefront can be detected. A collection of such diffuse projections at various angles can be processed to obtain an image of the scattering medium, although with great difficulty and of limited quality. Inversion of the diffusion process to form a crude image of the diffusion medium is called *diffusion tomography*. Though diffusion tomography

---

[15] Such as a hand or a head.

is limited by severe sensitivity and resolution problems, it may be important in some medical applications.

The theory of diffusion tomography is much more difficult to formulate than is the theory of diffraction tomography. The diffusion mechanism can be motivated by reference to the same arrangement of multiple screens as shown in Figure 6.15. In the case of diffusion, however, the screens are not attenuating screens; they are scattering screens. Every infinitesimal region of each screen multiplies the incoming wave by a phase term and scatters it both in the forward and in the backward directions. Eventually a propagating wave will be incident on both sides of each screen, and will be scattered with random phase shifts in all directions. To analyze this situation in the limit as the number of screens goes to infinity and the spacing goes to zero is quite difficult.

A different model of diffusion can be set up in terms of rays that are randomly bent upon passing through a screen. The collection of these rays forms an intensity field that is measured by a sensor array. Again, the analysis of this model of diffusion is quite difficult. The ray paths describe a random walk analogous to the particle walks under brownian motion. The particle density under brownian motion obeys a diffusion equation.

The fundamentals of the diffusion mechanism are not described here. Without any justification, diffusion is defined as the response to the equation

$$\nabla^2 \rho(x,y,z) - k^2 \rho(x,y,z) = 0.$$

This is a time-invariant homogeneous form of the *diffusion equation*. The constant k in this context is called the *diffusion coefficient*. The value of k depends on the diffusion medium. To solve the diffusion equation, notice that except for the sign of k^2, the diffusion equation is the same as the wave equation given by

$$\nabla^2 s(x,y,z) + k^2 s(x,y,z) = 0,$$

which describes diffraction. In the wave equation, the constant k is the wave number. The wave equation corresponds to the Huygens–Fresnel point-spread function of free space

$$h_d(x,y) = \frac{1}{2\pi} \left[\frac{-jdk}{(d^2+x^2+y^2)} + \frac{d}{(d^2+x^2+y^2)^{3/2}} \right] e^{jk\sqrt{d^2+x^2+y^2}}.$$

This suggests that the point-spread function for the diffusion equation can be obtained simply by replacing k by jk. This posits that the point-spread function for diffusion is

$$h_d(x,y) = \frac{1}{2\pi} \left[\frac{dk}{(d^2+x^2+y^2)} + \frac{d}{(d^2+x^2+y^2)^{3/2}} \right] e^{-k\sqrt{d^2+x^2+y^2}}.$$

The diffusion point-spread function is a real, decaying exponential with decay constant $1/k$. Now, by using the two-dimensional Fourier transform pair

6.7 Diffusion Tomography

$$\left[\frac{b/2\pi}{(b^2+x^2+y^2)^{1/2}} + \frac{ab}{b^2+x^2+y^2}\right] e^{-2\pi a\sqrt{b^2+x^2+y^2}} \Leftrightarrow e^{-2\pi b\sqrt{a^2+f_x^2+f_y^2}}$$

we see that

$$H_d(f_x,f_y) = e^{-d\sqrt{k^2+(2\pi)^2(f_x^2+f_y^2)}}.$$

This is the two-dimensional Fourier transform of the diffusion wavefront in the x,y plane for which $z = d$. To extend it to a three-dimensional Fourier transform, replace d by z and define the function

$$H(f_x,f_y,z) = e^{-z\sqrt{k^2+(2\pi)^2(f_x^2+f_y^2)}}.$$

Now take the one-dimensional Fourier transform with respect to the single variable z by referring to the one-dimensional Fourier transform pair

$$e^{-at}, t \geq 0 \leftrightarrow \frac{1}{a-j2\pi f}.$$

Then

$$H(f_x,f_y,f_z) = \frac{1}{\sqrt{k^2+(2\pi)^2(f_x^2+f_y^2)} - j2\pi f_z}.$$

Moreover,

$$P(f_x,f_y,f_z) = R(f_x,f_y,f_z)H(f_x,f_y,f_z).$$

The two-dimensional Fourier transform, denoted $P_d(f_x,f_y)$ or $P(f_x,f_y)$, of the projection in the plane at $z = d$ is

$$P(f_x,f_y) = \int_{-\infty}^{\infty} P(f_x,f_y,f_z) e^{j2\pi f_z z} df_z \bigg|_{z=d}$$

$$= \int_{-\infty}^{\infty} R(f_x,f_y,f_z) \frac{1}{\sqrt{k^2+(2\pi)^2(f_x^2+f_y^2)} - j2\pi f_z} e^{j2\pi f_z d} df_z.$$

As was done for diffraction tomography, the formulation of diffusion tomography can be expressed in a form parallel to that of two-dimensional projection tomography in the x,z plane by eliminating the y coordinate. To do so, require that $s(x,y,z)$ is independent of y. Then suppress f_y from the situation and also from the notation to write

$$P(f) = \int_{-\infty}^{\infty} R(f,f_z) \frac{1}{\sqrt{k^2+4\pi^2 f^2} - j2\pi f_z} e^{j2\pi f_z d} df_z$$

as the Fourier transform of $p(x)$, the diffused projection onto the x axis at $z = d$. The projection, at any angle θ has Fourier transform

$$P_\theta(f) = \int_{-\infty}^{\infty} R(f\cos\theta - f_z\sin\theta, f\sin\theta + f_z\cos\theta) \frac{1}{\sqrt{k^2+4\pi^2 f^2} - j2\pi f_z} df_z$$

as a consequence of the rotation property of the Fourier transform.

6.8 Optical-Coherence Tomography

Optical-coherence tomography is an imaging modality based on coherence-gated lightwave echoes used to form high-resolution images of optical scatterers near the surface of a partially transparent object.[16] A typical application of optical-coherence tomography is the imaging of very small blood vessels within a few millimeters of the skin's surface. A resolution of several microns is practical.

Optical-coherence tomography has the unusual property that it uses an optical beam with intentionally poor spatial coherence because good spatial coherence would actually degrade the resolution. Optical-coherence tomography uses a low-coherence optical beam to illuminate the interior of a partially transparent object. The beam penetrates the object and is back-reflected from structural details within the object and emerges from the object as a reflected beam. A copy of the low-coherence optical beam is formed by splitting the original optical beam into an illumination beam and a reference beam. The reference beam is delayed (such as could be done by reflecting it from a mirror) so that it is time-synchronized with the return of the reflected beam. It may be required that the delay τ in the reference beam can be varied.

A simple modal of the medium is as a density of scatterers, $\rho(x,y,z)$, embedded in a homogeneous attenuating medium. Although the signal is attenuated by the medium, it may be appropriate to regard the scattering as weak. Then, with respect to scattering, a version of a Born approximation may be suitable. The attenuation in the illumination can be modeled explicitly as a simple function of depth and not dependent on the scattering objects. Because the signal is attenuated, the scatterers appear to be weaker with depth and so the image fades. The image can be amplified as a function of depth to compensate for the fading, although this amplifies the noise as well. The depth of satisfactory imaging is the depth at which the attenuated image becomes too close to the noise.

The optical features may be spatially fixed or moving. When some of the scattering elements are moving, the reflected signal from those elements will exhibit a doppler shift that can be detected as such. In this way, the techniques of coherence tomography can be augmented to observe motion. Blood flow in biological tissue is a good example of such an application.

The transmitted optical beam at complex baseband at $z = 0$ is

$$c(x, y, t) = e^{-j\theta(t)} A(x, y),$$

where $\theta(t)$ is a gaussian random process with zero mean and correlation function $\phi(\tau)$, and $A(x,y)$ describes the beam. To get good resolution in x and y, a spatially narrow beam is necessary. Usually the aperture is the tip of a thin tapered optical probe. To observe a range of x and y, the beam is spatially scanned by moving the aperture or by

---

[16] Despite its name, optical-coherence tomography does not use the methods of tomography. It might better be called optical-coherence imaging.

6.8 Optical-Coherence Tomography

using an array of parallel beams formed by an array of tapered probes. Because of the Born approximation, the spatial distribution, $A(x, y)$ remains constant with z.

A copy of the transmitted beam must be delayed to serve as a reference beam. The reference beam is

$$c'(x, y, t) = e^{-j\theta(t-\tau)} A(x, y).$$

The illumination after a propagation delay at depth z is

$$c(x, y, t) = e^{-j\theta(t-z/c)} A(x, y).$$

This beam is partially reflected by $\rho(x, y, z)$ and received after a second propagation delay.

The cumulative reflected beam from all reflectors as it appears at the receiver is

$$v(x, y, t) = \int_0^\infty A(x, y) \rho(x, y, z) e^{-j\theta(t-2z)} dz.$$

The reflected beam is processed by multiplying it by the reference beam, then integrating. Thus

$$r(x, y) = \int_0^T c'(x, y, t) v^*(x, y, t) dt$$

$$= \int_0^T \int_0^\infty A(x, y) \rho(x, y, z) \left[e^{j\theta(t-2z/c)} e^{-j\theta(t-\tau)} \right] dz dt.$$

The expectation can be evaluated by using Corollary 15.2.2 of Chapter 15 in the third line of the next expression to give

$$E[r(x, y)] = \int_0^T \int_0^\infty A(x, y) \rho(x, y, z) E\left[e^{j\theta(t-2z)} e^{-j\theta(t-\tau)} \right] dz dt$$

$$= \int_0^T \int_0^\infty A(x, y) \rho(x, y, z) e^{-E\left[\theta(t-z) - \theta(t-\tau)\right]^2 / 2} dz dt$$

$$= TA(x, y) \int_0^\infty \rho(x, y, z) e^{-\sigma_\theta^2 (1 - \phi(z-\tau))} dz.$$

Let

$$h(z) = e^{\sigma_\theta^2 (1 - \phi(z))}.$$

Then

$$E[r(x, y)] = A(x, y) \big[h(z) * \rho(x, y, z) \big].$$

When a single spatial sensor is used, as is the normal method, this is integrated spatially across the beam. Then

$$\rho(z) = \int_{-\infty}^\infty \int_{-\infty}^\infty A(x, y) \rho(x, y, z) dx dy,$$

and

$$\frac{1}{T} E[r(z)] = h(z) * \rho(z).$$

is the expected return at depth z within the beam of the aperture $A(x, y)$ as a function of depth z. To sense over an x, y region, the transducer can be spatially scanned as by moving it. Alternatively, a transducer can be an array of transmitting or receiving elements to cover a range of the x, y plane.

6.9 Evanescent Wave Tomography

Closely related to the mathematics of diffusion tomography is near-field imaging or evanescent-wave optical imaging. Near-field imaging is used to image features smaller than the optical wavelength of the illumination by using a wave field containing strong evanescent components. The methods of evanescent-wave optical imaging are used for near-field microscopy and are related to modern methods of optical lithography that operate beyond the Abbe resolution limit.

To generate the evanescent wave, the object is illuminated by passing light through a subwavelength aperture. The subwavelength aperture could be the tip of a tapered optical fiber. The light is passed through a fiber whose tip is tapered to less than a wavelength, and this produces an evanescent wave launched from the tip. The evanescent wave extends a few wavelengths beyond the actual tip. Because the evanescent wave decays quickly in space, the tip must be within a few wavelengths of the object being viewed. Only the region of the object surface near the tip is illuminated by the evanescent wave. To view the entire surface, the x, y plane must be scanned by moving the aperture formed by the tip of the fiber.

There is a strong analogy between diffusing waves and evanescent waves. To see the analogy is to notice that a homogeneous evanescent wave in the z direction does satisfy the wave equation, but with an imaginary wave number k instead of a real k. Replacing k by jk in the wave equation effectively changes the sign of k and becomes a time-invariant form of the diffusion equation. In the plane wave expansion formalism, the defusing waves may be seen to consist of nothing but evanescent values whereas the usual diffracting waves generally consist of both evanescent and propagating components.

6.10 Merging of Multiple Images

Tomography, as described in Section 6.1 of this chapter, reconstructs an image from a collection of its projections. Deconvolution, as described in Section 7.2, enhances the quality of a single image by suppressing the blur of a point-spread function. Midway between the topics of tomography and deconvolution is the topic of combining several noisy images, each of which has been blurred by a different point-spread function.

Let $h_\ell(x, y)$ for $\ell = 1, \ldots, L$ be a set of L point-spread functions, and let

$$p_\ell(x, y) = h_\ell(x, y) ** \rho(x, y) \qquad \ell = 1, \ldots, L.$$

for the ℓth blurred copy of $\rho(x, y)$. In the special case for which $L = 1$, this problem reduces to the problem of deconvolution. In the special case for which

$$h_\ell(x,y) = \delta(t\cos\theta_\ell - r\sin\theta_\ell) \qquad \ell = 1,\ldots,L$$

this reduces to the problem of tomography. Thus, a projection is a special form of a filtered image using the filter

$$h(x,y) = \delta(t\cos\theta - r\sin\theta).$$

Consequently, the task of merging multiple images can be seen either as a generalization of tomography or as a generalization of deconvolution.

The problem of estimating $\rho(x,y)$ is applicable, for example, to the case where several images of a common scene are available, perhaps formed from different perspectives. It may be that each image is good in one direction and poor in the other direction. It is possible to merge these several images to obtain a single improved image that combines the best features of each of the several images.

For example, two filtered images, $p_1(x,y)$ and $p_2(x,y)$, of a common scene, $\rho(x,y)$ are given by

$$p_1(x,y) = h_1(x,y) ** \rho(x,y),$$
$$p_2(x,y) = h_2(x,y) ** \rho(x,y),$$

where $h_1(x,y)$ is the point-spread function of the first filter and $h_2(x,y)$ is the point-spread function of the second filter. The task is to recover $\rho(x,y)$.

A more general task is to estimate $\rho(x,y)$ from noisy versions of the two filtered images given by

$$v_1(x,y) = h_1(x,y) ** \rho(x,y) + n_1(x,y),$$
$$v_2(x,y) = h_2(x,y) ** \rho(x,y) + n_2(x,y),$$

where $n_1(x,y)$ and $n_2(x,y)$ are independent real or complex two-dimensional zero-mean gaussian noise processes with known two-dimensional correlation functions.

The task of estimating $\rho(x,y)$ from $v_1(x,y)$ and $v_2(x,y)$ is not easy to formulate in the two-dimensional space domain. The situation is easier to treat in the two-dimensional Fourier frequency plane. Suppose that the two-dimensional signal $\rho(x,y)$ has a transform $R(f_x,f_y)$ with support contained within some large circle of the f_x,f_y plane, as shown in Figure 6.17. The filters $h_1(x,y)$ and $h_2(x,y)$ may have supports in the frequency domain covering this circle or may have smaller supports. The figure shows an exaggerated case in which much of the circle is not covered. Each low-resolution image provides spectral components only within two strips, covering more of frequency space than either individually, but not all of the relevant frequency space. The corners of the circle in the figure are not covered by either strip. Hence the original signal $\rho(x,y)$ cannot be fully reconstructed from only $p_1(x,y)$ and $p_2(x,y)$. Other information, such as the side condition that the image is real and nonnegative, may partially compensate for the missing spectral data.

In the absence of noise, the two blurred images in the frequency domain are

$$P_1(f_x,f_y) = H_1(f_x,f_y)R(f_x,f_y),$$
$$P_2(f_x,f_y) = H_2(f_x,f_y)R(f_x,f_y).$$

6 Tomographic Imaging Systems

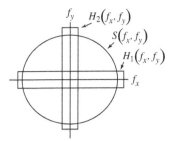

Figure 6.17 Illustrating the problem of image merging

Figure 6.17 illustrates how $P_1(f_x,f_y)$ and $P_2(f_x,f_y)$ may each contain part of the two-dimensional spectrum $R(f_x,f_y)$. Other parts of $R(f_x,f_y)$ may not appear in either $P_1(f_x,f_y)$ or $P_2(f_x,f_y)$.

The task is to estimate $R(f_x,f_y)$ from the combination of $P_1(f_x,f_y)$ and $P_2(f_x,f_y)$. A linear estimator will have the form

$$\widehat{R}(f_x,f_y) = G_1(f_x,f_y)P_1(f_x,f_y) + G_2(f_x,f_y)P_2(f_x,f_y),$$

where

$$G_1(f_x,f_y) = \frac{H_1^*(f_x,f_y)}{|H_1(f_x,f_y)|^2 + |H_2(f_x,f_y)|^2},$$

$$G_2(f_x,f_y) = \frac{H_2^*(f_x,f_y)}{|H_1(f_x,f_y)|^2 + |H_2(f_x,f_y)|^2}.$$

This gives the correct inverse at each point (f_x,f_y) for which either $H_1(f_x,f_y)$ or $H_2(f_x,f_y)$ is nonzero. When both terms are zero at frequency (f_x,f_y), then $R(f_x,f_y)$ cannot be determined from the given information.

In the presence of noise, the two-dimensional Wiener filter can be imitated to give

$$G_1(f_x,f_y) = \frac{H_1^*(f_x,f_y)}{|H_1(f_x,f_y)|^2 + |H_2(f_x,f_y)|^2 + N_1(f_x,f_y)},$$

$$G_2(f_x,f_y) = \frac{H_2^*(f_x,f_y)}{|H_1(f_x,f_y)|^2 + |H_2(f_x,f_y)|^2 + N_2(f_x,f_y)}.$$

When the noise is not well-modeled, the noise terms may be represented simply as C. Then the reconstruction becomes

$$\widehat{R}(f_x,f_y) = \frac{H_1^*(f_x,f_y)P_1(f_x,f_y) + H_2^*(f_x,f_y)P_2(f_x,f_y)}{|H_1(f_x,f_y)|^2 + |H_2(f_x,f_y)|^2 + C}.$$

The term C is an appropriate positive constant that protects against undue noise enhancement when $H_1(f_x,f_y)$ and $H_2(f_x,f_y)$ are both small.

Problems

6.1 Give two different functions $\rho(x,y)$ and $\rho'(x,y)$ that have identical projections on the x axis and identical projections on the y axis. Are the f_x and f_y slices of $R(f_x, f_y)$ and $R'(f_x, f_y)$ the same?

6.2 The projection of $\rho(x,y)$ at angle θ is

$$p_\theta(t) = \int_{-\infty}^{\infty} \rho(t\cos\theta - r\sin\theta, t\sin\theta + r\cos\theta) dr.$$

Show that for all $w(t)$,

$$\int_{-\infty}^{\infty} p_\theta(t) w(t) dt = \int_{-\infty}^{\infty}\int_{-\infty}^{\infty} \rho(x,y) w(x\cos\theta + y\sin\theta) dx dy,$$

provided the integrals exist.

6.3 Suppose that $\rho(x,y)$ separates as

$$\rho(x,y) = \rho_1(x)\rho_2(y).$$

Prove that when given the side information that $\rho(x,y)$ separates in this way, $\rho(x,y)$ is uniquely determined by two of its projections.

6.4 Prove that the generalized function $\delta(t)$ satisfies

$$\int_0^\pi \delta(x\cos\theta + y\sin\theta) d\theta = \frac{1}{\sqrt{x^2+y^2}}.$$

6.5 The circularly symmetric two-dimensional function $h(x,y)$ is defined in terms of the one-dimensional function $h(t)$ by $h(x,y) = h(\sqrt{x^2+y^2})$. Suppose that

$$\rho'(x,y) = h(x,y) ** \rho(x,y).$$

Show that the projections of $\rho(x,y)$ and $\rho'(x,y)$ satisfy

$$p'_\theta(t) = h(t) * p_\theta(t),$$

where $p_\theta(t)$ and $p'_\theta(t)$ are the projections of $\rho(x,y)$ and $\rho'(x,y)$ at angle θ.

6.6 Prove that

$$\int_0^{2\pi}\int_0^\infty f(r\cos\theta, r\sin\theta) r dr d\theta = \int_0^\pi \int_{-\infty}^\infty f(r\cos\theta, r\sin\theta) |r| dr d\theta.$$

6.7 Let

$$s(x,y) = \text{circ}(x - a, y)\text{circ}(x+a, y),$$

where a is less than one.

a. Describe the slices of the Fourier transform of $s(x,y)$.
b. Describe the slices of the Fourier transform of $s(x,y) ** s(x,y)$.

6.8 **a.** In n-dimensional euclidean space $\mathbb{R}^n$, what point of the $(n-1)$-dimensional hyperplane $\sum_i a_i x_i = c$ is closest to the point $x' = (x'_1, x'_2, \ldots, x'_n)$?

b. In n-dimensional euclidean space $\mathbb{R}^n$, what point of the $(n-2)$-dimensional hyperspace $\sum_i a_{i1} x_i = c_1$ and $\sum_i a_{i2} x_i = c_2$ is closest to the point $x' = (x'_1, \ldots, x'_n)$?

c. In n-dimensional euclidean space $\mathbb{R}^n$, what point of the $(n-r)$-dimensional manifold $\sum_i x_{i\ell} = c_\ell$ for $\ell = 1, \ldots, r$ is closest to the point $X' = (x'_1, \ldots, x'_n)$?

d. Give an alternating projection algorithm for inverting a sparse n by n system of linear equations that projects onto the intersection of r hyperplanes in each step. Must r divide n? What can be said about complexity?

6.9 Consider the two by two matrix

$$S = \begin{bmatrix} s_{11} & s_{12} \\ s_{21} & s_{22} \end{bmatrix}.$$

Given the "row projections" $p_j^{(R)} = \sum_i s_{ij}$ and the "column projections" $p_i^{(C)} = \sum_j s_{ij}$, can the elements of the matrix be recovered? How does this generalize to a three by three matrix? How does this change when diagonal projections are included?

6.10 (**Three-Dimensional Projection-Slice Theorem**) Let $\rho(x, y, z)$ have finite energy and Fourier transform $S(f_x, f_y, f_z)$. Let A be a unitary matrix defining a coordinate rotation:

$$\begin{bmatrix} t \\ r \\ q \end{bmatrix} = A \begin{bmatrix} x \\ y \\ z \end{bmatrix}.$$

The *projection* of $\rho(x, y, z)$ onto the t, r plane is defined as

$$p_A(t, r) = \int_{-\infty}^{\infty} \rho\left(A^T \begin{bmatrix} t \\ r \\ q \end{bmatrix}\right) dq,$$

where the integrand denotes that function of t, r, and q obtained by substituting for x, y, and z in $\rho(x, y, z)$, as indicated. Prove that

$$P_A(f_t, f_r) = R(A_{11} f_t + A_{21} f_r, A_{12} f_t + A_{22} f_r, A_{13} f_t + A_{23} f_r),$$

where $P_A(f_t, f_r)$ is the two-dimensional Fourier transform of $p_A(t, r)$.

6.11 A two-dimensional image of a three-dimensional scene can be constructed either as a slice or as a projection. Explain this statement. Under what circumstances is a set of projections sufficient to recover the three-dimensional scene?

6.12 Let $\rho(x, y, z)$ have finite energy and Fourier transform $R(f_x, f_y, f_z)$. As in Problem 6.10, let A be a unitary matrix defining a coordinate rotation:

$$\begin{bmatrix} t \\ r \\ q \end{bmatrix} = A \begin{bmatrix} x \\ y \\ z \end{bmatrix}.$$

The *conjection* of $\rho(x,y,z)$ onto the t axis is defined as

$$p_A(t) = \int_{-\infty}^{\infty}\int_{-\infty}^{\infty} \rho\left(A^T \begin{bmatrix} t \\ r \\ q \end{bmatrix}\right) drdq.$$

Prove that

$$P_A(f) = R(A_{11}f, A_{12}f, A_{13}f),$$

where $P_A(f)$ is the Fourier transform of $p_A(t)$.

6.13 A three-dimensional object, $\rho(x,y,z)$, is reduced to a stack of two-dimensional slices of that object by integrating over thin intervals in the z direction. Thus

$$\rho_k(x,y) = \int_{(k-\frac{1}{2})\Delta z}^{(k+\frac{1}{2})\Delta z} \rho(x,y,z)dz.$$

a. Show that this is equivalent to a convolution in the z direction with a rectangle function followed by sampling in z.
b. Describe tomographic images formed from projections of these slices.
c. Describe how to combine the images of the slices using Nyquist–Shannon interpolation to obtain the three-dimensional image. How does the result relate to $\rho(x,y,z)$?
d. Can the distortion be completely removed by deconvolution?
e. Repeat this analysis for the case in which the slices are defined by

$$\rho_k(x,y) = \int_{-\infty}^{\infty} e^{-(z-k\Delta z)^2/2\Delta z^2} \rho(x,y,z)dz.$$

6.14 Show that the complexity of filtered back-projection tomography is proportional to N^3. What part of the computations can be done while the data is still being collected? What computations can be done only after the full set of data is collected?

6.15 Suppose that the projections of a projection tomography system are contaminated by noise so that the actual measurements are

$$v_\theta(t) = p_\theta(t) + n_\theta(t),$$

where $n_\theta(t)$ is independent noise of power density spectrum $N(f)$ for each θ. For an image reconstructed using filtered back projection, how does the ramp filter affect the noise? How does the filtered noise affect the image?

6.16 A fan beam for projection tomography takes data at M uniformly spaced angles indexed by m. At each angle a fan of N uniformly spaced beams, indexed by n, measures projections.

a. What must be the relationship in the angular spacing $\Delta\phi$ of the fan origins and the spacing $\Delta\gamma$ of the beams within each fan so that the fan-beam data can be re-sorted into parallel beams?
b. Specify the jth parallel beam at the ith angle in terms of the nth fan beam at the mth angle.

6.17 Estimate how many hydrogen nuclei are in your body. Estimate how many hydrogen nuclei are in one of your major organs. Estimate how many hydrogen nuclei are in one cubic micron of your tissue. Repeat for one cubic nanometer. For a scale reference, the wavelength of visible light is about 500 nanometers. (Avogadro's number is approximately 6×10^{23}. A mole of water weighs about 18 grams.)

6.18 (Helical Scan Sensing) A helical scanner follows a helical path around a target collecting projections as it goes. A three-dimensional function $\rho(x,y,z)$ has Nyquist slices $\rho(x,y,k\Delta z)$ satisfying the Nyquist condition in the z direction. A helical-scan sensor takes projections

$$p_\theta(t) = \int_{-\infty}^{\infty} \rho(t\cos\theta - r\sin\theta, t\sin\theta + r\cos\theta, \theta/\Delta z).$$

a. Is this set of projections sufficient to recover the three-dimensional function $\rho(x,y,z)$?
b. Describe an interpolation algorithm.

6.19 Let $\rho(x_1, x_2, \ldots, x_n)$ denote an n-dimensional function of finite energy with Fourier transform $R(f_1, f_2, \ldots, f_n)$. Prove the n-dimensional generalization of the projection-slice theorem stated as follows:

Let x' denote a set of coordinates, given by $x = x'A$, where A is an orthogonal matrix, and let $f' = (f'_1, \ldots, f'_n)$ denote a set of frequency coordinates given by $f = f'A$. Then a *projection* onto an $(n-1)$-dimensional hyperplane is defined as

$$p_{x'_i}(x'_1, x'_2, \ldots, x'_{i-1}, x'_{i+1}, \ldots, x'_n) = \int_{-\infty}^{\infty} \rho(x'A) dx'_i.$$

The *slice* of the Fourier transform at f'_i is $R(f'_1, f'_2, \ldots, f'_{i-1}, 0, f'_{i+1}, \ldots, f'_n)$. Then the projection along x'_i has an $(n-1)$-dimensional Fourier transform equal to the slice of $R(f')$ at $f'_i = 0$.

Notes

The central theme of tomography is to use the projection-slice theorem to reconstruct images. This theorem was stated early by Radon in 1917 and was rediscovered independently in many fields. Bracewell (1956), and Bracewell and Riddle (1967), developed the idea for use in astronomy in connection with the inversion of multiple radio-telescope scans. DeRosier and Klug (1968) developed the idea in connection with the inversion of multiple electron microscope images. Cormack (1963) proposed using the method for the reconstruction of medical images from X-ray data.

Hounsfield (1972, 1973a, 1973b) built the first tomographic scanner for medical X-ray imaging. Systems for X-ray tomography are now in widespread use. These systems may image three-dimensional objects by imaging many two-dimensional cross sections. Algorithms that form three-dimensional images directly by using cone-beam illumination have been studied by Tuy (1983), Smith (1985), and others. A tutorial on algebraic reconstruction techniques was written by Gordon (1974).

The phenomenon of nuclear magnetic resonance was discovered in 1946 by Bloch. Thereafter, it was well understood that different tissues respond differently to magnetic excitation, but it was not immediately understood how to use this effect to form an image. Lauterbur (1973) contributed the idea of using gradient magnetic fields to encode the spatial information and to then recover the image from the received signal by tomographic processing. Moreover, Lauterbur was the first to demonstrate his technique. The important medical imaging technology of magnetic-resonance imaging grew out of Lauterbur's work. Phase encoding methods were introduced by Kumar, Welti, and Ernst (1975). Phase contrast methods were extended to magnetic resonance imaging by Moran (1982) and based on work in other fields, by Hahn (1960) and Singer (1971).

There have been many tutorial treatments of tomography such as those by Scudder (1978), and by Mersereau and Oppenheim (1974). The topic of sampling in polar coordinates for tomography was developed by Rattey and Lindgren (1981), following earlier work by Snyder and Cox (1977). An alternating-minimization algorithm for transmission tomography was proposed by O'Sullivan and Benac (2007). A tomographic formulation of the synthetic-aperture principle may be found in Munson, O'Brien, and Jenkins (1983). An alternative tomographic imaging radar, using multiple chirp slopes, was proposed by Bernfeld (1984), and by Feig and Grünbaum (1986). Three-dimensional tomographic formulations of radar have been discussed by Jakowitz and Thompson (1992), and by Coble (1992). The use of the Wigner distribution to obtain an alternative method of forming images from radar data–akin to tomography–was proposed by Barbarosa and Farina (1990), and by Chen (1994). The Lewis–Bojarski equation (Lewis, 1969; Bojarski, 1982) describes the relationship between inverse scattering and object shape, and is closely related to the projection-slice theorem.

Diffraction tomography can be traced back to the work of Wolf (1969). The paper by Mueller, Kaveh, and Wade (1979) was also influential. Techniques that solve the inverse scattering problem for diffusing waves were developed by Schotland (1997) based on work by Ishii, Leigh, and Schotland (1995). These methods can be modified to solve the inverse problem for the near-field diffraction as described in an article by Dines and Lytle (1979). A striking application is the article by Dziewonski and Woodhouse (1987). Optical-coherence tomography was discussed in a review article by Schmitt (1999).

7 Construction and Reconstruction of Images

The task of constructing or reconstructing an image of a scene when given a set of noisy data that is dependent on that scene is the task of image formation and is the subject of this book. Possibly some prior information about the scene is also given and may be helpful to that process. Image formation might also include the task of refining an existing image when given additional information about that image. In the latter case, the task of redefining or reconstructing an image might better be called *image restoration*.

The most familiar problem of image restoration starts with an image of a two-dimensional scene, but with the details of the image limited in some way. For example, the image of the scene may be blurred or poorly resolved. The signal-processing techniques of *deconvolution* or *deblurring* can enhance a poorly resolved image. When the blurring function is not known but must be inferred from the image itself, these techniques are called *blind deconvolution* or *blind deblurring*.

A more difficult task of image formation is the task of estimating an image of a scene from knowledge of some function of the scene. An instance of this task is the estimation of an image of a scene from only the magnitude of its two-dimensional Fourier transform when some side information about the scene is known. This process is called *phase retrieval* because the phase of the Fourier transform is implicit in the estimated image. In effect, the phase of the Fourier transform is inferred from the magnitude of the Fourier transform and the other side information.

Some topics of image estimation require their own chapters. One important topic deals with the estimation of an image from a set of its projections. This important topic of *tomography* was covered in Chapter 6. The study of likelihood methods for image formation is the subject of Chapter 8. The maximum-likelihood principle studied in that chapter is the basis of probabilistic methods that determine the image that best accounts for a set of fragmentary data. Such methods use raw data that may appear in a form that is quite different from the image.

7.1 The Inverse Problem

The task of describing a set of observations of an object when the observed object and the methods of observation are specified is called the *forward problem*. The task of forming an image of an observed object when given a set of observations of that object is called the *inverse problem*. The inverse problem is usually much harder than the forward problem. Indeed, it may be that the available observations are not sufficient to fully describe the object and the observation must be supplemented with side conditions based on experience and judgment.

The distinction between a direct problem and an inverse problem can be illustrated by the simple example of convolution.

$$r(t) = \int_{-\infty}^{\infty} h(t-\xi)s(\xi)d\xi.$$

Computing $r(t)$ from $s(t)$ and $h(t)$ is the direct problem. Computing $s(t)$ from $r(t)$ and $h(t)$ is the inverse problem. This instance of the inverse problem is called *deconvolution*. Usually this inverse problem of deconvolution is given only a noisy version of $r(t)$. Our interest in inverse problems extends well beyond this simple example to the much more difficult inverse problems of image formation from noisy data.

The direct problem is better stated as the task of describing the relationship between a set of observations and a large class of objects of a common form. For example, a cardiac ultrasound system is designed to observe any generic human heart, not simply a single specific heart that is already known. The point here is that the direct problem should not be defined too narrowly, yet should not be defined too broadly. Given a set of observations, the task of solving an inverse problem is to describe that object from the class of all objects of interest that could give rise to an observation of this kind.

Many inverse problems do not have a unique solution. Such problems are described as ill-posed or ill-conditioned. A problem is *well-posed* when it has a unique solution that varies continuously in the input data. A problem is *ill-posed* when it is not well-posed. A problem is *ill-conditioned* when small changes in the input lead to large changes in the output. This is only a subjective definition because the terms "small" and "large" have not been defined. A problem that is ill-posed need not be ill-conditioned. A problem that is ill-conditioned need not be ill-posed.

Any empirical remedy for an ill-posed problem is called *regularization*. An unregularized algorithm may put detail into a computed image that is unwarrented for that class of objects. The topic of regularization includes any method to avoid overfitting the data with a solution that has unwarrented complexity. Many methods of regularization are in use. One method is to include a penalty term in an optimization expression. A simple method of regularization is to choose the coarseness of the pixelization of a discrete image appropriate to the expected image. Pixels that are too small may invite overfitting the observed data with unwanted and unrealistic detail. Pixels that are too large will fail to reveal relevant detail. Methods of regularization are discussed in Section 8.11 of Chapter 8.

Some forward problems are linear. Some forward problems are nonlinear. A linear forward problem is one in which the observations are linear functions of the parameter being observed. A linear problem can be treated by the methods of linear algebra. This may require an appeal to the Nyquist–Shannon sampling theorem as needed to convert a continuous problem into a discrete problem, thereby converting integrations to summations, converting filters in continuous time into filters in discrete time, and converting Fourier transforms into discrete Fourier transforms. A linear problem such as linear filtering is then expressed in the form of a set of linear equations that can be expressed as a matrix equation. In such cases, the notion of an ill-posed problem may be related to the standard notions of singular matrices.

When a matrix is a square matrix of full rank, the number of equations may be the same as the number of unknowns and the problem is well-posed. When the number of equations is less than the number of unknowns, the problem is ill-posed. Regularization is the remedy when such a situation is anticipated. When the number of equations is more than the number of unknowns and the set does not have a consistent solution, as when there is noise in the measurements, a method of finding the most appropriate fit to these equations using other considerations is required.

These remarks and examples, however, are only motivational. Many imaging problems are large and have constraints. The analysis and the methods of solution for such problems can be complicated.

7.2 Deconvolution and Deblurring

The effect of passing a desired two-dimensional image through an unwanted two-dimensional point-spread function is a form of blurring. Blurring constitutes an elementary impairment of an image. An elementary task of image restoration is the suppression or reduction of blur.

Blurring of a One-Dimensional Signal

The one-dimensional signal $s(t)$ when passed through a known filter $h(t)$ is described by the one-dimensional convolution

$$r(t) = \int_{-\infty}^{\infty} h(t - \xi)s(\xi)\,d\xi.$$

The function $h(t)$ in this expression is a one-dimensional *linear time-invariant* filter. The notation $h(t)$ is used to label the filter itself and denotes the impulse response of that filter. The linear filter $h(t)$ is a *causal filter* when $h(t) = 0$ for t less than zero.

In general, a filter need not be time-invariant. A *time-variant* filter is written

$$r(t) = \int_{-\infty}^{\infty} h(t|\xi)s(\xi)\,d\xi,$$

where $h(t|\xi)$ is the time-varying response of the filter at time t to an impulse at time ξ.

Blurring of a Two-Dimensional Signal

The two-dimensional signal $s(x, y)$ when passed through the linear, two-dimensional space-invariant filter $h(x, y)$ is described by the two-dimensional convolution

$$r(x, y) = \int_{-\infty}^{\infty} \int_{-\infty}^{\infty} h(x - \xi, y - \eta)s(\xi, \eta)\,d\xi\,d\eta.$$

A linear, two-dimensional space-invariant filter is called a *point-spread function*. In general, a point-spread function need not be space-invariant. A space-varying point-spread function is written

$$r(x,y) = \int_{-\infty}^{\infty}\int_{-\infty}^{\infty} h(x,y|\xi,\eta)s(\xi,\eta)d\xi\,d\eta,$$

where $h(x,y|\xi,\eta)$ is the space-varying point-spread function of the two-dimensional filter.

Deconvolution

The inverse operation – the task of removing the effect of the convolution or blurring in whole or in part – is called *deconvolution* or *deblurring*. Deconvolution is the task of solving an integral equation of the form above[1] for the unknown function $s(x,y)$ when both the function $r(x,y)$ and the space-varying point-spread function $h(x,y|\xi,\eta)$ are given and where the integral is over some set known to be the domain of $s(x,y)$. The deblurring problem reduces to the deconvolution problem whenever the point-spread function depends only on the spatial difference between the points (x,y) and (ξ,η), not on each point separately, and the integration extends over infinite space. Thus two-dimensional deconvolution is the determination of the function $s(x,y)$ that satisfies the equation

$$r(x,y) = \int_{-\infty}^{\infty}\int_{-\infty}^{\infty} h(x-\xi, y-\eta)s(\xi,\eta)d\xi\,d\eta$$

when $r(x,y)$ and $h(x,y)$ are both known. The term "deconvolution" is natural because the given convolution

$$r(x,y) = h(x,y) ** s(x,y)$$

is to be undone.

The deconvolution problem is known to exhibit difficulty when a numerical solution is attempted. Small changes in $h(x,y)$ can result in large changes in the solution. Seemingly minor details in the numerical methods can have a large effect on the result. Deconvolution is an ill-conditioned problem.

The *inverse filter* for the two-dimensional filter $h(x,y)$, were it to exist, could be the filter $h^{-1}(x,y)$ such that

$$h^{-1}(x,y) ** h(x,y) = \delta(x,y),$$

where $\delta(x,y)$ is the two-dimensional delta function. Then the convolution

$$r(x,y) = h(x,y) ** s(x,y)$$

would be undone by

$$s(x,y) = h^{-1}(x,y) ** r(x,y).$$

---

[1] In the situation in which $s(x,y)$ is unknown, these are all examples of a Fredholm integral equation of the first kind. Many inverse problems reduce to the task of solving a Fredholm integral equation.

This simple statement may seem to settle the matter but the formula is unsatisfactory for many reasons. One difficulty is that $h^{-1}(x,y)$ does not exist as a proper function with finite energy. The inverse filter stated in the Fourier transform domain is

$$H^{-1}(f_x,f_y) = \frac{1}{H(f_x,f_y)}.$$

Clearly, even though the function $H(f_x,f_y)$ has finite energy, the function $H^{-1}(f_x,f_y)$ has infinite energy, which means that $h^{-1}(x,y)$ has infinite energy. Another difficulty is that $r(x,y)$ or $h(x,y)$ may not be precisely known and the simple deconvolution procedure can amplify errors. Moreover, when only the noisy signal

$$v(x,y) = (h(x,y) ** s(x,y)) + n(x,y)$$

is given and $h^{-1}(x,y)$ is known, the computation

$$h^{-1}(x,y) ** v(x,y) = s(x,y) + (h^{-1}(x,y) ** n(x,y))$$

would be unsatisfactory. Although the blur would be removed, the noise term, in general, would be made worse. The inverse filter would change noise of finite power into noise of infinite power which clearly is unacceptable.

A linear approach to deconvolution of a noisy signal is to find a filter, $g(x,y)$, that is a compromise between removing the blur and controlling the noise. When $v(x,y)$ is the input to the filter, the output of the filter is the estimate of $s(x,y)$ given by

$$\widehat{s}(x,y) = g(x,y) ** v(x,y)$$
$$= g(x,y) ** ((h(x,y) ** s(x,y)) + n(x,y)).$$

The noise $n(x,y)$ is modeled as stationary noise with zero mean and power density spectrum $N(f_x,f_y)$, where the power density spectrum is given by the Fourier transform of the autocorrelation function $\phi_n(\tau_x, \tau_y) = E[n(x,y)n^*(x+\tau_x, y+\tau_y)]$.

The signal $s(x,y)$ can be either deterministic or random. The case of a deterministic $s(x,y)$ leads to the whitened matched filter as discussed in Section 2.7. The case of a random $s(x,y)$ leads to the Wiener filter, which is treated in the forthcoming Theorem 7.2.1.

Whitened Matched Filter

Before stating the two-dimensional Wiener filter in Theorem 7.2.1, the whitened matched filter, given in Section 2.7 of Chapter 2, is restated in two dimension for contrast. A real or complex deterministic signal in real or complex noise of the simpler form

$$v(x,y) = s(x,y) + n(x,y)$$

is to be filtered in order to maximize the signal to noise ratio at the origin, where $s(x,y)$ is a *known* two-dimensional pulse of finite energy with two-dimensional Fourier transform $S(f_x,f_y)$, $h(x,y)$ is the known point-spread function, and $n(x,y)$ is zero-mean stationary noise with known power density spectrum denoted $\Phi_n(f_x,f_y)$ or $N(f_x,f_y)$.

The Fourier transform of the two-dimensional whitened matched filter for this noisy signal is

$$G(f_x, f_y) = \frac{S^*(f_x, f_y)}{\Phi_n(f_x, f_y)},$$

generalizing the case of a one-dimensional filter in Theorem 2.7.2 of Chapter 2. The two-dimensional matched filter $g(x, y)$ is the inverse Fourier transform of $G(f_x, f_y)$.

The Wiener Filter

Now turn to a different situation, replacing the finite-energy deterministic signal $s(x, y)$ with a stationary random process, also denoted $s(x, y)$. The problem is again written

$$v(x, y) = \big(h(x, y) ** s(x, y)\big) + n(x, y),$$

but now $s(x, y)$ is a two-dimensional stationary random process with zero mean and with power density spectrum $\Phi_s(f_x, f_y)$. The noise is a stationary random process with power density spectrum $\Phi_n(f_x, f_y)$, sometimes denoted $N(f_x, f_y)$.

A widely accepted criterion for choosing the filter $g(x, y)$ is to minimize the combined mean-square error from both the signal and the noise. The two-dimensional filter that minimizes the combined mean-square error is known as the (whitened) two-dimensional *Wiener filter*.[2]

Theorem 7.2.1 *The two-dimensional Wiener filter for the linear blurring function $h(x, y)$ is given by the point-spread function*

$$G(f_x, f_y) = \frac{H^*(f_x, f_y)\Phi_s(f_x, f_y)}{|H(f_x, f_y)|^2 \Phi_s(f_x, f_y) + \Phi_n(f_x, f_y)},$$

where $\Phi_s(f_x, f_y)$ and $\Phi_n(f_x, f_y)$ are the two-dimensional power density spectra of the signal and the noise, respectively.

Proof:
Only the case of a real signal in real noise is treated. The case of a complex signal in complex noise is similar.

The Wiener filter for real signals is derived by starting with the error signal which, for any point-spread function $g(x, y)$, is given by

$$\Delta(x, y) = s(x, y) - \big(g(x, y) ** v(x, y)\big),$$

where the received noisy signal is

$$v(x, y) = \big(h(x, y) ** s(x, y)\big) + n(x, y),$$

as stated in terms of the known point-spread function $h(x, y)$ and the stationary two-dimensional zero-mean noise $n(x, y)$.

---

[2] The Wiener filter in one dimension may be required to be a causal filter leading to a formulation in terms of the Wiener–Hopf equation. When causality is not required, the solution is different. The notion of a causal filter does not have a compelling generalization in two dimensions. The two-dimensional Wiener filter discussed here is a generalization of a noncausal one-dimensional Wiener filter.

The error signal $\Delta(x,y)$ is a real, stationary random process with zero mean and variance

$$\sigma^2 = \mathrm{E}\big[|\Delta(x,y)|^2\big]$$
$$= \mathrm{E}\big[s(x,y)^2\big] - 2\mathrm{E}\big[s(x,y)(g(x,y) ** v(x,y))\big] + \mathrm{E}\big[(g(x,y) ** v(x,y))^2\big].$$

The two-dimensional Wiener filter is the point-spread function $g(x,y)$ that minimizes this expression for σ^2.

To determine $g(x,y)$, substitute the expression for $v(x,y)$, expand the square, and take the expectation. Terms that are linear in $s(x,y)$ or in $n(x,y)$ have a zero expectation and need not be considered further. Only terms that are quadratic in $s(x,y)$ or $n(x,y)$ have a nonzero expectation.

The next step is to transform the expression for σ^2 into the frequency domain. This step is executed for the stationary random process $s(x,y)$, by using the two general expressions given by

$$\mathrm{E}\big[s(x,y)[u(x,y) ** s(x,y)]\big] = \int_{-\infty}^{\infty}\int_{-\infty}^{\infty} U(f_x,f_y)\Phi_s(f_x,f_y)\mathrm{d}f_x\mathrm{d}f_y$$

and

$$\mathrm{E}\big[(u(x,y) ** s(x,y))^2\big] = \int_{-\infty}^{\infty}\int_{-\infty}^{\infty} |U(f_x,f_y)|^2 \Phi_s(f_x,f_y)\mathrm{d}f_x\mathrm{d}f_y,$$

for any point-spread function $u(x,y)$ with Fourier transform $U(f_x,f_y)$. These two expressions are easily verified for any stationary random process $s(x,y)$ (see Problem 7.8).

With the aid of these two expressions, we have

$$\sigma^2 = \phi_s(0,0) - 2\int_{-\infty}^{\infty}\int_{-\infty}^{\infty} G(f_x,f_y)H(f_x,f_y)\Phi_s(f_x,f_y)\mathrm{d}f_x\mathrm{d}f_y$$
$$+ \int_{-\infty}^{\infty}\int_{-\infty}^{\infty} |G(f_x,f_y)|^2 |H(f_x,f_y)|^2 \Phi_s(f_x,f_y)\mathrm{d}f_x\mathrm{d}f_y$$
$$+ \int_{-\infty}^{\infty}\int_{-\infty}^{\infty} |G(f_x,f_y)|^2 \Phi_n(f_x,f_y)\mathrm{d}f_x\mathrm{d}f_y.$$

The *calculus of variations*[3] can now be used to find $G(f_x,f_y)$. To see how this works, we minimize instead the simpler one-dimensional expression

$$\sigma^2 = 2\int_{-\infty}^{\infty} G(f)A(f)\mathrm{d}f + \int_{-\infty}^{\infty} |G(f)|^2 B(f)\mathrm{d}f$$

over the choice of $G(f)$ with the condition that the corresponding functions in the time domain are real-valued, where $A(f) = -H(f)\Phi_s(f)$ and $B(f) = |H(f)|^2\Phi_s(f) + \Phi_n(f)$. The method of the calculus of variations replaces $G(f)$ by $G(f) + \epsilon\eta(f)$,

---

[3] Elementary calculus minimizes a *function* with respect to a *variable* by setting a derivative equal to zero. The calculus of variations minimizes an *expression* with respect to a *function* by generalizing this procedure.

7.2 Deconvolution and Deblurring

where $\eta(f)$ is an arbitrary function and ϵ is a scalar parameter. The derivative of the expanded expression with respect to ϵ set equal to zero is

$$\frac{\partial}{\partial \epsilon}\left[2\int_{-\infty}^{\infty}[G(f)+\epsilon\eta(f)]A(f)df + \int_{-\infty}^{\infty}|G(f)+\epsilon\eta(f)|^2 B(f)df\right]_{\epsilon=0} = 0.$$

This leads to

$$2\int_{-\infty}^{\infty}\eta(f)A(f)df + \int_{-\infty}^{\infty}G^*(f)\eta(f)B(f)df + \int_{-\infty}^{\infty}G(f)\eta^*(f)B(f)df = 0.$$

By Parseval's formula, the third term is real, so it may be replaced by its conjugate. Then the equality collapses to $2\int_{-\infty}^{\infty}[A(f)+G^*(f)B(f)]\eta(f)df = 0$. That is,

$$2\int_{-\infty}^{\infty}\left[-H(f)\Phi_s(f) + G^*(f)\left(|H(f)|^2\Phi_s(f)+\Phi_n(f)\right)\right]\eta(f)df = 0.$$

This holds for any arbitrary function $\eta(f)$, so the bracketed term must be zero. Otherwise, for each f, the arbitrary term $\eta(f)$ could be chosen so as to have the same sign as the bracketed term, thereby violating the equality. Setting the bracketed term equal to zero and solving for $G(f)$ gives

$$G(f) = \frac{H^*(f)\Phi_s(f)}{|H(f)|^2\Phi_s(f)+\Phi_n(f)}.$$

The same analysis in two dimensions gives the full expression for two-dimensional functions. It follows in the same way from the calculus of variations that

$$-H(f_x,f_y)\Phi_s(f_x,f_y) + G^*(f_x,f_y)\left[|H(f_x,f_y)|^2\Phi_s(f_x,f_y)+\Phi_n(f_x,f_y)\right] = 0.$$

The Wiener filter for two-dimensional functions that is given in the statement of the theorem now follows immediately. This completes the proof of the theorem. ∎

For white noise, the noise power density spectrum is constant. This is conventionally written as $\Phi_n(f_x,f_y) = N_0/2$. Therefore,

$$G(f_x,f_y) = \frac{H^*(f_x,f_y)\Phi_s(f_x,f_y)}{|H(f_x,f_y)|^2\Phi_s(f_x,f_y)+N_0/2}$$

is the Wiener filter for white noise. Furthermore, when $\Phi_s(f_x,f_y) = S_0/2$, both random processes are white, and

$$G(f_x,f_y) = \frac{H^*(f_x,f_y)}{|H(f_x,f_y)|^2+N_0/S_0}.$$

This version of the point-spread function is the basic form of the two-dimensional Wiener filter. This form of the Wiener point-spread function may be used in place of the Wiener point-spread function even when the signal and the noise are not white or are not known.

The Clean Algorithm

The two-dimensional Wiener filter makes use of the power density spectra of the noise process and the signal process but does not use any other properties of the image that might be known. The Wiener filter cannot create new frequency components in the image that are not already in the support of $H(f_x, f_y)$. Any linear filter can only amplify or attenuate the nonzero complex signal amplitudes at each frequency pair (f_x, f_y). For a point-spread function for which $H(f_x, f_y)$ is zero at some frequency pair (f_x, f_y), the output of the point-spread function is also zero at that frequency pair.

Methods of processing that incorporate known properties of the image in other ways are available. Any method that uses prior information to reconstruct frequencies that are not in the support of $H(f_x, f_y)$ is referred to as a *superresolution* method. Such methods are always nonlinear. Two important cases are discussed. The case in which the image is known or believed to have sparse support is discussed in this section. The case in which the image is known to be real and nonnegative is discussed in the next section. Both cases are nonlinear in the data.

The case of sparse support is the case in which the image is known to consist of a sparse arrangement of active regions of unknown size, shape, or location, and typically nonnegative. An image known to have sparse support is sometimes called a *mostly black image* because most of the image will be all black (or all white).

Whenever the scene $\rho(x, y)$ to be imaged is known to consist of a sparse set of small objects, or can be so modeled, the method of deconvolution known as the *Clean algorithm* may be appropriate. The clean algorithm is intuitively satisfying and useful, but it is developed as an empirical procedure. It is not supported by a basic principle of inference.

Suppose that the source $\rho(x, y)$ is a weighted sum of impulses of the form

$$\rho(x, y) = \sum_{\ell=1}^{L} a_\ell \delta(x - x_\ell, y - y_\ell),$$

where all amplitudes a_ℓ are positive and unknown. The locations (x_ℓ, y_ℓ) are also unknown. The number of terms L is finite but unknown as well. The noisy blurred image is given by

$$v(x, y) = \int_{-\infty}^{\infty} \int_{-\infty}^{\infty} h(x - \xi, y - \eta) \rho(\xi, \eta) \, d\xi \, d\eta + n(x, y),$$

where $h(x, y)$ is the point-spread function of the system. Because $h(x, y)$ may have undesirable sidelobes that create false objects or artifacts elsewhere in the image, it is called the *dirty point-spread function*. The goal of the clean algorithm is to replace $h(x, y)$ with the alternative point-spread function, $\widehat{h}(x, y)$, called the *clean point-spread function*.

The task is to estimate $\rho(x, y)$ when given the noisy blurred image $v(x, y)$ knowing that the noise term is small. The clean algorithm operates iteratively in a somewhat obvious way by estimating the terms $a_\ell \delta(x - x_\ell, y - y_\ell)$, one by one, starting with the largest and removing each term from $v(x, y)$ as it is estimated.

With the initialization $\widehat{s}(x,y) = 0$, the clean algorithm consists of the following steps.

Step 1

$$a = \max_{x,y} v(x,y),$$
$$(\widehat{x},\widehat{y}) = \mathrm{argmax}_{x,y} v(x,y)$$

Step 2

$$v(x,y) \leftarrow v(x,y) - ah(x-\widehat{x}, y-\widehat{y}),$$
$$\widehat{\rho}(x,y) \leftarrow \widehat{\rho}(x,y) + a\widehat{h}(x-\widehat{x}, y-\widehat{y})$$

Step 3
Either halt or return to Step 1.

The iterations are halted when they no longer improve the image. The clean point-spread function $\widehat{h}(x,y)$ could be a two-dimensional gaussian pulse or something similar. It is used to provide shape to the replacement impulses so that the image is more attractive. The width of the clean point-spread function might be chosen to match some measure of the width of the dirty point-spread function so as to prevent the image from displaying an unnatural appearance.

7.3 Deconvolution of Nonnegative Images

In problems of signal processing such as image formation, one may be asked to find a function in a space of real or complex finite-energy functions satisfying certain properties. In some cases, the correct formulation of this task is to find the function with the required properties consistent with the measured data that minimizes the euclidean distance between that function and some other function. In other cases, it is given that the desired function lies in the space of nonnegative real functions. Then the problem changes considerably. The euclidean distance is not the appropriate discrepancy measure in the space of nonnegative real functions.
Let

$$v(x,y) = \bigl(h(x,y) ** s(x,y)\bigr) + n(x,y)$$

be the blurred observation of a two-dimensional nonnegative finite-energy signal $s(x,y)$ in the presence of noise. A good estimate, $\widehat{s}(x,y)$, of the signal $s(x,y)$ removes as much of the blur and the noise as is appropriate. The matched filter is a real-valued inverting filter. The matched filter can produce negative values at its output so it is not suitable for problems having the prior condition that the image is nonnegative. Of course, one may proceed by first using the matched filter and then replacing negative values of the resulting function by zero. This procedure might give acceptable results, but it is a forced solution that does not satisfy any optimality criterion. Instead, an alternative procedure is described.

Any relationship between the blurred input signal and an estimated deblurred signal that satisfies the nonnegativity constraint must, of necessity, be nonlinear. This section will develop the *Richardson–Lucy algorithm*, which is an algorithm for computing a nonnegative signal or image estimate in the presence of noise. In the present section, a satisfying heuristic development is given for this algorithm. This development helps to show how the algorithm works. In Section 8.5, a formal development of the Richardson–Lucy algorithm is given starting from a more formal principal of inference.

Motivation

The motivation for the Richardson–Lucy algorithm is based on the Bayes formula of probability theory. Recall that a bivariate probability distribution can be decomposed into marginals and conditionals in two ways. These are $P_{jk} = p_j Q_{k|j}$ and $P_{jk} = q_k P_{j|k}$, where the marginals $p = [p_j]$ and $q = [q_k]$ are given by $p_j = \sum_k P_{jk}$ and $q_k = \sum_j P_{jk}$, and the conditionals $Q = [Q_{k|j}]$ and $P = [P_{j|k}]$ are given by $Q_{k|j} = P_{jk}/p_j$ and $P_{j|k} = P_{jk}/q_k$. Then the Bayes formula is

$$P_{j|k} = \frac{p_j Q_{k|j}}{\sum_i p_i Q_{k|i}}.$$

It is easy to compute the marginals and the conditionals from the joint distribution $P = [P_{jk}]$. It is difficult to compute a joint distribution $P = [P_{jk}]$ that has a specific marginal $q = [q_k]$ and a specific conditional $Q = [Q_{k|j}]$. This latter task has the nature of an inverse problem. The required solution need not exist. In general, any given probability vector q of length K and any given K by J probability transition matrix Q need not be compatible in the sense that a $[P_{jk}]$ need not exist that has both the marginal q and the conditional Q. Indeed, such pairs are rare. Even when there is a solution, the expression is not readily inverted to give an explicit expression for p when Q is not square.

In general, when the probability distribution q and the conditional probability distribution Q are both given, the expression $q_k = \sum_j p_j Q_{k|j}$ might not have a solution for a p with all nonnegative components.

Because $p_j = \sum_k q_k P_{j|k}$, the Bayes formula can be used to write

$$p_j = p_j \sum_k \frac{Q_{k|j}}{\sum_i p_i Q_{k|i}} q_k.$$

With an eye on this equality, the expression

$$p_j^{(r+1)} = p_j^{(r)} \sum_k \frac{Q_{k|j}}{\sum_i p_i^{(r)} Q_{k|i}} q_k$$

suggests itself as a way to iteratively compute the probability vector p satisfying $q_k = \sum_j p_j Q_{k|j}$ when a solution exists or to compute a best approximation when a solution

7.3 Deconvolution of Nonnegative Images

does not exist. Clearly, the iteration produces a sequence of probability vectors because when $p^{(r)}$ is a probability vector, then the $p_j^{(r+1)}$ are all nonnegative and the sum over j is equal to one, so $p^{(r+1)}$ is also a probability vector. To initialize the iterations, it is natural to choose the uniform distribution as the initial value $p^{(0)}$.

If the iteration does approach a limit point, then this limit point is clearly a probability vector and may be the desired p. A fixed point to which the iteration converges may exist even when the expression $q_k = \sum_j p_j Q_{k|j}$ does not have a probability vector p as a solution, as may be the case when q and Q are contaminated by measurement errors. In this case, the fixed point may be regarded in some sense as a best fit to the required conditions.

One can mimic this procedure to develop a similar iterative algorithm for the general problem of deblurring a nonnegative function. Suppose that a known nonnegative blurred function $v(t)$ satisfying

$$v(\xi) = \int_{-\infty}^{\infty} h(\xi|t)s(t)dt$$

is given, where the term $h(\xi|t)$ for each t is a known nonnegative function of ξ whose integral $\int_{-\infty}^{\infty} h(\xi|t)d\xi$ does not depend on t and where $s(t)$ is an unknown nonnegative function satisfying the expression to be computed from $v(\xi)$ and $h(\xi|t)$.

By rescaling $s(t)$, the integral of $v(\xi)$ is made equal to one. Because $h(\xi|t)$ does not depend on t, the integral of $h(\xi|t)$ is chosen to also equal one by rescaling, then imposing the constraint that $s(t)$ integrates to one.

The condition that characterizes this deblurring problem is that $s(t)$ is required to be real and nonnegative. Define

$$g(t|\xi) = \frac{h(\xi|t)s(t)}{\int_{-\infty}^{\infty} h(\xi|t)s(t)dt}.$$

The denominator on the right is equal to $v(\xi)$, so this expression can be rearranged as

$$h(\xi|t)s(t) = v(\xi)g(t|\xi).$$

But $h(\xi|t)$ integrates to one, which implies that

$$s(t) = \int_{-\infty}^{\infty} h(\xi|t)s(t)d\xi = \int_{-\infty}^{\infty} v(\xi)g(t|\xi)d\xi.$$

Therefore the unknown $s(t)$ satisfies the implicit expression

$$s(t) = s(t)\int_{-\infty}^{\infty} v(\xi)\frac{h(\xi|t)}{\int_{-\infty}^{\infty} h(\xi|t)s(t)dt}d\xi.$$

With this equality as the motivation, the Richardson–Lucy algorithm is now defined as the iteration

$$s^{(r+1)}(t) = s^{(r)}(t)\int_{-\infty}^{\infty} v(\xi)\frac{h(\xi|t)}{\int_{-\infty}^{\infty} h(\xi|t)s^{(r)}(t)dt}d\xi.$$

At iteration r, the estimate $s^{(r)}(t)$ is known and the new estimate $s^{(r+1)}(t)$ is computed. It is clear that when $s^{(r)}(t)$ is nonnegative, $s^{(r+1)}(t)$ is also nonnegative. The integral of $s^{(r)}(t)$ is easily seen to be $\int_{-\infty}^{\infty} v(\xi)d\xi$, which is independent of r. If $s^{(r)}(t)$ does converge to a fixed point, that fixed point may be the required solution to the integral equation. When it does not solve the integral equation, that fixed point might be regarded as the best fit to the specified conditions. Section 8.5 formally justifies this statement.

The same discussion applies in two dimensions. For nonnegative *two-dimensional* functions, the Richardson–Lucy algorithm can be written immediately as

$$s^{(r+1)}(x,y) = s^{(r)}(x,y) \int_{-\infty}^{\infty} \int_{-\infty}^{\infty} v(\xi,\eta) \frac{h(\xi,\eta|x,y)}{\int_{-\infty}^{\infty} \int_{-\infty}^{\infty} h(\xi,\eta|x,y)s^{(r)}(x,y)dxdy} d\xi d\eta,$$

because the above discussion can be easily repeated in two dimensions.

The two-dimensional Richardson–Lucy algorithm has been introduced here in an informal manner. In Section 8.5 of Chapter 8, a formal procedure for developing this algorithm is developed, as well as a formal statement regarding convergence, even in the presence of noise.

7.4 Diffractive Lenses

An ideal lens has a phase shift $(2\pi/\lambda)(x^2+y^2)/2f_\ell$ at the point (x,y). A thin lens, as shown in Figure 4.5, is an approximation to an ideal lens. A thin lens is an example of a *refractive lens*.

An ideal lens requires a phase shift between 0 and 2π, whereas a thin lens has phase shifts in excess of 2π. This motivates the following definition. A *diffractive lens* has a phase shift of $(2\pi/\lambda)(x^2+y^2)/2f_\ell$ (modulo 2π) at the point (x,y). The modulo 2π operation is the reason that a diffractive lens differs from a refractive lens.

A thin refractive lens and a thin diffractive lens provide similar approximations to an ideal lens at the design wavelength λ. However, the two lenses do not behave the same under a change in the wavelength. A diffractive lens is designed for a specific wavelength and is less tolerant of a change in the wavelength. It is better to describe a thin lens as introducing the delay Δ rather than a phase shift. The phase shift $2\pi\Delta/\lambda$ modulo 2π operation, which converts the residual thickness to distance, is different at different wavelengths.

Zone Plates

A diffractive lens can be approximated by quantizing the allowed phase shifts to a discrete set of values. The extreme case is a binary quantizer. This leads to the following definitions of the Fresnel zone plate and the Fresnel phase plate.

A *Fresnel zone plate* is the two-dimensional weighting function,

$$t(x,y) = \tfrac{1}{2}\left[1 + \text{sgn}(\cos\alpha(x^2+y^2)\right]\text{circ}(x/a,y/a),$$

7.4 Diffractive Lenses

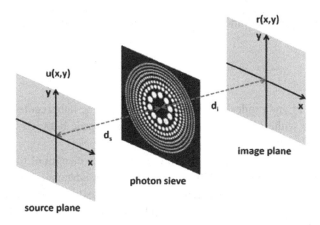

Figure 7.1 The photon sieve

taking values only in the binary set $\{0, 1\}$. By using the Fourier series expansion,

$$1 + \operatorname{sgn}(\cos 2\pi t) = \sum_{n=-\infty}^{\infty} \frac{\sin(\pi n/2)}{\pi n} e^{j2\pi nt},$$

with $2\pi t = \cos\alpha(x^2 + y^2)$, the Fresnel zone plate $t(x, y)$ behaves as the composite of ideal lenses with different focal lengths (see Problem 4.14).

A *Fresnel phase plate* is the two-dimensional weighting function,

$$t(x, y) = e^{j(\pi/2)[1+\operatorname{sgn}(\cos\alpha(x^2+y^2))]} \operatorname{circ}(x/a, y/a),$$

taking values only in the bipolar set $\{-1, +1\}$. As for the Fresnel zone plate, the Fresnel phase plate also behaves as the composite of multiple ideal lenses with different focal lengths.

The Photon Sieve

A *photon sieve* is an alternative kind of lens for the focusing of light by using the diffraction and interference of waves.[4] It consists of an opaque screen containing a suitable pattern of small holes. Instead of using a difference set to define the pattern of holes as discussed in the next section, the photon sieve is motivated by the Fresnel zone plate. A photon sieve is a *modified* binary Fresnel zone plate using small holes in place of the Fresnel rings. The rings of the Fresnel zone plate are each replaced by a very large number of holes, usually circular holes, as shown in Figure 7.1.

A simple way to define a photon sieve is to replace each ring of a Fresnel zone plate by a ring of circular holes sized to fit that ring. Let d_ℓ be the width of the ℓth ring, which is the difference between the outer and inner radii of that ring. Let r_i be

---

[4] Despite the exotic name, a photon sieve is conventionally analyzed using the diffraction of waves. Photons, as such, do not play a role in the theory and need not be mentioned.

the average of the outer and inner radii. The ℓth ring of width d_ℓ and radius r_ℓ then becomes

$$s_\ell(x,y) = \sum_i \text{circ}\left(\frac{x - x_{i\ell}}{d_\ell}, \frac{y - y_{i\ell}}{d_\ell}\right),$$

where $x_{i\ell}^2 + y_{i\ell}^2 = r_\ell^2$ and i indexes the holes in that ring. The holes centered $(x_{i\ell}, y_{i\ell})$ are equally spaced within the ring with sufficient guard space between the holes to maintain structural integrity.

When a small hole is comparable or smaller than the wavelength of the lightwave being imaged, the scalar theory of diffraction is no longer adequate. Then the more powerful vector theory of diffraction must be used. The scalar theory of diffraction given in Section 5.9 is adequate when the number of Fresnel rings is not too large because then the holes in the outer rings are not too small.

7.5 Coded Aperture Imaging

An electromagnetic wave in the optical frequency band can be focused into an image of the source of that wave by the use of a lens. The higher-energy photons of X-rays or gamma rays, however, interact with all species of atoms and so cannot penetrate any material. This means that there is no material that might be used to make a lens. Because a transparent material is not available at these frequencies, a conventional transmission lens cannot be constructed for X-rays or for gamma rays.[5] An alternative method of focusing uses reflection lenses at high angles of incidence. This requires the precision of a small fraction of a wavelength in the construction of the reflecting surface. Not only is it expensive but for wavelengths approaching 0.1 nanometer, impractical as well. An alternative way of forming an image is to use a Fresnel zone plate. For X-rays, the circular zones of a conventional Fresnel lens would have a width on the order of a nanometer or less, and thickness much smaller. It is not practical to construct a Fresnel zone plate or a photon sieve at X-ray or gamma-ray frequencies.

A *coded aperture* is an alternative imaging method that uses a pattern of pinholes satisfying the geometrical optics approximation. When an aperture containing a simple pinhole is placed in the path of the radiation, each ray, or incident photon, follows a straight line through the pinhole and onto a screen behind the pinhole as asserted by the geometrical optics approximation. The density distribution forming an image on the screen behind the single pinhole is an inverted copy of the radiation intensity distribution emitted by the source. The disadvantage of the pinhole camera used for X-ray imaging has the same disadvantage as does the pinhole camera used for optical imaging: it has poor sensitivity because only the area of the pinhole aperture captures radiation. This means that a long exposure time may be necessary in order to collect enough X-ray energy and the long exposure time is not acceptable in most

---

[5] X-rays have wavelengths in the range 10 picometers to 10 nanometers corresponding to frequencies in the range 3×10^{19} hertz to 3×10^{16} hertz.

7.5 Coded Aperture Imaging

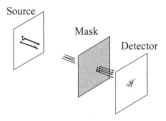

Figure 7.2 A coded aperture

applications. An alternative is to use multiple pinholes, which collectively comprise the aperture.

An array of multiple pinholes is shown in Figure 7.2. The arrangement of pinholes must be designed to facilitate the purpose of image formation. The active area of an array of N pinholes is N times the area of a single pinhole, so the aperture captures N times as much energy. Coded apertures for X-ray or gamma-ray telescopes may use more than 30,000 pinholes, thereby increasing the amount of captured light energy by this factor.

The wavelength of an X-ray or a gamma ray is so small that the geometrical optics approximation is quite adequate for distance scales somewhat larger than molecular sizes. Thus we model a pinhole camera at X-ray frequencies accurately as a pinhole through which X-ray radiation passes. In the geometrical optics approximation, a non-coherent source of intensity density $\rho(x,y)$ at distance d to the left of a single pinhole at the origin produces an image at distance d' to the right of the pinhole given by

$$r(x,y) = \rho\left(-\frac{x}{M}, -\frac{y}{M}\right),$$

where M is the ratio of the distances. To simplify the discussion, the image magnification M and the sign inversion can be suppressed from the equations by changing notation. Let $s(x,y) = \rho(-x/M, -y/M)$ denote the source after rescaling. A pinhole at the origin then produces the image $r(x,y) = s(x,y)$, while a pinhole at position (ξ, η) produces the image

$$r(x,y) = s(x - \xi, y - \eta).$$

This expression has the form of the convolution

$$r(x,y) = \delta(x - \xi, y - \eta) ** s(x,y)$$

of the source $s(x,y)$ with a single impulse at the location (ξ, η). An arrangement of L pinholes (represented as impulses) at locations (ξ_ℓ, η_ℓ) is given by

$$g(x,y) = \sum_{\ell=1}^{L} \delta(x - \xi_\ell, y - \eta_\ell),$$

where (ξ_ℓ, η_ℓ) is the location of the ℓth pinhole for $\ell = 1, \ldots, L$. This arrangement of L pinholes produces the image

7 Construction and Reconstruction of Images

$$r(x,y) = g(x,y) ** s(x,y)$$
$$= \sum_{\ell=1}^{L} s(x - \xi_\ell, y - \eta_\ell).$$

When additive sensor noise is included in the description, the noisy convolution

$$v(x,y) = \big(g(x,y) ** s(x,y)\big) + n(x,y)$$

is the received image.

The source $s(x,y)$ is not easily recognized in the recorded image $v(x,y)$ because the many pinholes cause the recorded image to consist of many overlapping copies of $s(x,y)$. The source $s(x,y)$ must be estimated by processing $v(x,y)$ using some form of deconvolution to partially remove $g(x,y)$. The pinhole pattern is chosen to enable a satisfactory deconvolution.

The quality of the deconvolved image depends on the pinhole pattern $g(x,y)$. The transform $|G(f_x,f_y)|^2$ should be nonzero for all (f_x,f_y) because otherwise, $G(f_x,f_y)S(f_x,f_y)$ would be zero and $S(f_x,f_y)$ could not be recovered at that spatial frequency.

One deconvolution method for coded-aperture imaging is to cascade a second two-dimensional binary pinhole array $h(x,y)$ with $g(x,y)$, where $h(x,y)$ is chosen so that $h(x,y) ** g(x,y)$ approximates an impulse and the largest sidelobe of $h(x,y) ** g(x,y)$ is small. One instance of this method is to choose $h(x,y) = g(-x,-y)$ and to choose $g(x,y)$ to be a binary pattern whose autocorrelation function $\phi(x,y) = g(x,y) ** g(-x,-y)$ has small sidelobes and so approximates an impulse.

To this purpose, it is convenient to constrain the pinholes of a coded aperture, as described next, to be a subset of points of $\mathbb{Z}^2$. The selected pinholes are centered within the selected squares of a checkerboard grid.

Definition 7.5.1 *A constrained m by m' coded aperture is an arrangement of black squares on an m by m' uniform array of unit squares.*

The squares that are not black can be regarded as white squares. The black squares locate the pinholes against a white background. A binary indicator function, $g(x,y)$, defined on the coded aperture is equal to one on the black squares or on a circle centered within that square and is equal to zero on the white squares. The task is to design the coded aperture so that $g(x,y)$ has a satisfactory autocorrelation function $\phi(x,y) = g(x,y) ** g(-x,-y)$.

One systematic method of design of the constrained array is based on the notion of a cyclic difference set.

Definition 7.5.2 *A cyclic difference set modulo an odd integer n and cyclic difference d is the set of nonnegative integers smaller than n such that every integer from 1 to n − 1 occurs the same number of times as the cyclic difference modulo n equal to d.*

For example, the set $\{1,2,3,5,6,9,11\}$ is a cyclic difference set modulo 15 with cyclic difference three. To check that this is so, notice that, modulo-15, the difference 1

7.5 Coded Aperture Imaging

occurs three times (as $2-1$, $3-2$, and $6-5$), the difference 2 occurs three times (as $3-1$, $5-3$, and $11-9$), the difference 3 occurs three times (as $5-2$, $6-3$, and $9-6$), the difference 4 occurs three times (as $5-1$, $6-2$, and $9-5$), the difference 5 occurs three times (as $6-1$, $11-6$, and $1-11$), difference 6 occurs three times (as $9-3$, $11-5$, and $2-11$), and so forth.

Any cyclic difference set modulo n for which n is the product of two coprime factors can be used to construct a coded aperture as follows: Let $n = mm'$ where m and m' are coprime. Simply number the squares of an m by m' array by the integers running down the extended diagonal. Because m and m' are coprime, the extended diagonal passes through every cell of the array, and every cell on the extended diagonal is assigned, in order, a unique integer from 1 to n. Those squares labeled with elements of the cyclic difference set are the holes of the coded aperture. Let $g_{ii'} = 1$ when (i, i') is an element of the coded aperture. Otherwise, let $g_{ii'} = 0$.

The correlation function is

$$\phi_{rr'} = \sum_{i'=0}^{m-1} \sum_{i'=0}^{m'-1} g_{ii'} g_{i+r,i'+r'},$$

where $g_{i+r,i'+r'} = 0$ when either $i + r$ is larger than one or $i' + r'$ is larger than one. This is the convolution $g(x, y) * * g(-x, -y)$, where $g(x, y) = \sum_i \sum_{i'} x^i y^{i'}$.

Because $g_{ii'}$ is equal to either zero or one, $\phi_{rr'}$ simply counts the number of places in which $g_{ii'}$ and $g_{i+r,i'+r'}$ are both one. The main lobe ϕ_{00} of the autocorrelation function is equal to the number of elements in the cyclic difference set. Moreover, no sidelobe has a value larger than d.

To show the assertion, refer to the *cyclic* correlation function

$$\phi^{\circ}_{rr'} = \sum_{i=0}^{m-1} \sum_{i'=0}^{m'-1} g_{ii'} g_{((i+r)),((i'+r'))},$$

where $((i + r))$ denotes $i + r$ modulo m and $((i' + r'))$ denotes $i' + r'$ modulo m'. This wraparound is equivalent to a linear correlation with the array g periodically repeated in two dimensions. Then the product $g_{ii'} g_{((i+r)),((i'+r'))}$ is always defined, and

$$\phi^{\circ}_{rr'} = \begin{cases} n & \text{when } r = 0 \pmod{m} \text{ and } r' = 0 \pmod{m'}, \\ d & \text{otherwise,} \end{cases}$$

where d is the number of times that each difference occurs in the difference set. The acyclic correlation function then satisfies

$$\phi_{rr'} = n \text{ when } (r, r') = (0, 0),$$
$$\phi_{rr'} \leq d \text{ otherwise.}$$

Example

A small example is the three by five coded aperture with $n = 7$ shown in Figure 7.3. This example is based on the cyclic difference set $\{1, 2, 3, 5, 6, 9, 11\}$. Starting in the upper left corner, the integers from 1 to 15 are written down the extended diagonal of

Figure 7.3 A difference-set coded aperture

a three by five array. Because three and five are coprime, every location of the array will be filled before the extended diagonal returns to the first square. The squares indexed by the elements of the cyclic difference set $\{1, 2, 3, 5, 6, 9, 11\}$ are shaded as in Figure 7.3. The shaded squares locate the holes of the coded aperture.

The cyclic correlation function always equals three except for the point $(r', r'') = (0, 0)$ because the underlying cyclic difference set has all cyclic differences equal to three. Therefore we know that the function

$$\phi(x, y) = g(x, y) ** g(-x, -y)$$

is equal to 7 at the origin and has no sidelobe larger than 3. This means that the cascade of pinhole arrays $g(x, y)$ and $g(-x, -y)$ produces an image of $s(x, y)$ scaled by an amplitude factor proportional to 7. The image is marred by several extraneous ghost images coming through the sidelobes. These ghost images are scaled in amplitude but no ghost image is scaled by a factor larger than 3. ∎

By constructing a cyclic difference set modulo a large integer n, say on the order of 30,000, one can construct a correlation function $\phi(x, y)$ with a peak amplitude n and with very small sidelobes compared to the mainlobe. Then $\phi(x, y)$ approximates an impulse and the estimated image

$$\widehat{s}(x, y) = g(-x, -y) ** [g(x, y) ** s(x, y)]$$
$$= \phi(x, y) ** s(x, y)$$

will be acceptable.

7.6 Phase Retrieval

Just as the one-dimensional signal $s(t)$ is uniquely determined in signal space by its one-dimensional Fourier transform $S(f)$, so too the two-dimensional signal $s(x, y)$ is uniquely determined in image space by its two-dimensional Fourier transform $S(f_x, f_y)$. This statement follows from the existence of the inverse Fourier transform. When some part of $S(f_x, f_y)$ is not known, however, the signal $s(x, y)$ is not uniquely determined unless that incomplete knowledge of the Fourier transform can be replaced in some way by side information.

In many instances of image formation, the magnitude of the two-dimensional Fourier transform can be measured but the phase cannot be measured. In such an application, prior side information such as the knowledge that $s(x, y)$ is real and nonnegative

7.6 Phase Retrieval

may be known and that side information may completely or partially compensate for the missing phase of the Fourier transform. Any computational algorithm that uses side information to recover the signal when the phase of the Fourier transform is unknown is called a *phase-retrieval algorithm* because recovering the signal $s(x,y)$ is equivalent to recovering the missing phase of the Fourier transform $S(f_x, f_y)$. Phase-retrieval algorithms, either in two dimensions or in three dimensions, are important in crystallography, optics, microscopy, and many other fields.

The Gerchberg–Saxton algorithm and the Fienup algorithm for phase retrieval are described in this section. An alternative algorithm for phase retrieval that may have better convergence properties is the Schulz–Snyder algorithm. The Schulz–Snyder algorithm is developed in Section 8.6 of Chapter 8 from a formal principle of inference after the method of alternating minimization is described. Chapter 9 discusses phase-retrieval methods developed specifically for sampled Fourier transforms.

The Gerchberg–Saxton Algorithm

One instance of phase retrieval arising in the space domain is the task of estimating the phase of a complex optical image $s_d(x,y)$ in one x,y plane at $z = d$ when given the magnitude $|s_d(x,y)|$ in that plane and also the magnitude $|s_{d'}(x,y)|$ in another x,y plane at $z = d'$, where the second image is a diffracted version of the first. The side information $|s_{d'}(x,y)|$ is obtained by probing the optical field magnitude in that second x,y plane and asserting the Huygens–Fresnel point-spread function connecting that plane to the image plane.[6]

The *Gerchberg–Saxton algorithm* for phase retrieval is an ad hoc iterative algorithm for retrieving the phase of a complex-valued wavefront in one plane, such as the image plane, from observations of that wavefront intensity in two different planes, one of which may be the image plane.

The common instance of the Gerchberg–Saxton algorithm retrieves the phase of the wavefront in the image plane from observations of the intensity in both that image plane and the far-field plane. The wavefronts in these two planes are related by the two-dimensional Fourier transform.

Let $s(x,y)$ and $S(f_x, f_y)$ be a two-dimensional Fourier transform pair denoted as

$$s(x,y) \Leftrightarrow S(f_x, f_y).$$

Given the two magnitudes, $|s(x,y)|$ and $|S(f_x, f_y)|$, the task is to find the complex image $s(x,y)$. Of course, once $s(x,y)$ is known, $S(f_x, f_y)$ is known as well.

This instance of the Gerchberg–Saxton algorithm alternates the two steps of computing tentative versions of the two-dimensional Fourier transform and the inverse two-dimensional Fourier transform. The first step computes the Fourier transform $\widehat{S}(f_x, f_y)$ from $\widehat{s}(x,y)$ using only the phase of $\widehat{S}(f_x, f_y)$, attaching it to the known magnitude $|S(f_x, f_y)|$. The second step computes the inverse two-dimensional Fourier

---

[6] Another application of this relationship is to calibrate the phase-adjustment of an optical system, such as is done in the Webb space telescope.

transform $\widehat{s}(x,y)$ using only the phase of the new $\widehat{s}(x,y)$, attaching that phase to the known magnitude $|s(x,y)|$. The algorithm is initialized with $s^{(0)}(x,y) = |s(x,y)|$.

Step 1

$$\widehat{s}^{(r)}(x,y) \Rightarrow \widehat{S}^{(r)}(f_x,f_y) = |\widehat{S}^{(r)}(f_x,f_y)|e^{j\hat{\theta}^{(r)}(f_x,f_y)}$$

$$\widehat{S}^{(r+1)}(f_x,f_y) = |S(f_x,f_y)|e^{j\hat{\theta}^{(r)}(f_x,f_y)}$$

Step 2 Halt or increment r.

Step 3

$$\widehat{S}^{(r)}(f_x,f_y) \Rightarrow \widehat{s}^{(r)}(x,y) = |\widehat{s}^{(r)}(x,y)|e^{j\hat{\theta}^{(r)}(x,y)}$$

$$\widehat{s}^{(r)}(x,y) = |s(x,y)|e^{j\hat{\theta}^{(r)}(x,y)}$$

Step 4 Return to Step 1.

The algorithm halts when it reaches a stationary point or when time expires.

The Gerchberg–Saxton algorithm can also be used, in principle, for any two planes related by a point-spread function $h(x,y)$ such as the Huygens point-spread function. The images $s_1(x,y)$ and $s_2(x,y)$ in the two planes are related by $s_2(x,y) = h(x,y) * * s_2(x,y)$. Now $S(f_x,f_y)$ is replaced by the convolution $s_1(x,y) = h(x,y) * * s_0(x,y)$. Step 1 of the algorithm is a convolution of $s_0(x,y)$ with point-spread function $h(x,y)$. Step 2 of the algorithm is a deconvolution of the point-spread function $h(x,y)$.

The Fienup Algorithm

The *Fienup algorithm*, shown in Figure 7.4, is an ad hoc iterative phase-retrieval algorithm for recovering the real nonnegative image $s(x,y)$ from the magnitude $|S(f_x,f_y)|$ of its Fourier transform using side information such as the side information that $s(x,y)$ is real and nonnegative, or that the support of $s(x,y)$ is within a stated region of the x,y plane. Noise is not considered. Section 8.6 of Chapter 8 discusses a probabilistic formulation of a more general problem in which the data consists of noisy measurements of the image magnitude.

The Fienup phase-retrieval algorithm is a heuristic algorithm that is not derived from an underlying fundamental principle of inference. The Fienup algorithm is not derived by minimizing an objective function. Nevertheless, in many situations it is a useful method for recovering $s(x,y)$ from the magnitude $|S(f_x,f_y)|$ of the Fourier transform $S(f_x,f_y)$.

An arbitrary image, $s_0(x,y)$, is chosen to initialize the estimate $\widehat{s}(x,y)$. At each iteration, a new estimate, $\widehat{s}(x,y)$, is computed from the previous estimate as follows. A signal in the frequency domain, $S'(f_x,f_y)$, is formed by combining the given magnitude $|S(f_x,f_y)|$ with the phase given by the phase of the Fourier transform of $\widehat{s}(x,y)$. That is, $S'(f_x,f_y) = |S(f_x,f_y)|e^{j\hat{\theta}(f_x,f_y)}$ where $\widehat{S}(f_x,f_y)e^{j\hat{\theta}(f_x,f_y)}$ is the Fourier transform of $\widehat{s}(x,y)$. The inverse Fourier transform of $S'(f_x,f_y)$ becomes the new estimate $\widehat{s}(x,y)$ insofar as it meets the constraining side conditions.

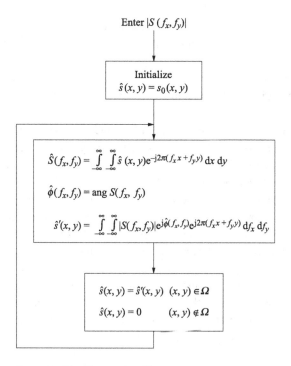

Figure 7.4 The Fienup algorithm

To impose a support constraint, let Ω be the set of (x, y) for which $s'(x, y)$ satisfies the image support constraints that comprise the side condition. This side condition could be that the set Ω is known to contain the support of $s(x, y)$ and so $s'(x, y)$ must be zero outside this set. The new estimate $\widehat{s}(x, y)$ is given by

$$\widehat{s}(x,y) = \begin{cases} s'(x,y) & (x,y) \in \Omega, \\ 0 & \text{otherwise.} \end{cases}$$

Other constraints, such as the constraint that $s(x, y)$ is real and nonnegative, can be handled in similar ways as may be appropriate. The iterations are repeated until a halt criterion is realized.

One wants to describe the performance of the Fienup algorithm in terms of the euclidean distance from the true image. This euclidean distance would be given by

$$d(\widehat{s}, s) = \int_{-\infty}^{\infty} |s(x,y) - \widehat{s}(x,y)|^2 \mathrm{d}x\mathrm{d}y$$
$$= \int_{-\infty}^{\infty} |S(f_x, f_y) - \widehat{S}(f_x,f_y)|^2 \mathrm{d}f_x\mathrm{d}f_y.$$

However, the true image $s(x, y)$ is not known, so the euclidean distance $d(\widehat{s}, s)$ cannot be computed. Instead, the empirical euclidean distance

$$\widehat{d}(\widehat{s}, s) = \int_{-\infty}^{\infty} (|S(f_x,f_y)| - |\widehat{S}(f_x,f_y)|)^2 \mathrm{d}f_x\mathrm{d}f_y$$

could be computed as a means of monitoring the iterations. The goal then is to find an image $\widehat{s}(x,y)$ satisfying any known constraints for which the empirical euclidean distance $\widehat{d}(\widehat{s},s)$ is zero or as small as possible. A *stagnation point* of the algorithm is a stable point of the algorithm at which $\widehat{d}(\widehat{s},s)$ is nonzero. Such stagnation points are not uncommon in applications of the Fienup algorithm.

There are variations of the Fienup algorithm that may be useful. These variations involve heuristic changes to the update rule that may speed convergence or impose other forms of side information as may be available. These can be suitably incorporated into the iteration process.

Although the Fienup algorithm is widely used and is often successful, no general proof of convergence is known or asserted. Indeed, there is no evident reason even to suppose that the first iteration always improves the image. It is only an empirical observation that an iteration usually improves the image. Furthermore, the algorithm is flawed by the stagnation problem. The iterated image may eventually no longer change noticeably with subsequent iterations even though various error measures may be converging to values strictly larger than zero. It then requires manual intervention by the user to force the algorithm away from a point of stagnation. Thus there is art in using the algorithm.

Conditions for Continuous Phase Retrieval

The task of phase retrieval is to recover $s(x,y)$ from the magnitude $|S(f_x,f_y)|$ of its Fourier transform when given that $s(x,y)$ satisfies certain known side conditions. The function $s(x,y)$ can be written as

$$s(x,y) = \int_{-\infty}^{\infty} |S(f_x,f_y)| e^{j\theta(f_x,f_y)} e^{j2\pi(f_x x + f_y y)} df_x df_y$$

where $\theta(f_x,f_y)$ is the unknown phase function. Certainly, the unknown phase function may be chosen in an infinite number of ways which means that an infinite number of $s(x,y)$ are associated with each $|S(f_x,f_y)|$. However, relatively benign side conditions on $s(x,y)$ usually suffice to make the choice of phase functions essentially unique. Only those phase functions $\theta(f_x,f_y)$ that give a function $s(x,y)$ consistent with the known side conditions can be used.

A common side condition is that $s(x,y)$ is real and nonnegative. This section will argue that for many cases this side information can partially substitute for the missing phase information, at least for the discretized two-dimensional version of the problem.

This side information that $s(x,y)$ is real and nonnegative is not completely equivalent to the missing phase information because some information is irretrievably lost. Specifically, translation of the real and nonnegative image $s(x,y)$ gives another real and nonnegative image $s(x-a,y-b)$, which, in the transform domain, corresponds to a simple multiplication of $S(f_x,f_y)$ by a complex exponential to give $S(f_x,f_y)e^{-j2\pi(f_x a + f_y b)}$. The complex exponential is discarded by the absolute value operation, which means that the image can be recovered from $|S(f_x,f_y)|$ only up to translation. Similarly, the mirror image of $s(x,y)$ is $s(-x,-y)$, which has the Fourier transform $S^*(f_x,f_y)$. This conjugation is lost when taking the absolute value of the Fourier transform. This means that $s(x,y)$ cannot be distinguished from $s(-x,-y)$. Knowing that $s(x,y)$ is real and nonnegative does not resolve this ambiguity. Thus we can hope to solve the phase-retrieval problem by asserting nonnegativity only up to translation and mirror image. Fortunately, these ambiguities are often regarded as acceptable.

A more egregious failure occurs whenever $s(x,y)$ can be decomposed as a convolution of two terms

$$s(x,y) = \int_{-\infty}^{\infty}\int_{-\infty}^{\infty} s_1(\xi,\nu)s_2(x-\xi,y-\nu)d\xi d\nu$$

This is because the convolution $s(x,y) = s_1(x,y) ** s_2(x,y)$ in the space domain implies the product $|S(f_x,f_y)| = |S_1(f_x,f_y)||S_2(f_x,f_y)|$ in the frequency domain. But this product is also the magnitude of the Fourier transform of the function $s_1(x,y) ** s_2(-x,-y)$. Consequently, the magnitude data cannot distinguish between $s_1(x,y) ** s_2(x,y)$ and $s_1(x,y) ** s_2(-x,-y)$. These are very different, and this ambiguity is not acceptable.

This is more egregious when $s(x,y)$ can be decomposed as a convolution of several nonnegative functions, such as $s_1(x,y) ** s_2(x,y) ** \cdots ** s_N(x,y)$. Any $s_\ell(x,y)$ can be replaced by $s_\ell(-x,-y)$ without changing the magnitude $|S(f_x,f_y)|$ and most such replacements give an incorrect image. These ambiguities are intrinsic. Fortunately, this kind of intrinsic ambiguity need not be a significant concern because it is rare for a real, nonnegative bivariate function to decompose as a convolution of two real bivariate functions.

The problem of phase retrieval can be restated entirely in the space domain by first noting that computation of $|S(f_x,f_y)|^2$ from $|S(f_x,f_y)|$ is trivial. The Fourier transform relationship

$$s(x,y) ** s(-x,-y) \Leftrightarrow |S(f_x,f_y)|^2$$

means that the autocorrelation function

$$\phi(x,y) = s(x,y) ** s(-x,-y)$$

can be easily computed as the inverse Fourier transform of $|S(f_x,f_y)|^2$. Thus the problem of computing $s(x,y)$ from the magnitude $|S(f_x,f_y)|$ is equivalent to the problem of computing $s(x,y)$ from its two-dimensional autocorrelation function $s(x,y) ** s(-x,-y)$. In this sense, phase retrieval can be seen as a special case of the problem of blind deconvolution which is discussed in Section 7.7.

This behavior of the two-dimensional Fourier transform with regard to phase retrieval is quite different from that behavior of the one-dimensional Fourier transform. In the one-dimensional case, the real nonnegative function $s(t)$ is not well identified by the magnitude of its Fourier transform $|S(f)|$. This striking contrast is related to the fact that every polynomial in one variable (or degree two or more) can be factored (in the complex field), whereas most polynomials in two variables cannot be so factored.

The phase-retrieval problem in one dimension is equivalent to the problem of finding $s(t)$ when given its autocorrelation function $s(t) * s(-t)$. These are equivalent because when $|S(f)|$ is known, then $|S(f)|^2$ is also known, and so, by the convolution theorem, $s(t) * s(-t)$ is easily computed and so is known as well. To demonstrate the nonuniqueness of phase retrieval in one dimension, note that restricting the function $s(t)$ to zero outside of a specified interval does not adequately compensate for the missing phase information because there are multiple functions $s(t)$ on a given finite support with the spectrum magnitude $|S(f)|$. Requiring the function to be real and nonnegative can reduce this ambiguity.

Conditions for Discrete Phase Retrieval

When $s(t)$ has a finite support, the dual of the sampling theorem applied in the frequency domain says that $S(f)$ can be recovered from its frequency-domain samples provided the frequency-domain samples are spaced sufficiently closely to avoid time-domain aliasing. This means that the time-domain function $s(t)$ must be time-constrained. Let $s(t) = 0$ for $|t| \geq T$. Then, by the dual of the sampling theorem, $S(f)$ must be sampled at a rate of $2T$ samples per hertz. Moreover, $S(f)$ goes to zero faster than $1/f$ as f goes to infinity because the integral of $|S(f)|^2$ is finite. To reconstruct $s(t)$ to within any given precision, only a finite number of samples of $S(f)$ are needed.

The z transform of the vector $\mathbf{s} = (s_0, \ldots, s_{n-1})$ is defined as the polynomial

$$\tilde{s}(z) = \sum_{i=0}^{n-1} s_i z^i.$$

The factorization of this monovariate polynomial is straightforward in principle. The fundamental theorem of algebra says that a polynomial of degree $n-1$ has $n-1$ zeros in the complex plane, some possibly repeated. This asserts the factorization

$$\tilde{s}(z) = \tilde{s}_0 \prod_{i=0}^{n-1} \left(1 - \frac{z}{a_i}\right)$$

over the complex numbers, and

$$\tilde{s}(z)\tilde{s}^*(z^*) = |\tilde{s}_0|^2 \prod_{i=0}^{n-1} \left(1 - \frac{z}{a_i}\right)\left(1 - \frac{z}{a_i^*}\right).$$

Because $S_k = \tilde{s}(e^{-j2\pi k/n})$, the term $|S_k|^2$ is obtained by setting $z = e^{-j2\pi k/n}$. When a_i is replaced by its conjugate a_i^*, the term $\tilde{s}(z)$ changes, but the product $\tilde{s}(z)\tilde{s}^*(z^*)$ does not change. Hence $|S_k|^2$ does not change. Because there are n such a_i, there are 2^n possibilities for the polynomial $s(z)$ that are consistent with the $|S_k|^2$. Consequently, there are 2^n different signal vectors consistent with $|\tilde{S}_k|$. This means that there are multiple solutions to the one-dimensional phase-retrieval problem. In some cases, some of these 2^n signal vectors can be similar (or identical) because two zeros of $\tilde{s}(z)$ are close together.

The discussion cannot be extended to two dimensions because there is no such fundamental theorem of algebra for bivariate polynomials. One may introduce an arbitrary polynomial, $\widetilde{s}(z,w)$, in two variables but this polynomial will rarely factor.

Because the phase retrieval problem is equivalent to solving the autocorrelation

$$\phi(x,y) = s(x,y) **s(-x,-y)$$

for $s(x,y)$, the phase retrieval problem is a variant of blind demodulation and our discussion of that topic can be adapted to this topic. In fact, a surprisingly small amount of side information may suffice to allow the reconstruction of a discrete two-dimensional image.

7.7 Blind Image Deconvolution

It is common experience that a projected optical image can be focused even when neither the projected image nor the blurring function is known. This everyday instance of blind image focusing motivates the topic of *blind deconvolution* that is studied in this section.[7] This is the task of removing the blur from an image that is caused by an unknown linear filter.

This section studies algorithms for blind deconvolution that remove an unknown space-invariant blur from a blurred two-dimensional image. Blind deconvolution is a more difficult version of the deconvolution problem because neither the blurring function nor the image is known.

Blind deconvolution of a two-dimensional noiseless blurred image computes $s(x,y)$ from $v(x,y)$ given the convolution

$$v(x,y) = \int_{-\infty}^{\infty}\int_{-\infty}^{\infty} h(x-\xi, y-\nu)s(\xi,\nu)\,d\xi\,d\nu$$

even though $h(x,y)$ is unknown. In the frequency domain, the two-dimensional convolution becomes

$$V(f_x,f_y) = H(f_x,f_y)S(f_x,f_y).$$

The task in the frequency domain that is equivalent to blind deconvolution is to find $H(f_x,f_y)$ and $S(f_x,f_y)$ when given a noisy version of the product $H(f_x,f_y)S(f_x,f_y)$. Clearly, this problem is not well-formulated. Without some form of regularization, there can be many solutions (perhaps solutions only within some numerical tolerance). Moreover, it may be difficult to tell which factor is the filter and which factor is the signal.

---

[7] The focusing of an optical image motivates the topic of this section, but in this case, the "deblurring" takes place within the optical field prior to converting the observed image into an intensity image in the detector, so it is not the same as the problem considered herein. It is mentioned only as an analogy. Moreover, this is a parametric form of blind equalization in that only the focal length of a lens can be changed to suppress blur.

A serious ambiguity would occur when $H(f_x,f_y)$ or $S(f_x,f_y)$ can be factored as the product of two or more terms. For example, $V(f_x,f_y)$ might have the form

$$V(f_x,f_y) = H_1(f_x,f_y)H_2(f_x,f_y)S_1(f_x,f_y)S_2(f_x,f_y).$$

These factors can be grouped into two terms in several incorrect ways. Suitable constraints or conditions must be imposed so that factorization of this kind almost never happens. A remedy for such ambiguity is regularization. An important form of regularization is a bandwidth constraint on $H(f_x,f_y)$ and $S(f_x,f_y)$ implying a bandwidth constraint on $V(f_x,f_y)$ as well. A bandwidth constraint is a suitable constraint for this purpose as will be described.

The difference in the nature of one-dimensional blind deconvolution and two-dimensional blind deconvolution can be understood by comparing two statements:

- Monovariate polynomials always factor.
- Bivariate polynomials rarely factor.

These statements will be discussed later in the section.

Blind Deconvolution in One Dimension

Before discussing blind deconvolution in two dimensions, blind deconvolution in one dimension is discussed. A fatal weakness of blind deconvolution in one dimension will be described. This weakness rarely happens in two dimensions.

The task of blind deconvolution in one dimension is to compute $s(t)$ from $v(t)$, where

$$v(t) = \int_{-\infty}^{\infty} h(t-\xi)s(\xi)d\xi,$$

even though $h(t)$ is unknown. In the frequency domain, this becomes

$$V(f) = H(f)S(f).$$

The equivalent task in the frequency domain is to find $H(f)$ and $S(f)$ when given the product $H(f)S(f)$. This is not a well-formulated problem. Without some form of regularization, there are many solutions. A bandwidth constraint is one form of regularization that, in itself, is not satisfactory. In one dimension, this problem does not have a satisfactory solution as it is given.

For a simple example, let $H(f)S(f) = e^{-\alpha t^2}$. There are many factorizations of $e^{-\alpha t^2}$. Thus, the terms $H(f) = e^{-\beta t^2}$ and $S(f) = e^{-\gamma t^2}$ are a valid factorization for any β and γ satisfying $\beta + \gamma = \alpha$. There are uncountably many such factorizations of $e^{-\alpha t^2}$ with this form or with other forms. Nonnegativity and support constraints on $s(t)$ are of no help in resolving this intrinsic ambiguity.

One way of reducing the number of ambiguities is by a bandwidth constraint on both $H(f)$ and $S(f)$. This can be written $H(f) = H(f)\text{rect}(f/B)$ and $S(f) = S(f)\text{rect}(f/B)$. This means that $V(f)$ can be sampled at a rate of $2B$ samples per second. With this form of regularization, blind deconvolution need be studied only for discrete sequences. As will be seen, however, for one-dimensional discrete sequences, this form of regularization by itself is not enough. Intrinsic ambiguities remain.

With additive noise included in the statement, the one-dimensional problem is based on the expression

$$v(t) = (h(t) * s(t)) + n(t).$$

The noise $n(t)$ is unknown, but in the usual case, the mean is zero and the correlation function is known. The task is to find $h(t)$ and $s(t)$ such that their convolution is a good fit to $v(t)$. Again, this task in one dimension is not well-behaved.

Blind Deconvolution in Two Dimensions

In two dimensions, the situation is much different. The noise-free blurred function in the frequency domain is $V(f_x,f_y) = H(f_x,f_y)S(f_x,f_y)$. When $V(f_x,f_y)$ is required to satisfy a two-dimensional bandwidth constraint, it may be written as $V(f_x,f_y) = V(f_x,f_y)\text{rect}(f_x/A,f_y/B)$. The sampling theorem in two dimensions converts this frequency-domain function $V(f_x,f_y)$ to a two-dimensional discrete sequence.

A two-dimensional convolution with additive noise has the form

$$v(x,y) = \int_{-\infty}^{\infty}\int_{-\infty}^{\infty} h(x-\xi, y-\eta)s(\xi,\eta)d\xi\,d\eta + n(x,y)$$
$$= (h(x,y) ** s(x,y)) + n(x,y).$$

The noise $n(x,y)$ is a two-dimensional zero-mean random process and the realization of the noise is not known but the two-dimensional correlation function of the noise is known. Side conditions on $s(x,y)$ may be known as well. The task is to estimate the signal $s(x,y)$ from the noisy observation $v(x,y)$ and perhaps to also estimate the point-spread function $h(x,y)$. It may be difficult to determine which factor in the convolution is the signal and which factor is the filter.

Conditions for Blind Image Deconvolution

The problem of blind deconvolution of two-dimensional signals is quite different from the problem of blind deconvolution of one-dimensional signals. Simple side conditions can suffice to make the blind deconvolution problem in two dimensions solvable. One important side condition is a bandwidth constraint imposed on $S(f_x,f_y)$ and $H(f_x,f_y)$ so that $s(x,y)$ and $h(x,y)$ may be sampled at a rate satisfying a Nyquist condition. This condition is important because the discrete version of two-dimensional convolution is highly constrained but, absent a Nyquist condition, the continuous version is not similarly constrained. Another common side condition is a nonnegativity constraint on the image $s(x,y)$, on the point-spread function $h(x,y)$, or on both. There may also be a boundedness constraint on the supports of $s(x,y)$ and $h(x,y)$. Such a constraint is that $s(x,y)$ can be nonzero only when $(x,y) \in \Omega_1$ for some support region Ω_1, and $h(x,y)$ can be nonzero only when $(x,y) \in \Omega_2$ for some support region Ω_2.

Even with such side conditions, the notion of a solution must be relaxed to allow for certain kinds of ambiguity in blind deconvolution. The reason is that for any real numbers a, b, and c,

$$h(x,y) ** s(x,y) = ch(x-a, y-b) ** c^{-1}s(x-a, y-b),$$

so we cannot hope to recover a, b, and c without additional information. Thus one must be content to recover $h(x, y)$ and $s(x, y)$ only up to amplitude and translation. It is also clear that a blind deconvolution algorithm cannot determine which of the two functions $h(x, y)$ and $s(x, y)$ is the image and which is the filter.

A more egregious ambiguity arises when either $h(x, y)$ or $s(x, y)$ can be decomposed as the convolution of two terms, such as

$$h(x, y) = h_1(x, y) \ast\ast h_2(x, y),$$

or

$$s(x, y) = s_1(x, y) \ast\ast s_2(x, y),$$

possibly with some approximation error. When both are true, this becomes

$$h(x, y) \ast\ast s(x, y) = h_1(x, y) \ast\ast h_2(x, y) \ast\ast s_1(x, y) \ast\ast s_2(x, y).$$

Now there are $2^4 - 2$ ways to partition the four terms on the right into two groups, excluding the two uninteresting partitions in which all terms are in the same group. In general, when $h(x, y)$ can be decomposed as an m-fold convolution and $s(x, y)$ as an n-fold convolution, there will be $2^{m+n} - 2$ nontrivial ways to decompose the convolution $h(x, y) \ast\ast s(x, y)$ and each decomposition corresponds to another solution, all but two of which are false solutions. Such ambiguities are intrinsic and can be resolved only by extrinsic side information.

A blurred version of a discrete image is the discrete convolution $v = h \ast\ast s$, which is described by

$$v_{ij} = \sum_k \sum_\ell h_{i-k, j-\ell} s_{k\ell}.$$

Our goal is to show that the two two-dimensional arrays h and s cannot, in general, be decomposed as $h_1 \ast\ast h_2$ and $s_1 \ast\ast s_2$. The terminology of polynomials is used for this development.

Irreducible Bivariate Polynomials

An n by n array $p = [p_{ij}], i = 0, \ldots, n - 1; j = 0, \ldots, n - 1$ can be represented by the bivariate polynomial $p(u, v) = \sum_i \sum_j p_{ij} u^i v^j$. Then $p(u, v) = p_1(u, v) p_2(u, v)$ if and only if $p = p_1 \ast\ast p_2$. To show that it is unusual for a two-dimensional array p to decompose as $p = p_1 \ast\ast p_2$, refer to the well-known fact that it is rare for a bivariate polynomial to factor. Just as a finite number of curves cannot fill the plane $\mathbb{R}^2$, and a finite number of surfaces cannot fill the space $\mathbb{R}^3$, so too a finite number of lower-dimensional hypersurfaces cannot fill $\mathbb{R}^n$. This statement can be made convincing by elementary arguments.

Consider a two by two array corresponding to the bivariate polynomial

$$p(x, y) = \sum_{i=0}^{1} \sum_{j=0}^{1} p_{ij} x^i y^j$$
$$= p_{11}(xy + ax + by + c).$$

The leading coefficient p_{11} has been factored out, so it is enough to study only the factorizability of the monic bivariate polynomial[8]

$$p(x,y) = xy + ax + by + c.$$

The set of three coefficients (a,b,c) define a point in three-dimensional euclidean space $\mathbb{R}^3$. Whenever $p(x)$ factors nontrivially, it factors as

$$\begin{aligned} p(x,y) &= (x+b)(y+a) \\ &= xy + ax + by + ab. \end{aligned}$$

Thus, $ab = c$ whenever $p(x,y)$ factors. The point $(a,b,ab) \in \mathbb{R}^3$ lies on a two-dimensional surface in $\mathbb{R}^3$ defined by the equation $ab = c$. For a point (a,b,c) that is not on this two-dimensional surface within $\mathbb{R}^3$, the polynomial $p(x,y)$ does not factor. Because an arbitrary point (a,b,c) almost surely does not satisfy $ab = c$, an arbitrary point in $\mathbb{R}^3$ almost always corresponds to a polynomial that does not factor. Moreover, an arbitrary point (a,b,c) is almost always the center of a ball of small radius ϵ such that every point in this ball in $\mathbb{R}^3$ corresponds to a polynomial that does not factor.

In a similar way, consider a real monic bivariate polynomial $p(x,y)$ of degree two in x and degree two in y of the form

$$p(x,y) = x^2 + axy + by^2 + cx + dy + e.$$

This polynomial has five free coefficients so it may be represented by a point in $\mathbb{R}^5$. The only nontrivial factorization into monic polynomials has the form

$$\begin{aligned} p(x,y) &= (x + Ay + B)(x + Cy + D) \\ &= x^2 + (A+C)xy + ACy^2 + (B+D)x + (AD+BC)y + BD \\ &= x^2 + axy + by^2 + cx + dy + e. \end{aligned}$$

The four free parameters (A,B,C,D) define a bivariate polynomial $p(x,y)$ that factors. These four free parameters define a four-dimensional hypersurface in $\mathbb{R}^5$ that leaves most of $\mathbb{R}^5$ empty. Only the points of $\mathbb{R}^5$ in the hypersurface specified in this way by the four parameters (A,B,C,D) correspond to a polynomial of this form that can be factored. Other points, not on one of these hypersurfaces, correspond to polynomials that do not factor. To provide some noise tolerance, include all points within ϵ of the hypersurface. Even so, almost all of $\mathbb{R}^5$ is empty.

For large n, this argument becomes even more compelling. A monic n by n real bivariate polynomial has $(n+1)^2 - 1$ real coefficients. The array of these coefficients can be regarded as a point in $\mathbb{R}^{n^2+2n}$. An n by n bivariate polynomial might factor, for example, as the product of two $n/2$ by $n/2$ bivariate polynomials when n is even. This replaces $n^2 + 2n$ coefficients with $2(n/2+1)^2 - 2 = n^2/2 + 2n$ coefficients. The set of $n/2$ by $n/2$ bivariate polynomials corresponds to a $n^2/2+2n$-dimensional hypersurface in $\mathbb{R}^{n^2+2n}$ occupying very little of $\mathbb{R}^{n^2+2n}$. Any factorization of the n by n bivariate

---

[8] A polynomial for which the coefficient of the term of highest order (requiring a definition of order) is a one is called a *monic polynomial*.

polynomial into an $n-i$ by $n-j$ bivariate polynomial times a i by j bivariate polynomial has fewer free coefficients and there are at most $n^2/4$ such other cases corresponding to hypersurfaces in $\mathbb{R}^{n^2+2n}$ of smaller dimension. The set of such hypersurfaces is sparse in $\mathbb{R}^{n^2+2n}$. Most points of $\mathbb{R}^{n^2+2n}$ correspond to bivariate polynomials that do not factor, even after allowing small perturbations of the coefficients.

An Algorithm for Blind Image Deconvolution

In general, the observed signal is contaminated by noise. Then

$$v(x,y) = \bigl(h(x,y) ** s(x,y)\bigr) + n(x,y)$$

is the observed signal where $h(x,y)$ and $s(x,y)$ are unknown and $n(x,y)$ is known two-dimensional zero-mean covariance-stationary noise with known covariance function $\phi(\tau,\nu)$. The task is to find both $h(x,y)$ and $s(x,y)$ so that $h(x,y) ** s(x,y)$ is the best fit to $v(x,y)$. This fit may be in the sense of the euclidian distance. A straightforward ad hoc procedure for blind deconvolution, constructed heuristically as a simple iteration, is described here to introduce the problem. A proof of convergence of this heuristic procedure is not given and none is claimed. Instead, in Section 8.4, a formal procedure is formulated and a better algorithm is derived.

Given the noise-free expression

$$v(x,y) = h(x,y) ** s(x,y)$$

with $v(x,y)$ known, the goal is to find the two nonnegative unknown functions $h(x,y)$ and $s(x,y)$ whose Fourier transforms both have bounded domains. The equivalent equation in the Fourier transform domain is

$$V(f_x,f_y) = H(f_x,f_y)S(f_x,f_y),$$

so the task is to factor $V(f_x,f_y)$ into the terms on the right. Because $V(f_x,f_y)$ is specified to be a product, the factorization does exist. Therefore the only concern is that it might not be unique. This is only a minor concern when the problem is appropriately discretized.

The nature of the problem suggests, absent of any theory, the following naive iterative procedure appropriately regularized as by discretization of the continuous functions. The iteration proceeds in the transform domain by choosing some initial nonnegative function for $h(x,y)$, and then alternating the following two steps with appropriate Fourier transforms and inverse Fourier transforms computed as implied.

Step 1
Let

$$\widehat{S}(f_x,f_y) = \begin{cases} \frac{V(f_x,f_y)}{H(f_x,f_y)} & \text{for } (f_x,f_y) \in \Omega_2, \\ 0 & \text{otherwise.} \end{cases}$$

Then
$$s(x,y) = \begin{cases} \widehat{s}(x,y) & \text{for } (x,y) \text{ such that } \widehat{s}(x,y) \geq 0, \\ 0 & \text{otherwise.} \end{cases}$$

Step 2
Let
$$\widehat{H}(f_x,f_y) = \begin{cases} \frac{V(f_x,f_y)}{S(f_x,f_y)} & \text{for } (f_x,f_y) \text{ satisfying } (f_x,f_y) \in \Omega_1, \\ 0 & \text{otherwise.} \end{cases}$$

Then
$$h(x,y) = \begin{cases} \widehat{h}(x,y) & \text{for } (x,y) \text{ satisfying } \widehat{h}(x,y) \geq 0, \\ 0 & \text{otherwise.} \end{cases}$$

With both $s(x,y)$ and $h(x,y)$ initialized as real functions, and Ω_1 and Ω_2 symmetric about the origin, the iterates $s(x,y)$ and $h(x,y)$ remain real. The iterations may be halted if and when the squared error

$$d^2(v, h**s) = \int_{-\infty}^{\infty}\int_{-\infty}^{\infty} [v(x,y) - (h(x,y)**s(x,y))]^2 dxdy$$

falls below some desired value. Otherwise, the iterations are halted when an iteration counter exceeds some specified value.

7.8 Optical Imaging from Point Events

Modern physics tells us that electromagnetic signals can exhibit either the behavior of waves or the behavior of particles. Weak signals are treated semiclassically as a stream of particles called *photons*. The detection of weak incident light can be thought of as the detection of a stream of photons that arrive intermittently at a photodetector. Each arriving photon is detected by means of an array of photoelectron-conversion sensors. To form an image,[9] a large number of photon detections must be collected at each cell, or pixel, of the array. When the number of photons is large, the law of large numbers asserts that the number of photons collected in each pixel of the array is nearly equal to the intensity of the image in that pixel, so it is sufficient to accumulate photoelectron conversion events in each pixel with a photodetector array (or with a photographic film). When the light intensity is strong or the exposure time is long, the number of received photons in each cell of the photodetector will be very large and the quantized nature of light can be ignored. However, this requires that the situation be stable for the duration of the collection process.

When either the propagating medium is fluctuating or the imaging sensor is shaking during the time it takes to collect the photoconversion events, the conventional

---

[9] Another example of imaging from multiple point events that would be spoiled by random motion is positron-emission tomography (PET), which is an important modality of medical imaging. Positron-emission tomography is discussed in Section 6.5.

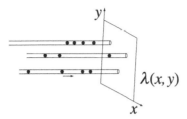

Figure 7.5 Illustrating spatially distributed poisson arrivals

optical image becomes blurred and possibly rendered useless. In the presence of such disturbances, the conventional ideas of optical image formation are unsatisfactory as such.

The algorithms developed in this section compute images directly from a large file of individual detected photon locations on the sensor array. In this way, the "camera" takes the form of a computational algorithm that processes a data file of measurements of individual photons. As many as tens of millions of photons – or more – and their parameters would need to be recorded, then processed, to form an image of an object by such an algorithm.

Let

$$u(x,y) = h(x,y) ** \rho(x,y)$$

be the optical signal arising from the object $\rho(x,y)$ as received at an array of photodetectors. Suppose that the incident light is so weak that the signal must be regarded as a stream of individual photons. The stream of photons at an infinitesimal detector cell of size $dxdy$ at location x, y is a poisson process with the infinitesimal arrival rate $\lambda(x,y)dxdy$ proportional to the intensity in that cell $u(x,y)dxdy$.

Figure 7.5 shows a way of visualizing[10] photons arriving at the cells of the sensor array as comprising individual poissonian streams. This illustration is not meant to be understood literally as a physical model of photon travel. It is only meant to clarify the meaning of $\lambda(x,y)$.

By ignoring the x, y coordinate, the full x, y array of poissonian streams can be pooled to form a single poissonian stream. This is because the merge of several poissonian streams is again a poissonian stream. Collectively, the combination of the multiple processes forms a poisson process with the combined arrival rate λ given by

$$\lambda = \int_{-\infty}^{\infty}\int_{-\infty}^{\infty} \lambda(x,y)dxdy.$$

This is the overall arrival rate of photoconversions at the photodetector array without regard for the x, y coordinate of the photoconversion. The ratio of $\lambda(x,y)$ to λ is the fraction expected in an infinitesimal cell at spatial coordinates (x,y), so this ratio can be interpreted as a probability density function. Given that a photon has arrived, the

---

[10] This semiclassical model of a photon as a particle on a well-defined path is adequate for the needs of this topic.

position of arrival on the x, y plane is a random variable with the probability density function

$$p(x,y) = \frac{\lambda(x,y)}{\lambda}.$$

In turn, this expression can be related to the intensity by

$$p(x,y) = \frac{u(x,y)}{u},$$

where

$$u = \int_{-\infty}^{\infty}\int_{-\infty}^{\infty} u(x,y) \mathrm{d}x\mathrm{d}y.$$

The distribution of photons may be impaired by uncompensated motion as shown in Figure 7.6. Two time-dependent impairments to the image-forming process are shown. These impairments are time-varying fluctuations in the propagation medium, such as may be due to atmospheric turbulence or by random jitter of the recording surface. In the absence of such impairments, the history of photon arrivals would be a list of position-of-arrival and time-of-arrival measurements of the form

$$\begin{array}{ccc} x_1 & y_1 & t_1 \\ x_2 & y_2 & t_2 \\ x_3 & y_3 & t_3 \\ & \vdots & \end{array}$$

In the presence of impairments, a photon arriving at time t is displaced from its correct position in the image by $\Delta x(t)$ and $\Delta y(t)$. Because of impairments, the observed history is the list

$$\begin{array}{ccc} x_1 + \Delta x(t_1) & y_1 + \Delta y(t_1) & t_1 \\ x_2 + \Delta x(t_2) & y_2 + \Delta y(t_2) & t_2 \\ x_3 + \Delta x(t_3) & y_3 + \Delta y(t_3) & t_3 \\ & \vdots & \end{array}$$

where $\Delta x(t)$ and $\Delta y(t)$ are unknown position offsets.

Figure 7.7 shows the actual and independently offset positions of the first five photons for the case where $\Delta x(t)$ and $\Delta y(t)$ change significantly from photon to photon so there is little of value in that history. In contrast, Figure 7.8 shows the same five photons for the case where $\Delta x(t)$ and $\Delta y(t)$ remain constant over the interval during which those five photons arrive. Whereas Figure 7.7 suggests that the image is irretrievably blurred or lost. Figure 7.8 suggests a situation where useful information remains in the data, at least over short time intervals. Although the absolute position of a photon arrival is not correct, the position difference between two photons is correct. This suggests that the data can be preprocessed to remove the effect of the unknown translation, such as by subtracting the position of the first photoconversion from the

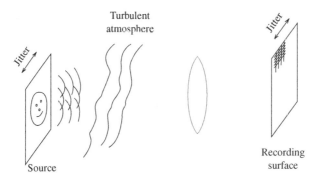

Figure 7.6 Some time-varying impairments

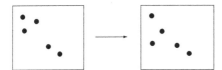

Figure 7.7 Illustrating actual and independently offset photon arrivals

Figure 7.8 Illustrating actual and collectively offset photon arrivals

remaining four. This process, known as *position differencing* (or, whimsically, *photon differencing*) reduces the five-photon data set to the form of four position differences

$$
\begin{array}{ccc}
x_2 - x_1 & y_2 - y_1 & t_2 \\
x_3 - x_1 & y_3 - y_1 & t_3 \\
x_4 - x_1 & y_4 - y_1 & t_4 \\
x_5 - x_1 & y_5 - y_1 & t_5.
\end{array}
$$

The probability density function of a sum is given by the convolution of the probability density functions of the summands. The probability density function of the difference $(u, v) = (x - x', y - y')$ is given by a convolution together with a sign reversal. Thus

$$p_D(u, v) = p(x, y) * * p(-x, -y).$$

This is the two-dimensional autocorrelation function of $p(x, y)$ and is proportional to the autocorrelation function of the image $u(x, y)$.

The imaging procedure is as follows. Given x, y, and t for each entry on a list of nN detected photoconversions, partition the data into N batches, each of n consecutive photoconversions. The batch size n is chosen small enough so that $\Delta x(t)$ and $\Delta y(t)$ can be adequately approximated as constant within each batch. Within each batch, subtract the position of the first photon detection event from the positions of the remaining $n-1$ detection events. Then the task is to estimate the image from the sequence of position differences.

This will be explained further for the case in which n is equal to two. The position offset is regarded as (almost) constant over the time interval between two photon arrivals. In this case, partition the $2N$ data points into N pairs of data points, compute pairwise position differences, and form a histogram[11] of the N position differences given by $v = N(x,y)/N$. By the law of large numbers, $N(x,y)/N$ converges to

$$p_D(x,y) = p(x,y) ** p(-x,-y),$$

which is proportional to the autocorrelation function of the image. For large N, the term $N(x,y)/N$ is nearly equal to $p_D(x,y)$. The two-dimensional Fourier transform of $N(x,y)/N$ is proportional to the magnitude-squared of the Fourier transform of the image

$$P_D(f_x,f_y) = |P(f_x,f_y)|^2,$$

and so

$$|P(f_x,f_y)| = \sqrt{P_D(f_x,f_y)}.$$

Image recovery from $|P(f_x,f_y)|$, for large enough N, is the task of phase retrieval, as described in Section 7.6. The Fienup algorithm, described in Section 7.6, is one algorithm that can be used to recover the image from the magnitude of its Fourier transform. The Schulz–Snyder algorithm, described in Section 8.6, is an alternative algorithm perhaps with better convergence behavior.

7.9 Imaging of a Semitransparent Bulk Material

An inhomogeneous semitransparent material is a bulk object that allows an illuminating signal to penetrate into it and scatters or back-reflects a small amount of the illuminating signal from details of the material as the unscattered signal continues to penetrate deeper into the material. The illuminating signal is thereby slowly attenuated.

Three instances of imaging bulk semitransparent material are described in this section. Each instance uses a simple geometrical optics model for wave propagation. The mathematical model in the three cases is similar, so the three instances are similar in form though they are very different as presented to the user. The three instances treated here are: *optical coherence microscopy*, *ultrasound imaging*, and *seismic imaging*. Each of the three instances launches a wave into a bulk material and processes the echo using a geometrical optics model of wave propagation under the Born approximation. Diffraction of the wave is usually not important in these three instances and is not considered in this section.

---

[11] On the continuous x,y plane, $N(x,y)$ is a scatter diagram of impulses. To speak of a histogram, the plane must be discretized into a grid of pixels (as it always is in practice), and N should be much larger than the number of pixels. For a 512 or 512 array, N should be in the tens of millions, or more.

In each case, the signal is attenuated within the material and becomes weaker as it reaches deeper into the material. The Born approximation posits that the attenuation is weak and can be ignored to first order. The Born approximation simplifies the analysis when it applies. Even so, as the signal is attenuated, the image deeper into the material fades. This can be partly compensated by amplification of the image as a function of depth, though noise and imperfections are amplified as well.

Optical Coherence Tomography

Optical coherence tomography[12] is a technique for obtaining images of moderate resolution of the immediate subsurface of a translucent or partially transparent material. A beam of light (typically in the near infrared) penetrating a surface, such as the skin of a medical patient, is attenuated by absorption and scattering by the details of small substructures that lie slightly below the surface. A small portion of the waveform is scattered back to the sensor or sensor array. The portion that is observed by the sensor or sensor array is used to form a cross-sectional image at each depth thereby forming, in principle, a three-dimensional image. The accessible depth is limited by the attenuation of the material being imaged.

Only the back-reflected light is used for imaging. Most of the light is not back-reflected. Most of the light is scattered into multiple directions and some of that scattered light is then rescattered. In this way much of the light scattered by the details of the material becomes converted into a background noise that may partially mask the desired back-reflected signal.

A long, properly encoded waveform, even though hidden in noise when its reflection is received by the sensor, becomes the input to a matched filter and can be well above the noise at the output of the matched filter. The matched filter for $s(t)$ is $s^*(-t)$. With $v(t) = s(t) + n(t)$ at the filter input, the filter output is the convolution $v(t) * s(-t)$. This filter output can be computed more directly by computing the correlation of $v(t)$ with $s(t)$. It is only necessary to save a copy of the transmitted signal $s(t)$ for use in the receiver (as in a delay line). Thus, the matched filter itself would be represented by the stored copy of the transmitted waveform. The output of the matched filter in the absence of noise is the autocorrelation of the transmitted waveform. The correlation of the reflected signal with a replica $s^*(t)$ is equivalent to a matched filter.

When the transmitter and the receiver are colocated, the waveform need not be permanently stored. The transmitted waveform can be saved as needed for computing the correlation with the received signal. Even a noise waveform can serve as the transmitted waveform. This is a common method for optical coherence imaging, recognizing that matched-filtering and correlation have the same mathematical expression. A low-coherence noisy waveform is transmitted. A copy of the noisy waveform is temporally stored, or delayed, and then correlated with the echo signal. This extracts the echo from the background noise.

---

[12] Despite its name, optical coherence tomography is not an instance of tomography of the form discussed in Section 6.10. It does not depend on the viewing of projections in the sense of the Radon transform. It does, however, share the property of producing cross-sectional images.

A wideband noise waveform has a correlation function with a sharp impulse-like autocorrelation. The echo signal is

$$r(x, y, ct) = \int \rho(x, y, ct - \xi) s(\xi) d\xi + n(x, y, ct),$$

where the integration is over the illuminated object. For each pair of coordinates x, y, this is a copy of $\rho(x, y, z)$ blurred in the z direction by the correlation function $\phi(z)$. The variables x and y are as seen by the transducer aperture, which is similar to an antenna aperture. The x and y coordinates of $\rho(x, y, z)$ can be explored by scanning the transducer.

Ultrasound Imaging

An ultrasound wave is an acoustic wave at a frequency larger than 20 kilohertz. An ultrasound waveform provides both spatial and temporal information regarding an object. An ultrasound waveform can be used for detection or for imaging in a way that is similar to the way that an electromagnetic waveform is used for detection or for imaging.

Much of the theory for radar detection and imaging given in Chapters 11 and 12 can be adapted for sonar detection and imaging using ultrasound, though with many differences. Ultrasound propagation and reflection in other materials such as biological tissue provides observable spatial and temporal signals that can be used for purposes such as medical diagnostics or for the inspection and grading of farm products.

The ultrasound propagation medium is not uniform and the propagation speed of an ultrasound wave is much smaller than the propagation speed of an electromagnetic wave. Depending on the specific host material, ultrasound travels in soft material such an air, water, or body tissue at approximately 1500 meters per second, and at least double this speed in hard material such as bone. Thus, a round trip in soft tissue over a distance of 7.5 centimeters takes about 100 microseconds. A pulse repetition frequency of from one to ten thousand pulses per second results in little or no range aliasing.

An image formed using a train of ultrasound pressure pulses as the insonification[13] is called an *ultrasound image* or a *sonogram*. Ultrasound in clinical applications is an acoustic wave at frequencies in the range from 1 to 18 megahertz. Higher acoustic frequencies have the advantage of better resolution. Higher acoustic frequencies have the disadvantage of stronger attenuation. Higher frequencies have less depth of penetration. Lower frequencies can image at deeper distances into the subject.

When a random semitransparent scattering material is insonified, the reflection from common internal scattering structures often behaves much like a noncoherent source.

An ultrasound imaging system consists of a transducer to launch a sound wave and the signal processing that forms the image. A transducer array consists of a one-dimensional or two-dimensional array of elements. A two-dimensional array might

---

[13] An object or scene is *illuminated* by light and *insonified* by sound.

use a square aperture about two centimeters on a side with transducer elements a few millimeters on a side. A 15 megahertz ultrasound wave traveling at a speed of 1500 meters per second has a wavelength of 0.1 millimeters. This would mean that an array element would have dimensions in the tens of wavelengths.

Medical ultrasound imaging may be used to penetrate human tissue to a depth of about ten or more centimeters. The ultrasound echo signal can be time-gated to form image slices at each depth. The signal is weakly attenuated by the host material and is scattered by various structures and boundaries. Some is reflected back to the source. The Born approximation may apply. A range of ten centimeters is about fifty times the size of a typical aperture element. Because this distance is well short of the far field, it suggests that the illumination function of an aperture element can be designed so that the transmit beam remains substantially confined. For this reason, it is common to describe the array as producing a collimated bundle of thin beams. The subject medium can scatter the signal in many directions. Thus, for the various purposes of signal processing, the elements of the array can be activated one-at-a-time, or collectively, as may be appropriate.

Geophysical Imaging

Acoustic waves propagated through the earth and reflected by internal details reveal much about the internal structure of our planet. Such waves, called *seismic waves*, can be used as the insonification to form an image of the subsurface or to reveal prominent details of the subsurface from its reflections. These details can be used to explore the earth's interior or to build a digital model of the substructure of a local geophysical region.

A seismic wave is an acoustic wave that propagates within a geophysical body. A seismic wave is generated by a natural or an artificial energy source. Most often a seismic wave is understood to mean a wave from a natural source. Nevertheless, the term seismic imaging may also refer to imaging of the local geology with an artificial energy source.

Acoustic waves can also be generated and transmitted locally for purposes of geological exploration. *Multipath* echos are not uncommon in seismic signals. A transmitted signal may travel an indirect path involving several reflections before returning to the site of the transmitter. This suggests that model-based methods of image reconstruction may be appropriate.

7.10 Near-Field Microscopy

Conventional microscopy based on the simple lens and the Fresnel approximation is subject to the Abbe diffraction limit on resolution. Near-Field microscopy is an instance of scanning microscopy that bipasses the limits of conventional optical imaging. Other such instances of scanning microscopy include microscopy based on electron tunneling, microscopy based on infrared absorption of various atomic species, and microscopy based on evanescent waves.

Each such instance of scanning microscopy moves an ultrasmall tip of a probe, or an array of such ultrasmall tips, across an x, y surface, mapping the interaction between the tip or tips and the surface as a function of x and y. The resolution of the image depends on the size of the tip. The map of this interaction is an image of some property of the surface that enables this interaction. The resulting image is of interest only when this interaction describes a property of interest of the surface.

Only imaging based on evanescent waves is described here. This instance of near-field microscopy is a variation of microscopy that bypasses the diffraction limit on resolution by exploiting the properties of evanescent waves. Evanescent waves attenuate rapidly with distance as measured by the number of wavelengths so the imaged object must be within a few wavelengths of the source of the evanescent wave. Otherwise, the signal becomes too weak to detect.

The evanescent wave[14] is formed by passing a planar wavefront through an aperture whose diameter is smaller than the wavelength of the illuminating wavefront. This results in an evanescent wave on the outer side of the aperture that will illuminate a target object at a small distance from the aperture. The reflection from the object is then detected. The optical resolution is determined by the size of the subwavelength aperture. A resolution on the order of a few percent of the optical wavelength can be obtained in this way. Because the target must be close to the aperture longitudinally, the aperture must be scanned across the x, y plane of a large scene in order to image the entire scene.

Evanescent waves can be described through the mathematics as a consequence of the Huygens–Fresnel principle. Evanescent waves can also be described through the physics as a consequence of simple experiments. Consider light that is completely reflected internally at a face of a prism. The geometrical optics view is that the reflected light is completely contained within the prism. There is an abrupt transition of the light intensity to zero at the boundary. However, experiments show that a second prism that is closer than a wavelength to the first prism will disturb the light within the first prism, even though the two prisms do not touch.[15] This phenomenon is explained physically by positing the existence of an evanescent wave on the outer surface of the first prism which is disturbed by the presence of the second prism.

7.11 Electron-Beam Microscopy

Optical microscopy in the visible band is limited by the wavelength of visible light. Optical microscopy can be extended into the ultraviolet frequencies with the aid of a ultraviolet photosensor but resolution is still limited by that ultraviolet wavelength. To obtain better resolution, a beam of electrons can be used as an alternative source of

---

[14] The existence of evanescent waves can be described formally as a consequence of Maxwell's equations. More generally, the evanescent waves can be described as a consequence of the wave equation. Instead, evanescent waves and Huygens principle are described in Section 4.2 using the two-dimensional Fourier transform applied to the superposition of plane waves.

[15] This behavior, which suggests the existence of evanescent waves, was first observed by Newton.

illumination in place of a beam of light. This imaging modality is known as electron-beam microscopy. An electron beam is used in place of a lightwave in order to obtain smaller resolution, perhaps on the order of 0.1 to 0.01 nanometers. This is because the intrinsic weight of an electron leads to waves of higher frequencies than are practical with lightwaves.

Just as a beam of light has both a wave nature and a particle nature, a beam of electrons has both a particle nature and a wave nature. In the case of light, the wave nature is its most evident manifestation. The particle nature of light is not evident in most common situations. In the case of electrons, the particle nature is its most evident manifestation. The wave nature of an electron beam is not evident in most common situations. The wave nature can be regarded as an emergent property of a beam of electrons rather than a property of an individual electron. These electron waves are called *de Broglie waves*.[16]

The particle of light has a mass that is directly proportional to the frequency and inversely proportional to the wavelength. A very small wavelength requires massive photons that would be difficult to create. Moreover, the light beam would be so coarsely quantized that the diffraction pattern becomes revealed only after long exposure times. Electrons, on the other hand, have a rest mass and a de Broglie wavelength inversely proportional to momentum.

A beam of electrons has the properties of a wave and can be focused by a properly shaped magnetic field imitating the function of an optical lens. Because it is a wave, an electron beam displays the properties of diffraction and interference.

There are two instances of electron-beam microscopy, called *electron reflection microscopy* and *electron transmission microscopy*. Both electron reflection microscopy and electron transmission microscopy are of interest. Electron reflection microscopy is a technique in which the target is a surface of interest from which a beam of electrons is reflected. The reflected beam is used to form an image of that surface. Electron transmission microscopy is a technique in which the target is an ultrathin material through which the electron beam is passed. In either case, diffraction of the resulting de Broglie wave is used to form an image of the target.

7.12 Spectral Imaging

The human eye perceives a mixture of blue light and yellow light as quite similar to monochromatic green light, perhaps even indistinguishable from green light. Yet viewing light with that color mixture through a few sharp color filters will reveal that these are very different. Moreover, light with that color mixture reflected from a blue

---

[16] The wavelength of an electron beam is called the *de Broglie wavelength*. The de Broglie wavelength is given by

$$\lambda = \frac{h}{p} = \frac{h}{mv},$$

where h is the Planck constant and p is the momentum of the electrons.

surface presents differently than reflected monochromatic green light from that blue surface.

One can say that the human eye undersamples the Nyquist sampling rate of the color spectrum, and so mixes the colors in the perceived lightwave signal.

The visible light spectrum consists of electromagnetic waves with frequencies in the terahertz range, usually described in terms of wavelength rather than frequency. The visible light spectrum consists of wavelengths from about 380 nanometers to about 700 nanometers. The various wavelengths are also called colors, but the color scale is subjective and without numerical precision. Various wavelengths or mixtures of wavelengths are given names as colors.

An object can reflect ambient light of a certain color only to the extent that the corresponding wavelength is contained within that ambient light and the object is responsive to that wavelength. White light is light that is flat over the visible light spectrum. The spectral energy density of white light is constant over wavelength λ in the visible band. White light contains all wavelengths equally; it has all colors and any color can be reflected from an illuminated object. A spectrum that is approximately constant may still be regarded as white light.

The human eye observes the color spectrum through three overlapping spectral filters that together cover the band of wavelengths called the color spectrum. These three color filters are the red filter $h_r(\lambda)$, the green filter $h_g(\lambda)$, and the blue filter $h_b(\lambda)$. The output of each filter is a sample of the light energy observed by that filter. Treating the color filters as simple linear filters, the output samples[17] s_r, s_g, and s_b are given by

$$s_r = \int_{\lambda_{min}}^{\lambda_{max}} h_r(\lambda)\rho(\lambda)d\lambda \quad s_g = \int_{\lambda_{min}}^{\lambda_{max}} h_g(\lambda)\rho(\lambda)d\lambda \quad s_b = \int_{\lambda_{min}}^{\lambda_{max}} h_b(\lambda)\rho(\lambda)d\lambda,$$

where each integral is over the visible spectrum going from λ_{min} to λ_{max}.

The perceived color depends on the relative values of the three samples s_r, s_g, and s_b observed by the three color filters. Any two wavelength spectra of incident light that produce the same triplet (s_r, s_g, s_b) will be perceived by the individual as the same color. Such an ambiguity may be seen as a form of aliasing.

The optical field $\rho(x, y, \lambda)$ reaching an aperture, such as the eye, at position (x, y) usually has a dependence on λ that is not well resolved by the three color filters. To fully observe the color spectrum, the required Nyquist rate on the λ axis may be rather high. The color spectrum $\rho(\lambda)$ is a nonnegative function of λ, and that spectrum does itself have a Fourier transform $R(f_\lambda)$. In the absence of spectral lines, the width of $R(f_\lambda)$ describes the necessary separation of samples on the wavelength (color) axis needed to observe the full spectrum without the aliasing of colors. In the presence of spectral lines, a Nyquist sampling approach would need to be very dense and so is not appropriate.

---

[17] By varying across the viewed image, they are better written as $s_r(x, y)$, $s_g(x, y)$, and $s_b(x, y)$.

A multispectral image of $s(x, y, \lambda)$ is, in general, a three-dimensional data cube with the axes labeled x, y, and λ. The first two axes describe spatial coordinates; the third axis describes wavelength (color). A multispectral imager may fill the data cube by scanning in the λ direction, forming an (x, y) image at each λ. Or the multispectral imager may fill the data cube by scanning in the y direction, forming an (x, λ) image at each y. More generally, computational methods may be used to produce the entire data cube at once as a three-dimensional image from a set of indirect data.

Problems

7.1 Define and derive the *multichannel Wiener filter* to treat the situation in which $s(x, y)$ is observed through two filters

$$v_1(x, y) = [h_1(x, y) * * s(x, y)] + n_1(x, y),$$
$$v_2(x, y) = [h_2(x, y) * * s(x, y)] + n_2(x, y),$$

where $n_1(x, y)$ and $n_2(x, y)$ are independent noise processes with the power density spectra $\Phi_1(f_x, f_y)$ and $\Phi_2(f_x, f_y)$, respectively, and the signal is a random process with the power density spectrum $\Phi_s(f_x, f_y)$.

7.2 By discussion of the following two examples, or other examples, show that knowing $|S(f)|$ may not be enough to determine the one-dimensional function $s(t)$ uniquely (up to translation and sign reversal) even with the side information that $s(t)$ is known to be real and nonnegative.

a. Sketch $s(t)$ and $|S(f)|$ when

$$s(t) = (1 \pm \epsilon \cos 2\pi f_0 t)\text{sinc}^2\left(\frac{t}{T_1}\right).$$

b. Sketch both versions of $s(t)$ and $|S(f)|$ when

$$s(t) = r(t) * r(t/2),$$

where $r(t) = \text{rect}(t) \pm \frac{1}{2}\text{rect}(t - 1)$.

7.3 Given that $s(t)$ has a contiguous closed support, can the support of $s(t)$ be determined from the support of $s(t) * s(-t)$? Given that $s(x, y)$ has contiguous closed support, can the support of $s(x, y)$ be determined from the support of $s(x, y) * * s(-x, -y)$? (A two-dimensional support is *contiguous* when any two points of the support can be connected by a curve contained within that support.)

7.4 The arrival times of a stream of photons comprise a poisson random process.

a. What is the probability density function on the waiting time, starting at time zero, for the arrival of the first photon?
b. What is the probability density function on the waiting time, starting at time zero, for the arrival of two photons?
c. What is the probability density function on the additional waiting needed before the second photon arrives after the first photon arrives?

7.5 **a.** Suppose that
$$h(x,y) * * s(x,y) = e^{-\pi(x^2+y^2)},$$
and one asks what are $h(x,y)$ and $s(x,y)$? Is there a unique answer to this question? Is the side information that $h(x,y)$ and $s(x,y)$ are both real and nonnegative helpful?

b. Suppose that the side conditions
$$S(f_x,f_y) = 0 \text{ for } f_x^2 + f_y^2 \geq 10,$$
$$H(f_x,f_y) = 0 \text{ for } f_x^2 + f_y^2 \geq 10$$
are given. Does the problem now have a solution? Does the problem have an approximate solution? What criteria should be used to define an approximate solution?

c. Numerically find an approximate solution.

7.6 **a.** In the set of monovariate polynomials with real coefficients, every polynomial that is irreducible over the reals has a degree not larger than two. What is the smallest number of irreducible factors (with real coefficients) that a real-valued polynomial of degree $2m$ can have? Give bounds on the number of ways in which a real-valued monovariate polynomial of degree ten can be factored into two real-valued monovariate polynomials.

b. Given that a real-valued monovariate polynomial of degree $2m$ has m real-valued irreducible factors, is it possible, in general, to perturb each coefficient by an amount less than ϵ so that the number of real-valued irreducible factors can be increased?

7.7 Let
$$s(x,y) = \text{rect}(x/a)\text{rect}(y/b)$$
with the Fourier transform $S(f_x,f_y)$. Let
$$s'(x,y) = \text{rect}(x/a)\text{rect}(y/b)\cos(2\pi x/A)$$
with the Fourier transform $S'(f_x,f_y)$. Show that, in some cases, when given $|S(f_x,f_y)|$ and the initial estimate $s(x,y) = s'(x,y)$, the Fienup iteration does not change the estimate $s(x,y)$. From this observation, give an explanation of the empirical fact that the Fienup algorithm sometimes produces false stripes as artifacts on a computed image.

7.8 For the covariance-stationary random process $s(x,y)$ with the two-dimensional covariance function $\phi(\xi,\eta) = E[s(x,y)s(x+\xi,y+\eta)]$, prove the following two statements for any point-spread function $u(x,y)$.

a. Let $\phi_1 = E[s(x,y)(u(x,y) * * s(x,y))]$. Then
$$\phi_1 = \int_{-\infty}^{\infty}\int_{-\infty}^{\infty} u(\xi,\eta)\phi_s(\xi,\eta)\,d\xi\,d\eta$$
$$= \int_{-\infty}^{\infty}\int_{-\infty}^{\infty} U(f_x,f_y)\Phi_s(f_x,f_y)\,df_x\,df_y.$$

b. Let $\phi_2 = E\big[(u(x,y) ** s(x,y))^2\big]$. Then

$$\phi_2 = \int_{-\infty}^{\infty}\int_{-\infty}^{\infty}\left[\int_{-\infty}^{\infty}\int_{-\infty}^{\infty} u(\xi,\eta)u(\xi-u,\eta-v)\,d\xi\,d\eta\right]\phi(u,v)\,du\,dv$$

$$= \int_{-\infty}^{\infty}\int_{-\infty}^{\infty} |U(f_x,f_y)|^2 \Phi_s(f_x,f_y)\,df_x\,df_y.$$

7.9 The deconvolution problem is usually complicated by the fact that the blurred image is truncated by the finite support of the recording frame. Therefore only

$$r(x,y) = \text{rect}\left(\frac{x}{L_1},\frac{y}{L_2}\right)\Big(h(x,y) ** s(x,y)\Big)$$

is known, rather than $h(x,y) ** s(x,y)$. What then is the known signal in the frequency domain? How does this change the problem of deconvolution? Under which circumstances is the Wiener filter still appropriate?

7.10 A diffractive lens defined for wavelength λ has a phase shift that is given by $(2\pi/\lambda)(x^2+y^2)/2f_\ell$ (modulo 2π). How does this lens perform as an imager for other wavelengths? Would a change in the definition to $(2\pi/\lambda)(x^2+y^2)/2f_\ell$ (modulo 4π) make the lens more tolerant to a change in wavelength?

7.11 Estimate how many holes are included in the innermost ring of a photon sieve to be used for wavelength λ.

7.12 A given image, $s_\lambda(x,y)$, is a function of position (x,y) for the color corresponding to wavelength λ.

a. Under what conditions can the image $s_\lambda(x,y)$ be represented by three "color samples" $s_{\lambda_1}(x,y)$, $s_{\lambda_2}(x,y)$, and $s_{\lambda_3}(x,y)$?
b. How does this change when the filtered image $s'_\lambda(x,y)$ is sampled, where $s_{i,\lambda}(x,y) = s_\lambda(x,y) * h_i(\lambda)$?
c. When there are two such systems of filters that each use somewhat different color filters $h_i^{(1)}(\lambda)$ and $h_i^{(2)}(\lambda)$ for $i =$ red, green, blue, can the two reconstructions of $s_\lambda(x,y)$ agree?

7.13 The whitened matched filter minimizes the probability of detection error of a pulse in stationary gaussian noise. Does this mean that the whitened matched filter minimizes the expected value of the difference between the noisy peak output of a filter and the noise-free peak output of that filter?

7.14 A discrete array $s = [s_{ij}]$ consists of 512 by 512 randomly chosen eight-bit numbers. How many such arrays are there? What can be said about the fraction of these arrays that can be decomposed as the convolution of two smaller arrays (within the precision of the fixed-point numbers)? How does this relate to the factorizability of bivariate polynomials?

7.15 Describe the iterations of the Gerchberg–Saxton algorithm under the following circumstances. The computations are to be cropped to a finite L by L region for some large L.

a. Given that $|s(x,y)| = |\text{rect}(x,y)|$ and $|S(f_x,f_y)| = |\text{sinc}(f_x,f_y)|$.
b. Given that $|s(x,y)| = |\text{sinc}(x,y)|$ and $|S(f_x,f_y)| = |\text{rect}(f_x,f_y)|$.
c. Given that $|s(x,y)| = |\text{rect}(x-a,y-b)|$ and $|S(f_x,f_y)| = |\text{sinc}(f_x,f_y)|$.

Notes

The Fienup algorithm (1978) can be seen as a variation of the Gerchberg–Saxton algorithm (Gerchberg and Saxton, 1972). Both of these algorithms are improvised without a formal underlying foundation. Schulz and Snyder (1992) proposed an alternative algorithm for phase retrieval. The Schulz–Snyder algorithm was given a foundation and derived as an alternating-minimization algorithm by Oktem and Blahut (2011). Phase retrieval was recognized early as an important problem early on in the field of X-ray crystallography, leading to methods such as were developed by Hauptman and Karle (1953) and by many others as are appropriate for applications in that field. Bruck and Sodin (1979), and later Bates (1984), argued in favor of the essential uniqueness of phase retrieval. By introducing the abstraction of *zero sheets*, Izraelevitz and Lim (1987) proposed a noniterative, computational algorithm that explicitly computes the image from the zeros of a two-dimensional polynomial. This idea was also discussed by Lane, Fright, and Bates (1987). Such abstract approaches have not yet found useful consequences. A more recent approach is to craft a likelihood function based on a probabilistic formulation of the problem, as is discussed in Chapter 8, thereby suppressing explicit reference to phase retrieval. The image is obtained directly by maximizing a loglikelihood function. With this approach, the phase of the Fourier transform need not be mentioned.

Wiener (1949) introduced the study of optimal filtering via the least-squares error criteria. Both causal and noncausal formulations of the problem are called Wiener filters, including problems without noise, and problems with constraints on the filter such as constraints on its support. The Richardson–Lucy iterative method of deblurring was developed to recover nonnegative images of astronomical objects from noisy blurred images (Richardson, 1972; Lucy, 1974). This method was independently rediscovered by Shepp and Vardi (1982) in the context of medical imaging.

Coded apertures are now widely used in X-ray and gamma-ray astronomy, and in other fields. An early treatment of coded apertures was given by Dicke (1968). A survey of cyclic difference sets was given by Helleseth and Kumar (1998).

The topic of blind deconvolution arose independently in many fields and a large literature now exists for this topic. Lane and Bates (1987) used ideas similar to those used in the study of phase retrieval to better understand blind deconvolution, arguing from the nature of zero sheets in four-dimensional space that blind deconvolution for discrete signals is essentially unique. Ghiglia, Romero, and Mastin (1996) demonstrated that zero-sheet methods are sensitive to noise. Ayers and Dainty (1982) studied iterative methods for blind convolution.

Nayar and Nakagawa (1994) studied the relationship between focusing and shape estimation. The Clean algorithm was introduced by Högbom (1974) for applications in radio astronomy. The performance of the Clean algorithm and its relationship to least-squares was studied by Schwarz (1978, 1979).

The statistics of speckle was studied by Goodman (1976) and by Tur, Chin, and Goodman (1982). Speckle in ultrasound systems was studied by Burckhardt (1978), who established that a slight change in perspective would give independent random speckle, thereby making possible the smoothing of speckle by averaging. In astronomy, the term "speckle" is used to describe the appearance of short-exposure images through a turbulent atmosphere. Labeyrie (1970) pointed out that multiple, short-exposure images collectively retain diffraction-limited information and that suitable phase-retrieval algorithms can recover the diffraction-limited image. Synge (1928) had earlier introduced the method of near-field scanning optical microscopy, which obtains superresolution by using observations at multiple positions through a subwavelength aperture.

8   Likelihood and Information Methods

A broader view of the topic of image formation is introduced in this chapter. The notion of the "correct" image is replaced by the notion of the "best" image. Rather than formulating the task as one of finding the single correct image, the task of image formation is viewed as the task of choosing the best image from the set of all possible images. The chosen image is the one that best accounts for a given set of data, usually a noisy and incomplete set of data. The set of possible images may be the set of all real-valued two-dimensional functions on a given support, or the set of all nonnegative real-valued two-dimensional functions on that given support. For this approach to be followed, one must define criteria upon which to decide which image best accounts for a given set of data. The class of information-theoretic criteria comprise a powerful class of optimality criteria. These are optimality criteria based on a probabilistic formulation of the image-formation problem.

Many imaging techniques are built on the idea of estimating a value separately for each cell of the image. Within each cell, the data is processed to estimate the signal within that cell without consideration of the signal in other cells. This criterion does have some intuitive appeal and leads to relatively straightforward and satisfactory computational procedures. This often-used criterion of separately estimating the signal for each cell cannot be defended at a fundamental level because other structure, such as dependence between neighboring cells, is ignored. Fundamental approaches can be developed by maximizing more global performance measures such as the likelihood function or the entropy of the full image. These methods, in principle, give better images and may be preferred to the more conventional methods. However, maximum-entropy and maximum-likelihood algorithms are complicated and nonlinear, and their behavior may be unsatisfactory in the presence of modal uncertainty.

The maximum-likelihood principle is a general information-theoretic principle for problems of decision and estimation. The maximum-likelihood principle is formulated for the decision problem using the criterion of minimizing the probability of error. The principle is formulated later for the estimation problem with the justification of minimizing the error variance.

The maximum-likelihood principle will be developed first for the case of a finite set of measurements, both discrete measurements and continuous measurements. Later, to apply the maximum-likelihood principle to waveforms, the continuous-time waveform $v(t)$ is approximated by a finite set of discrete-time samples to which the maximum-likelihood principle is applied. The maximum-likelihood principle for the waveform measurement $v(t)$ is then extended to the limit as the number of samples of $v(t)$ goes to infinity. In turn, this leads to the maximum-likelihood approach to image formation as an estimation problem.

8.1 Likelihood Functions and Decision Rules

The M-ary hypothesis-testing problem is the task of deciding in favor of one hypothesis from a set of M hypotheses indexed by $m = 0, \ldots, M-1$, one and only one of which is true. Corresponding to the mth hypothesis H_m is the probability distribution function $p_m(x)$. When given a measurement of the random variable X, one of the M hypotheses is to be selected as the hypothesis whose probability distribution produced that measurement. The natural rule is to choose the hypothesis for which the probability of decision error is as small as possible.

An enlarged version of this problem considers a block of independent measurements $(x_1, \ldots, x_n)$ of blocklength n occurring with all components of the same block based on the same probability distribution function $p_m(x_1, \ldots, x_n) = \prod_{\ell=1}^{n} p_m(x_\ell)$ corresponding to hypothesis H_m.

Given the finite block of data $(x_1, \ldots, x_n)$, the task is to decide which probability distribution function was used to generate that data. A *decision rule* is a function of the data, denoted $\widehat{m}(x_1, \ldots, x_n)$, for which $\widehat{m}$ is the estimated value of m when given the data vector $(x_1, \ldots, x_n)$. A *decision error* occurs when the mth probability density function $p_m(x)$ was used to generate the data block and the decision $\widehat{m}(x_1, \ldots, x_n)$ is not equal to m. Given that m is true, the conditional probability of error is denoted $p_{e|m}$. The average probability of decision error $p_e = \left(\sum_m p_{e|m} \right)/M$ is minimized over all decision rules when each m is equally likely to be true before the measurement is made. This condition means that each of M hypotheses occurs with the prior probability $1/M$.

When block probability distribution $p_m(x_1, \ldots, x_n)$ is under discussion, the index m is considered fixed and refers to hypothesis H_m. The data values $x_1, \ldots, x_n$ are realizations of the real-valued random variables $X_1, \ldots, X_n$. In this role as a function of $x_1, \ldots, x_n$ for a fixed m, the function $p_m(x_1, \ldots, x_n)$ is called a *probability distribution function*.[1] However, when the block measurement $x_1, \ldots, x_n$ is known but m is unknown, the function $p_m(x_1, \ldots, x_n)$ is viewed as a function of m for the fixed data vector $x_1, \ldots, x_n$. In this role, $p_m(x_1, \ldots, x_n)$ is known as a *likelihood function* and – reversing the notation of $p(x_1, \ldots, p_n | m)$ – is denoted $L(m|x_1, \ldots, x_n)$ or simply $L(m)$, where $L(m|x_1, \ldots, x_n) = p_m(x_1, \ldots, x_n)$. Because of its important role in the forthcoming Theorem 8.1.1, the likelihood function $L(m) = p_m(x_1, \ldots, x_n)$ arises frequently. Often it is more convenient to work with the likelihood function in the form of its logarithm. Then $\log L(m)$ is called the *loglikelihood function* and is denoted $\Lambda(m)$ or $\Lambda(m|x_1, \ldots, x_n)$, where $\Lambda(m|x_1, \ldots, x_n) = \log p_m(x_1, \ldots, x_n)$.

---

[1] Probability distribution function is the general term. For a continuous random variable, the probability distribution function is called a *probability density function*. For a discrete random variable, it is called a *probability mass function*.

8.1 Likelihood Functions and Decision Rules

The optimum decision rule for the M-ary hypothesis-testing problem is the decision rule with the smallest average probability of error p_e. The following theorem states that the optimum decision rule is the maximum-likelihood decision rule.

Theorem 8.1.1 *For the M-ary decision problem with equiprobable hypotheses:*

(i) Given the random block measurement $(X_1,\ldots,X_n)$ with realization $(x_1,\ldots,x_n)$, the optimum decision rule is

$$\widehat{m}(x_1,\ldots,x_n) = \underset{m}{\operatorname{argmax}}\, \Lambda(m|x_1,\ldots,x_n),$$

where the argmax on the right side denotes that m for which the function $p_m(x_1,\ldots,x_n)$ is largest for the data vector $x_1,\ldots,x_n$.

(ii) Moreover, when $X_1,\ldots,X_n$ is a gaussian vector random variable for each m with independent components of identical variance σ^2 and componentwise means $c_{m,1},\ldots,c_{m,n}$, the optimum decision rule is to choose that m for which the euclidean distance $\sum_{\ell=1}^{n}(x_\ell - c_{m,\ell})^2$ is smallest.

(iii) Moreover, when $\sum_{\ell=1}^{n} c_{m,\ell}^2$ is independent of m, the optimum decision rule is to choose that m for which the correlation $\sum_{\ell=1}^{n} x_\ell c_{m,\ell}$ is largest.

Proof:
Let $\mathcal{U}_m$ be the set of data vectors $(x_1,\ldots,x_n)$ for which the decision is that the mth probability density function was used to generate the memoryless block $(x_1,\ldots,x_n)$. Given that m is true, the probability of decision error is

$$p_{e|m} = \int_{\mathcal{U}_m^c} p_m(x_1,\ldots,x_n)dx_1\ldots dx_n.$$

The average probability of decision error over the M equiprobable hypotheses is

$$p_e = \sum_{m=0}^{M-1} \frac{1}{M} \int_{\mathcal{U}_m^c} p_m(x_1,\ldots,x_n)dx_1\ldots dx_n$$

$$= \sum_{m=0}^{M-1} \frac{1}{M}\left[1 - \int_{\mathcal{U}_m} p_m(x_1,\ldots,x_n)dx_1\ldots dx_n\right].$$

To minimize p_e, the measurement $(x_1,\ldots,x_n)$ is assigned to the set $\mathcal{U}_m$ for which $p_m(x_1,\ldots,x_n)$ is larger than $p_{m'}(x_1,\ldots,x_n)$ for all $m' \neq m$. A tie for the largest probability can be broken by any arbitrary rule such as break a tie in favor of the smallest index. This concludes the proof of the first statement of the theorem.

Suppose that for each m, $(x_1,\ldots,x_n)$ is a gaussian vector random variable with independent components of identical variances σ^2 and means $c_{m,1},\ldots,c_{m,n}$. Then

$$p_m(x_1,\ldots,x_n) = \prod_{\ell=1}^{n} \frac{1}{\sqrt{2\pi\sigma^2}} e^{-(x_\ell - c_{m,\ell})^2/2\sigma^2}$$

$$= \frac{1}{(\sqrt{2\pi\sigma^2})^n} e^{-\sum_\ell (x_\ell - c_{m,\ell})^2/2\sigma^2}.$$

The maximum-likelihood decision rule decides on that m for which $p_m(x_1,\ldots,x_n)$ is largest. To maximize $p_m(x_1,\ldots,x_n)$ over m, one should minimize $\sum_\ell (x_\ell - c_{m,\ell})^2$ over m. This completes the proof of the second statement of the theorem.

The third statement of the theorem follows from expanding the square and noting that the term $\sum_\ell x_\ell^2$ is independent of m and, by the condition of the third statement, the M terms $\sum_\ell c_{m,\ell}^2$ are independent of m as well. This completes the proof of the theorem. ∎

More generally, the index m can be replaced by the parameter γ. This parameter could be a scalar γ, a vector $\boldsymbol{\gamma}$, or an image also denoted $\boldsymbol{\gamma}$. The probability density function $p(x_1,\ldots,x_n|\boldsymbol{\gamma})$ is conditional on the parameter $\boldsymbol{\gamma}$. The parameter $\boldsymbol{\gamma}$ is not yet specified to be a random variable so there need not be a prior on $\boldsymbol{\gamma}$. Possibly, some components of the vector or image $\boldsymbol{\gamma}$ are discrete parameters and some components of $\boldsymbol{\gamma}$ are continuous parameters.

The loglikelihood function

$$\Lambda(\boldsymbol{\gamma}) = \log p(x_1,\ldots,x_n|\boldsymbol{\gamma})$$

is defined as a function of the scalar or vector γ.

The loglikelihood function may be of interest only for the purpose of finding where it achieves its maximum, while the value of the maximum itself may be of less interest. The value at which the maximum occurs is written

$$\widehat{\boldsymbol{\gamma}} = \underset{\gamma}{\operatorname{argmax}} \, \Lambda(\boldsymbol{\gamma}).$$

Constants that are added to or multiply the loglikelihood function do not affect the location of the maximum, so it is common practice to suppress such constants when they occur by redefining $\Lambda(\boldsymbol{\gamma})$. For example, the gaussian distribution with the known variance σ^2 and the unknown mean $\bar{x}$ has the loglikelihood function

$$\Lambda(\bar{x}) = -\log\sqrt{2\pi\sigma^2} - (x-\bar{x})^2/2\sigma^2,$$

as defined. For economy of notation, the known constants can be discarded to write

$$\Lambda(\bar{x}) = -(x-\bar{x})^2.$$

This abbreviated function might better be called a *likelihood statistic* when one wants a reminder that it is not precisely the loglikelihood function. A likelihood statistic is an example of a sufficient statistic. In general, a *statistic* is any function of the received data, and a *sufficient statistic* is a statistic that is equivalent to a likelihood function for the purpose at hand. No relevant information is missing from a sufficient statistic.

Problems in which the number of measurements n tends toward infinity are now considered. In such a case, the limit of $\Lambda(\boldsymbol{\gamma})$ as n goes to infinity may be infinite for all (or many) values of $\boldsymbol{\gamma}$. Then it would be meaningless to deal with the maximum over $\boldsymbol{\gamma}$ of the limit of $\Lambda(\boldsymbol{\gamma})$. The value of the maximum is not of immediate interest. The value of $\boldsymbol{\gamma}$ where the maximum occurs is more of interest or, better, the limit as n

goes to infinity of the sequence of values of γ that achieve the maximum for each n. To be precise, let

$$\hat{\gamma} = \lim_{n\to\infty} \operatorname*{argmax}_{\gamma} \log p_n(x_1,\ldots,x_n|\gamma),$$

provided the limit makes sense. The more immediate definition

$$\hat{\gamma} = \operatorname*{argmax}_{\gamma} \lim_{n\to\infty} \log p_n(x_1,\ldots,x_n|\gamma)$$

may be nonsensical because the limit may diverge to negative infinity for all data vectors x.

To avoid divergence of the limit, a normalized version of the loglikelihood function can be defined, which can be done in several ways. One choice is to define

$$\Lambda(\gamma) = \log[B_n p(x_1,\ldots,x_n|\gamma)],$$

where B_n is any convenient function of n, independent of γ, that will cause $\Lambda(\gamma)$ to have a finite limit. Alternatively, one can use a form called the *loglikelihood ratio*, also denoted by the symbol Λ and given by the general form

$$\Lambda(\gamma,\gamma') = \log \frac{p(x_1,\ldots,x_n \mid \gamma)}{p(x_1,\ldots,x_n|\gamma')},$$

or alternatively,

$$\Lambda(\gamma) = \log \frac{p(x_1,\ldots,x_n|\gamma)}{p(x_1,\ldots,x_n)},$$

where $p(x_1,\ldots,x_n)$ is a convenient reference probability distribution, perhaps corresponding to noise only. The purpose of introducing either form of the loglikelihood ratio is to have a form that remains finite as n goes to infinity. The maximum-likelihood estimate is then that value of γ maximizing $\Lambda(\gamma)$ or for which $\Lambda(\gamma,\gamma')$ is nonnegative for all γ'.

Likelihood of Waveforms

Now consider the loglikelihood function of the waveform $v(t)$ given by

$$v(t) = c(t,\gamma) + n(t),$$

where γ is an unknown vector parameter and $n(t)$ is additive white gaussian noise with the power density spectrum $N_0/2$. The loglikelihood function for this waveform is written formally as

$$\Lambda(\gamma) = -\frac{1}{N_0} \int_{-\infty}^{\infty} |v(t) - c(t,\gamma)|^2 dt,$$

with the superfluous constant N_0 retained so that the expression has a dimensionless form.

As written, the integral is infinite because over infinite time the noise has infinite energy. Therefore the formula with infinite limits can be understood only symbolically.

The formula is derived only for a finite observation time and only for white noise. Nonwhite noise can be converted to white noise by the use of a whitening filter, but this then strains the condition of a finite observation interval. Nevertheless, the use of a whitening filter is accepted under the condition that the observation interval is long in comparison with the response time of the whitening filter. A more elegant and more formal treatment of the case of nonwhite noise uses the methods of functional analysis to reach the same conclusion rigorously.

So that Theorem 8.1.1 can be used, the waveform $v(t)$ is replaced by a finite-dimensional vector of samples having independent noise. To ensure that the noise samples are independent is why $v(t)$ is required to have been passed through a whitening filter to whiten the noise.[2]

Now an expression is derived for the limit of $\Lambda(\gamma)$ as n goes to infinity for a waveform in additive white gaussian noise.

Proposition 8.1.2 *For the complex baseband signal $c(t,\gamma)$ parameterized by the vector γ and received as*

$$v(t) = c(t,\gamma) + n(t)$$

in the interval $[-T_0/2, T_0/2]$, where $n(t)$ is complex white gaussian noise with power density spectrum $N_0/2$, the integral

$$\Lambda(\gamma) = -\frac{1}{N_0} \int_{-T_0/2}^{T_0/2} |v(t) - c(t,\gamma)|^2 dt$$

is the loglikelihood function for the vector parameter γ.

Proof:
Consider the Fourier series expansion of

$$v(t) = c(t,\gamma) + n(t)$$

restricted to the interval $[-T_0/2, T_0/2]$. Then $v(t) = \sum_k V_k e^{j2\pi k/T_0}$, where the V_k are the complex Fourier expansion coefficients of $v(t)$. The expansion coefficients form a countably infinite set of independent, complex gaussian random variables. The mean values of the V_k are the Fourier coefficients C_k of $c(t,\gamma)$ on the interval $[-T_0/2, T_0/2]$. Because the gaussian noise is white, the random variables V_k are independent. The probability density function on the block $(V_{-K}, \ldots, V_K)$ is

$$p(V_{-K}, \ldots, V_K | c(t,\gamma)) = (2\pi\sigma^2)^{-(2K+1)} \prod_{k=-K}^{K} e^{-|V_k - C_k|^2/2\sigma^2},$$

and $\sigma^2 = N_0/2$. This probability distribution is conditional on the expected complex value $c(t,\gamma)$ of the received signal, which depends on the vector parameter γ. For a truncated set of complex samples,

---

[2] A whitening filter must respect the requirements of the Nyquist–Shannon sampling theorem. To properly reconcile conflicting requirements, a formal theory uses Karhunen–Loeve expansion coefficients in place of Fourier expansion coefficients.

$$\Lambda(\gamma) = \log\left((2\pi\sigma^2)^{2K+1} p(V_{-K}, \ldots, V_K | c(t, \gamma))\right)$$

is used as the loglikelihood function. The factor $(2\pi\sigma^2)^{2K+1}$ multiplying the probability is included as a normalizing constant. With this choice of normalizing constant, the loglikelihood statistic is

$$\Lambda(\gamma) = -\sum_{k=-K}^{K} |V_k - C_k|^2 / N_0.$$

Now let K go to infinity and use the energy theorem for the Fourier series[3] to write

$$\lim_{K \to \infty} \Lambda(\gamma) = -\frac{1}{N_0} \int_{-T_0/2}^{T_0/2} |v(t) - c(t, \gamma)|^2 dt,$$

which completes the proof of the proposition. ∎

8.2 The Maximum-Likelihood Principle

The estimation of an image from noisy data can be treated by the maximum-likelihood principle. Let v denote a finite set of noisy data samples, and let γ represent the image to be estimated. The data set v is fixed after the measurement takes place. From this fixed data vector, the image γ is to be estimated.

Let $p(v|\gamma)$ be the probability density function on the space of possible data sets v given that the image is γ. The loglikelihood function is

$$\Lambda(\gamma|v) = \log p(v|\gamma)$$

or, for brevity, simply $\Lambda(\gamma)$. The maximum-likelihood estimate of the image is defined as

$$\widehat{\gamma} = \operatorname*{argmax}_{\gamma} \Lambda(\gamma|v).$$

The maximum-likelihood image $\widehat{\gamma}$ is defined as the image γ that maximizes the loglikelihood function $\Lambda(\gamma|v)$.

Estimation of Variance

For an easy example of the use of the maximum-likelihood principle, consider a single random variable V measuring a zero-mean random gaussian signal S whose variance

---

[3] The energy theorem for the Fourier series

$$\int_{-T_0/2}^{T_0/2} |v(t)|^2 dt = \sum_{k=-\infty}^{\infty} |V_k|^2$$

is a special case of Parseval's identity for the Fourier series

$$\int_{-T_0/2}^{T_0/2} v(t) u^*(t) dt = \sum_{k=-\infty}^{\infty} V_k U_k^*.$$

σ_s^2 is unknown in independent zero-mean random additive gaussian noise N whose variance σ_n^2 is known. A measurement realization v is given by the scalar equation

$$v = s + n,$$

where s is a realization of the gaussian signal, and where n is a realization of the gaussian noise.

The elementary task is to estimate the signal variance σ_s^2 when given the measurement v. The estimator consists of an expression of the form

$$\widehat{\sigma_s^2} = \widehat{\sigma_s^2}(v, \sigma_n^2),$$

which takes the measurement v into the estimate $\widehat{\sigma_s^2}$.

Because the random variable V is the sum of two independent gaussian random variables, it is also a gaussian random variable. It has the probability density function

$$p(v) = \frac{1}{\sqrt{2\pi(\sigma_s^2 + \sigma_n^2)}} e^{-v^2/2(\sigma_s^2 + \sigma_n^2)}.$$

The loglikelihood function is the logarithm of the density function regarded as a function of the unknown σ_s^2. Ignoring additive constants, the loglikelihood is

$$\Lambda(\sigma_s^2) = -\frac{1}{2}\log_e(\sigma_s^2 + \sigma_n^2) - \frac{v^2}{2(\sigma_s^2 + \sigma_n^2)},$$

where the additive term $\log\sqrt{2\pi}$ has been ignored because it does not affect the subsequent maximization. The maximum-likelihood estimate $\widehat{\sigma_s^2}$ is found by setting the derivative with respect to σ_s^2 equal to zero. Thus

$$-\frac{1}{2(\sigma_s^2 + \sigma_n^2)} + \frac{v^2}{2(\sigma_s^2 + \sigma_n^2)^2} = 0,$$

which reduces to

$$\sigma_s^2 + \sigma_n^2 = v^2.$$

Checking the second derivative shows that this is a maximum. Because σ_s^2 must be nonnegative, we conclude that the estimate

$$\widehat{\sigma_s^2} = \max\left[0, v^2 - \sigma_n^2\right]$$

is the maximum-likelihood estimate of σ_s^2.

Estimation of Mean and Variance

Suppose now that the mean $\bar{v}$ and the variance σ_s^2 are to be estimated by making N independent measurements of the random variable s in the presence of memoryless additive gaussian noise. The N measurements are

$$v_i = s_i + n_i \qquad i = 1, \ldots, N,$$

where the s_i and the n_i are independent, zero-mean, gaussian random variables. The n_i all have the known variance σ_n^2. The s_i all have the unknown mean $\bar{v}$ and the unknown variance σ_s^2. The signal variance σ_s^2 is to be estimated from the N measurements.

The loglikelihood function is the simple sum

$$\Lambda(\sigma_s^2) = \sum_{i=1}^{N} \log\left[\frac{1}{\sqrt{2\pi(\sigma_s^2+\sigma_n^2)}} e^{-v_i^2/2(\sigma_s^2+\sigma_n^2)}\right].$$

To find the maximum, set the derivatives with respect to $\bar{v}$ and σ_s^2 both equal to zero. This gives the two expressions

$$0 = \sum_{i=1}^{N} (v_i - \bar{v})$$

and

$$0 = \sum_{i=1}^{N}\left[-\frac{1}{2(\sigma_s^2+\sigma_n^2)} + \frac{(v_i-\bar{v})^2}{2(\sigma_s^2+\sigma_n^2)^2}\right]$$

$$= -\frac{N}{2(\sigma_s^2+\sigma_n^2)} + \frac{\sum_{i=1}^{N}(v_i-\bar{v})^2}{2(\sigma_s^2+\sigma_n^2)^2}.$$

These reduce to the following estimates of the mean and the variance:

$$\widehat{\bar{v}} = \frac{1}{N}\sum_{i=1}^{N} v_i,$$

$$\widehat{\sigma_s^2} = \max\left[0, \frac{1}{N}\sum_{i=1}^{N}(v_i-\bar{v})^2 - \sigma_n^2\right].$$

The conclusion states that when $N = 1$, the estimated mean $\widehat{\bar{v}}$ is the measurement itself and the estimated variance σ_s^2 is zero.

This maximum-likelihood estimate of variance is similar to the case of the maximum-likelihood estimate of the variance σ_s^2 with a single measurement but with the single squared measurement v^2 replaced by the average of the squares $\sum_i (v_i - \widehat{\bar{v}})^2/N$ after the estimated mean is removed.

8.3 Alternating Maximization-Maximization

Many maximization problems, including many problems of maximum-likelihood image formation, cannot be solved analytically. For such problems, numerical methods are indicated. However, these problems are usually very large and often cannot be readily solved by direct numerical methods either. Fortunately, for some problems, it has been possible to devise a mixture of analytical and numerical methods that take the form of a partial analytical solution that can be numerically iterated. This is the method of alternating maximization (or to be more descriptive, alternating maximization-maximization) or of alternating minimization. The method requires the formulation of two maximization operations that can be alternated. The method of alternating minimization is the same except for a simple sign change. A typical alternating maximization algorithm is one based on the convex decomposition lemma.

8 Likelihood and Information Methods

The Richardson–Lucy algorithm of Section 8.5 and the Schulz–Snyder algorithm of Section 8.6 are such cases. Rather than giving an example here of alternating maximization, the Richardson–Lucy algorithm and the Schulz–Snyder algorithm are forward-referenced as two examples.

When two maximization operations are not available to form an alternating maximization algorithm, the more general method of alternating expectation-maximization may be appropriate. This alternative method alternates an expectation step and a maximization step. The method of alternating expectation-maximization is illustrated in this section by means of an example. The theory of alternating expectation-maximization is described in general terms in Section 8.9.

Variance Estimation

Alternating expectation-maximization is introduced by a simple example. The example uses the method in the form of alternating expectation-minimization to form an algorithm for estimating the variance of a gaussian signal in gaussian noise. Of course, an iterative algorithm is not needed for this problem because the problem can be solved analytically, as is done in Section 8.2. To get an iterative algorithm for this simple example, ignore the availability of an analytical solution and proceed with contrived ignorance.

The iterative algorithm is shown in Figure 8.1. The iterations shown in the figure converge to the maximum-likelihood estimate. Although an explicit solution is available for this problem, Figure 8.1 and its derivation are included here to demonstrate a method of treating more complicated problems that do not have an analytical solution. During every iteration nonnegativity and support are automatically inherited from the previous iteration as is typical in such an algorithm. The iterative algorithm is computationally tractable and is easy to implement.

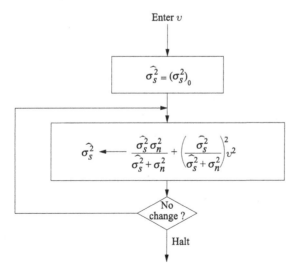

Figure 8.1 Iterative algorithm for estimating variance of signal in noise

8.3 Alternating Maximization-Maximization

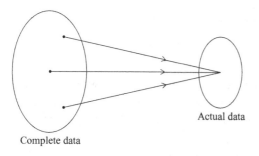

Figure 8.2 The complete data space

The received signal is $v = s + n$, where s is an unknown random signal with a zero-mean gaussian probability density function of *unknown* variance σ_s^2 and n is an independent noise of *known* variance σ_n^2. From v and σ_n^2, the maximum-likelihood estimate of σ_s^2 is to be computed.

To form an algorithm, the actual data set is augmented with additional data as though that additional data were also measured, though it is not. Instead, that missing data is estimated. The additional data is introduced so that the maximum-likelihood estimate can be found analytically from the data set when augmented by this additional data. Specifically, we act as though s and n have been individually measured and analytically solve the problem under this condition. The artificial data set (s, n) is conventionally called the *complete data*. The measured data $v = s + n$ is then conventionally called the *actual data set* or the *incomplete data set*.

Figure 8.2 shows how the actual data space $v = s + n$ fits into the complete data space (s, n). Another possibility that could be chosen for the complete data space is the data set (v, s). Either choice for the complete data space would be suitable for this example. Here (s, n) is chosen as the complete data set that is used to derive the iterative algorithm for estimating the variance.

The complete data space (s, n) is a bivariate random variable with the probability density function

$$p(s, n) = \left(\frac{1}{\sqrt{2\pi\sigma_s^2}} e^{-s^2/2\sigma_s^2}\right) \left(\frac{1}{\sqrt{2\pi\sigma_n^2}} e^{-n^2/2\sigma_n^2}\right),$$

which means that the loglikelihood function for the complete data space is

$$\Lambda(\sigma_s^2) = -\tfrac{1}{2} \log \sigma_s^2 - \frac{s^2}{2\sigma_s^2} - \tfrac{1}{2} \log \sigma_n^2 - \frac{n^2}{2\sigma_n^2}.$$

Then σ_s^2 is chosen to maximize $\Lambda(\sigma_s^2)$.

- The maximization step is derived as follows. Differentiate with respect to σ_s^2 to show that the maximum occurs at the point at which

$$-\frac{1}{2\sigma_s^2} + \frac{s^2}{2(\sigma_s^2)^2} = 0.$$

Solving this for σ_s^2 gives the estimate

$$\widehat{\sigma_s^2} = s^2.$$

This value would provide the desired estimate of σ_s^2 when s^2 is known. But s^2 is not known. Only $v = s + n$ is known. It must be represented by a suitable expected value of s^2. The expected value is based on the estimate $\widehat{\sigma_s^2}$ of σ_s^2.

- The expectation step is derived as follows. Suppose that σ_s^2 is known and evaluate the expectation

$$\widehat{s^2} = \mathrm{E}[s^2|v, \sigma_s^2, \sigma_n^2].$$

But the variance σ_s^2 is not known. Only an estimate of σ_s^2 is known. This is the estimate $\widehat{\sigma_s^2}$ computed in the maximization step. It will be used in this expression as though it were the true variance σ_s^2.

The expectation is provided by the following proposition.

Proposition 8.3.1 Let $v = s + n$ where s and n are independent gaussian random variables with the variances σ_s^2 and σ_n^2, respectively. Then

$$\mathrm{E}\left[s^2|v, \sigma_s^2, \sigma_n^2\right] = \frac{\sigma_s^2 \sigma_n^2}{\sigma_s^2 + \sigma_n^2} + \left(\frac{\sigma_s^2}{\sigma_s^2 + \sigma_n^2}\right)^2 v^2.$$

Proof:
Let

$$\widehat{s^2} = \mathrm{E}[s^2|v, \sigma_s^2, \sigma_n^2] = \int_{-\infty}^{\infty} s^2 p(s|v) ds,$$

where $p(s|v) = p(s|v, \sigma_s^2, \sigma_n^2)$ is the appropriate gaussian probability density function conditioned on v, σ_s^2, and σ_n^2. For brevity of notation, $p(s|v, \sigma_s^2, \sigma_n^2)$ is abbreviated $p(s|v)$ with the conditioning terms σ_s^2 and σ_n^2 implied within $p(s|v)$, but not explicitly written there. Also let

$$\widehat{s} = \mathrm{E}[s|v, \sigma_s^2, \sigma_n^2] = \int_{-\infty}^{\infty} s p(s|v) ds.$$

By the standard properties of moments,

$$\widehat{s^2} = \mathrm{E}\left[(s - \widehat{s})^2 + 2s\widehat{s} - \widehat{s}^2\right]$$
$$= \mathrm{E}\left[(s - \widehat{s})^2\right] + \widehat{s}^2.$$

To evaluate the two terms on the right, $p(s|v)$ needs to be computed first. By the Bayes rule,

8.3 Alternating Maximization-Maximization

$$p(s|v) = \frac{p(s,v)}{p(v)} = \frac{p(s)p(v|s)}{p(v)}$$

$$= \frac{\frac{1}{\sqrt{2\pi\sigma_s^2}}e^{-s^2/2\sigma_s^2}\frac{1}{\sqrt{2\pi\sigma_n^2}}e^{-(v-s)^2/2\sigma_n^2}}{\frac{1}{\sqrt{2\pi(\sigma_s^2+\sigma_n^2)}}e^{-v^2/2(\sigma_s^2+\sigma_n^2)}}$$

$$= \frac{1}{\sqrt{2\pi}}\sqrt{\frac{\sigma_s^2+\sigma_n^2}{\sigma_s^2\sigma_n^2}}e^{-\left(s-\frac{\sigma_s^2}{\sigma_s^2+\sigma_n^2}v\right)^2/2\left(\sigma_s^2\sigma_n^2/(\sigma_s^2+\sigma_n^2)\right)}.$$

This expression is now in the standard form of a gaussian distribution, so the conditional mean and the conditional variance can be recognized as

$$E[s|v,\sigma_s^2] = \frac{\sigma_s^2}{\sigma_s^2+\sigma_n^2}v$$

$$E[(s-\widehat{s})^2|v,\sigma_s^2] = \frac{\sigma_s^2\sigma_n^2}{\sigma_s^2+\sigma_n^2}.$$

Therefore, because $\widehat{s^2} = E[(s-\widehat{s})^2|v,\sigma_s^2] + \widehat{s}^2$, we conclude that

$$\widehat{s^2} = \frac{\sigma_s^2\sigma_n^2}{\sigma_s^2+\sigma_n^2} + \left(\frac{\sigma_s^2}{\sigma_s^2+\sigma_n^2}\right)^2 v^2,$$

as was to be proved. ∎

Now an iteration to compute the maximum-likelihood estimate of σ_s^2 can be given. From any estimate $\widehat{\sigma_s^2}$ of σ_s^2 compute an estimate $\widehat{s^2}$ of s^2 using Proposition 8.3.1. Then use $\widehat{s^2}$ to form the new estimate $\widehat{\sigma_s^2}$ of σ_s^2 as given by Proposition 11.3.1. Then repeat.

By combining the maximization step with the expectation step, the recursion

$$\sigma_s^2 \leftarrow \frac{\sigma_s^2\sigma_n^2}{\sigma_s^2+\sigma_n^2} + \left(\frac{\sigma_s^2}{\sigma_s^2+\sigma_n^2}\right)^2 v^2$$

is obtained as shown in Figure 8.1. Insert σ_s^2 on the right side to produce a new σ_s^2. Then repeat. The final estimate $\widehat{\sigma_s^2}$ is the limit point of this recursion. This is the point at which the two sides are equal. Thus

$$\widehat{\sigma_s^2} = \frac{\widehat{\sigma_s^2}\sigma_n^2}{\widehat{\sigma_s^2}+\sigma_n^2} + \left(\frac{\widehat{\sigma_s^2}}{\widehat{\sigma_s^2}+\sigma_n^2}\right)^2 v^2,$$

which can be reduced to the direct solution given in Section 8.2.

There is one more small point to be made. The right side of the recursion can be reworked to give the form

$$\sigma_s^2 \leftarrow \sigma_s^2 - \left(\frac{\sigma_s^2}{\sigma_s^2+\sigma_n^2}\right)^2(\sigma_s^2+\sigma_n^2-v^2),$$

which is more attractive because it is in the form of a main term and a correction term that goes to zero. When the recursion is written in this form, the limit

$$\sigma_s^2 = \max\left[0, v^2 - \sigma_n^2\right]$$

is easy to see. This, indeed, is the analytical solution obtained previously.

Because this problem can be solved analytically, this recursive algorithm is not needed. The reason for describing a recursive algorithm for this simple problem is given only to provide a simple example of a technique that is useful in complicated problems that cannot be solved analytically.

8.4 Other Principles of Inference

A set of noisy or incomplete data can be interpreted only when one has a model that relates that data to the underlying problem. Sampling in time, for example, represents a continuous-time function by its values at a discrete set of time instants. The samples are useful for recovering the continuous-time signal only if one accepts a model that imposes appropriate conditions on $s(t)$. One condition is the Nyquist condition, which is that the transform $S(f)$ equals zero for $|f|$ greater than $1/2T$, where T is the sampling interval. Then, using the Nyquist–Shannon interpolation formula, $s(t)$ can be recomputed for any t from the samples. By requiring $S(f)$ to satisfy such a condition, we are imposing a prior model onto the data that the sampled data by itself does not justify. Sometimes, an adequate model of this kind is not available; then the model must be construed by using some general principle of inference.

The maximum-likelihood principle is a popular principle of inference. The maximum-likelihood principle is discussed in some detail in the previous section. This principle requires that a conditional probability distribution on the data is known. This conditional probability distribution $p(x|\gamma)$ depends on the unknown parameter γ. Then, when x is measured, the maximum-likelihood estimate of γ is formed as $\widehat{\gamma} = \mathrm{argmax}_\gamma \Lambda(\gamma|x)$, where $\Lambda(\gamma|x) = \log p(x|\gamma)$.

In this section, other principles of inference are introduced. While these other principles are quite different from the maximum-likelihood principle, it is shown later that some can be reinterpreted to be seen as equivalent in some sense to the maximum-likelihood principle.

The maximum-likelihood principle assumes only that γ is unknown. It takes no position regarding whether γ itself is a random variable, only that the data is random and conditional on γ. However, when γ itself is a random variable with a known prior $p(\gamma)$, then one can write the Bayes formula

$$p(\gamma|x) = \frac{p(x|\gamma)p(\gamma)}{p(x)}$$

and use the estimate $\widehat{\gamma} = \mathrm{argmax}_\gamma p(\gamma|x)$. This alternative estimator is known as the *maximum-posterior estimator* or the *bayesian estimator*.

Maximum Entropy Principle

The *Jaynes maximum-entropy principle* is a principle of data reduction which says that when reducing a set of data into the form of an underlying model, one should be maximally noncommittal with respect to the missing data. The measure of uncertainty to be used is a function known as the entropy. The *Shannon entropy* of the probability distribution p is defined as

$$H(p) = -\sum_{j=0}^{J-1} p_j \log p_j.$$

The log base can be either 2, in which case the entropy is measured in units of *bits*, or the log base can be e, in which case the entropy is measured in units of *nats*. The maximum-entropy principle says that to estimate a probability distribution, p, satisfying certain known constraints on p such as

$$\sum_j p_j f_j = t,$$

then, of those distributions that are consistent with the constraints, one should choose as the estimate of p the probability distribution that has the maximum value of the entropy. The constraints are imposed to ensure that the choice satisfies various known conditions regarding the measurement data.

By using the inequality $\log_e x \leq x - 1$, it is easy to show that

$$H(p) \leq \log J,$$

and this inequality holds with equality if and only if $p_j = 1/J$ for all j.

Distance in Probability Space

The inequality $H(p) \leq \log J$ suggests that the difference $\log J - H(p)$ can be defined as a measure of distance from p to the equiprobable distribution $1/J$. Because $\log J = \sum_j p_j \log J$, this leads to the expression

$$\sum_{j=0}^{J-1} p_j \log \frac{p_j}{1/J} = \log J - H(p)$$

as a measure of the distance from the probability distribution p to the equiprobable distribution.

More generally, the *Kullback distance* (also called the *relative entropy*),[4] is defined as

$$D(p\|q) = \sum_{j=0}^{J-1} p_j \log \frac{p_j}{q_j}$$

as a measure of the "distance" from probability distribution q to probability distribution p, with both distributions on an alphabet of size J. The Kullback distance (or the

---

[4] The Kullback distance and the relative entropy are the same function. The first term is used when geometrical reasoning is appropriate. The second term is used when analytical reasoning is appropriate.

relative entropy) is nonnegative and equals zero if and only if p is equal to q. This is expressed by the *Gibbs inequality*

$$D(p\|q) \geq 0.$$

The Kullback distance from probability distribution p to probability distribution q is not the same as the Kullback distance from probability distribution q to probability distribution p. The Kullback distance is not symmetric and it does not satisfy the triangle inequality. It is not a metric. Nevertheless, the Kullback distance $D(p\|q)$ plays a role in the space of probability distributions on J symbols similar to the role that euclidian distance plays in the euclidean space $\mathbb{R}^n$. To emphasize a geometric interpretation, it is convenient to refer to the function $D(p\|q)$ as the Kullback distance.

The Kullback distance (or the relative entropy) can also be defined for probability density functions. The appropriate discrepancy measure on the space of probability density functions is the Kullback distance, which is a function on that space given by

$$D(p(x)\|q(x)) = \int_{-\infty}^{\infty} p(x)\log\frac{p(x)}{q(x)}dx.$$

As is the case for probability mass functions, the Kullback distance is an appropriate way to measure the separation between two probability density functions. It is not invariant under the interchange of $p(x)$ and $q(x)$.

The Kullback distance can be generalized to the space of nonnegative real functions. This discrepancy measure is the *Csiszár distance*. For continuous functions, the Csiszár distance is defined as

$$D(a(x)\|b(x)) = \int_{-\infty}^{\infty} a(x)\log_e\frac{a(x)}{b(x)}dx - \int_{-\infty}^{\infty} [a(x) - b(x)]dx,$$

where $a(x)$ and $b(x)$ are elements of the space of nonnegative real functions. A similar definition holds in the discrete case. The Csiszár distance is an appropriate way to measure the separation between two nonnegative functions. It plays a role in the space of nonnegative real functions analogous to the role played by euclidean distance in the space of real functions. It is not invariant under the interchange of $a(x)$ and $b(x)$. The Csiszár distance reduces to the Kullback distance when $a(x)$ and $b(x)$ are probability distributions.

Minimum Relative Entropy Principle

The *Kullback minimum-relative-entropy principle* is a principle of data reduction which says that when reducing a set of data to an estimate of a probability distribution p on a discrete set in the presence of a prior probability distribution q on that same discrete set, one should choose the admissible probability distribution p that minimizes $D(p\|q)$. This is the admissible distribution p that is closest to q in the Kullback distance. When q itself is admissible, then p should be chosen equal to q,

The admissible probability distributions are those that satisfy appropriate side conditions when side conditions are imposed. A side condition may take the form of an equality constraint, such as a constraint of the form

$$\sum_j p_j t_j = T$$

or an inequality constraint of the form

$$\sum_j p_j t_j \geq T.$$

A single equality constraint is treated by introducing lagrange multipliers λ and s, then minimizing the augmented object function as follows

$$\frac{\partial}{\partial p_j} \left[\sum_j p_j \log \frac{p_j}{q_j} + \lambda \left(\sum_j p_j - 1 \right) + s \left(\sum_j p_j t_j - T \right) \right] = 0,$$

which leads to

$$p_j = \frac{q_j e^{-st_j}}{\sum_j q_j e^{-st_j}},$$

where the lagrange multiplier s is to be chosen so that $\sum_j p_j t_j = T$ and λ so that $\sum_j p_j = 1$.

The procedure is essentially the same when, instead of only one constraint, there are many constraints of the form

$$\sum_j p_j t_{j\ell} = T_\ell \qquad \ell = 1, \ldots, L.$$

Then the solution is

$$p_j = \frac{q_j e^{-\sum_\ell s_\ell t_{j\ell}}}{\sum_j q_j e^{-\sum_\ell s_\ell t_{j\ell}}},$$

where s_ℓ is a lagrange multiplier corresponding to the ℓth constraint chosen so that the ℓth constraint is satisfied.

When the constraints are not consistent, it is not possible to choose the lagrange multipliers to satisfy all of the constraints. In this case, the problem has a solution only if some of the constraints can be relaxed to inequality constraints.

Convex Decomposition Lemma

Before ending this section, a lemma is developed that is useful to lift a certain kind of problem to a larger setting where it may be easier to solve. This lemma deals with non-negative functions that can be normalized so that they integrate to one. Accordingly, it is suggestive to use the notation of probability theory. Let $q(y|\gamma)$ be a probability density function[5] dependent on the parameter γ. The density function $q(y|\gamma)$ can be written in the form

$$q(y|\gamma) = \int_{-\infty}^{\infty} p(x|\gamma) Q(y|x) dx.$$

---

[5] The same discussion applies to discrete random variables and probability mass functions.

8 Likelihood and Information Methods

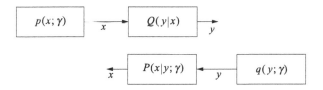

Figure 8.3 The forward channel and the backward channel

Thus, $q(y|\gamma)$ can be understood as the output of the channel $Q(y|x)$ when $p(x|\gamma)$ is the input. This situation also suggests that a "backwards channel," as shown in Figure 8.3, can be defined by the Bayes formula as

$$P(x|y|\gamma) = \frac{p(x|\gamma)Q(y|x)}{\int_{-\infty}^{\infty} p(x'|\gamma)Q(y|x')dx'}.$$

It may be desired to work with the logarithm of such an expression. The denominator then leads to the logarithm of an integral which is difficult to work with. The following lemma expresses the logarithm of an integral as the integral of a logarithm, but at the cost of a gratuitous minimization.

Lemma 8.4.1 (Convex Decomposition Lemma)

$$-\log \int_{-\infty}^{\infty} p(x|\gamma)Q(y|x)dx = \min_{P(x|y)} \int_{-\infty}^{\infty} P(x|y) \log \frac{P(x|y)}{p(x|\gamma)Q(y|x)} dx$$

Proof:
To minimize the right side by choice of the conditional probability density, introduce the lagrange multiplier λ. The augmented objective function to be minimized is

$$\int_{-\infty}^{\infty} P(x|y) \log \frac{P(x|y)}{p(x|\gamma)Q(y|x)} dx + \lambda \left(\int_{-\infty}^{\infty} P(x|y)dx - 1 \right).$$

Now use the calculus of variations to "differentiate," in effect, with respect to the term $P(x|y)$. To anticipate the conclusion of the derivation formally set the derivative with respect to $P(x|y)$ equal to zero to give

$$\log \frac{P(x|y)}{p(x|\gamma)Q(y|x)} + 1 + \lambda = 0,$$

where λ is to be chosen so that $P(x|y)$ is a probability density function. This means that

$$\widehat{P}(x|y) = \frac{p(x|\gamma)Q(y|x)}{\int_{-\infty}^{\infty} p(x'|\gamma)Q(y|x')dx'}$$

is a stationary point of the objective function.

The Gibbs inequality states that

$$\sum_{x,y} P(x,y) \log \frac{P(x,y)}{\widehat{P}(x,y)} \geq 0$$

with equality only if $\widehat{P}(x,y) = P(x,y)$. This shows that $\widehat{P}(x,y)$ achieves a minimum. The proof is complete. ∎

A proof using only the Gibbs inequality would be more concise but would be less informative.

8.5 Recovery of Nonnegative Signals

The recovery of a nonnegative image from a nonnegative blurred reproduction

$$v(x,y) = h(x,y) ** s(x,y)$$

was studied in Section 7.3. This is the task of solving such an equation for $s(x,y)$ when given the nonnegative terms $v(x,y)$ and $h(x,y)$. Noise may be present but noise is not considered here. The Richardson–Lucy algorithm was introduced in Section 7.3 as an iterative algorithm motivated by a heuristic informal development. Now, in this section, the Richardson–Lucy algorithm is developed using the more formal information-theoretic methods.

The Kullback distance is a measure of the difference between two nonnegative functions that integrate to one. The Kullback distance is used in this section to give a formal development of the Richardson–Lucy algorithm and to show that this algorithm converges. The discussion considers nonnegative functions that integrate to the same value. That value can be normalized to one. Probability distributions are such nonnegative functions that integrate to one.

The task is as follows: Given the nonnegative functions $h(x|y)$ and $v(x)$ such that $\int_{-\infty}^{\infty} v(x)dx = 1$ and $\int_{-\infty}^{\infty} h(x|y)dx = 1$ for each y, find a nonnegative function, $s(y)$, that integrates to one such that the quantity

$$\widehat{v}(x) = \int_{-\infty}^{\infty} h(x|y)s(y)dy$$

minimizes the Kullback distance $D(v(x) \| \widehat{v}(x))$ over all such $s(y)$. That is, the function

$$D\left(v(x) \middle\| \int_{-\infty}^{\infty} h(x|y)s(y)dy\right) = \int_{-\infty}^{\infty} v(x) \log \frac{v(x)}{\int_{-\infty}^{\infty} h(x|y)s(y)dy} dx$$

is to be minimized by the choice of $s(y)$. The function that achieves the minimum is written as

$$\widehat{s}(y) = \operatorname*{argmin}_{s(y)} \int_{-\infty}^{\infty} v(x) \log \frac{v(x)}{\int_{-\infty}^{\infty} h(x|y)s(y)dy} dx.$$

This minimization problem can be studied by temporarily ignoring the condition that $s(y)$ is nonnegative and introducing the Lagrange multiplier λ to constrain the integral of $s(y)$ to equal one. Then minimize the augmented object function

$$\int_{-\infty}^{\infty} v(x) \log \frac{v(x)}{\int_{-\infty}^{\infty} h(x|y)s(y)dy} dx + \lambda \left[\int_{-\infty}^{\infty} s(y)dy - 1\right],$$

as by using the calculus of variations.

The calculus of variations states that because $s(y)$ is arbitrary, it can be replaced by the alternative arbitrary function $s(y)+\epsilon\eta(y)$, where ϵ is an arbitrary scalar parameter and $\eta(y)$ is an arbitrary real function. Then set the derivative with respect to ϵ equal to zero at the point where ϵ is equal to zero. This is

$$\frac{\partial}{\partial \epsilon}\left[\int_{-\infty}^{\infty} v(x)\log\frac{v(x)}{\int_{-\infty}^{\infty} h(x|y)[s(y)+\epsilon\eta(y)]dy}dx + \lambda\int_{-\infty}^{\infty}(s(y)+\epsilon\eta(y))dy\right]_{\epsilon=0} = 0,$$

which leads to the equation

$$\int_{-\infty}^{\infty}\left[\frac{\int_{-\infty}^{\infty} h(x|y)\eta(y)dy}{\int_{-\infty}^{\infty} h(x|y)s(y)dy}v(x)\right]dx - \lambda\int_{-\infty}^{\infty}\eta(y)dy = 0,$$

where $\eta(y)$ is arbitrary. By pulling $\eta(y)$ out of both terms, this is rewritten as

$$\int_{-\infty}^{\infty}\left[\int_{-\infty}^{\infty}\frac{h(x|y)}{\int_{-\infty}^{\infty} h(x|y)s(y)dy}v(x)dx - \lambda\right]\eta(y)dy = 0.$$

This equation must hold for every function $\eta(y)$. Unless the bracketed term is zero, the left side could be made nonzero by choosing the sign of $\eta(y)$ to be ± 1 according to the sign of the bracketed term. Thus we conclude that the bracketed term must be zero and

$$\int_{-\infty}^{\infty}\frac{h(x|y)}{\int_{-\infty}^{\infty} h(x|y)s(y)dy}v(x)dx = \lambda.$$

To evaluate the constant λ, multiply both sides by $s(y)$ and integrate with respect to y, which leads to the conclusion that λ is equal to one.

Our development is incomplete because it ignores the requirement that $s(x)$ is nonnegative. A complete development leads to the statement that

$$\int_{-\infty}^{\infty}\frac{h(x|y)}{\int_{-\infty}^{\infty} h(x|y)s(y)dy}v(x)dx \leq 1$$

with equality at all y for which $s(y)$ is nonzero. This statement, known as a Kuhn–Tucker condition, is a necessary condition on any $s(x)$ that achieves the desired minimum. Moreover, because the objective function is a convex function, any local minimum is a global minimum, so the Kuhn–Tucker condition is both a necessary and a sufficient condition.

In particular, by multiplying through by $s(y)$, the Kuhn–Tucker condition becomes the statement that for all y,

$$s(y) = s(y)\int_{-\infty}^{\infty}\frac{h(x|y)}{\int_{-\infty}^{\infty} h(x|y)s(y)dy}v(x)dx.$$

By using the right side to create a new left side, this expression suggests the Richardson–Lucy algorithm. This is the iteration

$$s^{(r+1)}(y) = s^{(r)}(y)\int_{-\infty}^{\infty}\frac{h(x|y)}{\int_{-\infty}^{\infty} h(x|y)s^{(r)}(y)dy}v(x)dx.$$

The following theorem shows that this iteration converges to the desired $s(y)$.

8.5 Recovery of Nonnegative Signals

Theorem 8.5.1 (Richardson–Lucy Algorithm) *The iterated sequence*

$$s^{(r+1)}(y) = s^{(r)}(y) \int_{-\infty}^{\infty} \frac{h(x|y)}{\int_{-\infty}^{\infty} h(x|y)s^{(r)}(y)dy} v(x)dx$$

satisfies

$$\lim_{r \to \infty} s^{(r)}(y) = \underset{s(x)}{\operatorname{argmin}} D(v(x) \| \widehat{v}(x)),$$

where

$$\widehat{v}(x) = \int_{-\infty}^{\infty} h(x|y)\widehat{s}(y)dy.$$

Proof:
To prove convergence of the Richardson–Lucy algorithm, write the iteration as

$$s^{(r+1)}(y) = s^{(r)}(y)f^{(r)}(y),$$

where

$$f^{(r)}(y) = \int_{-\infty}^{\infty} \frac{h(x|y)}{\int_{-\infty}^{\infty} h(x|y)s^{(r)}(y)dy} v(x)dx.$$

Next, let

$$D_{\min} = \min_{s(y)} D\left(v(x) \middle\| \int_{-\infty}^{\infty} h(x|y)s(y)dy\right),$$

and let

$$D^{(r)} = D(v(x) \| v^{(r)}(x)),$$

where $v^{(r)}(x) = \int_{-\infty}^{\infty} h(x|y)s^{(r)}(y)dy$.

The proof first observes that the $D^{(r)}$ are monotonically decreasing and nonnegative so the sequence of the $D^{(r)}$ must converge, then observes that the limit of the sequence of $D^{(r)}$ must be $D_{\min}$. Finally, the proof notes that the $s^{(r)}(x)$ themselves must converge to the required function.

Recognizing that the second line in the chain of deductions below has the form $\log \int_{-\infty}^{\infty} p(y)f(y)dy$ permits Jensen's inequality for a concave function[6] to be applied to reach the third line below. With $s^{(r+1)}(y) = s^{(r)}(y)f^{(r)}(y)$ and

$$v^{(r)}(x) = \int_{-\infty}^{\infty} h(x|y)s^{(r)}(y)dy$$

used in the second line, this leads to the following chain of deductions:

$$D^{(r)} - D^{(r+1)} = \int_{-\infty}^{\infty} v(x) \log \frac{\int_{-\infty}^{\infty} h(x|y)s^{(r+1)}(y)dy}{v^{(r)}(x)} dx$$

$$= \int_{-\infty}^{\infty} v(x) \log \int_{-\infty}^{\infty} \left[\frac{h(x|y)s^{(r)}(y)}{\int_{-\infty}^{\infty} h(x|y)s^{(r)}(y)dy}\right] f^{(r)}(y)dy dx$$

---

[6] Jensen's inequality states that $f(E[x]) \geq E[f(x)]$ for any concave function $f(x)$.

$$\geq \int_{-\infty}^{\infty} v(x) \int_{-\infty}^{\infty} \frac{h(x|y)s^{(r)}(y)}{\int_{-\infty}^{\infty} h(x|y)s^{(r)}(y)dy} \log f^{(r)}(y) dy dx$$

$$= \int_{-\infty}^{\infty} s^{(r)}(y) f^{(r)}(y) \log \frac{s^{(r+1)}(y)}{s^{(r)}(y)} dy.$$

But $s^{(r)}(y) f^{(r)}(y) = s^{(r+1)}(y)$. Therefore

$$D^{(r)} - D^{(r+1)} \geq D\big(s^{(r)}(y) \| s^{(r+1)}(y)\big)$$

$$\geq 0.$$

Therefore $D^{(r+1)} \leq D^{(r)}$ with equality only when $s^{(r)}(y) = s^{(r+1)}(y)$. This means that the sequence of terms $D^{(1)}, D^{(2)}, D^{(3)}, \ldots$ is decreasing.

Because the terms of the sequence are nonnegative, the sequence of terms has a limit. Because no alternative possibility presents itself, one may expect that the limit is $D_{\min}$. The proof of this statement, however, is not trivial. To find the limit, write the above inequality as

$$\int_{-\infty}^{\infty} v(x) \log \frac{v^{(r+1)}(x)}{v^{(r)}(x)} dx \geq D\big(s^{(r+1)}(y) \| s^{(r)}(y)\big).$$

Summing both sides over r (and reversing the direction of the inequality) gives

$$\sum_{r=0}^{R} D\big(s^{(r+1)}(y) \| s^{(r)}(y)\big) \leq \sum_{r=0}^{R} \int_{-\infty}^{\infty} v(x) \log \frac{v^{(r+1)}(x)}{v^{(r)}(x)} dx$$

$$= \int_{-\infty}^{\infty} v(x) \log \frac{v^{(R+1)}(x)}{v^{(0)}(x)} dx$$

$$= \int_{-\infty}^{\infty} v(x) \log \frac{v(x)}{v^{(0)}(x)} dx - \int_{-\infty}^{\infty} v(x) \log \frac{v(x)}{v^{(R+1)}(x)} dx$$

$$\leq \int_{-\infty}^{\infty} v(x) \log \frac{v(x)}{v^{(0)}(x)} dx,$$

where the second inequality follows from the Gibbs inequality. The final term on the right side is a constant, independent of R. The sum on the left side is smaller for any value of R. This means that the summands $D\big(s^{(r+1)}(y) \| s^{(r)}(y)\big)$ on the left side must go to zero as r goes to infinity.

Because $D\big(s^{(r+1)}(y) \| s^{(r)}(y)\big)$ goes to zero, the sequence of $s^{(r)}(y)$ must converge. The Kuhn–Tucker conditions are satisfied in the limit, so the convergence of the $s^{(r)}(y)$ must be to $\hat{s}(y)$. This completes the proof of the theorem. ∎

For the one-dimensional, time-invariant blur function $h(t)$, the integrals become convolutions and the Richardson–Lucy algorithm has the form

$$s^{(r+1)}(t) = s^{(r)}(t) \left[\frac{h(t)}{h(t) * s^{(r)}(t)} * v(-t) \right],$$

requiring two convolutions in each iteration.

A similar development gives the corresponding algorithm in two dimensions. In the case of a two-dimensional, space-invariant point-spread function, the blurred function is given by the convolution

$$v(x,y) = h(x,y) ** s(x,y).$$

The Richardson–Lucy algorithm for a two-dimensional, space-invariant point-spread function takes the special form given in the following theorem.

Theorem 8.5.2 *The iterated sequence*

$$s^{(r+1)}(x,y) = s^{(r)}(x,y)\left[\frac{v(x,y)}{s^{(r)}(x,y) ** h(x,y)} ** h(-x,-y)\right]$$

satisfies

$$\lim_{r\to\infty} s^{(r)}(x,y) = \operatorname*{argmin}_{s(x,y)} D(v(x,y)\|\widehat{v}(x,y)),$$

where

$$\widehat{v}(x,y) = \int_{-\infty}^{\infty}\int_{-\infty}^{\infty} h(x-\xi, y-\eta)\widehat{s}(\xi,\eta)d\xi\, d\eta.$$

Proof:
The proof of the two-dimensional case is the same as the proof of the one-dimensional case. ∎

8.6 Recovery of Missing Phase

A phase retrieval algorithm recovers the missing phase of a two-dimensional Fourier transform $S(f_x,f_y)$ when given its magnitude $|S(f_x,f_y)|$, possibly using the side condition that $s(x,y)$ is real and nonnegative. The topic of phase retrieval is discussed in Section 7.6 and the discussion is now continued using the tools of this chapter. To this purpose, the Schulz–Snyder algorithm will be derived as an alternating minimization algorithm. The Schulz–Snyder algorithm is an iterative algorithm for recovering the phase of the two-dimensional Fourier transform of a nonnegative image $s(x,y)$ from the magnitude $|S(f_x,f_y)|$ of its Fourier transform.

The Schulz–Snyder algorithm recovers the phase of a discretized n by n representation of $|S(f_x,f_y)|$ sampled on an n by n grid. A sampled image represented by an array even as small as sixteen pixels by sixteen pixels is represented as a point in a very large vector space. In this small case, the point is in a space of dimension 256. Examples of minimization in a vector space of dimension two or three do not well-illustrate the task of minimization of a function in a large vector space where the local minima can be many and unexpected.

The squared-magnitude of the Fourier transform $|S(f_x,f_y)|^2$ has the autocorrelation function $\phi(x,y) = s(x,y) ** s(-x,-y)$ as its inverse two-dimensional

Fourier transform.[7] To obtain $\phi(x,y)$ from $|S(f_x,f_y)|^2$ is a straightforward inverse Fourier transform. Thereafter, the Schulz–Snyder iterations are entirely in the two-dimensional space domain. This is unlike the Fienup algorithm, which alternates between the frequency domain and the space domain.

The task of phase retrieval is equivalent to the task of recovering $s(x,y)$ from $\phi(x,y)$. The goal is to find a nonnegative function $\widehat{s}(x,y)$ whose autocorrelation function $\widehat{\phi}(x,y) = \widehat{s}(x,y) ** \widehat{s}(-x,-y)$ is equal to $\phi(x,y)$. Recognizing that there may be noise or other impairments, the goal becomes to find $\widehat{\phi}(x,y)$ as close as possible to $\phi(x,y)$ in terms of some measure of closeness. A natural choice for a measure of distance for the nonnegative functions $\phi(x,y)$ and $\widehat{\phi}(x,y)$ would be the Csiszár distance. Instead, $\phi(x,y)$ can be normalized so that it sums to one. This means that the underlying unknown $s(x,y)$ also sums to one. The normalization factor can be set aside at the start and recovered as a final step.

Because $\widehat{s}(x,y)$ is an estimate of $s(x,y)$, it is appropriate to also require that $\widehat{s}(x,y)$ sums to one. This means that both correlation functions $\phi(x,y)$ and $\widehat{\phi}(x,y)$ are also normalized so that their sums are equal to one. The Csiszár distance then reduces to the familiar Kullback distance $D_K(p(x)\|q(x))$. Accordingly, the problem is formulated in terms of the Kullback distance using the properties of probability distributions.

Formulation

The formulation of the problem of phase retrieval given the autocorrelation function $\phi(x,y)$ is given by

$$\widehat{s}(x,y) = \underset{s'(x,y)}{\operatorname{argmin}}\, D_K(\phi(x,y)\|\phi'(x,y)),$$

where $\phi'(x,y) = s'(x,y) ** s'(-x,-y)$ and $s'(x,y)$ ranges over the elements of the space of two-dimensional probability distributions on a pair of random variables X and Y, each of size n. This is the discrete probability space

$$\mathbb{P}^{n \times n} = \{s(x,y) \mid s(x,y) \geq 0,\ \sum_x \sum_y s(x,y) = 1\}.$$

The correlation functions $\phi(x,y)$ and $\widehat{\phi}(x,y)$ also have the form of probability distributions in that each is nonnegative and sums to one.

By using the Kullback distance as the objective function, the Schulz–Snyder algorithm builds the nonnegativity constraint into the heart of the algorithm. In contrast, the Fienup algorithm does not impose nonnegativity implicitly though it may impose nonnegativity by touching-up the result of each iteration.

---

[7] Because the algorithm is to be implemented in the discrete-signal domain, the Nyquist sampling condition applies. The variables x, y, f_x, f_y are to be understood as taking discrete values, equally spaced. A straightforward application of the sampling theorem (without using the special form imposed by the structure of $\phi(x,y)$) requires that the samples in the frequency domain must be sufficient to support the convolution $\phi(x,y) = s(x,y) ** s(-x,-y)$. Then the sampling rate would be twice that of $s(x,y)$.

A limit point $\widehat{\phi}(x,y)$ of the Schulz–Snyder algorithm is a point at which $D_K(\phi(x,y)\|\widehat{\phi}(x,y))$ is invariant under the Schulz–Snyder iteration. However, as a function of $s'(x,y)$, the objective function is not convex in $s'(x,y)$, as discussed below. Therefore, no statement is made asserting that a convergent of the algorithm must be a desired solution. It could be a local minimum or a saddle point.

Inspection of the Schulz–Snyder algorithm shows that the value of the sum of the components of the iterate $s^{(r)}(x,y)$ is preserved during the iterations of the algorithm. The function $s^{(r)}(x,y)$ is initialized at $r=0$ so that it sums to one. The computed autocorrelation function $\phi^{(r)}(x,y)$ then sums to one as well. Inspection also shows that the algorithm preserves the zeros of $s^{(r)}(x,y)$. Therefore prior information regarding the support of $s(x,y)$ is readily incorporated into the initialization by setting $s(x,y)$ equal to zero for points (x,y) known to be outside of the support.

Theorem 8.6.1 (Schulz–Snyder Algorithm) *For any two-dimensional autocorrelation function of the form $\phi(x,y) = \sum_{x,y} s(x,y)s(x+x',y+y')$ of an n by n array $s(x,y)$ that sums to one, the iteration*

$$s^{(r+1)}(x,y) = s^{(r)}(x,y)\left[\frac{\phi(x,y)+\phi(-x,-y)}{2\phi^{(r)}(x,y)} **s^{(r)}(x,y)\right],$$

where $\phi^{(0)}(x,y) = 1/n$ and $\phi^{(r)}(x,y) = s^{(r)}(x,y)**s^{(r)}(-x,-y)$, results in a nonincreasing sequence of nonnegative real functions $s^{(r)}(x,y)$ converging to a local minimum or a saddle point of the Kullback distance $D_K(\phi(x,y)\|\phi'(x,y))$ over the set of all $\phi'(x,y)$ of magnitude one.

Proof:
For brevity of notation, the proof of the theorem is given for the one-dimensional version of the algorithm. The two-dimensional version that is stated in the theorem is proved in the same way. The one-dimensional version of the algorithm is

$$s^{(r+1)}(x) = s^{(r)}(x)\left[\frac{\phi(x)+\phi(-x)}{2\phi^{(r)}(x)}*s^{(r)}(x)\right]$$

$$= s^{(r)}(x)\sum_\xi \frac{\phi(\xi)+\phi(-\xi)}{2\phi^{(r)}(\xi)}s^{(r)}(x+\xi),$$

where $\phi(x) = s(x)*s(-x)$ is the known autocorrelation function and $s(x)$ is unknown. The second line of the expression for $s^{(r+1)}(x)$ simply opens the convolution using ξ as the displacement of x as is needed for the form of a convolution.

The proof of the theorem is based on the convex decomposition lemma. To this purpose, use Jensen's inequality to write

$$\sum_x P(x|\xi)\log\frac{s(x)s(x+\xi)}{P(x|\xi)} \leq \log\sum_x P(x|\xi)\frac{s(x)s(x+\xi)}{P(x|\xi)}$$

$$= \log\sum_x s(x)s(x+\xi) = \log\phi(x),$$

where the minimum occurs with equality when
$$P(x|\xi) = \frac{s(x)s(x+\xi)}{\sum_x s(x)s(x+\xi)}.$$

Rewrite this inequality as
$$-\log \widehat{\phi}(x) = \operatorname*{argmin}_{P(x|\xi)} \sum_x P(x|\xi) \log \frac{P(x|\xi)}{s(x)s(x+\xi)}.$$

This expression lifts the one-dimensional function $\widehat{\phi}(x)$ by expressing it as a minimum of a function in two variables. Substitute this expression into the expression for the Kullback distance to lift the minimum of the Kullback distance to a double minimum:

$$\min_{s(x)} D(\phi(x)\|\widehat{\phi}(x)) = \min_{s(x)} \sum_\xi \phi(\xi) \log \phi(\xi) - \phi(\xi) \log \widehat{\phi}(\xi)$$

$$= \min_{s(x)} \min_{P(x|\xi)} \sum_\xi \left[\phi(\xi) \sum_x P(x|\xi) \log \frac{\phi(\xi) P(x|\xi)}{s(x)s(x+\xi)} \right].$$

The two minima are evaluated separately, then combined to obtain the alternating minimization algorithm.

To minimize the inner sum over all possible $P(x|\xi)$ for fixed ξ and fixed $s(x)$, introduce a lagrange multiplier λ to constrain $P(x|\xi)$ to be a probability distribution. Then set to zero the derivative of the objective function with respect to the xth component of $P(x|\xi)$ as follows:

$$\frac{\partial}{\partial P(x|\xi)} \left[\sum_x P(x|\xi) \log \frac{\phi(\xi) P(x|\xi)}{s(x)s(x+\xi)} + \lambda \sum_x (P(x|x_i) - 1) \right] = 0$$

$$\log \frac{\phi(\xi) P(x|\xi)}{s(x)s(x+\xi)} - 1 + \lambda = 0.$$

Choosing λ so that $P(x|\xi)$ sums to one, this reduces to
$$P(x|\xi) = \frac{s(x)s(x+\xi)}{\phi(\xi)}.$$

This conclusion agrees with the earlier statement following from Jensen's inequality. Because $P(x|\xi)$ is nonnegative and sums to one, it is a conditional probability distribution.

To minimize the outer sum over all possible $s(x)$ for fixed ξ and fixed $P(x|\xi)$, introduce a lagrange multiplier λ to constrain $s(x)$ to be a probability distribution. Then set to zero the derivative of the objective function with respect to the xth component of $s(x)$ as follows:

$$\frac{\partial}{\partial s(x)} \left[\sum_y \phi(\xi) \sum_x P(x|\xi) \log \frac{\phi(\xi) P(x|\xi)}{s(x)s(x+\xi)} - \lambda(s(x) - 1) \right] = 0,$$

which reduces to
$$\sum_y \phi(\xi) P(x|\xi) \left[\frac{1}{s(x)} + \frac{1}{s(x+\xi)} \right] = \lambda.$$

Therefore

$$\sum_y \phi(\xi)\left[P(x|\xi) + P(x - \xi|\xi)\right] = \lambda s(x).$$

The lagrange multiplier λ in this expression is equal to two, which is obtained by summing with respect to x. Therefore

$$s(x) = \frac{\sum_\xi \phi(\xi)\left(P(x|\xi) + P(x - \xi|\xi)\right)}{2}.$$

The alternating minimization algorithm now follows by iteratively alternating these two expressions computing $P(x|\xi)$ from $s(x)$ and $s(x)$ from $P(x|\xi)$ as given by

$$\cdots \to s(x) \to P(x|\xi) \to s(x) \to P(x|\xi) \to s(x) \to \cdots.$$

Reducing this two-part iteration to a simple iteration by merging the two equations into one equation results in the Schulz–Snyder iterative algorithm. ∎

Summing the second line in the first equation of the proof over x and then noting that $\sum_x s^{(r)}(x) s^{(r)}(x+\xi) = \phi^{(r)}(\xi)$ shows that the sum $\sum_x s^{(r+1)}(x)$ remains unchanged during an iteration. This means that $\sum_x \phi^{(r)}(x)$ also remains unchanged. This validates the assertion made earlier that the algorithm preserves the sum on x.

The Schulz–Snyder algorithm is easy to implement. It is computationally tractable and can be applied to any nonnegative real image. Moreover, the nonnegativity constraint, the support constraint, and the sum constraint are automatically inherited at every iteration from the previous iteration. The algorithm should be initialized with a nonnegative function whose support includes the support of the intended image and excludes anything known to be not in the support of that intended image. The algorithm never converts a zero value to a nonzero value. It is simple to include a support constraint in the Schulz–Snyder algorithm simply by constraining the initialization according to the support constraints.

The objective function that is minimized to derive the Schulz–Snyder algorithm is not convex, so, as for other phase retrieval algorithms, convergence to the global minimum cannot be asserted. Because the algorithm is based on an objective function, the iterations ensure that the objective function is never increasing. This means that the algorithm can converge only to a local minimum or to a saddle point. Convergence to an acceptable image appears to be more likely for an image that has considerable detail. Periodic substructures within the image, however, can confuse the algorithm.

Moreover, more than one function can have the same autocorrelation function. Therefore, $s(x - a, y - b)$ is also a solution to the problem, as is $s(-x, -y)$. However, the convex combination $\lambda s(x, y) + \bar{\lambda} s(x - a, y - b)$ has the Fourier transform magnitude given by

$$\left|\lambda S(f_x, f_y) + \bar{\lambda} S(f_x, f_y) e^{j2\pi(af_x + bf_y)}\right|^2 = \left|(\lambda^2 + 2\lambda\bar{\lambda}\cos(2\pi(af_x + bf_y)) + \bar{\lambda}^2)\right|$$
$$\left|S(f_x, f_y)\right|^2 \leq \left|S(f_x, f_y)\right|^2.$$

The inequality is strict for most values of f_x and f_y. This means that he convex combination of $s(-x,-y)$ and $s(x-a, y-b)$ does not have the required Fourier transform magnitude $|S(f_x,f_y)|$ for all f_x and f_y.

Specifically, when $\hat{s}(x,y)$ achieves a minimum of the objective function, then $\hat{s}(x-a, y-b)$ achieves that same minimum as well, but the convex combination $\lambda \hat{s}(x,y) + \bar{\lambda}\hat{s}(x-a, y-b)$ is not a solution to the problem because it does not have Fourier magnitude $|S(f_x,f_y)|$. The set of $S(f_x,f_y)$ satisfying the magnitude constraint is not convex. Ergo, the problem is not convex.

8.7 Blind Recovery of Nonnegative Signals

The task of two-dimensional blind deconvolution of nonnegative signals is to estimate the nonnegative functions $h(x,y)$ and $s(x,y)$ satisfying the equation

$$v(x,y) = h(x,y) ** s(x,y)$$

when given the realization $v(x,y)$, perhaps with additive noise $n(x,y)$ included. One procedure for estimating $h(x,y)$ and $s(x,y)$ was given in Section 7.6 without formal justification. This section suggests a somewhat more formal procedure for estimating $h(x,y)$ and $s(x,y)$.

The likelihood function for blind deconvolution in gaussian noise is

$$\Lambda\big(h(x,y), s(x,y)\big) = \int_{-\infty}^{\infty}\int_{-\infty}^{\infty} \big(v(x,y) - \big(h(x,y) ** s(x,y)\big)\big)^2 dxdy.$$

When the condition that the functions are nonnegative is not imposed, the maximum-likelihood principle would be used to write

$$\big(\hat{h}(x,y), \hat{s}(x,y)\big) = \underset{h(x,y), s(x,y)}{\operatorname{argmin}} \int_{-\infty}^{\infty}\int_{-\infty}^{\infty} \big(v(x,y) - \big(h(x,y) ** s(x,y)\big)\big)^2 dxdy$$

as the minimum of euclidean distance between $v(x,y)$ and $h(x,y) ** s(x,y)$. However, the solutions $h(x,y)$ and $s(x,y)$, in general, would take negative values and hence would not satisfy the nonnegativity constraint.

As an alternative to maximizing the likelihood, the Kullback distance given by

$$D\big(v(x,y) \| h(x,y) ** s(x,y)\big) = \int_{-\infty}^{\infty}\int_{-\infty}^{\infty} v(x,y) \log \frac{v(x,y)}{h(x,y) ** s(x,y)} dxdy$$

is minimized over both $h(x,y)$ and $s(x,y)$. The Kullback distance is a measure of discrepancy for nonnegative functions. The optimality criterion is to choose the pair of functions that minimizes the Kullback distance.

The minimum Kullback distance is then achieved by the pair $h(x,y)$ and $s(x,y)$ that minimizes the right side of this equation. This is a generalized version of the problem leading to the Richardson–Lucy algorithm in Section 8.5. In this case, however, two functions $s(x,y)$ and $h(x,y)$ must be found so as to minimize the objective function. This leads to two iterates in the generalized version of the Richardson–Lucy algorithm.

The Richardson–Lucy deconvolution algorithm for either $s(x,y)$ or $h(x,y)$ when the other is known would consist of either of the following two iterations:

$$s^{new}(x,y) = s^{old}(x,y)\left[\frac{v(x,y)}{s^{old}(x,y)**h(x,y)}**h(-x,-y)\right],$$

$$h^{new}(x,y) = h^{old}(x,y)\left[\frac{v(x,y)}{h^{old}(x,y)**s(x,y)}**s(-x,-y)\right].$$

Either iteration by itself is an iteration of the Richardson–Lucy algorithm.

Algorithm
The pair of alternating iterations

$$s^{(r+1)}(x,y) = s^{(r)}(x,y)\left[\frac{v(x,y)}{s^{(r)}(x,y)**h(x,y)}**h(-x,-y)\right],$$

$$h^{(r+1)}(x,y) = h^{(r)}(x,y)\left[\frac{v(x,y)}{h^{(r)}(x,y)**s(x,y)}**s(-x,-y)\right]$$

satisfies

$$\lim_{r\leftrightarrow\infty}(s^{(r)}(y), h^{(r)}(y)) = \underset{h(x,y),s(x,y)}{\operatorname{argmin}}\ D\big(v(x,y)\big\|\tilde{v}(x,y)\big).$$

Convergence of this ad hoc procedure is not claimed here.

The algorithm iterates both $s(x,y)$ and $h(x,y)$ concurrently. A slightly different version of the algorithm updates $s(x,y)$ and $h(x,y)$ alternately. Alternate one step of the iteration for $s(x,y)$ with one step of the iteration for $h(x,y)$ using the most recent iterate for the other function each time.

8.8 Differencing Methods in Photon Imaging

To develop the maximum-likelihood image for a given application, the probability distribution function $p(v|y)$ must be known. An algorithm for finding the argmax of the loglikelihood $\Lambda(y)$ can then follow. Our example in this section uses the method of expectation-maximization to derive an iterative algorithm to compute a maximum-likelihood image.

Section 7.8 describes a method to estimate an image in weak light by forming histograms of photon position differences, then using a general phase-retrieval procedure. That procedure is now reconsidered using the maximum-likelihood principle to obtain an estimator of an image.

Suppose that $2N$ photons, each with a position error that changes slowly in time, have been detected, as described in Section 7.8. The raw data consists of the observation of $2N$ space-time data points of the form $(x_i + \Delta x(t_i), y_i + \Delta y(t_i), t_i)$ for $i = 1,\ldots,2N$. The error terms $\Delta x(t_i)$ and $\Delta y(t_i)$ are slowly varying functions of time that change insignificantly during the interval between two photoconversions, but do change significantly over the full time of data collection. To remove the effect

8 Likelihood and Information Methods

of the position error in the spatial coordinates, the data is preprocessed by replacing the $2N$ spatial coordinates by N pairs of position differences $(x_{2\ell}-x_{2\ell-1}, y_{2\ell}-y_{2\ell-1})$. This step removes the unknown position displacement from the data set, and is given without further justification.

The random times t_i at which a photon arrives seem to have little to do with the unknown spatial distribution $\lambda(x,y)$ and it may seem that t_i need not be recorded. However, the integral of $\lambda(x,y)$ does depend on the average arrival rate, so at least the total collection interval $T = t_{2N} - t_1$ should be retained. A middle position regarding the t_i that allows the loglikelihood function to work out cleanly is taken. The arrival times are preprocessed to retain in the data set the time interval between the first of the photons in each pair of photons. The preprocessed data set now has the form

$$(x_1 - x_2, y_1 - y_2, t_3 - t_1)$$
$$(x_3 - x_4, y_3 - y_4, t_5 - t_3)$$
$$(x_5 - x_6, y_5 - y_6, t_7 - t_5)$$
$$\ldots,$$

with the time intervals defined so that they sum to $t_{2N} - t_1$. The data set consists of the photon-pair position differences and the time intervals.

Retaining the arrival times in the data set in the form of intervals yields an analytically tractable form for the likelihood function. The arrival times are a poisson random process with the parameter λ, so the probability density function $P_{w_n}(\tau)$ on the pairwise interval[8] is

$$P_{w_2}(\tau) = \lambda^2 \tau e^{-\lambda \tau}.$$

The term λ^2 cancels a term λ^{-2} that arises later in the likelihood function from the spatial data.

The probability density function for the entire set of preprocessed data can now be stated. Because the N differences $(x_{2\ell} - x_{2\ell-1}, y_{2\ell} - y_{2\ell-1}, t_{2\ell+1} - t_{2\ell-1})$ for $\ell = 1, \ldots, N$ are independent, the probability density function is a product. Thus the likelihood function is that same product.

Let $\boldsymbol{u}, \boldsymbol{v}$, and $\boldsymbol{\tau}$ be the vectors with the ℓth components $u_\ell = x_{2\ell} - x_{2\ell-1}$, $v_\ell = y_{2\ell} - y_{2\ell-1}$, and $\tau_\ell = t_{2\ell+1} - t_{2\ell-1}$, respectively. Then the loglikelihood function is

$$\Lambda(\lambda(x,y)) = \log p((\boldsymbol{u}, \boldsymbol{v}, \boldsymbol{\tau}) | \lambda(x,y))$$
$$= \log \prod_{\ell=1}^{N} P_D(u_\ell, v_\ell | \lambda(x,y)) P_{w_2}(\tau_\ell),$$

---

[8] The probability density function on the nth occurrence time for the homogeneous poisson process is

$$P_{w_n}(\tau) = \frac{(\lambda \tau)^{n-1}}{(n-1)!} \lambda e^{-\lambda \tau}.$$

where

$$P_D(u,v|\lambda(x,y)) = \frac{1}{\lambda^2}\int_{-\infty}^{\infty}\int_{-\infty}^{\infty} \lambda(x,y)\lambda(x+u,y+v)\mathrm{d}x\mathrm{d}y,$$

as described in Section 7.8, and $P_{w_2}(\tau)$ is as given previously. Consequently,

$$\Lambda(\lambda(x,y)) = \sum_{\ell=1}^{N}\log\left[\lambda^2\tau_\ell e^{-\lambda\tau_\ell}\frac{1}{\lambda^2}\int_{-\infty}^{\infty}\int_{-\infty}^{\infty}\lambda(x,y)\lambda(x+u_\ell,y+v_\ell)\mathrm{d}x\mathrm{d}y\right]$$

$$= \sum_{\ell=1}^{N}\log\tau_\ell - \lambda\sum_{\ell=1}^{N}\tau_\ell + \sum_{\ell=1}^{N}\log\int_{-\infty}^{\infty}\int_{-\infty}^{\infty}\lambda(x,y)\lambda(x+u_\ell,y+v_\ell)\mathrm{d}x\mathrm{d}y,$$

where $\lambda = \int_{-\infty}^{\infty}\int_{-\infty}^{\infty}\lambda(x,y)\mathrm{d}x\mathrm{d}y$. The first term is now suppressed because it does not depend on $\lambda(x,y)$ and so it does not affect the location of the maximum. Replace the sum of intervals in the second term by T and replace λ by its definition as an integral of $\lambda(x,y)$. The task is now to find $\widehat{\lambda}(x,y)$ satisfying

$$\operatorname*{argmin}_{\lambda(x,y)}\left[-T\int_{-\infty}^{\infty}\int_{-\infty}^{\infty}\lambda(x,y)\mathrm{d}x\mathrm{d}y + \sum_{\ell=1}^{N}\log\int_{-\infty}^{\infty}\int_{-\infty}^{\infty}\lambda(x,y)\lambda(x+u_\ell,y+v_\ell)\mathrm{d}x\mathrm{d}y\right],$$

where N can be very large. Finding the function $\widehat{\lambda}(x,y)$ that achieves the minimum is a formidable computational task. Fortunately, the method of alternating expectation-maximization can be used to derive an iterative algorithm for its solution. Before developing the algorithm, the next theorem states a necessary condition on $\lambda(x,y)$ to achieve the maximum. A formal proof of the algorithm based on the method of expectation-maximization appears in the next section.

Theorem 8.8.1 *A necessary condition for $\lambda(x,y)$ to maximize the photon difference loglikelihood function is*

$$1 = \frac{1}{T}\sum_{\ell=1}^{N}\frac{\lambda(x-u_\ell,y-v_\ell) + \lambda(x+u_\ell,y+v_\ell)}{\int_{-\infty}^{\infty}\int_{-\infty}^{\infty}\lambda(x,y)\lambda(x+u_\ell,y+v_\ell)\mathrm{d}x\mathrm{d}y}.$$

Proof:
To find this condition on $\lambda(x,y)$, replace $\lambda(x,y)$ by $\lambda(x,y) + \epsilon\eta(x,y)$ and use the methods of variational calculus. For any function $\eta(x,y)$ that is well-behaved, $\lambda(x,y)$ maximizes $\Lambda(\lambda(x,y))$ only when

$$\frac{\partial}{\partial\epsilon}\Lambda(\lambda(x,y) + \epsilon\eta(x,y))\Big|_{\epsilon=0} = 0$$

for any $\eta(x,y)$, where

$$\Lambda(\lambda(x,y)) = -T\int_{-\infty}^{\infty}\int_{-\infty}^{\infty}\lambda(x,y)\mathrm{d}x\mathrm{d}y + \sum_{\ell=1}^{N}\log\int_{-\infty}^{\infty}\int_{-\infty}^{\infty}\lambda(x,y)\lambda(x+u_\ell,y+v_\ell)\mathrm{d}x\mathrm{d}y.$$

Substituting $\lambda(x,y) + \epsilon\eta(x,y)$ in place of $\lambda(x,y)$ and differentiating with respect to ϵ gives

$$0 = -T \int_{-\infty}^{\infty}\int_{-\infty}^{\infty} \eta(x,y)\mathrm{d}x\mathrm{d}y$$
$$+ \sum_{\ell=1}^{N} \frac{\int_{-\infty}^{\infty}\int_{-\infty}^{\infty} [\lambda(x,y)\eta(x+u_\ell,y+v_\ell) + \eta(x,y)\lambda(x+u_\ell,y+v_\ell)]\mathrm{d}x\mathrm{d}y}{\int_{-\infty}^{\infty}\int_{-\infty}^{\infty} \lambda(x,y)\lambda(x+u_\ell,y+v_\ell)\mathrm{d}x\mathrm{d}y}$$

at $\epsilon = 0$. Make a change in variables to write

$$\lambda(x,y)\eta(x+u_\ell,y+v_\ell) = \lambda(x-u_\ell,y-v_\ell)\eta(x,y).$$

The integrand can now be expressed as a multiple of the arbitrary function $\eta(x,y)$ as follows:

$$0 = \int_{-\infty}^{\infty}\int_{-\infty}^{\infty} \left[-T + \sum_{\ell=1}^{N} \frac{\lambda(x-u_\ell,y-v_\ell) + \lambda(x+u_\ell,y+v_\ell)}{\int_{-\infty}^{\infty}\int_{-\infty}^{\infty} \lambda(x,y)\lambda(x+u_\ell,y+v_\ell)\mathrm{d}x\mathrm{d}y} \right] \eta(x,y)\mathrm{d}x\mathrm{d}y.$$

The standard method of variational calculus now asserts that the term multiplying $\eta(x,y)$ must equal zero for a $\lambda(x,y)$ that maximizes $\Lambda(\lambda(x,y))$. Otherwise, a function, $\eta(x,y)$, could be constructed making the integral larger than zero. The necessary condition follows and the proof of the theorem is complete. ∎

It is instructive to relate the condition of Theorem 8.8.1 first to the phase-retrieval problem and then to the iterative computational algorithm that is derived below.

To connect the condition to the phase retrieval problem, notice that, by the law of large numbers, the histogram of photoconversions converges with shrinking bin size to $NR(u,v)$, where

$$R(u,v) = \int_{-\infty}^{\infty}\int_{-\infty}^{\infty} \lambda(x,y)\lambda(x+u,y+v)\mathrm{d}x\mathrm{d}y,$$

and T will be approximately $2N\lambda$. Therefore, with $\widehat{\lambda}(x,y)$ denoting the estimated distribution, the condition on $\widehat{\lambda}(x,y)$ is approximated by

$$1 = \frac{1}{2\lambda} \int_{-\infty}^{\infty}\int_{-\infty}^{\infty} \frac{\widehat{\lambda}(x-u,y-v) + \widehat{\lambda}(x+u,y+v)}{\widehat{R}(u,v)} R(u,v)\mathrm{d}u\mathrm{d}v.$$

Choosing $\widehat{\lambda}(x,y)$ to equal $\lambda(x,y)$ leads to $\widehat{R}(u,v) = R(u,v)$ and this equation is satisfied. Thus solving the equation for $\lambda(x,y)$ is equivalent to finding $\lambda(x,y)$ when given the autocorrelation $R(x,y)$ of $\lambda(x,y)$. This is the problem of phase retrieval. Evidently, in this problem, a phase-retrieval algorithm approximates a maximum-likelihood solution when the data set is very large, but fails to approximate a maximum-likelihood solution when the data is not sufficient to closely estimate the autocorrelation function.

To restate the necessary condition on $\lambda(x,y)$ given in Theorem 8.8.1 in a form that suggests an iterative algorithm, multiply that equation by $\lambda(x,y)$. Then

$$\lambda(x,y) = \frac{\lambda(x,y)}{T} \sum_{\ell=1}^{N} \frac{\lambda(x-u_\ell, y-v_\ell) + \lambda(x+u_\ell, y+v_\ell)}{\int_{-\infty}^{\infty}\int_{-\infty}^{\infty} \lambda(x,y)\lambda(x+u_\ell, y+v_\ell)\mathrm{d}x\mathrm{d}y}.$$

This equation suggests an iterative procedure to compute $\lambda(x,y)$. Using the left side as an updated iterative based on the right side will give the iteration

$$\lambda^{(r+1)}(x,y) = \frac{\lambda^{(r)}(x,y)}{T} \sum_{\ell=1}^{N} \frac{\lambda^{(r)}(x-u_\ell, y-v_\ell) + \lambda^{(r)}(x+u_\ell, y+v_\ell)}{\int_{-\infty}^{\infty}\int_{-\infty}^{\infty} \lambda^{(r)}(x,y)\lambda^{(r)}(x+u_\ell, y+v_\ell)\mathrm{d}x\mathrm{d}y}.$$

When the sequence $\lambda^{(r)}(x,y)$ defined here converges to a limit point, that limit point is an extremum (in fact, it is a maximum). The algorithm follows from this discussion. It can be derived formally by using the method of alternating expectation-maximization.

8.9 Alternating Expectation-Maximization

The maximum-likelihood principle is a well-accepted principle of estimation theory for estimating a vector parameter γ. However, for many problems, finding the maximum of the likelihood function $\Lambda(\gamma)$ is not analytically tractable. Maximum-likelihood problems in image formation can be complicated. For example, an image γ to be estimated may consist of an array of 500 by 500 pixels. Then $\Lambda(\gamma)$ is a function of 500^2 variables, where the array γ can be represented as a point in 500^2-dimensional space. The method of alternating expectation-maximization,[9] constructs an iterative algorithm that determines a maximum-likelihood estimate of the array γ from measured data. In the usual formulation, the method evaluates an expectation, then computes a maximization, and then repeats these two steps until a convergence condition or a halting condition is reached. In most cases, after the equations of the expectation steps and the maximization steps are formulated, they can be merged into a single equation for the combined iteration. Examples of iterative algorithms derived by alternating expectation-maximization are given in Sections 8.3 and 11.8. Here these methods are recapitulated in a more formal discussion.

The development of the method of alternating expectation-maximization takes place in the context of two data spaces. The space in which the available data is defined is called the *actual data space* or the *incomplete data space*, the latter name referring to the fact that the actual data is not computationally tractable for determining an answer. A larger and somewhat arbitrary data space in which a more tractable set of data can be considered is called the *complete data space*.

The complete data must be estimated from the solution of the problem by means of an expectation. This means that an estimate of the complete data must be computable from the tentative solution and a tentative solution can be computable from the complete data. This suggests an iteration of alternating steps.

---

[9] Expectation-maximization is a method for constructing algorithms rather than itself an algorithm itself. Each algorithm constructed by this method is a computational procedure for a unique application and is specific to that application.

Were the complete data known, it would be analytically tractable to maximize the likelihood function. But an estimate of the complete data available from a previous iteration can be used for this purpose. In this way, an iterative algorithm is derived. First, estimate the loglikelihood function from the currently computed value of the unknown parameter γ and the data in the actual data space. Then maximize the likelihood to find a new γ, which is a tractable computation in the complete data space.

To use the expectation-maximization method, one must specify the complete data space. A suitable choice depends on an understanding of the problem. There may be several reasonable choices for the complete data space that work, but one choice may be considerably easier computationally or analytically. The formalism of the expectation-maximization method may allow some general statements about convergence to be made.

Covariance Estimation

A simple scalar instance of the estimation of the variance by the method of alternating expectation-maximization is described in Section 8.3. Now the more difficult vector version of this problem is described, and is then treated by the method of alternating expectation-maximization. The vector measurement is

$$v = s + n,$$

where s and n are real, zero-mean, gaussian vector random variables with covariance matrices $\boldsymbol{\Sigma}_s$ and $\boldsymbol{\Sigma}_n$, respectively.

The probability density functions of the random vectors s and n are[10]

$$p(s) = \frac{1}{\sqrt{\det[2\pi \boldsymbol{\Sigma}_s]}} e^{-\frac{1}{2}s^\dagger \boldsymbol{\Sigma}_s^{-1} s} \qquad p(n) = \frac{1}{\sqrt{\det[2\pi \boldsymbol{\Sigma}_n]}} e^{-\frac{1}{2}n^\dagger \boldsymbol{\Sigma}_n^{-1} n}.$$

Because the covariance matrices add under the addition of independent random vectors $v = s + n$, the probability distribution on the measurement v is

$$p(v|\boldsymbol{\Sigma}_s) = \frac{1}{\sqrt{\det[2\pi (\boldsymbol{\Sigma}_s + \boldsymbol{\Sigma}_n)]}} e^{-\frac{1}{2}v^\dagger (\boldsymbol{\Sigma}_s + \boldsymbol{\Sigma}_n)^{-1} v},$$

where the unknown $\boldsymbol{\Sigma}_s$ is explicitly mentioned in the notation. The loglikelihood statistic is

$$\Lambda(\boldsymbol{\Sigma}_s) = \log p(v|\boldsymbol{\Sigma}_s)$$
$$= -\tfrac{1}{2} \log \det(\boldsymbol{\Sigma}_s + \boldsymbol{\Sigma}_n) - \tfrac{1}{2} v^\dagger (\boldsymbol{\Sigma}_s + \boldsymbol{\Sigma}_n)^{-1} v,$$

where the nonrelevant constants have been suppressed. The function $\Lambda(\boldsymbol{\Sigma}_s)$ is to be minimized by the choice of the nonnegative-definite correlation matrix $\boldsymbol{\Sigma}_s$. Reversing the signs, the expression for the estimate of $\boldsymbol{\Sigma}_s$ is

---

[10] The dagger † denotes transpose for real vectors and matrices and denotes conjugate transpose for complex vectors and matrices.

$$\Lambda(\widehat{\boldsymbol{\Sigma}_s}) = \underset{\boldsymbol{\Sigma}_s}{\operatorname{argmin}} \left[\tfrac{1}{2} \log \det(\boldsymbol{\Sigma}_s + \boldsymbol{\Sigma}_n) + \tfrac{1}{2} v^\dagger (\boldsymbol{\Sigma}_s + \boldsymbol{\Sigma}_n)^{-1} v \right].$$

Some eigenvalues of the minimizing $\boldsymbol{\Sigma}_s$ may be zero. This means that the minimum must be replaced by the infimum when searching over only positive-definite matrices.

The task is to estimate the covariance matrix $\boldsymbol{\Sigma}_s$ from the measurement vector v when $\boldsymbol{\Sigma}_n$ is known. It may not be analytically tractable to maximize this loglikelihood function with respect to $\boldsymbol{\Sigma}_s$ by the elementary method of setting partial derivatives to zero. Instead, the maximization problem is solved here by an iterative computational algorithm.

The Complete Data

To set up the iterative algorithm, first replace the actual data $v = s + n$ with the complete data (s, n). Because s and n are independent, the likelihood of the complete data is a product and the loglikelihood $\Lambda(\boldsymbol{\Sigma}_s)$ is

$$\Lambda(\boldsymbol{\Sigma}_s) = \log[p(\boldsymbol{\Sigma}_s | \boldsymbol{\Sigma}_n, v, n)]$$
$$= \log \left[\left(\frac{1}{\sqrt{\det 2\pi \boldsymbol{\Sigma}_s}} e^{-\frac{1}{2} s^\dagger \boldsymbol{\Sigma}_s^{-1} s} \right) \left(\frac{1}{\sqrt{\det 2\pi \boldsymbol{\Sigma}_n}} e^{-\frac{1}{2} n^\dagger \boldsymbol{\Sigma}_n^{-1} n} \right) \right].$$

The complete-data loglikelihood $\Lambda(\boldsymbol{\Sigma}_s)$ is to be maximized by the choice of the matrix $\boldsymbol{\Sigma}_s$ subject to the condition that $\boldsymbol{\Sigma}_s$ is positive definite.

Maximization

To derive the maximization step, discard terms that do not depend on $\boldsymbol{\Sigma}_s$. The maximization step thereby reduces to

$$\widehat{\boldsymbol{\Sigma}}_s = \underset{\boldsymbol{\Sigma}_s}{\operatorname{argmax}} \left[-\log \det \boldsymbol{\Sigma}_s - s^\dagger \boldsymbol{\Sigma}_s^{-1} s \right]$$
$$= \underset{\boldsymbol{\Sigma}_s}{\operatorname{argmin}} \left[\log \det \boldsymbol{\Sigma}_s + s^\dagger \boldsymbol{\Sigma}_s^{-1} s \right],$$

where the second line follows by reversing signs. This minimization uses the following proposition.

Proposition 8.9.1 *Let A and B be positive-definite matrices. Then the minimum of the function*

$$f(A) = \log \det A + \operatorname{trace} A^{-1} B$$

occurs when A is equal to B.

Proof:
To minimize $f(A) = \log \det A + \operatorname{trace} A^{-1} B$, let $C = A^{-1} B$. Then

$$f(B) - f(A) = \log \det B + \operatorname{trace} B^{-1} B - \log \det A - \operatorname{trace} A^{-1} B$$
$$= \log \det C + \operatorname{trace} I - \operatorname{trace} C.$$

But det C is the product of the eigenvalues of C and trace C is the sum of the eigenvalues of C. The matrix C is full rank because both A and B, as positive-definite matrices, are full rank. Therefore, all eigenvalues c_ℓ of C are positive. Then

$$f(B) - f(A) = \sum_\ell \left[\log c_\ell - c_\ell + 1\right].$$

The inequality $\log x \leq x - 1$ shows that $\log c_\ell \geq c_\ell - 1$ for each term of the sum. Therefore, $f(B) - f(A) \geq 0$ and this term is equal to zero when A is equal to B. ∎

To use this proposition for the problem at hand, use the general relationship $x^\dagger A x = \text{trace} A x x^\dagger$ to write

$$\widehat{\Sigma}_s = \underset{\Sigma_s}{\text{argmin}} \left[\log \det \Sigma_s + \text{trace}\left(\Sigma_s^{-1} s s^\dagger\right)\right].$$

The proposition does not immediately apply because $B = s s^\dagger$ is not full rank. However there is always a full rank matrix arbitrarily close to $s s^\dagger$ to which the lemma does apply. In this sense, the lemma applies as well to $s s^\dagger$. Consequently, Proposition 8.9.1 leads to the conclusion that

$$\Sigma_s = s s^\dagger$$

is the maximum-likelihood estimate of Σ_s in terms of s, although s is unknown. This is the maximization step required by the alternating expectation-maximization algorithm.

Estimation

To derive the expectation step, the expression

$$E[s s^\dagger] = E[s s^\dagger \mid v, \Sigma_s, \Sigma_n]$$

must be evaluated as the estimate of $s s^\dagger$ when v, Σ_s, and Σ_n are given. This expectation $E[s s^\dagger]$ then serves in place of the unknown $s s^\dagger$ needed by the maximization algorithm. This evaluation of the expectation is provided by the corollary to the following theorem.

Theorem 8.9.2 *Let $v = s + n$ where s and n are independent, zero-mean, gaussian, real vector random variables with covariance matrices Σ_s and Σ_n, respectively. Then $p(s|v)$ is a multivariate gaussian probability density function with mean*

$$\bar{s} = \left(\Sigma_s^{-1} + \Sigma_n^{-1}\right)^{-1} \Sigma_n^{-1} v$$

and covariance matrix

$$\Sigma_{s|v} = \left(\Sigma_s^{-1} + \Sigma_n^{-1}\right)^{-1}.$$

Proof:
The Bayes formula is

$$p(s|v) = \frac{p(s) p(v|s)}{p(v)}.$$

8.9 Alternating Expectation-Maximization

Because all terms on the right are multivariate gaussian probability density functions, the term on the left is also a multivariate gaussian probability density function. It must have the form

$$p(s|v) = \frac{1}{\sqrt{\det(2\pi \Sigma_{s|v})}} e^{-\frac{1}{2}(s-\bar{s})^\dagger \Sigma_{s|v}^{-1}(s-\bar{s})}$$

with covariance matrix $\Sigma_{s|v}$ and mean $\bar{s}$. To find expressions for $\Sigma_{s|v}$ and $\bar{s}$, it is enough to inspect only the terms in the exponent of the Bayes formula for $p(s|v)$. The exponential terms combine to form the exponential term of $p(s|v)$ as

$$p(s|v) \sim \frac{e^{-\frac{1}{2}s^\dagger \Sigma_s^{-1} s} e^{-\frac{1}{2}(v-s)^\dagger \Sigma_n^{-1}(v-s)}}{e^{-\frac{1}{2}v^\dagger (\Sigma_s + \Sigma_n)^{-1} v}}.$$

Gather all the exponents together and identify terms. Identifying terms with those of the multivariate gaussian distribution for $p(s|v)$ gives the mean of $p(s|v)$ as

$$\bar{s} = (\Sigma_s^{-1} + \Sigma_n^{-1})^{-1} \Sigma_n^{-1} v,$$

and the variance of $p(s|v)$ is

$$\Sigma_{s|v} = (\Sigma_s^{-1} + \Sigma_n^{-1})^{-1},$$

as was to be proved. ∎

Corollary 8.9.3

$$E[ss^\dagger | v, \Sigma_s, \Sigma_n] = (\Sigma_s^{-1} + \Sigma_n^{-1})^{-1} + \bar{s}\bar{s}^\dagger.$$

Proof:

$$E[ss^\dagger | v, \Sigma_s, \Sigma_n] = E[(s-\bar{s})(s-\bar{s})^\dagger | v] + \bar{s}\bar{s}^\dagger$$
$$= (\Sigma_s^{-1} + \Sigma_n^{-1})^{-1} + \bar{s}\bar{s}^\dagger,$$

as was to be proved. ∎

The Iterative Algorithm

The maximization step and the expectation step can now be combined to describe the iterative algorithm. The maximization step is

$$\Sigma_s \leftarrow E[ss^\dagger | v, \Sigma_s, \Sigma_n]$$

and the expectation step, as given in Theorem 8.9.2 and Corollary 8.9.3, is

$$E[ss^\dagger | v, \Sigma_s, \Sigma_n] = (\Sigma_s^{-1} + \Sigma_n^{-1})^{-1} + \bar{s}\bar{s}^\dagger,$$

where

$$\bar{s} = (\Sigma_s^{-1} + \Sigma_n^{-1})^{-1} \Sigma_n^{-1} v.$$

Combining the two steps gives the iteration

$$\Sigma_s \leftarrow (\Sigma_s^{-1} + \Sigma_n^{-1})^{-1} + (\Sigma_s^{-1} + \Sigma_n^{-1})^{-1}\Sigma_n^{-1}vv^\dagger\Sigma_n^{-1}(\Sigma_s^{-1} + \Sigma_n^{-1})^{-1}.$$

This iteration can be manipulated into the alternative form

$$\Sigma_s \leftarrow \Sigma_s - \Sigma_s(\Sigma_s + \Sigma_n)^{-1}(\Sigma_s + \Sigma_n - vv^\dagger)(\Sigma_s + \Sigma_n)^{-1}\Sigma_s.$$

The alternative form may be preferred because it shows the iteration by means of a main term and a correction term on the right. This correction term is zero when

$$\Sigma_s = vv^\dagger - \Sigma_n.$$

While this is a sufficient condition for a fixed point, it may not be a necessary condition for a fixed point. For Σ_s to be a fixed point of the iteration, it is sufficient for the entire correction term to be zero. However, if $vv^\dagger - \Sigma_n$ is the only fixed point, then it must be the convergent of the iteration and so is the Σ_s that maximizes the likelihood.

8.10 Notions of Equivalence

It is natural to now ask how the maximum-likelihood principle is related both to the minimum-relative-entropy principle and to its special case of the maximum-entropy principle. Often the probability distribution that maximizes the likelihood also minimizes the relative entropy. In this sense, the maximum-likelihood principle and the minimum-relative-entropy principle (or the maximum-entropy principle) might seem to be equivalent. Although these principles seem to be different and are justified by different arguments, there are indeed similarities at a deeper level. We first show that the minimum-relative-entropy principle and the maximum-likelihood principle are connected by the law of large numbers and, to this extent, are equivalent.

Example
To anticipate the conclusion that is reached below, consider a gaussian density function as a prior distribution $q(x)$ on the random variable X. When given only the side information that x is contained in an interval of width x' as shown in Figure 8.4, how should the prior distribution be modified to form the posterior distribution $p(x)$. The minimum relative entropy principle says that $p(x)$ should be chosen to minimize $D(p(x)\|q(x))$ given by

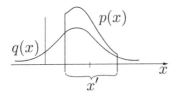

Figure 8.4 A prior and a posterior

8.10 Notions of Equivalence

$$D(p(x)\|q(x)) = \int_{-\infty}^{\infty} p(x)\log\frac{p(x)}{q(x)}dx$$
$$= \int_{x_0}^{\infty} p(x)\log\frac{p(x)}{q(x)}dx.$$

To find the minimum, introduce the lagrange multiplier λ and write

$$\widehat{p}(x) = \underset{p(x)}{\text{argmin}}\left(\int_{-x_0}^{\infty} p(x)\log\frac{p(x)}{q(x)}dx + \lambda\left(\int_{-x_0}^{\infty} p(x)dx - 1\right)\right).$$

The argmin can be found by the usual methods of the calculus of variations. This leads to the conclusion that $p(x) = Kq(x)$ for x in the given interval where K is chosen so that $\int_{x_0}^{\infty} p(x)dx = 1$.

This conclusion is illustrated in Figure 8.4. The figure illustrates this conclusion for a scalar measurement on the real line. For x in the given interval, the density function $p(x)$ is a scaled version of the density function $q(x)$. Elsewhere, $p(x)$ is zero. ∎

Let X be a discrete random variable described by the prior probability vector $\mathbf{q}$ so that X takes value x_j with probability q_j. The realization x of the random variable X is the unknown. Let y be the actual data obtained by the measurement described by $y = h(x)$. The data y is related to the unknown x by the defining function $y = h(x)$. The task is to estimate the value of x when given the measurement y. Because $h(x)$ is a many-to-one function, the measurement y does not determine a unique x, but only a set of possible x. Let $p(x)$ be a posterior probability distribution on x. Then we have the constraint on $p(x)$ that

$$\sum_{x:h(x)=y} p(x) = 1.$$

This follows because the posterior probability distribution is given y knowing that $h(x) = y$.

The following theorem states that when given a prior on a random variable and a noiseless measurement dependent on that random variable, the relative entropy is smallest when $p(x)$ is chosen equal to zero on all x that are not consistent with that measurement and $p(x)$ is proportional to $q(x)$ for all other x.

Theorem 8.10.1 *Let $\mathbf{q}$ be a prior on a set $\mathcal{X}$ and let $\mathbf{p}$ be any probability distribution on $\mathcal{X}$ that is nonzero only on those x satisfying $h(x) = y$. When measurement $y = h(x)$ is given, the relative entropy $D(\mathbf{p}\|\mathbf{q})$ is minimized over such $\mathbf{p}$ by*

$$p(x) = \frac{q(x)}{\sum_{x:h(x)=y} q(x)}.$$

for all x such that $h(x) = y$, and $p(x)$ equals zero for all other x.

Proof:
Let $\mathcal{X}'$ be the set of x satisfying the constraint $h(x) = y$. Then

$$\widehat{p}(x) = \underset{p(x)}{\operatorname{argmax}} \Big[\sum_{x \in \mathcal{X}'} p(x) \log \frac{p(x)}{q(x)} + \lambda \Big(\sum_{x \in \mathcal{X}'} p(x) - 1 \Big) \Big].$$

Setting the partial derivatives equal to zero shows that $p(x)/q(x)$ must be a constant for $x \in \mathcal{X}'$. The constraint requires that $p(x) = 0$ for $x \notin \mathcal{X}'$. The statement of the theorem follows. ∎

The significance of the theorem is that the $p(x)$ minimizing the relative entropy is the same as the conditional on x when given y. To verify this, recall that

$$q(x)p(y|x) = p(y)q(x|y).$$

When measurement $y = h(x)$ is given, then $p(y|x) = 1$ for that y and $p(y|x) = 0$ for all other y. Because $p(y) = \sum_{x:h(x)=y} q(x)$, we conclude that

$$q(x|y) = \frac{q(x)}{\sum_{x:h(x)=y} q(x)}.$$

This is the same $p(y)$ as is given in Theorem 8.10.1. Once again, as in the example of Figure 8.4, the probability distribution $q(x)$, after imposing the condition that $h(x) = y$, is replaced by a new probability $p(x)$ which is simply a multiple of $q(x)$ on that interval and zero elsewhere.

Motivation for Gaussian Distribution

A second notion of equivalence is an implicit equivalence between probability distributions and distance measures. A common practice is to use euclidean distance to measure the discrepancy between two vectors or two functions. This practice is regarded as quite reasonable. Another common practice is to use a gaussian probability density function as a surrogate probability density function when the true probability density function is not known. This is also regarded as reasonable. It may be partially justified by the loose statement that many probability densities are indeed gaussian, often as a consequence of the central limit theorem. It might also be justified by the heuristic statement that this choice is maximally noncommittal or by the formal statement that the gaussian probability density function maximizes the entropy for a specified mean and variance.

A more subtle justification is now provided, one that may be more compelling. Accepting the minimum euclidean distance as a criterion of estimation optimality, which is a widely accepted criteria, is equivalent to assuming a gaussian probability density function. A variant of this fact appears in Section 8.11. That section shows that using a quadratic penalty for regularization is equivalent to assuming a gaussian prior.

Suppose that a polynomial $y = \sum_{i=0}^{I-1} a_i x^i$ is given where the coefficients are unknown and are to be estimated from a set of noisy measurements $\{(x_j, y_j)\}$, where $y_j = \sum_{i=0}^{I-1} a_i x_j^i + n_i$, and where the n_i are random variables with unknown joint distributions. Suppose that for want of a better model, one assumes that the random variables are gaussian, with independent noise components of equal variance. Then the maximum of the loglikelihood function is the same as the minimum euclidean distance. Thus, choosing to minimize euclidean distance is equivalent to choosing a memoryless gaussian distribution.

Motivation for Poisson Distribution

Statements can be made for the poisson distribution that are parallel to the statements made for the gaussian distribution. It is common practice to use a poisson probability distribution for a point process when the true probability distribution is not known. In particular, accepting the Csiszár distance as an appropriate measure of distance in the space of nonnegative functions is tantamount to accepting a poisson distribution as appropriate for the data.

The Csiszár distance is written

$$D(y(x)\|\lambda(x)) = \sum_x \left[y(x)\log_e \frac{y(x)}{\lambda(x)} + \lambda(x) - y(x)\right]$$

$$= \sum_x \left[y(x)\log_e y(x) - y(x)\right] - \sum_x \left[y(x)\log_e \lambda(x) - \lambda(x)\right].$$

Then

$$e^{-D(y(x)\|\lambda(x))} = \frac{1}{Z}\prod_x e^{-\lambda(x)}\lambda(x)^{y(x)},$$

where $Z = [e/y(x)]^{-y(x)}$ does not depend on $\lambda(x)$. This formula is to be compared with the poisson distribution

$$p_\lambda(y) = \frac{e^{-\lambda}\lambda^y}{y!},$$

which has the same dependence on λ. Essentially the same formula arises in two different ways. The Csiszár distance is appropriate when treating nonnegative functions, in general, in a nonprobabilistic space. The poisson distribution is appropriate for point process in an probabilistic setting. Because the same formulas arise, the same computational procedures may apply. For example, the Richardson–Lucy algorithm, in addition to minimizing the Csiszár distance, can be seen as an algorithm of the expectation-maximization type.

8.11 Regularization

Inverse problems that are formulated on a continuous space are usually ill-posed. Inverse problems that are formulated on a finite-dimensional parameter space are usually well-posed, though possibly *ill-conditioned*, meaning that the solution varies widely, though continuously, with small changes in the data. Ill-posed or ill-conditioned behavior is treated by using various methods for regularizing the problem. The process of *regularization* of a problem is the process of restricting the class of acceptable solutions. The purpose of regularization is to replace an ill-posed problem, such as a problem on a continuous parameter space, by a well-posed problem usually on a finite-dimensional space as may be obtained by sampling. For example, the regularization may require the solution to satisfy certain smoothness conditions, and these required smoothness conditions become a part of the problem.

In an infinitely large space, there can be an infinite number of solutions that satisfy a finite set of data. Regularization can be described as a way of prioritizing these solutions so that only one preferred solution is obtained.

To this point, recall that the maximum-likelihood principle is a powerful principle of inference. The method of expectation-maximization provides a powerful tool to implement this maximum-likelihood principle. As is often the case with powerful tools, the tool may be too strong for the problem at hand. Consequently, maximum-likelihood image formation has a potential flaw that is subtle. An image is usually expected to have a satisfactory combination of crispness and smoothness, and less detail in an image is usually expected to be more likely than more detail, but this expectation is difficult to quantify. Whereas many computational problems improve as the variables are more finely quantized, maximum-likelihood imaging can degrade because the algorithm will create unnatural details in the image to account for small insignificant details in the data. The task of augmenting the computation with prior expectations in order to impose some degree of order or smoothness in the image is the process of regularization. When the prior expectation is in the form of a prior probability distribution, this becomes maximum-posterior estimation as discussed in Section 8.4.

An image on continuous space $\mathbb{R}^2$ has an infinite number of degrees of freedom and cannot be determined by a finite set of data. An elementary and commonly used way to regularize an image-formation computation is by two-dimensional sampling, commonly known as *pixelization*. The size of the pixels should be small enough to obtain a good resolution, yet not so small that artificial details are bothersome in the image.

Tikhonov Regularization

A standard method of regularization is the method of *penalties*. This method adds an empirical penalty term to the negative loglikelihood and then finds the minimum of this sum. Thus, the constrained maximum-likelihood estimate, when expressed as a minimization, is

$$\widehat{\gamma} = \underset{\gamma}{\operatorname{argmin}}[-\Lambda(\gamma) + c(\gamma)],$$

where $-\Lambda(\gamma)$ is the negative of the loglikelihood and $c(\gamma)$ is the empirical nonnegative strictly increasing penalty term. The penalty term means that low-energy states are preferred to high-energy states.

A quadratic constraint – such as an energy constraint – is often considered to be a good choice for a penalty constraint. Thus, the maximum-likelihood estimate under the quadratic penalty constraint is

8.11 Regularization

$$\widehat{\gamma} = \underset{\gamma}{\operatorname{argmin}} \left[\Lambda(\gamma) + \alpha^2 \|\gamma\|^2 \right],$$

where $\alpha^2 \|\gamma\|^2$ is the quadratic penalty term, written in terms of a constant α^2 and the energy in the image, denoted $\|\gamma\|^2$. The use of a quadratic penalty function is called *Tikhonov regularization*.

Example

An elementary example of Tikhonov regularization is for the analytical solution of least-squares problem. Suppose that y is linearly related to x in the form $y = ax + b$, but the parameter a, and perhaps b, is not known. A set of n noisy points of the curve (x_ℓ, y_ℓ) is given. When fit by the straight line $y = ax$ with b set equal to zero, each noisy point has a squared error of the form $(xa - y)^2$. The squared error in vector notation is $(a\mathbf{x} - \mathbf{y})^T(a\mathbf{x} - \mathbf{y})/n$. Therefore,

$$\widehat{a} = \underset{a}{\min}(a\mathbf{x} - \mathbf{y})^T(a\mathbf{x} - \mathbf{y}) + \lambda a^2.$$

Under differentiation with respect to a, this results in $2\mathbf{x}^T(\mathbf{x}a - \mathbf{y}) + 2\lambda = 0$. Therefore,

$$a = (\mathbf{x}^T\mathbf{x} + \lambda)^{-1}(\mathbf{x}^T\mathbf{y}).$$

Now allow b to be nonzero. Then

$$(\widehat{a}, \widehat{b}) = \underset{(a,b)}{\min}(a\mathbf{x} + b - \mathbf{y})^T(a\mathbf{x} + b - \mathbf{y}) + \lambda a^2.$$

Setting partial derivatives with respect to a and to b equal to zero gives two equations which can be solved to give the estimated straight line $y = \widehat{a}x + \widehat{b}$.

This instance of regularization uses the quadratic energy term to constrain the energy in a. The problem is posed without mention of a noise model.

Example

Another example of Tikhonov regularization is the estimation of the value of a vector $\mathbf{x}$ of blocklength n when given a vector measurement $\mathbf{y}$ of blocklength n satisfying

$$\mathbf{y} = c\mathbf{x} + \mathbf{z},$$

where $\mathbf{z}$ is a gaussian random vector of zero mean and covariance matrix $\mathbf{\Sigma}$. The conditional gaussian probability density function is

$$p(\mathbf{y}|\mathbf{x}) = \frac{1}{\sqrt{\det(2\pi\mathbf{\Sigma})}} e^{-\frac{1}{2}(\mathbf{y}-c\mathbf{x})^\dagger \mathbf{\Sigma}^{-1}(\mathbf{y}-c\mathbf{x})}.$$

For the unknown vector $\mathbf{x}$, the (negative) loglikelihood function is

$$\Lambda(\mathbf{x}) = (\mathbf{y} - c\mathbf{x})^\dagger \mathbf{\Sigma}^{-1}(\mathbf{y} - c\mathbf{x}).$$

Then, under a quadratic penalty constraint $\alpha^2 |\mathbf{x}|^2$,

$$\widehat{\mathbf{x}} = \underset{\mathbf{x}}{\operatorname{argmin}} \left[(\mathbf{y} - c\mathbf{x})^\dagger \mathbf{\Sigma}^{-1}(\mathbf{y} - c\mathbf{x}) + \alpha^2 |\mathbf{x}|^2 \right].$$

This maximum likelihood estimate for a deterministic vector parameter x under a quadratic penalty constraint is identical to a bayesian estimator with a gaussian prior, which arises in a much different way.

The bayesian estimator is discussed in Section 8.4. For this estimator, consider x to be a random vector with an independent gaussian distribution of variance σ^2 as a prior. For this formulation,

$$\hat{x} = \mathrm{argmin}_x \big[- \log p(x|y) \big]$$
$$= \mathrm{argmin}_x \big[- \log p(y|x) p(x) \big]$$
$$= \mathrm{argmin}_x \big[(y - cx)^\dagger \mathbf{\Sigma}^{-1} (y - cx) + \sigma^2 \|x\|^2 \big],$$

which is the same as before with the variance σ^2 replacing the free parameter α^2. Hence, choosing a quadratic penalty function gives the same estimate as was obtained by placing a prior on x so, in the sense of this observation, the two estimators are equivalent.

Grenander Sieves

The method of sieves can be used to suppress the artifacts of an ill-posed or ill-conditioned problem. One can say in such a problem that the search space, such as the space of functions, is too large. The method constrains the set of allowed images to be elements of a chosen subset of smooth images called a *Grenander sieve*. The sieve is defined as a sequence of sets of functions that depends on the size of the data set. For a larger data set, the sieve contains more functions. Because the size of the sieve is allowed to grow with the number of data measurements, it is hoped that the image estimate converges to the true image as the size of the data set goes to infinity.

Problems

8.1 Prove that the Kullback distance

$$D_K \big(p(x) \| q(x) \big) = \int_{-\infty}^{\infty} p(x) \log \frac{p(x)}{q(x)} dx$$

is convex in the first variable and convex in the second variable.

8.2 **a.** Prove that $\log_e x \leq x - 1$.

b. Prove that the Csiszár distance satisfies

$$D_C \big(a(x) \| b(x) \big) \geq 0$$

for any nonnegative functions $a(x)$ and $b(x)$.

8.3 A term of the form $\log \left(\sum_i q_i \right)$ is often difficult to handle analytically because the sum is inside the logarithm. Prove that, for a set of q_i that are all positive,

$$- \log \left(\sum_{i=1}^n q_i \right) = \min_{p \in \mathbb{P}^n} \sum_{i=1}^n p_i \log \frac{p_i}{q_i},$$

where p is a probability vector. This formula allows the sum to be pulled outside the logarithm at the cost of a superfluous minimization. This expression is a discrete form of the convex decomposition lemma.

8.4 A poisson process is a point process with the probability distribution

$$p_n = \frac{(\lambda T)^{n(T)}}{n(T)!} e^{-\lambda T},$$

where $n(T)$ is the number of events detected in the observation interval T. When λ is unknown and $n(T)$ is observed, p_n becomes the likelihood function

$$L(\lambda) = e^{\Lambda(\lambda)} = \frac{(\lambda T)^{n(T)}}{n(T)!} e^{-\lambda T}$$

in the parameter λ. The unknown λ can be estimated from the data $n(T)$ by maximizing this function.

Two independent poisson processes have the arrival rates a and b. The arrival rate b is known and the arrival rate a is unknown. The points of the two processes are pooled to form the single poisson process with the arrival rate $a+b$. The unlabeled occurrence times $(t_1, t_2, t_3, \ldots, t_N)$ of the pooled process comprise the observed data.

a. Find analytically the maximum-likelihood estimator of the unknown parameter a when given the observed data $(t_1, t_2, t_3, \ldots, t_N)$, and the parameter b is known.
b. Use the method of alternating expectation-maximization to construct an iterative algorithm to estimate a. What is the limit point of this iteration?

8.5 Let v be a gaussian block random variable of blocklength n with probability density function

$$p(v) = \frac{1}{\sqrt{\det 2\pi \Sigma}} e^{-\frac{1}{2}(v-\bar{v})^\dagger \Sigma^{-1}(v-\bar{v})}.$$

Calculate the mean $E[v]$ and the covariance matrix $E[vv^T] - E[v]E[v^T]$. The notation suggests the answer but the choice of notation is not a proof.

8.6 A maximum-likelihood estimate is known to be asymptotically unbiased in the size or quality of the data set. This means that the expected error goes to zero as the amount of data goes to infinity. Show that the problem of estimating the variance of a single gaussian random variable, measured in gaussian noise, is asymptotically unbiased with respect to the signal-to-noise ratio. (It will be helpful to recall that the square of a zero-mean, gaussian random variable with variance σ^2 is an exponential random variable with mean σ^2.)

8.7 The discrete noisy deblurring problem is the task of estimating a *deterministic* image s when given a noisy blurred copy of the image, possibly a complex image, as given by

$$v = Hs + n,$$

where n is a white gaussian noise vector with the covariance matrix $\Sigma_n = N_0 I$ and H is a known blurring matrix.

a. Show that the loglikelihood function for this problem is

$$\Lambda(s) = 2\text{Re}[n^\dagger \Sigma_n^{-1} Hs] - s^\dagger H^\dagger \Sigma_n^{-1} Hs.$$

b. Give an algorithm for computing the maximum-likelihood deblurred image $\hat{s}$.

8.8 Prove that the Csiszár distance $d_C(a(x)\|b(x))$ is nonnegative and is equal to zero if and only if $a(x)$ is equal to $b(x)$.

8.9 Show that the Csiszár distance does not satisfy the triangle inequality.

8.10 The discrete noisy deblurring problem is the task of estimating a *random* image, s, when given a noisy blurred image, possibly complex copy of the image, given by

$$v = Hs + n,$$

where s is a zero-mean, gaussian random process with covariance matrix Σ_s, n is white gaussian noise with the covariance matrix $\Sigma_n = N_0 I$, and H is a known blurring matrix.

a. Show that the loglikelihood function for this problem is

$$\Lambda(\Sigma_s) = -\log\det(H^\dagger \Sigma_s H + N_0 I) - v^\dagger(H^\dagger \Sigma_s H + N_0 I)^{-1} v.$$

b. Give an algorithm for computing the maximum-likelihood deblurred image, when given the constraint that Σ_s is diagonal.

8.11 a. Prove the *Pinsker inequality* for the relative entropy (in nats)

$$D(p_1\|p_2) \geq |p_1 - p_2|^2$$

for both discrete and continuous probability distributions.

b. Prove the inequality

$$D(p_1(x)\|p_2(x)) \geq \frac{1}{2}\int_{-\infty}^{\infty} \left(\sqrt{p_1(x)} - \sqrt{p_2(x)}\right)^2 dx.$$

8.12 a. Let

$$C = \begin{bmatrix} A & 0 \\ 0 & B \end{bmatrix},$$

where A and B are square matrices. Show that $\det C = \det A \det B$.

b. Let $v = s + n$ where s and n are gaussian vector random processes with covariance matrices Σ_s and Σ_n, respectively. Find the joint probability density function of (s, v) by evaluating $p(s)p(v|s)$.

c. Verify this conclusion by computing

$$\Sigma = \begin{bmatrix} E[ss^\dagger] & E[sv^\dagger] \\ E[vs^\dagger] & E[vv^\dagger] \end{bmatrix}$$

directly.

8.13 Let x and y each be a gaussian vector random variable of blocklength n with the joint probability density function

$$p(x,y) = \frac{1}{\sqrt{\det(2\pi \Sigma)}} e^{-\frac{1}{2}(x,y)\Sigma^{-1}(x,y)^\dagger},$$

where $\Sigma = \begin{bmatrix} \Sigma_x & R_{xy} \\ R_{xy} & \Sigma_y \end{bmatrix}$ is defined by $\Sigma = E\begin{bmatrix} \begin{bmatrix} x \\ y \end{bmatrix} \begin{bmatrix} x & y \end{bmatrix} \end{bmatrix}$. From $p(x,y)$, find expressions for the marginals $p(x)$ and $p(y)$ and the conditionals $p(x|y)$ and $p(y|x)$.

8.14 Prove the general matrix relation

$$x^\dagger A x = \text{trace} A x x^\dagger,$$

where A is a square matrix and x is a vector of compatible size.

8.15 Develop the alternative forms

$$\Sigma_s \leftarrow \Sigma_s - (\Sigma_s^{-1} + \Sigma_n^{-1})^{-1}\Sigma_n^{-1}(\Sigma_s + \Sigma_n - vv^\dagger)\Sigma_n^{-1}(\Sigma_s^{-1} + \Sigma_n^{-1})^{-1}$$

$$\Sigma_s \leftarrow \Sigma_s - \Sigma_s(\Sigma_s + \Sigma_n)(\Sigma_s + \Sigma_n - vv^\dagger)(\Sigma_s + \Sigma_n)^{-1}\Sigma_s$$

for the iterative algorithm for estimating covariance matrices given in Section 8.9.

8.16 Let $v = s + n$ where s and n are gaussian vector random variables with zero mean and with covariance matrices Σ_s and Σ_n, respectively. Find the conditional density function

$$p_{s|v}(s|v) = \frac{1}{\sqrt{\det(2\pi \Sigma_{s|v})}} e^{-\frac{1}{2}(s-\bar{s})^\dagger \Sigma_{s|v}^{-1}(s-\bar{s})}$$

from the Bayes formula.

8.17 Prove the matrix identities for compatible and invertible matrices, where Σ_s and Σ_n are covariance matrices.

(a) $\Sigma_s \Gamma^\dagger (\Gamma \Sigma_s \Gamma^\dagger + \Sigma_n)^{-1} = (\Gamma^\dagger \Sigma_n^{-1} \Gamma + \Sigma_s^{-1})^{-1} \Gamma^\dagger \Sigma_n^{-1}$
(b) $(\Sigma_s^{-1} + \Gamma^\dagger \Sigma_n^{-1} \Gamma)^{-1} = \Sigma_s - \Sigma_s \Gamma (\Gamma^\dagger \Sigma_s \Gamma + \Sigma_n)^{-1} \Gamma^\dagger \Sigma_s$

8.18 Develop the Richardson–Lucy algorithm with the Csiszár distance in place of the Kullback distance.

8.19 The circular two-dimensional gaussian density function

$$q(x,y) = \frac{1}{2\pi\sigma^2} e^{-(x^2+y^2)/2\sigma^2}$$

is given as the prior density function. When given the side information that $x = y$, the posterior distribution $p(x)$ is given by

$$p(x) = \frac{q(x,x)}{\int_{-\infty}^{\infty} q(x,x)dx}.$$

a. What is the posterior distribution $p(x) = \int q(x,y)dy$?
b. Does this differ from the marginal distribution $p(x)$?

c. Does this differ from the conditional distribution $p(x|y) = q(x,y)/\int q(x,y)dy$?
d. Repeat for the prior distribution

$$q(x,y) = \frac{1}{2\pi\sigma_1\sigma_2}e^{-(x^2/2\sigma_1^2)+(y^2/2\sigma_2^2)}.$$

8.20 The fixed-point conditions for blind deconvolution can be developed by the calculus of variations. Replace $h(x,y)$ and $s(x,y)$ by $h'(x,y) = h(x,y) + \epsilon\eta(x,y)$ and $s(x,y) + \epsilon'\eta'(x,y)$, respectively. Then the relative entropy is

$$D(v\|h' **s') = \int_{-\infty}^{\infty}\int_{-\infty}^{\infty} v(x,y)\log\frac{v(x,y)}{h'(x,y)**s'(x,y)}dxdy.$$

Now, establish the iteration by setting the partial derivatives with respect to ϵ and ϵ' equal to zero at the point at which both ϵ and ϵ' are equal to zero.

8.21 Verify convergence of the iterative algorithm for blind deconvolution given in Section 8.7. Is there a unique fixed point? Must the algorithm converge to the correct solution?

8.22 An *Ali–Silvey distance* between two probability distributions is defined for any nondecreasing function $f(\cdot)$ and any convex function $c(\cdot)$ as

$$d_{AS}(p_0,p_1) = f(\mathrm{E}[c(L(x))]),$$

where the function $L(x)$ denotes the likelihood ratio $p_1(x)/p_0(x)$.
Show that the Kullback relative entropy is an Ali–Silvey distance.

Notes

The maximum-likelihood principle is a long-standing and time-honored principle of statistics, having been introduced into statistics by Fisher (1925). Similarly, the entropy is an important quantity in the literature of mathematical informatics, having been introduced into information theory by Shannon (1948) and into quantum information theory by von Neumann (1932). An alternative statement of the entropy already had a prominent and long-standing place within the subject of statistical physics. Although expressed very differently and within a different setting, the alternative statement of the entropy in statistical physics is essentially equivalent.

The relative entropy was introduced into the statistics literature by Kullback (1959) and studied by Shore and Johnson (1980) and by Csiszár (1991). In his work, Csiszár argued that the only reasonable choices for distance measures are the quadratic distance in euclidian space, the Csiszár distance in the space of discrete nonnegative vectors, and the Kullback distance or relative entropy in probability space. Miller and Snyder (1987) observed that maximum likelihood and minimum relative entropy in many situations give the same answer and, in this sense, are equivalent. The important method of expectation-maximization for developing computational algorithms to obtain maximum-likelihood images was presented by Dempster, Laird, and Rubin (1977). Convergence of such algorithms was established under general conditions by Wu (1983). A stronger statement regarding convergence under certain special conditions was established by Vardi, Shepp, and Kaufman (1985).

The Richardson–Lucy algorithm (Richardson, 1972; Lucy, 1974) for deblurring nonnegative images was first developed in a heuristic manner, without any formal motivation or proof, and was then justified experimentally. The Richardson–Lucy algorithm was rediscovered in a probabilistic formulation by Shepp and Vardi (1982) in the context of medical imaging. It was given a formal framework based on minimizing the relative entropy, by Snyder, Schulz, and O'Sullivan (1992). Similar ideas were reached independently by Vardi and Lee (1993). O'Sullivan (1998, 2002) presented the formal development of the Richardson–Lucy algorithm as an alternating minimization algorithm. The blind version of the deconvolution algorithm along lines similar to the Richardson–Lucy algorithm was discussed by Ayers and Dainty (1982). By introducing the convex decomposition lemma, O'Sullivan (2002) completed the embedding of alternating minimization and alternating maximization algorithms for likelihood problems into the literature of information theory. The maximum-entropy principle was formulated by Jaynes (1957) based on the observation that any probability density function that is consistent with a set of constraints but has entropy smaller than the maximum entropy is atypical of the data. The Schulz–Snyder algorithm (1992) for phase retrieval was shown by Oktem and Blahut (2011) to be an alternating minimization algorithm for minimizing the Kullback distance between the measured autocorrelation and the computed autocorrelation of an image. Convergence of that algorithm was studied by Choi, Lanterman, and Raich (2006).

Imaging from weak light has been studied by Snyder and Schulz (1990). Schulz and Snyder (1992) developed a maximum-likelihood iteration for the photon-differencing problem by using the method of expectation-maximization. The topic of imaging from point event data was first formulated for emission tomography by Rockmore and Macovski (1967) using the maximum-likelihood principle, but their formulation was intractable because they had no way to compute the maximum-likelihood estimates. A computationally tractable form was developed by Shepp and Vardi (1982), and by Lange and Carson (1984), when they applied the method of expectation-maximization to this problem. Politte and Snyder (1991) described two algorithms based on two choices of the complete data space that can be developed by using the expectation-maximization method.

Regularization is often needed to make an ill-posed computational problem tractable or to prevent overfitting a set of data. An important contribution came from Tikhonov (1963). The method of sieves was suggested by Grenander (1981) as a general tool of statistical inference. The method of sieves was applied to the problem of noncoherent imaging by Snyder and Miller (1985).

9 Diffraction Imaging Systems

Because of their very small wavelengths (usually less than a nanometer), X-rays can be used to obtain valuable information about the structure of very small objects. The diffraction of X-rays by a crystal can provide information about the lattice structure of the crystal, the irregularities in the lattice structure, the shape of the molecules making up the crystal, and the thermal motion of those molecules.

The molecules of a crystal are arranged in a regular lattice with a spacing between them that is comparable to the wavelength of X-rays. An incident X-ray propagates as an electromagnetic wave satisfying the wave equation and is diffracted by the crystal. The diffracted wave forms sharp grating lobes in the far-field diffraction region because of the periodic structure of the crystal. The distribution of these grating lobes in solid angle depends on the lattice structure of the crystal, while the relative amplitudes of the many grating lobes depends on the electron structure of the molecules making up the crystal. The widths and the fluctuations of the grating lobes depend on the irregularities and strength in the crystal structure.

This chapter provides an introduction to crystallographic image formation while respecting much of the terminology and orientation of the other chapters of the book. The chapter begins with a discussion of the three-dimensional Fourier transform. Then the far-field diffraction of a plane wave by a three-dimensional crystal is described by using the three-dimensional Fourier transform. Methods of imaging the elements of the crystal are described in the latter part of the chapter.

Topics of mathematics such as the Fourier transform, lattice theory, and even the Karle–Hauptman inequality are important to crystallography but do form topics in mathematics that stand separate from crystallography. For full generality, these are described herein using the language of mathematics rather than the language of crystallography.

9.1 The Three-Dimensional Fourier Transform

A real or complex function, $s(x,y,z)$, of three variables x, y, and z is called a *three-dimensional function* or a *three-dimensional signal*. In many applications, $s(x,y,z)$ is a real-valued function. Often $s(x,y,z)$ is nonnegative and so it is a real-valued function that can be visualized as the density distribution of a diffuse cloud. The spatially

9.1 The Three-Dimensional Fourier Transform

varying density of the diffuse cloud at the point (x, y, z) is given by the function $s(x, y, z)$. In the general case, $s(x, y, z)$ is a complex function of x, y, and z.

Given the three-dimensional function $s(x, y, z)$, possibly complex, whose energy

$$E_p = \int_{-\infty}^{\infty}\int_{-\infty}^{\infty}\int_{-\infty}^{\infty} |s(x,y,z)|^2 dxdydz$$

is finite, the three-dimensional Fourier transform of $s(x, y, z)$ is a function of (f_x, f_y, f_z) defined as[1]

$$S(f_x,f_y,f_z) = \int_{-\infty}^{\infty}\int_{-\infty}^{\infty}\int_{-\infty}^{\infty} s(x,y,z)e^{-j2\pi(f_x x + f_y y + f_z z)} dxdydz$$

and is denoted by a triply shafted arrow

$$s(x,y,z) \iff S(f_x,f_y,f_z).$$

It is apparent that the inverse three-dimensional Fourier transform of $S(f_x, f_y, f_z)$ is given by

$$s(x,y,z) = \int_{-\infty}^{\infty}\int_{-\infty}^{\infty}\int_{-\infty}^{\infty} S(f_x,f_y,f_z)e^{j2\pi(f_x x + f_y y + f_z z)} df_x df_y df_z.$$

This is an immediate generalization of the inverse one-dimensional Fourier transform.

Other properties of the three-dimensional Fourier transform can be obtained easily as generalizations of familiar properties of the one-dimensional Fourier transform. Some of the basic properties are as follows:

1. Linearity: For any constants a and b, possibly complex,

$$as_1(x,y,z) + bs_2(x,y,z) \iff aS_1(f_x,f_y,f_z) + bS_2(f_x,f_y,f_z).$$

2. Sign reversal:

$$s(-x,y,z) \iff S(-f_x,f_y,f_z),$$
$$s(x,-y,z) \iff S(f_x,-f_y,f_z),$$
$$s(x,y,-z) \iff S(f_x,f_y,-f_z).$$

3. Conjugation:

$$s^*(x,y,z) \iff S^*(-f_x,-f_y,-f_z),$$
$$s^*(-x,-y,-z) \iff S^*(f_x,f_y,f_z).$$

When $s(x, y, z)$ is real, the transform satisfies $S^*(-f_x, -f_y, -f_z) = S(f_x, f_y, f_z)$.

4. Scaling: For any real nonzero constants a, b, and c,

$$s(ax,by,cz) \iff \frac{1}{|abc|}S\left(\frac{f_x}{a},\frac{f_y}{b},\frac{f_z}{c}\right).$$

---

[1] The three-dimensional variable (f_x, f_y, f_z) is said to be an element of the three-dimensional *frequency domain* or *frequency space*. Frequency space is also called *reciprocal space*, *dual space*, or *k-space*.

5. Origin translation: For any real constants a, b, and c,
$$s(x-a, y-b, z-c) \iff S(f_x, f_y, f_z)e^{-j2\pi(af_x+bf_y+cf_z)}.$$

6. Modulation: For any real constants a, b, and c,
$$s(x, y, z)e^{j2\pi(ax+by+cz)} \iff S(f_x-a, f_y-b, f_z-c).$$

7. Convolution:
$$g(x,y,z) ***\ h(x,y,z), \iff G(f_x,f_y,f_z)H(f_x,f_y,f_z),$$
where $***$ denotes a three-dimensional convolution given by
$$\int_{-\infty}^{\infty}\int_{-\infty}^{\infty}\int_{-\infty}^{\infty} g(\xi, \eta, \zeta)h(x-\xi, y-\eta, z-\zeta)d\xi d\eta d\zeta.$$

8. Product:
$$g(x,y,z)h(x,y,z) \iff G(f_x,f_y,f_z) *** H(f_x,f_y,f_z).$$

9. Parseval's formula:
$$\int_{-\infty}^{\infty}\int_{-\infty}^{\infty}\int_{-\infty}^{\infty} g(\mathbf{x})h^*(\mathbf{x})dxdydz = \int_{-\infty}^{\infty}\int_{-\infty}^{\infty}\int_{-\infty}^{\infty} G(\mathbf{f})H^*(\mathbf{f})df_x df_y df_z,$$
where $\mathbf{x} = (x,y,z)$ and $\mathbf{f} = (f_x, f_y, f_z)$.

10. Energy relation:
$$\int_{-\infty}^{\infty}\int_{-\infty}^{\infty}\int_{-\infty}^{\infty} |s(x,y,z)|^2 dxdydz = \int_{-\infty}^{\infty}\int_{-\infty}^{\infty}\int_{-\infty}^{\infty} |S(f_x,f_y,f_z)|^2 df_x df_y df_z.$$

11. Coordinate transformation: For any transform pair $s(x,y,z) \iff S(f_x,f_y,f_z)$ and for any three by three invertible matrix A,
$$s\left(A^T \begin{bmatrix} x \\ y \\ z \end{bmatrix}\right) \iff \frac{1}{\det A} S\left(A^{-1} \begin{bmatrix} f_x \\ f_y \\ f_z \end{bmatrix}\right).$$

12. Real part of transform: For any transform pair $s(x,y,z) \iff S(f_x,f_y,f_z)$,
$$s(x,y,z) + s^*(-x,-y,-z) \iff 2\text{Re}[S(f_x,f_y,f_z)],$$
which, for symmetric $s(x,y,z)$, reduces to $\text{Re}[s(x,y,z)] \iff \text{Re}[S(f_x,f_y,f_z)]$.

13. Friedel's property: A Fourier transform $S(f_x,f_y,f_z)$ is real if and only if
$$s(x,y,z) = s^*(-x,-y,-z).$$
A signal $s(x,y,z)$ is real if and only if
$$S(f_x,f_y,f_z) = S^*(-f_x,-f_y,-f_z).$$

14. Image doubling property: The signal $s(x,y,z)$ with Fourier transform $S(f_x,f_y,f_z)$ satisfies
$$2\text{Re}[S(f_x,f_y,f_z)] \iff s(x,y,z) + s^*(-x,-y,-z).$$

9.2 Transforms of Some Useful Functions

Many elementary three-dimensional functions can be defined as immediate generalizations of elementary one-dimensional or two-dimensional functions. Such functions are described in this section along with their Fourier transforms. Some of these Fourier transforms are obtained by a straightforward application of the separation of variables. Other elementary three-dimensional functions, such as the sphere and the helix, are also defined here. Fourier transforms of the latter two elementary functions must have their own derivations.

A list of three-dimensional Fourier transforms of various useful functions is given in Table 9.1. Some of these Fourier transform pairs are derived later in this section.

The *three-dimensional rectangle function* is defined as

$$\text{rect}(x,y,z) = \text{rect}(x)\text{rect}(y)\text{rect}(z).$$

The *three-dimensional sinc function* is defined as

$$\text{sinc}(f_x,f_y,f_z) = \text{sinc}(f_x)\text{sinc}(f_y)\text{sinc}(f_z).$$

The Fourier transform of the three-dimensional rectangle function is easily derived by a separation of variables as

$$\text{rect}(x,y,z) \iff \text{sinc}(f_x,f_y,f_z).$$

The scaled rectangle function

$$s(x,y,z) = \text{rect}(ax, by, cz)$$

has the three-dimensional Fourier transform

$$S(f_x,f_y,f_z) = \frac{1}{abc}\text{sinc}\left(\frac{f_x}{a}, \frac{f_y}{b}, \frac{f_z}{c}\right),$$

Table 9.1 A table of three-dimensional Fourier transform pairs

$s(x,y,z)$	$S(f_x,f_y,f_z)$
$\text{rect}(x,y,z)$	$\text{sinc}(f_x,f_y,f_z)$
$\text{sinc}(x,y,z)$	$\text{rect}(f_x,f_y,f_z)$
$\delta(x,y,z)$	1
1	$\delta(f_x,f_y,f_z)$
$\text{cyln}(x,y,z)$	$\text{jinc}(f_x,f_y)\text{rect}(f_z)$
$\text{sphr}(x,y,z)$	$\text{tinc}\sqrt{f_x^2+f_y^2+f_z^2}$
$\text{comb}_N(x,y,z)$	$\text{dirc}_N(f_x,f_y,f_z)$
$\text{comb}(x,y,z)$	$\text{comb}(f_x,f_y,f_z)$
$\text{comb}(x)$	$\text{comb}(f_x)\delta(f_y,f_z)$
$\text{helx}(x,y,z)$	$\sum_n J_n(2\pi f)e^{-jn\psi}\delta(n-2\pi f_z)$

which immediately follows from the scaling property of the one-dimensional Fourier transform.

The *three-dimensional impulse function* is defined by

$$\delta(x,y,z) = \delta(x)\delta(y)\delta(z).$$

The separation of variables immediately yields

$$\delta(x,y,z) \iff 1$$

as a three-dimensional Fourier transform pair. The right side is equal to one for all (f_x, f_y, f_z).

The three-dimensional gaussian pulse is defined by

$$e^{-\pi(x^2+y^2+z^2)} = e^{-\pi x^2} e^{-\pi y^2} e^{-\pi z^2}.$$

The separation of variables immediately leads to the Fourier transform pair

$$e^{-\pi(x^2+y^2+z^2)} \iff e^{-\pi(f_x^2+f_y^2+f_z^2)}.$$

The *three-dimensional comb function* is defined by

$$\text{comb}(x,y,z) = \text{comb}(x)\text{comb}(y)\text{comb}(z).$$

The comb function can be visualized as an infinitely small, infinitely dense grain at each point of the integer lattice $\mathbb{Z}^3$. The three-dimensional comb function satisfies the Fourier transform

$$\text{comb}(x,y,z) \iff \text{comb}(f_x, f_y, f_z).$$

The *three-dimensional finite comb function* is given by

$$\text{comb}_{N_x N_y N_z}(x,y,z) = \text{comb}_{N_x}(x)\text{comb}_{N_y}(y)\text{comb}_{N_z}(z).$$

The *three-dimensional dirichlet function* is given by

$$\text{dirc}_{N_x N_y N_z}(x,y,z) = \text{dirc}_{N_x}(x)\text{dirc}_{N_y}(y)\text{dirc}_{N_z}(z).$$

This immediately leads to the Fourier transform pair

$$\text{comb}_{N_x N_y N_z}(x,y,z) \iff \text{dirc}_{N_x N_y N_z}(f_x, f_y, f_z).$$

The *three-dimensional cylinder function* is given by

$$\text{cyln}(x,y,z) = \text{circ}(x,y)\text{rect}(z),$$

which is defined in terms of the two-dimensional circle function $\text{circ}(x,y)$ and the one-dimensional rectangle function $\text{rect}(z)$. The three-dimensional Fourier transform of $\text{cyln}(x,y,z)$ is

$$S(f_x, f_y, f_z) = \text{jinc}\left(\sqrt{f_x^2 + f_y^2}\right)\text{sinc}(f_z),$$

which is easily derived by a separation of variables.

9.2 Transforms of Some Useful Functions

The *three-dimensional sphere function* is defined by

$$\text{sphr}(x,y,z) = \begin{cases} 1 & \text{when } \sqrt{x^2+y^2+z^2} \leq 1/2, \\ 0 & \text{otherwise.} \end{cases}$$

Because the sphere function[2] sphr(x,y,z) is spherically symmetric, its Fourier transform must also be spherically symmetric. It can depend only on the sum $f_x^2+f_y^2+f_z^2$. The three-dimensional Fourier transform of sphr(x,y,z) evaluated at $f_x=0$, $f_y=0$, $f_z=f$ is

$$S(0,0,f) = \iiint_{\sqrt{x^2+y^2+z^2}\leq 1/2} e^{-j2\pi fz}\,dxdydz.$$

Expressed in spherical coordinates, this integral is

$$S(0,0,f) = \int_0^{1/2}\int_0^\pi \int_0^{2\pi} e^{-j2\pi fr\cos\phi} r^2 \sin\phi\,d\psi\,d\phi\,dr.$$

The triple integral is evaluated by elementary methods to give

$$S(0,0,f) = \frac{\pi}{2}\frac{\sin\pi f - \pi f\cos\pi f}{(\pi f)^3}.$$

Accordingly, the *tinc function* (or the *tinc pulse*) is defined as

$$\text{tinc}(t) = \frac{\sin\pi t - \pi t\cos\pi t}{2\pi^2 t^3},$$

by analogy with sinc(t) and jinc(t).

The tinc pulse should be contrasted with the sinc pulse and the jinc pulse. The tinc function is compared to the jinc function and the sinc function in Figure 9.1. For large t, the zeros of tinc(t) approach the half odd integers. The magnitudes of the sidelobes of tinc(t) decay with t as t^{-2}. The value of the tinc pulse at the origin is tinc$(0) = \pi/6$, which is the volume of a sphere of diameter one. The values of the jinc pulse and the sinc pulse at the origin are $\pi/4$ and 1, respectively.

The *three-dimensional tinc function* is given by

$$\text{tinc}(x,y,z) = \text{tinc}\sqrt{x^2+y^2+z^2}$$

Because the sphere function satisfies the one-dimensional Fourier transform property $S(0,0,f) = \text{tinc}(f)$, it follows that

$$\text{sphr}(x,y,z) \iff \text{tinc}\sqrt{f_x^2+f_y^2+f_z^2}$$

and

$$\text{tinc}(x,y,z) \iff \text{sphr}\sqrt{f_x^2+f_y^2+f_z^2}$$

---

[2] The sphere function is defined to have a unit diameter rather than a unit radius so that it is consistent with the rectangle function in one dimension, which has a unit diameter. The sphere function includes the interior although a sphere is sometimes defined as only including the surface points.

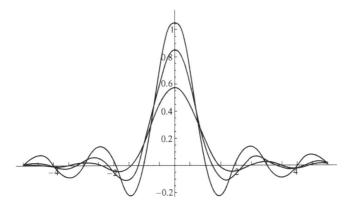

Figure 9.1 Comparison of the sinc, jinc, and tinc functions

are three-dimensional Fourier transform pairs.

Any three-dimensional function $s(x,y,z)$ can have its Fourier transform cropped by a sphere in the frequency domain. Two such cases are described by the three-dimensional Fourier transform pairs

$$s(x,y,z) * * * \text{tinc}(x,y,z) \iff S(f_x,f_y,f_z)\text{sphr}\sqrt{f_x^2+f_y^2+f_z^2}$$

and

$$s(x,y,z) * * * \text{tinc}(x,y,z)e^{j\pi z} \iff S(f_x,f_y,f_z)\text{sphr}\sqrt{f_x^2+f_y^2+(f_z-1/2)^2}.$$

The latter sphere is offset in the z direction from the origin by the radius of the sphere.

The next basic function is the helix function. The helix does not have a close analog in lower dimensions. An infinite helical curve is given parametrically by

$$x = a\cos\psi,$$
$$y = a\sin\psi,$$
$$z = b\psi,$$

where the parameter ψ ranges from $-\infty$ to ∞. A *finite* helical curve with T turns is obtained when ψ ranges from 0 to $2\pi T$.

The *three-dimensional helix function* is given by

$$\text{helx}(x,y,z) = \delta(x-\cos z)\delta(y-\sin z)$$
$$\doteq \delta(x-\cos z, y-\sin z).$$

An ideal infinite helix can be conceptualized as an infinitely thin, infinitely dense solid lying along a helical curve. The helix is written concisely in terms of a two-dimensional delta function but care must be taken in interpreting the formal properties of the delta function.

The helix function has radius equal to one and period equal to one. A helix of radius a and period b is given by

$$\text{helx}\left(\frac{x}{a},\frac{y}{a},\frac{z}{b}\right) = \delta\left(\frac{x}{a} - \cos\frac{z}{b}\right)\delta\left(\frac{y}{a} - \sin\frac{z}{b}\right).$$

Because the helix is periodic in the z direction, the Fourier transform[3] must be discrete in the f_z variable – that is, a Fourier series in z.

Theorem 9.2.1 *The three-dimensional Fourier transform of the helix function is given by*

$$S(f_x,f_y,f_z) = \sum_{n=-\infty}^{\infty} J_n\left(2\pi\sqrt{f_x^2 + f_y^2}\right) e^{jn\tan^{-1}(f_x/f_y)} \delta\left(f_z - \frac{n}{2\pi}\right),$$

where $J_n(a)$ is the nth-order Bessel function of the first kind.

Proof:

$$S(f_x,f_y,f_z) = \int_{-\infty}^{\infty}\int_{-\infty}^{\infty}\int_{-\infty}^{\infty} s(x,y,z) e^{-j2\pi(f_x x + f_y y + f_z z)} dxdydz$$

$$= \int_{-\infty}^{\infty}\int_{-\infty}^{\infty}\int_{-\infty}^{\infty} \delta(x - \cos z)\delta(y - \sin z) e^{-j2\pi(f_x x + f_y y + f_z z)} dxdydz.$$

The formal properties of the delta function are used to evaluate the x and y integrations. Executing the x integration across its delta function replaces x by $\cos z$. Executing the y integration across its delta function replaces y by $\sin z$. Therefore

$$S(f_x,f_y,f_z) = \int_{-\infty}^{\infty} e^{-j2\pi(f_x \cos z + f_y \sin z + f_z z)} dz.$$

Now let

$$f_x = f\sin\gamma,$$
$$f_y = f\cos\gamma.$$

Then

$$S(f_x,f_y,f_z) = \int_{-\infty}^{\infty} e^{-j2\pi(f\sin(z+\gamma) + f_z z)} dz.$$

To evaluate the integral, the following expansion[4] is used:

---

[3] Using prior knowledge of the constituent molecular fragments of the DNA molecule, Watson and Crick constructed their famous model of the DNA molecule after they (and others) observed that the diffraction pattern of DNA displays the characteristic pattern of the Fourier transform of a helical structure.

[4] To derive this identity, recall that any periodic $s(t)$ satisfies the Fourier series relationships

$$s(t) = \sum_{n=-\infty}^{\infty} S_n e^{jnt} \quad S_n = \frac{1}{2\pi}\int_{-\pi}^{\pi} s(t) e^{-jnt} dt.$$

In our case, $s(t) = e^{-ja\sin t}$, and

$$S_n = \frac{1}{2\pi}\int_{-\pi}^{\pi} e^{-ja\sin t} e^{-jnt} dt,$$

which is the definition of the Bessel function $J_n(a)$.

$$e^{-ja\sin t} = \sum_{n=-\infty}^{\infty} J_n(a)e^{jnt}$$

with $a = 2\pi f$ and $t = z + \gamma$. Then

$$S(f_x,f_y,f_z) = \sum_{n=-\infty}^{\infty} J_n(2\pi f) \int_{-\infty}^{\infty} e^{jn(z+\gamma) - j2\pi f_z z} dz$$

$$= \sum_{n=-\infty}^{\infty} J_n(2\pi f) e^{jn\gamma} \int_{-\infty}^{\infty} e^{-j2\pi (f_z - \frac{n}{2\pi})z} dz$$

$$= \sum_{n=-\infty}^{\infty} J_n(2\pi f) e^{jn\gamma} \delta\left(f_z - \frac{n}{2\pi}\right),$$

as was to be proved. ∎

The summation in Theorem 9.2.1 is the sum of the infinite set of two-dimensional functions,

$$S_n(f_x,f_y) = J_n\left(2\pi \sqrt{f_x^2 + f_y^2}\right) e^{jn\tan^{-1}(f_x/f_y)},$$

with each term multiplied by an impulse in f_z. For each integer n, the two-dimensional term $S_n(f_x,f_y)$ lies in the plane at $f_z = \frac{n}{2\pi}$. These planes are sometimes called *layers* or, when viewed in cross section, *layer lines*.

The layer $S_0(f_x,f_y)$, considered in isolation, is the Fourier transform of an infinite, hollow cylindrical surface. This conforms to the heuristic view that an infinite cylinder is a simplified approximation to an infinite helix. By filtering out all other layers of the one-dimensional Fourier transform, the infinite helix is converted to an infinite cylinder.

9.3 Sampling in Three Dimensions

The Nyquist–Shannon sampling theorem is given in one dimension in Chapter 2 and in two dimensions in Chapter 3. The Nyquist–Shannon sampling theorem in three dimensions given here is a straightforward generalization.

Theorem 9.3.1 (Nyquist–Shannon Sampling Theorem) *The three-dimensional finite-energy signal $s(x,y,z)$ whose transform $S(f_x,f_y,f_z)$ satisfies $S(f_x,f_y,f_z) = 0$ for $|f_x| > 1/2A$, $|f_y| > 1/2B$, and $|f_z| > 1/2C$ can be recovered from its three-dimensional Nyquist samples at $(x,y,z) = (\ell_x A, \ell_y B, \ell_z C)$, where $(\ell_x, \ell_y, \ell_z) \in \mathbb{Z}^3$, by the three-dimensional Nyquist–Shannon interpolation formula*

$$s(x,y,z) = \sum_{\ell_x=-\infty}^{\infty} \sum_{\ell_y=-\infty}^{\infty} \sum_{\ell_z=-\infty}^{\infty} s_{\ell_x \ell_y \ell_z} \mathrm{sinc}\left(\frac{x}{A} - \ell_x\right) \mathrm{sinc}\left(\frac{y}{B} - \ell_y\right) \mathrm{sinc}\left(\frac{z}{C} - \ell_z\right).$$

The development of the three-dimensional sampling theorem is anticipated by first reviewing the corresponding development of the one-dimensional Fourier transform.

One-Dimensional Sampling

The Nyquist–Shannon sampling theorem in one dimension, as described in Chapter 2, can be crisply summarized using the generalized Fourier transform pair

$$\big(s(t)\text{comb}(t)\big) * \text{sinc}(t) \leftrightarrow \big(S(f) * \text{comb}(f)\big)\text{rect}(f).$$

On each side of the expression, the inner term (the term in parentheses) describes the sampling operation. The outer term describes the interpolation operation. This is a Fourier transform pair, so the left side is equal to $s(t)$ whenever the right side is equal to $S(f)$. But the right side is equal to $S(f)$ whenever the support of $S(f)$ is contained within the support of $\text{rect}(f)$. The right side is equal to $S(f)$ because only the central image of $S(f)*\text{comb}(f)$ is supported by $\text{rect}(t)$. The other images of $S(f)$ appearing in $S(f) * \text{comb}(f)$ are not supported by $\text{rect}(f)$. Therefore, under this support condition, the left side is equal to $s(t)$ because the right side is equal to $S(f)$. On the left side, the convolution of the samples given by $s(t)\text{comb}(t)$ with $\text{sinc}(t)$ results in an interpolation that undoes the sampling operation.

Otherwise, when the support of $S(f)$ is larger than the support of $\text{rect}(f)$, there will be aliasing. Then portions of other images of $S(f)$ extend beyond the unit interval and contaminate the central image.

As written, the interpolation formula is an infinite sum requiring an infinite number of samples. When the set of samples is symmetrically truncated to n samples, the Fourier transform pair becomes

$$\big(s(t)\text{comb}_n(t)\big) * \text{sinc}(t) \leftrightarrow \big(S(f) * \text{dirc}_n(f)\big)\text{rect}(f).$$

Each image of the spectrum $S(f)$ then becomes unavoidably blurred by the dirichlet function. This blur is made negligible by making n large.

Sampling has a dual formulation. The dual inverts time and frequency to state that a time-limited function $s(t)$ can be represented by an infinite number of samples of $S(f)$ on the frequency axis. Let $s(t)$ be any complex function of time supported on the unit interval centered on the origin. Then the Fourier transform pair

$$\big(S(f)\text{comb}(f)\big) * \text{sinc}(f) \leftrightarrow \big(s(t) * \text{comb}(t)\big)\text{rect}(t)$$

implies that $S(f)$ can be recovered from its uniformly spaced samples in frequency by sinc interpolation in the frequency domain.

There is another important observation regarding only the sampled data with the interpolation step suppressed. The Fourier transform expression

$$s(t)\text{comb}(t) \leftrightarrow S(f) * \text{comb}(f),$$

relates the time-domain samples to a representation in the frequency domain. The right side is a periodic waveform in the frequency domain with unit period centered on the origin of the frequency axis. Every unit interval centered on an integer consists of one period of the waveform. This periodicity holds even when there is aliasing because the aliasing respects the periodicity. The images of $S(f)$ become folded onto each unit interval in the same way. The samples s_ℓ of the time-domain sequence of samples

$s(t)\mathrm{comb}(t) = \sum_{-\infty}^{\infty} s_\ell \delta(t-\ell)$ can then be seen as the coefficients of the Fourier series describing the periodic waveform $S(f) * \mathrm{comb}(f)$. As such, the samples of the Fourier series can be inverted to recover $S(f)$ by

$$S(f) = \sum_{\ell=-\infty}^{\infty} s_\ell e^{-j2\pi\ell}$$

$$= \sum_{\ell=-\infty}^{\infty} \left[s_{R,\ell} \cos(2\pi\ell) - s_{I,\ell} \sin(2\pi\ell)\right],$$

as asserted by the standard properties of the Fourier series.

With time t and frequency f interchanged, the Fourier transform pair is

$$s(t) * \mathrm{comb}(t) \leftrightarrow S(f)\mathrm{comb}(f).$$

This expression is the dual of the earlier expression. Now the frequency-domain function $S(f)$ is sampled and the infinite sequence of time-domain copies of $s(t)$ are nonoverlapping provided $s(t)$ is contained in the unit interval. This condition can be described as $s(t)\mathrm{rect}(t) = s(t)$.

Three-Dimensional Sampling

The Nyquist–Shannon sampling theorem in three dimensions is a straightforward generalization of the Nyquist–Shannon sampling theorem in one dimension. It is enough to describe sampling on the lattice $\mathbb{Z}^3$. Sampling on any other lattice is described by using the scaling and coordinate conversion properties of the Fourier transform that are given in Section 9.1.

The sampling theorem for $\mathbb{Z}^3$ follows from the three-dimensional Fourier transform pair

$$\bigl(s(\mathbf{x})\mathrm{comb}(\mathbf{x})\bigr) * * * \mathrm{sinc}(\mathbf{x}) \iff \bigl(S(\mathbf{f}) * * * \mathrm{comb}(\mathbf{f})\bigr)\mathrm{rect}(\mathbf{f}),$$

where $\mathbf{x} = (x,y,z)$ and $\mathbf{f} = (f_x,f_y,f_z)$. The samples of $s(x,y,z)$ given by $s(\mathbf{x})\mathrm{comb}(\mathbf{x})$ are values on the points of the unit three-dimensional lattice $\mathbb{Z}^3$. To avoid aliasing requires that the support of the transform $S(f_x,f_y,f_z)$ lies inside the unit cube centered at the origin of the frequency domain.

For the needs of this chapter,[5] the dual of the sampling theorem is the version that is the more important. The appropriate relationship is the dual expression

$$\bigl(S(\mathbf{f})\mathrm{comb}(\mathbf{f})\bigr) * * * \mathrm{sinc}(\mathbf{f}) \iff \bigl(s(\mathbf{x}) * * * \mathrm{comb}(\mathbf{x})\bigr)\mathrm{rect}(\mathbf{x}),$$

which simply interchanges the roles of the signal domain and the frequency domain. As before $\mathbf{x} = (x,y,z)$ and $\mathbf{f} = (f_x,f_y,f_z)$.

Moreover, with the interpolation operation set aside, the expression reduces to

$$S(f_x,f_y,f_z)\mathrm{comb}(f_x,f_y,f_z) \iff s(x,y,z) * * * \mathrm{comb}(x,y,z).$$

---

[5] In the study of diffraction, the grating lobes of the diffraction pattern sample the function $S(f_x,f_y,f_z)$ in frequency space.

9.3 Sampling in Three Dimensions

The left side is periodic in all three dimensions. When there is no aliasing, each three-dimensional unit period, or unit cell, is a translate of $s(x, y, z)$. When there is aliasing, each cell contains a superposition of the center of $s(x, y, z)$ and the tails of neighboring copies of $s(x, y, z)$. The samples

$$S(f_x, f_y, f_z) \mathrm{comb}(f_x, f_y, f_z) = \sum_{\ell_x} \sum_{\ell_y} \sum_{\ell_z} s_{\ell_x \ell_y \ell_z} \delta(x - \ell_x, y - \ell_y, z - \ell_z)$$

are the coefficients of the three-dimensional Fourier series describing the periodic three-dimensional waveform $s(x, y, z) * * * \mathrm{comb}(x, y, z)$.

The dual of the three-dimensional Nyquist–Shannon sampling theorem states that when the support of $s(x, y, z)$ lies inside a unit cube, then the sampling operation in the frequency domain

$$S'(f_x, f_y, f_z) = \mathrm{comb}(f_x, f_y, f_z) S(f_x, f_y, f_z)$$

can be inverted in the signal domain by

$$s(x, y, z) = \mathrm{rect}(x, y, z) s'(x, y, z),$$

corresponding to the interpolation formula

$$S(f_x, f_y, f_z) = \mathrm{sinc}(f_x, f_y, f_z) * * * S'(f_x, f_y, f_z)$$

in the frequency domain.

However, the values of the set of *magnitude* samples $|S(\ell_x, \ell_y, \ell_z)|$ do not determine $s(x, y, z)$ in the absence of side information. The observations are samples of the magnitude $|S(f_x, f_y, f_z)|$ of $S(f_x, f_y, f_z)$, and not samples of $S(f_x, f_y, f_z)$ itself. When the samples of $|S(f_x, f_y, f_z)|$ are squared, they become the samples of $|S(f_x, f_y, f_z)|^2$, which is the Fourier transform of the autocorrelation $s(x, y, z) * * * s(-x, -y, -z)$.

The case in which the support of $s(x, y, z)$ is contained in a cell of size L on each side can now be treated by using the scaling property of the Fourier transform. Then it is sufficient to know $S(f_x, f_y, f_z)$ for values of (f_x, f_y, f_z) given by the points of the cubic lattice with spacing L^{-1}. These three-dimensional Nyquist samples are the points of frequency space with coordinates $(\ell_x/L, \ell_y/L, \ell_z/L)$ where ℓ_x, ℓ_y, and ℓ_z are integers. The complex values of the set of samples $S(\ell_x/L, \ell_y/L, \ell_z/L)$ completely determines $s(x, y, z)$.

More general sampling lattices are easily treated using the properties of the three-dimensional Fourier transform. A different scale factor can be used on each of the three coordinate axes. Moreover, the theory applies to any lattice $M\mathbb{Z}^3$ where M is a full-rank three by three matrix. The three-dimensional Fourier transform property for any pair is

$$s(M^T x) \iff \frac{1}{\det M} S(M^{-1} f),$$

where $x=(x,y,z)$ and $f=(f_x,f_y,f_z)$. This expression establishes the relationship between a lattice in image space and the reciprocal lattice in frequency space. It is straightforward to embed the matrices M and M^{-1} into the equations described above.

9.4 Diffraction by a Three-Dimensional Object

The diffraction pattern of a three-dimensional object is a consequence of the scattering of a coherent wave. The density of scattering centers is described by the three-dimensional function $\rho(x,y,z)$. The real nonnegative function $\rho(x,y,z)$ may be thought of as the electron density distribution within the molecule being imaged. An estimate of the electron density $\rho(x,y,z)$ amounts to an estimate of the structure of the molecule.[6]

The idealized scattering arrangement is as follows. A plane wave passes through a partially transparent object, $\rho(x,y,z)$. Within the infinitesimal volume centered at the point (x,y,z) is the infinitesimal scatterer $\rho(x,y,z)\mathrm{d}x\mathrm{d}y\mathrm{d}z$. When the scatterer is illuminated with the plane wave, it coherently reradiates, or scatters, the spherical wavelet $\rho(x,y,z)\mathrm{d}x\mathrm{d}y\mathrm{d}z$ in all directions. As the incident wave sweeps through the partially transparent object $\rho(x,y,z)$, each infinitesimal volume reradiates some of the signal that is incident on that infinitesimal volume. The fraction of the incident signal that is reradiated is negligible, a condition known as the *Born approximation*. The Born approximation means that the incident wave maintains its complex amplitude as it sweeps through the scattering object. This approximation allows us to model the scattered radiation from two identical infinitesimal cells at different depths within the scatterer as equal except for the phase of the incident wave. The Born approximation also allows us to ignore multiple scattering, meaning that the amount of scattered energy that is subsequently rescattered is inconsequential.

In this section, $\rho(x,y,z)$ is an arbitrary scattering function of three-dimensional image space. Because the scattered wave is observed in the far field, the Fraunhofer approximation is implicit in this discussion of diffraction. The next section considers the scattering object to be a three-dimensional array, such as a periodic crystal, also with the signal scattered into the far field.

Consider the two points $(0,0,0)$ and (x,y,z) within the scatterer $\rho(x,y,z)$ and the two lines from these points to a common point in the far field. The Fraunhofer approximation amounts to the approximation that these two lines are parallel. When ϕ is zero, both lines to the far field are parallel to the z axis.

First restrict the situation to two dimensions as shown in Figure 9.2. Then find the difference in the lengths of two paths to a common point in the far field. The point in the far field is in the x,z plane at angle ϕ from the z axis. One path to this point is the line originating at the point $(0,0,0)$. The other path is the line originating at the point $(x,0,z)$. For a point whose path from the origin to a point in the far field has length $c\tau$,

---

[6] The coherent scattering of an X-ray by the electron cloud is known as Thomson scattering.

9.4 Diffraction by a Three-Dimensional Object

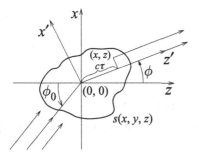

Figure 9.2 Illustrating the phase delay

the path from the point $(x, 0, z)$ to that point has length $c\tau - z$. The path difference is z. This means that the phase difference is $2\pi f_0 z/c$. For a nonzero value of the angle ϕ, rotate the coordinate system to the new coordinate system $(x', 0, z')$ with the z' axis in the direction of propagation. Then

$$z' = x \sin\phi + z \cos\phi,$$

and z' is the path difference with respect to a scattering element at the origin. Hence the phase difference from a scattering element due to the path difference is

$$\Delta\theta = \frac{2\pi f_0}{c}(x \sin\phi + z \cos\phi).$$

There will also be a contribution due to the path difference in the illumination to the scattering object. The incident angle ϕ_0 is defined so that this contribution has a negative sign. Otherwise, the form of the path difference is the same,

$$\Delta\theta = \frac{-2\pi f_0}{c}(x \sin\phi_0 + z \cos\phi_0).$$

The contribution to the signal in the far field at angle ϕ from the scattering cell at $(x, 0, z)$ is

$$dv = \rho(x, y, z) e^{-j2\pi (f_0/c)[x \sin\phi + z \cos\phi - x \sin\phi_0 - z \cos\phi_0]} dx dy dz.$$

The total signal in the far field at angle ϕ is the superposition of all of these contributions

$$v(\phi) = \int_{-\infty}^{\infty} \int_{-\infty}^{\infty} \int_{-\infty}^{\infty} \rho(x, y, z) e^{-j(2\pi/\lambda)\left(x(\sin\phi - \sin\phi_0) + z(\cos\phi - \cos\phi_0)\right)} dx dy dz.$$

This integral is now recognized to have the form of a Fourier transform. Therefore,

$$v(\phi) = R\left(\frac{\sin\phi - \sin\phi_0}{\lambda}, 0, \frac{\cos\phi - \cos\phi_0}{\lambda}\right).$$

is the diffraction pattern, where $R(f_x, f_y, f_z)$ denotes the Fourier transform of $\rho(x, y, z)$.

In general, to concisely specify an arbitrary direction, let s_i and s_o be unit vectors along the incident direction and the scattered direction, respectively. Let i_x, i_y, and i_z be unit vectors along the x, y, and z axes, and let

$$\mathbf{x} = x\mathbf{i}_x + y\mathbf{i}_y + z\mathbf{i}_z.$$

Then
$$v(\phi, \psi) = \int_{-\infty}^{\infty}\int_{-\infty}^{\infty}\int_{-\infty}^{\infty} \rho(x,y,z) e^{-j(2\pi/\lambda)\mathbf{x}\cdot(\mathbf{s}_i - \mathbf{s}_o)} dx\,dy\,dz,$$

where
$$\mathbf{x} \cdot (\mathbf{s}_i - \mathbf{s}_o) = x(\cos\psi \sin\phi - \cos\psi_0 \sin\phi_0)$$
$$+ y(\sin\psi \sin\phi - \sin\psi_0 \sin\phi_0) + z(\cos\phi - \cos\phi_0).$$

Therefore, with
$$f_x = \frac{1}{\lambda} i_x \cdot (\mathbf{s}_i - \mathbf{s}_o),$$
$$f_y = \frac{1}{\lambda} i_y \cdot (\mathbf{s}_i - \mathbf{s}_o),$$
$$f_z = \frac{1}{\lambda} i_z \cdot (\mathbf{s}_i - \mathbf{s}_o),$$

the received signal in the far field is expressed concisely as
$$v(\phi, \psi) = R(f_x, f_y, f_z)$$
$$= R\left(\frac{\cos\psi \sin\phi - \cos\psi_0 \sin\phi_0}{\lambda}, \frac{\sin\psi \sin\phi - \sin\psi_0 \sin\phi_0}{\lambda}, \frac{\cos\phi - \cos\phi_0}{\lambda}\right).$$

This important conclusion states that the ideal X-ray diffraction pattern can be expressed simply as the three-dimensional Fourier transform of the scattering density. Computing the diffraction pattern of $\rho(x, y, z)$ then reduces to the task of computing the three-dimensional Fourier transform of $\rho(x, y, z)$.

The *scattering function* is a function of spherical angle defined as
$$A(\phi, \psi) = \frac{1}{\lambda^3}$$
$$R\left(\frac{\cos\psi \sin\phi - \cos\psi_0 \sin\phi_0}{\lambda}, \frac{\sin\psi \sin\phi - \sin\psi_0 \sin\phi_0}{\lambda}, \frac{\cos\phi - \cos\phi_0}{\lambda}\right).$$

As defined, the scattering function is normalized so that the scaling property of the Fourier transform makes $A(\phi, \psi)$ into a function of d/λ where d is a scaling parameter. Thus, under scaling of the wavelength by d,
$$A(\phi, \psi) = \left(\frac{d}{\lambda}\right)^3$$
$$R\left(\frac{\cos\psi \sin\phi - \cos\psi_0 \sin\phi_0}{\lambda/d}, \frac{\sin\psi \sin\phi - \sin\psi_0 \sin\phi_0}{\lambda/d}, \frac{\cos\phi - \cos\phi_0}{\lambda/d}\right).$$

The scattering function is defined so that it is unchanged when d and λ are changed in the same proportion. Finally,
$$v(\phi, \psi) = \lambda^3 A(\phi, \psi)$$

is the scattered signal.

Example

Under the Born approximation, the scattering function from the interior of a sphere of diameter d at wavelength λ is

$$A(\phi, \psi) = \left(\frac{d}{\lambda}\right)^3 \text{tinc}\left(\frac{\cos\psi\sin\phi - \cos\psi_0\sin\phi_0}{\lambda/d}, \frac{\sin\psi\sin\phi - \sin\psi_0\sin\phi_0}{\lambda/d}, \frac{\cos\phi - \cos\phi_0}{\lambda/d}\right).$$

This expression describes the scattering from a spherical ball of unit diameter for which all points in the interior of the ball are scatterers. Because of symmetry, there is no loss of generality in choosing $\psi_0 = \phi_0 = 0$. Then

$$A(\phi, \psi) = \left(\frac{d}{\lambda}\right)^3 \text{tinc}\left(\frac{\cos\psi\sin\phi}{\lambda/d}, \frac{\sin\psi\sin\phi}{\lambda/d}, \frac{\cos\phi - 1}{\lambda/d}\right)$$

$$= \left(\frac{d}{\lambda}\right)^3 \text{tinc}\left(\frac{\sqrt{2 - 2\cos\phi}}{\lambda/d}\right).$$

Therefore the signal in the far field at solid angle ϕ, ψ is

$$v(\phi, \psi) = d^3 \text{tinc}\left(\frac{\sqrt{2 - 2\cos\phi}}{\lambda/d}\right),$$

which is independent of ψ.

Example

One way to identify an inverse Fourier transform in simple cases is by pattern recognition. One may recognize some characteristic details of the magnitude of a Fourier transform. This requires familiarity with some elementary cases of the three-dimensional Fourier transform. For example, one may be familiar with the magnitude of the three-dimensional Fourier transform of the simple helix, an instance of great historical significance.

Theorem 9.4.1 *The scattering function of the ideal infinite helix illuminated along the longitudinal axis is*

$$A(\phi, \psi) = \frac{a^2 b}{\lambda^3} \sum_{n=-\infty}^{\infty} J_n\left(2\pi \frac{a}{\lambda}\sin\phi\right) e^{jn(\psi + \pi/2)} \delta\left(\frac{\cos\phi - 1}{\lambda} + \frac{n}{2\pi b}\right).$$

Proof:
This formula follows easily from the relationship

$$A(\phi, \psi) = \frac{1}{\lambda^3} R\left(\frac{\cos\psi\sin\phi}{\lambda}, \frac{\sin\psi\sin\phi}{\lambda}, \frac{\cos\phi - 1}{\lambda}\right)$$

by substitution of the usual trigonometric form of the direction cosines into the Fourier transform of the helix in place of f_x, f_y, and f_z. ∎

Because of the delta function, the scattering function given in Theorem 9.4.1 is zero everywhere except on a finite set of values of ϕ given by

$$\cos\phi_n = 1 - \frac{n}{2\pi}\frac{\lambda}{b},$$

This expression has solutions not larger than 90° only for $n = 0, \ldots, \lfloor 2\pi b/\lambda \rfloor$. Thus, except for a finite number of values of $\cos\phi_n$, the function $A(\phi, \psi)$ is zero. For each ϕ_n with $n = 0, \ldots, \lfloor 2\pi b/\lambda \rfloor$, the scattering function is

$$A(\phi_n, \psi) = \frac{a^2 b}{\lambda^3} J_n\left(2\pi \frac{a}{\lambda}\sin\phi_n\right) e^{jn(\psi + \pi/2)}.$$

This scattering function can be visualized as a finite set of coaxial cones, the nth such cone is at angle ϕ_n and has an amplitude

$$|A(\phi_n, \psi)| = \frac{a^2 b}{\lambda^3}\left|J_n\left(2\pi \frac{a}{\lambda}\sin\phi_n\right)\right|,$$

which is determined as a function of angle ϕ_n in terms of the nth-order Bessel function, where $\cos\phi_n = 1 - n(\lambda/2\pi b)$.

A helix of finite length L can be described as an infinite helix multiplied by the rectangle function $\mathrm{rect}(z/L)$ in the z direction. Accordingly, the Fourier transform of the infinite helix is convolved with a sinc function along the z axis. This broadens each layer of the Fourier transform into a sinc function and these sinc functions partially overlap. This means that each of the coaxial cones forming $|A(\phi_n, \psi)|$ is now diffuse in ϕ. For a helix that is made sufficiently long, this broadening of the layers is negligible.

Finally, for an ideal N_x by N_y by N_z array of finite-length helices, the function $A(\phi, \psi)$ is sampled in three dimensions by multiplying it by a three-dimensional dirichlet function. This means that the values of the magnitude $|A(\phi_n, \psi)|$ can be observed only at the grating lobes of the dirichlet function. When the available grating lobe samples have the Bessel function amplitudes of a helix, it may be reasonable to hypothesize that $\rho(x, y, z)$ is a helix even when the samples are sparse and insufficient to do the inversion computationally.

9.5 Image Formation from Diffraction Data

A full measurement of the transform $R(f_x, f_y, f_z)$ of $\rho(x, y, z)$ for all values of f_x, f_y, f_z, were such a measurement obtainable, would reduce the task of estimating the scattering density $\rho(x, y, z)$ to the task of computing the inverse three dimensional Fourier transform. However, for small wavelengths such as X-ray wavelengths, only the absolute value of the diffraction can be measured. The phase of the diffraction cannot be measured. Moreover, the magnitude cannot be measured for large values of $\sqrt{f_x^2 + f_y^2 + f_z^2}$.

An X-ray sensor can measure only the magnitude of the far-field diffraction pattern.[7] It cannot measure the phase. This is because the wavelength of an X-ray may be on the order of a tenth of a nanometer or less, and the many and various spatial

---

[7] In the field of X-ray crystallography, the diffraction pattern is called the *visibility function* and the autocorrelation function is called the *Patterson map*.

factors that affect the phase of the measurement cannot be controlled to an accuracy of a small percentage of this small wavelength.

Moreover, the absolute value of the observed signal describes $|R(f_x,f_y,f_z)|$ for only some values of f_x, f_y, f_z. Then the task is to estimate $\rho(x,y,z)$ when given the magnitude $|R(f_x,f_y,f_z)|$ of its Fourier transform at only some of the points (f_x,f_y,f_z) of frequency space.

The general signal-processing task is to recover the scattering function $\rho(x,y,z)$ from the three-dimensional Fourier transform magnitude $|R(f_x,f_y,f_z)|$ using the side condition that $\rho(x,y,z)$ is real and nonnegative. This is the task known as *phase retrieval*. By simple squaring, the measurement $|R(f_x,f_y,f_z)|$ can be regarded as a measurement of the squared function $|R(f_x,f_y,f_z)|^2$, which is the Fourier transform of the three-dimensional autocorrelation function $\rho(x,y,z) \ast\ast\ast \rho(-x,-y,-z)$.

The two-dimensional phase-retrieval problem is discussed in Sections 7.6 and 7.8 of Chapter 7, and in Section 8.6 of Chapter 8. The three-dimensional phase retrieval problem is discussed in Section 9.9 of this chapter. As indicated therein, the prior information that $\rho(x,y,z)$ is real and nonnegative may partially compensate for the fact that the phase of its Fourier transform $R(f_x,f_y,f_z)$ is not known. Other prior information may also be used when available.

For a cubic array, the magnitude of $R(f_x,f_y,f_z)$ for an array can be measured only for those values of $f_x, f_y,$ and f_z of the form

$$f_x = (\cos\psi\sin\phi - \cos\psi_0\sin\phi_0)/\lambda,$$
$$f_y = (\sin\psi\sin\phi - \sin\psi_0\sin\phi_0)/\lambda,$$
$$f_z = (\cos\phi - \cos\phi_0)/\lambda,$$

corresponding to the grating lobes of the crystal array. To determine the values of f_x, f_y, and f_z at which the function $R(f_x,f_y,f_z)$ can be observed by an appropriate choice of $\phi, \psi, \phi_0,$ and ψ_0, first consider $\phi_0 = \psi_0 = 0$. Then the equations become

$$\lambda f_x = \cos\psi\sin\phi,$$
$$\lambda f_y = \sin\psi\sin\phi,$$
$$\lambda f_z + 1 = \cos\phi.$$

Thus

$$\lambda^2 f_x^2 + \lambda^2 f_y^2 + (\lambda f_z + 1)^2 = 1.$$

This is the equation of a sphere in frequency space,[8] called an *Ewald sphere*. The radius of the Ewald sphere is $1/\lambda$ and its center is at $f_z = -\lambda^{-1}$. The Ewald sphere is a surface that passes through the origin. All points (f_x,f_y,f_z) on the Ewald sphere can be observed by the choice of ϕ and ψ. Moreover, the center of this sphere can be relocated by the choice of ϕ_0 and ψ_0 to any point at distance λ^{-1} from the origin, thereby forming another Ewald sphere, again passing through the origin. The union

---

[8] Frequency space is also called reciprocal space or *k*-space.

of the surfaces of all Ewald spheres forms the set of all points in the interior of a data sphere of radius $2\lambda^{-1}$ centered at the origin. Thus, by the choice of ϕ_0, ψ_0, ϕ, and ψ, any spatial frequency point (f_x, f_y, f_z) within the interior of a larger sphere of radius $2\lambda^{-1}$ is visible. We conclude that the observable data is given by $|R(f_x, f_y, f_z)|$ for all f_x, f_y, f_z that satisfy $f_x^2 + f_y^2 + f_z^2 \leq 2\lambda^{-1}$. This larger ball in frequency space is the observable *Ewald data ball* of radius $2\lambda^{-1}$ in frequency space. The size of the Ewald data ball is determined by the choice of the wavelength λ. To measure large spatial frequencies corresponding to small details in $\rho(x, y, z)$, a small value of λ is needed. Were the complex values $R(f_x, f_y, f_z)$ including phase fully observed only within the data ball, but not elsewhere, the resulting image $\rho'(x, y, z)$ would be blurred because of the size of the data ball. The blurred image would be

$$\rho'(x, y, z) = \text{tinc}\left(\frac{x}{r}, \frac{y}{r}, \frac{z}{r}\right) * * * \rho(x, y, z),$$

where $r = 2\lambda^{-1}$ is the radius of the Ewald data ball. By making λ smaller, the tinc function can be made narrower by increasing the radius $2\lambda^{-1}$ of the Ewald ball.

To discretize the problem, the sampling theorem, or in this case, the dual of the sampling theorem, can be used to determine a sufficient set of discrete samples of $|R(f_x, f_y, f_z)|$. It may be impractical to fully observe all discrete points within the entire Ewald data ball because this requires full angular coverage of both the illumination and the scattered wave. When angular coverage is incomplete, the estimate of $\rho(x, y, z)$ must be computed with only partial Fourier magnitude data within the Ewald data ball.

Finally, although the theory of diffraction from a single molecule is valid, it would be experimentally unrealistic to measure diffraction from a single molecule. Among other reasons, the signal would be far too weak. Instead, massive numbers of identical molecules in the form of a crystal provide the diffraction data. The basic crystal has the form of an array.

9.6 Lattice Structure of Arrays

The geometric structure of a crystal can be described as a manifestation of the mathematical structure of a lattice. An array based on a lattice forms a tessellation of euclidean space $\mathbb{R}^3$ such as by parallelepipeds. A parallelogram is a tessellate that tiles the plane $\mathbb{R}^2$. A parallelepiped is a tessellate that tiles three-space $\mathbb{R}^3$. The corner points of the parallelepiped tessellation form a mathematical *lattice*, denoted Λ. A three-dimensional lattice is defined by any three nonplanar vectors in $\mathbb{R}^3$. The lattice is isomorphic to $\mathbb{Z}^3$. A *primitive cell* of the lattice is defined by the eight points forming a parallelepiped. Rotation of the coordinate system is not considered relevant though it is common to align one vector along the x axis and one vector in the x, y plane. These tessellates in three-space describe the cells of a crystal.

In the sense of the lattice isomorphism, all lattices in $\mathbb{R}^3$ are equivalent. This means that one lattice can be morphed into another. However, the primitive cells of some special lattices fail to display the significant symmetry of that lattice. A more delicate

Figure 9.3 Two Bravais Lattices

classification recognizes special symmetries of those lattices and declares that there are fourteen types of symmetry in the lattices of $\mathbb{R}^3$. To recognize this special symmetry, a cell of a lattice is described by a larger set of points in $\mathbb{R}^3$ called a *Bravais cell*. Some Bravais cells are described by nine or more points of the lattice even though the primitive cells of these lattices are described by only eight points.

The notion of a Bravais cell can be described in terms of the lattices of the plane $\mathbb{R}^2$. There are five Bravais lattice cells in $\mathbb{R}^2$. A lattice in $\mathbb{R}^2$ is defined by two non-parallel vectors of length a and b from the origin and at angle α. These lattices may be classified by $a = b$ versus $a \neq b$ and by $\alpha = 90°$ versus $\alpha \neq 90°$. These conditions define all four cases of primitive cells in $\mathbb{R}^2$. However, one of these cases has a special subcase. The special subcase is the lattice with $a = \sqrt{2}b$ and $\alpha = 45°$. This is commonly described as a face-centered square described by five points of the lattice, as shown on the left side of Figure 9.3. It is the fifth of the five Bravais lattices in two dimensions. The face-centered square is a Bravais cell which for this special case provides an alternative to the notion of a primitive cell. The symmetry for this particular lattice is so special that it is declared to be one of the Bravais lattice cells and represented by the face-centered square.

In three dimensions, there are fourteen Bravais lattices. Of these, seven are basic three-dimensional lattices based on the relative lengths and the angles of the three vectors defining the lattice. There are seven additional special Bravais lattices based on specific angles and lengths that invoke additional symmetry. These special lattices are described as side-centered, face-centered, or body-centered versions of one of the seven basic Bravais lattices. For example, the three vectors $i_x, (i_y+i_z)/\sqrt{3}, (i_y-i_z)/\sqrt{3}$ define the *body-centered* cubic Bravais lattice in three dimensions, which is shown on the right side of Figure 9.3. The corresponding primitive cell is not the cubic cell. Every point of the body-centered cubic lattice can be reached by integer combinations of the three basis vectors. The body-centered cubic Bravais lattice in the figure is unique up to scaling.

Some structures, such as a lattice or molecule, have the property of centrosymmetry. Some do not. A structure is *centrosymmetric* when there is at least one special point of the structure called the *center of symmetry*. A center of symmetry is a point of the structure such that for every point of the structure at (x, y, z) relative to the center of symmetry, there is another point at $(-x, -y, -z)$. This can describe the center of symmetry of a molecule, which is a point of the molecule such that for every point of the molecule at (x, y, z) relative to the center of symmetry there is another point of the molecule at $(-x, -y, -z)$.

9.7 Diffraction by Three-Dimensional Arrays

Whereas a single molecule with electron density $\rho(x,y,z)$ is physically too small to scatter observable waves, a regular array of such molecules is large enough to scatter observable waves. A crystal $\rho_{array}(x,y,z)$ in the form of an N_x by N_y by N_z regular array is a very large array when N_x, N_y, and N_z are very large. In a crystal these numbers are huge, typically in the millions. Then the scattered waves at discrete angles are very large and can be observed. We only need to find an expression for the scattered wave in terms of the electron density $\rho(x,y,z)$ of the single molecule and the electron density $\rho_{array}(x,y,z)$ of the entire crystal. This is the forward problem. The inverse problem is to invert that expression to find $\rho(x,y,z)$ as a function of the observable properties of the scattered wave. These are properties that depend on $\rho_{array}(x,y,z)$. Any method of solving this inverse problem for a model of $\rho(x,y,z)$ from measurements of these properties is a crystallographic image formation algorithm.

An ideal crystal is a three-dimensional array of identical molecular elements, as shown in Figure 9.4. An elementary instance of a crystal consists of an array of identical molecules in a cubic array of crystal cells. Even a very large molecule,[9] such as a complex organic molecule with hundreds of nonhydrogen atoms, can often be made to form crystals. By forming a crystal and using it as a three-dimensional diffraction grating, one has a method for probing the crystal structure or for forming a three-dimensional model of the molecule. The wavelength of the incident radiation must be comparable to the crystal spacing and the details of the molecule element, which makes X-rays a natural choice to use for most crystallography because a wavelength somewhat smaller than the crystal spacing is appropriate to produce grating lobes. Wavelengths of 0.7 or 1.5 angstroms are typical.

From a mathematical point of view, the analysis of a crystallographic diffraction pattern is an exercise in the theory of the three-dimensional Fourier transform. Indeed, the discussion of the forward problem in this section may be viewed as an extension of the mathematics describing antenna arrays, which was studied in Section 5.2, from two dimensions into three dimensions.

Each three by three full-rank matrix M specifies a three-dimensional lattice Λ. The lattice Λ is defined as the set of points $x = (x,y,z)$ given by

$$x = M\ell,$$

where $\ell = (\ell_x, \ell_y, \ell_z)$ ranges over all three-dimensional vectors with integer components. The integer-valued vectors $\ell = (\ell_x, \ell_y, \ell_z)$ are the elements of the lattice $\mathbb{Z}^3$.

An elementary instance of a three-dimensional lattice is a cubic lattice. In this case, M is a diagonal matrix. Setting M equal to the identity matrix recovers the lattice $\mathbb{Z}^3$ and consists of all points of euclidean space $\mathbb{R}^3$ with integer-valued components. For a general matrix M, the lattice Λ may be succinctly denoted as $M\mathbb{Z}^3$.

---

[9] It is a matter for others to discuss whether a particular molecule can be made to form a crystal, perhaps within a suitable bath.

9.7 Diffraction by Three-Dimensional Arrays

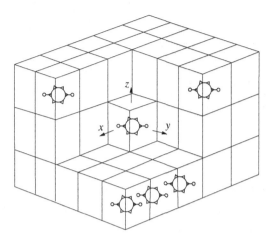

Figure 9.4 A crystal as a periodic array

For any M, the lattice $M^{-1}\mathbb{Z}^3$ is called the *reciprocal lattice* of $M\mathbb{Z}^3$. The function comb(Mx) defines a set of impulses on the points of the lattice $M\mathbb{Z}^3$. Then the Fourier transform pair

$$\text{comb}(Mx) \iff \frac{1}{\det M}\text{comb}(M^{-1}f)$$

describes a set of impulses on the points of the reciprocal lattice $M^{-1}\mathbb{Z}^3$.

An infinitely large ideal crystal with a lattice structure is defined as a three-dimensional convolution of the element $\rho(x, y, z)$ with a three-dimensional array of impulses on the lattice points[10]

$$\rho_{\text{array}}(x, y, z) = \rho(x, y, z) \ast \ast \ast \sum_{\ell_x}\sum_{\ell_y}\sum_{\ell_z} \delta(x - M\ell)$$

$$= \sum_{\ell_x}\sum_{\ell_y}\sum_{\ell_z} \rho(x - M\ell),$$

where $\rho(x - M\ell)$ denotes the translation of $\rho(x, y, z)$ to the lattice point $M\ell$.

A finite crystal $\rho_{\text{array}}(x, y, z)$ is obtained when the lattice points are restricted to a finite region of space. A finite crystal on a rectangular N_x by N_y by N_z lattice can be written using ρ_{array} as

---

[10] Only crystals whose cells correspond to the cells of a lattice are considered herein. Three-dimensional Fourier transform properties then apply directly. Such crystals satisfy the *centrosymmotric property*, which is the statement that there is a point of the cell designated as coordinate $(0, 0, 0)$ called the symmetric center of the cell such that whenever the point (x, y, z) is in the cell, the point $(-x, -y, -z)$ is also in the cell. Not every tessellation of $\mathbb{R}^3$ has the centrosymmetric property. Similarly, an element of the array can itself have the centrosymmetric property. Then there is a point of the element, designated as coordinate $(0, 0, 0)$ called the symmetric center of the element. A cube has the centrosymmetric property. It follows from Friedel's property that the Nyquist samples $R_{\text{array}}(f_x, f_x, f_z)$ of $\rho(x, y, z)$ are real if and only if $\rho(x, y, z)$ has a symmetric center.

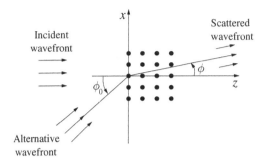

Figure 9.5 A two-dimensional crystal

$$\rho_{\text{array}}(x,y,z) = \sum_{\ell_x=0}^{N_x-1}\sum_{\ell_y=0}^{N_y-1}\sum_{\ell_z=0}^{N_z-1} \rho(x - d_x\ell_x, y - d_y\ell_y, z - d_z\ell_z).$$

Figure 9.4 shows an embodiment of this equation in which multiple copies of the molecule $\rho(x,y,z)$ form the crystal $\rho_{\text{array}}(x,y,z)$.

The Fourier transform of $\rho_{\text{array}}(x,y,z)$ follows immediately. It is

$$R_{\text{array}}(f_x,f_y,f_z) = R(f_x,f_y,f_z)\text{dirc}_{N_xN_yN_z}(d_xf_x, d_yf_y, d_zf_z).$$

The grating lobes are

$$R_{\text{array}}(\ell_x/d_x, \ell_y/d_y, \ell_z/d_z) = R(\ell_x/d_x, \ell_y/d_y, \ell_z/d_z)\text{dirc}_{N_xN_yN_z}(\ell_x, \ell_y, \ell_z),$$

where $\text{dirc}_{N_xN_yN_z}(\ell_x, \ell_y, \ell_z)$ is the three-dimensional dirichlet function. The grating lobes are called *structure factors* in the field of crystallography.

The complex amplitudes $R(\ell_x/d_x, \ell_y/d_y, \ell_z/d_z)$ are the Nyquist samples of one cell of the three-dimensional array, such as a crystal. These are the Nyquist samples of $\rho(x,y,z)$ provided $\rho(x,y,z)$ is fully contained within a cell of the crystal. Only a subset of these Nyquist samples can be observed by the methods of diffraction crystallography, and only their magnitudes.

The diffraction of a wavefront by an ideal finite rectangular array is described next. Figure 9.5 shows a cross section of the array in two dimensions. Specifically, the figure shows the cross section lying in the x, z plane. As compared to the time at which the wave passes the origin, a point on a wave moving along the z axis reaches the position with coordinate z with the time delay $\tau = z/c$. To derive the diffraction pattern directly, start with the simple geometry of Figure 9.5 and calculate the signal scattered in the x, z plane at the angle ϕ from the z axis by a monochromatic incident plane wave that moves along the z axis. Then examine the scattering in an arbitrary direction.

This analysis of scattering from a three-dimensional array is not different from the analysis in Section 9.4 in which $v(\phi, \psi)$ was obtained in terms of $R(f_x,f_y,f_z)$ for scattering from an object $\rho(x,y,z)$, so there is no need to repeat it. Thus, the array $v(\phi, \psi)$ can be described in terms of $R_{\text{array}}(f_x,f_x,f_z)$. The diffraction pattern for the complete

9.7 Diffraction by Three-Dimensional Arrays

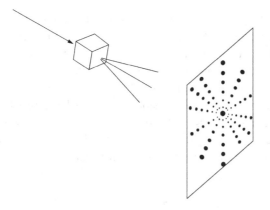

Figure 9.6 Illustrating formation of a Laue scattering pattern

three-dimensional rectangular array can now be written in terms of the Fourier transform $R(f_x, f_y, f_z)$ of a single element of the array multiplied by a dirichlet function on each axis. Thus,

$$v(\phi, \psi) = R_{\text{array}} \left(\frac{\cos\psi \sin\phi - \cos\psi_0 \sin\phi_0}{\lambda}, \frac{\sin\psi \sin\phi - \sin\psi_0 \sin\phi_0}{\lambda}, \frac{\cos\phi - \cos\phi_0}{\lambda} \right),$$

where

$$\ell_x = (\cos\psi \sin\phi - \cos\psi_0 \sin\phi_0)/\lambda,$$
$$\ell_y = (\sin\psi \sin\phi - \sin\psi_0 \sin\phi_0)/\lambda,$$
$$\ell_z = (\cos\phi - \cos\phi_0)/\lambda.$$

Although the grating lobes are described in the spherical coordinates ψ and ϕ, they are indexed by $\ell = (\ell_x, \ell_y, \ell_z)$ and can be so-labeled.

This signal $v(\phi, \psi)$ is large only when the three-dimensional dirichlet function is large. This occurs when $(d_x f_x, d_y f_y, d_z f_z) = (\ell_x, \ell_y, \ell_z)$. At other values of ϕ and ψ, the scattered signal $v(\phi, \psi)$ has negligibly small values. The magnitude of the scattering function $R(f_x, f_y, f_z)$ can be measured only on the grating lobes where it is proportional to $|R(\ell_x/d_x, \ell_y/d_y, \ell_z/d_z)|$. This set of amplitude-modulated grating lobes is called the *Laue scattering pattern* of the array. An illustration of a Laue scattering pattern is shown in Figure 9.6.

Grating lobes of the dirichlet function are found if and only if the following three equations, known as the *Bragg–Laue equations* for a three-dimensional array, such as a rectangular crystal, are satisfied:

$$(s_i - s_0) \cdot i = \ell_x \lambda / d_x,$$
$$(s_i - s_0) \cdot j = \ell_y \lambda / d_y,$$
$$(s_i - s_0) \cdot k = \ell_z \lambda / d_z,$$

where ℓ_x, ℓ_y, and ℓ_z are integers. Measuring the angles of the grating lobes[11] determines the terms on the left sides of the Bragg–Laue equations. From these equations, d_x, d_y, and d_z can then be computed. For an array that is not a rectangular lattice, the Bragg–Laue equations take a more general form as anticipated by the coordinate transformation properties of the three-dimensional Fourier transform.

The left side of each of the Bragg–Laue equations cannot be larger than two because all terms on the left are unit vectors, and a dot product of two unit vectors has a magnitude not larger than one. This gives the maximum values for the integers ℓ_x, ℓ_y, and ℓ_z. The grating lobes can be observed for all integers satisfying

$$\ell_x \le 2d_x/\lambda,$$
$$\ell_y \le 2d_y/\lambda,$$
$$\ell_z \le 2d_z/\lambda.$$

Moreover, because the vector magnitude $\|s_i - s_0\|$ is not larger than two, the inequality

$$\lambda\sqrt{(\ell_x/d_x)^2 + (\ell_y/d_y)^2 + (\ell_z/d_z)^2} \le 2$$

must also be satisfied. For a cubic lattice, $(\lambda/d)\sqrt{\ell_x^2 + \ell_y^2 + \ell_z^2} \le 2$. This leads to the notion of the large Ewald sphere.

So that grating lobes are formed, the wavelength λ must be chosen smaller than twice the array spacing. Otherwise, no positive integer will satisfy the Bragg–Laue inequalities. All samples are then outside of the Ewald sphere and are not visible. This can be compared to the notion of an evanescent wave in optics. To illustrate this point, the Bragg–Laue equalities are written as

$$\cos\psi \sin\phi = \ell_x \lambda/d_x + \cos\psi_0 \sin\phi_0,$$
$$\sin\psi \sin\phi = \ell_y \lambda/d_y + \sin\psi_0 \sin\phi_0,$$
$$\cos\phi = \ell_z \lambda/d_z + \cos\phi_0.$$

The Bragg–Laue equations specify a relationship between the angles of incidence and the angles of reflection. Both sets of angles, (ϕ, ψ) and (ϕ_0, ψ_0), can be varied to observe the grating lobes. However, the grating lobes are determined by the array and are fixed. Changing the viewing angles only changes which grating lobes are observed.

A grating lobe exists only when the magnitude of the right side of each equation is not larger than one. Otherwise, for some triples of integers (ℓ_x, ℓ_y, ℓ_z), the equations will fail to have a solution in ψ and ϕ.

When the solution exists, it has the form

$$\psi = \tan^{-1} \frac{(\ell_x \lambda/d_x) + \cos\psi_0 \sin\phi_0}{(\ell_y \lambda/d_y) + \sin\psi_0 \sin\phi_0},$$

$$\phi = \tan^{-1} \frac{[[\quad]^2 + [\quad]^2]^{\frac{1}{2}}}{(\ell_z \lambda/d_z) + \cos\psi_0},$$

---

[11] The Bragg–Lane grating lobes, as limited by the Ewald sphere, provide Nyquist samples at the Nyquist sampling rate.

where, for each grating lobe in the equation for ϕ, the open brackets contain the terms in the numerator and the denominator, respectively, of the equation for ψ. The width of an ideal grating lobe in each direction is inversely proportional to N_x, N_y, and N_z. These are extremely large integers for a typical crystal. Therefore, from a mathematical point of view, the grating lobes would be exquisitely thin. However, from a physical point of view, the spacings in a crystal are not exact and, because of thermal noise, are fluctuating in time. This causes the actual grating lobes to be much wider than they would be in an idealized array. Furthermore, the X-ray beam is not strictly monochromatic. The incident X-ray beam will contain a range of wavelengths, and different wavelengths will have grating lobes at slightly different angles. This again broadens or moves the observed grating lobes. For this reason, the X-ray beam may be filtered to limit its bandwidth.

9.8 Sampling of an Array

An array of identical scattering elements located on the points of a lattice[12] forms a three-dimensional diffraction grating. The diffraction pattern of the array is dominated by the grating lobes. The magnitudes of these grating lobes comprise the only part of the diffraction pattern that is observable. The complex amplitudes of the grating lobes can be viewed as an instance of three-dimensional sampling of the frequency-domain function $R(f_x,f_y,f_z)$. Each grating lobe of the diffraction pattern has a complex amplitude comprising a sample of $R(f_x,f_y,f_z)$.

The discussion of diffraction by an array begins with the full set of complex samples on the grating lobes, but then recognizes that the grating lobes outside of the Ewald data ball are not visible in the diffraction pattern. Only grating lobes that are within an Ewald data ball are visible. Moreover, only the magnitude of the samples on the grating lobes can be measured. Thus, only the magnitudes of those grating lobes that are inside the Ewald data ball can be measured. These facts are discussed as straightforward mathematical consequences of the three-dimensional Fourier transform and the spherical geometry.

To establish the sufficiency of the samples, were they known, the Bragg–Laue equations must be tested against the Nyquist–Shannon sampling theorem. The conclusion is that the grating lobes underlying the Bragg–Laue equations and the Nyquist–Shannon sampling theorem are in exact correspondence. This correspondence is because both are explained as mathematical consequences of the same three-dimensional dirichlet function or the three-dimensional comb function.

Knowing the complex samples of $R(f_x,f_y,f_z)$ at all points of the lattice $\mathbb{Z}^3$ allows $\rho(x,y,z)$ to be fully reconstructed by means of the Nyquist–Shannon sampling theorem in three dimensions. However, for a cubic lattice of size d, only the samples $R(f_x,f_y,f_z)\text{comb}(f_x,f_y,f_z)$ that are within the Ewald data ball of radius $r = 2d/\lambda$ can be

---

[12] An array of identical objects such as molecules on a lattice is a mathematical model of a crystal.

observed. This means that the array of samples is cropped as indicated by multiplying the sampled array with the function $\text{sphr}(f_x, f_y, f_z)$ as follows.

The commutativity of multiplication shows that

$$\left[R(f)\text{comb}(f)\right]\text{sphr}(rf) = \left[R(f)\text{sphr}(rf)\right]\text{comb}(f),$$

where $f = (f_x, f_y, f_z)$. This means that missing the complex samples outside of the Ewald data ball is equivalent to passing the scattering function such as the electron density $\rho(x, y, z)$ through the point-spread $\text{tinc}(x/r, y/r, z/r)$ function as

$$\widehat{\rho}(x, y, z) = \rho(x, y, z) * * * \text{tinc}(x/r, y/r, z/r),$$

where r is the radius of the Ewald data ball. This reconstruction would be possible when all of the complex values of the grating lobes are observed. The reconstruction may not do as well when the sampled data values are reduced to only their magnitudes.

Knowing the magnitude $|R(f_x, f_y, f_z)|$ of $R(f_x, f_y, f_z)$ is equivalent to knowing the squared magnitude $|R(f_x, f_y, f_z)|^2$, which is the Fourier transform of the autocorrelation $\rho(x, y, z) * * * \rho(-x, -y, -z)$. Because the autocorrelation of a function supported on a unit cube is supported on a cube of size two, the Nyquist samples of $R(f_x, f_y, f_z)$ are not the Nyquist samples of $|R(f_x, f_y, f_z)|^2$. This is because the support of $|R(f_x, f_y, f_z)|^2$ is the support of $R(f_x, f_y, f_z)$ convolved with the reversal of that support.[13] This seems to say that the sampling rate must be larger for $|R(f_x, f_y, f_z)|^2$, which is the case when side information is ignored. The Nyquist sampling rate is a necessary condition only in the absence of side information. Here the side information is that the inverse Fourier transform is an autocorrelation function. Moreover, for many applications, $\rho(x, y, z)$ is real and nonnegative. Of course, it is not enough to just know side information; it must also be used and this requires a practical algorithm.

Practical algorithms for this problem are hard to come by because, as is the case in other problems of phase retrieval, convexity arguments are not available. When the function $\rho(x, y, z)$ has Fourier transform magnitude $|R(f_x, f_y, f_z)|$, then the function $\rho(-x, -y, -z)$ and the functions $\rho(x - a, y - b, z - c)$ for any a, b, and c also have the Fourier transform magnitude $|R(f_x, f_y, f_z)|$. However, convex combinations of these solutions are not themselves solutions. Nevertheless, all of these solutions can be considered equivalent, to be disambiguated later. For most problems of interest under a nonnegative constraint, the solution is unique up to this equivalence and can be easily verified by solving the forward problem. Practical algorithms for phase retrieval of nonnegative functions from the magnitude of the Fourier transform use iterative trial and error aided by side information and partial analytic expressions. The Fienup algorithm and the Schulz–Snyder algorithm are studied in earlier chapters. Methods developed for the needs of crystallography are studied in this chapter.

---

[13] In general, the *support* of the convolution of two three-dimensional functions is equal to the set sum of the supports of the two functions. The *set sum* of two sets $\mathcal{A}$ and $\mathcal{B}$ is the set
$\mathcal{A} + \mathcal{B} = \{x + y : x \in \mathcal{A}, y \in \mathcal{B}\}$.

9.9 Image Formation from Array Diffraction

The tasks of diffraction crystallography are varied. The simplest task of diffraction crystallography is to determine the lattice structure of the crystal from the diffraction data. The harder task is to determine the structure of the individual molecule, as by forming an image of $\rho(x,y,z)$ from the diffraction data, which is the topic studied in this section. This is a primary method of determining the three-dimensional structure of a molecule.

The diffracted X-ray pattern from a single molecule, were this measurement possible to implement, would be far too weak to extract from the noise. Instead, an immense number of copies of the molecule arranged in a crystal, when such a crystal can be formed, may collectively give a sufficiently strong signal in the form of the diffraction grating lobes. The diffraction pattern in three-dimensional frequency space from an array on the elementary lattice $\mathbb{Z}^3$ is

$$R_{\text{array}}(f_x,f_y,f_z) = R(f_x,f_y,f_z)\text{dirc}_{N_x}(d_x f_x)\text{dirc}_{N_y}(d_y f_y)\text{dirc}_{N_z}(d_z f_z),$$

where

$$f_x = (\cos\psi \sin\phi - \cos\psi_0 \sin\phi_0)/\lambda,$$
$$f_y = (\sin\psi \sin\phi - \sin\psi_0 \sin\phi_0)/\lambda,$$
$$f_z = (\cos\phi - \cos\phi_0)/\lambda,$$

and the grating lobes satisfy the Bragg–Laue equations. The signal on the grating lobes is strong enough to be measured. In contrast, scattering in directions other than the grating lobe directions, the signal is not strong enough to be measured.

Accordingly, a difficult computational task of diffraction crystallography is the general inverse problem of forming an estimated image of the element scattering function $\rho(x,y,z)$ from the signal scattered by a crystal of such elements. Only the magnitudes of the grating lobes can be measured. The magnitude of the scattered signal along the grating lobes is proportional to $|R(\ell_x/d_x, \ell_y/d_y, \ell_z/d_z)|$. These are the scaled magnitudes of the Nyquist samples[14] of the Fourier transform of $\rho(x,y,z)$. In principle, were all complex Nyquist samples in Fourier space measured satisfying the Nyquist condition, they could be inverted to find $\rho(x,y,z)$. Moreover, not all of these grating lobes can be observed. Only those samples inside of the Ewald sphere are visible. The other is invisible. This means that the resolution is limited as is described by convolution of $\rho(x,y,z)$ with the tinc function. Because only the magnitude of a Nyquist sample, not the phase, can be measured, and only on the grating lobes, the diffraction data is not invertible as given. In general, the inversion of the Nyquist magnitude samples could use some form of prior knowledge that $\rho(x,y,z)$ is real and nonnegative or knowledge of the atomic species making up the molecule.

---

[14] The signal on the grating lobes form the Nyquist samples or the *structure factors* of $\rho(x,y,z)$.

Direct construction of $\rho(x,y,z)$ from the magnitude $|R(f_x,f_y,f_z)|$ requires recovery of the phase of $R(f_x,f_y,f_z)$. In general, $|R(f_x,f_y,f_z)|^2 = (\text{Re}[R(f_x,f_y,f_z)])^2 + (\text{Im}[R(f_x,f_y,f_z)])^2$. In the simplest case, when the imaginary part is equal to zero, $|R(f_x,f_y,f_z)| = |\text{Re}[R(f_x,f_y,f_z)]|$. In this case, only the sign of $|R(f_x,f_y,f_z)|$ needs to be recovered. This is so when there is a choice of origin for which $|R(f_x,f_y,f_z)|$ is real. Recovery of the sign implicitly recovers the origin at a center of symmetry.

The function $R(f_x,f_y,f_z)$ is real for some choice of origin if and only if there is a center of symmetry such that $\rho(x,y,z) = \rho(-x,-y,-z)$. A centrosymmetric molecule $\rho(x,y,z)$ with the origin at the center of symmetry satisfies the Friedel property. This property states that the Fourier transform of $\rho(x,y,z)$ is real and so the structure factors are real because the structure factors are samples of $R(f_x,f_y,f_z)$. A molecule that is not centrosymmetric can be modeled as centrosymmetric by attaching a virtual mate or 'twin' $\rho(-x,-y,-z)$ with respect to *any* point which is then designated as the *new* origin $(0,0,0)$. Then

$$R_{\text{twin}}(f_x,f_y,f_z) = R(f_x,f_y,f_z) + R^*(-f_x,-f_y,-f_z)$$

Any point can be selected as the center of symmetry. The virtual paired molecule is then positioned accordingly with respect to this point. Equivalently, the structure factors of $\rho(x,y,z) + \rho(-x,-y,-z)$ can be declared to be real and the center of symmetry is then implicit as is the location of the virtual twin molecule. The two images can be described as twin images or double images. They are isomers, only one of which is correct.

Reference to the proposition shows that recovery of only the signs of the real part of $R(f_x,f_y,f_z)$ allows the twin images to be recovered by the inverse three-dimensional Fourier transform. However, only the magnitude of $R(f_x,f_y,f_z)$ is known, but the magnitude could suffice as a possible first approximation of a structured search. Although the signs are not known, the expressions of Section 9.11 may be used to recover some signs. Knowing that $\rho(x,y,z)$ is nonnegative, a reconstruction of the twin pair of $\rho(x,y,z)$ and $\rho(-x,-y,-z)$ can be teased from these conditions.

9.10 Model Construction from Diffraction Data

For small molecules, the number of atoms of each atomic species contained in the molecule is usually known, and the electron density is approximately the same function $a_i(x,y,z)$ for all atoms of the ith atomic species. This is not exact because the electron density $a_i(x,y,z)$ may be influenced by nearby atoms within the molecule.

For a molecule $\rho(x,y,z)$ that consists of a moderate number of known atoms, it may be sufficient to determine only the location of each known atom within the molecule. To this purpose, it may be possible to form an image by working with a parametric model of the form

$$\rho(x,y,z) = \sum_{\ell=1}^{n} a_\ell(x - x_\ell, y - y_\ell, z - z_\ell),$$

where $a_\ell(x,y,z)$ is the model of the electron density for the ℓth atom with $\ell = 1,\ldots,n$ indexing the n individual atoms in order by weight. Possibly, an atom $a_\ell(x,y,z)$ is modeled simply as a weighted impulse, an approximation called the *atomicity assumption*. The coordinates (x_ℓ, y_ℓ, z_ℓ) then are the unknowns. There are $3n$ unknowns in the parametric model. When the number of atoms is not too large, there may be more than $3n$ samples of $|R(f_x,f_y,f_z)|$. This means that one can expect that $\rho(x,y,z)$ can be computed from the magnitude data on the grating lobes by computing the unknown parameters. To do so, one needs a systematic way of working through the data.

The autocorrelation function

$$\phi(x,y,z) = \rho(x,y,z) *** \rho(-x,-y,-z)$$

has the Fourier transform $|R(f_x,f_y,f_z)|^2$, which is straightforward to compute by squaring the observed magnitude data $|R(f_x,f_y,f_z)|$. The support of $\phi(x,y,z)$ is twice as big as the support of $\psi(x,y,z)$. This means that the observed Bragg–Laue samples of $|R(f_x,f_y,f_z)|$ are not spaced closely enough to be Nyquist samples of $|R(f_x,f_y,f_z)|^2$. It is not straightforward to compute a sufficient set of samples of the autocorrelation function $\phi(x,y,z)$ from $|R(f_x,f_y,f_z)|$. Algorithms that do not recognize and use the fact that $\phi(x,y,z)$ is an autocorrelation function may not apply.

For a molecule with n atoms, the autocorrelation function has the form

$$\phi(x,y,z) = \sum_{\ell=1}^{n} a_\ell(x-x_\ell, y-y_\ell, z-z_\ell) *** \sum_{\ell'=1}^{n} a_{\ell'}(-x-x_{\ell'}, -y-y_{\ell'}, -z-z_{\ell'})$$

$$= \sum_{\ell=1}^{n} \sum_{\ell'=1}^{n} a_\ell(x-x_\ell, y-y_\ell, z-z_\ell) *** a_{\ell'}(-x-x_{\ell'}, -y-y_{\ell'}, -z-z_{\ell'}).$$

This is a parametric model. The n coordinate points (x_ℓ, y_ℓ, z_ℓ) for $\ell = 1, \ldots, n$ are the unknowns. There are n^2 terms here, so it is reasonable to examine an expression in this form when the number of atoms n is not large, possibly to a few hundred atoms.

Were the $a_\ell(x,y,z)$ unit impulses, then $\phi(x,y,z)$ would consist of a sum of impulses of the form

$$\phi(x,y,z) = \sum_{\ell=1}^{n} \sum_{\ell'=1}^{n} a_\ell a_{\ell'} \delta(x-(x_\ell-x_{\ell'}), y-(y_\ell-y_{\ell'}), z-(z_\ell-z_{\ell'})).$$

The computational task is then to determine the individual terms in this expression. Each term of $\phi(x,y,z)$ is located at a pairwise difference between two molecules. From this information, one could determine all pairwise separations in three dimensions and then infer the molecular structure. For $a_\ell(x,y,z)$ that have approximate dimensions comparable to the atomic spacing and can be resolved, it may be possible to force useful inferences by proceeding along these lines.

The molecule might contain only a small number of different atomic species and only one or two instances of the largest atomic species. The largest atom has an electron density, $a_1(x,y,z)$, that may be known. The largest peaks in $\phi(x,y,z)$ that are not at the origin, provided they can be resolved, correspond to pairwise differences in the positions of the large atoms of the same atomic species, or to the pairwise difference in the position of a large atom and a next smaller atom.

Because the origin of the coordinate system is arbitrary, it can be chosen so that $x_1 = y_1 = z_1 = 0$. When there is only a single largest atom, the autocorrelation function is written

$$\phi(x,y,z) = a_1(x,y,z) *** a_1(x,y,z)$$
$$+ \sum_{\ell=2}^{n} a_\ell(x - x_\ell, y - y_\ell, z - z_\ell) *** a_1(x,y,z)$$
$$+ a_1(x,y,z) *** \sum_{\ell'=2}^{n} a_{\ell'}(-x - x_{\ell'}, -y - y_{\ell'}, -z - z_{\ell'})$$
$$+ \sum_{\ell=2}^{n} \sum_{\ell'=2}^{n} a_\ell(x - x_\ell, y - y_\ell, z - z_\ell) *** a_{\ell'}(-x - x_{\ell'}, y - y_{\ell'}, z - z_{\ell'}).$$

The first term involves no displacement terms, and so is not useful. The last term will be small when the other a_ℓ are small in comparison to a_1. The two middle terms will have peaks both at the locations of individual atoms and at the mirror images of these locations provided the individual atoms can be resolved. In this way, one may obtain initial estimates $(\widehat{x}_\ell, \widehat{y}_\ell, \widehat{z}_\ell)$ for $\ell = 2, \ldots, n$. The estimated molecule then is

$$\widehat{\rho}(x,y,z) = \sum_{\ell=1}^{n} a_\ell(x - \widehat{x}_\ell, y - \widehat{y}_\ell, z - \widehat{z}_\ell).$$

This initial solution can be refined by various iterative techniques that use the difference between $|\widehat{R}(f_x,f_y,f_z)|$ and $|R(f_x,f_y,f_z)|$ as an error to be reduced by adjusting the solution.

Such methods of parametric model fitting may require some manual intervention in the computations. A skilled crystallographer might construct a good model of a small molecule in this way. Indeed, such parametric methods are responsible for some early successes in diffraction imaging of moderate-sized molecules.

9.11 Direct Methods from Diffraction Data

Imaging methods that solve the inverse problem from diffraction data by first recovering the phase of the Fourier transform $R(f_x,f_y,f_z)$ directly from the magnitude $|R(f_x,f_y,f_z)|$ are called *direct methods*. These methods are based on the fact that the unknown $\rho(x,y,z)$ is known to be real and nonnegative, which knowledge can compensate for the missing phase. This means that there are constraining relationships between the phases of the Fourier coefficients. Direct methods exploit these relationships.

To show the need for direct methods, notice that even for the case of a centrosymmetric $\rho(x,y,z)$, in which all phases of the Fourier coefficients are real can be computationally difficult. When the Fourier coefficients are real, all phases are either 0 or $\pi/2$. A set of n such phases can be described as a binary vector of length n. But it can be formidable to recover these n phases by undirected search when n is equal to several hundred. For $n = 100$, there would be 2^{100} binary words to try corresponding to the 100 unknown phases. For complex Fourier coefficients, the situation is much more difficult. Clearly, for general values of the phase, an unstructured search is intractable. Direct methods must be structured.

9.11 Direct Methods from Diffraction Data

Two formulas that are useful in devising a direct computational estimation procedure are the *Sayre formula* and the *Karle–Hauptman formula*. These formulas can be used to obtain constraints in the frequency domain that are helpful in many cases of computational crystallographic image formation even when the number of particles n is rather large.

These formulas are used as aides in searching for the phase function $\theta(x,y,z)$ so that the function $|R(x,y,z)|e^{j\theta(x,y,z)}$ then inverts to a real nonnegative function $\rho(x,y,z)$. As stated, each formula is valid only under certain symmetry conditions. Nevertheless, each can often be used as an approximation in the absence of the symmetry conditions.

Center of Symmetry

Many of the inequalities of this section posit that the Nyquist samples are real numbers. Then when given the magnitudes of the frequency-domain Nyquist samples that are known to be real, it is only necessary to determine the signs of these real-valued Nyquist samples. The intrinsic side information available to determine the signs is that the space-domain function $\rho(x,y,z)$ is real and nonnegative.

The basic case consists of a real, nonnegative $\rho(x,y,z)$ satisfying the symmetry property

$$\rho(x,y,z) = \rho(-x,-y,-z)$$

arranged in the infinite array defined by the elements of $\mathbb{Z}^3$ centered symmetrically on the origin. This symmetry condition in the frequency domain is

$$R(f_x,f_y,f_z) = R^*(f_x,f_y,f_z),$$

which means that $R(f_x,f_y,f_z)$ is real.

Even when the support of $\rho(x,y,z)$ extends beyond the unit cell, the symmetry is maintained and the Nyquist samples are real, though they are aliased. When the center of $\rho(x,y,z)$ is away from the origin of the reference lattice, there is a translate of the lattice in which $\rho(x,y,z)$ is centered. It is enough to change the center of the mathematical reference lattice accordingly. Treating the Nyquist samples as real numbers and determining only the signs implicitly chooses the origin of that lattice as the center of symmetry.

The Harker–Kasper Inequality

The Harker–Kasper inequality is one instance of a large class of inequalities of this kind. It is a precursor for the more elaborate instances that follow. As is appropriate to this chapter, the Harker–Kasper inequality is described in three dimensions, though it holds for functions of any dimension.

The Harker–Kasper inequality is an elementary inequality regarding the Nyquist spectral samples of a real, nonnegative function. The Nyquist samples are the Fourier coefficients of the three-dimensional periodic waveform. The magnitudes of the Nyquist samples are known as structure factors in the field of crystallography.

Proposition 9.11.1 (Harker–Kasper) *The frequency-domain Nyquist samples of the nonnegative real-valued centrosymmetric function $\rho(x,y,z)$ satisfy the inequality*

$$|R(i)|^2 \leq R(0)\tfrac{1}{2}|R(0) + R(2i)|,$$

where $\mathbf{0} = (0,0,0)$, $\mathbf{i} = (i,j,k)$ is any triple index, and $2\mathbf{i} = (2i,2j,2k)$.

Proof:
Because $\rho(x,y,z)$ is centrosymmetric, all Fourier samples are real. Because $R(i)$ is real, it is given by

$$|R(i)|^2 = \left| \int_{-\infty}^{\infty}\int_{-\infty}^{\infty}\int_{-\infty}^{\infty} \rho(x,y,z)\cos\left(2\pi(i_x x + i_y y + i_z z)\right)dxdydz \right|^2.$$

The squared magnitude then satisfies the following chain of inferences (the Schwarz inequality is used in the second line):

$$|R(i)|^2 = \left| \int_{-\infty}^{\infty}\int_{-\infty}^{\infty}\int_{-\infty}^{\infty} \left(\sqrt{\rho(x,y,z)}\right)\left(\sqrt{\rho(x,y,z)}\cos\left(2\pi(i_x x + i_y y + i_z z)\right)\right)dxdydz \right|^2$$

$$\leq \int_{-\infty}^{\infty}\int_{-\infty}^{\infty}\int_{-\infty}^{\infty} \rho(x,y,z)dxdydz \int_{-\infty}^{\infty}\int_{-\infty}^{\infty}\int_{-\infty}^{\infty} \rho(x,y,z)$$

$$\left(\cos\left(2\pi(i\cdot x)\right)\right)^2 dxdydz$$

$$= R(0)\int_{-\infty}^{\infty}\int_{-\infty}^{\infty}\int_{-\infty}^{\infty} \tfrac{1}{2}\rho(x,y,z)\Big[1 + \cos\left(2\pi(2i\cdot x)\right)\Big]dxdydz$$

$$\leq R(0)\tfrac{1}{2}(R(0) + R(2i)),$$

where $x = (x,y,z)$, This completes the proof of the proposition. ∎

The Harker–Kasper inequality is occasionally useful to determine the sign of a structure factor. For example, when a centrosymmetric function with the origin at the center of symmetry has $|R(i)^2| = 0.5$, $R(0) = 1$, and $|R(2i)| = 0.8$, then $R(2i)$ must have a positive sign.

Inspection of the proof reveals that the condition that the sample $R(2i)$ is real is not used, nor does the proof refer to other Fourier components. This suggests the following corollary

Corollary 9.11.2 *Any two frequency-domain Nyquist samples of the real nonnegative function $\rho(x,y,z)$ of the form $R(i)$ and $R(2i)$ are related by*

$$|R(i)|^2 \leq R(0)\tfrac{1}{2}|1 + \text{Re}[R(2i)]|$$

whenever $R(i)$ is real.

Proof:
The proof is nearly the same as the proof of Proposition 9.11.1 ∎

Finally, note that $R(i)$ can always be made real by a translation of the origin. This leads to the next corollary.

9.11 Direct Methods from Diffraction Data

Corollary 9.11.3 *Any two frequency-domain Nyquist samples of the real nonnegative function $\rho(x,y,z)$ of the form $R(i)$ and $R(2i)$ are related by*

$$|R(i)|^2 \leq R(0)\tfrac{1}{2}\big|1 + \text{Re}[R(2i)e^{-j2\theta(i)}]\big|,$$

where $R(i) = |R(i)|e^{j\theta(i)}$.

Proof:
A suitable translation of the origin multiplies $R(i)$ by $e^{-j\theta(i)}$ and multiplies $R(2i)$ by $e^{-j2\theta(i)}$. Corollary 9.11.2 then applies. ∎

The Sayre Formula

The Sayre formula, given next in Theorem 9.11.4, is a statement regarding the three-dimensional Fourier transform $R(f_x,f_y,f_z)$ of $\rho(x,y,z)$ for the case in which n identical and essentially nonoverlapping atoms (or substructures) given by $a(x,y,z)$ comprise the molecule $\rho(x,y,z)$ as given by

$$\rho(x,y,z) = \sum_\ell a(x - x_\ell, y - y_\ell, z - z_\ell).$$

The Sayre formula is often stated in the concise form

$$R(f_x,f_y,f_z) \ast\ast\ast\ R(f_x,f_y,f_z) = R(f_x,f_y,f_z)B(f_x,f_y,f_z),$$

where the function $B(f_x,f_y,f_z)$ depends only on the element $a(x,y,z)$ and is independent of the number of elements n. In this form, the expression suggests that the support of $R(f_x,f_y,f_z) \ast\ast\ast\ R(f_x,f_y,f_z)$ must be equal to the support of $R(f_x,f_y,f_z)$, which can be so only when the support is infinite. This is in accord with the observation that $\rho(x,y,z)$ is finite.

Because the expression is asserted to be independent of n, the function $B(f_x,f_y,f_z)$ apparently can be found from the case with $n = 1$. Thus

$$A(f_x,f_y,f_z)B(f_x,f_y,f_z) = A(f_x,f_y,f_z) \ast\ast\ast\ A(f_x,f_y,f_z),$$

so that

$$B(f_x,f_y,f_z) = \frac{A(f_x,f_y,f_z) \ast\ast\ast\ A(f_x,f_y,f_z)}{A(f_x,f_y,f_z)}.$$

Because $a(x,y,z)$ is known, the transform $A(f_x,f_y,f_z)$ is known. Therefore $B(f_x,f_y,f_z)$ is known, as well.

The Sayre formula is motivated by considering the one-dimensional pulse train $\rho(t)$ consisting of identical nonoverlapping pulses $a(t)$ given by

$$\rho(t) = \sum_{\ell=1}^{n} a(t - \ell T).$$

Convolving $a(t)$ with the square of $\rho(t)$ gives

9 Diffraction Imaging Systems

$$a(t) * \rho(t)^2 = a(t) * \left[\sum_{\ell=1}^{n} a(t - \ell T) \sum_{\ell'=1}^{n} a(t - \ell' T)\right]$$

$$= \int_{-\infty}^{\infty} a(t - \xi) \sum_{\ell=1}^{n} a(\xi - \ell T) \sum_{\ell'=1}^{n} a(\xi - \ell' T) d\xi$$

$$= \int_{-\infty}^{\infty} a(t - \xi) \sum_{\ell=1}^{n} \sum_{\ell'=1}^{n} a(\xi - \ell T) a(\xi - \ell' T) d\xi.$$

For pulses that do not overlap, the double sum reduces to a single sum:

$$a(t) * \rho^2(t) = \int_{-\infty}^{\infty} a(t - \xi) \sum_{\ell=1}^{n} a^2(\xi - \ell T) d\xi$$

$$= \sum_{\ell=1}^{n} \int_{-\infty}^{\infty} a(t - \xi) a^2(\xi - \ell T) d\xi$$

$$= \sum_{\ell=1}^{n} \int_{-\infty}^{\infty} a(t - \xi - \ell T) a^2(\xi) d\xi$$

$$= \int_{-\infty}^{\infty} a^2(\xi) \sum_{\ell=1}^{n} a(t - \xi - \ell T) d\xi$$

$$= a^2(t) * \rho(t).$$

In the frequency domain, this statement becomes

$$A(f)\Big(R(f) * R(f)\Big) = R(f)\Big(A(f) * A(f)\Big).$$

The statement holds for a train of nonoverlapping pulses even when the pulse train is not periodic.

The proof of the following theorem gives the Sayre formula, which explicitly refers to identical nonoverlapping atoms. The condition of nonoverlapping means that the three-dimensional supports of the individual atoms are nonintersecting, but those three-dimensional supports can be in any geometrical arrangement. The formula in the theorem is written concisely using the summary notation $f = (f_x, f_y, f_z)$.

Theorem 9.11.4 *The Fourier transform $R(f)$ of the molecule $\rho(x, y, z)$ composed of n nonoverlapping identical atoms, $a(x, y, z)$, satisfies the Sayre formula*

$$R(f)[A(f) * * * A(f)] = A(f)[R(f) * * * R(f)],$$

where $A(f)$ is the Fourier transform of $a(x, y, z)$.

Proof:
In the space domain, the corresponding formula that is to be proved is

$$a(x, y, z) * * * \big(\rho(x, y, z)\big)^2 = \rho(x, y, z) * * * \big(a(x, y, z)\big)^2,$$

as follows from the convolution theorem. By using the summary notation $x = (x, y, z)$, this formula can be concisely written as

$$a(x) *** \rho(x)^2 = \rho(x) *** a(x)^2$$

Using this notation, the term on the right is developed as follows:

$$a(x) *** \rho(x)^2 = \int_{-\infty}^{\infty}\int_{-\infty}^{\infty}\int_{-\infty}^{\infty} a(x - \xi) \sum_{\ell=1}^{n} a(\xi - x_\ell) \sum_{\ell'=1}^{n} a(\xi - x_{\ell'}) d\xi$$

$$= \sum_{\ell=1}^{n}\sum_{\ell'=1}^{n} \int_{-\infty}^{\infty}\int_{-\infty}^{\infty}\int_{-\infty}^{\infty} a(x - \xi) a(\xi - x_\ell) a(\xi - x_{\ell'}) d\xi,$$

where $d\xi = d\xi_x d\xi_y d\xi_z$. The condition that the atoms do not overlap means that $a(\xi - x_\ell)a(\xi - x_{\ell'}) = 0$ unless ℓ is equal to ℓ'. Hence the equation collapses to

$$a(x) *** \rho(x)^2 = \sum_{\ell=1}^{n} \int_{-\infty}^{\infty}\int_{-\infty}^{\infty}\int_{-\infty}^{\infty} a(x - \xi) a^2(\xi - x_\ell) d\xi$$

$$= \sum_{\ell=1}^{n} \int_{-\infty}^{\infty}\int_{-\infty}^{\infty}\int_{-\infty}^{\infty} a(x - x_\ell - \xi) a^2(\xi) d\xi$$

$$= \int_{-\infty}^{\infty}\int_{-\infty}^{\infty}\int_{-\infty}^{\infty} a^2(\xi) \sum_{\ell=1}^{n} a(x - x_\ell - \xi) d\xi$$

$$= a(x)^2 *** \rho(x),$$

as was to be proved. ■

Of course, there are not many applications in which a target molecule has only identical nonoverlapping atoms. However, the formula extends to a variety of other cases. The Sayre formula applies as well to identical nonoverlapping molecules or identical nonoverlapping groups of atoms. Because the formula is robust, it may be useful as a first approximation and not the final answer even when the atoms are intersecting or are not identical.

The Karle–Hauptman Formula

The Karle–Hauptman formula is a statement that a matrix of a certain kind has a nonnegative determinant. Before describing the general form of the Karle–Hauptman formula, an elementary instance of the formula is described. This is the simple statement that for any real and nonnegative one-dimensional function $s(t)$, the Fourier transform $S(f)$ satisfies the elementary inequality

$$\det \begin{bmatrix} S(0) & S(\Delta) \\ S(-\Delta) & S(0) \end{bmatrix} \geq 0,$$

where Δ is any value of f. To verify this inequality, write the determinant as

$$S(0)^2 - S(\Delta)S(-\Delta) = \int_{-\infty}^{\infty} s(t)dt \int_{-\infty}^{\infty} s(t')dt' - \int_{-\infty}^{\infty} s(t)e^{-j2\pi \Delta t}dt \int_{-\infty}^{\infty} s(t')e^{j2\pi \Delta t'}dt'.$$

The second term n on the right is real because $s(t)$ is real and nonnegative and is invariant under conjugation. But then that second term can be written

$$\int_{-\infty}^{\infty}\int_{-\infty}^{\infty} s(t)s(t')e^{-j2\pi\Delta(t-t')}dtdt' = \int_{-\infty}^{\infty}\int_{-\infty}^{\infty} s(t)s(t')\cos 2\pi\Delta(t-t')dtdt'$$

$$\leq \int_{-\infty}^{\infty}\int_{-\infty}^{\infty} s(t)s(t')dtdt',$$

and the assertion follows.

This simple two by two matrix inequality for the one-dimensional Fourier transform holds as well for the n by n three-dimensional Fourier transform. Using the triple index $i = (i, j, k)$, the definition of the Karle–Hauptman matrix is as follows.

Definition 9.11.5 *A general n by n Karle–Hauptman matrix for a three-dimensional function $s(x, y, z)$ is a matrix of frequency-domain Nyquist samples of $S(f_x, f_y, f_z)$ using the abbreviated notation $S(\ell i)$ for any triple index $i = (i, j, k)$ and*

$$M = \begin{bmatrix} S(0i) & S(1i) & S(2i) & \cdots & S((n-1)i) \\ S(1i)^* & S(0i) & S(1i) & \cdots & S((n-2)i) \\ S(2i)^* & S(1i)^* & S(0i) & \cdots & S((n-3)i) \\ \vdots & \vdots & \vdots & \vdots & \vdots \\ S((n-1)i)^* & S((n-2)i)^* & S((n-3)i)^* & \cdots & S(0i) \end{bmatrix},$$

where $\ell i = (\ell i, \ell j, \ell k)$ and for any n such that the indices, regarded cyclically, do not repeat, The ℓk element of matrix M is $S(\ell i, \ell j, \ell k)$. ∎

The Karle–Hauptman matrix[15] M is hermitian. As for any hermitian matrix, the determinant of the matrix M is real. This is because the matrix satisfies $M^* = M^T$, which implies that

$$(\det M)^* = \det M^* = \det M^T = \det M.$$

Moreover, the forthcoming Theorem 9.11.6 states that the determinant of M is nonnegative.

Before stating the theorem, a factorization of the Karle–Hauptman matrix into the product of two matrices is developed. To simplify the discussion, this factorization of the Karle–Hauptman matrix will be developed for a function $S(i)$ in one dimension. The development for a function $S(i, j, k)$ in three dimensions is the same.

For the discretized signal $s = (s_0, s_1, \ldots, s_{m-1})$, where s_i is real and nonnegative with discrete Fourier transform $S(i)$, let $B_{ik} = \sqrt{s_i}e^{-j2\pi ik/m}$. Then define the matrix of such terms,

---

[15] The Karle–Hauptman matrix is an instance of a *Toeplitz matrix*. Any matrix $A = [A_{ik}]$ whose values depend on only $i - k$, and not on i and k individually, is called a Toeplitz matrix.

$$B = \begin{bmatrix} B_{00} & B_{01} & B_{02} & \cdots & B_{0\ell} \\ B_{10} & B_{11} & B_{12} & \cdots & B_{1\ell} \\ B_{20} & B_{21} & B_{22} & \cdots & B_{2\ell} \\ \vdots & \vdots & \vdots & \vdots & \vdots \\ B_{\ell 0} & B_{\ell 1} & B_{\ell 2} & \cdots & B_{\ell\ell} \end{bmatrix}$$

for any i. The ik term of the matrix is M, then

$$\sum_i B_{ik} B^*_{ik'} = \sum_i \sqrt{s_i} e^{-j2\pi ik/m} \sqrt{s_i} e^{-j2\pi ik'/m}$$
$$= \sum_i s_i e^{-j2\pi(k-k')i/m}$$
$$= S(k-k').$$

Now let $k' = k - \ell'i$ for $\ell' = 0, \ldots, \ell$, where i is now a new index. Then $\sum_i B_{ik} B^*_{ik'} = S(\ell'i)$. This describes a typical term of the matrix product of the matrices B and $B^\dagger$. Therefore, the matrix product is

$$BB^\dagger = \begin{bmatrix} S(0i) & S(1i) & S(2i) & \cdots & S(\ell i) \\ S(1i)^* & S(0i) & S(1i) & \cdots & S((\ell-1)i) \\ S(2i)^* & S(1i)^* & S(0i) & \cdots & S((\ell-2)i) \\ \vdots & \vdots & \vdots & \vdots & \vdots \\ S(\ell i)^* & S((\ell-1)i)^* & S((\ell-2)i)^* & \cdots & S(0i) \end{bmatrix},$$

where $B^\dagger$ denotes the complex transpose. The right side is equal to M, which means that the one-dimensional Karle–Hauptman matrix factors as $M = BB^\dagger$. The three-dimensional Karle–Hauptman matrix factors in the same way.

Theorem 9.11.6 (Karle–Hauptman) *A Karle–Hauptman matrix M for the real non-negative function $s(x,y,z)$ satisfies*

$$\det M \geq 0.$$

Proof:
For brevity, an n by n one-dimensional version of this theorem using the single variable t is proved. The n by n two-dimensional version using (x,y) and the n by n three-dimensional version using (x,y,z) are proved in the same way.

The Karle–Hauptman matrix is now shown to be always positive-semidefinite by using the factorization $M = BB^\dagger$. Then

$$aMa^\dagger = aBB^\dagger a^*$$
$$= \sum_i \sum_{i'} a_i \sum_k B_{ik} B_{i'k} a_{i'}$$
$$= \sum_k \left(\sum_i a_i B_{ik}\right)\left(\sum_{i'} a_{i'} B_{i'k}\right)^*$$
$$= \left\|\sum_i a_i B_{ik}\right\|^2 \geq 0,$$

as was asserted. This completes the proof of the theorem. ∎

The Karle–Hauptman inequality is used to find relationships between real Fourier components that help to determine the phases of those components. Because there are a large number of matrices that can be extracted from the Karle–Hauptman matrix inequality, it may be possible to determine the signs or phases of many components. A sign, once determined, can then be used in subsequent trials.

The simplest instance states that the determinant of any three by three Karle–Hauptman matrix is nonnegative. This may constrain the sign of one of the terms in that equation. For example, the three by three determinant

$$\det \begin{bmatrix} 1 & S(i) & S(2i) \\ S(i^*) & 1 & S(i) \\ S(2i^*) & S(i^*) & 1 \end{bmatrix} \geq 0,$$

evaluates as

$$1 - 2|S(i)|^2 - |S(2i)|^2 + S(i)^2 S(2i^*) + S(i^*)^2 S(2i) \geq 0.$$

When all terms are real, this reduces to

$$[1 - S(2i)][1 + S(2i) - 2S(i)^2] \geq 0.$$

The first term is nonnegative. Therefore, the second term must also be nonnegative. Consequently,

$$S(i)^2 \leq \tfrac{1}{2}[1 + S(2i)],$$

which is the Harker–Kasper inequality derived earlier.

Phase Extension

The need for solutions to the phase recovery problem has led to the development of inexact mathematical methods that can be justified only by success in practical applications. The tangent formula provides such a method. For a centrosymmetric molecule, the Karle–Hauptman inequality may be a useful tool under the approximation that the value of $R(f_x, f_y, f_z)$ is real. For a noncentrosymmetric molecule, this is not an appropriate method. Instead, it may be more productive to treat $R(f_x, f_y, f_z)$ as complex and seek to recover the phase angles using the Karle–Hauptman tangent or phase extension formula. The tangent formula is an inexact formula that allows the phase of one grating lobe or structure factor to be approximated from the phases of two other grating lobes whose phases are known.

With h defining the trivariate index $h = (i, j, k)$, the Karle–Hauptman tangent formula is the tangent formula

$$\langle \tan \phi_h \rangle = \frac{\sum_h R_{h'} R_{h-h'} \sin(\phi_{h'} + \phi_{h-h'})}{\sum_h R_{h'} R_{h-h'} \cos(\phi_{h'} + \phi_{h-h'})},$$

where ϕ_h is the phase of R_h. The notation $\langle \ \rangle$ on the left indicates that this is an approximation.

The phase extension formula computes the phase of one grating lobe (or structure factor) from the known or estimated phases of two other grating lobes. It can then be iterated to compute other phase angles of other grating lobes. In this way a small number of candidate phases can be extended to many phases. With prior knowledge of the molecular constituents possibly incorporated, this can be repeated until a meaningful solution emerges. The tangent formula is not exact and the phases on the right side are rarely known exactly, so the tangent formula for the phase extension must be used with great care.

The tangent formula is a heuristic formula that is not exact and is not justified by a proper mathematical derivation. Nevertheless, it has proven to be very useful to move a structured search in the right direction. The Sayre formula

$$R(f) = \frac{A(f)}{A(f) * * * A(f)} [R(f) * * * R(f)]$$

holds for a molecule composed of identical atoms $a(x,y,z)$. For a general molecule, it does not apply. Nevertheless, it is used as a guide to suggest the heuristic expression

$$R(f) = B[R(f) * * * R(f)]$$

as an approximate fit to the general molecule.

The sampled form of the version of this expression in one dimension replaces $R(f)$ with $R_k e^{-\phi_k}$. Then

$$R_k e^{-\phi_k} \approx \sum_{k'} R_{k'} R_{k-k'} e^{-\phi_{k'}} e^{-\phi_{k-k'}}.$$

The real and imaginary parts are

$$R_k \cos(\phi_k) \approx \sum_{k'} R_{k'} R_{k-k'} \cos(\phi_{k'} + \phi_{k-k'}),$$

$$R_k \sin(\phi_k) \approx \sum_{k'} R_{k'} R_{k-k'} \sin(\phi_{k'} + \phi_{k-k'}),$$

which leads to the tangent formula.

9.12 Diffraction from Fiber Arrays

Some molecules cannot be made to form a regular crystal. The standard techniques of diffraction imaging cannot be used to form images for such molecules. However, even the molecules in a large disordered collection will individually diffract X-rays, and the cumulative diffraction pattern does depend in some way on the shape of the individual molecules even though these molecules are not in an orderly array. When there are a large number of molecules, individually disordered, the expected diffraction pattern is the statistical average of the individual diffraction patterns, randomly oriented and positioned. Though the information in this diffraction pattern of a disordered array of molecules is far weaker than the information in the diffraction pattern of an ordered array, there still may be some useful information.

The nature of the disorder can take various forms, and each form requires individual analysis. One very disordered case that might be considered consists of identical molecules strewn randomly at the points of a three-dimensional Poisson process, which is a random set of locations in three dimensions. In addition, each molecule has a random orientation that is described by a rotation matrix that is uniformly distributed and independent of the rotation matrix defining the orientation of any other molecule. A variation of this is a powder in which each grain of the powder is itself a small crystal. These grains, though themselves regular crystals individually, are completely disordered in both position and rotation. Each grain is randomly oriented in angle, so each has a diffraction pattern that is randomly oriented. The diffraction pattern of the powder is a composite of the diffraction patterns of the individual grains. Evidently, the diffraction pattern has the form of a series of concentric shells because the only frequency-domain dependence that is not averaged out is the dependence in the radial direction.

A case of some importance in practice is a regular array of molecules, but with the orientation of each molecule within the array uniformly and independently distributed over all rotations in three-dimensional space. To suggest the possibilities, only a simple idealized case is discussed. This idealized case has each molecule randomly rotated about only the z axis in an otherwise regular array. These molecules are said to be cylindrically disordered. This is called a *fiber array*.

Let $\rho(x, y, z)$ be any function, and let

$$\rho_\psi(x, y, z) = \rho(x \cos \psi - y \sin \psi, x \sin \psi + y \cos \psi, z)$$

be that function rotated by random angle ψ about the z axis. This function has the Fourier transform

$$R_\psi(f_x \cos \psi - f_y \sin \psi, f_x \sin \psi + f_y \cos \psi, f_z).$$

Define the fiber array as

$$p(x, y, z) = \sum_{\ell_x=0}^{N-1} \sum_{\ell_y=0}^{N-1} \sum_{\ell_z=0}^{N-1} \rho_{\psi_{\ell_x \ell_y \ell_z}}(x - \ell_x, y - \ell_y, z - \ell_z),$$

where the $\psi_{\ell_x \ell_y \ell_z}$ are independent random variables denoting angles that are uniformly distributed on the interval $[0, 2\pi)$ With this model, each molecule is at a lattice point indexed by the integers ℓ_x, ℓ_y, and ℓ_z. This molecule has a random rotation about the z axis, denoted by $\psi_{\ell_x \ell_y \ell_z}$, but is not rotated otherwise. The scattering function is given in terms of the Fourier transform

$$P(f_x, f_y, f_z) = \sum_{\ell_x=0}^{N-1} \sum_{\ell_y=0}^{N-1} \sum_{\ell_z=0}^{N-1} R_{\psi_{\ell_x \ell_y \ell_z}}(f_x, f_y, f_z) e^{-j2\pi(\ell_x f_x + \ell_y f_y + \ell_z f_z)}.$$

The ensemble average of this random function $P(f_x, f_y, f_z)$ is the expectation

$$E[P(f_x, f_y, f_z)] = \sum_{\ell_x=0}^{N-1} \sum_{\ell_y=0}^{N-1} \sum_{\ell_z=0}^{N-1} E[R_{\psi_{\ell_x \ell_y \ell_z}}(f_x, f_y, f_z)] e^{-j2\pi(\ell_x f_x + \ell_y f_y + \ell_z f_z)}.$$

The expectation term is $E[|R_\psi(f_x,f_y,f_z)|]$, for every value of indices ℓ_x, ℓ_y, and ℓ_z. It can be moved outside the summation signs. Then the sums can be executed and the residual phase term suppressed by choice of origin, leading to

$$E[P(f_x,f_y,f_z)] = E[R_\psi(f_x,f_y,f_z)]\text{dirc}_N(f_x,f_y,f_z),$$

where

$$E[R_\psi(f_x,f_y,f_z)] = \frac{1}{2\pi}\int_0^{2\pi} R(f_x\cos\psi - f_y\sin\psi, f_x\sin\psi + f_y\cos\psi, f_z)d\psi$$

is the expectation, where $R_\psi(f_x,f_y,f_z)$ is $R(f_x,f_y,f_z)$ after a rotation by ψ about the z axis.

A similar analysis applied to the squared magnitude of $P(f_x,f_y,f_z)$ gives

$$E[|P(f_x,f_y,f_z)|^2] = E[|R_\psi(f_x,f_y,f_z)|^2]\text{dirc}_N(f_x,f_y,f_z)$$

for the expected intensity.

Example

Consider an array of helices, each helix randomly rotated around the z axis by a uniformly distributed angle. A finite helix of length L has the Fourier transform

$$R(f_x,f_y,f_z) = \sum_{n=-\infty}^{\infty} J_n\left(2\pi\sqrt{f_x^2+f_y^2}\right) e^{-jn\tan^{-1}(f_x/f_y)} \text{sinc}(L(f_z - n/2\pi)),$$

and $R_\psi(f_x,f_y,f_z)$ is $R(f_x,f_y,f_z)$ under a random rotation about the z axis. The complex exponential contained within $R_\psi(f_x,f_y,f_z)$ can be written

$$e^{-jn\tan^{-1}\left(\frac{f_x\cos\psi - f_y\sin\psi}{f_x\sin\psi + f_y\cos\psi}\right)} = e^{-jn\tan^{-1}(f_x/f_y)}e^{jn\psi}.$$

The expectation operator applied to this expression leads to the term $E[e^{jn\psi}]$, which is zero except for the term with $n = 0$. Consequently,

$$E[R_\psi(f_x,f_y,f_z)] = \text{sinc}(Lf_z)J_0\left(2\pi\sqrt{f_x^2+f_y^2}\right),$$

and

$$E[P(f_x,f_y,f_z)] = \text{sinc}(Lf_z)J_0\left(2\pi\sqrt{f_x^2+f_y^2}\right)\text{dirc}_N(f_x,f_y,f_z).$$

This is the same as the diffraction pattern for an array of cylindrical shells. By allowing each helix in the array to have an arbitrary angle about the z axis, the details of the helix have been averaged into a cylindrical shell. This is the mean of the ensemble of random variables.

The variance of the ensemble can be evaluated in a similar way. It is, approximately,

$$E[|P(f_x,f_y,f_z)|^2] = \sum_{n=-\infty}^{\infty} J_n^2(2\pi f)\text{sinc}^2(L(f_z - n/2\pi))\text{dirc}_N(f_x,f_y,f_z),$$

where $f^2 = f_x^2 + f_y^2$. This is an approximation because the interaction between the sinc sidelobes coming off the individual layers has been ignored as negligible.

9.13 Diffraction from Excited Arrays

The individual molecules in a crystal array always have thermal energy which causes these molecules and their constituent atoms to randomly vibrate. This vibration leads to time-varying irregularities in the crystal spacing that affects the diffraction pattern. Observation of this effect in the diffraction pattern provides an indirect means of observing the vibration energy and measuring the amount of irregularity in the crystal. Moreover, even when there is no thermal vibration, the molecules are not placed perfectly at the lattice points. There are imperfections in the spacing and the orientation. Because the diffraction pattern is a Fourier transform, these imperfections can be studied as the effect of irregularities on the Fourier transform of an array.

Thermal energy of the individual atoms of a molecule must be consistent with the thermal energy of other atoms of that molecule. This may sometimes help to identify an atom within a molecule when several possibilities are otherwise suggested.

The irregularities in the positions of the elements in an array manifest themselves primarily as phase errors in the terms making up the grating lobes. The effect of phase errors on the Fourier transform due to irregularities in the spacings of an array is studied for the general case in Chapter 15. In this section, the conclusions of Chapter 15 are referenced to describe the effect that the vibration and imperfections of a crystal have on the diffraction pattern.

Proposition 15.3.1 of Chapter 15 describes the expected Gabor bandwidth of a one-dimensional pulse $s(t)$ contaminated by phase noise as given by $s(t)e^{j\theta(t)}$. This expression for the expected Gabor bandwidth is

$$E[B_G^2(v)] = B_G^2(s) + \int_{-\infty}^{\infty} f^2 \Phi(f) df,$$

where $B_G^2(s)$ is the Gabor bandwidth of the individual pulse $s(t)$, and $\Phi(f)$ is the power density spectrum of the phase noise. When the term $B_G^2(s)$ on the right is small, it may be adequate to use the approximation

$$E[B_G^2(v)] \approx \int_{-\infty}^{\infty} f^2 \Phi(f) df$$

as a simple way to estimate the bandwidth of the phase noise. To discuss this broadening of the Gabor bandwidth and the relationship to crystal diffraction, the time variable t is replaced by the three space variables x, y, and z. Because there are three space variables, the situation is somewhat more elaborate but not more complicated. Replacing the simple variable time by the three space variables does not change the nature of the conclusion. It is enough to consider only one space variable. The Gabor bandwidth B_G^2 is determined by the width of a one-dimensional grating lobe, which is negligibly small when the number of cells of the crystal is large. Thus, by observing the Gabor bandwidth of the main lobe of the diffraction pattern, one can estimate the bandwidth of the spatial phase noise. The variance of position errors in the elements of the crystal can be estimated by describing the effect of position error on the phase error, then estimating the three-dimensional bandwidth of the phase error from the Gabor bandwidths of the main lobe of the diffraction pattern.

There is more to say because the array may also exhibit thermal vibration. This is because there are two consequences of thermal vibration of an array: Vibration affects both the space structure and the time structure of the grating lobes. The vibration will affect the space structure because the irregularities in the space structure of the diffracted wave cause the grating lobes to be widened. The vibration will affect the time structure because then the spatial irregularities are time-varying, so the structure of the grating lobes will display time fluctuations. By measuring the shape of the grating lobes as a function of time, it is possible to make inferences about the thermal fluctuations of the crystal and the strength of its bonds.

Problems

9.1 (Coordinate Transformation) Prove the three-dimensional Fourier transform relationship

$$S\left(A^T \begin{bmatrix} x \\ y \\ z \end{bmatrix}\right) \iff \frac{1}{|A|} S\left(A^{-1} \begin{bmatrix} f_x \\ f_y \\ f_z \end{bmatrix}\right),$$

where $|A|$ is the determinant of A. State and prove the coordinate rotation property in three dimensions by specializing $|A|$ to an eulerian rotation matrix.

9.2 Find the three-dimensional Fourier transforms of $\mathrm{rect}(x,y,z)$ and $\mathrm{cyln}(x,y,z)$.

9.3 Let $s(x,y,z)$ be a three-dimensional function of finite energy.

a. Show that the three-dimensional Fourier transform $S(f_x,f_y,f_z)$ of $s(x,y,z)$ satisfies the property

$$s^*(x,y,z) \iff S^*(-f_x,-f_y,-f_z).$$

b. Show that

$$s(-x,-y,-z) \iff S(-f_x,-f_y,-f_z).$$

c. What is the inverse Fourier transform of $\mathrm{Re}[S(f_x,f_y,f_z)]$?

9.4 A three-dimensional function $s(x,y,z)$ of finite energy is to be estimated from its Fourier transform $S(f_x,f_y,f_z)$ which is known only at those points within the sphere

$$f_x^2 + f_y^2 + f_z^2 \le k^2.$$

a. Suppose that $s(x,y,z)$ is computed by simply setting $S(f_x,f_y,f_z)$ equal to zero at all f_x,f_y,f_z outside this sphere and computing the three-dimensional inverse Fourier transform of this cropped data set. How is $s(x,y,z)$ degraded?

b. Propose other ways to infer values of $S(f_x,f_y,f_z)$ outside this sphere. Discuss the merits of these proposals.

9.5 Show that $\mathrm{tinc}(0) = \pi/6$, first by using l'Hôpital's rule, and then by using a Taylor series expansion.

9.6 a. Given a general nonrectangular lattice,
$$v = Mi,$$
defined by matrix M, where $i \in \mathbb{Z}^3$, find the diffraction pattern for a crystal defined by this lattice and the scattering function $\rho(x, y, z)$.

b. State the general form of the Bragg–Laue equations for the nonrectangular lattice Mi.

9.7 a. The autocorrelation of a circle function is the two-dimensional *circular hat function*, denoted chat(x, y), and defined as
$$\text{chat}(x, y) = \text{circ}(x, y) * * \text{circ}(x, y).$$
Find the Fourier transform of cyln$(x, y, z) * * *$ cyln(x, y, z).

b. Find the general features of the autocorrelation function of a hexagonal molecule made of cylindrical atoms. That is, find the autocorrelation function of a nonoverlapping hexagonal arrangement of six cylinder functions in the x, y plane and oriented in the z direction.

c. Find the intensity of the scattering function of a hexagonal arrangement of six nonoverlapping circle (or cylinder) functions in the x, y plane. Give a "signature" or "template" for recognizing a hexagonal subsection within a molecule by recognizing its scattering intensity within a more complicated Fourier transform.

9.8 Given the three-dimensional Fourier transform pair
$$s(x, y, z) \iff S(f_x, f_y, f_z),$$
what is the inverse Fourier transform of $\text{Re}[S(f_x, f_y, f_z)]$? That is, complete the three-dimensional Fourier transform pair
$$? \iff \text{Re}[S(f_x, f_y, f_z)].$$
Suppose that $S(f_x, f_y, f_z)$ is real. What does this imply about $s(x, y, z)$?

9.9 The three-dimensional generalization of the two-dimensional circular hat function chat(x, y) is called the *three-dimensional circular hat function* and is defined as
$$\text{chat}(x, y, z) = \text{sphr}(x, y, z) * * * \text{sphr}(x, y, z).$$
Give a closed form expression for the three-dimensional circular hat function chat(x, y, z). What is the three-dimensional Fourier transform of chat(x, y, z)?

9.10 Describe the function
$$s(x, y, z) = \text{circ}(x, y)\text{comb}(z)$$
and find its Fourier transform.

9.11 a. Prove the Nyquist–Shannon sampling theorem for three-dimensional sampling on a cubic lattice.

b. Let $s(x,y,z)$ be a function whose Fourier transform $S(f_x,f_y,f_x)$ is zero for all points satisfying $f_x^2 + f_y^2 + f_z^2 > \frac{1}{4}$. Derive a tinc-interpolation formula for samples of $s(x,y,z)$ on a cubic lattice. Why is tinc interpolation better than three-dimensional sinc interpolation?

c. The *grocer's lattice* in three dimensions is familiar from the commonplace packing of spheres. Define this lattice as

$$\begin{bmatrix} x \\ y \\ z \end{bmatrix} = M \begin{bmatrix} i_x \\ i_y \\ i_z \end{bmatrix}$$

for an appropriate matrix M. Use this lattice to define a sampling pattern for a function, $s(x,y,z)$, with the Fourier transform satisfying $S(f_x,f_y,f_z) = 0$ for $f_x^2 + f_y^2 + f_z^2 > \frac{1}{4}$. By how much is the sampling density improved compared to cubic sampling? Give an interpolation formula.

d. Do grocers use the grocer's lattice to save space or for another reason?

9.12 Denote the three rows of the three by three full-rank matrix M by the vectors a, b, and c, respectively.

a. Show that the three rows of the reciprocal matrix M^{-1} can be written using the vector cross product and the vector dot product as

$$a^* = \frac{b \times c}{a \cdot (b \times c)} \quad b^* = \frac{c \times a}{a \cdot (b \times c)} \quad c^* = \frac{a \times b}{a \cdot (b \times c)}.$$

b. Using standard properties of the vector cross product, verify that $MM^{-1} = I$ based on the above equations.

9.13 a. Show that the following

$$s(x) + s(-x) \leftrightarrow 2\text{Rm}[S(f)]$$

is a Fourier transform pair, where $x = (x,y,z)$ and $f = (f_x,f_y,f_z)$.

b. Show that the following

$$s(x - \tfrac{1}{2}) + s(-x - \tfrac{1}{2}) \leftrightarrow 2\text{Re}[S(f)\cos(\pi f)]$$

is a Fourier transform pair where $\tfrac{1}{2} = (\tfrac{1}{2},\tfrac{1}{2},\tfrac{1}{2})$. Comment about the case $s(x) = s(-x)$.

c. Explain the distinction between $s(-x+a)$ and $s(-x-a)$, where $a = (a,b,c)$.

d. Find the Fourier transforms of the two expressions $s(x-a)+s(-x-a)$ and $s(x-a)+s(-x+a)$. Which of these is the transform of $2\text{Re}[S(f)]$?

9.14 Derive the identity

$$e^{-ja\sin t} = \sum_{n=-\infty}^{\infty} J_n(a) e^{-jnt}.$$

9.15 The ideal finite helix is helx(x, y, z)rect(z/L). Find its three-dimensional Fourier transform.

9.16 Consider the ideal infinite helix with $b = \beta/2$ and $a = \beta$. Sketch the scattering function $A(\phi, \psi)$ for the helix illuminated along the longitudinal axis. Sketch the scattering function for the helix illuminated normal to the longitudinal axis. How does the scattering function change when the infinite helix is truncated to a helix of length 100β?

9.17 For fixed x and y, expand helx(x, y, z) in a Fourier series in the form

$$\text{helx}(x, y, z) = \sum_{n=-\infty}^{\infty} s_n(x, y) e^{jnz}.$$

Compute the Fourier transform of helx(x, y, z) starting from this expansion. What is $s_n(x, y)$ for each n and what is its two-dimensional Fourier transform?

9.18 a. Prove the three-dimensional Fourier transform pair

$$\text{sinc}(z) \iff \delta(f_x, f_y)\text{rect}(f_z),$$

where $\delta(f_x, f_y)$ is the two-dimensional delta function.

b. Show that

$$\delta(x, y)\text{rect}(z) *** \text{helx}(x, y, z) = \text{ring}(x, y, z),$$

where ring(x, y, z) is the infinite cylindrical surface ring(x, y) for all z.

c. Use the convolution theorem to find the Fourier transform of cyln(x, y, z).

9.19 How does the Fourier transform of helx(x, y, z) change when the helix is rotated around the z axis by $90°$? How does the Fourier transform of helx(x, y, z) change when the helix is translated along the z axis by $\pi/4$?

9.20 The discontinuous helix, denoted dhlx(x, y, z), is a set of discrete points uniformly spaced along the helix. It can be defined as

$$\text{dhlx}(x, y, z) = \text{helx}(x, y, z)\text{comb}\left(\frac{z}{a}\right),$$

where a is a parameter that determines the spacing. Find the Fourier transform of the discontinuous helix. What happens when $a = 1$?

9.21 Let $S(f_x, f_y, f_z)$ be the three-dimensional Fourier transform of the real nonnegative function $s(x, y, z)$. The Nyquist samples of $S(f_x, f_y, f_z)$ are denoted $S(\ell_x, \ell_y, \ell_z)$, where $(\ell_x, \ell_y, \ell_z) \in \mathbb{Z}^3$.

a. Prove that the Nyquist sample $S(0, 0, 0)$ is real and positive.
b. Prove that the origin of image space can be chosen so that the six samples $S(\pm 1, 0, 0)$, $S(0, \pm 1, 0)$, and $S(0, 0, \pm 1)$ are all real.
c. Can the origin be chosen so that the eight points $S(\pm 1, \pm 1, \pm 1)$ are all real?
d. Under what conditions on $s(x, y, z)$ are all Nyquist samples $S(\ell_x, \ell_y, \ell_z)$ real?

9.22 Let M be a full-rank three by three real matrix

a. Write out the equation for the Ewald data ellipsoid corresponding to the three-dimensional lattice $M\mathbb{Z}^3$.
b. Use the coordinate transformation properties of the three-dimensional Fourier transform to describe how the Fourier transform of the sphere function changes when the sphere function is transformed into an ellipsoid function by the matrix M.

9.23 (**Sayre formula**) The Sayre formula can be concisely written as

$$R(f) = \frac{1}{B(f)} R(f) * R(f),$$

$$B(f) = \frac{1}{A(f)} A(f) * A(f).$$

Suppose that $R(f)$ and $A(f)$ can be approximated by the sum of a finite number of impulses

$$R(f) = \sum_k R_k \delta(f - f_k).$$

Show that the Sayre formula reduces to

$$R_k = \frac{1}{B_k} \sum_h R_h R_{k-h}.$$

What is B_k?

9.24 (**Structure Invariants**) Let $R(f_x, f_y, f_z) = |R(f_x, f_y, f_z)| e^{j\phi(f_x, f_y, f_z)}$ be the Fourier transform of $\rho(x, y, z)$. Let $R_0(f_x, f_y, f_z) = |R_0(f_x, f_y, f_z)| e^{j\phi_0(f_x, f_y, f_z)}$ be the Fourier transform of $\rho(x - x_0, y - y_0, z - z_0)$.

a. Prove that the sum

$$\phi_0(f_x, f_y, f_z) + \phi_0(f_x', f_y', f_z') + \phi_0(f_x'', f_y'', f_z'')$$

does not depend on (x_0, y_0, z_0) provided

$$(f_x, f_y, f_z) + (f_x', f_y', f_z') + (f_x'', f_y'', f_z'') = (0, 0, 0).$$

b. Generalize this statement to a sum of N terms.
c. Explain how the structure invariants might be useful in an attempt to recover phase of $R(f_x, f_y, f_z)$.

9.25 Wall paint is colored by microscopic color particles suspended within a transparent liquid that dries to a solid. Suppose that light is scattered (in the Born approximation) from all atoms in the interior of each color particle and that the sum of the volumes of all color particles is constrained. Use the formula for the three-dimensional Fourier transform of a sphere to explain how the diameter of these particles should be chosen to maximize the hiding power of the paint (at wavelength λ). In general terms, how should the diameter be chosen to maximize the hiding power when the scattering is only from the surface of each particle?

Notes

X-ray diffraction was introduced into crystallography by Laue in 1912 to demonstrate the wave properties of X-rays. It was quickly realized that X-rays provide a way to probe the structure of crystallized molecules. The two Braggs, father and son, extensively studied (1929, 1942) the problem of reconstruction of the scattering distribution from the scattered waves. This was followed by the widespread realization that the shape of a molecule could be deduced from the X-ray diffraction pattern. Early methods of image formation described the molecule being imaged as composed of a finite set of discrete points representing atoms. Accordingly, a finite system of equations could be set up to locate those atoms. O'Neill and Walter (1963) recognized that side conditions such as knowing the atomic constituents of a molecule may compensate for the missing phase.

Duane (1925) first introduced the Fourier transform into the field of crystallography. This had immediate consequences. Patterson (1935) realized that the autocorrelation function of the scattering function could be recovered as the inverse Fourier transform of the diffraction intensity. For sparse molecules, the autocorrelation function, provides all pairwise interatomic distances allowing many small molecules to be described. Sayre (1952a, 1952b) made the observation that the Nyquist spacing for a function with the support of the autocorrelation function is half the Nyquist spacing for the original function. However, the side information that the function is an autocorrelation function may partially offset this condition.

Harker and Kasper (1948), Karle and Hauptman (1950), and Sayre (1952a, 1952b) introduced various direct constraints on the phase of the Fourier transform for special cases based on the magnitude of the Fourier transform. Hauptman and Karle (1953) developed direct methods for recovering the scattering function of small molecules based on the structure invariants and a least-squares formulation. Such direct methods use prior knowledge of the number and kinds of atoms comprising the molecule. Bricogne (1984) proposed the use of statistical methods to solve the phase-retrieval problem for large molecules. The classic paper of Cochran, Crick, and Vand (1952) discussed the Fourier transform of the helix as an important signature in studying many complex molecules such as polymers. Watson and Crick (1953) used this observation to great advantage in deducing that the structure of a DNA strand is a double helix. A survey of the methods of diffraction imaging is provided by Millane (1990).

10 The Woodward Ambiguity Function

A two-dimensional radar can be described, within some approximations, as a device for forming a two-dimensional convolution of the reflectivity density function $\rho(\tau, \nu)$, expressed in delay τ and doppler ν coordinates,[1] of an illuminated scene with a two-dimensional function $\chi(\tau, \nu)$ that is determined by the transmitted radar waveform $s(t)$. The two-dimensional function $\chi(\tau, \nu)$, also expressed in τ, ν coordinates, is called an *Woodward ambiguity function*. The two-dimensional output $s(\tau, \nu)$ of the convolution

$$s(\tau, \nu) = \chi(\tau, \nu) ** \rho(\tau, \nu),$$

is a summary depiction of the radar image under the usual approach to radar processing. Thus, when described in x, y coordinates, the ambiguity function $\chi(x, y)$ plays the role of a point-spread function through which the actual scene $\rho(x, y)$ is passed to form the image $s(x, y)$. Accordingly, the waveform $s(t)$ should be designed so as to have an attractive ambiguity function $\chi(\tau, \nu)$ according to the needs of the application.

A radar uses the delay and the doppler of the received echo waveform as a means of obtaining information about a target scene. This requires the use of waveforms that are carefully designed to provide adequate resolution and to avoid ambiguity. The major analytical tool used to design such waveforms is the ambiguity function. The ambiguity function is a two-dimensional function $\chi(\tau, \nu)$ in τ, ν coordinates defined as a functional of the one-dimensional waveform $s(t)$, where τ represents delay and ν represents doppler. Every one-dimensional waveform of energy E_p is associated with a two-dimensional ambiguity function $\chi(\tau, \nu)$ of energy E_p^2. The ambiguity function provides considerable insight into the performance of the waveform $s(t)$.

The Woodward ambiguity function is introduced formally in this chapter as a mathematical object and its mathematical properties are stated and proved. Some examples are given. Chapter 11 studies the performance of imaging radars using the ambiguity function as the primary tool, identifying the coordinates of the ambiguity function with the delay and the doppler of an echo. Chapter 12 studies the performance of search radars using the ambiguity function as the primary tool, using the τ, ν coordinates of the ambiguity function to measure the range and the radial velocity as indicated by the delay and the doppler.

---

[1] The relationship between the τ, ν coordinates and the x, y coordinates is described below.

10.1 Theory of the Ambiguity Function

Every finite energy pulse or waveform $s(t)$, whether real or complex, is associated with a complex function of two variables called the ambiguity function of that pulse. The ambiguity function captures many of the performance properties of a radar or sonar system that uses that pulse. It is used as a means of understanding and evaluating the usefulness of the pulse $s(t)$ for a particular radar or sonar application.

Definition 10.1.1 *The ambiguity function of the finite energy pulse $s(t)$ is a complex function of two variables given by*

$$\chi(\tau,\nu) = \int_{-\infty}^{\infty} s(t+\tau/2)s^*(t-\tau/2)e^{-j2\pi\nu t}dt.$$

The ambiguity function provides a mapping from the set of complex-valued, finite-energy functions of one variable into the set of complex-valued, finite-energy functions of two variables.[2] This mapping is denoted

$$s(t) \to \chi(\tau,\nu).$$

The ambiguity function is a complex function with a real part and an imaginary part. Thus,

$$\chi(\tau,\nu) = \chi_R(\tau,\nu) + j\chi_I(\tau,\nu).$$

The magnitude of the ambiguity function $|\chi(\tau,\nu)|$ is called the *ambiguity surface*.
An alternative formulation of the definition is

$$\chi(\tau,\nu) = \int_{-\infty}^{\infty} \left(s(t+\tau/2)e^{-j\pi\nu t}\right)\left(s(t-\tau/2)e^{j\pi\nu t}\right)^* dt.$$

The alternative form shows that $\chi(\tau,\nu)$ can be thought of as the correlation between the waveform $s(t)$ shifted both in time and in frequency and the same waveform $s(t)$ shifted in both time and in frequency in the opposite directions. The ambiguity function is the response of a matched filter to the pulse $s(t+\tau/2)e^{-j\pi\nu t}$.

An asymmetric form of the ambiguity function, which is slightly different but essentially equivalent to the symmetric form, is given by the definition

$$\chi'(\tau,\nu) = \int_{-\infty}^{\infty} s(t)s^*(t-\tau)e^{-j2\pi\nu t}dt.$$

While the asymmetric form might appear simpler at the onset, the symmetric form is chosen for Definition 10.1.1 because it simplifies the appearance and the proof of

---

[2] Formally, the ambiguity function $\chi(\tau,\nu)$ is a map from one-dimensional signal space $L^2(\mathbb{C})$ into two-dimensional signal space $L^2(\mathbb{C}^2)$. This is denoted $\chi(L^2(\mathbb{C})) \subset L^2(\mathbb{C}^2)$.
One-dimensional or two-dimensional signal space is the set of monovariate or bivariate complex signals of finite energy under the condition that two signals whose difference has zero energy are considered to be equivalent signals and are regarded as the same element of that signal space. The condition of equivalence is a transitive relationship. Two signals that are equivalent to the same signal are equivalent to each other, so the signal space is well-defined.

10.1 Theory of the Ambiguity Function

many later statements. Either form of the ambiguity function can be used without comment depending on its convenience. The two forms are related by a linear phase term. This can be seen by the change in variables, $t' = t + \tau/2$, as follows:

$$\chi(\tau, \nu) = \int_{-\infty}^{\infty} s(t + \tau/2) s^*(t - \tau/2) e^{-j2\pi \nu t} dt$$

$$= e^{j\pi \tau \nu} \int_{-\infty}^{\infty} s(t + \tau/2) s^*(t - \tau/2) e^{-j2\pi \nu (t + \tau/2)} dt$$

$$= e^{j\pi \tau \nu} \int_{-\infty}^{\infty} s(t') s^*(t' - \tau) e^{-j2\pi \nu t'} dt'$$

$$= e^{j\pi \tau \nu} \chi'(\tau, \nu).$$

In particular, notice that

$$|\chi(\tau, \nu)| = |\chi'(\tau, \nu)|.$$

Both forms have the same ambiguity surface.

By setting $\nu = 0$ or $\tau = 0$, the ambiguity function is reduced to other well-known functions. By setting $\nu = 0$ in $\chi(\tau, \nu)$, the ambiguity function is reduced to the autocorrelation function of the pulse $s(t)$. Thus,

$$\chi(\tau, 0) = \phi(\tau).$$

By setting $\tau = 0$ in $\chi(\tau, \nu)$, the ambiguity function is reduced to the Fourier transform of the square of the pulse

$$\chi(0, \nu) = \int_{-\infty}^{\infty} |s(t)|^2 e^{-j2\pi \nu t} dt.$$

By setting both $\tau = 0$ and $\nu = 0$, the value of the ambiguity function is the energy of the pulse

$$\chi(0, 0) = \int_{-\infty}^{\infty} |s(t)|^2 dt$$

$$= E_p.$$

The value of the ambiguity function at the origin is equal to the energy in the pulse.

Theorem 10.1.2 *Suppose that*

$$s(t) \rightarrow \chi(\tau, \nu).$$

Then

(i) $s(t - \Delta) \rightarrow e^{-j2\pi \nu \Delta} \chi(\tau, \nu)$,
(ii) $s(t) e^{j2\pi ft} \rightarrow e^{-j2\pi f\tau} \chi(\tau, \nu)$,
(iii) $s(at) \rightarrow |a|^{-1} \chi(a\tau, \nu/a)$,
(iv) $s(t) e^{j\pi \alpha t^2} \rightarrow \chi(\tau, \nu - \alpha \tau)$.

Proof:
We provide a proof only of item (iv). Let $s'(t) = s(t)e^{j\pi\alpha t^2}$. Then

$$\chi_{s'}(\tau,\nu) = \int_{-\infty}^{\infty} s'(t+\tau/2)s'^*(t-\tau/2)e^{-j2\pi\nu t}dt$$

$$= \int_{-\infty}^{\infty} s(t+\tau/2)s^*(t-\tau/2)e^{j2\pi\alpha\tau t}e^{-j2\pi\nu t}dt$$

$$= \chi_s(\tau, \nu - \alpha\tau),$$

which proves the statement of item (iv). ∎

The statement in the fourth line of the theorem is known as the *quadratic-phase property*.

Theorem 10.1.3 *Let $S(f)$ be the Fourier transform of $s(t)$. The ambiguity function can be written*

$$\chi(\tau,\nu) = \int_{-\infty}^{\infty} S(f+\nu/2)S^*(f-\nu/2)e^{j2\pi f\tau}df.$$

Proof:
Write the ambiguity function in the form

$$\chi(\tau,\nu) = \int_{-\infty}^{\infty} \left(s(t+\tau/2)e^{-j\pi\nu t}\right)\left(s(t-\tau/2)e^{j\pi\nu t}\right)^* dt.$$

The first term has the transform $S(f+\nu/2)e^{j\pi(f+\nu/2)\tau}$. The second term has the transform $S(f-\nu/2)e^{-j\pi(f-\nu/2)\tau}$. Then, by Parseval's formula,

$$\chi(\tau,\nu) = \int_{-\infty}^{\infty} \left(S(f+\nu/2)e^{j\pi(f+\nu/2)\tau}\right)\left(S(f-\nu/2)e^{-j\pi(f-\nu/2)\tau}\right)^* df$$

$$= \int_{-\infty}^{\infty} \left(S(f+\nu/2)e^{j\pi f\tau}\right)\left(S(f-\nu/2)e^{-j\pi f\tau}\right)^* df,$$

as was to proved. ∎

It is immediately apparent from Theorem 10.1.3 that the ambiguity function can also be written

$$\chi(\tau,\nu) = e^{j\pi\nu\tau}\int_{-\infty}^{\infty} S(f+\nu)S^*(f)e^{j2\pi f\tau}df.$$

Corollary 10.1.4 (Duality) *Suppose that $s(t)$ with Fourier transform $S(f)$ has the ambiguity function $\chi(\tau,\nu)$. Then $S(t)$ with Fourier transform $s(-f)$ has the ambiguity function $\chi(-\nu,\tau)$.*

Proof:
The proof follows directly from Theorem 10.1.3. ∎

The ambiguity function $\chi(\tau,\nu)$ is a two-dimensional function. It has a two-dimensional inverse Fourier transform. This is described using the following definition.

Definition 10.1.5 *The Wigner function of the finite-energy pulse $s(t)$ is*

$$\Xi(f,t) = \int_{-\infty}^{\infty} s(t+\tau/2)s^*(t-\tau/2)e^{j2\pi\tau f}\,d\tau.$$

This expression in the definition superficially resembles the expression defining $\chi(\tau,\nu)$, but it is actually quite different because the integration is in the variable τ rather than the variable t.

Theorem 10.1.6 *The two-dimensional inverse Fourier transform of the ambiguity function of the finite-energy pulse $s(t)$ is the Wigner function of that pulse.*

Proof:
The two-dimensional inverse Fourier transform of $\chi(\tau,\nu)$ is

$$\int_{-\infty}^{\infty}\int_{-\infty}^{\infty} \chi(\tau,\nu)e^{j2\pi(\nu t+\tau f)}\,d\nu d\tau = \int_{-\infty}^{\infty}\left[\int_{-\infty}^{\infty}\chi(\tau,\nu)e^{j2\pi\nu t}\,d\nu\right]e^{j2\pi\tau f}\,d\tau$$

$$= \int_{-\infty}^{\infty} s(t+\tau/2)s^*(t-\tau/2)e^{j2\pi\tau f}\,d\tau.$$

The right side is the Wigner function, as was to be proved. ∎

The statement of the theorem is written symbolically as

$$\Xi(t,f) \leftrightarrow \chi(\tau,\nu).$$

Thus, for each finite energy pulse $s(t)$, the Wigner function of $s(t)$ and the Woodward function of $s(t)$ are a two-dimensional Fourier transform pair.

10.2 Ambiguity Functions of Some Simple Pulses

There are a few simple pulses whose ambiguity functions are convenient to compute in closed form. Some of these functions are listed in Table 10.1. The ambiguity function of many other pulses must be computed numerically.

Table 10.1 A table of ambiguity functions

$s(t)$	$\chi(\tau,\nu)$
$\text{rect}(t/T)$	$(T-|\tau|)\text{sinc}[\nu(T-|\tau|)]\text{rect}(\tau/2T)$
$\text{sinc}(t/T)$	$T(1-T|\nu|)\text{sinc}[(1-T|\nu|)(\tau/T)]\text{rect}(T\nu/2)$
$\text{chrp}_\alpha(t)\text{rect}\left(\frac{t}{T}\right)$	$(T-|\tau|)\text{sinc}[(\nu-\alpha\tau)(T-|\tau|)]\text{rect}(\tau/2T)$
$e^{-\pi t^2}$	$\sqrt{\frac{1}{2}}e^{-\pi(\tau^2+\nu^2)/2}$
$e^{-\pi(t/T)^2}$	$\sqrt{\frac{1}{2}}Te^{-\pi((\tau/T)^2+(T\nu)^2)/2}$

The ambiguity function of the simple rectangular pulse $s(t) = \text{rect}(t/T)$ is easy to compute. Because the term $s(t+\tau/2)s^*(t-\tau/2)$ equals one for $|t| \leq (T-|\tau|)/2$, and otherwise equals zero, the product is

$$s(t+\tau/2)s^*(t-\tau/2) = \text{rect}\left(\frac{t}{T-|\tau|}\right),$$

which is zero for $|\tau|$ larger than T. Consequently,

$$\chi(\tau,\nu) = \begin{cases} (T-|\tau|)\text{sinc}(\nu(T-|\tau|)) & \text{for } |\tau| \leq T, \\ 0 & \text{for } |\tau| > T. \end{cases}$$

Notice that for $\nu = 0$, $\chi(\tau,0)$ takes the form of a triangular pulse in τ,

$$\chi(\tau,0) = \begin{cases} T-|\tau| & |\tau| \leq T, \\ 0 & |\tau| > T, \end{cases}$$

and for $\tau = 0$, $\chi(0,\nu)$ takes the form of a sinc pulse in ν,

$$\chi(0,\nu) = T\text{sinc}(\nu T).$$

The zeros of $\chi(\tau,\nu)$ define curves in the τ,ν plane. These are the *zero curves* $\nu(T-|\tau|) = k$ for $k = \pm 1, \pm 2, \ldots$.

The ambiguity function of the gaussian pulse $s(t) = e^{-\pi t^2}$ is derived as follows:

$$\chi(\tau,\nu) = \int_{-\infty}^{\infty} s(t+\tau/2)s^*(t-\tau/2)e^{-j2\pi\nu t}dt$$

$$= \int_{-\infty}^{\infty} e^{-\pi(t+\tau/2)^2} e^{-\pi(t-\tau/2)^2} e^{-j2\pi\nu t}dt$$

$$= e^{-\pi\tau^2/2} \int_{-\infty}^{\infty} e^{-2\pi t^2} e^{-j2\pi\nu t}dt.$$

The integration now has the form of a Fourier transform of a gaussian pulse. Therefore,

$$\chi(\tau,\nu) = e^{-\pi\tau^2/2}\sqrt{\tfrac{1}{2}}e^{-\pi\nu^2/2}$$

$$= \sqrt{\tfrac{1}{2}}e^{-\pi(\tau^2+\nu^2)/2}.$$

Thus, the ambiguity function of a gaussian pulse is a two-dimensional gaussian pulse.

Another important pulse is a chirp pulse. Any pulse whose instantaneous frequency varies linearly across the duration of the pulse is called a *linear frequency-modulated pulse*. Such a pulse has quadratically increasing phase. A linear frequency-modulated pulse whose amplitude is constant is also called a *chirp pulse* or a *quadratic-phase pulse*. The chirp pulse has strong properties that are useful for a variety of purposes. The *finite chirp pulse* is defined in the complex baseband form as

$$s(t) = e^{j\pi\alpha t^2}\text{rect}(t/T),$$

which is nonzero only for $|t| \leq T/2$. The passband form of the finite chirp pulse is

$$\tilde{s}(t) = \cos(2\pi f_0 t + \pi\alpha t^2)\text{rect}(t/T).$$

The "instantaneous frequency" of the chirp pulse is

$$f(t) = f_0 + \alpha t.$$

Usually αT is small compared to f_0.

The ambiguity function of the finite chirp pulse is easily derived by starting with the ambiguity function of the simple rectangular pulse,

$$\chi(\tau,\nu) = (T - |\tau|)\operatorname{sinc}(\nu(T - |\tau|)) \qquad |\tau| \leq T.$$

The quadratic-phase property for ambiguity functions, given in Theorem 10.1.2, says that when

$$s(t) \to \chi(\tau,\nu),$$

then

$$s(t)e^{j\pi\alpha t^2} \to \chi(\tau, \nu - \alpha\tau).$$

Therefore, the ambiguity function for the finite chirp pulse is

$$\chi(\tau,\nu) = \begin{cases} (T - |\tau|)\operatorname{sinc}((\nu - \alpha\tau)(T - |\tau|)) & |\tau| \leq T, \\ 0 & |\tau| > T. \end{cases}$$

The correlation function of the finite chirp pulse is now easily obtained by setting $\nu = 0$,

$$\phi(\tau) = \chi(\tau, 0)$$
$$= (T - |\tau|)\operatorname{sinc}(\alpha\tau(T - |\tau|)) \qquad |\tau| \leq T.$$

10.3 More Properties of the Ambiguity Function

The ambiguity function has a tidy set of interlocking properties. It is unusual for an arbitrary two-dimensional function to satisfy these properties, and so, in the set of all possible two-dimensional finite-energy functions, ambiguity functions are uncommon. Usually, in radar design problems, one wants an ambiguity surface $|\chi(\tau,\nu)|$ or ambiguity function $\chi(\tau,\nu)$ with certain properties, and one must work backwards to find a signal $s(t)$ that has this ambiguity function. However, this tends to be an ill-posed problem. No satisfactory techniques are available for finding the waveform corresponding to a surface, nor is a satisfactory set of practical rules known for determining whether a desired ambiguity surface is, in fact, an ambiguity surface. A waveform, $s(t)$, that gives rise to a desired ambiguity surface need not exist.

A proposed ambiguity function $\chi(\tau,\nu)$ can be tested, in principle, by Corollary 10.1.4 to see whether its inverse Fourier transform has the factored form $s(t + \tau/2)s^*(t - \tau/2)$ required by Property 6 of Section 10.5. For most $\chi(\tau,\nu)$, the inverse Fourier transform can only be performed numerically and is subject to numerical loss of precision, so that factorization test may be unsatisfactory.

The main properties of the ambiguity function are given next. Some of these properties are similar to properties of the Fourier transform.

10 The Woodward Ambiguity Function

Property 1 (Symmetry Property)

$$\chi(\tau, \nu) = \chi^*(-\tau, -\nu).$$

Proof:
The proof follows immediately from the definition.

Property 2 (Maximum Property)
The ambiguity function is real and positive at the origin, and equal to E_p. The largest value of the ambiguity surface is always at the origin,

$$|\chi(\tau, \nu)| \leq \chi(0, 0) = E_p,$$

with strict inequality for all (τ, ν) other than $(0, 0)$.

Proof:
At the origin,

$$\chi(0, 0) = \int_{-\infty}^{\infty} |s(t)|^2 dt = E_p.$$

A straightforward application of the Schwarz inequality gives

$$|\chi(\tau, \nu)|^2 = \left| \int_{-\infty}^{\infty} \left(s(t + \tau/2)e^{-j\pi \nu t}\right)\left(s(t - \tau/2)e^{j\pi \nu t}\right)^* dt \right|^2$$

$$\leq \int_{-\infty}^{\infty} \left|s(t + \tau/2)e^{-j\pi \nu t}\right|^2 dt \int_{-\infty}^{\infty} \left|s(t - \tau/2)e^{j\pi \nu t}\right|^2 dt$$

$$= \int_{-\infty}^{\infty} |s(t)|^2 dt \int_{-\infty}^{\infty} |s(t)|^2 dt = E_p^2.$$

The Schwarz inequality is strict unless the two functions in the first line are equal. ∎

Property 3 (Volume Property)

$$\int_{-\infty}^{\infty} \int_{-\infty}^{\infty} |\chi(\tau, \nu)|^2 d\tau d\nu = |\chi(0, 0)|^2 = E_p^2.$$

Proof:
Recall that $\chi(\tau, \nu)$ can be expressed in either of the following two ways:

$$\chi(\tau, \nu) = e^{j\pi \tau \nu} \int_{-\infty}^{\infty} s(t) s^*(t - \tau) e^{-j2\pi \nu t} dt,$$

or

$$\chi(\tau, \nu) = e^{j\pi \tau \nu} \int_{-\infty}^{\infty} S(f + \nu) S^*(f) e^{j2\pi f \tau} df.$$

Therefore, we have the fourfold integral

$$\int_{-\infty}^{\infty}\int_{-\infty}^{\infty} |\chi(\tau,\nu)|^2 d\tau d\nu$$

$$= \int_{-\infty}^{\infty}\int_{-\infty}^{\infty}\int_{-\infty}^{\infty}\int_{-\infty}^{\infty} s(t)s^*(t-\tau)S^*(f+\nu)S(f)e^{-j2\pi(\nu t+f\tau)} dt df d\tau d\nu.$$

Now identify groups of terms as Fourier transforms:

$$\int_{-\infty}^{\infty} s^*(t-\tau)e^{-j2\pi f\tau} d\tau = S^*(f)e^{-j2\pi ft}$$

and

$$\int_{-\infty}^{\infty} S^*(f+\nu)e^{-j2\pi \nu t} d\nu = s^*(t)e^{j2\pi ft}.$$

This gives

$$\int_{-\infty}^{\infty}\int_{-\infty}^{\infty} |\chi(\tau,\nu)|^2 d\tau d\nu = \int_{-\infty}^{\infty}\int_{-\infty}^{\infty} |s(t)|^2 |S(f)|^2 dt df$$
$$= [\chi(0,0)]^2,$$

as was to be proved. ∎

Property 3 gives a strong and important condition necessary for a function to be an ambiguity function. The volume under the surface $|\chi(\tau,\nu)|^2$ is equal to the square of the value of $|\chi(\tau,\nu)|$ at the origin. When one wants the surface $|\chi(\tau,\nu)|$ to have a narrow main lobe centered at the origin, there must be an excess of volume that cannot fit under this narrow main lobe, and so that volume must appear somewhere else in the τ,ν plane. Perhaps, away from the main lobe which is at the origin, $|\chi(\tau,\nu)|$ is small and the excess volume is more or less uniformly distributed across a large region of the τ,ν plane, or perhaps the excess volume is concentrated in one or more large lobes, called *sidelobes*, at places other than the origin. Such extraneous sidelobes are called ambiguities.[3]

10.4 Shape and Resolution Parameters

The shape of the main lobe of the ambiguity surface $|\chi(\tau,\nu)|$ of $s(t)$ near the origin determines, in part, the performance of a system such as a radar that uses the waveform $s(t)$. The curvature of the function $|\chi(\tau,\nu)|$ at the origin in both directions determines the accuracy of estimates of position and velocity. The width of the main lobe in both directions determines the ability of the radar to resolve two closely spaced targets. This section relates the shape of the main lobe of $\chi(\tau,\nu)$ and of $|\chi(\tau,\nu)|$ to the properties of the waveform $s(t)$.

---

[3] In order to study these extraneous sidelobes of pulse trains, Woodward introduced and named the ambiguity function as such in his 1953 book. However, the true role of the ambiguity function is much broader. A better name would have been the "Woodward function," which might be preferred. There is nothing ambiguous about the ambiguity function.

The *shape parameters* are parameters that describe the shape of the main lobe of $\chi(\tau,\nu)$ at the peak. They are the coefficients of a quadratic surface fit to the ambiguity surface $|\chi(\tau,\nu)|$ near the origin as described by the second partial derivative of $|\chi(\tau,\nu)|$. To relate the partial derivatives of the ambiguity surface to the properties of the pulse $s(t)$, it is convenient to begin with the alternative form of the ambiguity function given by

$$\chi'(\tau,\nu) = \int_{-\infty}^{\infty} s(t)s^*(t-\tau)e^{j2\pi\nu t}dt.$$

This alternative form has the same ambiguity surface as $\chi(\tau,\nu)$.

The next theorem uses the following terms:

$$\bar{t} = \frac{1}{E_p}\int_{-\infty}^{\infty} t|s(t)|^2 dt,$$

$$\overline{t^2} = \frac{1}{E_p}\int_{-\infty}^{\infty} t^2|s(t)|^2 dt,$$

$$\overline{tf} = \frac{1}{E_p}\frac{j}{2\pi}\int_{-\infty}^{\infty} ts(t)\dot{s}^*(t)dt,$$

$$\bar{f} = \frac{1}{E_p}\frac{j}{2\pi}\int_{-\infty}^{\infty} s(t)\dot{s}^*(t)dt = \frac{1}{E_p}\int_{-\infty}^{\infty} f|S(f)|^2 df,$$

$$\overline{f^2} = \frac{1}{E_p}\int_{-\infty}^{\infty} f^2|S(f)|^2 df = \frac{1}{E_p}\frac{-1}{4\pi^2}\int_{-\infty}^{\infty} |\dot{s}(t)|^2 dt,$$

which are first defined in Section 2.5.

Theorem 10.4.1 *When first and second derivatives of $\chi'(\tau,\nu)$ exist at the origin, then near the origin*

$$\chi(\tau,\nu) = \chi(0,0)\left(1 + j2\pi\nu\bar{t} - 2\pi^2\nu^2\overline{t^2} + j2\pi\tau\bar{f} - 2\pi^2\tau\overline{f^2} - 4\pi^2\tau\nu\overline{tf}\right)$$

up to the terms of second order in τ and ν.

Proof:
Expand $\chi(\tau,\nu)$ in a Taylor series as follows:

$$\chi(\tau,\nu) = \chi(0,0) + \tau\frac{\partial\chi}{\partial\tau} + \nu\frac{\partial\chi}{\partial\nu} + \frac{1}{2}\tau^2\frac{\partial^2\chi}{\partial\tau^2} + \tau\nu\frac{\partial^2\chi}{\partial\tau\partial\nu} + \frac{1}{2}\nu^2\frac{\partial^2\chi}{\partial\nu^2} + \cdots.$$

This expansion exists whenever the partial derivatives exist. (Notice, however, that the ambiguity function of a square pulse fails to have a first partial derivative with respect to τ.)

Now substitute the Taylor series

$$s(t-\tau) = s(t) - \tau\dot{s}(t) + \tfrac{1}{2}\tau^2\ddot{s}(t) + \cdots,$$

and the series

$$e^{j2\pi\nu t} = 1 + j2\pi\nu t - 2\pi^2\nu^2 t^2 + \cdots,$$

into the definition of $\chi'(\tau,\nu)$. Then, up to terms of second order,

$$\begin{aligned}
\chi'(\tau,\nu) &= \int_{-\infty}^{\infty} s(t)s^*(t-\tau)e^{j2\pi\nu t}dt \\
&\approx \int_{-\infty}^{\infty} \left(|s(t)|^2 - \tau s(t)\dot{s}^*(t) + \frac{\tau^2}{2}s(t)\ddot{s}^*(t)\right)\left(1 + j2\pi\nu t - 2\pi^2\nu^2 t^2\right)dt \\
&= \int_{-\infty}^{\infty} \left(|s(t)|^2 + j2\pi\nu t|s(t)|^2 - 2\pi^2\nu^2 t^2|s(t)|^2 - \tau s(t)\dot{s}^*(t)\right. \\
&\quad \left. - j\tau 2\pi\nu t s(t)\dot{s}^*(t) + \frac{\tau^2}{2}s(t)\ddot{s}^*(t)\right)dt \\
&= \chi(0,0)\left(1 + j2\pi\nu\overline{t} - 2\pi^2\nu^2\overline{t^2} + j\tau 2\pi\overline{f} - 2\pi^2\tau^2\overline{f^2} - 4\pi^2\tau\nu\overline{tf}\right),
\end{aligned}$$

as was to be proved. ∎

Theorem 10.4.2 *When $\chi(\tau,\nu)$ has first and second derivatives at the origin, then near the origin*

$$|\chi(\tau,\nu)| = E_p\left(1 - 2\pi^2(T_G^2\nu^2 + 2T_G B_G\rho\tau\nu + B_G^2\tau^2)\right)$$

up to the terms of second order in τ and ν, where T_G, B_G, and ρ are the Gabor parameters of pulse $s(t)$.

Proof:
The shape of the ambiguity surface $|\chi(\tau,\nu)|$ is closely related to the shape of $\chi(\tau,\nu)$. The shape of the squared ambiguity surface near the origin is given by

$$|\chi(\tau,\nu)|^2 = \chi(0,0)^2 \left|1 + j2\pi\nu\overline{t} - 2\pi^2\nu^2\overline{t^2} + j2\pi\tau\overline{f} - 2\pi^2\tau\overline{f^2} - 4\pi^2\tau\nu\overline{tf}\right|^2.$$

Up to the terms of second order, this becomes

$$|\chi(\tau,\nu)| = E_p\left(1 - 4\pi^2\nu^2(\overline{t^2} - \overline{t}^2) - 4\pi^2\tau^2(\overline{f^2} - \overline{f}^2) - 8\pi^2\tau\nu\text{Re}[\overline{tf} - \overline{t}\overline{f}]\right)^{1/2}.$$

Consequently, because $\sqrt{1+2x^2} = 1 + x^2$ up to the terms of second order, we have

$$\begin{aligned}
|\chi(\tau,\nu)| &= E_p\left(1 - 2\pi^2\nu^2(\overline{t^2} - \overline{t}^2) - 2\pi^2\tau^2(\overline{f^2} - \overline{f}^2) - 4\pi^2\tau\nu\text{Re}[\overline{tf} - \overline{t}\overline{f}]\right) \\
&= E_p\left(1 - 2\pi^2\nu^2 T_G^2 - 2\pi^2\tau^2 B_G^2 - 4\pi^2\tau\nu T_G B_G\rho\right)
\end{aligned}$$

up to terms of second order, where T_G is the Gabor timewidth of the pulse $s(t)$, B_G is the Gabor bandwidth of the pulse $s(t)$, and

$$T_G B_G \rho = \text{Re}[\overline{tf} - \overline{t}\overline{f}]$$

is the Gabor skew parameter of the pulse $s(t)$. The Gabor parameters are external descriptors of the waveform $s(t)$ in the sense that they can be obtained from the ambiguity surface $|\chi(\tau,\nu)|$ without knowledge of the details of $s(t)$.

The reciprocal of the Gabor timewidth of the pulse measures the width of the quadratic approximation to the main lobe of the ambiguity surface in the ν direction.

The reciprocal of the Gabor bandwidth measures the width of the quadradic approximation to the main lobe of the ambiguity surface in the τ direction. The shape of the main lobe can be seen more completely by describing the intersection of the main lobe with a horizontal plane. When the plane is set high enough, it slices through the main lobe just below the peak. The shape of the cut describes the shape of the main lobe near the peak. Specifically, choose a convenient constant, C, and set

$$|\chi(\tau,\nu)| = E_p(1 - 2\pi^2 C).$$

Within the quadratic approximation to $\chi(\tau,\nu)$, this becomes the equation of the following ellipse in τ and ν,

$$\tau^2 B_G^2 + 2\tau\nu T_G B_G \rho + \nu^2 T_G^2 = C.$$

As a function of τ and ν, this is known as the *uncertainty ellipse*. The uncertainly ellipse provides a summary description of the main lobe of the ambiguity function. The constant C has no special importance. Only the shape of the uncertainty ellipse is of interest.

Suppose the pulse $s(t)$ with uncertainty ellipse

$$B_G^2 \tau^2 + T_G^2 \nu^2 = C$$

is given. Then, by the quadratic-phase property, the chirp pulse $s(t)e^{j\pi\alpha t^2}$ has the uncertainty ellipse

$$B_G^2 \tau^2 + T_G^2 (\nu - \alpha\tau)^2 = C,$$

which can be rewritten as

$$(B_G^2 + \alpha^2 T_G^2)\tau^2 - 2\alpha T_G^2 \nu\tau + T_G^2 \nu^2 = C.$$

Consequently, when the pulse $s(t)$ has the Gabor bandwidth B_G, then the pulse $s(t)e^{j\pi\alpha t^2}$ has the Gabor bandwidth $\sqrt{B_G^2 + \alpha^2 T_G^2}$ and has a nonzero skew parameter. The uncertainty ellipses for the pulses $s(t)$ and $s(t)e^{j\pi\alpha t^2}$ are shown in Figure 10.1.

The Gabor bandwidth, the Gabor timewidth, and the Gabor skew parameter ρ describe the shape and width of the main lobe of the ambiguity function near the maximum. These parameters are relevant to statements about errors in estimates of τ or ν. The Gabor parameters can also be used as a measure of resolution, although the curvature at the peak need not give a good description of the width of the main lobe for statements about resolution. Other resolution criteria may measure the width of the main lobe in a way that may be more appropriate to a given task. For example, the ambiguity function $\chi(\tau,\nu)$ has a Woodward resolution in the τ direction given by

$$\Delta\tau = \frac{\int_{-\infty}^{\infty} |\chi(\tau,0)|^2 d\tau}{|\chi(0,0)|^2} = \frac{1}{E_p^2} \int_{-\infty}^{\infty} |\phi(\tau)|^2 d\tau,$$

where $\phi(\tau) = \chi(\tau,0)$.

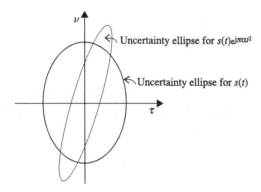

Figure 10.1 Two uncertainty ellipses

10.5 More Properties of the Ambiguity Function

The properties of the ambiguity function that are now given in this section are more technical than those given in Section 10.3. Because these properties, especially Property 7, may be of independent mathematical interest, the ambiguity function $\chi(\tau,\nu)$ is here called the *Woodward function*. As such, $\chi(\tau,\nu)$ might be better defined as the Fourier transform of the Wigner function $\Xi(f,t)$ as given by Theorem 10.1.6. Thus,

$$\Xi(f,t) \Leftrightarrow \chi(\tau,\nu)$$

could be used as an alternative definition of $\chi(\tau,\nu)$. The Wigner function and the Woodward function are a two-dimensional Fourier transform pair.

The volume property of $\chi(\tau,\nu)$ given as Property 3 of Section 10.3 is a special case of the even stronger property given next as Property 4. This generalization of the volume property states that the squared ambiguity surface equals its own two-dimensional Fourier transform, but with a sign reversal.

Property 4 (Ambiguity Surface Squared Property)

$$\int_{-\infty}^{\infty}\int_{-\infty}^{\infty} |\chi(\tau,\nu)|^2 e^{-j2\pi(\xi\tau+\eta\nu)} d\tau d\nu = |\chi(-\eta,\xi)|^2.$$

Proof:
The proof is similar to the proof of Property 3. Again, by substitution of the relationships

$$\chi(\tau,\nu) = e^{j\pi\tau\nu} \int_{-\infty}^{\infty} s(t)s^*(t-\tau)e^{-j2\pi\nu t} dt,$$

$$\chi(\tau,\nu) = e^{j\pi\tau\nu} \int_{-\infty}^{\infty} S(f+\nu)S^*(f)e^{j2\pi f\tau} df,$$

the double integral expands into the four-fold integral

$$\int_{-\infty}^{\infty}\int_{-\infty}^{\infty}|\chi(\tau,\nu)|^2 e^{-j2\pi(\xi\tau+\eta\nu)}d\tau d\nu$$

$$=\int_{-\infty}^{\infty}\int_{-\infty}^{\infty}\int_{-\infty}^{\infty}\int_{-\infty}^{\infty} s(t)s^*(t-\tau)S^*(f+\nu)S(f)e^{-j2\pi(\tau(f+\xi)+\nu(t+\eta))}dt df d\tau d\nu.$$

To evaluate two of these four integrations, isolate the τ integral and the ν integral and recognize each as a Fourier transform. The τ integral, in isolation, is evaluated using the change in variables $\tau' = t - \tau$ to give

$$\int_{-\infty}^{\infty} s^*(t-\tau)e^{-j2\pi\tau(f+\xi)}d\tau = e^{-j2\pi t(f+\xi)} \int_{-\infty}^{\infty} s^*(\tau')e^{+j2\pi\tau'(f+\xi)}d\tau'$$

$$= e^{-j2\pi t(f+\xi)}S^*(f+\xi).$$

The ν integral, in isolation, is evaluated using the change in variables $\nu' = f + \nu$ to give

$$\int_{-\infty}^{\infty} S^*(f+\nu)e^{-j2\pi\nu(t+\eta)}d\nu = e^{+2\pi f(t+\eta)} \int_{-\infty}^{\infty} S^*(\nu')e^{-j2\pi\nu'(t+\eta)}d\nu'$$

$$= e^{+j2\pi f(t+\eta)}s^*(t+\eta).$$

The four-fold integral now becomes

$$\int_{-\infty}^{\infty}\int_{-\infty}^{\infty}|\chi(\tau,\nu)|^2 e^{-j2\pi(\xi\tau+\eta\nu)}d\tau d\nu$$

$$=\int_{-\infty}^{\infty} s(t)s^*(t+\eta)e^{-j2\pi\xi t}dt \int_{-\infty}^{\infty} S(f)S^*(f+\xi)e^{j2\pi f\eta}df$$

$$=\int_{-\infty}^{\infty} s(t)s^*(t-\eta')e^{-j2\pi\xi t}dt \int_{-\infty}^{\infty} S(f)S^*(f+\xi)e^{-j2\pi f\eta'}df,$$

where the second line follows by setting $\eta' = -\eta$. The first integral on the right evaluates to $\chi(\eta',\xi)e^{-j\pi\eta'\xi}$. The second integral on the right evaluates to $\chi^*(\eta',\xi)e^{j\pi\eta'\xi}$. Therefore,

$$\int_{-\infty}^{\infty}\int_{-\infty}^{\infty}|\chi(\tau,\nu)|^2 e^{-j2\pi(\xi\tau+\eta\nu)}d\tau d\nu = |\chi(-\eta,\xi)|^2,$$

as was to be proved. ∎

By replacing η by $-\eta$, Property 4 can be stated in the alternative form

$$\int_{-\infty}^{\infty}\int_{-\infty}^{\infty}|\chi(\tau,\nu)|^2 e^{-j2\pi(\xi\tau-\eta\nu)}d\tau d\nu = |\chi(\eta,\xi)|^2.$$

Property 5 (Width Property)

$$\int_{-\infty}^{\infty}|\chi(\tau,\nu)|^2 d\tau = \int_{-\infty}^{\infty}|\chi(\tau,0)|^2 e^{-j2\pi\nu\tau}d\tau,$$

$$\int_{-\infty}^{\infty}|\chi(\tau,\nu)|^2 d\nu = \int_{-\infty}^{\infty}|\chi(0,\nu)|^2 e^{-j2\pi\nu\tau}d\nu.$$

Proof:
To derive the first expression, let $\xi = 0$ and replace η by τ on the right side of the preceding equation. The second equation is derived in a similar way. ∎

Property 6 (Inverse Property)

$$\int_{-\infty}^{\infty} \chi(\tau,\nu) e^{j2\pi\nu t} d\nu = s(t+\tau/2)s^*(t-\tau/2),$$

$$\int_{-\infty}^{\infty} \chi(\tau,\nu) e^{-j2\pi f\tau} d\tau = S(f+\nu/2)S^*(f-\nu/2).$$

Proof:
The first expression is the inverse Fourier transform of the defining equation for $\chi(\tau,\nu)$ with τ fixed. The second expression is the inverse Fourier transform of the expression in Theorem 10.1.3 with ν fixed. ∎

The first line of Property 6 immediately leads to the expression

$$s(\tau) = \frac{1}{s^*(0)} \int_{-\infty}^{\infty} \chi(\tau,\nu) e^{j\pi\nu\tau} d\nu$$

simply by replacing t by $\tau/2$. The same expression, with τ replaced by t, can be used with any bivariate function $f(\tau,\nu)$ to produce a monovariate function $s(t)$ as

$$s(t) = \frac{1}{s^*(0)} \int_{-\infty}^{\infty} f(t,\nu) e^{j\pi\nu t} d\nu.$$

The Woodward function of this $s(t)$, however, recovers $f(\tau,\nu)$ if and only if $f(\tau,\nu)$ is itself a Woodward function $\chi(\tau,\nu)$.

In principle, $s(t)$ can be recovered from its Woodward function $\chi(\tau,\nu)$ by the integral given above. However, the inverse formula may be unsatisfactory for numerical computation. It can give a poorly behaved numerical computation because small changes in the Woodward function $\chi(\tau,\nu)$ can cause large changes in $s(\tau)$.

Property 6, and its consequences, suggest that Woodward functions are scarce in the signal space of two-dimensional functions because most two-dimensional functions, $f(\tau,t)$, cannot be factored in the form $f(\tau,t) = s(t+\tau/2)s^*(t-\tau/2)$. This motivates Property 7, which is stated next and is proved in Theorem 10.10.1 of Section 10.10.

Property 7 (Lines Property)
The three Woodward functions $\chi_a(\tau,\nu)$, $\chi_b(\tau,\nu)$, and $\chi_c(\tau,\nu)$ of pulses $s_a(t)$, $s_b(t)$, and $s_c(t)$ can satisfy the convex combination

$$\lambda \chi_a(\tau,\nu) + \bar{\lambda} \chi_b(\tau,\nu) = \chi_c(\tau,\nu),$$

if and only if $\chi_b(\tau,\nu)$ is a constant multiple of $\chi_a(\tau,\nu)$.

Property 7 says that for any two Woodward functions $\chi_a(\tau,\nu)$ and $\chi_b(\tau,\nu)$, the convex combination[4] $\lambda\chi_a(\tau,\nu) + \overline{\lambda}\chi_b(\tau,\nu)$ for $0 < \lambda < 1$ is a Woodward function if and only if $\chi_a(\tau,\nu) = A\chi_b(\tau,\nu)$, where A is a complex constant.

10.6 Ambiguity Function of a Pulse Train

A long-duration waveform can be formed by periodically repeating a suitable short-duration waveform. In this context, the elemental waveform is called a *pulse*, although that elemental waveform itself may be complicated. A pulse train, then, is a waveform consisting of a finite or infinite number of nonoverlapping pulses. This section studies pulse trains with N uniformly spaced, identical copies of pulse $s(t)$. Then, with the pulse train beginning at time zero, the pulse train is

$$p(t) = \sum_{n=0}^{N-1} s(t - nT_r),$$

or, with the pulse train centered at time zero, it is

$$p(t) = \sum_{n=0}^{N-1} s\big(t - nT_r + \tfrac{1}{2}(N-1)T_r\big),$$

where T_r is the pulse repetition interval and the pulse $s(t)$ has a width smaller than T_r. The ambiguity function of the pulse train $p(t)$, which is denoted $\chi_p(\tau,\nu)$, is now related to the ambiguity function of the pulse $s(t)$, which is denoted $\chi_s(\tau,\nu)$.

Theorem 10.6.1 *The pulse train*

$$p(t) = \sum_{n=0}^{N-1} s\big(t - nT_r + \tfrac{1}{2}(N-1)T_r\big)$$

has the ambiguity function

$$\chi_p(\tau,\nu) = \sum_{n=-(N-1)}^{N-1} \chi_s(\tau - nT_r, \nu)\mathrm{dirc}_{N-|n|}\nu T_r,$$

where $\chi_s(\tau,\nu)$ is the ambiguity function of the pulse $s(t)$.

Proof:
It is convenient to work with the pulse train in the form

$$p(t) = \sum_{n=0}^{N-1} s(t - nT_r),$$

and later slide the pulse train to the left to center it on the origin. By definition,

---

[4] In any vector space, the *convex combination* of a set of points is a linear combination in which all coefficients are nonnegative and sum to one.

10.6 Ambiguity Function of a Pulse Train

$$\chi_p(\tau,\nu) = \int_{-\infty}^{\infty} \sum_{n=0}^{N-1} s\left(t - nT_r + \tfrac{1}{2}\tau\right) \sum_{m=0}^{N-1} s\left(t - mT_r - \tfrac{1}{2}\tau\right)^* e^{-j2\pi\nu t} dt$$

$$= \sum_{n=0}^{N-1}\sum_{m=0}^{N-1} \int_{-\infty}^{\infty} s\left(t + \tfrac{1}{2}\tau - nT_r\right) s^*\left(t - \tfrac{1}{2}\tau - mT_r\right) e^{-j2\pi\nu t} dt.$$

Replace t by $t + \frac{m+n}{2}T_r$

$$\chi_p(\tau,\nu) = \sum_{n=0}^{N-1}\sum_{m=0}^{N-1} e^{-j2\pi\frac{m+n}{2}\nu T_r} \int_{-\infty}^{\infty} s\left(t + \frac{\tau}{2} + \frac{m-n}{2}T_r\right) s^*$$

$$\left(t - \frac{\tau}{2} - \frac{m-n}{2}T_r\right) e^{-j2\pi\nu t} dt$$

$$= \sum_{n=0}^{N-1}\sum_{m=0}^{N-1} e^{-j2\pi\frac{m+n}{2}\nu T_r} \chi_s\left(\tau + (m-n)T_r, \nu\right).$$

The sum over the N by N array is indexed along rows and columns by n and m. To proceed, the two-index sum is rearranged by indexing along the subdiagonals.

Given any N by N matrix, A, one can sum the N^2 elements by first summing down the main diagonal, then summing all subdiagonals above the main diagonal and all subdiagonals below the main diagonal. This leads to the identity

$$\sum_{n=0}^{N-1}\sum_{m=0}^{N-1} A_{nm} = \sum_{n=0}^{N-1} A_{nn} + \sum_{n=1}^{N-1}\sum_{k=0}^{N-1-n} A_{k(k+n)} + \sum_{n=1}^{N-1}\sum_{k=0}^{N-1-n} A_{(k+n)k}.$$

The first term is a sum down the main diagonal, the second term is the sum of all minor diagonals in the upper triangular matrix, and the third term is the sum of all minor diagonals in the lower triangular matrix. The first two terms can be combined to give

$$\sum_{n=0}^{N-1}\sum_{m=0}^{N-1} A_{nm} = \sum_{n=0}^{N-1}\sum_{k=0}^{N-1-n} A_{k(k+n)} + \sum_{n=1}^{N-1}\sum_{k=0}^{N-1-n} A_{(k+n)k}.$$

Replace n by $-n$ in the second term on the right

$$\sum_{n=0}^{N-1}\sum_{m=0}^{N-1} A_{nm} = \sum_{n=0}^{N-1}\sum_{k=0}^{N-1-n} A_{k(k+n)} + \sum_{n=-1}^{-N+1}\sum_{k=0}^{N-1-n} A_{(k-n)k}.$$

In the case of the theorem, the elements of matrix A are

$$A_{nm} = e^{-j2\pi\frac{m+n}{2}\nu T_r} \chi_s\left(\tau + (m-n)T_r, \nu\right).$$

Then

$$A_{k(k+n)} = e^{-j2\pi\frac{2k+n}{2}\nu T_r} \chi_s(\tau + nT_r, \nu),$$

and

$$A_{(k-n)k} = e^{-j2\pi\frac{2k+|n|}{2}\nu T_r} \chi_s(\tau + nT_r, \nu).$$

Now combine the two sums as follows:

$$\chi_p(\tau,\nu) = \sum_{n=0}^{N-1}\sum_{m=0}^{N-1} A_{nm}$$

$$= \sum_{n=-(N-1)}^{N-1} \sum_{k=0}^{N-1-|n|} e^{-j2\pi \frac{2k+|n|}{2}\nu T_r} \chi_s(\tau+nT_r,\nu)$$

$$= \sum_{n=-(N-1)}^{N-1} e^{-j\pi|n|\nu T_r}\chi_s(\tau+nT_r,\nu) \sum_{k=0}^{N-1-|n|} e^{-j2\pi k\nu T_r}.$$

Finally, the sum on k is executed using the relationship

$$\sum_{k=0}^{N-1} e^{-j2\pi Ak} = e^{-j\pi(N-1)A}\mathrm{dirc}_N A$$

to obtain

$$\chi_p(\tau,\nu) = \sum_{n=-(N-1)}^{N-1} e^{-j\pi|n|\nu T_r}\chi_s(\tau+nT_r,\nu) e^{-j\pi(N-1-|n|)\nu T_r}\mathrm{dirc}_{N-|n|}\nu T_r$$

$$= e^{-j\pi\nu T_r(N-1)} \sum_{n=-(N-1)}^{N-1} \chi_s(\tau-nT_r,\nu)\mathrm{dirc}_{N-|n|}\nu T_r.$$

When the pulse train is centered by redefining the time origin, the phase term drops out, and the proof is complete. ∎

The statement of Theorem 10.6.1 does not require that the pulse $s(t)$ has a simple structure. The pulse may have a complex structure. For example, $s(t)$ may itself be a train of pulses with its own pulse repetition interval. The formula for the pulse train ambiguity function may then be embedded into itself to describe a pulse train of pulse trains.

Theorem 10.6.1 is basic to understanding the performance of any pulse train. Each summand is a product of two terms. For each n from $-(N-1)$ to $+(N-1)$, the first term of each summand in Theorem 10.6.1 is the ambiguity function of a single pulse translated along the τ axis by nT_r, where T_r is the pulse repetition interval. The second term is a dirichlet function due to the pulse train. For fixed n, the dirichlet function has grating lobes at those values of ν that are integer multiples of $1/T_r$.

Let $s(t)$ be a pulse whose ambiguity function $\chi_s(\tau,\nu)$ has a main lobe described by an uncertainty ellipse as shown in Figure 10.2. The uniform pulse train $p(t) = \sum_{n=0}^{N-1} s(t-nT_r+\frac{1}{2}(N-1)T_r)$ has an ambiguity function $\chi_p(\tau,\nu)$ described by Theorem 10.6.1. The theorem asserts that $\chi_s(\tau,\nu)$ will be repeated $2N-1$ times along the τ axis. Each such copy of $\chi_s(\tau,\nu)$ is multiplied by a dirichlet function.

The structure of the ambiguity function of the pulse train $p(t)$ can be portrayed by combining this uncertainty ellipse for each value of the delay offset nT_r with the

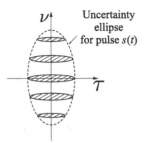

Figure 10.2 The formation of doppler grating lobes

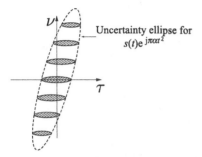

Figure 10.3 The formation of doppler grating lobes – chirped pulse

grating lobes of the dirichlet function. Then, for the term $\chi_p(\tau, \nu)$ one obtains the depiction shown in Figure 10.2.

The main lobe of $\chi_s(\tau, \nu)$ repeats in $\chi_p(\tau, \nu)$ whenever τ is a multiple of T_r, and so T_r is called the *delay ambiguity*. Similarly, the dirichlet function has grating lobes in the ν direction that repeat for ν a multiple of $1/T_r$. The ν separation between grating lobes is called the *doppler ambiguity*. Generally, one wants both the delay ambiguity and the doppler ambiguity to be large. But these goals conflict. Hence a compromise between the delay ambiguity and the doppler ambiguity is always necessary with a uniform pulse train.

For the real pulse $s(t)$, consider the uncertainty ellipse of the chirp pulse $s(t)e^{j\pi\alpha t^2}$ which follows from the quadratic-phase property. This uncertainty ellipse of $\chi_s(\tau, \nu)$, shown in Figure 10.3, has poor resolution along the line of the major axis. The ambiguity function of the pulse train, however, is relatively narrow in both the τ direction and the ν direction. Thus, a pulse train of chirp pulses can be used to give good resolution in both the τ direction and the ν direction even though the pulse is wide in comparison to the width of the ambiguity function in that direction.

10.7 Ambiguity Function of a Costas Pulse

A uniform pulse train has ambiguities, or false peaks, in both the τ direction and the ν direction. Whenever these ambiguities are not acceptable, a waveform must be designed with the periodicity suppressed so that the ambiguities do not occur. There are many ways to suppress the ambiguities while still maintaining the basic structure of a pulse train. One can use an irregular spacing of the pulses

$$p(t) = \sum_{n=0}^{N-1} s(t - \sum_{\ell=1}^{n} T_\ell),$$

where T_ℓ is the ℓth pulse spacing; or one can use an irregular pattern of the phase shifts of the pulses

$$p(t) = \sum_{n=0}^{N-1} s(t - nT_r) e^{-j\theta_n},$$

where θ_n is the phase angle of the nth pulse; or one can use an irregular pattern of the frequency shifts of the pulses

$$p(t) = \sum_{n=0}^{N-1} s(t - nT_r) e^{-j2\pi \Omega_n t},$$

where Ω_n is the frequency shift of the nth pulse.

Each of these approaches can be developed in a variety of ways in hopes of obtaining a satisfactory ambiguity function. One such waveform is studied in this section. This waveform is based on the method of using a regular pattern of pulses modified by an irregular pattern of frequencies. Let

$$p(t) = \sum_{n=0}^{N-1} s\left(t - nT_r + \tfrac{1}{2}(N-1)T_r\right) e^{-j2\pi \Omega_n t},$$

where the nth frequency is

$$\Omega_n = \frac{\theta_n}{T_r},$$

and the θ_n are a permutation of the integers

$$\{\theta_0, \theta_1, \ldots, \theta_{N-1}\} = \{1, 2, 3, \ldots, N\}.$$

The goal is to choose the "firing sequence" $\{\theta_0, \theta_1, \ldots, \theta_{N-1}\}$ so that the ambiguity surface has a sharp peak at the origin.

Costas Pulse
One example of such a waveform is known as a *Costas pulse*. Let $s(t) = \text{rect}(t/T_r)$ and $N = 4$. Then choose $(\theta_0, \theta_1, \theta_2, \theta_3) = (2, 4, 3, 1)$. The frequency sequence is $\Omega_n = \theta_n/T_n$ for $n = 0, \ldots, 3$. The Costas pulse

$$p(t) = \sum_{n=0}^{3} \text{rect}\left((t - nT_r)/t_r\right) e^{-j2\pi \Omega_n t}$$

is described as a "frequency hopping" pattern of four "subpulses."

The firing sequence (2, 4, 3, 1) can be used to describe the following four by four array:

$$A = \begin{bmatrix} 0 & 0 & 0 & 1 \\ 1 & 0 & 0 & 0 \\ 0 & 0 & 1 & 0 \\ 0 & 1 & 0 & 0 \end{bmatrix}.$$

The nth column of the array has a single one in the row corresponding to the nth term of the firing sequence. The array A has the property that any translated copy of A overlaid on A itself has only a single one in common with A.

This means that, for any given pair of integers, (r, s), there can be at most one other pair, (m, n), such that both $n - m = r$ and $\Omega_n - \Omega_m = s$. This requirement motivates the following definition.

Definition 10.7.1 *An N by N Costas array, $A = [A_{ij}]$, is a column permutation of an N by N identity matrix such that*

$$\sum_{i=0}^{N-1-r} \sum_{j=0}^{N-1-s} A_{ij} A_{i+r, j+s} \leq 1 \quad \text{for} \quad (r, s) \neq (0, 0).$$

A larger Costas pulse of blocklength 10 is defined by the firing sequence 2, 4, 8, 5, 10, 9, 7, 3, 6, 1. This Costas pulse has ten subpulses. A much larger Costas pulse – perhaps with hundreds of subpulses – can be found that has an ambiguity function with a sharp and a large central peak at the origin and a small irregular background elsewhere on the τ, ν plane. Such a Costas pulse can be described by a large Costas array.

Central Peak

The ambiguity function of a Costas pulse, denoted $\chi_p(\tau, \nu)$, can be written with the summations and the integration interchanged as follows:

$$\chi_p(\tau, \nu) = \sum_{n=0}^{N-1} \sum_{m=0}^{N-1} \int_{-\infty}^{\infty} s\left(t + \frac{\tau}{2} - nT_r\right) e^{-j2\pi \Omega_n (t + \tau/2)}$$

$$s^*\left(t - \frac{\tau}{2} - mT_r\right) e^{j2\pi \Omega_m (t - \tau/2)} e^{-j2\pi \nu t} dt,$$

here written with the first subpulse centered at the origin.

Replace t with $t + \frac{m+n}{2} T_r$ and identify $\chi_s(\tau, \nu)$ as the ambiguity function of the subpulse $s(t)$. The same method used in the proof of Theorem 10.6.1 leads to the expression

$$\chi_p(\tau, \nu) = \sum_{n=0}^{N-1} \sum_{m=0}^{N-1} e^{-j2\pi \frac{m+n}{2} (\nu T_r + \Omega_n - \Omega_m)} e^{-j2\pi (\Omega_n + \Omega_m) \frac{\tau}{2}}$$

$$\times \chi_s(\tau + (m - n)T_r, \nu + \Omega_n - \Omega_m).$$

The terms with $m = n$ form the central peak. These terms are the "central" terms. The terms with $m \neq n$ form the background. These terms are the "self-noise" terms. The two kinds of terms will be considered separately, denoting their subsums as $\chi^{(1)}(\tau, \nu)$ and $\chi^{(2)}(\tau, \nu)$. Then

$$\chi_p(\tau, \nu) = \chi_p^{(1)}(\tau, \nu) + \chi_p^{(2)}(\tau, \nu).$$

The first term provides the desired impulse-like central peak. The second term provides a low-level background, or pedestal, that is needed to satisfy the volume property of the ambiguity function.

Because $m = n$ and $\Omega_m = \Omega_n$ for the central term, the argument of the pulse ambiguity function $\chi_s(\tau, \nu)$ is the same in all terms of the sum, so the expression for $\chi_p^{(1)}(\tau, \nu)$ becomes

$$\chi_p^{(1)}(\tau, \nu) = \chi_s(\tau, \nu) \left[\sum_{n=0}^{N-1} e^{-j2\pi n \nu T_r} e^{-j2\pi \Omega_n \tau} \right] e^{j\pi(N-1)\nu T_r},$$

where now the extra phase term is included on the right to center the Costas pulse on the origin.

The central term $\chi_s^{(1)}(\tau, \nu)$ cannot be readily evaluated for all τ and ν. Therefore, it is evaluated only along the τ axis and the ν axis. To evaluate $\chi_p^{(1)}(\tau, \nu)$ along the ν axis, set $\tau = 0$, leading to

$$\chi_p^{(1)}(0, \nu) = \left[\chi_s(0, \nu) \sum_{n=0}^{N-1} e^{-j2\pi n \nu T_r} \right] e^{j\pi(N-1)\nu T_r}$$
$$= \chi_s(0, \nu) \mathrm{dirc}_N \nu T_r$$
$$= T \frac{\sin \pi \nu T}{\pi \nu T_r} \frac{\sin N \pi \nu T_r}{\sin \pi \nu T_r}.$$

For $T_r = T$, this becomes

$$\chi_p^{(1)}(0, \nu) = NT \mathrm{sinc}\, \nu NT,$$

where T is the duration of a subpulse and NT is the duration of the entire Costas pulse. This is the same as the ambiguity function of a unit amplitude pulse of duration NT.

To evaluate the function $\chi_p^{(1)}(0, \nu)$ along the τ axis, set $\nu = 0$. Then

$$\chi_p^{(1)}(\tau, 0) = \left[\chi_s(\tau, 0) \sum_{n=0}^{N-1} e^{-j2\pi \Omega_n} \right] e^{j\pi(N-1)\tau/T}$$
$$= \left[\chi_s(\tau, 0) \sum_{n=0}^{N-1} e^{-j2\pi n\tau/T} \right] e^{j\pi(N-1)\tau/T}$$

because the set of Ω_n is a permutation of the set of n/T:

$$\chi_p^{(1)}(\tau, 0) = (T - |\tau|) \frac{\sin N \pi \tau / T}{\sin \pi \tau / T} \qquad |\tau| \leq T$$
$$\approx (T - |\tau|) \mathrm{dirc}(\tau N/T) \qquad |\tau| \leq T.$$

In contrast, the ambiguity function of a single pulse of duration T is $(T - |\tau|)$. The main lobe in the τ direction is due to the main lobe of the dirichlet function. The first zero is at $\tau = T/N$, which can be used as the half-width of $\chi_p^{(1)}(\tau, 0)$. Approximating the dirichlet function as the sinc function near the origin, the factor $\text{sinc}(\tau N/T)$ is the Fourier transform of a pulse of width T/N.

The first grating lobe of the dirichlet function is at $\tau = T$, which is where the amplitude term satisfies $T - |\tau| = 0$. This means that there are no grating lobes.

Thus, in summary, the central peak of $\chi_p^{(1)}(\tau, 0)$ has height N. The width of the central peak in the ν direction is as if there were a single pulse of duration NT. The width of the central peak in the τ direction is as though there were a single pulse of duration T/N.

Background Pedestal

The large central spike sits on a background "pedestal" of small amplitude due to the "self-noise" term given by

$$\chi^{(2)}(\tau, \nu) = \sum_{n=0}^{N-1} \sum_{m \neq n} e^{-j2\pi \frac{m+n}{2}(\nu T_r + \Omega_n - \Omega_m)} e^{-j2\pi(\Omega_n + \Omega_m)\frac{\tau}{2}}$$

$$\times \chi_s(\tau + (m-n)T_r, \nu + \Omega_n - \Omega_m).$$

The inequality $|\sum_i z_i| \leq \sum_i |z_i|$ for any set of complex numbers ensures that this term satisfies

$$|\chi^{(2)}(\tau, \nu)| \leq \sum_{n=0}^{N-1} \sum_{m \neq n} |\chi_s(\tau + (m-n)T, \nu - (\Omega_n - \Omega_m))|.$$

The m, n term of the summation has its peak at τ of the form $(n-m)T$ and at ν of the form $(\Omega_n - \Omega_m)$. For a Costas pulse, two such peaks never occur at the same place. Were only the peaks of $|\chi_s(\tau, \nu)|$ present, the background pedestal would satisfy a maximum value of one. However, each $\chi_s(\tau, \nu)$ does have sidelobes which contribute, so only an approximation of the form $|\chi^{(2)}(\tau, \nu)| \lesssim 1$ can be asserted without a deeper analysis.

Welch–Costas Array

General rules for constructing or classifying all possible Costas arrays are not known, but constructions are known for special cases. A simple construction is available when $N+1$ is a prime. The construction uses modulo-p arithmetic and an integer π between zero and p that has the properties that $\pi^{p-1} = 1$ modulo p and $\pi^n \neq 1$ modulo p for $n < p$. The integer π is then said to have the multiplicative order $p - 1$ modulo p. Such an element π of multiplicative order $p - 1$ modulo p is called a *primitive element* of modulo-p arithmetic. When p is a prime, such a primitive element always exists. For example, 7 is a prime, so there must be a primitive element for modulo-seven arithmetic. One way of finding it is by trial and error. First, try $\pi = 2$ to see whether 2 is primitive. Thus, $\pi = 2$, $\pi^2 = 4$, and $\pi^3 = 1$. Because $\pi^4 = \pi$, succeeding powers of π will only repeat this cycle, so $\pi = 2$ is not primitive. Next, try $\pi = 3$ to see whether 3 is primitive. Thus, modulo seven, $\pi = 3$, $\pi^2 = 2$, $\pi^3 = 6$, $\pi^4 = 4$, $\pi^5 = 5$, and $\pi^6 = 1$. Therefore, 3 is a primitive element.

Definition 10.7.2 *Let p be a prime and let $N = p - 1$. An N by N Welch–Costas array is the N by N array*

$$A_{ij} = \begin{cases} 1 & j = \pi^i, \\ 0 & j \neq \pi^i, \end{cases}$$

for $i = 1, \ldots, N$, where π is a primitive element of the integer modulo p arithmetic system. ∎

For example, with $p = 7$ and primitive element $\pi = 3$, the firing sequence is 3, 2, 6, 4, 5, 1.

Theorem 10.7.3 *A Welch–Costas array is a Costas array.*

Proof:
For any fixed r and s, the theorem fails only when distinct integer pairs (i, j) and (i', j') exist such that

$$j = \pi^i \quad j' = \pi^{i'},$$
$$j + s = \pi^{i+r} \quad j' + s = \pi^{i'+r},$$

where, without loss of generality, j' can be required to be less than j. Multiply the upper two equations by π^r, and then eliminate π^{i+r} from each pair of equations to obtain

$$j + s = j\pi^r \quad j' + s = j'\pi^r.$$

Subtracting these two equations gives

$$j - j' = (j - j')\pi^r.$$

Then, either $j - j' = 0$ or $\pi^r = 1$. But any $p - 1$ successive powers of π are distinct and $r < p - 1$, so the second condition cannot be satisfied. Consequently, $j = j'$ and the theorem is proved. ∎

10.8 The Cross-Ambiguity Function

The ambiguity function $\chi(\tau, \nu)$ measures the similarity of the pulse $s(t)$ to time-delayed and frequency-shifted versions of itself. To measure the similarity of pulse $s_1(t)$ to time-delayed and frequency-shifted versions of a *different* pulse $s_2(t)$, the cross-ambiguity function is defined and studied in this section.

Definition 10.8.1 *Let $s_1(t)$ and $s_2(t)$ be two finite energy pulses. The two-dimensional function*

10.8 The Cross-Ambiguity Function

$$\chi_{12}(\tau,\nu) = \int_{-\infty}^{\infty} s_1(t+\tau/2)s_2^*(t-\tau/2)e^{-j2\pi\nu t}dt$$

is called the cross-ambiguity function of $s_1(t)$ with $s_2(t)$.

A slightly different version of the cross-ambiguity function is defined in the asymmetric form

$$\chi'_{12}(\tau,\nu) = \int_{-\infty}^{\infty} s_1(t)s_2^*(t-\tau)e^{-j2\pi\nu t}dt.$$

The asymmetric form of the cross-ambiguity function has a mathematical structure similar to another concept. This is the modification of the Fourier transform given by

$$S_\tau(f) = \int_{-\infty}^{\infty} s(t)g(t-\tau)e^{-j2\pi f t}dt,$$

which, as such, goes by the name of the *short-time Fourier transform*. Although the cross-ambiguity function and the short-time Fourier transform have the same mathematical form, the motivation in the two cases is different and the theory is developed in different directions. The cross-ambiguity function is thought of as "comparing" $s_1(t)$ with $s_2(t)$. The short-time Fourier transform is thought of as weighting $s(t)$ by the sliding "window" $g(t)$ and forming the Fourier transform for each such windowed signal.

Theorem 10.8.2 *The cross-ambiguity function can be written as*

$$\chi_{12}(\tau,\nu) = \int_{-\infty}^{\infty} S_1(f+\nu/2)S_2^*(f-\nu/2)e^{j2\pi f\tau}df.$$

Proof:
This proof is similar to the proof of Theorem 10.1.3. ∎

The cross-ambiguity function $\chi_c(\tau,\nu)$ has other properties that parallel the properties of the ambiguity function. They are proved in the same way.

Property 1 (Antisymmetry)

$$\chi_{12}(\tau,\nu) = \chi_{21}^*(-\tau,-\nu).$$

Property 2 (Origin)

$$\chi_{12}(0,0) = \int_{-\infty}^{\infty} s_1(t)s_2^*(t)dt \le \sqrt{E_{p_1}E_{p_2}},$$

where E_{p_1} and E_{p_2} denote the energy of $s_1(t)$ and $s_2(t)$, respectively.

Property 3

$$\int_{-\infty}^{\infty}\int_{-\infty}^{\infty} |\chi_{12}(\tau,\nu)|^2 d\tau d\nu = E_{p_1}E_{p_2}.$$

Example

An important instance of the cross-ambiguity function occurs when $s_1(t)$ is a time-delayed and frequency-shifted version of $s_2(t)$. Suppose that

$$s_2(t) = s(t),$$
$$s_1(t) = s(t - \tau_0)e^{j2\pi\nu_0 t}.$$

Then

$$\chi_{12}(\tau, \nu) = \int_{-\infty}^{\infty} s_1(t + \tau/2) s_2^*(t - \tau/2) e^{-j2\pi\nu t} dt$$
$$= \int_{-\infty}^{\infty} s(t - \tau_0 + \tau/2) s^*(t - \tau/2) e^{-j2\pi(\nu-\nu_0)t} dt.$$

Let $t = t' + \tau_0/2$, so that

$$\chi_{12}(\tau, \nu) = e^{-j2\pi(\nu-\nu_0)\tau_0/2} \int_{-\infty}^{\infty} s\left(t + \frac{\tau - \tau_0}{2}\right) s^*\left(t - \frac{\tau - \tau_0}{2}\right) e^{-j2\pi(\nu-\nu_0)t} dt$$
$$= e^{-j\pi(\nu-\nu_0)\tau_0} \chi(\tau - \tau_0, \nu - \nu_0).$$

The term $e^{-j\pi(\nu-\nu_0)\tau_0}$ has no effect on the magnitude. Thus, the magnitude is

$$|\chi_{12}(\tau, \nu)| = |\chi_{12}(\tau - \tau_0, \nu - \nu_0)|.$$

Moreover, whenever $(\nu - \nu_0)\tau_0$ is small, the *radar imaging approximation* is

$$\chi_{12}(\tau, \nu) \approx \chi(\tau - \tau_0, \nu - \nu_0),$$

in which the small phase term on $\chi_{12}(\tau, \nu)$ is neglected.

Example

For another example, suppose that

$$s_2(t) = s(t),$$
$$s_1(t) = s(t - \tau_0)e^{j2\pi\nu_0 t} + s(t - \tau_1)e^{j2\pi\nu_1 t}.$$

Then

$$\chi_{12}(\tau, \nu) = e^{-j\pi(\nu-\nu_0)\tau_0} \chi(\tau - \tau_0, \nu - \nu_0) + e^{-j\pi(\nu-\nu_1)\tau_1} \chi(\tau - \tau_1, \nu - \nu_1).$$

Whenever both $(\nu - \nu_0)\tau_0$ and $(\nu - \nu_1)\tau_1$ are small, the radar imaging approximation gives

$$\chi_{12}(\tau, \nu) \approx \chi(\tau - \tau_0, \nu - \nu_0) + \chi(\tau - \tau_1, \nu - \nu_1).$$

When the main lobes of the two translates of $\chi(\tau, \nu)$ are resolved, they can be observed individually in $\chi_{12}(\tau, \nu)$ and the two peaks provide estimates of the positions (τ_0, ν_0) and (τ_1, ν_1) of those peaks.

10.9 The Sample Cross-Ambiguity Function

For a complex baseband representation $s(t)$ of the transmitted signal and a complex baseband representation $v(t)$ of the received echo signal, the cross-ambiguity function between the transmitted signal and the received signal has the form

$$\chi_c(\tau, \nu) = \int_{-\infty}^{\infty} v(t) s^*(t - \tau) e^{-j2\pi \nu t} dt.$$

The cross-ambiguity function is useful in many ways and can be described as a matched filter and can be motivated in various other ways. Figure 10.4 shows the underlying nature of the cross-ambiguity function. The incoming radar signal $v(t)$ from a point reflector is a time-delayed and frequency-shifted replica in noise of the reference waveform $s(t)$. The time delay enters because of the total propagation delay from the transmitter of the waveform to the receiver of the reflected waveform. The frequency shift enters because of the total relative motion from the source of the waveform to the receiver of the reflected waveform. The processing task is to match the received signal $v(t)$ to a copy of the transmitted signal $s(t)$ provided as a reference signal. As shown in Figure 10.4, the reference is delayed, frequency shifted, and multiplied with the received signal. When the amplitude, reference time delay, and frequency shift match those of the received signal, the product of the two is real, positive, and integrates to the energy in the received signal (accounting for different scaling).

The cross-ambiguity function can be seen as the output of a matched filter for a noisy delayed and frequency-shifted copy of $s(t)$. The delayed complex baseband pulse $s(t - \tau_0)$ received in additive white noise, where τ_0 is a known constant, is

$$v(t) = s(t - \tau_0) + n(t).$$

The matched filter

$$g(t) = s^*(-t - \tau_0)$$

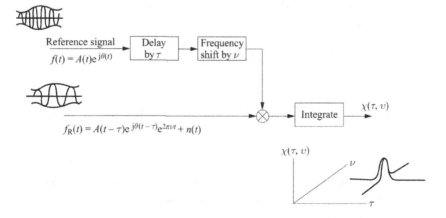

Figure 10.4 Principle of coherent processing

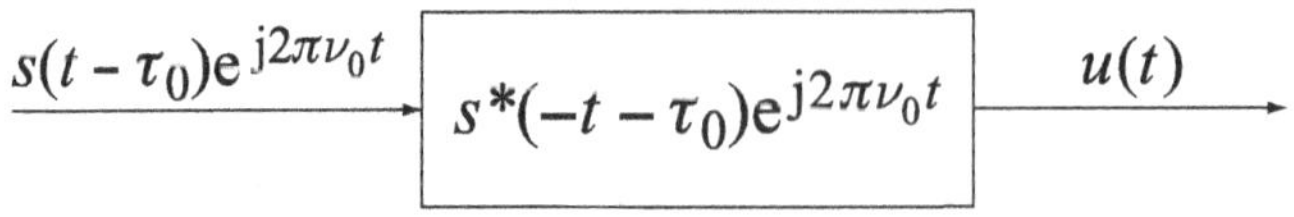

10 The Woodward Ambiguity Function

$$s(t-\tau_0)e^{j2\pi\nu_0 t} \longrightarrow \boxed{s^*(-t-\tau_0)e^{j2\pi\nu_0 t}} \xrightarrow{u(t)}$$

Figure 10.5 Matched filter with time and frequency offset

will maximize the signal-to-noise ratio at time zero. Then

$$u(0) = \int_{-\infty}^{\infty} v(t)s^*(t-\tau_0)dt$$

is the output of the matched filter at time zero.

More generally, let the received signal be the pulse $s(t-\tau_0)e^{j2\pi\nu_0 t}$ with both a known delay, τ_0, and a known frequency offset, ν_0. Then

$$v(t) = s(t-\tau_0)e^{j2\pi\nu_0 t} + n(t)$$

is observed in additive white noise $n(t)$. The matched filter for this case is

$$g(t) = [s(-t-\tau_0)e^{-j2\pi\nu_0 t}]^*$$
$$= s^*(-t-\tau_0)e^{j2\pi\nu_0 t}.$$

The output of the matched filter at time zero is

$$u(0) = \int_{-\infty}^{\infty} v(\xi)g(-\xi)d\xi$$
$$= \int_{-\infty}^{\infty} v(t)s^*(t-\tau_0)e^{-j2\pi\nu_0 t}dt.$$

The matched filter for this case is shown in Figure 10.5. The value $u(0)$ is the value of the cross-ambiguity function at $\tau = \tau_0$ and $\nu = \nu_0$. Thus, the matched filter output can be obtained from the computed cross-ambiguity function. To emphasize that the input $v(t)$ is the measured signal including noise, this instance is called the *sample cross-ambiguity function*. The sample cross-ambiguity function is random because the noise is random.

Now suppose that the values of τ_0 and ν_0 are unknown. Both values are to be estimated from an observation of $v(t)$. The matched filter

$$g(t) = s^*(-t-\tau)e^{-j2\pi\nu t}$$

maximizes the output when $\tau = \tau_0$ and $\nu = \nu_0$. The output of the filter is the value

$$\chi_c(\tau,\nu) = \int_{-\infty}^{\infty} v(t)s^*(t-\tau)e^{-j2\pi\nu t}dt$$

at the location $\tau = \tau_0$ and $\nu = \nu_0$. No linear functional of $v(t)$ can have a larger signal-to-noise ratio at $\tau = \tau_0$ and $\nu = \nu_0$ than the sample cross-ambiguity function. The expected value of $\chi_c(\tau,\nu)$ is

10.9 The Sample Cross-Ambiguity Function

$$E[\chi_{12}(\tau,\nu)] = \int_{-\infty}^{\infty} E[v(t)]s^*(t-\tau)e^{-j2\pi\nu t}dt$$

$$= \int_{-\infty}^{\infty} s(t-\tau_0)s^*(t-\tau)e^{j2\pi\nu_0 t}e^{-j2\pi\nu t}dt$$

$$= e^{-j2\pi(\nu-\nu_0)\tau_0}\chi(\tau-\tau_0, \nu-\nu_0),$$

which has its maximum magnitude at $\tau = \tau_0$ and $\nu = \nu_0$. Consequently, τ_0 and ν_0 can be estimated by computing $\chi_c(\tau,\nu)$ and finding the values of τ and ν at which the maximum occurs.

The sample cross-ambiguity function is linear in the received signal. Suppose that the received signal consists of the sum of two copies of the known pulse $s(t)$ in the form

$$v(t) = \rho_0 s(t-\tau_0)e^{j2\pi\nu_0 t} + \rho_1 s(t-\tau_1)e^{j2\pi\nu_1 t} + n(t),$$

where ρ_0 and ρ_1 are amplitude parameters. In this case, the optimality property of the matched filter can only be regarded as a heuristic estimator because the matched filter was not developed for the sum of two signals. However, it does apply for each signal in the absence of interference from the other. Although there is no statement of optimality, we are free to process the composite signal $v(t)$ using a cross-ambiguity calculation. Compute the sample cross-ambiguity function

$$\chi_c(\tau,\nu) = \int_{-\infty}^{\infty} v(t)s^*(t-\tau)e^{-j2\pi\nu t}dt$$

when it maximizes the signal-to-noise ratio for each pulse individually. The expected value is

$$E[\chi_c(\tau,\nu)] = \rho_0 e^{-j2\pi(\nu-\nu_0)\tau_0}\chi(\tau-\tau_0, \nu-\nu_0) + \rho_1 e^{-j2\pi(\nu-\nu_1)\tau_1}\chi(\tau-\tau_1, \nu-\nu_1).$$

The expected magnitude $|E[\chi_c(\tau,\nu)]|$ has a peak at $(\tau,\nu) = (\tau_0,\nu_0)$ and a peak at $(\tau,\nu) = (\tau_1,\nu_1)$ when the peaks are well-resolved and sidelobes can be neglected. Then each signal appears in noise uncontaminated by the other signal, so the matched filter is effectively optimal for each received pulse.

In the general case, suppose that $v(t)$ arises as a continuum of copies of $s(t)$ in the form

$$v(t) = \int_{-\infty}^{\infty}\int_{-\infty}^{\infty} \rho(\tau',\nu')s(t-\tau')e^{j2\pi\nu' t}d\tau'd\nu' + n(t)$$

for some function, $\rho(\tau',\nu')$, possibly complex. The optimality property that led to the matched filter cannot be stretched far enough to serve in this case. Nevertheless, the matched filter can still be used. The sample cross-ambiguity function

$$\chi_c(\tau,\nu) = \int_{-\infty}^{\infty} v(t)s^*(t-\tau)e^{-j2\pi\nu t}dt$$

has an expected value

$$E[\chi_c(\tau,\nu)] = \int_{-\infty}^{\infty}\int_{-\infty}^{\infty} \rho(\tau',\nu')e^{-j2\pi(\nu-\nu')\tau'}\chi(\tau-\tau', \nu-\nu')d\tau'd\nu'.$$

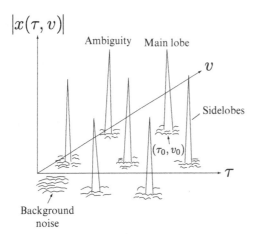

Figure 10.6 Illustrating the sample cross-ambiguity surface

The radar imaging approximation, that $(\nu - \nu')\tau'$ is small, when satisfied says that the exponential is approximately equal to one. This leads to the two-dimensional convolution

$$E[\chi_c(\tau, \nu)] = \chi(\tau, \nu) ** \rho(\tau, \nu)$$

called the *radar imaging equation*. Thus, the result of the cross-ambiguity computation has an easily interpreted form. This convenient interpretation provides another partial justification for using the sample cross-ambiguity function.

The computation of the sample ambiguity function is robust and computationally tractable. Moreover, to compute the sample cross-ambiguity function, a prior model of the statistics of $\rho(\tau, \nu)$ is not needed.

Figure 10.6 shows a sketch of the sample cross-ambiguity function of a uniform pulse train waveform for the case where the reflectivity density $\rho(\tau, \nu)$ is the simple offset impulse, $\delta(\tau - \tau_0, \nu - \nu_0)$ and the received signal is contaminated by additive noise. Only one peak is the main lobe. The other peaks are ambiguities of the pulse train.

10.10 Ambiguity Functions in Signal Space

Complex signal space $L^2(\mathbb{C})$ consists of all complex functions of one variable of finite energy. Two-dimensional complex signal space $L^2(\mathbb{C}^2)$ consists of all complex functions of two variables of finite energy. An ambiguity function is a complex function of two variables of finite energy and so it is an element of two-dimensional complex signal space. The operation of computing the ambiguity function maps elements of one-dimensional complex signal space $L^2(\mathbb{C})$ into two-dimensional complex signal space $L^2(\mathbb{C}^2)$. The image of the ambiguity-function operator is denoted $\chi(L^2(\mathbb{C})) \subset L^2(\mathbb{C}^2)$. The subset $\chi(L^2(\mathbb{C}^2))$ is the set of all two-dimensional functions

10.10 Ambiguity Functions in Signal Space

in complex two-dimensional signal space that are ambiguity functions. Each element of $\chi(L^2(\mathbb{C}))$ is an ambiguity function of an element of $L^2(\mathbb{C}^2)$.

The following theorem states that the convex combination of two ambiguity functions can never be an ambiguity function unless the two are related by a scalar multiplier.[5]

Theorem 10.10.1 *The three ambiguity functions $\chi_a(\tau,\nu)$, $\chi_b(\tau,\nu)$, and $\chi_c(\tau,\nu)$ of pulses $s_a(t)$, $s_b(t)$, and $s_c(t)$ satisfy the convex combination*

$$\lambda \chi_a(\tau,\nu) + \overline{\lambda}\chi_b(\tau,\nu) = \chi_c(\tau,\nu)$$

only when $\chi_b(\tau,\nu)$ is a scalar multiple of $\chi_a(\tau,\nu)$.

Proof:
The proof makes use of the Schwarz inequality in the following square-root form. This form states that for any two real or complex pulses $r(t)$ and $s(t)$ of finite energy,

$$\left| \int_{-\infty}^{\infty} r(t)s^*(t)\,dt \right| \leq \left(\int_{-\infty}^{\infty} |r(t)|^2\,dt \int_{-\infty}^{\infty} |s(t)|^2\,dt \right)^{1/2},$$

with equality if and only if $r(t)$ is a real or complex multiple of $s(t)$. That is, $r(t) = As(t)$ for some complex constant A.

The ambiguity function satisfies $\chi(0,0) = E_p$. Therefore, the energies of the three ambiguity functions asserted in the theorem statement satisfy

$$\lambda E_{p,a} + \overline{\lambda} E_{p,b} = E_{p,c}.$$

The volume property of the ambiguity function states that the energy of the convex combination $\chi_c(\tau,\nu)$ is

$$\int_{-\infty}^{\infty} \int_{-\infty}^{\infty} |\chi_c(\tau,\nu)|^2\,d\tau\,d\nu = |\chi_c(0,0)|^2 = E_{p,c}^2.$$

Evaluate this integral for the convex combination $\lambda \chi_a(\tau,\nu) + \overline{\lambda}\chi_b(\tau,\nu)$ by first opening the square and distributing the integral, and proceeding as follows:

$$E_{p,c}^2 = \int_{-\infty}^{\infty} \int_{-\infty}^{\infty} |\lambda \chi_a(\tau,\nu) + \overline{\lambda}\chi_b(\tau,\nu)|^2\,d\tau\,d\nu$$

$$= \lambda^2 E_{p,a}^2 + 2\lambda\overline{\lambda}\,\mathrm{Re}\left[\int_{-\infty}^{\infty} \int_{-\infty}^{\infty} \chi_a(\tau,\nu)\chi_b^*(\tau,\nu)\,d\tau\,d\nu\right] + \overline{\lambda}^2 E_{p,b}^2$$

$$\leq \lambda^2 E_{p,a}^2 + 2\lambda\overline{\lambda}\left|\int_{-\infty}^{\infty} \int_{-\infty}^{\infty} \chi_a(\tau,\nu)\chi_b^*(\tau,\nu)\,d\tau\,d\nu\right| + \overline{\lambda}^2 E_{p,b}^2$$

$$\leq \lambda^2 E_{p,a}^2 + 2\lambda\overline{\lambda}\left[\int_{-\infty}^{\infty} \int_{-\infty}^{\infty} |\chi_a(\tau,\nu)|^2\,d\tau\,d\nu\right]$$

---

[5] In brief, there are no lines in the image set $\chi(L^2(\mathbb{C}^2))$ except for the "vertical" lines of the form $\{A\chi(\tau,\nu), A \in \mathbb{C}\}$. This property is preserved by the Fourier transform, which means that it holds as well for the set of Wigner functions. The set of Wigner functions $\Xi(L^2(\mathbb{C}^2))$ in $L^2(\mathbb{C}^2)$ has no nontrivial lines.

$$\times \int_{-\infty}^{\infty}\int_{-\infty}^{\infty} |\chi_b(\tau,\nu)|^2 d\tau\, d\nu \Big]^{1/2} + \bar{\lambda}^2 E_{p,b}^2$$
$$= \lambda^2 E_{p,a}^2 + 2\lambda\bar{\lambda}(E_{p,a}^2 E_{p,b}^2)^{1/2} + \bar{\lambda}^2 E_{p,b}^2$$
$$= (\lambda E_{p,a} + \bar{\lambda} E_{p,b})^2,$$

where the second inequality is the Schwarz inequality. But this expression can hold with equality only when the two inequalities are satisfied with equality. This is so only when $\chi_a(\tau,\nu)$ and $\chi_b(\tau,\nu)$ are scalar multiples of each other. In this case, the first inequality is also satisfied with equality. Both inequalities are satisfied with equality for ambiguity functions that are equal except for scaling. This completes the proof of the theorem. ∎

Problems

10.1 Let $s(t)$ be the complex pulse defined as follows:

$$s(t) = \begin{cases} 1-j & -T \le t \le 0, \\ 1+j & 0 \le t \le T, \\ 0 & \text{otherwise.} \end{cases}$$

Compute and sketch $\chi(\tau,\nu)$.

10.2 Let $s(t)$ be a rectangular pulse of width T_1. Let $s'(t)$ be a pulse train consisting of N copies of pulse $s(t)$ with pulse $s(t)$ repeated every T_2 seconds. Let $s''(t)$ be a pulse train consisting of R copies of pulse train $s'(t)$ with pulse train $s'(t)$ repeated every T_3 seconds.

Find $\chi(\tau,\nu)$ for the pulse train of pulse trains $s''(t)$.

10.3 Compute the ambiguity function of the triangular pulse

$$s(t) = \begin{cases} 1-|t| & |t| \le 1, \\ 0 & \text{otherwise.} \end{cases}$$

10.4 Derive expressions for the ambiguity functions of $s(t)\cos\pi\alpha t^2$ and $s(t)\sin\pi\alpha t^2$ in terms of the ambiguity function of $s(t)$.

10.5 The *Barker-coded pulse* of length 7, abbreviated as (+ + + − − + −), is sketched as

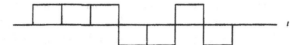

That is, the pulse consists of three counts of +1, followed by two counts of −1, followed by one count of +1, and followed by one count of −1.

a. Compute the autocorrelation function of this pulse. What is the ratio between the main lobe and the largest sidelobe?
b. Compute and display the ambiguity function of this Barker pulse.
c. Repeat for the Barker-coded pulse of length 13:

$$(+, +, +, +, +, -, -, +, +, -, +, -, +).$$

10.6 a. Compute and graph the ambiguity function of the rectangle function rect(t).
b. Compute and graph the ambiguity function of the chirp function rect(t)(e)$^{j\alpha t^2}$.

10.7 The *Golay-coded pulse pair* $(s_1(t), s_2(t))$ (also called *complimentary codes*) of blocklength 4 is denoted $((+ + +-), (+ + --+))$. Each pulse consists of four contiguous and identical square pulses with positive or negative sign as denoted.

This pair of pulses has the property that the sum of their two autocorrelation functions has no sidelobes.

a. Sketch the two pulses.
b. Compute the two autocorrelation functions of these two pulses. Compute and sketch the sum of the two autocorrelation functions.
c. Compute and display the ambiguity functions of these two pulses. Compute and display the sum of the two complex ambiguity functions.
d. Define the pulse

$$s(t) = s_1(t - T/2) + s_0(t + T/2)$$

for T large. Does $s(t)$ have correlation sidelobes? Does the ambiguity function of $s(t)$ violate the volume property?
e. Repeat for the Golay-coded pulse pair of blocklength 20, given by

$$(+ + + + + - + - - + + + - - + + + - + -)$$
$$(+ - + - - - + + - - - + - - + - + + + +).$$

10.8 A uniform pulse train has N pulses with pulse centers separated in time by a multiple of T_r. The nth pulse of the pulse train is stepped in complex frequency by the nth multiple of Ω. That is,

$$P(t) = \sum_{n=0}^{N-1} p(t - nT_r) e^{-j2\pi n\Omega t}.$$

a. Use the quadratic-phase property of ambiguity functions to find the pulse-train ambiguity function $\chi_P(\tau,\nu)$ in terms of the pulse ambiguity function $\chi_p(\tau,\nu)$. How should Ω and T_r be related to make this work out simply?

b. How should $p(t)$ be defined so that $P(t)$ is itself a simple chirp pulse? Verify that $\chi_P(\tau,\nu)$, derived in part a, does reduce to the ambiguity function of a chirp pulse.

10.9 Derive the ambiguity function of the passband pulse

$$s(t) = s_R(t)\cos 2\pi f_0 t - s_I(t)\sin 2\pi f_0 t$$

in terms of the ambiguity functions of the modulation components $s_R(t)$ and $s_I(t)$ and their cross-ambiguity function. How does the ambiguity function of the complex baseband pulse

$$s(t) = s_R(t) + js_I(t)$$

relate to the ambiguity function of the passband pulse?

10.10 The *cross-ambiguity function* of the two pulses $s_1(t)$ and $s_2(t)$ is defined as

$$\chi_{12}(\tau,\nu) = \int_{-\infty}^{\infty} s_1(t+\tau/2) s_2^*(t-\tau/2) e^{-j2\pi\nu t} dt.$$

a. Prove that

$$s_1(t) = \frac{1}{s_2^*(0)} \int_{-\infty}^{\infty} \chi_{12}(t,\nu) e^{j\pi\nu t} d\nu.$$

b. Prove that unless $s_1(t) = cs_2(t)$ for all t and for some constant c, the cross-ambiguity function is not equal to the ambiguity function $\chi(\tau,\nu)$ of any pulse $p(t)$.

10.11 Let

$$s(t) = \begin{cases} e^{-j2\pi\beta t} & 0 \le t \le T, \\ e^{+j2\pi\beta t} & -T \le t < 0, \\ 0 & \text{otherwise}. \end{cases}$$

Calculate the ambiguity function of $s(t)$. Is the ambiguity surface symmetric about the τ and ν axes?

10.12 Prove that the cross-ambiguity function satisfies

$$\int_{-\infty}^{\infty}\int_{-\infty}^{\infty} |\chi_{12}(\tau,\nu)|^2 d\tau d\nu = \int_{-\infty}^{\infty} |s_1(t)|^2 dt \int_{-\infty}^{\infty} |s_2(t)|^2 dt.$$

10.13 a. Show that the ambiguity function of the pulse

$$r(t) = s(t-\Delta)e^{-j(\alpha t+\beta)}$$

satisfies

$$|\chi_r(\tau,\nu)| = |\chi_s(\tau,\nu)|.$$

b. Let $p(t)$ be a pulse that is equal to zero for $|t|$ larger than $T/2$. Define

$$s(t) = p(t) + p(t - 2T),$$
$$r(t) = p(t) - p(t - 2T).$$

Show that $|\chi_r(\tau, \nu)| = |\chi_s(\tau, \nu)|$.

c. Conclude that the set of all pulses corresponding to a given ambiguity surface can be rather large.

10.14 Prove that for any two ambiguity functions $\chi_1(\tau, \nu)$ and $\chi_2(\tau, \nu)$, the sum $a\chi_1(\tau, \nu) + b\chi_2(\tau, \nu)$ is an ambiguity function if and only if $\chi_1(\tau, \nu) = c\chi_2(\tau, \nu)$ for some constant c. (In the language of geometry, the space of ambiguity functions contains no "lines" except those through the origin.)

10.15 a. Prove the stronger form of the uncertainty principle given by

$$T_G^2 B_G^2 (1 - \rho^2) \geq \frac{1}{(4\pi)^2}.$$

b. Prove that the uncertainty ellipse

$$B_G^2 \tau^2 + 2 B_G T_G \rho \tau \nu + T_G^2 \nu^2 = \frac{1}{4\pi^2}$$

has an area not larger than one and is equal to one if and only if the pulse is gaussian.

10.16 Prove that, when the uncertainty ellipse

$$B_G^2 \tau^2 + T_G^2 \nu^2 = \frac{1}{4\pi^2}$$

of pulse $s(t)$ has area A, the uncertainty ellipse of $s(t) e^{j\pi \alpha t^2}$ also has area A.

10.17 Let

$$s(t) = a(t) e^{j\theta(t)},$$

where $a(t)$ is a real, nonnegative function.

a. Show that

$$a(t)^2 = \int_{-\infty}^{\infty} \chi(0, \nu) e^{j 2\pi \nu t} d\nu.$$

b. Show that when $a(t) = 1$,

$$\dot{\theta}(t) = -j \int_{-\infty}^{\infty} \frac{\partial \chi(0, \nu)}{\partial \tau} e^{j 2\pi \nu t} d\nu.$$

10.18 Find a "firing sequence" for a constant amplitude waveform of duration $16T$ based on a 16 by 16 Costas array. Describe the uncertainty ellipse of this waveform. Describe approximately the relationship between the sidelobes and the mainlobe.

10.19 Let $s'(t) = h(t) * s(t)$ for some impulse response $h(t)$. Express $\chi'(\tau, \nu)$ in terms of $\chi(\tau, \nu)$ and $H(f)$.

10.20 Find the Wigner function of the pulse $\text{rect}(t/T)$.

10.21 Show that the pair of conditions

(i) $\int_{-\infty}^{\infty} \int_{-\infty}^{\infty} |s(x,y)|^2 dx dy = C$
(ii) $\int_{-\infty}^{\infty} \int_{-\infty}^{\infty} s(x,y) e^{-j2\pi(x\xi - y\eta)} dx dy = s(\xi, \eta)$

is satisfied by an infinite number of $s(x,y)$. What is the nature of such an $s(x,y)$?

10.22 **a.** Suppose that $s(t)$ has the Fourier transform $S(f)$ and $s'(t)$ has the Fourier transform $S'(f) = S(f) e^{j\pi \beta f^2}$. Prove that when

$$s(t) \to \chi(\tau, \nu),$$

then

$$s'(t) \to \chi(\tau + \beta \nu, \nu).$$

b. Using part (a) and the quadratic phase property of the ambiguity function, prove that when $\chi(\tau, \nu)$ is an ambiguity function, then for any ϕ,

$$\chi'(\tau, \nu) = \chi(\tau \cos \phi - \nu \sin \phi, \tau \sin \phi + \nu \cos \phi),$$

is also an ambiguity function.

10.23 Let $\Xi_s(f, t)$ denote the Wigner function of the pulse $s(t)$. Prove the following:

a. $s(t - t_0) e^{-j2\pi f_0 t}$ has Wigner function $\Xi_s(f - f_0, t - t_0)$;
b. $\sqrt{\gamma} s(\gamma t)$ has Wigner function $\Xi_s(f/\gamma, \gamma t)$;
c. $s(t) e^{j\pi \alpha t^2}$ has Wigner function $\Xi(f - \alpha t, t)$;
d. $s(t) r(t)$ has Wigner function $\Xi_s(f, t) *_f \Xi_r(f, t)$, where $*_f$ is some modified notion of convolution.

10.24 (Apodization) To reduce the doppler sidelobes of a sample ambiguity function, the definition is modified to the alternative form

$$\chi_c(\tau, \nu) = \int_{-\infty}^{\infty} h(t) s(t) v^*(t - \tau) e^{-j2\pi \nu t} dt,$$

where $h(t)$ is a function called an *apodizing window*. Let $s(t) = v(t) = \text{rect}\left(\frac{t}{T}\right)$, and choose the window

$$h(t) = \cos \pi \frac{t}{T} \qquad |t| \leq T/2.$$

a. By introducing $h(t)$, what happens to the first sidelobe?
b. By introducing $h(t)$, what happens to main lobe?
c. By introducing $h(t)$, what happens to the output noise?

10.25 An ambiguity function can be defined for a function of two variables. Let $s(x,y)$ be a bivariate function whose energy $E_p = \int_{-\infty}^{\infty} \int_{-\infty}^{\infty} |s(x,y)|^2 dx dy$ is finite. Define the *spatial ambiguity function*,

$$\chi(\tau_x, \tau_y, \nu_x, \nu_y) = \int_{-\infty}^{\infty} \int_{-\infty}^{\infty} s\left(x - \frac{\tau_x}{2}, y - \frac{\tau_y}{2}\right) s^*\left(x + \frac{\tau_x}{2}, y + \frac{\tau_y}{2}\right) e^{-j2\pi(\nu_x x + \nu_y y)} dx dy.$$

What is the ambiguity function of $s(x,y)e^{j\pi\alpha(x^2+y^2)}$ expressed in terms of the ambiguity function of $s(x,y)$?

10.26 The spatial ambiguity function of the two-dimensional pulse $s(x,y)$ is the function

$$\chi(\tau_x, \tau_y, \nu_x, \nu_y) = \int_{-\infty}^{\infty} \int_{-\infty}^{\infty} s\left(x - \frac{\tau_x}{2}, y - \frac{\tau_y}{2}\right) s^*\left(x + \frac{\tau_x}{2}, y + \frac{\tau_y}{2}\right) e^{-j2\pi(\nu_x x + \nu_y y)} dx dy.$$

a. Is there a quadratic-phase property for the spatial ambiguity function?
b. Suppose that $s(x,y)$ is propagated under Fresnel diffraction. How does the spatial ambiguity function change?
c. Show that the spatial autocorrelation function and the power density spectrum can be recovered from $\chi(\tau_x, \tau_y, \nu_x, \nu_y)$. Describe the "interchange" of these two quantities under Fresnel propagation.
d. Suppose that $s(x,y)$ is passed through an ideal lens. How does the spatial ambiguity function change?
e. How does the spatial ambiguity function change under the cascade of three operations: Fresnel diffraction, followed by an ideal lens, followed by Fresnel diffraction? Does this provide an alternative proof of the lens law?

Notes

The important role that the ambiguity function plays in the design of radar waveforms and radar systems was recognized as early as 1953 by Woodward. The role of complex functions in signal analysis had been recognized earlier by Ville (1948). The basic theorems about the ambiguity function were developed by Siebert (1956, 1958), and also by Lerner (1958), Wilcox (1960), and Price and Hofstetter (1965). Sussman (1962) studied the synthesis of a waveform that has an ambiguity function approximating a desired function in the least-squares sense. Klauder (1960) related the ambiguity function to the Wigner distribution, a function that is important within the subject of quantum mechanics. Papoulis (1974) described the role of the ambiguity function in Fourier optics. The inverse problem of determining whether an arbitrary complex function of two variables is an ambiguity function is difficult, as is the problem of finding a pulse, $s(t)$, that corresponds to a given ambiguity surface. This problem was studied by DeBuda (1970). The deep algebraic properties of the mapping that takes $s(t)$ into its ambiguity function $\chi(\tau, \nu)$ were studied by Auslander and Tolimieri (1985).

The uncertainty principle for waveforms was discussed by Gabor (1946), and by Kay and Silverman (1957). The uncertainty ellipse was named by Helstrom (1960). Additional properties of the ambiguity function were described by Rihaczek (1965) and Grünbaum (1984). The term "chirp" was attributed to Oliver by Klauder, Price, Darlington, and Albersheim (1960). Costas (1984) introduced the Costas array as one way of designing a frequency-hopping pulse with a good ambiguity function. Golomb and Taylor (1984) surveyed the known methods for constructing Costas arrays. Efficient digital computation of the ambiguity function of an arbitrary waveform was studied by Tolimieri and Winograd (1985).

11 Radar Imaging Systems

A conventional radar consists of a transmitter that illuminates a region of interest, a receiver that collects the signal reflected by objects in that region of interest, and a processor that extracts information of interest from the received signal. A radar processor consists of a preprocessor, a detection and estimation function, and a postprocessor. This chapter is concerned with the preprocessor, which is an essentially linear stage of processing at the front end of the processing chain. In the conventional preprocessor, the signal is extracted from the noise, and the entire signal reflected from each resolution cell is gathered into a single statistic. The preprocessor resolution cells could be range cells, doppler cells, or range/doppler cells. The cells might be annotated with additional data such as the direction of arrival of the signal in that cell.

An *imaging radar* uses the output of the preprocessor to form an image of the observed scene. A *search radar* detects targeted objects in a scene and makes estimates regarding those objects such as their position or velocity. The postprocessor refines postdetection data such as by establishing track histories on detected targets or by making other inferences about the detected targets.

The radar preprocessor can often be described as the computation – or a part of the computation – of a sample cross-ambiguity function, though possibly that computation is greatly simplified and not easily recognized as such. The output of this preprocessor can be described in a compact way under several easily satisfied approximations. This output can be described as the two-dimensional convolution of the reflectivity density $\rho(x,y)$ of the radar scene and the ambiguity function $\chi(x,y)$ of the transmitted waveform where the τ and ν coordinates of the ambiguity function are both expressed in terms of x and y coordinates of the scene as $\rho(\tau(x,y), \nu(x,y))$. Thus, the ambiguity function plays the role of a two-dimensional point-spread function through which the scene $\rho(x,y)$ is viewed. This unifying description of the preprocessor output is powerful and it makes many aspects of the radar performance clearly evident. This formulation is most suitable in the usual case in which rectangular coordinates are appropriate to the situation. An alternative description of radar imaging based on the projection-slice theorem, as given in Chapter 6, may be more suitable in the uncommon situation in which polar coordinates are more appropriate.

11.1 The Received Signal

Figure 11.1 shows an elementary block diagram of a radar system. The baseband pulse $s(t)$ is "up-converted" to a center frequency f_0 thereby forming the passband pulse $\tilde{s}(t)$ that is transmitted by the antenna. The signal transmitted at the spherical angle (ϕ, ψ) is given by $E_t(\phi, \psi)\tilde{s}(t)$, where the antenna pattern $E_t(\phi, \psi)$ of the transmitting antenna is a function of the angles ϕ and ψ. The signal reflected from the reflector is proportional to[1] $\rho E_t(\phi, \psi)\tilde{s}(t - R_1/c)$, where the parameter ρ or $\rho(x,y)$ is the *reflectivity* or *reflectivity density* of the reflector or scene. The reflectivity is usually dependent on the incident angle and on the scattered angle of that reflected signal. The echo signal that is then intercepted by the receiving antenna is

$$\tilde{v}(t) = E_r(\phi', \psi')\rho E_t(\phi, \psi)\tilde{s}(t - (R_1 + R_2)/c) + n(t).$$

where $n(t)$ is additive noise and $E_r(\phi', \psi')$ is the antenna pattern of the receiving antenna as a function of the angles ϕ' and ψ', with these angular coordinates distinguished by primes. The received passband echo signal $\tilde{v}(t)$ is "down-converted" to the complex baseband representation $v(t)$. The signal-processing task is to process the received baseband signal $v(t)$ to obtain information about the reflector or reflectors such as position, velocity, size, or even detailed shape.

An electromagnetic wave travels through space at a finite speed. In free space, it travels at the free space speed of light c (approximately 3×10^8 meters per second). An electromagnetic wave consists of a propagating vector field. Because the electric field is a vector, it has an orientation, called the *polarization* of the wave, that in free space is orthogonal to the direction of propagation but is otherwise arbitrary. Because the polarization is another degree of freedom of the radar signal, it may sometimes carry useful information. Other kinds of waves, such as sonar, seismic, and acoustic waves do not have a polarization. For the most part, polarization is considered only occasionally in this book.

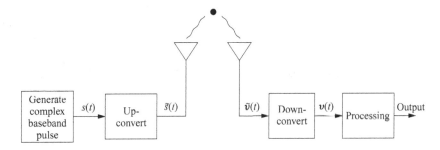

Figure 11.1 Elementary radar block diagram

---

[1] A proportionality constant associated with spherical wave propagation is usually suppressed in a discussion of this kind and is treated separately in a power budget calculation.

11.1 The Received Signal

The propagation of electromagnetic waves through free space and the interaction of these waves with reflecting objects and with antennas is described using Maxwell's equations. These phenomena are studied in the science of electrodynamics. This section adopts the simplified point of view of signal processing, modeling the signal at the antenna input or at the receiver input to include all effects that are relevant to the processing of the signal but ignoring any detail of the electromagnetic phenomena giving rise to these effects that is irrelevant. Thus, for the purposes of this chapter, it suffices to treat an antenna as a linear element joining free space to the transmitter or receiver with an antenna pattern that is a function of the angle of arrival as studied in Chapter 5.

The general nature of the received signal must be partly known. Were the signal completely unknown, a receiver could not be intelligently designed. On the other hand, were the signal completely known, there would be no point in receiving it. The received signal must conform to an appropriate prior parametric model. A common parametric model has the form

$$v(t) = s(t, \gamma) + n(t),$$

where the noise-free complex received signal $s(t, \gamma)$ is dependent on the parameter γ and $n(t)$ is additive random noise usually arising in the receiver electronics. The parameter γ might be rather simple, such as a single real number or a vector of real numbers, or might be as elaborate as a two-dimensional image, or perhaps something even more complicated. For image formation, which is the topic of interest here, the parameter γ usually denotes a two-dimensional function $\rho(x, y)$ of the spatial coordinates x and y describing the reflectivity density of the scene.

Signal Models

The *passband signal* $\tilde{s}(t)$ is transmitted at one point. The passband signal $\tilde{s}'(t)$ is received at a distant point at distance R from the transmitter. The received signal is

$$\tilde{s}'(t) = a\tilde{s}(t - R/c).$$

The *complex passband representation* of the transmitted passband signal $\tilde{s}(t)$ is $s(t)e^{-j2\pi f_0 t}$ where, in this representation, $s(t)$ is a complex signal. The real part operator reduces this complex representation to the real passband signal $\tilde{s}(t)$ by $\tilde{s}(t) = \text{Re}[s(t)e^{-j2\pi f_0 t}]$. The complex passband representation of the signal received at a point at distance R is

$$s'(t)e^{-j2\pi f_0 t} = as(t - R/c)e^{-j2\pi f_0(t - R/c)}.$$

The *complex baseband representation* of the received waveform is formed by removing the carrier $e^{-j2\pi f_0 t}$. Accordingly, the complex baseband representation is

$$s'(t) = as(t - R/c)e^{j2\pi f_0 R/c}.$$

This form of the received signal at complex baseband with both the modulation delay and the carrier delay explicitly displayed is the most convenient form for our purposes.

The received complex baseband signal $s'(t)$ differs from the transmitted complex baseband signal $s(t)$ by virtue of the amplitude attenuation a, the time delay R/c, and the phase shift $2\pi f_0 R/c$. The attenuation a may be a predictable consequence of the antenna gain pattern and the inverse square-law attenuation experienced by the propagating signal, as well as of amplification in the receiver. With the amplitude a suppressed from the expression, the abbreviated received signal is written

$$s'(t) = s(t - R/c)e^{j2\pi f_0 R/c}.$$

In this way, the amplitude is often omitted from the equations in studies of the signal processing because it plays no essential role other than establishing the gain needed in the receiver. The amplitude term can be re-inserted later when desired. This amplitude term is important, however, in Chapter 12 when discussing estimation accuracy and probability of error of a signal in noise.

A radar system makes use of echoes, so the range between the transmitter and the reflector is considered and also the range between the reflector and the receiver. Thus, the transmitted signal, $s(t)$, travels the distance R_1 to a reflector and scatters a fraction ρ in the direction of the receiver. The echo then travels the distance R_2 to the receiver. Then

$$s'(t) = \rho s(t - (R_1 + R_2)/c)e^{j2\pi f_0(R_1+R_2)/c}$$

is the complex baseband signal at that receiver with the attenuation term a suppressed.

The parameter ρ is called the *reflectivity*. The parameter ρ is included explicitly because, in most problems of interest, the reflectivity is an unknown parameter. Observation of the unknown reflectivity ρ is the usual purpose of a radar. In general, the reflectivity ρ from a single point object is a complex number because there can be a phase shift during the process of reflection. The *radar cross section* of the reflector is defined as $\sigma = |\rho|^2$.

Monostatic Reflection

In a monostatic radar system, the transmitter and the receiver are colocated, so $R_1 = R_2 = R$. In a monostatic radar, the time delay is $2R/c$, and the constant ρ represents the fraction of the signal incident on the reflector that is reflected back toward the receiver. The received signal at complex baseband is

$$s'(t) = \rho s(t - 2R/c)e^{j4\pi f_0 R/c},$$

with the attenuation from all other causes suppressed. The signal attenuation is not relevant to the imaging discussion here but is of major importance for studying the signal-to-noise ratio later.

Reflectivity

A physical object has spatial extent, and also has an appearance that depends on the aspect from which it is viewed. Suppose that a sinusoidal signal, such as an electromagnetic wave, is incident on a surface. A portion of this signal is reflected back in the direction of incidence, and a portion is scattered into other directions. Reflection from the real object can be quite complicated. It may be beyond our abilities to completely predict the reflection by even fairly simple objects, although numerical computational methods may give satisfactory results. One can sometimes postulate a parametric model for reflection with parameters that can be filled in by measurement or computation. For our purposes, we consider only ideal reflectors that, though they may be of complex shape, reflect signals instantly with no internal memory. Thus, the possibility of certain artifacts that sometimes occur in a radar return signal, such as those due to multiple reflections or electrical currents induced in the reflector, is ignored. In general, the reflectivity is dependent on carrier frequency f_0. More generally, the reflectivity varies with time because the aspect angle, and therefore the scattering interaction with the reflector varies as the radar or the reflector moves or rotates.

When the spatial details of a reflecting scene are comparable to the resolution of the imaging system, the reflectivity may be replaced by a continuous spatial function $\rho(x,y)$. Let $s(t)\Delta x \Delta y$ be an incident complex signal on a small cell of area $\Delta x \Delta y$ centered at that point of the scene with spatial coordinates (x, y). The reflected complex signal from that cell is an attenuated and phase-shifted copy of the incident signal. The reflected signal in a direction of interest from a small cell centered at (x, y) reflects a portion of the incident signal as described by the factor $\rho(x,y)$ called the *reflectivity density*. The reflectivity density is complex, in general, because the reflection may include a phase shift and is dependent on the aspect angle.

The *radar cross-section density* at the point (x, y) is a real, positive function given by

$$\sigma(x,y) = |\rho(x,y)|^2.$$

The radar cross section σ of the object with reflectivity density $\rho(x,y)$ is the integral of $|\rho(x,y)|^2$ over the object or perhaps the square of the integral of ρ over the object depending on the nature of that object. The reflectivity density $\rho(x,y)$ and the radar cross-section density $\sigma(x,y)$ also depend on the angle of incidence of the incident radiation, as measured by the angular coordinates ϕ, ψ. This dependence is made explicit by writing $\rho(x,y,\phi,\psi)$ and $\sigma(x,y,\phi,\psi)$.

Bistatic Reflection

One may also be interested in applications in which the transmitter and the receiver are located in different places. One reason might be so that a receiver near a radar target can be passive and covert, while the transmitter, though active, can be far away. This is called a *bistatic radar* and is illustrated in Figure 11.2. Then one is interested in the signal from direction (ϕ, ψ) that is reflected into direction (ϕ', ψ'). In this way, the *bistatic* (or bidirectional) *reflectivity density* $\rho(x,y,\phi,\psi,\phi',\psi')$ is also defined and the bistatic radar cross-section density is $\sigma(x,y,\phi,\psi,\phi',\psi')$.

The following usual properties of the bistatic cross section may help in the understanding of the bistatic case:

- For sufficiently smooth and perfectly conducting bodies, in the limit of small wavelength, the bistatic cross section is approximately the monostatic cross section at

11 Radar Imaging Systems

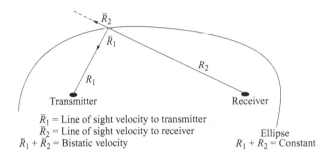

Figure 11.2 Geometry of a bistatic radar

the bisector of the bistatic angle between the direction to the transmitter and the direction to the receiver.
- The bistatic cross section is unchanged under an interchange of the positions of the transmitter and receiver.

Polarization

The reflectivity depends on the polarization of the incident signal. One may consider the vertical and horizontal components of the polarization of the incident signal, and also the vertical and horizontal components of the polarization of the reflected signal. Consequently, there are four reflectivity density functions. These can be arranged in a matrix of functions,

$$\rho = \begin{bmatrix} \rho_{11}(x,y) & \rho_{12}(x,y) \\ \rho_{21}(x,y) & \rho_{22}(x,y) \end{bmatrix}.$$

Similarly, for the radar cross-section density,

$$\sigma = \begin{bmatrix} \sigma_{11}(x,y) & \sigma_{12}(x,y) \\ \sigma_{21}(x,y) & \sigma_{22}(x,y) \end{bmatrix}.$$

This matrix is known as the *scattering matrix*. Each element of the scattering matrix may also be written as a function of ϕ and ψ for a monostatic radar, and as a function of ϕ, ψ, ϕ', and ψ' for a bistatic radar. These may reveal different attributes of the reflecting scene. Usually, only the strongest or most important element of the scattering matrix is observed and the others are ignored.

Delay and Doppler

The location of the transmitter, the reflector, or the receiver may change with time. The distance

$$R(t) = \sqrt{(x'(t) - x(t))^2 + (y'(t) - y(t))^2 + (z'(t) - z(t))^2}$$

from the time-dependent location $(x'(t), y'(t), z'(t))$ at the time t to the time-dependent location $(x(t), y(t), z(t))$ is the distance that the signal $s(t)$ must travel.

11.1 The Received Signal

The complex baseband signal at the reflector is the signal $s(t - R(t)/c)e^{j2\pi f_0 R(t)/c}$, which is delayed in time by $R(t)/c$ and phase-shifted by $2\pi f_0 R(t)/c$. In a monostatic radar system, the transmitter and the receiver are at the same place. Then the monostatic return seen at the receiver is $s'(t) = s(t - 2R(t)/c)e^{j4\pi f_0 R(t)/c}$.

In a bistatic radar system, the transmitter, the reflector, and the receiver are at different points. The transmitter is at location $(x'(t), y'(t), z'(t))$, the echo is from an object at location $(x(t), y(t), z(t))$, and the receiver is at location $(x''(t), y''(t), z''(t))$. This leads to two range expressions:

$$R_1(t) = \sqrt{(x'(t) - x(t))^2 + (y'(t) - y(t))^2 + (z'(t) - z(t))^2},$$

$$R_2(t) = \sqrt{(x''(t) - x(t))^2 + (y''(t) - y(t))^2 + (z''(t) - z(t))^2}.$$

The received complex baseband signal in the bistatic case is

$$s'(t) = s(t - (R_1(t) + R_2(t))/c)e^{j2\pi f_0 (R_1(t) + R_2(t))/c}.$$

The received complex baseband signal in the monostatic case is

$$s'(t) = s(t - 2R(t)/c)e^{j4\pi f_0 R(t)/c}.$$

The monostatic and bistatic cases are both described by the same equation,

$$s'(t) = s(t - \tau(t))e^{j2\pi f_0 \tau(t)},$$

where either $\tau(t) = 2R(t)/c$ or $\tau(t) = (R_1(t) + R_2(t))/c$.

Whenever there is motion in the monostatic system, $R(t)$ is not constant, and the received signal depends on the time-varying delay. Then $R(t)$ may be adequately described by the expression

$$R(t) = R_0 + \dot{R}_0 t + \tfrac{1}{2}\ddot{R}_0 t^2 + \delta R,$$

where δR is a small correction term. Indeed, $R(t) = R_0 + Vt$ may be an adequate approximation, with V for velocity replacing $\dot{R}_0$. In the case of a two-way delay under this approximation, the received complex baseband signal is

$$s'(t) = s(t - 2(R + Vt)/c)e^{j4\pi f_0 (R + Vt)/c}.$$

The term $2f_0 V/c$ in the exponent is called the *doppler shift*, or simply the *doppler*, and the term $2f_0 R/c$ is called the *phase shift*. These are the contributions coming from the passband signal. The modulating pulse $s(t)$ undergoes a range delay and a dilation of the time axis, either compression or expansion, depending on the sign of V. In sonar applications, this dilation of the time axis may be important. In radar applications, the velocity V is small compared to c so that the dilation of $s(t)$ is usually negligible.

By ignoring the dilation term Vt/c within the pulse $s(t)$, the pulse $s'(t)$ can be expressed more compactly as the complex baseband signal

$$s'(t) = s(t - \tau_0)e^{j2\pi\nu_0 t}e^{j2\pi\theta},$$

where τ_0 is a modulation delay, ν_0 is a frequency offset, and θ is a phase offset. The same equation can be used for the bistatic radar case by setting $\tau_0 = (R_1 + R_2)/c$

and $\nu_0 = f_0(\dot{R}_1 + \dot{R}_2)/c$, and for the monostatic radar case by setting $\tau_0 = 2R/c$ and $\nu_0 = 2V/c$.

The dilation of the time axis may appear in the carrier as a frequency shift that is quite noticeable. Typically, in radar systems, V/c is in the range of 10^{-6} to 10^{-7}. Then, for f_0 in the VHF band (above 30×10^6 hertz), the doppler shift may be tens of hertz or more.

In most applications, the elementary narrowband delay-doppler approximation based on $R(t) \approx R_0 + \dot{R}_0 t$ is entirely adequate. In other applications, it is not. Section 11.2 studies the processing of a signal under the delay-doppler approximation. Section 11.3 studies the focusing and motion compensation corrections for the terms $\ddot{R}_0$ and δR that are sometimes needed to better describe $R(t)$.

Specular and Diffuse Reflection

Radar reflectors may have complicated structures and the many structures are different, each to each. A general yet flexible model of the reflectivity density is required. The reflectivity density $\rho(x, y, \theta, \phi)$ depends on the size, shape, and surface properties of the object or scene being illuminated. It depends on the angle of viewing and so changes with time when there is motion in the situation. When the object or scene has a surface or structure that is rough as compared to the wavelength of the illumination, $\rho(x, y, \theta, \phi)$ will change rapidly with movement in the situation. When the object or scene has a surface that is smooth as compared to the wavelength of the illumination, $\rho(x, y, \theta, \phi)$ will change only slowly with movement.

Two models of reflection are used, one deterministic and one probabilistic, by which the received signal can contain information about a scene. These are referred to as *specular reflection* and *diffuse reflection*, respectively. In the case of a specular reflector, $\rho(x, y)$ is modeled as an unknown deterministic function of the spatial coordinates x and y. In the case of a diffuse reflector, $\rho(x, y)$ is modeled as an unknown random function of the spatial coordinates with zero mean and an unknown variance[2] $\sigma^2(x, y)$ that is a function of the spatial coordinates (x, y).

Specular reflection from a planer surface can be described as reflection from a mirror. A wavefront incident on the planar surface at one angle is reflected at a corresponding angle. Reflection from a curved surface is more complicated. Treating the curved surface as a continuum of infinitesimal planes, a plane wave incident on the curved specular surface is reflected at a continuum of angles. A coherent incident wave retains its coherence upon reflection.

Diffuse reflection from a planar surface is scattered into a continuum of directions. The surface is rough and reflections from different minute regions have different phase delays. An incident waveform, whether spatially coherent or spatially noncoherent,

---

[2] The conventional notation for radar cross section is σ. The conventional notation for statistical variance is σ^2. The radar cross section is defined in terms of the reflectivity ρ as $\sigma = |\rho|^2$. The variance of the reflectivity is defined as $\sigma^2 = E[\rho - E[\rho]]^2]$. Both notational conventions are well established in the literature and both conventions will be respected herein, thereby accepting some risk of confusion.

becomes spatially noncoherent upon reflection. Constructive and destructive interference causes a temporally coherent reflection to display a speckle pattern. This speckle pattern changes with slight changes in the viewing angle.

Most of this chapter studies specular reflectors. Diffuse reflectors are studied only in Section 11.8. A unified statement of the two cases can be given by modeling the reflector $\rho(x,y)$ in each resolution cell as a gaussian random variable for which the variance is zero for a specular model and for which the mean is zero for the diffuse model. Reflector models in which the mean and the variance are both nonzero are not considered herein.

Typically, specular reflections are due to objects whose significant details are large in comparison to the wavelength of the illumination, while diffuse reflections are due to objects whose significant details are small in comparison to the wavelength. These are sometimes referred to, more simply, as *reflection* and *backscatter*, respectively, though these terms are not always distinguished and may be used interchangeably.

Signal Processing

For a single point reflector, the received signal is a single pulse in additive gaussian noise. To detect a pulse or waveform, the optimum detector is a matched filter followed by a threshold. When τ and ν are unknown, the matched-filter computation must be executed for all τ and ν. The output of this computation is the sample cross-ambiguity function. The expected image then consists of the sample cross-ambiguity function with the peak output located at the coordinates of that single point.

When there are two point reflectors sufficiently separated in range and doppler and not threatened by interference from major sidelobes, the image is the superposition of two copies of the ambiguity function corresponding to those two points. Some interference between the two superimposed copies of $\chi(\tau,\nu)$ may come from minor sidelobes of the ambiguity surface.

When there are n point reflectors or a continuum of point reflectors, The sample ambiguity function can generalize the above but the theoretical justification is less strong. The sample ambiguity function is computationally tractable and the radar imaging equation, to be defined in the next section, is an attractive formulation. Nevertheless, no claims are made regarding optimality of the sample ambiguity function. Such prior knowledge regarding the objects in the scene is not included in the computation of the sample ambiguity function. Resolution cells are regarded as independent and computed independently, whereas strong reflectors may occupy several adjacent cells. The methods of regularization can help to soften these limitations.

11.2 The Radar Imaging Equation

The ambiguity function, studied in Chapter 10 is used in this chapter as a description of the output of a conventional radar preprocessor in terms of two-dimensional filter theory. Thus, the noise-free value of the sample cross-ambiguity function is,

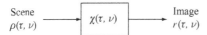

Figure 11.3 A filtering view of imaging

within some close approximations, described as a filtered version of the scene. This noise-free image is related to the reflectivity density $\rho(\tau,\nu)$ in range-doppler coordinates by a two-dimensional convolution of $\rho(\tau,\nu)$ with the ambiguity function $\chi(\tau,\nu)$ of the imaging waveform. This relationship is so useful that the radar imaging equation is given as a definition. Later it is shown, within some close approximations, that the output of a conventional radar preprocessor is described by this definition of the imaging equation.

Definition 11.2.1 *The radar imaging equation for the pulse $s(t)$ is the statement that the radar image $s(\tau,\nu)$ is the two-dimensional convolution*

$$s(\tau,\nu) = \chi(\tau,\nu) ** \rho(\tau,\nu)$$

of the reflectivity density $\rho(\tau,\nu)$ with the ambiguity function $\chi(\tau,\nu)$ of the pulse $s(t)$.

Both τ and ν can be expressed as functions $\tau(x,y)$ and $\nu(x,y)$ of x and y, so the imaging equation can be written instead as a function of the spatial variables x and y, now reusing the notation as $s(x,y)$, in the form

$$s(x,y) = \chi(\tau(x,y),\nu(x,y)) ** \rho(\tau(x,y),\nu(x,y))$$

by the substitutions $\tau = \tau(x,y)$ and $\nu = \nu(x,y)$.

The radar imaging equation is exact because it is a definition. The radar imaging equation is not an exact description of the result of the sample cross-ambiguity computation, but it is usually a close approximation. The radar imaging approximation results in the radar imaging equation. The radar imaging equation defines the radar image as the two-dimensional scene $\rho(\tau,\nu)$ seen through the two-dimensional filter whose point-spread function is $\chi(\tau,\nu)$. This filtering viewpoint is powerful because it submerges all the details of the processing into a simple formula that exposes the underlying limit of resolution. This is illustrated in Figure 11.3. Were $\chi(\tau,\nu)$ an impulse, the radar image $s(\tau,\nu)$ would be equal to $\rho(\tau,\nu)$. But $\chi(\tau,\nu)$ is never an impulse, so the convolution will blur the details of $\rho(\tau,\nu)$ by the main lobe of $\chi(\tau,\nu)$. Significant sidelobes of $\chi(\tau,\nu)$, if any, will lead to weak translated copies of $\rho(\tau,\nu)$ to be found as echos in $s(\tau,\nu)$. Based on this approximating formulation of the preprocessor as a two-dimensional filter of the scene, one can formulate some methods for the manipulation and processing of radar images by using two-dimensional Fourier transform techniques, but these methods are limited by the approximations needed to describe the output by the radar imaging equation.

Image of a Point Reflector

The relationship of the actual scene $\rho(x,y)$ to the output of the radar imaging equation is studied first for a set of point reflectors. The scene is

11.2 The Radar Imaging Equation

$$\rho(x,y) = \sum_i \rho_i \delta(x_i, y_i),$$

where ρ_i is the complex amplitude of the point reflector at coordinate (x_i, y_i). The received echo signal is

$$v(t) = \sum_i \rho_i s(t - \tau_i) e^{j2\pi \nu_i t},$$

where ρ_i is complex. The sample cross-ambiguity function for the set of point reflectors is

$$\chi_c(\tau, \nu) = \int_{-\infty}^{\infty} v(t) s^*(t - \tau) e^{-j2\pi \nu t} dt$$

$$= \sum_i \rho_i e^{-j\pi(\nu - \nu_i)\tau_i} \chi(\tau - \tau_i, \nu - \nu_i).$$

The radar approximation

$$\chi_c(\tau, \nu) = \sum_i \rho_i \chi(\tau - \tau_i, \nu - \nu_i)$$

applies whenever the terms $(\nu - \nu_i)\tau_i$ in the exponent are small.

Even when the terms $(\nu - \nu_i)\tau_i$ are not small, they have little effect on the magnitude of $\chi_c(\tau, \nu)$ whenever the points (τ_i, ν_i) are sparse and widely separated compared to the width of the main lobe of $\chi(\tau, \nu)$. This means that the approximation

$$|\chi_c(\tau, \nu)| \approx \sum_i |\rho_i \chi(\tau - \tau_i, \nu - \nu_i)|$$

$$\approx \sum_i |\rho_i| |\chi(\tau - \tau_i, \nu - \nu_i)|$$

is appropriate. The phase terms are now suppressed.

Imaging a Continuum of Reflectors

When there is a continuum of reflectors, the received signal is

$$v(t) = \int_{-\infty}^{\infty} \int_{-\infty}^{\infty} \rho(\tau', \nu') s(t - \tau') e^{j2\pi \nu' t} d\tau' d\nu'.$$

Then

$$\chi_c(\tau, \nu) = \int_{-\infty}^{\infty} \int_{-\infty}^{\infty} \rho(\tau', \nu') e^{-j\pi(\nu - \nu')\tau'} \chi(\tau - \tau', \nu - \nu') d\tau' d\nu'.$$

In a fine-resolution imaging application, $(\nu - \nu')$ may be a few hertz and τ' may be a few milliseconds. Therefore, $(\nu - \nu')\tau'$ is small for the region where the integral is significant. Then, again, the radar-imaging approximation gives

$$\chi_c(\tau, \nu) \approx \int_{-\infty}^{\infty} \int_{-\infty}^{\infty} \rho(\tau', \nu') \chi(\tau - \tau', \nu - \nu') d\tau' d\nu',$$

which is the imaging equation. Even when the radar imaging approximation fails far away from the origin, it does not spoil any of the general conclusions about the structure of $\chi_c(\tau, \nu)$, which are largely due to the main lobe of $\chi(\tau, \nu)$. However, the approximation may hinder an attempt to recover $\rho(\tau, \nu)$ from $s(\tau, \nu)$ by deconvolution methods.

Ambiguity Function as a Point-Spread Function

The imaging equation views the reflectivity density $\rho(\tau, \nu)$ as a two-dimensional signal that is the input to a two-dimensional filter with point-spread function $\chi(\tau, \nu)$. The image is the two-dimensional signal that is the output of the filter. One wants the point-spread function $\chi(\tau, \nu)$ to be a two-dimensional impulse so that the image, as given by the radar imaging equation, would be a scalar multiple of the reflectivity density function of the scene $\rho(\tau, \nu)$. However, no such ambiguity function with $\chi(\tau, \nu)$ equal to $\delta(\tau, \nu)$ exists. The resolution in the scene is limited by the resolution of the imaging point-spread function $\chi(\tau, \nu)$. The resolution in the range direction is determined by the τ width of the main lobe of $\chi(\tau, \nu)$; the resolution in the doppler direction is determined by the ν width of the main lobe of $\chi(\tau, \nu)$. The major sidelobes and the grating lobes as well as the minor lobes produce various processing artifacts in the image.

The radar imaging equation makes it easy to see the effect of the major sidelobes that arise in the ambiguity function of a pulse train. Such an ambiguity function has grating lobes in the doppler direction and delay ambiguities in the delay direction. To understand the effect of these ambiguities, approximate the grating lobes and the delay ambiguities as impulses. Then, the ambiguity function of a uniform pulse train is replaced by

$$\chi(\tau, \nu) \approx \sum_{i=-I}^{I} \sum_{j=-J}^{J} \delta\left(\tau - iT_r, \nu - j\frac{1}{T_r}\right),$$

which is a two-dimensional array of impulses. The image

$$r(\tau, \nu) = \rho(\tau, \nu) ** \chi(\tau, \nu)$$

then has the form of an array of copies of $\rho(\tau, \nu)$. When the scene $\rho(\tau, \nu)$ is confined to a rectangle, as shown in Figure 11.4, of width smaller than T_r in the τ direction and width smaller than T_r^{-1} in the ν direction, the copies do not overlap. There will be no image aliasing and the ambiguities will not be of major importance. This will require the reflecting elements to be contained within the T_r by T_r^{-1} rectangle in the τ, ν plane. Any reflectors outside of this rectangle will lead to aliasing. The scene could be limited to this rectangle by the beam of the transmit antenna or by the beam of the receive antenna, or it may be that the objects of the scene themselves are so confined.

The sampling interval T_r has to be large enough to encompass the delay spread of the scene without aliasing and the sampling frequency T_r^{-1} has to be large enough to encompass the doppler spread of the scene without aliasing. Making one of these terms larger makes the other smaller, so a pulse-train waveform must compromise in the choice of T_r. This compromise is at the heart of the design of pulse-train waveforms. In some cases, it may not be possible to satisfy simultaneously the constraints on both

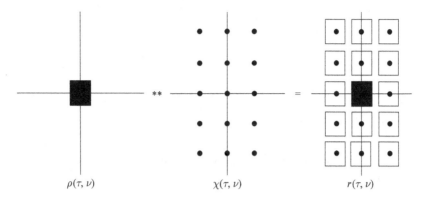

Figure 11.4 Illustrating an image without image aliasing

the delay and doppler. Then one must either abandon the use of a pulse waveform or accept the existence of artifacts in the image caused by aliasing.

11.3 Imaging Resolution

The quality of a radar image is determined in large part by the ambiguity function of the imaging waveform. The ambiguity function determines the imaging resolution permitted by the main lobe, describes false images due to major sidelobes, and predicts self-clutter due to minor sidelobes. The resolution can be understood by considering a point reflector at the center of a scene of interest. Choosing the center of the scene as the origin of the coordinate system, a point reflector $\rho(x,y) = \delta(x,y)$ at the origin of the x,y plane results in a complex image that is equal to the ambiguity function:

$$r(x,y) = \chi(\tau(x,y), \nu(x,y)).$$

Consequently, imaging performance is also determined by the ambiguity function expressed using the functions $\tau(x,y)$ and $\nu(x,y)$. The resolution of the image is determined by the width of the main lobe of $r(x,y)$. The arguments of the ambiguity function are related to x and y by

$$\tau(x,y) = \frac{2}{c}R(x,y),$$

$$\nu(x,y) = 2\frac{f_0}{c}\dot{R}(x,y).$$

To first order, the main lobe has a shape described by the uncertainty ellipse, which can be used as one descriptor of resolution.

Choose any measure of resolution, designating it by $\Delta\tau$ or $\Delta\nu$, respectively, in the τ and ν directions. The resolution $\Delta\tau$ and $\Delta\nu$ is related to the resolution in R or $\dot{R}$ by

$$\Delta\tau = \frac{2}{c}\Delta R,$$

$$\Delta\nu = 2\frac{f_0}{c}\Delta\dot{R}.$$

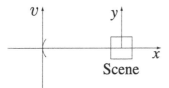

Figure 11.5 Sidelooking geometry

In turn, the right side of these equations can be related to the resolution in the x and y coordinate axes as follows:

$$\Delta\tau = \frac{2}{c}\left[\frac{\partial R}{\partial x}\Delta x + \frac{\partial R}{\partial y}\Delta y\right],$$

$$\Delta\nu = 2\frac{f_0}{c}\left[\frac{\partial \dot{R}}{\partial x}\Delta x + \frac{\partial \dot{R}}{\partial y}\Delta y\right].$$

For example, a common geometrical configuration known as a sidelooking radar is shown in Figure 11.5. The origin of the coordinate system is at the center of the scene. The velocity vector is parallel to the y coordinate axis. The radar antenna at time zero is on the x coordinate axis pointing in the direction toward the center of the scene and moving with velocity V. The relationships now take the simple forms

$$\tau(x,y) = \frac{2}{c}R(x,y,x',y')$$

$$\approx \frac{2}{c}(x-x'),$$

and

$$\nu(x,y) = 2\frac{f_0}{c}\dot{R}(x,y,x',y')$$

$$= 2\frac{f_0}{c}V\frac{y}{R}.$$

The earlier expressions then become

$$\Delta x = \frac{c}{2}\Delta\tau,$$

$$\Delta y = \frac{R}{V}\frac{c}{2f_0}\Delta\nu = \frac{\lambda R}{2V}\Delta\nu$$

for the spatial resolution of the sidelooking radar.

The reciprocal of the waveform bandwidth is often used as a measure of the delay resolution, and the reciprocal of the waveform timewidth (or duration) is often used as a measure of the doppler resolution. Thus,

$$\Delta\tau = \frac{1}{B} \qquad \Delta\nu = \frac{1}{T},$$

where B is an appropriate measure of the bandwidth of the transmitted waveform, and T is an appropriate measure of the timewidth.

For a waveform that is a coherently processed pulse train, the duration of the waveform may be long. The resolution in cross range can be defined as the first null in the ambiguity function in the ν direction, which is at $1/NT_r$. The range resolution and cross-range resolution are then expressed as

$$\Delta x = \frac{c}{2B} \qquad \Delta y = \frac{\lambda R}{2L},$$

where $L = VT = VNT_r$ is the distance that any point on the antenna travels in time T. It is natural to call L the "synthetic-aperture length." For this reason, the real antenna is visualized as sweeping out an interval of length L called the *synthetic aperture*.

11.4 Focusing and Motion Compensation

In some instances of an imaging radar, the delay and the doppler alone are not adequate to fully describe the significant effects on carrier phase due to a change in delay because of motion. Then the next term of the series expansion, which is the second time derivative of the delay, must be included. Although this term is somewhat different for every point of the scene, often it is sufficiently accurate to make the same correction for all points of a scene. Then this term can be treated simply as a phase correction that is made to the received signal for a reflection from the center of the scene.

The complex baseband signal received from a point reflector at the point (x, y) is

$$v(t) = s(t - \tau(x, y, t))e^{j2\pi f_0 \tau(x,y,t)},$$

where $\tau(x, y, t)$ is the path delay at time t. In the exponential term, $\tau(x, y, t)$ can be approximated up to the quadratic term in t as

$$\tau(x, y, t) \approx \tau_o(x, y) + \dot{\tau}_o(x, y)t + \ddot{\tau}_o(x, y)t^2.$$

The quadratic term of this approximation is to be included in the simplest possible way. Specifically, the simplest approximation in the exponential term that respects the quadratic term is

$$\tau(x, y, t) \approx \tau_o(x, y) + \dot{\tau}_o(x, y)t + \tfrac{1}{2}\ddot{\tau}_o(0, 0)t^2.$$

This quadratic correction term is independent of x and y. The quadratic term is written this way so that it need not be separately applied for each point (x, y) in the scene. To this purpose, note that the delay-doppler matched-filter response of the received signal $v(t)$ from a point reflector at (x, y) can be rearranged as follows:

$$r(x, y) = \int_{-\infty}^{\infty} v(t)s^*(t - \tau(x, y, t))e^{-j2\pi f_0 \tau(x,y,t)} dt$$

$$\approx e^{-j2\pi f_0 \tau_0} \int_{-\infty}^{\infty} \left[v(t)e^{-j\pi \ddot{\tau}_0(0,0)t^2} \right] s^*(t - \tau_0(x, y))e^{-j2\pi \dot{\tau}_0(x,y)t} dt.$$

This reduces to the computation of the cross-ambiguity function $\chi(\tau, \nu)$ of the "focused" waveform $v'(t)$ defined as

$$v'(t) = v(t)e^{-j\pi \ddot{\tau}_0(0,0)t^2}.$$

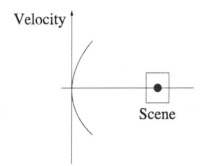

Figure 11.6 A constant-radius reference trajectory

The quadratic term is here described as a *focusing compensation* because it is incorporated into the computation simply by phase-shifting the received signal quadratically in time by the phase $\ddot{\tau}_0(0,0)t^2$. This correction alters the phase – up to quadratic terms – to make it appear as if the radar is following a circular (or even parabolic) path centered at the origin of the scene, as shown in Figure 11.6.

There may also be small, undesired irregularities in the radar path. To account for minor irregularities in the radar motion, it is useful to further modify the approximation to the form

$$\tau(x,y,t) = \tau_0'(x,y) + \dot{\tau}_0(x,y)t + \tfrac{1}{2}\ddot{\tau}_0(0,0)t^2 + \delta\tau(t).$$

Whereas the third term on the right accounts for the difference between a straight line and a circle, the fourth term on the right accounts for measured deviations in the motion of the moving radar from the reference straight line. The last term is referred to as the *motion compensation* term.

The focusing and motion compensation terms stabilize the signal from each reflecting element so that the reflecting element effectively has (nearly) constant coordinates in the τ, ν plane. The received signal actually is the composite of many echoes from many points, so the ideal compensation would be different for every point of the scene. However, when the scene is not too large, it is enough when the focusing and motion compensation terms can be approximated so that they are the same for every point in the scene.

The approximation of the focusing term as independent of x and y is now examined with the aid of Figure 11.7, which depicts an imaging radar following a straight line at a constant velocity V, and a depicted scene to be imaged. The range rate and range acceleration with respect to the origin are given by

$$\dot{R} = -V\sin\phi \qquad \ddot{R} = \frac{V^2}{R}\cos^2\phi.$$

The range rate $\dot{R}$ and the range acceleration $\ddot{R}$ result in a doppler ν and a doppler rate α given by

$$\nu = -\frac{2f_0}{c}V\sin\phi \qquad \alpha = \frac{2V^2}{\lambda R}\cos^2\phi.$$

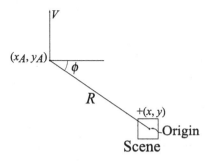

Figure 11.7 Elementary geometry

The doppler rate causes a quadratic-phase change, $\Delta\theta = \alpha t^2/2$ (in cycles). Over time T, this phase change is given by

$$\Delta\theta = \frac{(VT)^2}{\lambda R}\cos^2\phi.$$

The phase change is removed by the focusing compensation. For simplicity of exposition, only the case with $\phi = 0$ is considered. Then $\Delta\theta = (VT)^2/\lambda R$. The cross-range resolution Δy for a synthetic-aperture radar, given by

$$\Delta y = \frac{\lambda R}{2VT},$$

can be used to eliminate VT in the expression for $\Delta\theta$. Thus,

$$\Delta\theta = \frac{\lambda R}{4\Delta y^2}.$$

This correction is a function of range. When $\Delta\theta$ is corrected for only a nominal range R, the phase correction will be in error for values of range $R + \delta R$ as given by

$$\delta(\Delta\theta) = \frac{\lambda}{4\Delta y^2}\delta R.$$

This is the phase error due to a constant focusing correction for a reflector at a range $R + \delta R$, which is offset by δR compared to the range to the center of the scene.

Suppose that an uncompensated, quadratic-phase deviation of at most 1/8 cycle is deemed to be acceptable.[3] The degradation due to the doppler-rate correction is the value of δR satisfying the equation

$$\frac{1}{8} = \frac{\lambda}{4\Delta y^2}\delta R.$$

This gives the acceptable width in range

$$\delta R = \frac{1}{2}\frac{\Delta y^2}{\lambda}$$

as the half-width in the y direction of the imaged scene.

---

[3] This results in the equivalent of 0.5 dB loss in the signal-to-noise ratio. At 1/4 cycle, the loss would be about 2 dB. Such considerations are studied in Chapter 15.

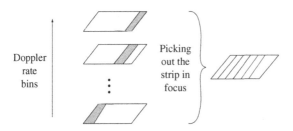

Figure 11.8 Conceptualizing multiple doppler-rate bins

To illustrate, suppose that the y resolution is Δy is 10 feet and λ is 0.1 foot, then the half-width of a satisfactory scene in the range direction is 500 feet. This does not mean that a larger scene cannot be processed in this way. It only means that the scene will be attenuated at the scene edges in the range direction. When the range spread of the scene is large, it is not possible to focus the entire scene at once. The remedy is to make the doppler-rate correction separately for each of several subscenes. Then the scene can be divided into range strips that are individually focused. This is expressed by saying there is more than one doppler-rate bin. Because the focusing takes place in the signal processing, it is straightforward to reprocess the same data many times for different subscenes, as shown in Figure 11.8. Each computation of a subscene uses a different quadratic-phase correction. Then the appropriate range strip is found in each sample cross-ambiguity function and assembled into the full scene.

11.5 Structure of Imaging Systems

A high-resolution imaging radar that moves during the duration of the waveform by a distance that is large compared to the size of its physical antenna is called a *synthetic-aperture radar*. This name refers to the notion that by moving the physical antenna, a larger antenna is synthesized. It is satisfying to describe an imaging radar using this idea of a synthetic aperture. The notion of a synthetic antenna aperture is appealing, and there is a long history of describing an imaging radar in this way. However, the analogy is flawed in several ways and can lead to false conclusions when it is taken literally.

A synthetic-aperture radar can be described simply as a single antenna element that is moved from position to position over a series of pulses. Neglecting the motion of the antenna during a single pulse, the antenna at each position is pulsed and the return at that position and the position itself are both recorded. The pulse returns of the recorded sequence are later coherently combined, each with an appropriate position-dependent phase shift.

A real array is different. A pulse is simultaneously transmitted through all elements of the array, and the reflected return is simultaneously received by all elements of that array. The synthetic aperture could be emulated by sequentially transmitting a sequence of identical pulses one at a time, one pulse from each real antenna in turn.

11.5 Structure of Imaging Systems

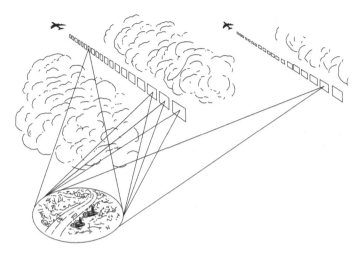

Figure 11.9 Bistatic imaging radar

The return received at that antenna is recorded and then the returns are coherently added together.

While a synthetic aperture is superficially similar to a real array, the differences are significant. A synthetic array can receive a return only at the same antenna element from which the pulse was transmitted. Any signal that might have been received at other antenna elements is lost because those other antenna elements do not exist. Hence the synthetic array processes less information than would a real array. The array element is moving while transmitting and receiving. Hence each echo pulse is individually doppler-shifted in the synthetic array but not in the real array. Because of these differences, the analogies between a synthetic array and a real array should not be pushed too far.

The simplest and most common arrangement for a synthetic-aperture imaging radar places the transmitter and the receiver on the same moving platform. This is called a *monostatic imaging radar*. A less common arrangement places the transmitter and the receiver at different locations, as shown in Figure 11.9. This is called a *bistatic imaging radar*. We will develop the ideas mainly for monostatic imaging systems. Bistatic imaging systems are a simple generalization and obey the same principles of image formation. The main difference is in the coordinate transformation taking τ, ν coordinates to rectangular x, y coordinates.

The transmitter or receiver of a synthetic-aperture radar system must be in motion with respect to the target scene. The two most popular arrangements are the *swath-mode imaging radar*, shown in Figure 11.10 and the *spotlight-mode imaging radar*, shown in Figure 11.11.

In the simplest case, a swath-mode imaging radar has a straight-line, constant-velocity trajectory. It illuminates and images a continuous strip that is parallel to the trajectory. The resolution is limited by the amount of time a reflecting element is illuminated. This time duration is limited by the beamwidth of the real antenna.

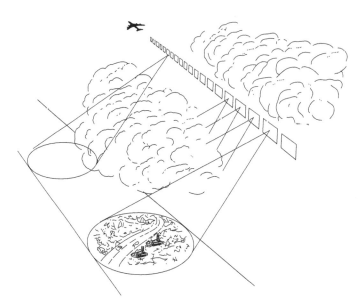

Figure 11.10 Swath-mode synthetic-aperture radar

Figure 11.11 Spotlight-mode, synthetic-aperture radar

A spotlight-mode imaging radar enhances resolution by keeping the antenna beam directed at the target area for a longer time by rotating an antenna or by phasing an array. In the extreme case, a spotlight-mode imaging radar has a constant-speed, circular trajectory centered on a small scene that is to be imaged. For long flight paths,

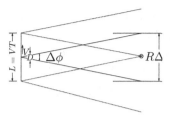

Figure 11.12 Swath-mode limit on illumination time

spotlight-mode imaging suggests a polar coordinate system, which suggests the methods of tomography for geometries in which the synthetic-aperture length L is long, as is treated in Section 11.9.

The synthetic-aperture image is created in the τ, ν plane. Because of changing geometry, objects do not normally have persistent τ, ν coordinates. To obtain good ν resolution, long waveforms are needed and, during the duration of this waveform, the doppler coordinate ν of a reflector will change. In general, objects will be out of focus or smeared because their τ and ν coordinates are changing during the exposure. This imposes the requirements of focusing and motion compensation that are studied in Section 11.4.

The cross-range resolution for a sidelooking synthetic-aperture radar was given in Section 11.2 as

$$\Delta y = \frac{\lambda R}{2L}.$$

For a uniform pulse train with pulse repetition interval T_r, the *synthetic-aperture length* is $L = V(N-1)T_r$, which is approximately the distance that the antenna moves during the duration of the pulse train.

The cross-range resolution of the synthetic aperture can be compared to the cross-range resolution of a real aperture of length L. For a real aperture that is uniformly illuminated, the antenna pattern is a sinc function. A real sidelooking aperture of length D that is uniformly illuminated has its first null at $\Delta\phi = \lambda/D$ and the cross-range resolution is

$$\Delta y = R\Delta\phi = \frac{\lambda R}{D}.$$

For a real antenna to have the same cross-range resolution as a synthetic aperture of length L, the real aperture must have a width $D = 2L$. A real aperture must be twice as long as a synthetic aperture to obtain the same cross-range resolution. As before, the notion of a "synthetic aperture" is seen to be not exact.

An implicit condition underlying the use of the ambiguity function of the imaging radar is that the antenna beam is directed at the scene throughout the duration of the waveform $s(t)$. When VT is large, this may become a limitation that cannot be ignored. Figure 11.12 shows a notional depiction of an swath-mode imaging radar with a staring antenna beam illuminating the scene. The width of the beam at range R is $R\Delta\phi$. When VT is smaller than $R\Delta\phi$, a scene element will be illuminated for the full waveform of duration T, and the antenna beamwidth need not be considered further. When VT is larger than $R\Delta\phi$, the beam moves off a target element during the waveform duration. Instead, a spotlight-mode imaging radar may be used, rotating the antenna beam to keep a scene illuminated.

A swath-mode imaging radar using a fully illuminated real antenna of width D has a main beam of width $\Delta\phi = \lambda/D$. The staring antenna beam of width $\Delta\phi$ illuminates a region that changes with time because of the motion of the transmitter. An element of the scene is fully illuminated only for a time T satisfying

$$VT = R\Delta\phi.$$

The cross-range resolution of a swath-mode imaging radar is

$$\Delta y = \frac{\lambda R}{2VT}.$$

The real aperture of the synthetic-aperture radar antenna has a width D. Then $\Delta\phi = \lambda/D$ and $VT = R\Delta\phi$, so the resolution becomes

$$\Delta y = \frac{D}{2}.$$

The cross-range resolution is limited to one-half of the physical size of a staring real antenna because the reflected signal from an object ceases when the beam of the staring antenna moves off of that object. The resolution cannot improve further. The remedy is to steer the beam.

Radar Astronomy

A much different instance of an imaging radar with a much different geometry is shown in Figure 11.13. This geometry occurs in the imaging of a distant space-based object, such as a planet or an asteroid, using the methods of radar astronomy with an earth-based radar. The basic concepts regarding the ambiguity function as a point-spread function in τ, ν coordinates are still appropriate for space-based objects. The main difference lies in the details of the geometric transformation from τ, ν coordinates to spherical coordinates describing the surface of the imaged object. The surface is nonplanar and the relative motion has a different form. For earth-based imaging of space-based objects, the doppler is due to the relative rotation of the planet or asteroid being imaged.

The rotation of the earth enters into the description of the received signal only through the motion at the point at which the radar is located. This results in a simple frequency offset in the transmitted and received waves, and is easily compensated. The rotation of the reflecting planet or asteroid, in contrast, produces different values of doppler for different reflecting elements at different points of that object. The surface of that planet or asteroid can be described in range/doppler coordinates.

Because points in different "hemispheres" of the object can have the same range/doppler coordinates, the processing output will be the superposition of

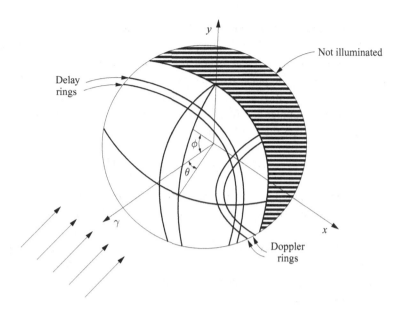

Figure 11.13 Delay coordinates on a distant sphere

two hemispheres. The image can be disambiguated by dual aperture interferometry, which is described in Section 11.7.

11.6 Computing the Cross-Ambiguity Function

Computation of the sample cross-ambiguity function

$$\chi_c(\tau,\nu) = \int_{-\infty}^{\infty} v(t)s^*(t-\tau)e^{-j2\pi\nu t}dt$$

from the received signal $v(t)$ and the transmitted signal $s(t)$ is a central task of radar signal processing. Depending on the requirements of the radar, the processing can be simple or substantial. The processing has been performed by simple analog circuits, either at baseband or at passband, or by a sophisticated optical processor, or by a special-purpose digital processor. One might choose the waveform $s(t)$ so as to keep the computations manageable, or one may develop fast computational algorithms that compute $\chi_c(\tau,\nu)$ for an arbitrary $s(t)$ chosen for its resolution or ambiguity performance.

Depending on the circumstances of the application, the computation of the sample cross-ambiguity function might be a massive computation for a received waveform that extends over many seconds, or it might be a simpler computation for a received pulse that extends only over a few milliseconds. The complexity is determined by the requirements of the application. When τ is set to a single fixed value, the structure of the computation has the form of a Fourier transform. When ν has a single fixed value, the structure of the computation has the form of the matched-filter convolution. The structure of a processing implementation, or the choice of waveform, may emphasize one of these details to simplify the computations. Simple radars, including the earliest radars, use a waveform, $s(t)$, that makes the computation trivial.

For a *pulse radar*, the pulse $s(t)$ may be chosen to be a short pulse of duration T such that $\nu T \ll 1$ for all ν of relevance. Then $\chi_c(\tau,\nu)$ does not depend on ν for the relevant values of doppler and it is sufficient to do the computations for $\nu = 0$. The computation now takes the simple form

$$\chi_c(\tau,0) = \int_{-\infty}^{\infty} v(t)s^*(t-\tau)dt,$$

which can be computed by passing $v(t)$ through a filter with the impulse response $s^*(-t)$. The ambiguity function then reduces to a matched filter.

For a *doppler radar*, the pulse $s(t)$ may be chosen to be a long pulse of narrow bandwidth B so that $B\tau \ll 1$ for all relevant τ. Then $\chi_c(\tau,\nu)$ does not depend on τ for the relevant values of range delay and it is sufficient to do the computation for $\tau = 0$. Then the computation is

$$\chi_c(0,\nu) = \int_{-\infty}^{\infty} v(t)s^*(t)e^{-j2\pi\nu t}dt.$$

For a rectangular pulse $s(t)$ of width T, this integral takes the simple form

$$\chi_c(0,\nu) = \int_{-T/2}^{T/2} v(t)e^{-j2\pi\nu t}dt.$$

In this case, $\chi_c(0,\nu)$ can be sampled by passing $v(t)$ through a bank of passband filters, as can be described by the Fourier transform

$$\chi_c(0,\nu_k) = \int_{-T/2}^{T/2} v(t)e^{-j2\pi\nu_k t}dt$$

for some discrete set of ν_k.

A waveform that has a resolution in both τ and ν is the pulse train

$$p(t) = \sum_{n=0}^{N-1} s(t-nT_r),$$

where $s(t) = 0$ for $|t| \geq T/2$. Then

$$\chi_c(\tau,\nu) = \int_{-\infty}^{\infty} \sum_{n=0}^{N-1} v(t)s^*(t-nT_r+\tau)e^{-j2\pi\nu t}dt.$$

Now move the sum outside the integral and make a change of variables in each summand so that $t - nT_r + \tau$ becomes replaced by t. This gives

$$\chi_c(\tau,\nu) = e^{j2\pi\nu\tau} \sum_{n=0}^{N-1} e^{-j2\pi\nu n T_r} \int_{-T/2}^{T/2} s^*(t)v(t+nT_r-\tau)e^{-j2\pi\nu t}dt.$$

11.6 Computing the Cross-Ambiguity Function

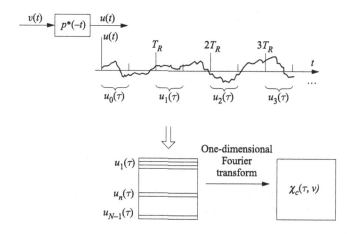

Figure 11.14 Computing the sample cross-ambiguity function

There are now n integrals, each from $-T/2$ to $T/2$, but with each segment of $v(t)$ in the integrand translated. Moreover, when T is chosen so that $\nu T \ll 1$ for all relevant ν, this takes the approximate form

$$\chi_c(\tau, \nu) = e^{j2\pi\nu\tau} \sum_{n=0}^{N-1} e^{-j2\pi\nu n T_r} \int_{-T/2}^{T/2} s^*(t) v(t + nT_r - \tau) dt.$$

With this approximation, each returned pulse is passed through a matched filter to provide an output, say $u_n(\tau)$. Then the sample cross-ambiguity function takes the form of a one-dimensional Fourier transform,

$$\chi_c(\tau, \nu) = e^{j2\pi\nu\tau} \sum_{n=0}^{N-1} u_n(\tau) e^{-j2\pi\nu n T_r},$$

for each value of τ. When T is small in comparison to the pulse repetition interval T_r, there are considerable computational savings by taking advantage of the nature of the pulse train. Figure 11.14 shows the computation of the ambiguity function $\chi_c(\tau, \nu)$ for a pulse train using the standard tools of signal processing. The output of a single matched filter is broken into intervals, each interval corresponding to one of the $u_n(\tau)$ for a single pulse of the pulse train. These segments of the matched-filter output are read as rows into a two-dimensional array. The one-dimensional Fourier transform of each column of the array then provides the desired sample cross-ambiguity function.

A popular pulse train is a train of chirped pulses. One reason for the popularity of chirped pulses is that for a pulse train of chirped pulses the computation of the cross-ambiguity function can be arranged to take the form of a two-dimensional Fourier transform. To do this, let $s(t) = e^{j\pi\alpha t^2}$ and write as before

11 Radar Imaging Systems

$$\chi_c(\tau,\nu) = e^{j2\pi\nu\tau} \sum_{n=0}^{N-1} e^{-j2\pi\nu nT_r} \int_{-T/2}^{T/2} e^{j\pi\alpha t^2} v(t+nT_r-\tau)dt$$

$$= e^{j2\pi\nu\tau} \sum_{n=0}^{N-1} e^{-j2\pi\nu nT_r} u_n(\tau),$$

where

$$u_n(\tau) = \int_{-T/2}^{T/2} e^{j\pi\alpha t^2} v(t+nT_r-\tau)dt.$$

For convenience, the time origin for the received signal is now chosen so that τ is zero at the center of the scene. With another change of variables such that $t - \tau$ is replaced by t, this becomes

$$u_n(\tau) = e^{j\pi\alpha\tau^2} \int_{-T/2+\tau}^{T/2+\tau} e^{j2\pi\alpha t\tau} \left[e^{j\pi\alpha t^2} v(t+nT_r) \right] dt.$$

Define the nth "dechirped" pulse as

$$v_n(t) = e^{j\pi\alpha t^2} v(t+nT_r).$$

Each received pulse is "dechirped" by multiplying it by $e^{j\pi\alpha t^2}$. This prepares the pulse for further processing and reduces the bandwidth considerably, thereby resulting in a practical sampling rate.

The processing of each pulse now takes the form of a Fourier transform of the dechirped pulse

$$u_n(\tau) = \int_{-T/2+\tau}^{T/2+\tau} e^{j2\pi\alpha t\tau} v_n(t) dt.$$

The limits of integration are offset by τ because, for a reflector with delay coordinate τ, the return occupies the (recentered) interval from $-T/2+\tau$ to $T/2+\tau$. To simplify the expression, note that τ is small compared to $T/2$ and so make the approximation that $T/2 + \tau \approx T/2$. The approximation in the limits of integration means that, for reflectors near the edge of the scene, a small part of the received signal is wasted so the image near the edge of the scene will be attenuated by a small amount in the proportion as τ is to T.

The computation of the sample cross-ambiguity function now takes the form

$$\chi_c(\tau,\nu) = \sum_{n=0}^{N-1} e^{j2\pi\nu nT_r} \int_{-T/2}^{T/2} e^{j2\pi\alpha t\tau} v_n(t) dt.$$

This computation is in the form of a two-dimensional Fourier transform of the train of dechirped pulses as shown in Figure 11.15. One axis is a discrete Fourier transform and one axis is a continuous Fourier transform.

The computation involves a sum on one axis, and an integration on the other. This computation may be approximated in either of two ways, either converting the sum

11.6 Computing the Cross-Ambiguity Function

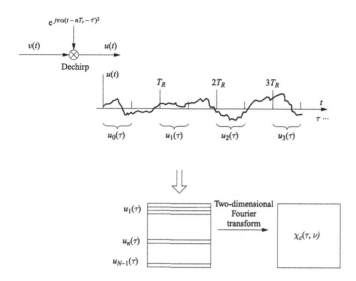

Figure 11.15 Receiver computations for a chirp pulse train

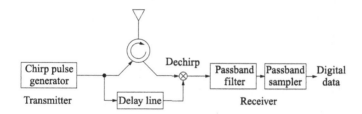

Figure 11.16 Block diagram of a chirp transmitter/receiver

to an integral or converting the integral to a sum. For an optical processor, the computation can be put in the form of a continuous two-dimensional Fourier transform by replacing the discrete Fourier transform with a continuous Fourier transform by interpolation. Then a lens can form the two-dimensional Fourier transform. This has been a powerful approach early on. For a digital processor, the computation can be put in the form of a discrete two-dimensional Fourier transform by replacing the continuous Fourier transform with a discrete Fourier transform by sampling. This has now become the preferred approach.

The Chirped Pulse

The quadratic-phase pulse is central in the study of optics in the form of the Fresnel approximation to the Huygens principle. It plays an important role in the study of diffraction and the behavior of the ideal lens. In radar, the quadratic-phase pulse is used as a wide bandwidth pulse for which the delay of a pulse can be converted to a frequency offset in the processing. The technique that transmits a chirped pulse and dechirps each pulse of the received signal is called *stretch* or *frequency compression*.

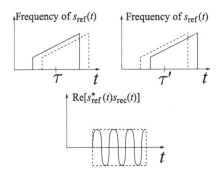

Figure 11.17 Frequency compression

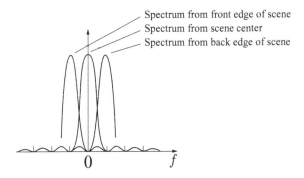

Figure 11.18 Illustrating the compressed signal

Figure 11.16 shows a block diagram of a radar that uses frequency compression. In this diagram, the transmitted pulse $s(t)$ is also delayed to match the propagation delay corresponding to the center of the scene. This becomes the reference chirp pulse

$$s_{\text{ref}}(t) = e^{j\pi\alpha t^2} \qquad |t| \leq T/2,$$

relative to the time origin of the received signal. The received echo pulse $s_{\text{rec}}(t)$ for a reflector with delay τ with respect to the scene center is

$$s_{\text{ref}}(t) = e^{j\pi\alpha(t-\tau)^2} \qquad |t-\tau| \leq T/2.$$

Then the *dechirped* signal is

$$s^*_{\text{ref}}(t)s_{\text{rec}}(t) = e^{j2\pi\alpha\tau t - j\pi\alpha\tau^2}$$

for $|t-\tau| \leq T/2$ and $|t| \leq T/2$. The dechirped signal takes the simple form of a sinusoid at the frequency $\alpha\tau$ with the phase shift $\alpha\tau^2/2$. The frequency $\alpha\tau$ of the dechirped signal depends on the range offset. Figure 11.17 illustrates how the process of dechirping changes a delay into a frequency offset that is proportional to range.

For a scene with many reflectors, the dechirped signal is a superposition of many such sinusoids, each corresponding to one of the reflectors. This is depicted in the frequency domain in Figure 11.18 for three such reflectors. The bandwidth of the dechirped signals depends on the range spread and on the pulse duration of the chirped pulse. The bandwidth can be far smaller than the bandwidth of the chirped pulse. Consequently, by placing the sampling after the dechirping, the sampling can use a much lower sampling rate than it would were the sampling to occur before the dechirping.

11.7 Dual Aperture Imaging

There are many ways by which multiple antenna apertures can be used to provide additional information to an imaging radar. Multiple antennas can be used with the methods of interferometry to make angle measurements to reflecting objects within the scene. Multiple antennas can also be used to suppress signals or echoes from other directions that interfere with a desired signal. Each range-doppler cell corresponds to an (x,y) point of a stationary scene and that point is at its own angle with respect to the antenna. Moving objects appear at a false point. So that the returns from different cells can be sorted and each detected moving object can be recognized as such and associated with its proper angle, the interferometry calculation must be placed after the calculation of the sample cross-ambiguity function. This requires the computation of a sample cross-ambiguity function for each of two antennas.

Another application of interferometry is to annotate a radar image with height information. The normal output image of an imaging radar is two-dimensional, presenting the image in x,y coordinates as though it were viewed from directly overhead. The height of objects in the scene is not directly evident in a single cross-ambiguity function. However, to enhance the image, an interferometer incorporated into the imaging radar with two antennas separated vertically can measure the height of individual reflectors.[4]

The raw signal at the radar antenna is the composite of the echoes from many reflecting elements. Some of these echoes are completely masked by stronger echoes. This is why interferometry must be performed after the computation of the sample cross-ambiguity function so that the echo signals are sorted into range-doppler cells.

Figure 11.19 shows the concept. Two antennas separated vertically are used. The signal is transmitted from one antenna, but the echo is received at both antennas. Each received echo signal is used to compute a sample cross-ambiguity function. Because the antennas are closely spaced in comparison to the range, both computed images will have nearly the same magnitude. The phase of the signal in each (τ, ν) cell, however, will be slightly different because of the slight difference in path length from the dominant reflecting element in that cell to each of the two antennas. This phase difference between the cross-ambiguity functions at range-doppler coordinates (τ, ν) is given by

$$\Delta\theta(\tau,\nu) = 2\pi \frac{d}{\lambda} \sin\phi(\tau,\nu),$$

where $\phi(\tau,\nu)$ is the declination angle to the dominant reflecting element with range-doppler coordinates (τ, ν). Hence at each (τ, ν) coordinate pair, $\phi(\tau,\nu)$ can be easily

---

[4] Interferometric imaging radar was use by the Apollo orbital vehicle to measure the depth of the dust layer overlaying the lunar rock layer. The early earth-based radar imaging of the surface of Venus used dual antennas to separate the upper and lower hemispheres of the planet.

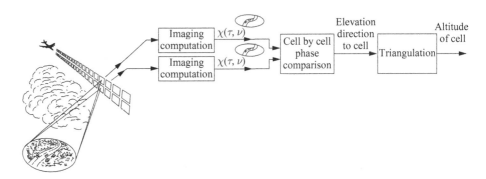

Figure 11.19 Imaging interferometry

computed from the phase difference $\Delta\theta(\tau,\nu)$ by inverting the expression. Then, because the distance to the reflecting cell is known from the measurement of τ, the height can be easily computed by elementary trigonometry in each (τ,ν) cell.

Another application of interferometry in imaging radar systems places two antennas side by side. Then the azimuth angle to each reflecting point can be computed from the two signals received at the two antennas by the same methods of interferometry. Although the azimuth angle could be computed from the doppler and the flight geometry, an accurate determination of azimuth would require that the radar velocity vector be precisely known and the reflector be stationary. In contrast, an interferometric measurement gives an independent measurement of azimuth with respect to the antenna baseline rather than to the velocity vector of the radar with respect to the reflector. This angular measurement may be more accurate. Moreover, the interferometric measurement gives an estimate of azimuth even for moving reflectors, whereas the doppler can be used to infer angle only when it is known that the object is not moving. By reconciling the interferometric measurement of azimuth to the measurement based on doppler, the motion of the reflector can be detected and corrected in the image.

One can also use two antennas to form an antenna pattern with a null in order to reject interference. The rotating sphere shown in Figure 11.13 to denote a planet or asteroid is an example of a situation with interference from another reflector that is spatially separated but has the same τ,ν coordinates. Each hemisphere has the same set of τ,ν coordinates. With a single antenna, the radar image will consist of the superposition of both hemispheres. With two antennas, the relative phase can be used to separate the images of the two hemispheres.

11.8 Radar Imaging of Diffuse Reflectors

A radar reflector may be modeled as a specular reflector or may be modeled as a diffuse reflector, as discussed in Section 11.1, or as a combination of the two. The specular reflection model is usually the more appropriate model at radar frequencies

11.8 Radar Imaging of Diffuse Reflectors

and so is the common model used at radar frequencies. The diffuse reflector model leads to an unconventional method of radar processing that is much different than the methods using the matched filter and the ambiguity function. It may be appropriate for high-resolution imaging of certain targets such as fields or forests. The imaging of diffuse targets is described here as an example of the method of alternating expectation-maximization.

A specular model for a radar reflector is one in which the surface of the object is smooth with respect to the radar wavelength. A diffuse model for a radar reflector is one in which the surface of the object is rough with respect to the radar wavelength. There are a large number of small reflectors in a typical resolution cell of a diffuse reflector such that no single reflecting element dominates that cell. Constructive and destructive interference from the multitude of reflectors in a cell creates a speckled pattern in the scene that appears as multiplicative noise. The amplitude and phase of the return from each individual resolution cell is random. A slight change in the imaging geometry, such as by viewing angle, can cause a large change in the complex amplitude of a cell and so a change in the speckled pattern. In such a model, the reflectivity density is denoted $\widetilde{\rho}(x,y)$ and modeled as a two-dimensional complex random process with zero mean and variance[5] $\sigma^2(x,y) = E[|\widetilde{\rho}(x,y)|^2]$. The expected value, or mean, of the reflectivity density $E[|\widetilde{\rho}(x,y)|]$ is posited to equal zero and is not estimated. Instead, the variance $\sigma^2(x,y)$ as a function of x and y is estimated. The effective reflectivity density of each cell may then be defined as $\rho(x,y) = \left[E[|\widetilde{\rho}(x,y)|^2]\right]^{1/2}$.

Texture

The diffuse reflectivity density $\widetilde{\rho}(x,y)$ may be interpreted as the product of a *texture function* $u(x,y)$ and a smooth reflectivity density $\rho(\tau,\nu)$. With this interpretation, the diffuse reflectivity density

$$\widetilde{\rho}(x,y,t) = u(x,y,t)\rho(x,y)$$

is observed, while $\rho(x,y)$ is of primary interest. The texture function $u(x,y,t)$ modeled as a complex, stationary, memoryless random field with zero mean and unit variance. The function $\widetilde{\rho}(x,y)$ has zero mean and variance $E[|\widetilde{\rho}(x,y)|^2]$. The reason for separating the texture function from the reflectivity density $\rho(x,y)$ would be so that the texture function is the random factor whereas $\rho(\tau,\nu)$ is a deterministic function. A particular realization of the $\widetilde{\rho}(x,y)$ is usually not regarded as an informative part of the reflectivity density, though the statistical parameters of the texture could be of separate interest when the roughness of the target surface is of interest.

Signal in Noise

Before treating the problem at hand, recall the problem of estimating the variance of a gaussian signal in gaussian noise that was discussed in Section 8.3. A measurement of the form

$$v = s + n$$

---

[5] Again, as noted in an earlier footnote, the clash in traditional notation is now more egregious. The variance, which is usually denoted σ^2, refers to the expected radar cross section $E[|\rho(x,y)|^2]$, which is usually denoted $\sigma(x,y)$ in the radar literature. Neither convention can be ignored. The clash must be tolerated.

is given, where s is a gaussian random variable of unknown variance σ_s^2 and n is a gaussian random variable of known variance σ_n^2. The task is to estimate the variance σ_s^2. The loglikelihood is

$$\Lambda(\sigma_s^2) = -\tfrac{1}{2}\log_e(\sigma_s^2 + \sigma_n^2) - \frac{v^2}{2(\sigma_s^2 + \sigma_n^2)}.$$

The estimate of the variance is

$$\widehat{\sigma_s^2} = \max\{0, v^2 - \sigma_n^2\}.$$

More generally, when there are M independent measurements, then the maximum-likelihood estimate of σ_s^2 is

$$\widehat{\sigma_s^2} = \max\{0, \widehat{v^2} - \sigma_n^2\},$$

where

$$\widehat{v^2} = \frac{1}{M}\sum_{\ell=1}^{M} v_\ell^2$$

is the average squared value.

Image Variance

The problem of interest is the problem of estimating the variance function $\sigma^2(x,y) = E[|\tilde{\rho}(x,y)|^2]$ of the diffuse reflectivity density from a reflected radar signal when the mean is zero. This is a more general version of the problem of estimating the variance of a single random variable. No simple analytic solution is evident for the maximum-likelihood estimate. Instead, an iterative computational algorithm is developed as an example of the method of alternating expectation-maximization. The algorithm suggests nontraditional, alternative approaches to the processing of radar signals to form images. The algorithm to be developed, however, is computationally intensive and does not have convergence guarantees. The method of development of this iterative algorithm is the main contribution of this section.

To regularize the problem and describe it in terms of matrix and vector operations, the diffuse reflectivity density $\rho(x,y)$ is discretized to form the m by m discrete array $\rho = [\rho_{i'i''}]$. The (i', i'')th element is the reflectivity $\rho_{i'i''}$, which is modeled as a zero-mean gaussian random variable of variance $\sigma_{i'i''}^2$ that is independent of all other elements. Regard the array of independent reflectivity samples as arranged into a single vector, denoted ρ, and regard the array of variances as arranged into a single vector of length $N = m^2$, denoted σ^2. The covariance matrix Σ_ρ of ρ is the diagonal matrix with the elements of the vector σ^2 on the diagonal. Thus, the model of the reflectivity is that every element of the reflectivity ρ has its own variance and these samples are independent. Because the sample values are defined as independent random variables, the covariance matrix Σ_ρ is diagonal.

The measured data is a vector v of length M given by

$$v = \Gamma^\dagger \rho + n,$$

where $\mathbf{\Gamma}$ is an M by N matrix and $\mathbf{n}$ is a white gaussian-noise vector with diagonal covariance matrix $N_0 \mathbf{I}$ and $\boldsymbol{\rho}$ is a gaussian random vector with diagonal covariance matrix $\boldsymbol{\Sigma}_\rho$. The measurement vector $\mathbf{v}$ has the form of a sum of a gaussian signal term and a gaussian noise term. Because of the matrix $\mathbf{\Gamma}$, this instance of an estimation task is more general than the instance studied in Section 8.3. Because $\boldsymbol{\rho}$ and $\mathbf{n}$ are both gaussian and independent, the measurement vector $\mathbf{v}$ is a gaussian random variable with the covariance matrix $\mathbf{K} = \mathbf{\Gamma}^\dagger \boldsymbol{\Sigma}_\rho \mathbf{\Gamma} + N_0 \mathbf{I}$.

By setting $\mathbf{s} = \mathbf{\Gamma}^\dagger \boldsymbol{\rho}$, this takes the form

$$\mathbf{v} = \mathbf{s} + \mathbf{n},$$

which is the form studied in Section 8.2. Moreover, $\mathbf{\Gamma}$ itself is not specified here to be invertible. The problem differs from the problem studied in Section 8.3 both because of the inclusion of the measurement matrix $\mathbf{\Gamma}$ and because $\boldsymbol{\Sigma}_\rho$ is constrained to be diagonal. For an invertible $\mathbf{\Gamma}$, using the methods of that section to compute $\boldsymbol{\Sigma}_s$ and then simply writing $\boldsymbol{\Sigma}_s = \mathbf{\Gamma}^\dagger \boldsymbol{\Sigma}_\rho \mathbf{\Gamma}$ need not produce a diagonal $\boldsymbol{\Sigma}_s$.

The probability density function for $\mathbf{v}$ is

$$p(\mathbf{v}) = \frac{1}{\sqrt{\det(2\pi \mathbf{K})}} e^{-\frac{1}{2}\mathbf{v}^\dagger \mathbf{K}^{-1} \mathbf{v}},$$

where $\det(2\pi \mathbf{K}) = (2\pi)^N \det \mathbf{K}$ and $\mathbf{K} = \mathbf{\Gamma}^\dagger \boldsymbol{\Sigma}_\rho \mathbf{\Gamma} + N_0 \mathbf{I}$. This leads to the loglikelihood function $\Lambda(\boldsymbol{\Sigma}_\rho) = -\log \det \mathbf{K} - \mathbf{v}^\dagger \mathbf{K}^{-1} \mathbf{v}$. The maximum-likelihood estimate then is

$$\widehat{\boldsymbol{\Sigma}}_\rho = \underset{\boldsymbol{\Sigma}}{\operatorname{argmax}} \left[-\log_e \det \mathbf{K} - \mathbf{v}^\dagger \mathbf{K}^{-1} \mathbf{v} \right],$$

where $\mathbf{K} = \mathbf{\Gamma}^\dagger \boldsymbol{\Sigma}_\rho \mathbf{\Gamma} + N_0 \mathbf{I}$ and the range of the argmax is over all $\boldsymbol{\Sigma}_\rho$ that are diagonal N by N nonnegative-definite matrices.

This argmax problem for $\boldsymbol{\Sigma}_\rho$ has no evident closed-form solution. Therefore, an iterative algorithm is developed using the method of alternating expectation-maximization. To form this algorithm, the actual data $\mathbf{v}$ is replaced by the complete data $(\boldsymbol{\rho}, \mathbf{n})$. The mapping from the complete data to the actual data is given by

$$\mathbf{v} = \mathbf{\Gamma}^\dagger \boldsymbol{\rho} + \mathbf{n}.$$

The probability density function for the complete data is

$$p(\boldsymbol{\rho}, \mathbf{n}) = \frac{1}{\sqrt{\det(2\pi \boldsymbol{\Sigma}_\rho)}} e^{-\frac{1}{2}\boldsymbol{\rho}^\dagger \boldsymbol{\Sigma}_\rho^{-1} \boldsymbol{\rho}} \frac{1}{\sqrt{\det(2\pi N_0 \mathbf{I})}} e^{-\frac{1}{2}\mathbf{n}^\dagger \boldsymbol{\Sigma}_n^{-1} \mathbf{n}}.$$

The loglikelihood statistic Λ_{cd} for the complete data is

$$\Lambda_{cd}(\boldsymbol{\Sigma}_\rho) = -\log_e \det \boldsymbol{\Sigma}_\rho - \boldsymbol{\rho}^\dagger \boldsymbol{\Sigma}_\rho^{-1} \boldsymbol{\rho},$$

where all irrelevant constants have been dropped. Because the unknown matrix $\boldsymbol{\Sigma}_\rho$ is a diagonal matrix, this reduces to

$$\Lambda_{cd}(\boldsymbol{\Sigma}_\rho) = -\sum_{i=1}^{I} \log_e \sigma_i^2 - \sum_{i=1}^{I} \frac{|\rho_i|^2}{\sigma_i^2}.$$

Suppose that the estimate Σ^{old} is given for the unknown diagonal matrix Σ_ρ. The expectation of the complete-data loglikelihood when given the actual data and the estimate Σ^{old} is

$$E\left[\Lambda_{cd}(\Sigma_\rho)|v, \Sigma^{\text{old}}\right] = -\sum_{i=1}^{N} \log \sigma_i^2 - \sum_{i=1}^{N} \frac{1}{\sigma_i^2} E\left[|\rho_i|^2 | v, \Sigma^{\text{old}}\right].$$

The expectation on the far right is an estimation-theoretic calculation of a standard form previewed in Section 8.3 and stated in Theorem 8.9.2 and in Corollary 8.9.3. This expectation step is stated as a general proposition using the notation $[\ ]_{ii}$ to designate the ith diagonal element of the matrix.

The expectation step is the content of the following proposition.

Proposition 11.8.1 *Let $v = \Gamma^\dagger \rho + n$, where ρ and n are independent, gaussian vector random variables with covariance matrices Σ_ρ and $N_0 I$, respectively. Then*

$$E\left[|\rho_i|^2 | v, \Sigma_\rho\right] = \sigma_i^2 - \sigma_i^4 \left[\Gamma K^{-1} \Gamma^\dagger - \Gamma K^{-1} S K^{-1} \Gamma^\dagger\right]_{ii},$$

where $K = \Gamma^\dagger \Sigma_\rho \Gamma + N_0 I$ and $S = vv^\dagger$.

Proof:
As in the proof of Theorem 8.9.2, the proof uses the Bayes formula

$$p(\rho|v) = \frac{p(\rho, v)}{p(v)} = \frac{p(\rho)p(v|\rho)}{p(v)}$$

to compute the probability density function $p(\rho|v)$ and to then recognize terms. Because $p(\rho|v)$ is gaussian, it has the form

$$p(\rho|v) = \frac{1}{\det(2\pi \Sigma_{\rho|v})} e^{-\frac{1}{2}(\rho-\bar{\rho})^\dagger \Sigma_{\rho|v}^{-1}(\rho-\bar{\rho})}.$$

Recall that

$$p(v) = \frac{1}{\sqrt{\det(2\pi K)}} e^{-\frac{1}{2}v^\dagger K^{-1} v} \qquad p(\rho) = \frac{1}{\sqrt{\det(2\pi \Sigma_\rho)}} e^{-\frac{1}{2}\rho^\dagger \Sigma_\rho^{-1} \rho}$$

and

$$p(v|\rho) = \frac{1}{\sqrt{\det(2\pi \Sigma_n)}} e^{-\frac{1}{2}(v-\Gamma\rho)^\dagger \Sigma_n^{-1}(v-\Gamma\rho)},$$

where $\Sigma_n = N_0 I$. Therefore, suppressing the proportionality constant gives

$$p(\rho|v) \sim e^{-\frac{1}{2}(\rho-\bar{\rho})^\dagger \Sigma_{\rho|v}^{-1}(\rho-\bar{\rho})}$$

$$\sim \frac{e^{-\frac{1}{2}\rho^\dagger \Sigma_\rho^{-1} \rho} e^{-\frac{1}{2}(v-\Gamma\rho)^\dagger \Sigma_n^{-1}(v-\Gamma\rho)}}{e^{-\frac{1}{2}v^\dagger (\Gamma^\dagger \Sigma_\rho \Gamma + \Sigma_n)^{-1} v}},$$

using Bayes formula for the second line.

Next, collect terms in the exponent to write

11.8 Radar Imaging of Diffuse Reflectors

$$(\rho - \bar{\rho})^\dagger \Sigma_{\rho|v}^{-1}(\rho - \bar{\rho}) = \rho^\dagger(\Sigma_\rho^{-1} + \Gamma^\dagger \Sigma_n^{-1}\Gamma)\rho$$
$$+ v^\dagger\left(\Sigma_n^{-1} - (\Gamma^\dagger \Sigma_\rho \Gamma + \Sigma_n)^{-1}\right)v$$
$$- v^\dagger \Sigma_n^{-1}\Gamma \rho - \rho^\dagger \Gamma^\dagger \Sigma_n^{-1} v.$$

By comparing both sides, the terms $\Sigma_{\rho|v}$ and $\bar{\rho}$ now can be readily recognized as

$$\Sigma_{\rho|v} = (\Sigma_\rho^{-1} + \Gamma^\dagger \Sigma_n^{-1}\Gamma)^{-1}$$

and, by identifying the last terms with $\Sigma_{\rho|v}^{-1}\bar{\rho}$,

$$\bar{\rho} = \Sigma_{s|v}\Gamma^\dagger \Sigma_n^{-1} v = (\Sigma_{\rho|v}^{-1} + \Gamma^\dagger \Sigma_n^{-1}\Gamma)^{-1}\Gamma^\dagger \Sigma_n^{-1} v.$$

Next, by using $\Sigma_s = \Gamma^\dagger \Sigma_\rho \Gamma$, these can be rewritten in an alternative form as

$$\Sigma_{\rho|v} = \Sigma_n \left(\Sigma_n^{-1} - (\Sigma_s + \Sigma_n)^{-1}\right) \Sigma_n,$$
$$\bar{\rho} = -\Sigma_s \Gamma^\dagger \left(\Gamma \Sigma_s \Gamma^\dagger + \Sigma_n\right)^{-1} v.$$

By the standard properties of moments,

$$E[\rho_i^2 | v, \Sigma_n] = E[|\rho_i - \bar{\rho}_i|^2 + 2\rho_i\bar{\rho}_i - \bar{\rho}_i^2 | v, \Sigma]$$
$$= E[|\rho_i - \bar{\rho}_i|^2 | v, \Sigma] + \bar{\rho}_i^2 = \Sigma_{\rho|v} + \bar{\rho}_i^2$$
$$= \left(\Sigma_\rho^{-1} + \Gamma^\dagger \Sigma_n^{-1}\Gamma\right)^{-1} + \left(\Sigma_\rho^{-1} + \Gamma^\dagger \Sigma_n^{-1}\Gamma\right)^{-1}$$
$$vv^\dagger \left(\Sigma_\rho^{-1} + \Gamma^\dagger \Sigma_n^{-1}\Gamma\right)^{-1}.$$

This completes the proof of the proposition. ∎

The iterative algorithm for imaging diffuse reflectivity can now be given. It is stated in the following proposition. Each iterate inherits nonnegativity from the previous iterate. The iterates enjoy the property that the likelihood is monotonically increasing.

Proposition 11.8.2 *The maximum-likelihood variance image $\widehat{\Sigma}_\rho$ is computed from $S = vv^\dagger$ as the limit of the iteration*

$$(\sigma_i^2)^{\text{new}} = (\sigma_i^2)^{\text{old}} - (\sigma_i^2)^{\text{old}} \left[\Gamma K^{-1}\Gamma^\dagger - \Gamma K^{-1} S K^{-1}\Gamma^\dagger\right]_{ii},$$

where

$$K = \Gamma^\dagger \Sigma_\rho^{\text{old}} \Gamma + N_0 I,$$

and Σ_ρ^{old} is the diagonal matrix with $(\sigma_i^2)^{\text{old}}$ as the ith diagonal element.

Proof.
The maximum-likelihood estimate from the complete data (ρ, n) is

$$\widehat{\sigma_i^2} = [\rho\rho^\dagger]_{ii}.$$

Because ρ is unknown, the right side of this equation is replaced by the most recent update for expectation $\mathrm{E}[\rho_i^2|\nu, \mathbf{\Sigma}_\rho]$. Now use the identity[6]

$$(\mathbf{\Sigma}_\rho^{-1} + \mathbf{\Gamma}^\dagger \mathbf{\Sigma}_n^{-1} \mathbf{\Gamma})^{-1} = \mathbf{\Sigma}_\rho - \mathbf{\Sigma}_\rho \mathbf{\Gamma}(\mathbf{\Gamma}^\dagger \mathbf{\Sigma}_\rho \mathbf{\Gamma} + \mathbf{\Sigma}_n)^{-1} \mathbf{\Gamma}^\dagger \mathbf{\Sigma}_\rho$$

to write

$$\mathbf{\Sigma}_{\rho|\nu} + \bar{\rho}\bar{\rho}^\dagger = \mathbf{\Sigma}_\rho - \mathbf{\Sigma}_\rho(\mathbf{\Gamma}K^{-1}\mathbf{\Gamma}^\dagger - \mathbf{\Gamma}K^{-1}S K^{-1}\mathbf{\Gamma}^\dagger)\mathbf{\Sigma}_\rho,$$

where $K = \mathbf{\Gamma}^\dagger \mathbf{\Sigma}_\rho \mathbf{\Gamma} + \mathbf{\Sigma}_n$ and $S = \nu\nu^\dagger$. Because $\mathbf{\Sigma}_\rho$ is constrained to be a diagonal matrix, the update for the new diagonal element is

$$\sigma_s^2 \leftarrow \sigma_s^2 - \sigma_s^2[\mathbf{\Gamma}K^{-1}\mathbf{\Gamma}^\dagger - \mathbf{\Gamma}K^{-1}SK^{-1}\mathbf{\Gamma}^\dagger]_{ii}\sigma_s^2,$$

as was to be proved. ∎

The conclusion of the proposition can be generalized. When there are multiple measurements ν_ℓ of the vector ν indexed by ℓ, the proposition reads the same except that $S = \sum_\ell \nu_\ell \nu_\ell^\dagger$. The problem can be further generalized by the inclusion of a specular component of the reflection represented by the mean of ρ. Then the reflectivity, as modeled, has both a mean and a variance.

To close the section, note that the iteration given in Proposition 11.8.2 immediately reveals the limiting point. There is no change during an iteration when

$$\mathbf{\Gamma}K^{-1}\mathbf{\Gamma}^\dagger = \mathbf{\Gamma}K^{-1}SK^{-1}\mathbf{\Gamma}^\dagger.$$

When $\mathbf{\Gamma}$ is invertible, this immediately reduces to

$$K^{-1} = K^{-1}SK^{-1},$$

which leads to the conclusion that $K = S$.

11.9 Coherent and Noncoherent Radar Tomography

Most of this chapter deals with radar image formation in the range-doppler rectangular coordinate system. This is the conventional approach. For a synthetic-aperture radar with a long aperture described in rectangular coordinates, the angle to a cell changes with motion of the radar and this requires a focusing correction. To treat this requirement in a different way, image formation for an imaging radar can be formulated in polar coordinates using the methods of tomography. This tomographic formulation further clarifies the theory of an imaging radar by exploring it from another

---

[6] This is an instance of the Woodbury matrix identity

$$(A + UCV)^{-1} = A^{-1} - A^{-1}U(C^{-1} + VA^{-1}U)^{-1}VA^{-1},$$

where A is an n by n matrix and C is a k by k matrix, and all inverses exist.

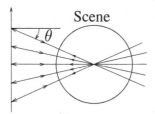

Figure 11.20 Tomographic depiction of synthetic-aperture radar

point of view.[7] It may be that the tomographic formulation is the superior point of view in instances in which the synthetic aperture is long enough so that a polar (r, θ) coordinate system fits the situation better than a rectangular (x, y) coordinate system.

We adopt a tomographic formulation using a chirp pulse train for which every pulse of the waveform is transmitted and received at its own viewing angle as shown in Figure 11.20. Each pulse can be processed individually to form an ambiguity function with poor doppler resolution oriented at one aspect angle of the scene. The batch of these projections is processed to form a tomographic image. In this way, the doppler variable used in rectangular coordinates is replaced by a changing angle in polar coordinates. An individual pulse is short, so the doppler shift on an individual pulse could be ignored. Although the radar antenna is physically moving during each pulse to obtain different viewing angles, this approximation regards the antenna as stationary during each pulse, but moving between pulses.

To fully process an individual pulse by computing a sample cross-ambiguity function on each pulse would give for each pulse

$$\chi_s(\tau, \nu) = \chi^{(p)}(\tau, \nu) ** \rho(\tau, \nu)$$

by the radar imaging approximation, where $\chi^{(p)}(\tau, \nu)$ is the ambiguity function of a single pulse. The delay τ and the doppler ν are in a coordinate system appropriate for that pulse. Because the pulse is short, $\chi^{(p)}(\tau, \nu)$ is wide in the ν direction, possibly wider than the illuminated scene. Then $\chi^{(p)}(\tau, \nu)$ can be approximated as independent of ν. Thus,

$$\chi^{(p)}(\tau, \nu) = \phi(\tau).$$

Therefore, the image computed for one pulse is

$$\chi_s(\tau, \nu) = \phi(\tau) ** \rho(\tau, \nu),$$

where $\phi(\tau)$ is the autocorrelation function of the pulse. When $\phi(\tau)$ does not depend on ν, the double convolution becomes a single convolution

$$\chi_s(\tau, \nu) = \phi(\tau) * \int_{-\infty}^{\infty} \rho(\tau, \nu) d\nu.$$

---

[7] The ambiguity function and the projection-slice theorem can now be seen as two ways of dealing with the geometric effect of a changing viewing angle – one in rectangular coordinates and one in polar coordinates.

The integral is a projection along the ν direction of the pulse. The left side is independent of ν, and is denoted by $p'(\tau)$. Thus,

$$p'(\tau) = \phi(\tau) * \int_{-\infty}^{\infty} \rho(\tau,\nu)d\nu.$$

This says that the projection $p(\tau)$ is convolved with the autocorrelation function of the illuminating pulse $s(t)$. In the frequency domain, using the projection-slice theorem, this becomes

$$P'(f) = |S(f)|^2 R(f,0),$$

where $R(f,0)$ is a slice of the Fourier transform of $\rho(\tau,\nu)$ and $|S(f)|^2$ is the Fourier transform of $\phi(\tau)$. For each pulse, the antenna has changed perspective on the scene, and the local τ,ν coordinate system has a τ axis from the antenna to the center of the scene. It can be regarded as a polar coordinate system with coordinates r and θ. In the r,θ polar coordinate system, centered at the center of a scene, the return from reflectors at range r after matched filtering that return is the complex pulse

$$p_\theta(r) = \phi(r) * \int_{-\infty}^{\infty} \rho(r,r')dr'.$$

Thus, the pulse transmitted at angle θ provides a projection at angle θ. These projections can be used to form an image by the methods of tomography.

For a chirp pulse, $e^{j\pi t^2}$ with Fourier transform $\frac{1+j}{\sqrt{2}}e^{-j\pi f^2}$, the power spectrum $|S(f)|^2 = \sqrt{2}$ simply one. However, the chirp pulse has a finite duration T. Thus, $s(t) = e^{j\pi t^2}\text{rect}(t/T)$ and

$$|S(f)|^2 = \left| \frac{1+j}{\sqrt{2}} e^{-j\pi f^2} * \text{sinc}(fT) \right|^2.$$

To better understand the processing for the chirp pulse it may be helpful to develop the processing equations directly. The radar return seen by a pulse $s(t)$ at angle θ is

$$v_\theta(t) = \int_{-\infty}^{\infty} p_\theta(r)s(t - 2(R+r)/c)dr,$$

where

$$p_\theta(r) = \int_{-\infty}^{\infty} \rho(r,r')dr'.$$

When the chirp pulse has a duration T that is long in comparison with the dispersion in delay across the scene, the end points of the pulse $s(t)$ can be approximated without causing significant error. Thus, the received pulse is

$$v_\theta(t) = \int_{-\infty}^{\infty} p_\theta(r)e^{-j[2\pi f_0(t-2(R+r)/c) + \pi\alpha(t-2(R+r)/c)^2]}dr.$$

Define the dechirped pulse as

$$c_\theta(t) = v_\theta(t)e^{j[2\pi f_0(t-2R/c) + \pi\alpha(t-2R/c)^2]}$$

$$= \int_{-\infty}^{\infty} p_\theta(r)e^{j[4\pi f_0 r/c + \pi\alpha(2rt/c - 4Rr/c^2 + 2r^2/c^2)]}dr.$$

This expression has the appearance of a Fourier transform except for the phase term that is quadratic in r. This term under the condition that the scene size is such that $2\alpha r^2/c^2$ is a negligible angle. Then

$$c_\theta(t) = \int_{-\infty}^{\infty} p_\theta(r) e^{j2\pi(2f_0+\alpha(t-\tau_o))r/c} dr,$$

where $\tau_o = 2R/c$. Now $c_\theta(t)$ can be recognized as a Fourier transform. Specifically,

$$c_\theta(t) = P_\theta(2f_0 + \alpha(t-\tau_o)/c),$$

and $c_\theta(t)$ is the Fourier transform of a projection. This means that a pulse transmitted at angle θ generates a slice at angle θ of the Fourier transform. A collection of such pulses for various θ gives a sequence of slices. These can be coherently combined as suggested by the methods of projection tomography.

Problems

11.1 **a.** Suppose that an imaging radar uses a real antenna of length d that is rigidly fixed so that its beam is perpendicular to the velocity vector. For a uniformly illuminated aperture of length d, what is the real antenna pattern in the ϕ direction?

b. Approximate the beam by an idealized uniform beam of angular width $\Delta\phi = \lambda/d$. That is,

$$A(\phi) = \begin{cases} 1 & \text{for } |\phi| \leq \frac{1}{2}(\lambda/d) \\ 0 & \text{otherwise.} \end{cases}$$

Within this approximation, find a limit in terms of d on the cross-range resolution of a synthetic-aperture imaging radar.

11.2 A bistatic imaging radar with its transmitter and receiver separated by 150 nautical miles (one nautical mile equals 6,080 feet) images a 1 nautical mile by 1 nautical mile scene that is centered at a point 100 nautical miles from the transmitter and 100 nautical miles from the receiver. Sketch the situation with the scene aligned with the local τ, ν coordinate system.

Does one nanosecond of τ resolution for the bistatic radar provide better or poorer spatial resolution than for the case of a monostatic imaging radar with similar parameters?

11.3 A waveform for a sidelooking imaging radar uses a uniform pulse train with rectangular chirped pulses of duration $T = 1$ microsecond and chirp rate α equal to 100 MHz/microsecond. The pulse repetition interval T_r is 1 millisecond. With $V = 1{,}000$ ft/sec and $f_0 = 10$ GHz, a scene centered at $R = 100$ nautical miles (one nautical mile equals 6,080 feet) is to be imaged.

How many pulses should be used so that the cross-range resolution is comparable to the range resolution?

Where are the ambiguities of a point reflector at the scene center?

What second derivative $\ddot{R}$ correction is needed? For how big a scene does this $\ddot{R}$ correction continue to apply?

How must the radar antenna illuminate the scene so that no ghosts are folded into the scene via the grating and range ambiguities?

Can the antenna be rigidly fixed at an angle with respect to the velocity vector, or must provision be made for rotating the antenna?

11.4 A real aperture of length L has half the resolution in cross range as does a synthetic aperture of length L. Suppose that a stationary real aperture consists of an array of elements with spacing of $L_r = VT_r$, where V is the velocity of the synthetic aperture and T_r is the pulse spacing of the synthetic-aperture waveform.

Describe how the real aperture can be used so that the resolution of the real aperture becomes the same as the resolution of a synthetic aperture of the same length. Does this have an effect on the signal-to-noise ratio?

11.5 A synthetic-aperture imaging radar is used to image a scene at a nominal angle of 45° from the radar velocity vector. Does the inferred position of the image depend on knowledge of the magnitude of the velocity vector? If so, how does the image change because of a 1% error in knowledge of the velocity magnitude? How does the image change because of an error in the velocity direction?

11.6 An optical beam is focused by using a quadratic-phase lens. A radar beam is focused by using a parabolic antenna. Can these two statements be reconciled? How?

11.7 Consider a coherent pulse train that consists of n identical, equispaced pulses.

a. What is the ambiguity surface for this pulse train?
b. Give an expression for the sum of the individual ambiguity surfaces of the individual pulses.
c. Compare these two cases.
d. A proposed imaging radar computes the sample cross-ambiguity function for each individual echo pulse, then adds the magnitudes. Is this a reasonable proposal? Why?

11.8 An aircraft moving along the y axis at a velocity of 1,000 ft/sec is crossing the x axis at $t = 0$. The aircraft carries a sidelooking, synthetic-aperture imaging radar that uses a carrier frequency of 10 GHz. Along the x axis is a perfectly straight highway on which two cars are moving at location $x = 100$ nm and $y = \pm 10$ ft (and within the beam): one moving with velocity equal to 100 ft/sec, and one moving with velocity equal to -100 ft/sec. Describe the position and appearance of these cars in the radar image for a waveform duration of 0.1 sec and a waveform bandwidth of 10 MHz.

11.9 An imaging radar at a carrier frequency of 10 GHz and a pulse repetition interval of 1 millisecond is carried by an aircraft with a velocity of 1,000 ft/sec. Describe the locus of ground points corresponding to each doppler grating lobe.

Describe an interferometric imaging radar that will reject large discrete objects from entering the image through a grating lobe by determining and cancelling the grating lobe in which the signal entered. Will this work for small distributed objects in the grating lobe? What advantage might this have over a scheme using a sharp antenna beam that illuminates only objects in the main doppler lobe?

11.10 Show that when a reflecting object viewed by a synthetic-aperture radar has a velocity in the range direction, it shifts the position of the imaged object in the cross-range direction, and when it has a velocity in the cross-range direction, it blurs the image of the object in the range direction.

11.11 A helicopter has a rotor with three uniformly spaced blades, each 10 feet long, rotating at 60 rpm. Each blade has a reflectivity density uniformly distributed along its length. Each blade has a radar cross section of one square foot, and the body of the helicopter has a radar cross section of 10 square feet. Does the image change when the blades are not rotating?

Describe the general appearance of the sample cross-ambiguity function for a helicopter illuminated by a radar with a 10 GHz carrier and a pulse train waveform consisting of 100 pulses of pulse width 1 microsecond spaced 100 microseconds apart. An understandable sketch will suffice.

11.12 Explain how an interferometric technique can be used to erase moving vehicles from a synthetic-aperture radar image. Explain how this technique can be used to relocate the detected positions of moving vehicles to their proper place in the scene. Select parameters for a sidelooking synthetic-aperture radar with an aircraft velocity of 300 meters per second and a range of 100 kilometers. What baseline distance should be used for the interferometer?

11.13 A common waveform for a synthetic-aperture radar is a uniformly spaced pulse train of chirp pulses. For a sidelooking antenna in uniform motion at velocity V with a pulse interval is T_r, what is the angle between the grating lobes? By spacing two antenna elements, design a two-element phased array for the real antenna that will have a real antenna null exactly on the first grating lobe of a detected reflector. How must T_r be controlled as a function of V to make this work? What happens to the second and third grating lobes?

11.14 Reorganize the computations of the algorithm for radar imaging of diffuse targets to improve computational efficiency. Approximately how many multiplications are required in each iteration?

11.15 Show that the doppler effect moves a spectroscopic line at wavelength λ_0 to a spectroscopic line at wavelength λ given by

$$\lambda = \lambda_0 \frac{1 + (V/c)\cos\phi}{\sqrt{1 + (V/c)^2}}$$

for a velocity vector with magnitude V and angle ϕ with respect to the line of sight where c is the speed of light. The denominator is significant only when the velocity V is large enough to produce relativistic effects. Can there be a "doppler-like" effect even when the motion is orthogonal to the line of sight?

Notes

The early history of imaging radar is described in the article by Sherwin, Ruina, and Rawcliff (1962) and in the article by Brown and Porcello (1969). This early work was dominated by the need to process massive amounts of data at a time when digital computers did not yet exist. The processing was made possible by means of the dechirping of chirp pulses to change the processing problem into the form of a two-dimensional Fourier transform that is suitable for optical processing using a lens. Later work, as by Brown and Fredericks (1969), studied various motion effects. The use of range-doppler sorting in radar astronomy, as by the ambiguity function, to separate radar returns from different features of a planet so as to form an image was suggested and later demonstrated by Green (1962). He described the magnitude image as a convolution of the radar cross section with the square of the ambiguity function, and realized that his ideas were essentially those of a synthetic-aperture radar applied to the radar astronomy problem. Tagfors and Campbell (1973) gave an early survey of radar astronomy. A radar system for astronomical imaging would be flawed because of the doppler ambiguity were it not for the use of interferometry to resolve this ambiguity, so it is not surprising that the incorporation of interferometry into synthetic-aperture radar first occurred in radar astronomy. It was proposed by Manasse (1959) and demonstrated by Campbell (1971).

Synthetic-aperture radar was originally conceived as a way to image a two-dimensional surface. Graham (1974) described the incorporation of interferometry into a synthetic-aperture radar as a way of annotating the image with altitude. A similar application to synthetic-aperture sonar was patented by Spiess and Anderson (1983). Further work on interferometric synthetic-aperture radar was reported by Zebker and Goldstein (1986), and by Hirasawa and Kobayashi (1986). A landmark paper by Munson, O'Brien, and Jenkins (1983) presented the then novel idea that, under the right conditions, synthetic-aperture radar can be given a tomographic formulation. The use of a tomographic formulation of imaging radar in planetary astronomy suggests the use of a three-dimensional projection-slice theorem with the constraint that the image lies on a surface, not necessarily a planer surface. This method of treating three-dimensional objects, motivated by the topic of imaging the moon and planets, was studied by Webb and Munson (1995), and by Webb, Munson, and Stacy (1998). A general formulation of a theory of radar imaging of three-dimensional objects was presented by Jakowatz and Thompson (1995).

The concept of a synthetic-aperture radar immediately suggests the notion of a synthetic-aperture sonar. Synthetic-aperture sonar was studied early on by Cutrona (1975) and followed by a patent by Gilmour (1978). High-resolution synthetic-aperture sonar using hydrophone arrays for beam steering has been developed for imaging the ocean floor as well as for other underwater applications.

Motivated by the particular characteristic of reflection at optical frequencies, Shapiro, Capron, and Harney (1981) proposed that lidar systems use a diffuse model of reflectivity in which the two-dimensional reflectivity density is the desired zero-mean random image. A similar model of diffuse reflectivity had been proposed by

Van Trees (1971) for microwave frequencies, with earlier work in this direction by Gaarder (1968). The use of maximum-likelihood methods for radar imaging of diffuse reflector scenes was proposed by Snyder, O'Sullivan, and Miller (1989). Iterative algorithms for this imaging problem, as formed by the expectation-maximization method, were studied by Moulin, O'Sullivan, and Snyder (1992), and by Lanterman (2000). This method of imaging the reflectivity variance is a radical departure from the more traditional methods for forming radar images.

12 Radar Search Systems

A radar processor consists of a preprocessor, a detection and estimation function, and a postprocessor. A received radar signal first encounters the preprocessor, which often can be viewed as the computation of the sample cross-ambiguity function, though perhaps in the suitably simplified form of a conventional matched filter. A search radar is one whose preprocessor output typically consists of isolated peaks that are examined to detect large objects in the environment and to estimate parameters associated with those objects. The topics of detection and estimation are studied here in the context of a radar search system but apply to many situations.

A reflecting object may be made up of many individual reflecting elements, such as corners, edges, and so on. When the resolution of the radar is coarse compared to the size of the individual reflecting elements, then the reflecting object may be regarded as a single point object and it appears as a single peak in the preprocessor output. The search radar detects that peak, and the delay and doppler coordinates of the peak form an estimate of the delay and doppler coordinates of the reflector considered as a single object.

When the resolution of the radar is fine compared to the size of an individual reflector, there will be structure in the processed image. Then the search radar begins to take on the character of an imaging radar. Thus, a somewhat loose distinction between a search radar and an imaging radar can be made based on the relationship between the resolution of the radar and the spacing between reflecting elements within a local scene. This chapter studies the tasks of detection and estimation as applied to the specialized problems of radar and sonar search systems. These tasks are special cases of the general theories of detection and estimation, which are well-developed topics in the subject of mathematical statistics.

12.1 The Radar Range Equation

For a search radar, the resolution of the radar is usually comparable to the size of the target objects. The reflectivity density $\rho(\tau, \nu)$ might then be modeled as a sparse set of discrete point reflectors, each point reflector corresponding to an object and represented by an impulse. The ith point reflector has a complex reflectivity, ρ_i, at the delay-doppler coordinates (τ_i, ν_i). Then the reflectivity density $\rho(\tau, \nu)$ is the sum of amplitude-weighted impulses given by

12.1 The Radar Range Equation

$$\rho(\tau,\nu) = \sum_i \rho_i \delta(\tau - \tau_i, \nu - \nu_i),$$

with one impulse at each reflecting object. By design of a search radar, the separation between any two impulses in the τ, ν plane is usually large in comparison to the resolution of the radar.

It may be more convenient to deal with the radar cross section of a reflector because it is more convenient to deal with the reflected signal energy rather than the reflected signal amplitude. Thus,

$$\sigma(\tau,\nu) = |\rho(\tau,\nu)|^2$$
$$= \sum_i \sigma_i \delta(\tau - \tau_i, \nu - \nu_i),$$

where $\sigma_i = |\rho_i|^2$ is the radar cross section of the ith target object. Because each object is modeled as a point target, it can be characterized simply by its radar cross section σ_i, which is a nonnegative real number in contrast to the complex reflectivity ρ_i.

The performance of a search radar depends in large measure on the amount of reflected energy E_{pr} in the pulse that reaches the receiver as compared to the internal thermal noise inevitably generated within the receiver.

In the case of a direct one-way, point-to-point link between a transmitter and a receiver, the equation that expresses the received pulse energy E_{pr} in terms of the transmitted pulse energy E_{pt} is

$$E_{pr} = \left(\frac{\lambda^2}{4\pi}G_r\right)\left(\frac{1}{4\pi R^2}\right) G_t E_{pt},$$

where G_t and G_r are the gains of the transmitting and receiving antennas in the direction of propagation, R is the distance from transmitter to receiver, and λ is the wavelength of the radiation. This energy expression is called the *Friis equation*.

In the case of a two-way monostatic or bistatic radar between transmitter and receiver, the equation that expresses the received pulse energy E_{pr} in terms of the transmitted pulse energy E_{pt} is

$$E_{pr} = \left(\frac{\lambda^2}{4\pi}G_r\right)\left(\frac{1}{4\pi R_r^2}\right) \sigma \left(\frac{1}{4\pi R_t^2}\right) G_t E_{pt},$$

where G_t and G_r are the gains of the transmitting and receiving antennas in the direction of propagation, σ is the radar cross section of the reflector, R_t and R_r are the distances from transmitter to reflector and from reflector to receiver, and λ is the wavelength of the radiation. This energy expression is called the *radar range equation*.[1]

The Friis equation and the radar range equation are written with the terms arranged to tell the story of the energy bookkeeping by starting at the right side. The transmitted

---

[1] Replacing E_{pr} and E_{pt} with P_r and P_t restates the radar range equation in terms of power rather than energy. The energy in a pulse is the power integrated over time. Energy is usually the more convenient quantity for our purposes.

energy E_{pt} appears at the output of the antenna as the effective radiated energy $G_t E_{pt}$. The energy spreads in spherical waves. The term $4\pi R_t^2$ is the area of a sphere of radius R_t. By dividing $G_t E_{pt}$ by this term, we have the energy per unit area that passes through a surface at the distance R_t from the transmitter. A reflector reradiates a portion of this energy in the direction of the receiver, as described by the radar cross section σ. Dividing the reflected energy by $4\pi R_r^2$ gives the energy per unit area that passes through a spherical surface at the distance R_r from the reflector. Of this reflected energy, a fraction is captured by the receiving antenna. The energy captured by the receiving antenna is determined by the effective area A_e of the receiving antenna. It is related to the antenna gain of the receiving aperture by antenna theory as

$$A_e = G_r \frac{\lambda^2}{4\pi}.$$

The product of all factors leads to the radar range equation. For a monostatic radar with $R_t = R_r = R$ and $G_t = G_r = G$, the equation reduces to the form

$$E_{pr} = G^2 \frac{\lambda^2 \sigma}{(4\pi)^3 R^4} E_{pt}.$$

The radar range equation is written in many alternative forms by grouping the terms differently, or by renaming groups of terms. Another commonly used form of the equation is

$$E_{pr} = G_r \left(\frac{\lambda^2}{4\pi R_r^2}\right) \left(\frac{\sigma}{4\pi \lambda^2}\right) \left(\frac{\lambda^2}{4\pi R_t^2}\right) G_t E_{pt}.$$

This rearrangement of terms normalizes range by wavelength and normalizes σ by the square of the wavelength.

12.2 Coherent Detection of a Pulse in Noise

The radar-detection problem in its most elementary form consists of the received signal $v(t)$, which is either the pulse $s(t)$ contaminated by the stationary additive noise $n(t)$ or is the noise $n(t)$ alone. In general, the pulse $s(t)$ is a complex pulse and $n(t)$ is a complex noise process. The noise process $n(t)$ is known by its mean and by its autocorrelation function $\phi(\tau)$. The pulse may be received in the form $s(t - \tau_0) e^{j(2\pi \nu_0 t + \theta_0)}$, in which form it includes a known arrival time delay, τ_0, a known doppler shift, ν_0, and a known phase shift, θ_0. When convenient to do so, the pulse $s(t)$ is easily redefined so it includes all of these known parameters. They do not need to be displayed explicitly.

The task is to detect the presence of the pulse $s(t)$. In this section, because the phase shift θ_0 is known, the task is called *coherent detection*. When the phase shift θ_0 is not known, as is the case studied in Section 12.5, the task is called *noncoherent detection*.

The detection of a real or complex pulse in noise is a special case of the general theory of binary hypothesis testing. In general, this problem of testing hypotheses consists of two hypotheses, called H_0 and H_1. One and only one of these two hypotheses is true, and the problem is to decide which hypothesis is true. Hypothesis H_0 is called

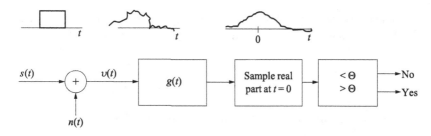

Figure 12.1 Detection of a pulse in noise

the *null hypothesis* and hypothesis H_1 is called the *alternative hypothesis*. They are defined as

$$H_0: v(t) = n(t) \quad \text{(noise only)},$$
$$H_1: v(t) = s(t) + n(t) \quad \text{(signal plus noise)},$$

where the signal $s(t)$ is a known real or complex pulse of finite energy E_p and the noise $n(t)$ is a real or complex, stationary, zero-mean random process. Complex noise is denoted $n(t) = n_R(t) + jn_I(t)$ whose real and imaginary noise components $n_R(t)$ and $n_I(t)$ are independent, identically distributed stationary, zero-mean random processes.

The real noise $n(t)$ has the autocorrelation function $\phi(\tau) = E[n(t)n(t+\tau)]$ and the power density spectrum $N(f)$, which is the Fourier transform of $\phi(\tau)$. Complex noise has the autocorrelation function $\phi(\tau) = E[n(t)n^*(t+\tau)]$ and the power density spectrum $N(f)$, which is the Fourier transform of random process $\phi(\tau)$. Commonly, $n(t)$ is real or complex *gaussian* noise,[2] but the gaussian property need not be imposed at this time. Only the second-order properties of $n(t)$ will be used initially.

The elementary detection problem is to decide between H_0 and H_1 based on an observation of $v(t)$. This amounts to formulating a rule that assigns every possible $v(t)$ to either of two disjoint sets, $\mathcal{U}_0$ or $\mathcal{U}_1 = \mathcal{U}_0^c$, that together include all possible measurements. When $v(t)$ is in $\mathcal{U}_0$, the decision is in favor of hypothesis H_0; that there is no pulse in the received signal. When $v(t)$ is in $\mathcal{U}_1$, the decision is in favor of hypothesis H_1; that there is the pulse $s(t)$ in the received signal. Of course, the sets $\mathcal{U}_0$ and $\mathcal{U}_1$ cannot be described by enumerating all of the $v(t)$ that belong to each set; those would be infinite lists. The two sets must be described by a rule called a *decision rule*. The decision rule will decide whether the observed $v(t)$ is in $\mathcal{U}_0$ or in $\mathcal{U}_1$. Thus, the decision rule will decide whether the received signal $v(t)$ consists of a pulse in noise or of noise only.

The decision rule that we choose in this case is to pass the received signal $v(t)$ through a filter with the impulse response $g(t)$, as shown in Figure 12.1, and then to test the amplitude of the real part at the output of the filter at $t=0$ by comparing that amplitude to a threshold, Θ, and deciding H_0 or H_1 based on whether the filter output is smaller or larger than Θ. Although such a filter must be causal, that causality

---

[2] Complex gaussian noise with independent, identically distributed components is called *circular* gaussian noise. This is because the bivariate probability distribution is unchanged under a coordinate rotation.

condition is not yet imposed. For an amplitude larger than the threshold at time zero, the decision is that the pulse is present in the received signal. Then the filter output is the convolution

$$u(t) = \int_{-\infty}^{\infty} v(\xi)g(t-\xi)d\xi.$$

The filter output at the sampling instant $t=0$ is equal to the correlation

$$u(0) = \int_{-\infty}^{\infty} g(-t)v(t)dt = \int_{-\infty}^{\infty} g(-t)(s(t) + u(t))dt.$$

For noise of power density spectrum $N(f)$, the filter $g(t)$ that maximizes the signal-to-noise ratio for pulse $s(t)$ at time zero is the whitened matched filter for $s(t)$ as was developed in Section 2.7.

It is not obvious at this point that maximizing the signal-to-noise ratio at the single time instant $t=0$ is the optimum thing to do for the detection problem. In general, it is not. It is shown at the end of the next section that for white gaussian noise, the matched filter followed by a threshold test is the optimum detection procedure. When the noise is not gaussian, the matched-filter detector is still a very practical and popular detector. Then it is optimal only in the restricted class of detectors that have the form of a linear filter followed by a threshold test.

To determine the performance of the matched-filter detector, two probabilities of error must be evaluated. For the detection problem, the probability that the real part of the filter output exceeds the threshold Θ when the input is noise only is called the *probability of false alarm* and is denoted α or p_{FA}. The probability that the filter output does not exceed the threshold when the input is signal plus noise is called the *probability of missed detection* and is denoted β or p_{MD}.

The matched filter collapses the received waveform $v(t)$ into the single number with real part $\text{Re}[u(0)]$. The probability distribution of $\text{Re}[u(0)]$ under each hypothesis must be determined.

Under the hypothesis H_0, the mean $E[\text{Re}[u(0)]]$ of $\text{Re}[u(0)]$ is zero. Under the hypothesis H_1, the mean $E[\text{Re}[u(0)]]$ of $\text{Re}[u(0)]$ is a real number described as follows:

$$E[\text{Re}[u(0)]] = \int_{-\infty}^{\infty} g(-t)s(t)dt$$
$$= \int_{-\infty}^{\infty} G(f)S(f)df$$
$$= \int_{-\infty}^{\infty} \frac{|S(f)|^2}{N(f)} df.$$

This follows from Section 2.7 where the whitened matched filter $G(f) = S^*(f)/N(f)$ is discussed showing that the mean of the output at time zero is always real. The signal power at the sampling instant is the square of this term.

Under either hypothesis, the variance per component when the noise is complex gaussian noise is the real number

12.2 Coherent Detection of a Pulse in Noise

$$N = \tfrac{1}{2}\text{var}[u(0)] = \text{var}\big[\text{Re}[u(0)]\big] = \sigma^2$$
$$= \int_{-\infty}^{\infty} N(f)|G(f)|^2 df$$
$$= \int_{-\infty}^{\infty} \frac{|S(f)|^2}{N(f)} df.$$

The signal-to-noise ratio is

$$\frac{S}{N} = \frac{\big(E[\text{Re}[u(0)]]\big)^2}{N} = \int_{-\infty}^{\infty} \frac{|S(f)|^2}{N(f)} df.$$

For white noise, the power density spectrum is $N(f) = N_0/2$ and the signal-to-noise ratio is

$$\frac{S}{N} = \frac{2E_p}{N_0}.$$

This statement holds for white noise with any noise probability density function.

To numerically evaluate the probability of false alarm p_{FA} and the probability of missed detection p_{MD}, the noise probability density function must be specified more fully. When the input to the matched filter is white gaussian noise, the output at $t = 0$ is a gaussian random variable, and the mean and variance under each hypothesis have been found. Therefore, for gaussian noise, the probability density functions on the output of the matched filter under hypotheses H_0 and H_1 are

$$p_0(x) = \frac{1}{\sqrt{2\pi}\sigma} e^{-x^2/2\sigma^2},$$
$$p_1(x) = \frac{1}{\sqrt{2\pi}\sigma} e^{-(x-A)^2/2\sigma^2},$$

where $A = E[\text{Re}[u(0)]]$. The probability of a false alarm and the probability of a missed detection are given by

$$p_{FA} = \int_{\Theta}^{\infty} p_0(x) dx,$$
$$p_{MD} = \int_{-\infty}^{\Theta} p_1(x) dx.$$

Make the change in the variables $z = \frac{x}{\sigma}$ in the first integral and $z = \frac{x-A}{\sigma}$ in the second integral. Using the symmetry of the gaussian function, write the integrals in terms of the following standard function:

$$Q(x) = \int_{x}^{\infty} \frac{1}{\sqrt{2\pi}} e^{-z^2/2} dz,$$

called the *error function*. Then

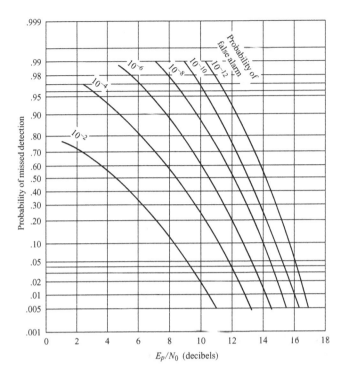

Figure 12.2 Performance of a coherent detector in gaussian noise

$$p_{FA} = Q\left(\frac{\Theta}{\sigma}\right),$$

$$p_{MD} = Q\left(\frac{A - \Theta}{\sigma}\right).$$

The error probabilities are also written in terms of the ratio E_p/N_0 as

$$p_{FA} = Q\left(\lambda\sqrt{2E_p/N_0}\right),$$

$$p_{MD} = Q\left((1-\lambda)\sqrt{2E_p/N_0}\right),$$

where $\lambda = \Theta/A$ is a parameter between $-\infty$ and ∞. Normally, λ is between 0 and 1. The expressions depend only on the ratio E_p/N_0 and the parameter λ. By crossplotting $p_{FA}(\lambda)$ and $p_{MD}(\lambda)$ for each value of λ, the parameter λ can be suppressed from the presentation.

The performance of the coherent detector of a pulse in gaussian noise is shown in Figure 12.2 with E_p/N_0 expressed in decibels. The performance of the coherent detector should be compared to the performance of the noncoherent detector that is analyzed in Section 12.5 and shown in Figure 12.7.

It is important to observe that for both the coherent detector and the noncoherent detector, the performance of the matched-filter detector in gaussian noise depends only on the ratio E_p/N_0 but not on the detailed structure of $s(t)$. Only the energy of the pulse matters. The shape of the pulse may be chosen for reasons of convenience or to satisfy other conditions.

12.3 The Neyman–Pearson Theorem

Now that the matched filter followed by a threshold has been studied as one kind of detector of a signal in noise, the discussion turns to the larger task of finding the optimal such detector of a signal in noise. This requires a statement of optimality, as given in this section. The conclusion is that when the noise is white gaussian noise, but not otherwise, the optimal detector is indeed the matched filter followed by a threshold as studied in Section 12.2.

When the noise is nonwhite gaussian noise, the optimal detector is the *whitened* matched filter followed by a threshold, as will be shown. When the noise is not gaussian, or when it is dependent on the signal, then the optimal detector does not have this structure of a filter followed by a threshold. Generally, it will have nonlinearities in addition to a threshold.

The Neyman–Pearson theorem is a general theorem in the subject of binary hypothesis testing. It deals with the case in which there is no prior probability on the validity of the two hypotheses, H_0 and H_1. In many applications, it may be unreasonable to insist that such a probability distribution could be known, even in principle. For example, one would not like to insist that a meaningful probability model could be constructed for the event that there will be an aircraft at a particular point of space at a given time. The Neyman–Pearson point of view sidesteps this question by refusing to introduce the notion of a prior probability on the two hypotheses.

To start out, suppose that the real signal $v(t)$ is summarized by the single number, called a *statistic* or *measurement*. To emphasize that it is a function of $v(t)$, the statistic x is denoted $\Lambda(v(t))$. Perhaps the statistic $\Lambda(v(t))$ is a sample of $v(t)$ at some time t_0, so that $\Lambda(v(t)) = v(t_0)$, or perhaps is a sample of a filtered version of $v(t)$. First, the detection rule given only the statistic x is determined. Then the method used to form the optimum statistic from the received signal $v(t)$ is developed.

The statistic $x = \Lambda(v(t))$ is a real random variable. Associated with each of the two simple hypotheses H_0 and H_1 is a probability distribution on the statistic x. When H_0 is true, then $p_0(x)$ is the probability density function on the statistic. When H_1 is true, then $p_1(x)$ is the probability density function on the statistic x.

As for the case of pulse detection, a simple measurement consists of an observation of one realization of the statistic x. For some values of x, the decision is that hypothesis H_0 is true. For other values of x, the decision is that hypothesis H_1 is true. A hypothesis-testing rule is a partition of the measurement space into two disjoint sets, $\mathcal{U}_0$ and $\mathcal{U}_1$. The decision is that H_0 or H_1 is true according to whether x is an element of $\mathcal{U}_0$ or of $\mathcal{U}_1$. Each partition is a different hypothesis-testing rule and, except for randomized rules, which are not considered, there are no other rules. The best hypothesis-testing rule should be used, but first the meaning of "best" rule should be stated.

Accepting hypothesis H_1 when H_0 is actually true is called a *type I error*[3] (or a *false alarm* or a *false positive*), and the probability of this event is denoted by α. Accepting hypothesis H_0 when H_1 is actually true is called a *type II error* (or a *missed detection* or a *false negative*), and the probability of this event is denoted by β. Obviously,

$$\alpha = \int_{\mathcal{U}_1} p_0(x) \qquad \beta = \int_{\mathcal{U}_0} p_1(x).$$

The problem is to specify $(\mathcal{U}_0, \mathcal{U}_1)$ so that α and β are as small as possible. This is not yet a well-defined problem because α generally can be made smaller by reducing $\mathcal{U}_1$, although β thereby is increased. The Neyman–Pearson point of view assumes that a maximum value of β is specified. The sets $\mathcal{U}_0$ and $\mathcal{U}_1$ must be determined to minimize α subject to the constraint that β is not larger than this specified maximum.

Theorem 12.3.1 (Neyman–Pearson Theorem) *For any real number Θ, let*

$$\mathcal{U}_0(T) = \{x : p_1(x) \leq p_0(x)e^{-\Theta}\},$$
$$\mathcal{U}_1(T) = \{x : p_1(x) > p_0(x)e^{-\Theta}\},$$

and let α^* and β^* be the probabilities of type I and type II error corresponding to this choice of decision regions. Suppose that α and β are the probabilities of type I and type II errors corresponding to some other choice of decision regions, and suppose that $\alpha < \alpha^*$. Then $\beta > \beta^*$.

Proof:
Let $\{(\mathcal{U}'_0, \mathcal{U}'_1)\}$ be any other decision procedure such that $\alpha < \alpha^*$. Define the indicator functions on x as: $\phi(x) = 1$ when $x \in \mathcal{U}_1(\Theta)$ and otherwise $\phi(x) = 0$; and $\phi'(x) = 1$ when $x \in \mathcal{U}'_1$ and otherwise $\phi'(x) = 0$. Then $(p_1(x) - p_0(x)e^{-\Theta})(\phi(x) - \phi'(x)) \geq 0$ for all x, which can be verified by separately examining the cases $x \in \mathcal{U}_0(\Theta)$ and $x \in \mathcal{U}_1(\Theta)$. Therefore,

$$\int_{-\infty}^{\infty} (p_1(x) - p_0(x)e^{-\Theta})(\phi(x) - \phi'(x)) dx \geq 0.$$

Expanding the product gives

$$\int_{-\infty}^{\infty} p_1(x)\phi(x)dx - \int_{-\infty}^{\infty} p_1(x)\phi'(x)dx \geq e^{-\Theta}\left[\int_{-\infty}^{\infty} p_0(x)\phi(x)dx - \int_{-\infty}^{\infty} p_0(x)\phi'(x)dx\right].$$

Hence

$$(1 - \beta^*) - (1 - \beta) \geq e^{-\Theta}(\alpha^* - \alpha).$$

By assumption $\alpha^* - \alpha > 0$, so it must hold that $\beta - \beta^* > 0$, as was to be proved. ∎

Notice that because the theorem does not explicitly define α^* as a function of β^* but rather expresses the optimum pair parametrically as $\alpha^*(\Theta)$ and $\beta^*(\Theta)$, many values

---

[3] The convention is unfortunate in that the errors are denoted type I and type II, while the hypotheses are subscripted 0 and 1.

12.3 The Neyman–Pearson Theorem

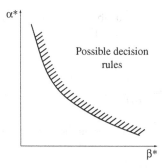

Figure 12.3 The region of possible decision rules

of the threshold Θ must be examined to find the smallest α^* satisfying the constraint on the maximum permissible β^*. An efficient procedure is to compute $\alpha^*(\Theta)$ and $\beta^*(\Theta)$ for a range of Θ, then to crossplot $\alpha^*(\Theta)$ and $\beta^*(\Theta)$ to construct a graph of α^* versus β^*.

Figure 12.3 presents a typical graph of α^* versus β^*. This graph provides a visual portrayal of the Neyman–Pearson theorem. The theorem says that every possible decision rule has a performance lying above or on the curve, but no decision rule has a performance lying below the curve. The Neyman–Pearson threshold rules have their performance lying on the curve.

The Neyman–Pearson decision regions can be rewritten in terms of the loglikelihood ratio, which is given by

$$\Lambda(x) = \log \frac{p_0(x)}{p_1(x)},$$

as the sets

$$\mathcal{U}_0(T) = \mathcal{U}_1(T)^c = \{x : \Lambda(x) \geq \Theta\}.$$

The decision is that H_0 is true when the loglikelihood ratio is at least as large as the threshold Θ. It is clear that $\Lambda(x)$ (or any monotonic function of $\Lambda(x)$) is the significant function in the Neyman–Pearson theorem rather than the two probability distributions individually.

Suppose that the single measurement x is replaced by a block of independent, identically distributed measurements. Consider the block as a vector of measurements of length n. Then write the vector measurement as an n-tuple of simple measurements,

$$v = (x_1, \ldots, x_n),$$

where x_ℓ is the value of the ℓth measurement. With the measurement in the form of blocks, the theory holds just as before. However, the vector structure permits the problem to be broken into pieces. The loglikelihood ratio for the vector v is

$$\Lambda(v) = \log \frac{p_0(v)}{p_1(v)}.$$

Because the measurements are independent, the probability of a block is the product of the probabilities of the individual measurements. This means that the logarithm becomes a sum, so that

$$\Lambda(v) = \sum_{\ell=1}^{n} \log \frac{p_{0\ell}(x_\ell)}{p_{1\ell}(x_\ell)},$$

where $p_{0\ell}(x)$ and $p_{1\ell}(x)$ are probability density functions on the ℓth component

$$\Lambda(v) = \sum_{\ell=1}^{n} \Lambda(x_\ell).$$

The loglikelihood ratio of a sum of independent measurements is the sum of the loglikelihood ratios of the individual measurements.

When $p_{0\ell}(x)$ is gaussian with zero mean and variance σ^2 and $p_{1\ell}(x)$ is gaussian with mean A_ℓ and variance σ^2, the loglikelihood ratio $\Lambda(x_\ell)$ is

$$\Lambda(x_\ell) = \frac{|x_\ell - A_\ell|^2}{2\sigma^2} - \frac{|x_\ell|^2}{2\sigma^2}$$

$$= -\frac{x_\ell A_\ell^* + x_\ell^* A_\ell}{2\sigma^2} + \frac{|A_\ell|^2}{2\sigma^2}.$$

The second term does not depend on the measurement x_ℓ. It is an uninformative constant that can be ignored, as can the denominator of the first term. Then, for independent components, it is enough to write

$$\Lambda(v) = \sum_{\ell=1}^{n} \text{Re}[x_\ell A_\ell^*].$$

When the finite block v is obtained by sampling the pulse $v(t) = s(t) + n(t)$ in white noise, this becomes

$$\Lambda(v(t)) = \sum_{\ell=1}^{L} \text{Re}[v(t_\ell)s(t_\ell)].$$

In the limit, this becomes

$$\Lambda(v(t)) = \int_{-\infty}^{\infty} \text{Re}[v(t)s(t)] dt.$$

Thus, for a gaussian noise source, the matched filter is not only optimal in the class of linear decision rules, it is optimal in general. The conclusion is that the threshold detector in Section 12.2 is optimal for gaussian noise because it is the loglikelihood detector.

When the noise is not white, it can be made white by a noise whitening filter. This means that the whitened matched filter followed by a threshold is optimal for stationary gaussian noise with power density spectrum $N(f)$.

12.4 Rayleigh and Ricean Probability Distributions

When the carrier phase is unknown, the detection statistic is the magnitude of the complex matched-filter output. This statistic corresponds to taking the square root of the sum of the squares of the real and imaginary components. When the input is a signal in additive white gaussian noise, the output of each component of the matched filter at the sample instant is a gaussian random variable defined in terms of two random variables, say X and Y. To evaluate the probability of error of noncoherent detection, the density function of the magnitude of complex gaussian noise is required. This is the density function of the square root of the sum of the squares of two independent gaussian random variables. It can be viewed as a transformation of the noise from rectangular coordinates (in-phase and quadrature components) to polar coordinates (amplitude and phase components).

Noise Only

For random variables X and Y that are independent, zero mean, and have equal variance, the two-dimensional gaussian probability density function is

$$p(x, y) = \frac{1}{2\pi\sigma^2} e^{-(x^2+y^2)/2\sigma^2},$$

as shown in Figure 12.4. The transformation from rectangular coordinates to polar coordinates is

$$r = \sqrt{x^2 + y^2},$$
$$\phi = \tan^{-1}\frac{x}{y}.$$

An expression for the probability density function of the amplitude r is required.

The probability of the measurement lying within any region $\mathcal{A}$ must be the same whether the measurement is expressed in rectangular coordinates or in polar coordinates. That is, for any region $\mathcal{A}$,

$$\int_{\mathcal{A}} p(r, \phi) dr d\phi = \int_{\mathcal{A}} \frac{1}{2\pi\sigma^2} e^{-(x^2+y^2)/2\sigma^2} dx dy,$$

where $p(r, \phi)$ is the probability density function in polar coordinates. On the right side, substitute

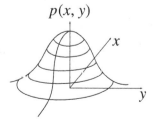

Figure 12.4 Two-dimensional gaussian probability density function

$$x^2 + y^2 = r^2,$$
$$dxdy = rd\phi dr.$$

This gives

$$\int_A p(r,\phi)drd\phi = \int_A \frac{1}{2\pi\sigma^2} e^{-r^2/2\sigma^2} rdrd\phi,$$

from which it follows that

$$p(r,\phi) = \frac{r}{2\pi\sigma^2} e^{-r^2/2\sigma^2} \qquad 0 \le \phi < 2\pi,$$

which is uniform in ϕ. Integrating over the range of ϕ multiplies $p(r,\phi)$ by 2π giving the marginal

$$p(r) = \frac{r}{\sigma^2} e^{-r^2/2\sigma^2} \qquad r \ge 0.$$

This probability density function is known as a *rayleigh probability density function*, and the corresponding random variable is called a *rayleigh random variable*. It is the probability density function for the envelope of unbiased, complex gaussian noise. A rayleigh random variable has mean $\sigma\sqrt{\pi/2}$ and variance $\sigma\sqrt{2}$.

To write the rayleigh probability density function in a standard form, define the rayleigh density function as

$$p_{ra}(z) = ze^{-z^2/2} \qquad z \ge 0.$$

Then

$$p(r) = \frac{1}{\sigma} p_{ra}\left(\frac{r}{\sigma}\right),$$

with the change of variables $z = r/\sigma$.

Signal Plus Noise

For a signal plus noise, a two-dimensional gaussian probability density function with the nonzero mean $(\bar{x}, \bar{y})$ must be used. When the mean is nonzero, the two-dimensional gaussian probability density function is

$$p(x,y) = \frac{1}{2\pi\sigma^2} e^{-\left((x-\bar{x})^2 + (y-\bar{y})^2\right)/2\sigma^2}.$$

Write the mean in the form

$$\bar{x} = A\cos\theta,$$
$$\bar{y} = A\sin\theta,$$

where θ is an unknown phase angle. Then the in-phase and quadrature components have the gaussian distributions

$$p_R(x) = \frac{1}{\sqrt{2\pi}\sigma} e^{-(x - A\cos\theta)^2/2\sigma^2},$$

$$p_I(y) = \frac{1}{\sqrt{2\pi}\sigma} e^{-(y - A\sin\theta)^2/2\sigma^2}.$$

12.4 Rayleigh and Ricean Probability Distributions

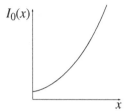

Figure 12.5 The modified Bessel function

As before, the transformation of variables gives

$$p(r, \phi) = \frac{r}{2\pi\sigma^2} e^{-(r^2 - 2Ar\cos(\theta-\phi) + A^2)/2\sigma^2}.$$

Integrating over ϕ gives

$$p(r) = \frac{r}{2\pi\sigma^2} e^{-r^2/2\sigma^2} e^{-A^2/2\sigma^2} \int_{-\pi}^{\pi} e^{Ar\cos(\theta-\phi)/\sigma^2} d\phi.$$

The integral is clearly independent of θ because the integral is periodic and extends over one period whatever the value of θ. The integral can be expressed in terms of the standard function $I_0(x)$ known as the *modified Bessel function* of the first kind and order zero.[4] The modified Bessel function $I_0(x)$ is shown in Figure 12.5. Now write

$$p(r) = \frac{r}{\sigma^2} e^{-(r^2 + A^2)/2\sigma^2} I_0\left(\frac{Ar}{\sigma^2}\right) \qquad r > 0.$$

To express this function in terms of a single parameter, let

$$p_{ri}(z, a) = z e^{-(z^2 + a^2)/2} I_0(az).$$

Therefore, setting

$$z = \frac{r}{\sigma} \quad \text{and} \quad a = \frac{A}{\sigma},$$

gives

$$p(r) = \frac{1}{\sigma} p_{ri}\left(\frac{r}{\sigma}, \frac{A}{\sigma}\right).$$

The probability density $p_{ri}(z, a)$ is known as a *ricean probability density function* with the parameter a. The corresponding random variable is called a *ricean random variable*. The parameter a defines a family of probability densities, one for each value of a, as shown in Figure 12.6. The ricean probability density function reduces to the rayleigh probability density function when a is equal to zero. When a is large, the ricean probability density function resembles a gaussian probability density function with a large mean. However, the ricean density is zero when its argument is negative, while the gaussian density is not zero for any value of its argument.

---

[4] The modified Bessel function of the first kind and order zero is defined by the integral

$$I_0(x) = \frac{1}{2\pi} \int_{-\pi}^{\pi} e^{x\cos\theta} d\theta.$$

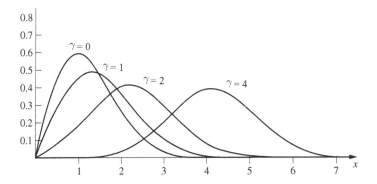

Figure 12.6 Ricean probability density functions

12.5 Noncoherent Detection of a Pulse in Noise

Binary hypothesis testing for pulse detection in gaussian noise is now treated for the case in which the pulse has an unknown phase, but is otherwise known. The two hypotheses are

$$H_0: v(t) = n(t),$$
$$H_1: v(t) = s(t)e^{-j\theta} + n(t).$$

This model corresponds to the usual radar detection problem in which a known passband pulse of unknown phase is received in noise. A pulse whose phase is unknown is called a *noncoherent pulse* or, better, a noncoherently received pulse. The noncoherent detector compares the *magnitude* of the matched-filter output to a threshold whereas the coherent detector compares the *real part* of the matched-filter output to a threshold. In other respects, the coherent detector and the noncoherent detector are the same. The difference in performance is seen by comparing Figures 12.2 and 12.7.

The analysis of the noncoherent detector exactly parallels the analysis of the coherent detector except that the magnitude function means that the gaussian densities are replaced by the ricean and rayleigh densities. These densities are

$$p_0(r) = \frac{r}{\sigma^2} e^{-r^2/2\sigma^2} \qquad r \geq 0,$$

$$p_1(r) = \frac{r}{\sigma^2} e^{-(r^2+A^2)/2\sigma^2} I_0\left(\frac{rA}{\sigma^2}\right) \qquad r \geq 0,$$

where $A = E[u(0)]$.

Accordingly, the log-likelihood ratio is

$$\log \frac{p_0(r)}{p_1(r)} = \frac{A^2}{2\sigma^2} - \log I_0\left(\frac{rA}{\sigma^2}\right).$$

The log-likelihood ratio is to be compared to a threshold Θ. Because the loglikelihood ratio is monotonic in r, however, it is enough to compare r itself directly to a threshold. This threshold is a different threshold, but it is also denoted Θ. As before, the

12.5 Noncoherent Detection of a Pulse in Noise

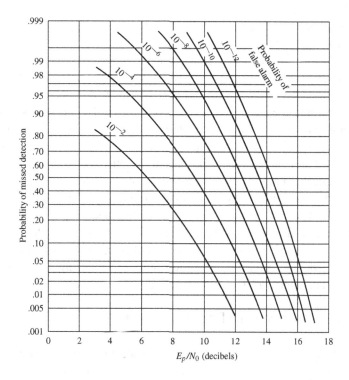

Figure 12.7 Performance of a noncoherent detector in gaussian noise

probabilities of false detection error and missed detection error are

$$\alpha = \int_{\Theta}^{\infty} p_0(r)dr,$$

$$\beta = \int_{-\infty}^{\Theta} p_1(r)dr.$$

From the theory of the matched filter, $(A/\sigma)^2 = 2E_p/N_0$ in unbiased white gaussian noise. Redefine the variable of integration as $z = r/\sigma$ and normalize the threshold as $\Theta' = \Theta/\sigma$. Then, for any Θ',

$$\alpha(\Theta') = \int_{\Theta'}^{\infty} z e^{-z^2/2} dz = e^{-\Theta'^2/2},$$

$$\beta(\Theta') = \int_0^{\Theta'} z e^{-(z^2 + 2E_p/N_0)/2} I_0\left(z\sqrt{2E_p/N_0}\right) dz,$$

where the lower limit in the second integral is set equal to zero because the integrand is zero for negative z.

These expressions now depend only on the ratio E_p/N_0 and the normalized threshold Θ'. By crossplotting $\alpha(\Theta')$ and $\beta(\Theta')$ for each value of Θ', the threshold Θ' can be suppressed from the presentation. The performance of the noncoherent detector in gaussian noise is shown in Figure 12.7. Comparing Figure 12.7 to Figure 12.2 shows that the noncoherent detector requires at most 1 dB more E_p/N_0 than the coherent detector in the region of high E_p/N_0.

Instead of detecting just one pulse, an entire pulse train can be detected as a whole. There are several cases. The pulse train can be fully coherent, or it can be internally coherent but noncoherent as a whole, or it can be fully noncoherent from pulse to pulse. By regarding a coherent pulse train itself as a pulse, the discussion in Section 12.2 applies without change. Both Figure 12.2 and Figure 12.7 apply just as before, but now E_p refers to the total energy in the pulse train. To compute the detection statistic for an internally *coherent* pulse train, filter each pulse with a matched filter and sample the complex value of each filter output, then sum the complex values and compare the magnitude of that sum to a threshold to make a detection decision. To compute the detection statistic for an internally *noncoherent* pulse train, filter each pulse with a matched filter and sample the complex value of each filter output, then sum the magnitudes of those complex values and compare that sum to a threshold to make a detection decision. The error probability of a internally noncoherent pulse train is studied in Section 14.1.

12.6 Arrival Time Estimation of a Baseband Pulse

A simple estimation problem is the estimation of the arrival time τ_0 of a real-valued baseband pulse $s(t)$ in noise. The received baseband waveform $v(t)$ is

$$v(t) = s(t - \tau_0) + n(t).$$

The noise $n(t)$ is real-valued zero-mean covariance-stationary noise with the known correlation function $\phi(\tau)$ and power density spectrum $N(f)$. The development herein is limited to the class of estimators called *linear estimators* because higher-order moments of the noise $n(t)$ might be unknown or not given and these higher-order moments are not needed for a linearized analysis. A linear estimator is determined using only the second-order properties of the noise. This means that to derive the optimum estimator, the noise $n(t)$ need not be gaussian noise. The performance, however, is evaluated only for gaussian noise.

The linear estimator passes $v(t)$ through the linear noncausal filter $g(t)$ and determines the delay estimate $\widehat{\tau}_0$ as the time when the output of the filter is a maximum. Without loss of generality, the noiseless output $g(t) * s(t)$ of the noncausal filter $g(t)$ can be constrained to take its maximum value at the origin. In the absence of noise and with input $s(t-\tau_0)$, the filter output is $g(t) * s(t-\tau_0)$. This means that $g(t) * s(t-\tau_0)$ has its maximum value at τ_0. The linear estimator chooses the time at which the maximum value of $g(t) * v(t)$ occurs as the estimate $\widehat{\tau}_0$. Because $v(t)$ contains noise, the estimate $\widehat{\tau}_0$ can be in error.

One wants to use the best, or optimum, estimator, and this requires a notion of optimality. A minimum-variance linearized estimator is one in which the filter $g(t)$ is chosen to minimize the variance of the error due to noise. Instead, the filter that maximizes the signal-to-noise ratio might be considered because this filter is already known. It is the whitened matched filter

12.6 Arrival Time Estimation of a Baseband Pulse

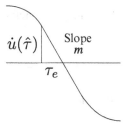

Figure 12.8 Linear Approximation to $\dot{u}(t)$

$$G(f) = \frac{S^*(f)}{N(f)},$$

which is defined in Theorem 2.7.2 of Section 2.7. When $N(f) = N_0/2$, this filter reduces to $G(f) = S^*(f)$. The factor $N_0/2$ in the denominator is neglected because it does not affect the signal-to-noise ratio at the filter output.

It turns out that the whitened matched filter does indeed minimize the error variance, as shown in the next proposition, so maximizing the signal-to-noise ratio is equivalent to minimizing noise variance. Moreover, Section 12.7, which follows, shows that for gaussian noise, this linearized estimator is asymptotically optimal over the set of all possible estimators, not just over the set of linearized estimators.

The estimate $\hat{\tau}_0$ of the delay τ_0 is taken at the peak of the filter output, here denoted $u(t)$. This is the value of τ where $\dot{u}(t)$, the time derivative of $u(t)$, is equal to zero. An asymptotic statement of the variance σ^2 of the delay error is found by using a linearized analysis based on the triangle shown in Figure 12.8. The linearized analysis is valid when the matched-filter output is sufficiently large compared to the noise. Otherwise, as suggested by Figure 12.8, second-order terms become important when the noise is large and the linearized approximation breaks down.

The linearized triangle of Figure 12.8 shows that the (negative) slope m of $\dot{u}(t)$ at $t = 0$ satisfies $m = \dot{u}(\hat{\tau})/\tau_e$, which means that the error τ_e in the estimate $\hat{\tau}$ is given by the linearized approximation $\tau_e = \dot{u}(\hat{\tau})/m$. It follows that the variance is

$$\sigma_\tau^2 \simeq \frac{\text{var}[\dot{u}(\hat{\tau}_0)]}{m^2}.$$

The denominator in this expression for σ_τ^2 is the (negative) slope m of $\dot{u}(t)$ at $\hat{\tau}_0$, which is the second derivative of $u(t)$:

$$m = \frac{d}{dt} E[\dot{u}(t)]_{t=\hat{\tau}_0}$$

$$= \frac{d}{dt} E\left[\int_{-\infty}^{\infty} \dot{g}(\xi) v(t-\xi) d\xi \right]_{t=\hat{\tau}_0} = \frac{d}{dt} \int_{-\infty}^{\infty} \dot{g}(\xi) s(t-\xi) d\xi \bigg|_{t=\hat{\tau}_0}$$

$$= \int_{-\infty}^{\infty} \dot{g}(\xi) \dot{s}(t-\xi) d\xi \bigg|_{t=\hat{\tau}_0}.$$

The Parseval formula at $t = 0$ then gives

$$m = \frac{d}{dt} E[\dot{u}(t)]_{t=\hat{\tau}_0} = (2\pi)^2 \int_{-\infty}^{\infty} f^2 S(f) G(f) df,$$

where $G(f)$ is the Fourier transform of $g(t)$.

The numerator in the expression for σ_τ^2 is the variance in $\dot{u}(t)$, which is due to the noise only. The variance is given by

$$\text{var}[\dot{u}(t)] = \int_{-\infty}^{\infty}\int_{-\infty}^{\infty} \dot{g}(\xi)\dot{g}(\xi')\text{E}[n(t-\xi)n(t-\xi')]\,d\xi\,d\xi'$$

$$= \int_{-\infty}^{\infty}\int_{-\infty}^{\infty} \dot{g}(\xi)\dot{g}(\xi')\phi(\xi-\xi')\,d\xi\,d\xi',$$

which is independent of t. Make the change in variables $\xi - \xi' = \eta$ to obtain

$$\text{var}[\dot{u}(t)] = \int_{-\infty}^{\infty} \phi(\eta)\int_{-\infty}^{\infty} \dot{g}(\xi)\dot{g}(\xi-\eta)\,d\xi\,d\eta.$$

Transform this formula into the frequency domain using the Parseval formula by noting that the convolution $\int_{-\infty}^{\infty} \dot{g}(\xi)\dot{g}(\xi-\eta)\,d\xi$ has the transform $(2\pi f)^2|G(f)|^2$. Then

$$\text{var}[\dot{u}(t)] = (2\pi)^2 \int_{-\infty}^{\infty} f^2 N(f)|G(f)|^2\,df.$$

Proposition 12.6.1 *The matched-filter estimator of real baseband pulse arrival time in additive stationary gaussian noise of power density spectrum $N(f)$ has, asymptotically, the smallest arrival-time error variance of any linear estimator. The error variance satisfies the approximation*

$$\sigma_{\tau_0}^2 \simeq \left[(2\pi)^2 \int_{-\infty}^{\infty} f^2 \frac{|S(f)|^2}{N(f)}\,df\right]^{-1},$$

where the approximation is asymptotically tight as the bracketed term increases without limit.

Proof:
Taking the ratio of $\text{var}[\dot{u}(t)]$ to m^2 gives

$$\sigma_{\tau_0}^2 = \frac{1}{(2\pi)^2} \frac{\int_{-\infty}^{\infty} f^2 N(f)|G(f)|^2\,df}{\left[\int_{-\infty}^{\infty} f^2 S(f)G(f)\,df\right]^2}.$$

The error variance is to be minimized by the choice of $G(f)$. To this purpose, the denominator is manipulated into a form that is appropriate for using the Schwarz inequality. This form involves the term $N^{1/2}(f)$ which is well-defined because $N(f)$ is real and positive. Then

$$\left|\int_{-\infty}^{\infty} f^2 S(f)G(f)\,df\right|^2 = \left|\int_{-\infty}^{\infty} \left(fG(f)N^{1/2}(f)\right)\left(\frac{fS(f)}{N^{1/2}(f)}\right)df\right|^2$$

$$\leq \int_{-\infty}^{\infty} f^2|G(f)|^2 N(f)\,df \int_{-\infty}^{\infty} f^2 \frac{|S(f)|^2}{N(f)}\,df$$

with equality if and only if

$$G(f) = \frac{S^*(f)}{N(f)}.$$

Therefore, for this choice of filter,

$$\sigma_{\tau_0}^2 \simeq \frac{\text{var}[\dot{u}(\tilde{\tau}_0)]}{m^2} = \frac{(2\pi)^2 \int_{-\infty}^{\infty} f^2 \frac{|S(f)|^2}{N(f)} df}{\left[(2\pi)^2 \int_{-\infty}^{\infty} f^2 \frac{|S(f)|^2}{N(f)} df\right]^2}.$$

The conclusion of the theorem follows. ∎

When the noise is gaussian, this estimator is asymptotically optimum in the signal-to-noise ratio as follows from the discussion in the next section. When the noise is nongaussian, the linear estimator is not the optimal estimator, but it is a valid estimator that depends only on the power density spectrum. Within the approximation of the linearized model for any covariance stationary process, the matched-filter estimator is the best *linear* estimator. When one knows only the power density spectrum of covariance stationary noise, but not higher-order moments, one chooses to use this estimator. It is robust. There may be a *nonlinear* estimator that is better when a better noise model is known, but the nonlinear estimator may be sensitive to errors in the model. It is usually not robust, meaning that it is sensitive to modeling errors.

Corollary 12.6.2 *The variance of the error in the matched-filter estimate of the arrival time of a real baseband pulse $s(t)$ in white gaussian noise is*

$$\sigma_{\tau_0}^2 \simeq \frac{1}{(2\pi)^2(2E_p/N_0)\overline{f^2}}$$

asymptotically in E_p/N_0, where

$$\overline{f^2} = \frac{1}{E_p} \int_{-\infty}^{\infty} f^2 |S(f)|^2 df.$$

Proof:
Setting $N(f) = N_0/2$ in Proposition 12.6.1 provides the proof of the corollary. ∎

When $\overline{f} = 0$, the statement of the corollary is that

$$\sigma_{\tau_0}^2 \lesssim \frac{1}{(2\pi)^2(2E_p/N_0)B_G^2},$$

where B_G is the Gabor bandwidth of the pulse $s(t)$. This expression is seen again in Proposition 12.8.1 where it describes the performance of the noncoherent estimate of pulse arrival time.

How should the estimator for a real baseband pulse be changed for a complex baseband pulse in circularly symmetric gaussian noise? The two noise components are independent. The real and complex components of the received signal can be processed separately and the two estimates weighted and averaged. But the signal output of the whitened matched filter for a complex pulse is real at the peak (and equal to E_p) and finding the peak of the real part is equivalent to the weighted average, so the estimator should find the peak of the real part of the output signal. Other than this consideration, the estimator for a known complex baseband pulse is the same as the estimator for a known real baseband pulse.

In the language of the ambiguity function, to estimate τ_0 from the received baseband signal $v(t) = s(t) + n(t)$, first compute the cross-ambiguity function $\chi_c(\tau, \nu)$ for $s(t)$ and $v(t)$ with $\nu = 0$ (the matched filter). Then find the peak of $\chi_c(\tau, 0)$. When $s(t)$ and $v(t)$ are complex, find the peak of $\text{Re}[\chi_c(\tau, 0)]$. Thus, the sample cross-ambiguity function $\chi_c(\tau, \nu)$ is a statistic for estimating τ_0 in the presence of stationary white noise. The next section shows that when the white noise is gaussian, this is asymptotically the minimum-variance estimator.

12.7 The Cramer–Rao Theorem

The Cramer–Rao theorem states an information-theoretic lower bound on the variance of any real-valued estimate $\widehat{\theta}$ of a parameter θ based on noisy data. For an unbiased estimate of the single parameter θ based on a random measurement X with probability distribution $q(x|\theta)$, one form of the Cramer–Rao lower bound is

$$\sigma_\theta^2 \geq \frac{1}{\mathrm{E}\big[(\partial/\partial\theta)\log q(x|\theta)\big]^2},$$

provided the derivative exists. An alternative form of the Cramer–Rao lower bound expressed in terms of the second derivative is

$$\sigma_\theta^2 \geq \frac{-1}{\mathrm{E}\big[(\partial^2/\partial\theta^2)\log q(x|\theta)\big]},$$

provided the second derivative exists.

An estimate $\widehat{\theta}$ is a function of the random measurement X. This means that the estimate $\widehat{\theta}(X)$ is a random variable because it is a function of the random variable X. As such, the estimate itself has a probability density function conditional on the parameter θ. In general, any function $y(x)$ of the random measurement X defines a random variable $Y(X)$ with a conditional probability density function denoted $p(y|\theta)$ which depends on $q(x|\theta)$.

The following theorem prepares for the Cramer–Rao theorem, given as the theorem immediately thereafter.

Theorem 12.7.1 *The variance σ_θ^2 of any unbiased estimate of θ satisfies*

$$\sigma_\theta^2 \int_{-\infty}^{\infty} q(x|\theta)\left(\frac{q(x|\gamma)}{q(x|\theta)} - 1\right)^2 dx \geq (\theta - \gamma)^2$$

for all θ and γ for which $q(x|\theta)$ and $q(x|\gamma)$ are defined.

12.7 The Cramer–Rao Theorem

Proof:
Let E_θ denote the expectation corresponding to the density $q(x|\theta)$. Then, using the Schwarz inequality in the second line, we have

$$\sigma_\theta^2 \left(E_\theta\left[1 - \frac{q(x|\gamma)}{q(x|\theta)}\right]^2\right) = \left(E_\theta\left[\widehat{\theta}(x) - E_\theta[\widehat{\theta}(x)]\right]^2\right)\left(E_\theta\left[1 - \frac{q(x|\gamma)}{q(x|\theta)}\right]^2\right)$$

$$\geq \left(E_\theta\left[(\widehat{\theta}(x) - E_\theta[\widehat{\theta}(x)])\left(1 - \frac{q(x|\gamma)}{q(x|\theta)}\right)\right]\right)^2$$

$$= \left(\int_{-\infty}^{\infty} q(x|\theta)\left(\widehat{\theta}(x) - \theta\right)\left(1 - \frac{q(x|\gamma)}{q(x|\theta)}\right)dx\right)^2$$

$$= \left(\int_{-\infty}^{\infty} (q(x|\theta) - q(x|\gamma))\left(\widehat{\theta}(x) - \theta\right)dx\right)^2$$

$$= (\theta - \gamma)^2.$$

The inequality holds between the first and the last terms. This is the statement of the theorem, so the proof is now complete. ∎

This theorem is now specialized to the Cramer–Rao theorem as follows.

Theorem 12.7.2 (Cramer–Rao) *Suppose that the conditional density function $q(x|\theta)$ is differentiable with respect to θ. The variance of an unbiased estimate of θ is bounded by*

$$\sigma_\theta^2 \geq \frac{1}{E_\theta\left[(\partial/\partial\theta)\log q(x|\theta)\right]^2}.$$

Proof:
Let $\gamma = \theta + \Delta$ so that $\gamma - \theta = \Delta$. Then rewrite Theorem 12.7.1 as

$$\sigma_\theta^2 \int_{-\infty}^{\infty} q(x|\theta)\left(1 - \frac{q(x|\theta + \Delta\theta)}{q(x|\theta)}\right)^2 dx \geq \Delta\theta^2.$$

It now follows that

$$\sigma_\theta^2 \geq \frac{1}{\int_{-\infty}^{\infty} q(x|\theta)\left[\frac{1}{\Delta\theta}\left(1 - \frac{q(x|\theta + \Delta\theta)}{q(x|\theta)}\right)\right]^2 dx}.$$

Taking the limit of this expression as $\Delta\theta$ goes to zero completes the proof of the theorem. ∎

The Cramer–Rao inequality can be weaker than its precursor given in Theorem 12.7.1. Nevertheless, the Cramer–Rao theorem is preferred because the probability density function appears inside a logarithm which makes the inequality easier to use. For a block of independent measurements, the block probability distribution is the product of the individual probability distributions so the Cramer–Rao inequality reduces to a sum of simple terms, one term for each measurement. In contrast, the inequality of Theorem 12.7.1, as such, is not additive for independent measurements. Indeed, the nonadditivity is the reason that the inequality of Theorem 12.7.1 can sometimes be stronger than the Cramer–Rao inequality.

An unbiased estimator that asymptotically satisfies the Cramer–Rao bound with equality is called an *efficient estimator*. An efficient estimator need not exist for a given problem.

Example
An example of an efficient estimator is the linear estimator of a gaussian random variable S in additive gaussian noise as described in Section 8.1. The linear estimator for S is described in Theorem 8.1.1 of that section. The variance of the estimate is

$$\sigma_{\hat{S}}^2 = \left(\frac{\sigma_s^2}{\sigma_s^2 + \sigma_n^2}\right)^2 \sigma_s^2.$$

A straightforward evaluation of the Cramer–Rao bound for a gaussian random variable shows that the variance of the linear estimator satisfies the Cramer–Rao bound with equality for gaussian random variables. Therefore, for a gaussian random variable, the linear estimator is an efficient estimator. In the case of a gaussian random variable, the linear estimator is an asymptotically optimal estimator in the sense of minimum error variance. For a nongaussian random variable, the Cramer–Rao bound does not assert this optimality.

Block Random Variables
The Cramer–Rao inequality continues to hold with the measurement X replaced by a vector of independent measurements. To generalize Theorem 12.7.2 to the case of block measurements, simply replace the distribution $q(x|\theta)$ with the distribution $\boldsymbol{q}(\boldsymbol{x}|\theta)$ of the block $\boldsymbol{x}$. That is, replace x with the block $\boldsymbol{x} = (x_1, \ldots, x_n)$ and replace $q(x|\theta)$ with $\boldsymbol{q}(\boldsymbol{x}|\theta)$.

For independent measurements, $q(x_1, \ldots, x_n|\theta) = \prod_{\ell=1}^{n} q_\ell(x_\ell|\theta)$ where the q_ℓ need not be the same. The logarithm of a product is the sum of the logarithms so when multiple independent measurements are available from which to estimate a parameter, the straightforward generalization of the following theorem can be used. It holds even when the independent measurements are not identical.

The theorem can be applied to dependent *gaussian* processes by diagonalizing the covariance matrix so that the components x'_i in the new coordinate system are independent.

Theorem 12.7.3 *The variance of an unbiased estimate of θ based on a set of n independent measurements satisfies*

$$\sigma_\theta^2 \geq \frac{1}{\mathrm{E}_\theta\left[\sum_{\ell=1}^{n}(\partial/\partial\theta)\log q_\ell(x_\ell|\theta)\right]^2},$$

where $q_\ell(x_\ell|\theta)$ is the conditional probability density function of the ℓth measurement x_ℓ.

Proof:
The logarithm of the probability distribution is additive for independent measurements. ∎

For n identically distributed measurements, the theorem reduces to

$$\sigma_\theta^2 \geq \frac{1}{n E_\theta \left[(\partial/\partial\theta) \log q(x|\theta) \right]^2}.$$

Example
The variance N of a zero-mean, stationary gaussian-noise process is to be estimated based on n independent observations of that random process. The logarithm of a zero-mean gaussian random variable of variance N is

$$\log_e q(x|N) = -\frac{1}{2} \log_e(2\pi N) - \frac{x^2}{2N}.$$

The derivative with respect to N is

$$\frac{\partial}{\partial N} \log_e q(x|N) = -\frac{1}{2N} + \frac{x^2}{2N^2}.$$

Using the Isserlis theorem[5] in the following string of equalities to set $E[x^4] = 3(E[x^2])^2$ leads to

$$E_N \left[\frac{\partial}{\partial N} \log q(x|N) \right]^2 = E_N \left[-\frac{1}{2N} + \frac{x^2}{2N^2} \right]^2$$

$$= E_N \left[\frac{1}{4N^2} - \frac{1}{N}\frac{x^2}{2N^2} + \frac{x^4}{4N^4} \right]$$

$$= \frac{1}{4N^2} - \frac{1}{2N^2} + 3\frac{1}{4N^2} = \frac{1}{2N^2},$$

where the last line asserts the Isserlis theorem. Therefore, for n independent measurements, the Cramer–Rao bound given in Theorem 12.7.2 states that

$$\sigma_N^2 \geq \frac{2N^2}{n}$$

for any unbiased estimator. The variance of any estimate of the noise power N approaches zero no faster than does the reciprocal of the number of measurements.

This bound is now compared to the maximum-likelihood estimate of N. For a single sample, this is

$$\widehat{N} = \underset{N}{\mathrm{argmax}} \left[\log p(x|N) \right]$$

$$= \underset{N}{\mathrm{argmax}} \left[-\frac{1}{2} \log 2\pi N - \frac{x^2}{2N} \right].$$

---

[5] The Isserlis theorem states that

$$E[X_1 X_2 X_3 X_4] = E[X_1 X_2] E[X_3 X_4] + E[X_1 X_3] E[X_2 X_4] + E[X_1 X_4] E[X_2 X_3]$$

for dependent gaussian random variables X_1, X_2, X_3, X_4. Therefore, $E[X^4] = 3[E[X^2]]^2$.

Setting the derivative with respect to N equal to zero leads to the equality $-1/2N + x^2/2N^2 = 0$ and so the estimate $\widehat{N} = x^2$ is the maximum-likelihood estimate. The mean of this estimate is $\overline{N} = E[\widehat{N}] = E[x^2] = N$. The variance of this estimate is $\sigma_{\widehat{N}}^2 = E[\widehat{N}^2] - \overline{N}^2 = E[x^4] - N^2 = 2N^2$ because $E[x^4] = 3N^2$ for a gaussian random variable. Accordingly, the maximum-likelihood estimate for n independent measurements has variance

$$\sigma_{\widehat{N}}^2 = \frac{2N^2}{n}.$$

This agrees with the Cramer–Rao bound for this problem. No estimator can have a variance asymptotically smaller than that asserted by the Cramer–Rao bound, so it can be concluded that the maximum-likelihood estimator is the minimum-variance unbiased estimator of N.

Random Processes

The Cramer–Rao inequality also holds with the measurement X replaced by a discrete or continuous random process. A gaussian random process is defined on discrete or continuous time, extending from negative infinity to positive infinity. A zero-mean gaussian process on discrete time is completely described by its (infinite) covariance matrix Σ. Now the x in $q(x|\theta)$ might refer to a sample sequence of the random process. The term $q(x|\theta)$ is not meaningful, as such, when x is an infinite sequence or waveform. It becomes meaningful only when the sequence or sampled waveform is truncated to a finite block. The Cramer–Rao theorem then applies to blocks of each finite size, and so applies in the limit as the blocklength goes to infinity.[6]

Example

A third example of an efficient estimator is the linear estimator of the time of arrival of a baseband pulse in white gaussian noise. The linear estimator for white noise, not necessarily gaussian noise, is described in Corollary 12.6.2 of Section 12.6. Now the gaussian noise $n(t)$ is a continuous random process. A straightforward evaluation of the Cramer–Rao bound shows that this variance of the linear estimator satisfies the Cramer–Rao bound with equality for gaussian random variables. Therefore, for a gaussian random variable, the linear estimator is an efficient estimator. In the case of a gaussian random variable, the linear estimator is an optimal estimator in the sense of minimum variance. For a nongaussian random variable, the Cramer–Rao bound does not assert optimality.

12.8 Noncoherent Estimation of Pulse Parameters

The search radar forms the estimates of the target parameters from the received echo signal. After a pulse or waveform is detected, the parameters such as the *time of*

---

[6] A formal treatment requires the introduction of Karhunen–Loeve functions and the Toeplitz distribution theorem. These tools are not discussed herein.

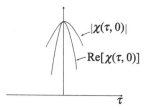

Figure 12.9 Illustrating the curvature of $\chi(\tau, 0)$

arrival or the *frequency of arrival* may be estimated. The noncoherent estimators for a complex baseband pulse are studied in this section. Errors in the conversion from the passband representation to the complex baseband representation are not included. The coherent estimators for a complex baseband pulse are not studied because it is not natural for the phase of the waveform to be known when the frequency or time of arrival is not known.

When represented at complex baseband, the received signal has the form

$$v(t) = s(t - \tau_0)e^{j2\pi\nu_0 t}e^{j\theta} + n(t),$$

where $s(t)$ is a known waveform with finite energy. The additive noise $n(t)$ is a stationary, complex random noise process

$$n(t) = n_R(t) + jn_I(t),$$

whose components $n_R(t)$ and $n_I(t)$ are independent with known and identical correlation functions $\phi(\tau)$ and power density spectrum $N(f)$. For white noise $N(f)$ is, by convention, equal to $N_0/2$. The delay parameter τ_0 and the doppler parameter ν_0 in the received signal $v(t)$ are unknown and are to be estimated from that $v(t)$.

The optimal estimator in additive white gaussian noise first computes the sample cross-ambiguity function. Were the phase known, the coordinates of the *real part* of the peak would be the asymptotically optimal estimates $(\widetilde{\tau}_0, \widetilde{\nu}_0)$ of the parameters (τ_0, ν_0). But the phase of the received signal is not known, so the coordinates of the *magnitude* of the peak are the asymptotically optimal estimates $(\widehat{\tau}_0, \widehat{\nu}_0)$ of the parameters (τ_0, ν_0). For noise that is nonwhite or nongaussian, computation of the sample cross-ambiguity function is not optimal for detection or estimation, though it is still a practical estimator.

The difference between the performance of the noncoherent estimator and a presumed coherent estimator in white noise can be attributed to the fact that the peak of the real part of the ambiguity function is sharper than the peak of the magnitude of that same ambiguity function, as illustrated in Figure 12.9, so the peak of the real part would be located more accurately in noise.

To estimate τ_0 when ν_0 is known, find the peak of $|\chi(\tau, \nu_0)|$ as a function of τ with ν_0 fixed. This is a noncoherent estimator of τ_0. To estimate ν_0 when τ_0 is known, find the peak of $|\chi(\tau_0, \nu)|$ as a function of ν with τ_0 fixed. This is a noncoherent estimator of ν_0. To simultaneously estimate τ_0 and ν_0, find the peak of $|\chi(\tau, \nu)|$ with respect to both variables. This is a noncoherent estimator of the pair τ_0, ν_0.

Estimation of Time of Arrival

The estimation of pulse parameters is noncoherent when the phase of the received pulse is not known, which is the natural case. This section shows that the asymptotic error variance of the noncoherent estimator in white gaussian noise of the arrival time of a passband pulse $s(t)$ whose frequency is known is

$$\sigma_{\tau_0}^2 \approx \frac{1}{(2\pi)^2(2E_p/N_0)B_G^2},$$

where B_G is the Gabor bandwidth of $s(t)$.

Proposition 12.8.1 *The variance of the error of the noncoherent estimate of pulse arrival time τ_0 based on finding the maximum of the ambiguity surface $|\chi(\tau, \nu - \nu_0)|$ with ν fixed at ν_0 is given by*

$$\sigma_{\tau_0}^2 = \frac{1}{(2\pi)^2(2E_p/N_0)B_G^2}$$

asymptotically for large E_p/N_0.

Proof:
The peak of any smooth function is found by setting the derivative of that function equal to zero. To determine the error in the estimate of the location of the peak due to small noise terms, use a quadratic approximation to the peak of the ambiguity surface $|\chi(\tau, \nu)|$. The partial derivatives are computed from the local approximation

$$|\chi(\tau,\nu)| \approx E_p\left(1 - 2\pi^2\nu^2T_G^2 - 4\pi^2\tau\nu T_G B_G \rho - 2\pi^2\tau^2 B_G^2\right),$$

as given in Section 10.4. This quadratic surface can be compared to the earlier quadratic surface

$$\chi(\tau,\nu) = E_p\left(1 + j2\pi\nu\bar{t} - 2\pi^2\nu^2\overline{t^2} - 4\pi^2\tau\nu\overline{tf} + j2\pi\tau\bar{f} - 2\pi^2\tau^2\overline{f^2}\right)$$

$$\text{Re}[\chi(\tau,\nu)] = E_p\left(1 - 2\pi^2\nu^2\overline{t^2} - 4\pi^2\tau\nu\overline{tf} - 2\pi^2\tau^2\overline{f^2}\right)$$

for the peak of the complex ambiguity function itself, where the first line is taken from Section 10.4 and the second line follows by dropping the imaginary part.

Comparing these expressions, the term multiplying τ^2 in the quadratic approximation to the real part is $\overline{f^2}$, whereas in the magnitude surface, it is B_G^2. Corollary 12.6.2 uses the quadratic approximation for the peak of $\chi(\tau,\nu)$ in its proof. To complete the proof of the proposition, it is only necessary to refer to the expression of Corollary 12.6.2 and change $\overline{f^2}$ to B_G^2. The proposition is thereby proved. ∎

Estimation of the Pulse Arrival Frequency

The asymptotic error variance of the noncoherent estimator in white gaussian noise of the arrival frequency of a passband pulse whose arrival time is known is

$$\sigma_{\nu_0}^2 \approx \frac{1}{(2\pi)^2(2E_p/N_0)T_G^2},$$

12.8 Noncoherent Estimation of Pulse Parameters

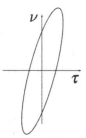

Figure 12.10 An uncertainty ellipse

where T_G is the Gabor timewidth of $s(t)$. A derivation of this expression is not given. The expression is suggested by analogy with Proposition 12.8.1 and can be justified by duality.

Estimation of Both Time and Frequency of Arrival

When both τ and ν are unknown and both are to be estimated, each estimate may be less accurate than it would be were the other parameter known. An instance of the uncertainty ellipse in Figure 12.10 shows that when ν is known, the Gabor bandwidth B_G appears in the equation for σ_τ^2. When ν is unknown, we may expect that the projection of the uncertainty ellipse onto the frequency axis replaces the role of B_G. To find the maximum excursion of the ellipse

$$\tau^2 B_G^2 + 2\tau\nu T_G B_G \rho + \nu^2 T_G^2 = 1$$

in the τ direction, rewrite it in the form

$$(B_G^2 - \rho^2 B_G^2)\tau^2 + (T_G\nu + \rho B_G \tau)^2 = 1.$$

The maximum value of τ occurs when ν is such that the second term on the left equals zero. Consequently, B_G^2 is replaced by $B_G^2 - \rho^2 B_G^2$ suggesting that

$$\sigma_\tau^2 = \frac{1}{(2\pi)^2(2E_p/N_0)B_G^2(1-\rho^2)}.$$

This is the statement of the corollary to the following proposition.

The proposition is evident when the uncertainty ellipse is diagonal because ρ is then zero. Otherwise, the proposition is proved by rotating into the principal coordinate system, then rotating back.

Proposition 12.8.2 *The covariance matrix for the error in a noncoherent estimate of τ_0 and ν_0 satisfies*

$$\Sigma = \frac{1}{(2\pi)^2(2E_p/N_0)} \begin{bmatrix} B_G^2 & \rho B_G T_G \\ \rho B_G T_G & T_G^2 \end{bmatrix}^{-1}$$

in the asymptotic limit of a large signal-to-noise ratio.

Proof:

The peak of the sample ambiguity surface $|\chi_c(\tau,\nu)|$ occurs where the partial derivatives are both equal to zero. In general, the τ and ν errors are dependent. Rotate the axes into a new coordinate system, so that the two components are independent. In the new coordinate system, the diagonal elements are the eigenvalues. These are denoted λ_1 and λ_2. They describe the curvature of the peak in the principal coordinate system. In that coordinate system the errors (up to first-order terms) in estimating the location of the peak on the two axes are independent. Referring to Proposition 12.8.1, the two variances of the errors are given by $\sigma_{\lambda_i}^2 = [(2\pi)^2(2E_p/N_0)\lambda_i]^{-1}$ for $i = 1, 2$. This is written in matrix form as

$$\Sigma' = \left[(2\pi)^2(2E_p/N_0)\begin{bmatrix} \lambda_1 & 0 \\ 0 & \lambda_2 \end{bmatrix}\right]^{-1},$$

or as

$$\begin{bmatrix} \lambda_1 & 0 \\ 0 & \lambda_2 \end{bmatrix}\Sigma' = \frac{1}{(2\pi)^2(2E_p/N_0)}.$$

Rotating back to the original coordinate system, this becomes

$$\begin{bmatrix} B_G^2 & \rho B_G T_G \\ \rho B_G T_G & T_G^2 \end{bmatrix}\Sigma = \frac{1}{(2\pi)^2(2E_p/N_0)}.$$

Moving the two by two matrix to the right side of the expression completes the proof. ∎

Corollary 12.8.3 *The variances for the delay and doppler errors in a joint noncoherent estimate of τ_0 and ν_0 satisfy*

$$\sigma_{\tau_0}^2 = \frac{1}{(2\pi)^2(2E_p/N_0)B_G^2(1-\rho^2)},$$

$$\sigma_{\nu_0}^2 = \frac{1}{(2\pi)^2(2E_p/N_0)T_G^2(1-\rho^2)}$$

in the asymptotic limit of a large signal-to-noise ratio.

Proof:

The inverse of the matrix appearing in the proposition is

$$\begin{bmatrix} B_G^2 & \rho B_G T_G \\ \rho B_G T_G & T_G^2 \end{bmatrix}^{-1} = \frac{1}{B_G^2 T_G^2(1-\rho^2)}\begin{bmatrix} T_G^2 & -\rho B_G T_G \\ -\rho B_G T_G & B_G^2 \end{bmatrix}.$$

The diagonal elements are $1/B_G^2(1-\rho^2)$ and $1/T_G^2(1-\rho^2)$ as given in the two expressions of the corollary. ∎

12.9 Clutter

The cumulative unwanted signal received by a radar due to reflections from background objects other than objects of interest is called *clutter*. The term usually carries the connotation that the clutter is caused by a dense set of small, unresolved or unresolvable reflecting elements, perhaps with moving elements such as raindrops, swaying tree branches, birds, or ocean waves; or perhaps fixed elements such as buildings or other structures.

Clutter can be treated simply as noise and can be processed as noise such as by computing the sample cross-ambiguity function of the received signal with the transmitted signal. The image, as described by the radar imaging equation, includes the clutter sources. Each small source of clutter contributes its own small peak to the image. In total, the clutter causes an interfering background in the sample cross-ambiguity function that the target main lobe must rise above.

The effect of clutter on the radar cross-ambiguity function might resemble thermal noise in that it creates a background intensity above which the main lobe of the sample cross-ambiguity function of an object of interest must rise so that it can be detected. Clutter is different from thermal noise, however, because clutter is dependent on the transmitted signal and has statistical properties that can be exploited to improve performance. Clutter has an internal structure that thermal noise does not have.

The quality of a radar in clutter is sometimes judged by the notion of the *subclutter visibility*. This imprecise term refers to the amount by which the maximum allowable clutter-to-signal ratio exceeds the maximum allowable noise-to-signal ratio. This is the amount by which the clutter alone can exceed the noise alone while keeping the performance required of the radar.

A more general approach to dealing with clutter is to treat the clutter signal as non-white gaussian noise and to design "whitening" filters into the system that improve performance by de-emphasizing spectral regions where the clutter is strong. This models the clutter in terms of the signal spectrum but ignores the dependence between the clutter and the signal during detection.

Clutter with Only Delay

The simplest form of clutter occurs when there is no motion between the radar and the sources of clutter. This clutter may be caused by a large number of echoes from small, scattered, stationary reflectors. To model this clutter, let ρ_i be a random set of independent complex random variables indexed by i describing the amplitudes of the independent clutter sources. Then $E[\rho_i \rho_j^*] = 0$ for $i \neq j$. The reflector with the reflected signal ρ_i has a delay, denoted τ_i. The clutter signal from a set of stationary reflectors indexed by i is

$$n(t) = \sum_i \rho_i s(t - \tau_i),$$

where $s(t)$ is the transmitted pulse. The τ_i are a set of time delays distributed irregularly within a region of the time axis and the ρ_i are random complex amplitudes.

Noise Model of Clutter

One way to deal with clutter is to treat it as equivalent nonwhite noise. When motion is insignificant, the sample correlation function $\phi_c(\tau)$ of the clutter $n(t)$ is

$$\phi_c(\tau) = \int_{-\infty}^{\infty} n(t) n^*(t-\tau) dt$$

$$= \sum_i \sum_j \rho_i \rho_j^* \int_{-\infty}^{\infty} s(t-\tau_i) s^*(t-\tau-\tau_j) dt.$$

The Fourier transform $\Phi_c(f)$ of the clutter correlation function $\phi_c(\tau)$ is obtained by using the convolution theorem in each term inside the summations to convert that convolution to a product. Then

$$\Phi_c(f) = \sum_i \sum_j \rho_i \rho_j^* [S(f) e^{j2\pi \tau_i}][S(f) e^{j2\pi \tau_j}]^*$$

$$= |S(f)|^2 \sum_i \sum_j \rho_i \rho_j^* e^{j2\pi(\tau_i - \tau_j)},$$

where $S(f)$ is the transform of the signal $s(t)$ on a long but finite duration observation interval. The double summation yields a real number because the ij term is the complex conjugate of the ji term. The clutter power density spectrum is the expectation

$$N(f) = E[\Phi_c(f)]$$

$$= |S(f)|^2 E\left[\sum_i \sum_j \rho_i \rho_j^* e^{j2\pi(\tau_i - \tau_j)}\right].$$

The reflector amplitudes are modeled as independent random variables, so this expression becomes $N(f) = \gamma |S(f)|^2$ for some real constant γ dependent on the random distribution of the reflectors. This means that the power density spectrum of the stationary clutter is proportional to the signal power spectrum.

Accordingly, the whitened matched filter is

$$G(f) = \frac{S^*(f)}{N(f)} = \frac{1}{\gamma S(f)},$$

because $N(f) = \gamma |S(f)|^2$. This filter is not realizable because $S(f)$ must go to zero as $|f|$ goes to infinity thereby implying that $|S(f)|^{-1}$ goes to infinity. Indeed, the output of the filter would be an impulse because $G(f)S(f)$ is the constant $1/\gamma$ independent of frequency. The output signal-to-noise ratio then would be

$$\frac{S}{N} = \int_{-\infty}^{\infty} \frac{|S(f)|^2}{N(f)} df$$

$$= \int_{-\infty}^{\infty} \frac{1}{\gamma} df,$$

which is infinite. This naive clutter model gets as much value from weak signals as it gets from strong signals. Accordingly, a whitened matched filter for this clutter model is not meaningful.

Urkowitz Filter

A better filter in the presence of clutter is the *Urkowitz filter*. The Urkowitz filter is an ad hoc filter that incorporates additive thermal noise that is independent of the signal. When additive thermal noise with spectrum $N(f)$ is included, the total power density spectrum is $\gamma|S(f)|^2 + N(f)$. Then the clutter-matched filter is

$$G(f) = \frac{S^*(f)}{\gamma|S(f)|^2 + N(f)}.$$

This filter behaves like the noise-whitened matched filter at frequencies where the thermal noise term $N(f)$ dominates the clutter term $\gamma|S(f)|^2$.

When there is motion in the situation because the radar is moving or the clutter reflectors are moving, each individual reflecting element creating the clutter has a doppler shift. Different reflecting elements have a different doppler shift causing the clutter term $\gamma|S(f)|^2$ to be slightly spread on the frequency axis.

$$n(t) = \sum_i \rho_i s(t - \tau_i) e^{j2\pi \nu_i t},$$

where τ_i and ν_i are the delay and doppler coordinates of the ith reflecting element. This again has the appearance of noise when the individual terms are not resolved. It can be treated by an ad hoc modification of the Urkowitz filter such as replacing $|S(f)|^2$ by the expectation of the spread of $|S(f)|^2$ given by $E[S(f + \nu_i)S^*(f - \nu_j)]$.

Clutter-Free Region

Because clutter is formed by copies of the transmitted waveform shifted in delay and doppler, the clutter will affect the sample cross-ambiguity function only in the range-doppler region corresponding to range and doppler contained in the clutter as modified by convolution with the ambiguity function of the transmitted waveform $s(t)$.

The sample cross-ambiguity function is described by the radar imaging equation $M(\tau, \nu) = \chi(\tau, \nu) * * \rho(\chi, \nu)$. An instance of this convolution is depicted in Figure 12.11. In this instance, the clutter reflectivity density $\rho_c(\tau, \nu)$ is deemed to be confined to a rectangular region as shown in the center of Figure 12.11. The radar waveform $s(t)$ is a periodic pulse train. The ambiguity function of $s(t)$ consists primarily of a main lobe and grating lobes as shown in idealized form on the left side of Figure 12.11. Between the lobes of $\chi(\tau, \nu)$ are the minor fluctuations of $\chi(\tau, \nu)$. This combination can result in a "clutter-free" region. In this region the clutter has little effect on the sample cross ambiguity function, as shown in Figure 12.11. The figure shows an example in which the clutter reflectivity density $\rho_c(\tau, \nu)$ is confined to small values of ν. A target of interest whose doppler coordinate lies in the clutter-free region of the image may not be significantly affected by the clutter.

512   12 Radar Search Systems

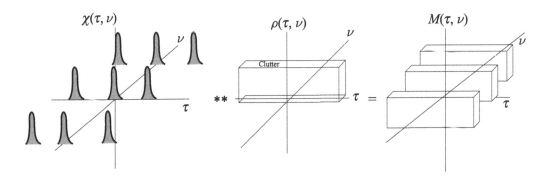

Figure 12.11 Illustrating the effect of clutter

12.10 Detection of Moving Objects

A *moving-target detection* radar is a radar whose purpose is to detect those elements in a scene that are in motion with respect to the background. In general, the radar task of detecting and locating moving objects in three-dimensional space is a problem with six unknowns consisting of three position coordinates and three velocity components. When restricted to the earth's surface, the problem has four unknowns consisting of two position coordinates and two velocity components. Some of these unknown parameters are visible in the sample cross-ambiguity function, others are visible in the antenna beam and can be measured by interferometry.

A moving-target detection radar may be stationary or it may itself be moving. For a stationary moving-target detection radar fixed on the ground, the targets are in motion with respect to the radar and exhibit a doppler based on the speed in the radial direction of the reflecting object. An object with a doppler shift can be detected when it can be separated from the clutter. When the radar itself is moving, doppler is due both to the radar itself and to the moving reflector. The clutter is spread in the doppler direction by the motion of the radar itself. The detection of a moving object and the estimation of the parameters of that object are more difficult.

A radar waveform that is designed to have an ambiguity function with good resolution in the doppler direction can be used to detect moving objects even though there may be stationary or slowly moving reflectors that are much larger. A moving-target detection radar is one that is so designed to detect the presence of moving objects in the presence of such interference by using a waveform with an appropriate ambiguity function. A moving-target detection radar detects moving reflectors and measures one or more components of the velocity vector, and may also measure the range and the angle from the radar to the moving reflector.

The ambiguity function can be used to study the effect of clutter on a linear receiver. The radar return from a moving reflector of interest is cluttered by echoes from stationary or slowly moving reflectors in the antenna main beam. The sample cross-ambiguity function computed from the received signal will be cluttered by the effect of the

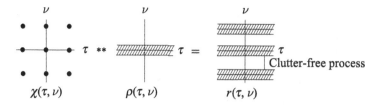

Figure 12.12 Plan view for narrowband clutter

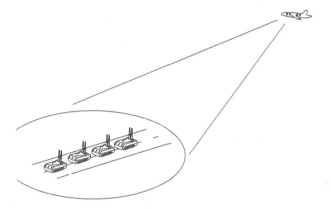

Figure 12.13 Moving-target detection from moving radar

echoes. The clutter may mask a signal of interest whose doppler is within the band of doppler values of the clutter.

The two-dimensional clutter reflectivity density $\rho(\tau, \nu)$ is (two-dimensionally) convolved with the two-dimensional ambiguity function of the radar waveform to produce the clutter that appears in the image. Figure 12.12 shows a plan view corresponding to Figure 12.11 which illustrates that when the clutter is confined to small values of ν, that clutter contaminates certain regions of the sample ambiguity function while leaving other regions essentially free of clutter. Moving objects that lie in the clutter-free region are readily detected when they are sufficiently strong with respect to thermal noise and with respect to the incidental clutter projected into the clutter-free region through the minor lobes of the ambiguity function.

For a radar that is moving, as shown in Figure 12.13, clutter suppression can be much more difficult because an object in the background, although stationary, is in relative motion with respect to the moving radar. Reflections from stationary objects have a doppler shift due to the motion of the radar and the clutter is not confined to small values of ν. The clutter-free regions shown in Figure 12.11 are reduced in size because of the motion of the radar itself. Whenever the doppler of the clutter is comparable to the doppler of the moving target, the doppler alone is not enough to separate the weaker targets from the clutter. Dual antennas used to suppress clutter can be a remedy for this limitation.

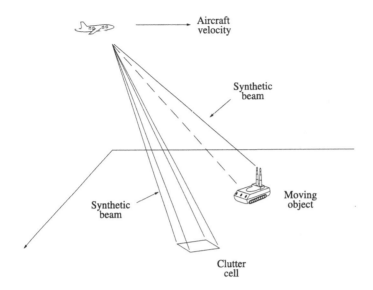

Figure 12.14 Conceptual beams with the same doppler

Clutter Nulling

A radar that is moving will result in doppler in the signal from stationary objects. Each fixed point on the ground has a radial velocity as seen by the moving radar. Choose the coordinate system so that the velocity V is in the positive x direction. Then the radial velocity of a fixed point at angle θ is $V \cos \theta$, resulting in a two-way doppler shift of $2f_0(V/c)\cos \theta$. A pair of antennas separated horizontally by distance d can be used to form a null in the antenna pattern at angle θ. This eliminates the clutter from that angle, but can also attenuate the signal from angles close to θ.

All reflectors with the same range R and the same relative velocity V_r fall into the same delay-doppler cell. Because moving vehicles are few compared to the number of cells, two moving vehicles rarely fall into the same delay-doppler cell. Hence, with rare exceptions, only one moving object and one stationary source of clutter will be found in each delay/doppler cell. The signal within a specified delay/doppler cell may be conceptualized as coming from the two virtual or synthetic beams as shown in Figure 12.14. The clutter beam is eliminated by the clutter-nulling operation.

The two signals in the same cell have the same delay, or range, and the same doppler, or relative velocity. For a radar with velocity V and a moving reflector with velocity V_r, the relative velocities $V_r^{(t)}$ and $V_r^{(c)}$ of the target and the clutter are

$$V_r^{(t)} = V \cos \theta^{(t)} + V_t \cos \phi,$$
$$V_r^{(c)} = V \cos \theta^{(c)},$$

respectively, where $\theta^{(t)}$ and $\theta^{(c)}$ are the angles from the radar velocity vector to the target and the clutter, respectively, and where ϕ is the angle from the target velocity vector to the radar. The two relative velocities $V_r^{(t)}$ and $V_r^{(r)}$ are equal.

An efficient way to null clutter is to recall that linear operations can be done in any order. First form the sample cross-ambiguity functions $\chi^{(1)}(\tau,\nu)$ and $\chi^{(2)}(\tau,\nu)$, one for each of two horizontally-separated antennas using a transmitted signal from only one antenna. Then form

$$\chi(\tau,\nu) = e^{+j\phi(\tau,\nu)/2}\chi^{(1)}(\tau,\nu) + e^{-j\phi(\tau,\nu)/2}\chi^{(2)}(\tau,\nu),$$

where $\phi(\tau,\nu)$ is the relative phase angle between the two antennas needed to null the stationary clutter that has those τ,ν coordinates. Then each cell of the composite function $\chi(\tau,\nu)$ has the clutter from a stationary object in its cell suppressed. Anything still in that cell is either noise or a moving object at a location different than that of the clutter.

Problems

12.1 **a.** Use a series expansion on a term of the form $\sqrt{(A+x)^2 + y^2}$ to explain why the ricean probability density function looks like a gaussian probability density function when A is sufficiently large.

b. The modified Bessel function of the first kind $I_0(x)$ can be approximated as

$$I_0(x) \approx \frac{e^x}{\sqrt{2\pi x}}$$

when x is large. Using this approximation as a starting point, show again that the ricean probability density function looks like a gaussian probability density function when A is large.

12.2 A noncoherent estimator of delay and doppler finds the coordinates at which the sample ambiguity surface achieves its maximum. A suggestion for an alternative approach is to first measure the phase at the peak of the sample ambiguity function, then to correct the ambiguity function for that phase and follow with a coherent estimator to estimate delay and doppler. Is this suggestion sound? Explain.

12.3 **a.** A pulse train consisting of N identical pulses is noncoherently received in white gaussian noise with a single random phase that is the same on all pulses of this pulse train. What is the Neyman–Pearson detector of the pulse train?

b. A pulse train consisting of N identical pulses is noncoherently received in white gaussian noise with the phase of each pulse random, independent, and uniformly distributed. What is the Neyman–Pearson detector of this pulse train?

c. A pulse train consisting of N real identical pulses is received in white gaussian noise, and the amplitude of each pulse is random, independent, and rayleigh-distributed. What is the Neyman–Pearson detector of this pulse train?

12.4 The *rayleigh probability density function* (or the rayleigh pulse) is defined as the radial marginal $p(r)$ of the two-dimensional gaussian probability density function

12 Radar Search Systems

$$p(x,y) = \frac{1}{2\pi} e^{-(x^2+y^2)/2}.$$

The *maxwellian probability density function* (or the maxwellian pulse) is defined as the radial marginal $p(r)$ of the three-dimensional gaussian probability density function

$$p(x,y,z) = \frac{1}{(2\pi)^{3/2}} e^{-(x^2+y^2+z^2)/2}.$$

Find expressions for the rayleigh probability density function and the maxwellian probability density function and their Fourier transforms.

12.5 Consider the trapezoidal pulse

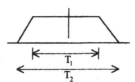

a. Given that $2E_p/N_0 = 10$ dB, $T_1 = 1$ microsecond and $T_2 = 1.1$ microseconds, what is the variance of the error of the best noncoherent estimator of the pulse arrival time.
b. Now let the pulse width be increased so that $T_1 = 2$ microseconds and $T_2 = 2.1$ microseconds. The value of $2E_p/N_0$ is unchanged. How will the error variance of the estimator change?

12.6 Give an example of a signal for which the largest Nyquist sample is smaller than the largest value of the signal. For the signal of this example, how does the probability of missed detection depend on the offset between the location of the maximum and the nearest sample. Give approximate numerical values for the probability of missed detection. Does testing each Nyquist sample against a threshold provide an optimal method of finding the threshold crossings of the waveform?

12.7 The whitened matched filter minimizes the probability of detection error of a pulse in stationary gaussian noise. Does this mean that the whitened matched filter minimizes the expected value of the error between the noisy peak output of a filter and the noise-free peak output of that filter?

12.8 A passband pulse at carrier frequency f_0 is

$$\tilde{s}(t) = s_R(t) \cos 2\pi f_0 t - s_I(t) \sin 2\pi f_0 t.$$

Let $f_0' = f_0 + \Delta f$. Then

$$\tilde{s}(t) = s_R'(t) \cos 2\pi f_0' t - s_I'(t) \sin 2\pi f_0' t,$$

where

$$s_R' = s_R(t) \cos 2\pi \Delta f t + s_I(t) \sin 2\pi \Delta f t,$$
$$s_I' = - s_R(t) \sin 2\pi \Delta f t + s_I(t) \cos 2\pi \Delta f t.$$

This means that with full mathematical precision, one can choose to regard $\tilde{s}(t)$ as the complex baseband pulse $s'_R(t) + s'_I(t)$ modulated onto carrier frequency f'_0.

Because the doppler frequency is $\nu = (V/c)f'_0$, this appears to say that the doppler can be increased simply by redefining the carrier frequency.

a. Explain why this reasoning, though in itself is mathematically correct, is not useful.
b. Does this interpretation permit a more accurate estimate of frequency of arrival of the pulse? Explain.

12.9 The Fourier transform of the waveform $s(t)$ is

$$S(f) = \text{rect}\left(\frac{f - f_0}{B}\right) + \text{rect}\left(\frac{f + f_0}{B}\right),$$

where f_0 is larger than B.

a. What is $\chi(\tau, 0)$?
b. Evaluate the variance of a noncoherent estimator of pulse arrival time in white gaussian noise based on the assumption that a linearized analysis in the vicinity of the main lobe is adequate.
c. With reference to the figure, use the fact that the matched-filter noise output when the input is white noise has the autocorrelation function $(N_0/2)\chi(\tau, 0)$. Give an expression for the probability that, in white gaussian noise, the first sidelobe on the right or the left of the main lobe is larger than the main lobe.
d. Noise may cause a small perturbation in the position of the main lobe. Noise may also cause a false detection of one of the principal sidelobes. Give a refined expression for the variance developed in part (b) based on both kinds of error.

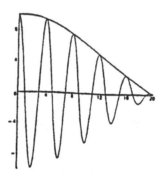

12.10 The digital computation of a sample cross-ambiguity function uses a discrete (τ, ν) grid consistent with the spatial frequency content of $\chi(\tau, \nu)$. To simplify the computation for estimating the arrival time of the pulse given in the previous problem, a two-step estimator, consisting of a coarse estimate followed by a fine estimate, is proposed. A coarse estimate of τ_0 is computed by using only that portion of $S(f)$ contained in a single frequency interval. Then $|\chi(\tau, 0)|$ is computed by using all of $S(f)$, but computing $|\chi(\tau, 0)|$ only in the vicinity of the coarse estimate.

Specifically, to obtain the coarse estimate, the received signal is first filtered by the filter $h(t)$, where $h(t)$ corresponds to $\text{rect}((f-f_0)/B)+\text{rect}((f+f_0)/B)$ in the frequency domain. What is the variance of the coarse estimate? Given that the noise is gaussian, at $E_p/N_0 = 10$ dB and $f_0/B = 10$, what is the probability that the coarse estimate is closer to the main lobe than to a sidelobe? Explain.

12.11 A radar with the carrier frequency f_0 and a narrow beam fixed at an up-looking angle of $45°$ is used to measure the altitude of satellites in low circular orbits, moving right to left.

Because the satellite is known to be in a circular orbit, its velocity and altitude are interrelated by the laws of mechanics. For h small compared to the earth's radius r_e, the relationship is approximated by

$$v(h) = v_0 \left(1 - \frac{1}{2}\frac{h}{r_e}\right).$$

a. Express this equation as a constraint in the τ, ν plane of the form

$$a\nu + b\tau = 1$$

for some appropriate constants, a and b. Show the location of a typical cross-ambiguity function by sketching the uncertainty ellipse of the cross-ambiguity function in the τ, ν plane.

b. Given that the uncertainty ellipse of the waveform $s(t)$ has the form

$$T_G^2 \nu^2 + B_G^2 \tau^2 = C,$$

and taking advantage of the constraining condition, give an expression for the variance of a noncoherent estimator of τ_0.

c. When the pulse $s(t)$ is replaced by $s(t)e^{j\pi \alpha t^2}$, how will the variance change? Is there a best choice for α?

12.12 A moving-target detection radar uses the waveform

$$s(t) = \sum_{n=0}^{N-1} p(t - nT_r).$$

A primitive form of a clutter-suppression algorithm passes the received echo $v(t)$ through the filter

$$g(t) = p(t) - p(t - T_r),$$

prior to detection.

a. What is the filter output due to stationary clutter?
b. What is the filter output due to a moving reflector?
c. Let E_p be the energy of pulse $p(t)$. In the presence of white gaussian noise of power density spectrum $N_0/2$, what is the signal-to-noise ratio of a moving reflector at the output of the filter?

12.13 a. Prove that, as long as enough energy exists in each segment so that the linearized analysis holds, with no loss in arrival-time estimation accuracy, a pulse train waveform can be chopped into segments, and the arrival time of the pulse train can be estimated by averaging the estimates of the arrival time of each segment.

(This justifies the common practice of estimating the arrival time of a pulse train by estimating the arrival time of each pulse individually, and then averaging after accounting for each pulse offset time. For time-of-arrival estimation, it is not necessary to maintain coherence in long-duration waveforms, provided the signal-to-signal ratio of the individual segments is sufficiently large.)

b. Prove that the arrival frequency of a pulse train can be estimated much more accurately than by estimating and averaging the arrival frequencies of the individual pulses. Estimating the arrival frequency of a pulse train by averaging estimates made on individual pulses will result in excessive degradation in accuracy.

12.14 A four-way hypothesis-testing problem with at most two pulses is given with $v(t) = a_1 s(t - \tau_1) + a_2 s(t - \tau_2) + n(t)$, where $a_1, a_2 \in \{0, 1\}$. Each pulse is either present or absent. Sketch the minimum-distance decision regions for the case in which $\text{Re}\left[\int s(t - \tau_1) s^*(t - \tau_2) dt\right]$ is negative. Give expressions for the conditional probabilities of error when $n(t)$ is white gaussian noise.

Notes

The detection of objects by means of reflected radio waves became an obvious possibility as soon as Heinrich Hertz demonstrated the transmission and reflection of radio waves in 1887. A formal theory of detection and estimation of radar signals came much later. Detection is an instance of the problem of the testing of binary hypotheses, as was studied by Neyman and Pearson (1933), and also by Zadah and Ragazzini (1952), Marcum (1960), Swerling (1957, 1960), and many others. Detection of a signal with an unknown delay or unknown frequency is a more difficult problem. By quantizing delay to one of M discrete values, the detection of a signal with an unknown arrival time becomes an M-ary detection problem. It was treated as such by Peterson and Birdsall (1953), and also by Middleton and Van Meter (1955). Selin (1965) employed an averaging technique to treat the detection of a signal with an unknown doppler.

Detection of multiple, overlapping echo signals, each with an unknown delay and an unknown doppler, is a yet more complicated detection problem. The resolution of multiple targets was treated by Helstrom (1960) using the method of maximum likelihood. A direct approach to the combined problem of multitarget detection and estimation of their parameters was due to Nilsson (1961), who introduced a hybrid loss function that assigned penalties to both detection and estimation errors; the loss was then minimized with respect to a posterior probability measure. The solution requires a computer to perform a multidimensional maximization; as the number of targets grows, the computational requirements grow in complexity and difficulty.

Detection of moving targets against a stationary background has a somewhat different history because of the problem of clutter. Moving-target detection had its origins during World War II. This early work was treated in the reports of Emslie (1946) and Emerson (1954). Detection in the presence of clutter has been studied from many points of view, as in the papers of Capon (1964); Manasse (1961); and Sekine, Ohtani, and Muska (1981). Urkowitz (1953) proposed the model of clutter as statistically independent of the signal, but showing the same spectrum.

13 Passive and Baseband Systems

A propagating medium may be teeming with a multitude of weak signals even when it appears superficially to be empty. For example, an acoustic medium such as the water in a lake may appear quite still, and yet it may contain numerous faint pressure waves originating in various submerged objects and reflecting off other submerged objects. A passive solar system may be able to intercept these waves and extract useful information from the raw received data. Indeed, these invisible pressure waves might be used to form images of submerged objects. Likewise, a seismographic sensor or an array of such sensors on the surface of the earth can measure tiny vibrations of the earth's surface and deduce the location of distant earthquakes and other geophysical disruptions or can form crude images of geological structures.

Even the electromagnetic environment in which we are immersed contains immense quantities of information. Some of these electromagnetic signals can be intercepted by suitable large apertures and formed into detailed images of far distant galaxies. No illumination is provided by the sensor, nor can illumination be provided in such an application. We need only gather the data with appropriate passive sensors and process that sensed data into the images that allow us to observe these galaxies.

Passive surveillance systems include systems for radio astronomy, seismic data analysis, electromagnetic surveillance, and sonar surveillance. A passive surveillance system collects a propagating signal that the system itself does not illuminate or stimulate. The signal originates in the environment, often in the scene or the object being observed. The information-bearing signals that are incident on a set of receivers are collected at one or more places within the propagation medium. From this set of received signals, contaminated by noise and perhaps by other forms of interference, parameters such as the locations of radiation sources or the properties of the propagation medium can be estimated. By fully processing radiation intercepted at decentralized locations over time, it may even be possible to obtain images of the radiation source itself or images of the environment surrounding that radiation source.

The functions of detection and estimation in a passive surveillance system are described primarily in the context of systems for the passive location of radar or sonar sources. The function of imaging in a passive surveillance system is discussed primarily in the context of radio astronomy.

13.1   Radio Astronomy

Our everyday environment is rich with invisible signals of many kinds, including electromagnetic waves in the radio bands. Some of these radio waves are emitted naturally by radio sources in distant galaxies. The task of radio astronomy is to detect and form images of the astronomical — actually extragalactic – sources of these electromagnetic waves. The distribution of the received signal coherently across a real or synthetic-aperture is used to form an image of these radiation sources. The methods used to form images from the received signals is the means by which galaxies are discovered and imaged.[1]

In addition to forming an image of the radiation intensity $\rho(x,y)$ from a galaxy, one can also image both the radial velocity and the temperature distribution of that radiation source by measurements of spectral lines. Perhaps the most important spectral line for this purpose is the hydrogen spectral line at 1.420405 GHz. The doppler effect will move the observed frequency (or wavelength) of this line by an amount proportional to the radial velocity. The doppler effect also causes this spectral line to be broadened by an amount proportional to the temperature at the source because of the thermal motion of the hydrogen atoms. By observing such details, an intensity image of a radiation source can be annotated with additional information, such as an estimated temperature map. Such refinements are not discussed herein. Only methods for the formation of the intensity image will be described.

Figure 13.1 shows several elements of an antenna array that forms a radio telescope. Each element of the array is a large, high-gain antenna. The antenna beams are very narrow and are steered so that all beams point in parallel to the same point at infinity. The wavelengths in the microwave band are large, which would lead to poor resolution for a single antenna or a small array of such antennas. To obtain good resolution in the usual way would require an impractically large array. Instead, the usual ways of forming a filled-aperture array are sidestepped in radio astronomy by using a sparse array combined with correlation methods. The image formation methods of radio astronomy rest on the important van Cittert–Zernicke theorem, which is given herein as Theorem 13.1.3. This theorem provides the theoretical validation of the massive radio telescopes now in widespread use. Before giving this theorem, a satisfying heuristic development of the methods of radio astronomy is given. These methods must be formally validated, however, by the van Cittert–Zernicke theorem.

Baseband Cross-Correlation

A radio telescope uses a pair of antennas to form a kind of correlation-computing interferometer. Consider any pair of antennas of the array, as shown in Figure 13.2. Suppose that there is only a single stationary random signal $s(t)$ with intensity ρ

---

[1] In recent years, radio astronomy has rivaled classical optical astronomy in importance. Some present-day radio telescopes, typically in or below the UHF band, use a sparse array of antennas that spans an entire continent.

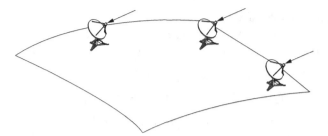

Figure 13.1 Radio astronomy

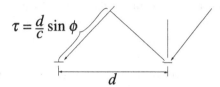

Figure 13.2 Illustrating the phase delay

coming at angle ϕ from a point source at infinity. With the phase reference chosen so that the phase delay at the first element is zero, the complex baseband signals received at the two antennas are

$$v_1(t) = s(t),$$
$$v_2(t) = s(t)e^{j2\pi(d/\lambda)\sin\phi}$$

during an observation interval of duration T. The transmitted signal $s(t)$ from the point source at infinity is spatially coherent across the array of antennas. For a signal $s(t)$ that is generated by a great many independent and random radiation sources in the resolution cell at angle ϕ, the central limit theorem implies that the signal from that cell is accurately modeled temporally as a complex, white stationary gaussian random process of mean zero and variance ρ equal to the signal power. The imaging task is to estimate the signal power ρ and the direction ϕ of that incoming signal. The standard procedure is to compute the sample cross-correlation, given by

$$\frac{1}{T}\int_0^T v_1(t)v_2^*(t)dt = e^{-j2\pi(d/\lambda)\sin\phi}\frac{1}{T}\int_0^T |s(t)|^2 dt$$
$$= \widehat{\rho}\, e^{-j2\pi(d/\lambda)\sin\phi},$$

where the estimate $\widehat{\rho}$ is defined as

$$\widehat{\rho} = \frac{1}{T}\int_0^T |s(t)|^2 dt.$$

The estimate $\widehat{\rho}$ of the intensity has the expected value ρ. The angle ϕ is determined by inverting the phase term $2\pi(d/\lambda)\sin\phi$.

Passband Cross-Correlation

An equivalent computation of this estimate can be formulated at passband, which is sometimes preferred. At passband, the received signals are

$$\tilde{v}_1(t) = A(t)\cos 2\pi(f_0 t + \theta),$$
$$\tilde{v}_2(t) = A(t)\cos 2\pi(f_0 t + \theta + (d/\lambda)\sin\phi).$$

Consequently, using a standard trigonometric identity leads to

$$2\int_0^T \tilde{v}_1(t)\tilde{v}_2(t)dt = \cos(2\pi(d/\lambda)\sin\phi)\int_0^T |s(t)|^2 dt,$$

where the negligible contribution to the integral of the double-frequency term at $2f_0$ has been neglected. When regarded as a function of the angle ϕ, the term $\cos(2\pi(d/\lambda)\sin\phi)$ is referred to as the *fringe pattern*. An observation of the phase of the fringe pattern provides a measurement of ϕ.

One way to measure the phase of the fringe pattern, is to phase-shift the first of the two passband inputs $\tilde{v}_1(t)$ by 90° to form $\tilde{v}'_1(t)$ and then compute

$$2\int_0^T \tilde{v}'_1(t)\tilde{v}_2(t)dt = \sin(2\pi(d/\lambda)\sin\phi)\int_0^T |s(t)|^2 dt.$$

The phase of the fringe is then computed as an arc-tangent of the ratio of this integral and the previous integral.

It does not matter to the theory whether the computation is at complex baseband or at passband. The difference is computational, not consequential. For our purposes, the simplicity of the complex baseband formulation is preferred.

Two Point Sources

Next suppose that there are two point sources at angles ϕ_1 and ϕ_2 that are each emitting an independent stationary gaussian random signal. These signals at complex baseband are denoted $s_1(t)$ and $s_2(t)$. The transmitted signals have zero means and have variances ρ_1 and ρ_2, respectively. The received signals are

$$v_1(t) = s_1(t) + s_2(t),$$
$$v_2(t) = s_1(t)e^{j2\pi(d/\lambda)\sin\phi_1} + s_2(t)e^{j2\pi(d/\lambda)\sin\phi_2}.$$

Now the computational task is to estimate the parameters ρ_1, ρ_2, ϕ_1, and ϕ_2. Again, the accepted procedure is to compute the sample cross-correlation.

$$\frac{1}{T}\int_0^T v_1(t)v_2^*(t)dt = \widehat{\rho}(\phi_1)e^{-j2\pi(d/\lambda)\sin\phi_1} + \widehat{\rho}(\phi_2)e^{-j2\pi(d/\lambda)\sin\phi_2} + n_e(t),$$

where

$$\widehat{\rho}(\phi_1) = \frac{1}{T}\int_0^T |s_1(t)|^2 dt,$$

$$\widehat{\rho}(\phi_2) = \frac{1}{T}\int_0^T |s_2(t)|^2 dt,$$

and the so-called *self-noise* term is

$$n_e(t) = e^{j2\pi(d/\lambda)\sin\phi_2} \frac{1}{T}\int_0^T s_1(t)s_2^*(t)dt + e^{j2\pi(d/\lambda)\sin\phi_1} \frac{1}{T}\int_0^T s_2(t)s_1^*(t)dt.$$

These two integrals both grow more slowly than linearly with time because the independent $s_1(t)$ and $s_2(t)$ are uncorrelated. For large enough T, the self-noise term can be treated as negligible in comparison with the other terms. This consideration is justified because the variance of the noise terms is inversely proportional to T.

Continuum of Infinitesimal Signals

Now consider a continuum of infinitesimal signals distributed in the angle ϕ. The infinitesimal signal at angle ϕ is denoted $s(t,\phi)d\phi$, where $s(t,\phi)$ is the signal density as a function of ϕ. For our needs, it is enough that the signal densities at different angles are uncorrelated. They are not required to be independent because only second-order terms are in play. Then we can write

$$E[s(t,\phi)s^*(t,\phi')] = \rho(\phi)\delta(\phi - \phi').$$

The received signals are

$$v_1(t) = \int_0^{2\pi} s(t,\phi)d\phi,$$

$$v_2(t) = \int_0^{2\pi} s(t,\phi)e^{j2\pi(d/\lambda)\sin\phi}d\phi.$$

The sample cross-correlation coefficient is computed from the received signals as the integral

$$\int_0^T v_1(t)v_2^*(t)dt = \int_0^T \int_0^{2\pi}\int_0^{2\pi} s(t,\phi)s^*(t,\phi')e^{-j2\pi(d/\lambda)\sin\phi'}d\phi d\phi' dt.$$

The expected value of the integral is

$$E\left[\int_0^T v_1(t)v_2^*(t)dt\right] = \int_0^T \int_0^{2\pi}\int_0^{2\pi} \rho(\phi)\delta(\phi - \phi')e^{-j2\pi(d/\lambda)\sin\phi'}d\phi d\phi' dt$$

$$= T\int_0^{2\pi}\int_0^{2\pi} \rho(\phi)\delta(\phi - \phi')e^{-j2\pi(d/\lambda)\sin\phi'}d\phi d\phi'$$

$$= T\int_0^{2\pi} \rho(\phi)e^{-j2\pi(d/\lambda)\sin\phi}d\phi.$$

Moreover, because ϕ is measured with respect to the antenna boresight, $\rho(\phi)$ is negligible except near $\phi = 0$, the approximation $\sin\phi \approx \phi$ allows this expectation to be written

$$\frac{1}{T}E\left[\int_0^T v_1(t)v_2^*(t)dt\right] \approx \int_{-\infty}^{\infty} \rho(\phi)e^{-j2\pi(d/\lambda)\phi}d\phi$$

$$= R(d/\lambda),$$

where $R(f)$ is the Fourier transform of $\rho(\phi)$. Consequently, within this approximation, the expected value of the cross-correlation coefficient is equal to the Fourier transform

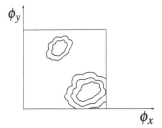

Figure 13.3 A density profile of radio sources

$R(f)$ at the single spatial frequency $f = d/\lambda$. For large T, the sample cross-correlation coefficient, computed from the observations, approaches its expectation which is the true cross-correlation coefficient. Therefore, the sample cross-correlation coefficient provides an estimate of the Fourier transform of $\rho(\phi)$ at the single spatial frequency $f = d/\lambda$.

To obtain acceptable accuracy in the estimate of $R(d/\lambda)$ for a continuum of sources, large values of T are needed to suppress the self-noise term. Then the notion of self-noise that must be suppressed becomes less useful. It is better to describe this as a task of reducing the error variance when computing a correlation coefficient from long samples of two dependent stationary random processes.

Sampling Fourier Space

An observation of a radio source is two-dimensional, as shown in Figure 13.3. When the scene is small and the z axis is pointed at the scene, the angular coordinates can be described by the small angles ϕ_x and ϕ_y. Then the source is described as the function $\rho(\phi_x, \phi_y)$ with Fourier transform $R(f_x, f_y)$. Just as before, each computed sample cross-correlation coefficient provides a good estimate of the two-dimensional Fourier transform at the single point (f_x, f_y) of frequency space provided T is sufficiently large. To form an image, the function $R(f_x, f_y)$ must be determined for many values (f_x, f_y) of frequency space. These Fourier samples must satisfy the Nyquist criterion in frequency space or the data set must be supplemented in some way by prior information.

Consider three colinear antennas with pairwise spacings d_1, d_2, and $d_1 + d_2$. By computing the correlation using two antennas at a time, three samples of the two-dimensional Fourier transform can be estimated. When the three antennas lie along the x axis, then the three values of the Fourier transform are $R(d_1/\lambda, 0)$, $R(d_2/\lambda, 0)$, and $R((d_1 + d_2)/\lambda, 0)$. When the three antennas are in a straight line at the angle ψ with respect to the x axis, then the three values of the Fourier transform are at $R(d_1 \cos\psi/\lambda, d_1 \sin\psi/\lambda)$, $R(d_2 \cos\psi/\lambda, d_2 \sin\psi/\lambda)$, and $R((d_1 + d_2)\cos\psi/\lambda, (d_1 + d_2)\sin\psi/\lambda)$. When the three antennas are not colinear, the three points of the Fourier transform are also not colinear.

With M antennas, the M received signals can be correlated two at a time to produce estimates of many samples of the Fourier transform. There are $M(M-1)/2$ antenna pairs. The arrangement of antenna placements within the array should be designed

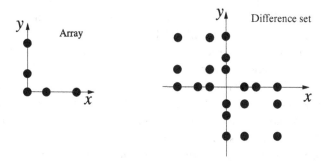

Figure 13.4 An array and its difference set

so that the pairwise vector differences are all different. Then $M(M-1)/2$ samples of $R(f_x, f_y)$ can be computed. Figure 13.4 shows an array of five antennas, as well as the twenty vector differences between pairs of antennas. These vector differences determine the twenty points of the f_x, f_y plane where the values of $R(f_x, f_y)$ are obtained. Because $\rho(x, y)$ is real, $R(f_x, f_y) = R(-f_x, -f_y)$, thus only ten of the samples of $R(f_x, f_y)$ are unique. This is in accord with the statement that $M(M-1)/2$ samples of the Fourier transform can be obtained.

The intensity distribution $\rho(x, y)$ is the inverse Fourier transform of $R(f_x, f_y)$. The twenty samples of $R(f_x, f_y)$ shown in the figure provide only incomplete knowledge of $\rho(x, y)$. To form a high-resolution image of $\rho(x, y)$, one needs many more samples of $R(f_x, f_y)$. A good image might need to use between 10^5 and 10^6 samples of $R(f_x, f_y)$. To obtain more samples by pairwise correlation would require something on the order of one thousand radio antennas, each consisting of a large and expensive high-gain dish. These samples need not be taken simultaneously. A galaxy changes on a much longer time scale than the time needed to make observations.

To obtain many Fourier samples using a modest number of radio antennas, one obtains the effect of many antennas by reusing a smaller number of antennas many times. One method is to mount one or more antennas on a track so that one or more antennas can be physically repositioned with respect to the other antennas after each set of data is collected. A second method is to allow the normal rotation of the earth to rotate the array with respect to the source. A new set of data is taken after the earth rotates a small amount, and this data set consists of samples of $R(f_x, f_y)$ in a new reference frame that is rotated with respect to the first reference frame. Such observation methods produce Fourier transform samples that are not on a rectangular grid. This means that computing the inverse Fourier transform and controlling sampling artifacts are not straightforward tasks.

Prior Information

Even with all of these methods for enlarging the data set, the number of measured components of the Fourier transform may still be smaller than the number of pixels in a high-resolution image. The Fourier samples may fail to satisfy the Nyquist condition. Moreover, there may be some regions of the Fourier plane that are only sparsely sampled, or not observed at all. For these reasons, methods have been developed that can supplement the measured data with prior knowledge about an image. Two simple prior conditions on the image are of great value. One is that the image is real and nonnegative; the other is that the object being imaged is compact and a large part of that image is black, a so-called *mostly black* image. The prior condition that the image is real and nonnegative is a precise condition. The prior condition that the image is mostly black requires some degree of judgment.

Various principles of inference are invoked to infer the missing Fourier data from the condition that the image is nonnegative. The *clean algorithm*, which is discussed in Section 7.2, uses the prior information that the image is mostly black. The *Richardson–Lucy algorithm*, which is discussed in Section 7.3, uses the prior information that the image is real and nonnegative.

The van Cittert–Zernike Theorem

The introduction to the topic of radio astronomy given above describes the relationship between the autocorrelation function and the image of the source in an informal way that provides a satisfying overview. A formal justification is given now based on the van Cittert–Zernike theorem,[2] which is stated below.

In the ideal case, a radio telescope fills an array in frequency space consisting of n by n pixels with n^2 spatial correlations. The image is the two-dimensional Fourier transform of this two-dimensional array. This method of forming an image is to be compared to a real uniformly illuminated square aperture. An L by L square aperture has the two-dimensional sinc function $\text{sinc}(L\alpha, L\beta)$ as its antenna pattern. This pattern can be steered by filling the aperture with an array to be scanned so as to form an image of the far field with the same resolution as would be obtained by using the full aperture to form a beam in the far field. These two methods both form an image of the same far field. The images are equivalent in the sense that, except for possible discretization effects, they differ only by the scaling by fixed constants. This equivalence anticipates the forthcoming van Cittert–Zernike theorem.

The van Cittert–Zernike theorem relates two stochastic quantities defined in terms of the same spatial random process. The two quantities are the *spatial coherence function* of the two-dimensional random process and the *spatial power density spectrum* of that process, both of which are defined in terms of the time-varying two-dimensional zero-mean, stationary random process $S(x, y, t)$. To emphasize that that this is a two-dimensional random variable in space that varies in time, it might better be written $S(x, y)(t)$. Although the time variable t is included, it has little to do with the theory of this section.

---

[2] The van Cittert–Zernike theorem was first developed in the field of noncoherent optics but is of fundamental importance in other instances of coherence. Indeed, Goodman states "by the van Cittert–Zernike theorem, which is undoubtedly one of the most important theorems of modern optics" (Goodman, 1985, p. 207).

The spatial coherence function and the spatial power density spectrum of the two-dimensional random process S(x, y, t) are defined separately. The theorem then states that the two quantities are a two-dimensional Fourier transform pair.[3]

Definition 13.1.1 *The spatial coherence function of the real two-dimensional, time stationary, random process* S(x, y, t) *is*

$$\phi(u, v) = E[S(x, y, t) | S(x+u, y+v, t)],$$

which is independent of time. ∎

The spatial coherence function $\phi(u, v)$ does not depend on the time variable t because the spatial random process is stationary in time. A spatial coherence function is also called a two-dimensional correlation function. A spatial random process with a finite spatial support has a spatial coherence function with a finite spatial support.

The random process S(x, y, t) does not have a Fourier transform with respect to its spatial coordinates because the Fourier transform of a random process is not defined. Moreover, a realization $s(x, y, t)$ may have infinite energy, which means that a realization of the random process does not have a Fourier transform either. Instead, the cropped version $s_{L \times L}(x, y, t) = s(x, y, t)\text{rect}(x/L, y/L)$ does have finite energy and so does have a Fourier transform $S_{L \times L}(f_x, f_y)$.

Definition 13.1.2 *The spatial power density spectrum is the limit*

$$S(f_x, f_y) = \lim_{L \to \infty} \frac{1}{4L^2} E[|S_{L \times L}(f_x, f_y, t)|^2],$$

where $S_{L \times L}(f_x, f_y, t)$ is the two-dimensional Fourier transform of $s_{L \times L}(x, y, t)$. ∎

The proof of the following theorem mimics, step-by-step, the proof of the Wiener–Khintchine theorem, which is given as Theorem 2.8.1 of Chapter 2. Because of the importance of the van Cittert–Zernike theorem, its proof is written out in detail,[4] although the formal justification of some technical steps involving limits is suppressed.

Theorem 13.1.3 (van Cittert–Zernike Theorem) *The spatial power density spectrum of a well-behaved stationary two-dimensional random process is equal to the Fourier transform of the spatial coherence function of that random process.*

---

[3] The random process with realization $s(x, y)$ is denoted S(x, y), using a sans serif S. The cropped realization $s_{L \times L}(x, y) = s(x, y)\text{rect}(x/L, y/L)$ has finite energy and so has Fourier transform $S_{L \times L}(f_x, f_y)$.

[4] The van Cittert–Zernike theorem has a structure that parallels the structure of the Wiener–Khintchine theorem. The Wiener–Khintchine theorem states that the correlation function $\phi(\tau)$ of a stationary random process and the power density spectrum $P(f)$ of that random process are a Fourier transform pair. As is the case for the Wiener–Khintchine theorem, the van Cittert–Zernike theorem is an exact mathematical statement. There are no approximations in the theorem itself, as sometimes claimed. The approximations come in the applications for which the various quantities or relationships might not exactly fit the conditions of the theorem.

Proof:

The proof consists of a long sequence of equalities using the abbreviated notation $\mathbf{x}$ to represent x, y and $d\mathbf{x}$ to represent $dxdy$ as well as $\mathbf{f} \cdot \mathbf{x}$ to represent $f_x x + f_y y$:

$$S(f_x, f_y) = \lim_{L \to \infty} \frac{1}{4L^2} E\left[|S_{L \times L}(f_x, f_y, t)|^2\right]$$

$$= \lim_{L \to \infty} \frac{1}{4L^2} E\left[\int_{-L}^{L}\int_{-L}^{L} s(\mathbf{x}, t) e^{j2\pi(\mathbf{f} \cdot \mathbf{x})} d\mathbf{x} \int_{-L}^{L}\int_{-L}^{L} s^*(\mathbf{x}', t) e^{-j2\pi(\mathbf{f} \cdot \mathbf{x}')} d\mathbf{x}'\right]$$

$$= \lim_{L \to \infty} \frac{1}{4L^2} E\left[\int_{-L}^{L}\int_{-L}^{L}\int_{-L}^{L}\int_{-L}^{L} s(\mathbf{x}, t) s^*(\mathbf{x}', t) e^{-j2\pi[\mathbf{f} \cdot (\mathbf{x} - \mathbf{x}')]} d\mathbf{x} d\mathbf{x}'\right]$$

$$= \lim_{L \to \infty} \frac{1}{4L^2} \int_{-L}^{L}\int_{-L}^{L}\int_{-L}^{L}\int_{-L}^{L} E\left[s(\mathbf{x}, t) s^*(\mathbf{x}', t)\right] e^{-j2\pi[\mathbf{f} \cdot (\mathbf{x} - \mathbf{x}')]} d\mathbf{x} d\mathbf{x}'$$

$$= \lim_{L \to \infty} \frac{1}{4L^2} \int_{-L}^{L}\int_{-L}^{L}\int_{-L}^{L}\int_{-L}^{L} \phi(\mathbf{x} - \mathbf{x}') e^{-j2\pi[\mathbf{f} \cdot (\mathbf{x} - \mathbf{x}')]} d\mathbf{x} d\mathbf{x}'$$

$$= \lim_{L' \to \infty} \frac{1}{4L'^2} \int_{-L'}^{L'}\int_{-L'}^{L'} \lim_{L \to \infty} \int_{-L}^{L}\int_{-L}^{L} \phi(\mathbf{x} - \mathbf{x}') e^{-j2\pi[\mathbf{f} \cdot (\mathbf{x} - \mathbf{x}')]} d\mathbf{x} d\mathbf{x}'$$

$$= \lim_{L' \to \infty} \frac{1}{4L'^2} \int_{-L'}^{L'}\int_{-L'}^{L'} \int_{-\infty}^{\infty}\int_{-\infty}^{\infty} \phi(\mathbf{x} - \mathbf{x}') e^{-j2\pi[\mathbf{f} \cdot (\mathbf{x} - \mathbf{x}')]} d\mathbf{x} d\mathbf{x}'$$

$$= \lim_{L' \to \infty} \frac{1}{4L'^2} \int_{-L'}^{L'}\int_{-L'}^{L'} \int_{-\infty}^{\infty}\int_{-\infty}^{\infty} \phi(\mathbf{x}) e^{-j2\pi[\mathbf{f} \cdot \mathbf{x}]} d\mathbf{x} d\mathbf{x}'$$

$$= \left[\int_{-\infty}^{\infty}\int_{-\infty}^{\infty} \phi(\mathbf{x}) e^{-j2\pi[\mathbf{f} \cdot \mathbf{x}]} d\mathbf{x}\right]\left[\lim_{L' \to \infty} \frac{1}{4L'^2} \int_{-L'}^{L'}\int_{-L'}^{L'} 1 \, d\mathbf{x}'\right]$$

$$= \int_{-\infty}^{\infty}\int_{-\infty}^{\infty} \phi(\mathbf{x}) e^{-j2\pi(\mathbf{f} \cdot \mathbf{x})} d\mathbf{x}$$

$$= \Phi(f_x, f_y).$$

This completes the proof of the theorem. ∎

The statement of the theorem, as appearing in the proof, can be summarized by stating that the expression

$$\phi(x, y) \Leftrightarrow S(f_x, f_y)$$

is a Fourier transform pair. This statement is *not* a definition of $S(f_x, S_y)$ Although the Fourier transform pair is sometimes said to be a definition, that approach would leave $S(f_x, S_y)$ without independent meaning connecting it to the random process. Because $S(x, y, t)$ is a random process, the power density spectrum $S(f_x, f_y)$ must have a suitable definition as such.

In the application to radio astronomy, $f_x = \alpha/\lambda$ and $f_y = \beta/\lambda$, where α and β are the direction cosines. Then

$$\phi(x, y) \Leftrightarrow S(\alpha/\lambda, \beta/\lambda).$$

This statement of the van Cittert–Zernicke theorem provides the formal validation of modern radio astronomy.

Hardlimiting

The computation of the large number of samples of a correlation function needed by a radio telescope is a massive task. The task of image formation in radio astronomy can be computation-limited or it can be communication-limited, or both. The sampled data gathered at remote antennas must be brought to a common location where the correlation function is computed. Most of the relevant information is contained in the first bit of each quantized sample. The second bit of each sample has much less information, and subsequent bits have even less. Additive noise reduces the information content of the second and subsequent bits even further. Hardlimiting of the received signal may be a suitable remedy in both cases. The hardlimiter reduces each sample of the received waveform to a single bit, thereby reducing both the communication burden and the computation burden.

As will be seen herein, the number of waveform samples entering a correlation estimate is much more important than the wordlength of the individual samples.

Early on, computation of the many correlation functions required for a radio astronomy image was a limiting factor and remains a burden. Also, modern and future radio telescopes have remote decentralized antennas. The communication of the large-bandwidth signals to a central processing site could become a limiting factor. Both of these limitations can be lessened by the fact that most of the useful information in a gaussian noise process is in the first bit. For the task of computing a cross-correlation function, the gaussian noise process can be approximated by its sign with only a predictable effect in the correlation function.

A *hardlimiter* is defined symmetrically as the function

$$\mathrm{hrd}(x) = \begin{cases} 1 & \text{if } x > 0, \\ 0 & \text{if } x = 0, \\ -1 & \text{if } x < 0, \end{cases}$$

or asymmetrically with the case $x = 0$ adjoined to either of the other two cases. The *hardlimited* reduction of signal $s(t)$ given by $\mathrm{hrd}(s(t))$, is called a hardlimited signal or a *clipped* signal. When $s(t)$ is sampled as the signal $s(t_k)$ for $k = 0, 1, 2, \ldots$, the signal $\mathrm{hrd}(s(t_k))$ for $k = 0, 1, 2, \ldots$ is a one-bit quantized signal.

Degradation in the estimated correlation function can be offset by integrating over a longer interval. That is,

$$\int_0^{T_1} s_a(t)s_b(t+\tau)dt \sim \int_0^{T_2} \mathrm{hrd}(s_a(t))\mathrm{hrd}(s_b(t+\tau))dt,$$

meaning that, but for scaling, the two sides are asymptotically equivalent under a suitable interpretation.

The following proposition uses a standard definite integral to lift the hardlimited signal to an expression that is more amenable to analysis.

13 Passive and Baseband Systems

Proposition 13.1.4 *The hardlimiter representation* hrd($s(t)$) *of* $s(t)$ *can be written*

$$\text{hrd}(s(t)) = \frac{1}{\pi} \int_{-\infty}^{\infty} \frac{\sin(\xi s(t))}{\xi} d\xi.$$

Proof:
The following widely tabulated standard integral

$$\int_{-\infty}^{\infty} \frac{\sin(\xi \alpha)}{\xi} d\xi = \begin{cases} \pi & \text{if } \alpha > 0, \\ 0 & \text{if } \alpha = 0, \\ -\pi & \text{if } \alpha < 0 \end{cases}$$

reduces to the definition of the hardlimiter. This comparison provides an immediate proof of the proposition. ∎

The functions of hardlimiting and sampling can be performed in either order. In each case, the result is a single bit per sample.

One may compute the sample covariance function of $s(t)$ or the sample covariance function of hrd($s(t)$). The first uses a multibit by multibit multiplier. The second uses a single bit by single bit multiplier. The following discussion derives the relationship between the two covariance functions,[5] so either covariance function can be expressed in terms of the other.

Theorem 13.1.5 (van Vleck–Middleton Theorem) *Let* $Y(t) = \text{hrd}(X(t))$, *where* $X(t)$ *is a zero-mean stationary gaussian random process with autocorrelation function* $\phi_{xx}(\tau)$. *Then the autocorrelation function* $\phi_{yy}(\tau)$ *of* $Y(t)$ *is*

$$\phi_{yy}(\tau) = \frac{2}{\pi} \sin^{-1}\left(\frac{\phi_{xx}(\tau)}{\phi_{xx}(0)}\right).$$

Proof
Rather than giving a proof of the van Vleck–Middleton theorem, as such, a general reformulation of that theorem is stated and proved instead. The proof of the van Vleck–Middleton theorem is a special case of the proof of the next theorem. ∎

Theorem 13.1.6 *Let* $Y_1(t) = \text{hrd}(X_1(t))$ *and* $Y_2(t) = \text{hrd}(X_2(t))$, *where* $X_1(t)$ *and* $X_2(t)$ *are zero-mean stationary gaussian random processes with cross-correlation function* $\phi_{X_1 X_2}$. *Then the cross-correlation of* $Y_1(t)$ *and* $Y_2(t)$ *is*

$$\phi_{y_1 y_2} = \frac{2}{\pi} \sin^{-1}\left(\frac{\phi_{x_1 x_2}}{\sqrt{\phi_{x_1 x_1} \phi_{x_2 x_2}}}\right).$$

---

[5] When a random process has a zero mean, the covariance function and the correlation function are the same thing.

13.1 Radio Astronomy

Proof:
The expression in Proposition 13.1.4 is rewritten here as

$$\mathrm{hrd}(x) = \frac{1}{2\pi j} \int_{-\infty}^{\infty} \frac{e^{j2\pi\xi x} - e^{-j2\pi\xi x}}{\xi} d\xi.$$

This is the form that is used twice in this proof. To begin

$$\phi_{y_1 y_2} = \mathrm{E}[Y_1(t) Y_2(t)]$$
$$= -\frac{1}{4\pi^2} \mathrm{E}\left[\int_{-\infty}^{\infty} \int_{-\infty}^{\infty} \left(e^{j2\pi\xi X_1(t)} - e^{-j2\pi\xi X_1(t)}\right)\left(e^{j2\pi\xi X_2(t)} - e^{-j2\pi\xi X_2(t)}\right) \frac{d\xi}{\xi} \frac{d\nu}{\nu}\right].$$

Expand the product in the integrand and distribute the expectation over the resulting four terms. The four terms differ only in the signs in the exponent. Theorem 15.2.1 of Chapter 15 states that

$$\mathrm{E}\left[e^{j(\pm\xi x_1(t) \pm \nu x_2(t))}\right] = e^{\xi^2 \pm 2\rho\xi\nu + \nu^2},$$

where $\rho = \mathrm{E}[X_1(t) X_2(t)]$. In the expression for $\phi_{y_1 y_2}$, there are four terms represented in the expression on the left and two terms represented in the expression on the right. This is because each term on the right occurs twice. Therefore, the four terms reduce to

$$\phi_{y_1 y_2} = -\frac{1}{2\pi^2} \int_{-\infty}^{\infty} \int_{-\infty}^{\infty} \left(e^{-(\xi^2 + 2\rho\xi\nu + \nu^2)/2} - e^{-(\xi^2 - 2\rho\xi\nu + \nu^2)/2}\right) \frac{d\xi}{\xi} \frac{d\nu}{\nu}$$
$$= \frac{1}{2\pi^2} \int_{-\infty}^{\infty} \int_{-\infty}^{\infty} e^{-(\xi^2 + \nu^2)/2} \left(e^{\rho\xi\nu} - e^{-\rho\xi\nu}\right) \frac{d\xi}{\xi} \frac{d\nu}{\nu}.$$

This can be collapsed even further. Regard this expression as the difference of two double integrals. Make the change $\xi' = -\xi$ in the second term with reference to the equality to make it identical to this first. To this purpose, note that with the substitution $x' = -x$, the generic integral

$$\int_{-\infty}^{\infty} e^{-x} \frac{dx}{x} = \int_{\infty}^{-\infty} e^{-x'} \frac{dx'}{x'} = -\int_{-\infty}^{\infty} e^{-x'} \frac{dx'}{x'}.$$

The second term of ϕ_{yy} now combines with the first term, giving

$$\phi_{yy} = \frac{1}{\pi^2} \int_{-\infty}^{\infty} \int_{-\infty}^{\infty} e^{-(\xi^2 + \rho\xi\nu + \nu^2)/2} \frac{d\xi}{\xi} \frac{d\nu}{\nu}.$$

It remains to evaluate this definite integral. Regard the integral as a function of ρ. Denote that function $I(\rho)$ and differentiate it with respect to ρ. Passing the differentiation under the integral sign gives

$$\frac{dI(\rho)}{d\rho} = \frac{1}{\pi^2} \int_{-\infty}^{\infty} \int_{-\infty}^{\infty} e^{-(\xi^2 + \nu^2)/2} e^{\rho\xi\nu} \frac{d\xi}{\xi} \frac{d\nu}{\nu}$$
$$= \frac{2}{\pi^2} \int_{-\infty}^{\infty} \int_{-\infty}^{\infty} e^{-(\xi^2 + 2\rho\xi\nu + \nu^2)/2} d\xi\, d\nu$$
$$= \frac{1}{\pi} \frac{1}{\sqrt{1 - \rho^2}}.$$

The last line follows because it is a standard integral of a two-dimensional gaussian pulse. Then

$$I(\rho) = \frac{2}{\pi} \int_0^\rho \frac{1}{\sqrt{1-x^2}} dx$$

$$= \frac{2}{\pi} \sin^{-1} \rho.$$

Setting $\rho = \phi_{x_1 x_2}/\sqrt{\phi_{x_1 x_1} \phi_{x_2 x_2}}$ completes the proof of the theorem. ∎

Corollary 13.1.7 *Let $Y(t) = \mathrm{hrd}(X(t))$, where $X(t)$ is a zero-mean stationary gaussian random process with covariance function $\phi_{xx}(\tau)$. Then the covariance function of $Y(t)$ is given by*

$$\phi_{yy}(\tau) = \frac{2}{\pi} \sin^{-1}\left(\frac{\phi_{xx}(\tau)}{\phi_{xx}(0)}\right).$$

Proof:
The corollary is a special case of the theorem with $X_1 = X(t)$ and $X_2 = X(t+\tau)$, as well as $Y_1 = Y(t)$ and $Y_2 = Y(t+\tau)$. ∎

13.2 Magnetic Anomaly Detection

A ferromagnetic metallic object distorts the ambient magnetic field in its vicinity. The earth has an ambient magnetic field and ferromagnetic objects distort that ambient field. Hence, one may detect a nearby metallic object by measuring anomalies in the earth's magnetic field. This means of detection is referred to as *magnetic anomaly detection*. Magnetic anomaly detection can be used for detecting submarines and land mines, as well as small objects buried in soil or walls. Magnetic anomaly detection has the advantage of being a passive method of detection relying only on the earth's natural magnetic field. An advantage is that soil, water, and most other materials are essentially transparent to a magnetic sensor. Magnetic anomaly detection has the disadvantages of poor sensitivity, poor resolution, and short range. An anomaly in the magnetic field attenuates rapidly with distance and the gradients in the magnetic field quickly become weak as the distance from the source increases. A magnetic anomaly detector works only over short distances.

Magnetic fields can also be created by electrical currents. By precisely measuring a very weak magnetic field structure in one region of space, one can hope to deduce something about the distribution of electrical currents in a nearby region of space. The general task is referred to as *magnetic imaging*. The topic of magnetic imaging is not discussed herein. Only the simpler task of magnetic anomaly detection is discussed.

This section addresses the processing techniques that can be used for magnetic anomaly detection to extract the desired information from the background magnetic field under ideal circumstances. The task of magnetic anomaly detection can be quite delicate, requiring precise measurements of the structure of the local magnetic field. The actual performance is limited by the error in the known local value of the earth's magnetic field, by sensor measurement bias, and by receiver noise. A magnetic anomaly detector can be degraded by ferromagnetic material within the host itself.

A magnetic field is a vector field. It is a vector function of space written as

$$H(x,y,z) = H_x(x,y,z)i_x + H_y(x,y,z)i_y + H_z(x,y,z)i_z,$$

where i_x, i_y, and i_z are unit vectors. The terms x, y, and z appear here in two roles. Within the subscripts, the terms x, y, and z denote the three directions of the three vector components. Within the parentheses, the terms x, y, and z denote a point in (x,y,z) space at which the magnetic field is designated. The three components of the magnetic field are three functions of space that together respect the laws of physics.

A magnetic field must satisfy Maxwell's equations. When there are no electrical currents or time-varying electrical fields, Maxwell's equations in free space reduce to

$$\nabla \cdot H = 0,$$
$$\nabla \times H = 0,$$

as expressed by the divergence and curl operators. Because magnetic monopoles do not exist, the simplest static magnetic field is a dipole field.[6]

The ambient magnetic field of the earth is a slowly varying function of space. As seen in a small geographic region, the local earth's magnetic field can be treated as a constant vector field denoted $H_0 = H_{0x}i_x + H_{0y}i_y + H_{0z}i_z$. Any local variations in the earth's normal magnetic field are minimal and can be accommodated in the analysis by an error term. The distortion in the magnetic field caused by a target object is here regarded as an additive field created by that object. The earth's magnetic field H_0 is regarded as a bias vector field to which is added the vector field $H(x,y,z)$ created by the target object in response to the earth's magnetic field. The total magnetic field is

$$H_t(x,y,z) = H_0 + H(x,y,z)$$

where the addition is vector addition.

Magnetometer

A *magnetometer* is a device for measuring properties of the total magnetic field $H_t(x,y,z)$ at location (x,y,z). The goal is to learn something about $H(x,y,z)$ in order to detect and locate nearby metallic objects. Three versions of a magnetometer are to be described. These are: a scalar magnetometer; a vector magnetometer; and a gradient magnetometer. For each version, the performance depends on both the magnitude of $H(x,y,z)$ and the direction of $H(x,y,z)$ with respect to the direction of the vector H_0.

---

[6] Every magnetic source begins to look like a magnetic dipole asymptotically with increasing distance because, except for the dipole term, all terms in an expansion decay more quickly than $1/R^3$ with increasing distance. The dipole field decreases as $1/R^3$ with increasing distance.

A *scalar magnetometer* measures only the magnitude $|H_t(x,y,z)|$ of $H_t(x,y,z)$ at the point (x,y,z). A scalar magnetometer can detect the presence of an object but does not locate that object. The magnitude of $H_t(x,y,z)$ depends strongly on the magnitude of $H(x,y,z)$ when the vector $H(x,y,z)$ is comparable to H_0. The magnitude $|H_t(x,y,z)|$ can be applied to a threshold Θ in order to detect a target. A target is detected when the inequality

$$|H_t(x,y,z)| \geq \Theta$$

is satisfied, where Θ is larger than $|H_0|$. The probability of a false alarm depends on the threshold Θ. The probability of a missed detection also depends on $|H(x,y,z)|$ which depends in turn on the magnitude and the direction of $H(x,y,z)$. The detection is formulated as a standard parametric problem of testing binary hypotheses in the presence of receiver noise and model bias. A scalar magnetometer can be used with simple thresholding to detect the existence of a nearby metallic object. The performance will depend on the angle between the vectors $H(x,y,z)$ and H_0 as well as the magnitude of each term.

A *vector magnetometer* measures the three components of the total magnetic field $H_t(x,y,z)$. When a global coordinate reference system is maintained so that the orientation of the vector H_0 is known, the bias field vector H_0 can be subtracted from the measured field vector to obtain $H(x,y,z)$. Then the direction to the nearby object can be computed directly from $H(x,y,z)$. When such an observation of direction is available at each of several decentralized locations, as when the magnetometer is moving, the magnetic object can be located by means of triangulation.

A *gradient magnetometer* (or a gradiometer) is a much more complicated instance of a magnetometer. It measures all, or most, of the nine partial derivatives of the magnetic field. An ideal gradient magnetometer measures

$$\Delta H = \begin{bmatrix} \frac{\partial H_x}{\partial x} & \frac{\partial H_x}{\partial y} & \frac{\partial H_x}{\partial z} \\ \frac{\partial H_y}{\partial x} & \frac{\partial H_y}{\partial y} & \frac{\partial H_y}{\partial z} \\ \frac{\partial H_z}{\partial x} & \frac{\partial H_z}{\partial y} & \frac{\partial H_z}{\partial z} \end{bmatrix}.$$

The nine matrix elements are the derivatives of each of the three magnetic field components with respect to each of the three spatial coordinates. To express this set of nine derivatives compactly, the three spatial coordinates (x,y,z) are denoted instead by (x_1,x_2,x_3). Then the nine derivatives are denoted $\partial H_k/\partial x_j$ for $k = 1,2,3$ and $j = 1,2,3$, and

$$\Delta H = \left[\frac{\partial H_i}{\partial x_j} \right].$$

The three by three matrix ΔH of partial derivatives must correspond to a field that satisfies Maxwell's equations.

Dipole Field

A magnetic dipole field is the simplest instance of a magnetic field. The magnetic field $H(x_1, x_2, x_3)$ for a magnetic dipole $m = (m_1, m_2, m_3)$ at the origin is

$$H(x_1, x_2, x_3) = \frac{3(m \cdot R)R}{R^5} - \frac{m}{R^3},$$

where $R = (x_1, x_2, x_3)$ and $R = |R|$. The magnitude of the dipole field decreases as $1/R^3$. Terms of a magnetic field expansion other than the dipole field decay more quickly than $1/R^3$. The inverse dependence of the field on R^3 means that even the magnetic dipole field will be weak at large R. This is why magnetic anomaly detection is useful only for short distances.

The derivatives of the dipole field equations can be expressed concisely in terms of the three direction cosines, here denoted α_i for $i = 1, 2, 3$, and satisfying $\alpha_i = x_i/R$. Differentiating the magnetic dipole field at the point (x_1, x_2, x_3) gives

$$\frac{\partial H_k}{\partial x_j} = -\sum_{i=1}^{3} M_i N_{ijk},$$

where $M_i = 3m_i/R^4$ for $i = 1, 2, 3$, and the nine quantities N_{ijk} form the third-rank tensor given by

$$N_{ijk} = 5\alpha_i \alpha_j \alpha_k - (\delta_{ki}\alpha_i + \delta_{kj}\alpha_j + \delta_{ji}\alpha_k).$$

These are the nine elements of a third-rank tensor.

Dipole Location

The more difficult problem of magnetic anomaly detection when there is a magnetic dipole at the origin is to compute both the distance to the dipole and the magnitude of the dipole from the matrix ΔH. This is the task of inverting the dipole gradient equations to express the position, the magnitude, and the direction of the dipole in terms of the elements of ΔH.

The problem of magnetic anomaly detection is shown in Figure 13.5. The direction to a magnetic dipole, m, at an unknown location, (x, y, z) is to be estimated when given the partial derivatives of the magnetic field H at the origin. The point dipole m and the location R are unknown. Each unknown vector has three unknown components, resulting in six scalar unknowns. At the origin, we have the three by three array of partial derivatives $\frac{\partial H_k}{\partial x_j}$ for $k = 1, 2, 3$ and $j = 1, 2, 3$. Maxwell's equations require that the trace of this matrix equals zero (because $\nabla \cdot B = 0$), and also that the matrix is symmetric. Therefore, there are only five independent measurements, and the six independent unknowns cannot be estimated. To obtain a full solution, additional information, such as the scalar range $|R|$ or the magnetic field magnitude $|H|$, must be available.

In detail, the task is to determine the three position coordinates (x_1, x_2, x_3) and the three dipole components (m_1, m_2, m_3) from measurements of the three by three matrix of the partial derivatives of H at the unknown point (x_1, x_2, x_3). Any two of the direction cosines are independent and imply the third because the three direction

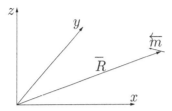

Figure 13.5 Location of an unknown magnetic dipole

cosines satisfy $\alpha_1^2 + \alpha_2^2 + \alpha_3^2 = 1$. The three scaled moments and any two of the three direction cosines constitute a set of five independent unknowns to be computed from the matrix of the partial derivatives of H. This solution is best obtained in two steps: first compute the direction cosines from ΔH, then compute the scaled moments from both ΔH and the direction cosines.

The first step in the inversion is to solve the equations for the scaled moments M_i so they can be eliminated. This is done with the aid of another third-rank tensor

$$\widetilde{N}_{jk\ell} = \frac{3}{2}\alpha_j\alpha_k\alpha_\ell - \frac{1}{2}(\delta_{j\ell}\alpha_k + \delta_{k\ell}\alpha_j),$$

defined in terms of the unknown direction cosines. The $\widetilde{N}_{jk\ell}$ provide an inverse for N_{ijk} in the sense that

$$\sum_j \sum_k N_{ijk}\widetilde{N}_{jk\ell} = \delta_{i\ell},$$

as can be checked easily. Combining the equations for the field derivative $\partial H_k/\partial x_j$ with $\widetilde{N}_{jk\ell}$ and summing yields

$$M_\ell = -\sum_j \sum_k \frac{\partial H_k}{\partial x_j}\widetilde{N}_{jk\ell},$$

which gives the three scaled moments in terms of the three direction cosines and the nine measurements. To eliminate the three scaled moments, substituting this equation for M_ℓ into the equations for the nine field derivatives. This yields

$$\frac{\partial H_k}{\partial x_j} = -\sum_i M_i \frac{\partial H_m}{\partial x}$$

$$= -\sum_\ell \sum_m \sum_i \widetilde{N}_{\ell m i} N_{ijk}\frac{\partial H_m}{\partial x}.$$

The scaled moments now have been eliminated in this set of equations. The term $\widetilde{N}_{\ell m i}N_{ijk}$ depends only on the direction cosines. Therefore, this system of equations relates the unknown direction cosines to the observed field derivatives. The solution of this system lies at the crux of the inversion problem. Once the direction cosines are known, the scaled moments M_i can be obtained from the earlier expression.

To solve this system of equations, the coordinate system is to be chosen so that the matrix of derivatives is a diagonal matrix. This entails no loss of generality because the general system of equations can always be rotated into a principal axis frame. Therefore,

$$\sum_{m=1}^{3} \frac{\partial H_k}{\partial x_m} \sum_{i=1}^{3} \tilde{N}_{mmi} N_{ijk} = \begin{cases} \frac{\partial H_k}{\partial x_k} & j = k, \\ 0 & j \neq k. \end{cases}$$

Because $\sum_i \alpha_i^2 = 1$, the sum on i can be collapsed as given by

$$\sum_{i=1}^{3} \tilde{N}_{mmi} N_{ijk} = \sum_{i=1}^{3} \left[\frac{3}{2} \alpha_m^2 \alpha_i - \delta_{mi} \alpha_m \right] \left[5\alpha_i \alpha_j \alpha_k - \delta_{ki} \alpha_j - \delta_{kj} \alpha_i - \delta_{ji} \alpha_k \right]$$

$$= -\frac{1}{2} \alpha_m^2 (\alpha_j \alpha_k + \delta_{kj}) + \alpha_m (\alpha_j \delta_{mk} + \alpha_k \delta_{mj})$$

$$= -\frac{1}{2} \alpha_m^2 (\alpha_j \alpha_k + \delta_{kj}) + \alpha_k \alpha_j \delta_{mk} + \alpha_j \alpha_k \delta_{mj}.$$

This reduces to

$$\sum_{i=1}^{3} \tilde{N}_{mmi} N_{ikk} = -\frac{1}{2} \alpha_m^2 (\alpha_k^2 + 1) + 2\alpha_m \alpha_k \delta_{mk} \qquad \text{when } j = k,$$

$$\sum_{i=1}^{3} \tilde{N}_{mmi} N_{ijk} = \left(-\frac{1}{2} \alpha_m^2 + \delta_{mk} + \delta_{mj} \right) \alpha_k \alpha_j \qquad \text{when } j \neq k,$$

as the equations to be solved.

13.3 Passive Location of Radiation Emitters

An image can be symbolic such as a diagram. A symbolic instance of an image is a map of a geographical region with objects of interest, such as radiating objects, marked on the map by special symbols. The symbols on such a map might show the estimated locations of radars in a military theater or these symbols might mark the estimated locations of active mobile telephones within a cell of a cellular base station to facilitate managing the network. To be marked on a map, the radiating objects must be located by monitoring their radiation intercepted at two or more sites.

The passive location of the emitter of received radiation is a task encountered in electromagnetic surveillance, passive sonar, and seismic data analysis. Passive location of radiating emitters detects and locates the emitters passively using only their received radiation. Range cannot be determined from an unknown signal that is passively received at a single site on a direct path from the transmitter. Passive detection and location of an object radiating a signal that is unknown to the location system requires that a direct-path signal be received at two or more locations. These measurements require two or more decentralized receivers that can intercept the same radiation from that emitter. Because the modulation on the signal is unknown, the measurements are differential measurements such as differential phase of arrival, differential time of arrival, or differential frequency of arrival. The first case can measure direction of arrival using two apertures at a single location. The other cases require two remote apertures at decentralized locations.

The content of the passively received waveform is unknown and that signal is regarded as a random noise-like waveform. Only the comparison of the signal as received in two apertures is considered relevant. The two apertures may be near to each other or may be far from each other. Phase difference measurements are appropriate for apertures that are close together. Time difference and frequency difference measurements are appropriate for apertures that are far from each other.

When two apertures are closely spaced, as described in Section 13.5, the direction of arrival can be measured by observing the difference in the carrier phase at the two apertures. This direction establishes a straight line of position on which the emitter is deemed to lie. In the usual case, the modulation is virtually the same in both apertures and a comparison of the time-of-arrival of the modulation is not informative regarding the location of the transmitter. At two closely spaced apertures, only the phase difference is informative and is informative only to estimate a direction of arrival.

When two apertures are distant from each other, there are several possibilities. For a narrowband signal, the comparison of the modulation delay at the two apertures may still be unhelpful because the time of arrival of a narrowband signal cannot be measured accurately. Then the aperture at each station can be split into two apertures and a direction-of-arrival measurement made at each location. This gives two lines of position, and the intersection of those two straight lines locates the source of radiation. The accuracy of this method depends on the accuracy of the phase measurement and the distances to the emitter.

In contrast, when the bandwidth of the modulation is sufficiently large, the time difference in the modulation received at the two apertures can be used to locate the radiation source. This is so when the spacing of the two apertures is large in comparison to c/B, where c is the speed of light and B is the modulation bandwidth.

The task of passive location is usually understood to employ receivers at several decentralized stations. At each of several widely separated and possibly moving receivers, as shown in Figure 13.6, an attenuated and delayed version of a signal, $s(t)$ is observed, but is contaminated by receiver noise. In a passive location system, the signal $s(t)$ itself is unknown. Only the received noisy version of the signal $s(t)$ is observed. The source of the signal is at an unknown location (x, y) and that location is to be estimated from the received signals at the several receiver locations. To this purpose, the several observed signals must be brought to a common location by means of relays.

Differential Delay

A formal statement of the problem for the case with receivers that are not moving is as follows. Given the received signals at baseband or passband

$$v_i(t) = s(t - R_i/c) + n_i(t) \quad i = 1, \ldots, I,$$

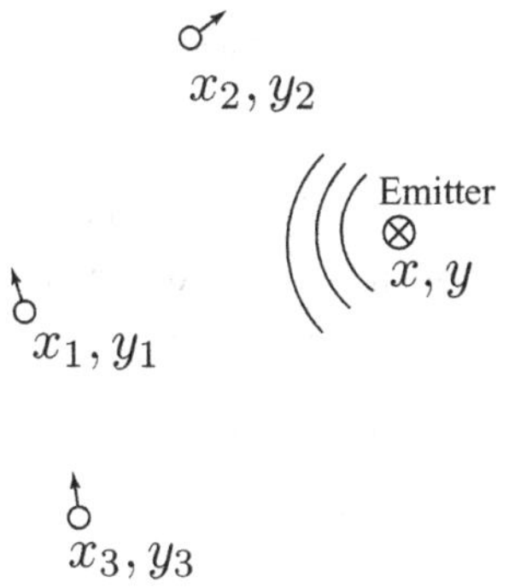

13.3 Passive Location of Radiation Emitters

Figure 13.6 Geometry for passive source location

or at complex baseband

$$\tilde{v}_i(t) = s(t - R_i/c)e^{2\pi f_0 t} + n_i(t)e^{2\pi f_0 t} \quad i = 1,\ldots,I,$$

form an estimate, $(\hat{x},\hat{y})$, of the source location (x,y). In a two-dimensional problem, the range R_i has the form

$$R_i = \sqrt{(x-x_i)^2 + (y-y_i)^2}.$$

The only case considered is the case in which the propagation velocity c is a known constant and the locations of the receivers (x_i, y_i) are known. A two-step approach to the task of location first forms estimates $\hat{\tau}_{ij}$ of the differential delays $\tau_{ij} = (R_i - R_j)/c$ as intermediate parameters. The estimates $\hat{\tau}_{ij}$ comprise the input to a geometric computation to determine the location from which the signal has emanated. The set of differential delays forms a statistic for the task of location.

The differential delay $\tau_{ij} = (R_i - R_j)/c$ defines a hyperbola with foci at the two receiver stations. With three decentralized receivers, there are three such differential delays and three such hyperbolas, though the errors are not independent.

Intersection of Hyperboli

Differential time of arrival defines a hyperbola. Two estimates $\hat{\tau}_{12}$ and $\hat{\tau}_{13}$ of differential time of arrival using three receivers defines two hyperboli with a common focus. Position $(\hat{x},\hat{y})$ can be estimated from the pair of estimates $\hat{\tau}_{12}$ and $\hat{\tau}_{13}$. This is the task of computing the intersection of two hyperboli with a common focus: Find (x,y) by solving the two equations

$$c\hat{\tau}_{12} = \sqrt{(x-x_2)^2 + (y-y_2)^2} - \sqrt{(x-x_1)^2 + (y-y_1)^2},$$
$$c\hat{\tau}_{13} = \sqrt{(x-x_3)^2 + (y-y_3)^2} - \sqrt{(x-x_1)^2 + (y-y_1)^2},$$

where (x_1,y_1), (x_2,y_2), and (x_3,y_3) are the three distinct foci of the two hyperboli. These equations can be algebraically solved for (x,y) using nothing more complicated than the rooting of a quadratic equation. First, temporally translate the coordinate system so that $x_1 = y_1 = 0$, and let $R = \sqrt{x^2 + y^2}$. The method of solution is to set up a quadratic equation for R and find its two roots. Once R is known, x and y can be computed easily.

We begin with a restatement of the problem. The statement with two equations in the two unknowns x and y is lifted to a problem with three equations in three unknowns $x, y,$ and R:

$$(R + c\tau_{12})^2 = (x_2 - x)^2 + (y_2 - y)^2,$$
$$(R + c\tau_{13})^2 = (x_3 - x)^2 + (y_3 - y)^2,$$
$$R^2 = x^2 + y^2.$$

Expand the first two equations and subtract the third equation from each to obtain

$$2Rc\tau_{12} + c^2\tau_{12}^2 = -2x_2 x - 2y_2 y + R_2^2,$$
$$2Rc\tau_{13} + c^2\tau_{13}^2 = -2x_3 x - 2y_3 y + R_3^2,$$

where

$$R_2^2 = x_2^2 + y_2^2,$$
$$R_3^2 = x_3^2 + y_3^2.$$

The solution of these equations for x and y in terms of R is given by

$$\begin{bmatrix} x \\ y \end{bmatrix} = \frac{1}{2} \begin{bmatrix} x_2 & y_2 \\ x_3 & y_3 \end{bmatrix}^{-1} \begin{bmatrix} -2Rc\tau_{12} - c^2\tau_{12}^2 - R_2^2 \\ -2Rc\tau_{13} - c^2\tau_{13}^2 - R_3^2 \end{bmatrix},$$

or as

$$\begin{bmatrix} Dx \\ Dy \end{bmatrix} = \frac{1}{2} \begin{bmatrix} y_3 & -y_2 \\ -x_3 & x_2 \end{bmatrix} \begin{bmatrix} -2Rc\tau_{12} - c^2\tau_{12}^2 - R_2^2 \\ -2Rc\tau_{13} - c^2\tau_{13}^2 - R_3^2 \end{bmatrix},$$

where $D = x_2 y_3 - x_3 y_2$.

This gives expressions for x and y that are linear in R. Substitute the expressions for x and y into the equation $R^2 = x^2 + y^2$ to obtain a quadratic equation in R which has two roots. One root is the desired R so it must be real and positive. The other must be real but can be either positive or negative. A negative root can be discarded. The positive roots give one or two possible values of R from which x and y are easily computed. When both roots of the quadratic equation are positive, then the pair of measurements (τ_{12}, τ_{13}) is explained by either of two (x, y) locations. In this case, there are two solutions resulting in an ambiguity.

The solution for the point (x, y) involves a matrix inverse that fails to exist whenever the three foci lie on a straight line. This special case can be subsumed in the same set of equations by restructuring the equations so that the determinant D is removed from the denominator of the equations. Let D denote the determinant $x_2 y_3 - y_2 x_3$. Then the above equation can be written in the form

$$\begin{bmatrix} Dx \\ Dy \end{bmatrix} = \begin{bmatrix} y_3 & -y_2 \\ -x_3 & x_2 \end{bmatrix} \begin{bmatrix} a_1 R + b_1 \\ a_2 R + b_2 \end{bmatrix},$$

for appropriate definitions of a_1, a_2, b_1, and b_2. Then the equation $D^2 R^2 = D^2 x^2 + D^2 y^2$ can be written as a quadratic in R even when $D = 0$. It always has two roots for R. By the nature of the problem, at least one root is positive. When one root is negative, it can be discarded. When both are positive, one root is the desired value of R. The other root is a false value of R leading to an ambiguous solution.

Having computed R using the quadratic equation, the equations reduce to a pair of linear equations in the unknown x and y and are readily solved provided D is nonzero. Otherwise, when D is zero, the expression $x^2 + y^2 = R^2$ must be used.

Differential Delay and Doppler

More generally, a signal transmitted at the center frequency f_0 by a radiation source, possibly itself moving, has undergone a time delay and a frequency shift when received at a moving point. The ith received signal can be written in complex form as

$$v_i(t) = s(t - R_i(t)/c) e^{j2\pi (f_0/c) \dot{R}_i(t) t} e^{j\theta_i} + n_i(t),$$

where $s(t)$ is the complex representation of the transmitted signal and $R_i(t)$, given by

$$R_i(t) = \sqrt{(x(t) - x_i(t))^2 + (y(t) - y_i(t))^2},$$

is the distance from the emitter to the receiver. The derivative $\dot{R}_i(t)$, given by

$$\dot{R}_i(t) = \frac{(x(t) - x_i(t))(\dot{x}(t) - \dot{x}_i(t)) + (y(t) - y_i(t))(\dot{y}(t) - \dot{y}_i(t))}{\sqrt{(x(t) - x_i(t))^2 + (y(t) - y_i(t))^2}},$$

is the rate of change of these distances and θ_i is the constant phase shift.

The only case treated here is the case in which the radiation source is not moving so $\dot{x}(t) = \dot{y}(t) = 0$. Only the receivers are moving. The received signal can also be written as

$$v_i(t) = s(t - \tau_i(t)) e^{j2\pi \nu_i(t)} e^{j\theta_i} + n_i(t),$$

where $\tau_i(t) = R_i(t)/c$ and $\dot{\tau}_i(t) = (f_0/c) \dot{R}_i(t)$. In the first approximation, $\tau_i(t)$ is taken to be the constant τ_i and $\nu_i(t)$ is taken to be the constant ν_i.

The estimator studied here is based on the two-input sample cross-ambiguity surface

$$|\chi_{ij}(\tau, \nu)| = \left| \int_{-\infty}^{\infty} v_i(t) v_j^*(t + \tau) e^{j2\pi \nu t} dt \right|.$$

The terms $v_i(t)$ and $v_j(t)$ in the integrand are measured at two decentralized receivers. Noise enters the computation through both $v_i(t)$ and $v_j(t)$. This differs from the problem studied in Chapter 12 where one of the signals is a stored replica of the transmitted signal and so does not introduce noise. Moreover, the signal $s(t)$ itself is random.

The differential delay and the differential doppler

$$\tau_{ij} = (R_i - R_j)/c,$$

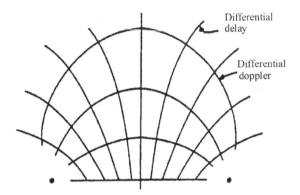

Figure 13.7 Lines of constant differential delay and differential doppler

$$\nu_{ij} = (\dot{R}_i - \dot{R}_j)f_0/c.$$

are defined as the intermediate variables from which the location is computed. In the absence of noise, the sample cross-ambiguity surface is

$$|\chi_{ij}(\tau,\nu)| = \left| \int_{-\infty}^{\infty} s(t-\tau_i)e^{j2\pi\nu_i t}s^*(t-\tau_j-\tau)e^{-j2\pi\nu_j t}e^{j2\pi\nu t}dt \right|$$

$$= \left| \int_{-\infty}^{\infty} s(t)s^*(t-\tau+\tau_{ij})e^{j2\pi(\nu-\nu_{ij})t}dt \right|,$$

which has a peak at $\tau = \tau_{ij} = \tau_i - \tau_j$, $\nu = \nu_{ij} = \nu_i - \nu_j$. Thus, by computing the sample cross-ambiguity surface from $v_i(t)$ and $v_j(t)$, we have a way of estimating τ_{ij} and ν_{ij}. The estimates $\hat{\tau}_{ij}$ and $\hat{\nu}_{ij}$ satisfy the two equations

$$\hat{\tau}_{ij} = (R_i(x,y) - R_j(x,y))/c,$$
$$\hat{\nu}_{ij} = (\dot{R}_i(x,y) - \dot{R}_j(x,y))f_0/c.$$

These expressions can be inverted to obtain position estimates $\hat{x}$ and $\hat{y}$. Each equation defines a curve in the x,y plane upon which the radiation source must lie. Figure 13.7 shows a family of differential delay curves in the x,y plane, each curve consisting of all points for which $\tau_{12} = (R_1(x,y) - R_2(x,y))/c$ is a constant. The figure also shows a family of differential doppler curves in the x,y plane, each consisting of all points for which $\nu_{12} = (\dot{R}_1(x,y) - \dot{R}_2(x,y))f_0/c$ is a constant. The loci of constant differential delay are easy to describe. They are hyperbola. The curves of constant differential doppler have no such simple description: they must be computed numerically. The highlighted curves correspond to a specific pair of measurements. The intersection of these two curves determines the location of the radiation source. When the curves have two intersections, the solution is ambiguous.

When there are three or more receivers, one can compute the sample cross-ambiguity surface for each pair. For example, when there are three receivers, one can compute the three magnitudes $|\chi_{12}(\tau,\nu)|$, $|\chi_{23}(\tau,\nu)|$, and $|\chi_{31}(\tau,\nu)|$. The peaks of these surfaces provide the differential delay estimates $\hat{\tau}_{12}$, $\hat{\tau}_{23}$, and $\hat{\tau}_{31}$ and the differential doppler estimates $\hat{\nu}_{12}$, $\hat{\nu}_{23}$, and $\hat{\nu}_{31}$. The differential delay coordinates are not independent because they satisfy

$$\tau_{12} + \tau_{23} + \tau_{31} = 0.$$

The estimates $\hat{\tau}_{12}$, $\hat{\tau}_{23}$, and $\hat{\tau}_{31}$, however, need not sum to zero because they contain errors.

When there is little or no motion, differential doppler is not meaningful for estimating position. Then the position estimate $(\hat{x},\hat{y})$ is found using only the differential delay estimates.

13.4 Estimation of Differential Parameters

Measured parameters such as phase, frequency of arrival, and time of arrival are not useful for locating the emitters when those parameters of the transmitted waveform are not known. Instead, one may estimate *differential* parameters using a common signal that is received at two or more subapertures. Measurement accuracy of these differential parameters is discussed in this section.

The estimation of the differential time-of-arrival is described first for a pair of complex baseband signals of the form

$$v_1(t) = s(t - \tau_1) + n_1(t),$$
$$v_2(t) = s(t - \tau_2) + n_2(t),$$

received at two decentralized points, where $s(t)$ is an unknown waveform with finite energy. In general, $s(t)$ is complex. Each noise, $n_1(t)$ or $n_2(t)$, is an independent, stationary complex gaussian noise process with identical and known correlation function $\phi(\tau)$ and power density spectrum $N(f)$.

The estimation problem is to determine the differential delay

$$\tau_{12} = \tau_1 - \tau_2,$$

from the received waveforms $v_1(t)$ and $v_2(t)$. The parameters τ_1 and τ_2 are unknown but fixed.

A coherent estimator of τ_{12} computes the sample cross-correlation function of $v_1(t)$ and $v_2(t)$. This can be described indirectly as the computation of the sample cross-ambiguity function

$$\chi_{12}(\tau,\nu) = \int_{-\infty}^{\infty} v_1(t) v_2^*(t-\tau) e^{-j2\pi\nu t} dt,$$

and finding the peak of the real part $\text{Re}[\chi_{12}(\tau,0)]$ with ν equal to zero. Because ν is set to zero, it is enough to compute the correlation

$$\chi_{12}(\tau,0) = \int_{-\infty}^{\infty} v_1(t) v_2^*(t-\tau) dt.$$

The estimate of the differential delay $\hat{\tau}_{ij}$ is the point at which $\chi(\tau,\nu)$ takes it peak.

More generally, the received signals may have the unknown phases θ_1 and θ_2 so that

$$v_1(t) = s(t-\tau_1) e^{j\theta_1} + n_1(t),$$

$$v_2(t) = s(t - \tau_2)e^{j\theta_2} + n_2(t).$$

Then a noncoherent estimator must be used. A noncoherent estimator of τ_{12} first computes the sample cross-ambiguity function of $v_1(t)$ and $v_2(t)$ given by

$$\chi_{12}(\tau, \nu) = \int_{-\infty}^{\infty} v_1(t) v_2^*(t - \tau) e^{-j2\pi \nu t} dt,$$

and then finds the peak of the magnitude function $|\chi_{12}(\tau, 0)|$ with $\nu = 0$ regarded as a function of τ. This reduces to a complex cross-correlator followed by a peak-magnitude detector.

More generally, the received signals also have unknown frequency offsets

$$v_1(t) = s(t - \tau_1) e^{j2\pi \nu_1 t} e^{j\theta_1} + n_1(t),$$
$$v_2(t) = s(t - \tau_2) e^{j2\pi \nu_2 t} e^{j\theta_2} + n_2(t).$$

A noncoherent estimator of τ_{12} and ν_{12} computes the sample cross-ambiguity function of $v_1(t)$ and $v_2(t)$ then finds the peak of $|\chi_{12}(\tau, \nu)|$ with respect to both τ and ν. The analysis of error is illustrated by the analysis of the error in the noncoherent estimation of differential pulse arrival time τ. Recall from the proof of Proposition 12.8.1 that the variance of the τ error can be written

$$\sigma_\tau^2 = \frac{\mathrm{E}\left[(\partial N_e(\tau, 0)/\partial \tau)^2\right]}{\left[\partial^2 |\chi(\tau, 0)|/\partial \tau^2\right]_{\tau=0}^2},$$

where $|\chi(\tau, 0)|$ and $N_e(\tau, 0)$ are determined from the first two terms of a series expansion of the sample cross-ambiguity surface. Thus,

$$|\chi_{12}(\tau, \nu)| \approx |\chi(\tau, \nu)| + N_e(\tau, \nu),$$

where

$$N_e(\tau, \nu) = \mathrm{Re}\left[\frac{\chi(\tau, \nu)}{|\chi(\tau, \nu)|}(N_R(\tau, \nu) + jN_I(\tau, \nu))\right].$$

The proof of Proposition 12.8.1 shows that the denominator in the expression for σ_τ^2 is equal to $[(2\pi)^2 1_G^2 E_p]^2$. Because the denominator does not depend on noise, it remains the same in the estimator of differential delay. The noise term, however, is not the same because noise enters the sample cross-ambiguity function in two ways.

To find the numerator, inspect $N(\tau, \nu)$ at $\nu = 0$,

$$N(\tau, 0) = \int_{-\infty}^{\infty} s(t) n_2^*(t) dt + \int_{-\infty}^{\infty} n_1(t) s^*(t) dt + \int_{-\infty}^{\infty} n_1(t) n_2^*(t) dt.$$

Whereas the analysis of the time-of-arrival error variance has one noise term at this point, the analysis of the differential time-of-arrival has three noise terms. The first two arise from signal times noise terms. The third of these is a noise-times-noise term and is incidental when noise is smaller than the signal. Thus, for small noise, the variance of the estimate of differential time of arrival is about twice as large as the estimate of

13.5 Estimation of Direction

time of arrival. When the noise is not small, there is an additional noise-only term. A formal analysis involving expectations of products of random processes is tedious.

13.5 Estimation of Direction

The interferometric method of using the relative phase at a pair of antennas or hydrophones to measure the direction of arrival, as discussed in Section 5.5, is used in a great variety of situations, both passive and active. In a situation in which several receiving apertures are closely spaced in comparison to the distance to the source of radiation, the incoming wave may be considered to be a plane wave, and the difference in the carrier phase at the two apertures is a statistic pertaining to the direction to that radiation source.

A system for the estimation of the direction of arrival from the differential phase at two or more apertures is called an interferometer. The interferometer is introduced and analyzed in Section 5.5 of Chapter 5, but is treated in that section without including noise. Now, in this section, the question of whether there is a better way of using the area occupied by the two apertures is investigated. Approaching the problem from its fundamentals deepens our understanding of an interferometer.

The discussion begins with a single two-dimensional contiguous aperture treated as a limited resource, asking how the signal in that aperture should be sampled and processed in order to best extract the direction of arrival. The distribution of the phase of an incoming waveform across an aperture depends on the direction of arrival of the waveform. The aperture can be divided into two or more subapertures to estimate the direction of arrival by a phase-comparison interferometer. This widely used approach is described in Section 5.5 without any statement of optimality.

Now, in this section, the study of elementary methods for the estimation of the direction of arrival of a waveform is continued. The error due to both external noise and noise introduced within the receivers are analyzed.

The simplest instance of the problem, as studied in Section 5.5, consists of partitioning the total available aperture into two disjoint subapertures of the same size with the two received complex-baseband signals in the two subapertures given by

$$v_1(t) = s(t) + n_1(t),$$
$$v_2(t) = s(t)e^{j\Delta\theta} + n_2(t),$$

where $n_1(t)$ and $n_2(t)$ are taken to be independent, complex, white gaussian-noise processes, each with variance σ^2, where $\Delta\theta$ is the difference in the carrier phase due to path length difference from a point in the far field to the two phase centers. The usual way to estimate $\Delta\theta$ when $s(t)$ is a known pulse (as when the received signal is a reflected version of the transmitted signal[7]) is to pass each received signal through a

---

[7] When $s(t)$ is an unknown signal, correlation methods are necessary, as studied in the next section.

filter matched to $s(t)$ and to sample the phase of the two filter outputs at the expected peak. The two complex samples at the expected peak of the matched-filter outputs are:

$$u_1 = A + n_1,$$
$$u_2 = Ae^{j\Delta\theta} + n_2,$$

where $n_1 = n_{1R} + jn_{1I}$ and $n_2 = n_{2R} + jn_{2I}$ are independent, complex gaussian random variables of variance σ^2 per component. The difference in the phase angles of u_1 and u_2 provides an estimate of $\Delta\theta$. Because of the noise terms, this estimate includes an error, denoted θ_e, which becomes larger as the angle from the boresight increases.

Proposition 13.5.1 *A differential-phase interferometer with two apertures of equal size with spacing d and with a known signal and independent, circularly symmetric zero-mean noise of variance σ_n^2 per component on each aperture has an error in the measurement of direction at angle ϕ with zero mean and variance asymptotically satisfying*

$$\sigma_\phi^2 = 2\left(\frac{\lambda}{2\pi d\cos\phi}\right)^2 \frac{\sigma_n^2}{A^2},$$

as the signal-to-noise ratio goes to zero, where A is the signal strength and λ is the wavelength.

Proof:
The phase-difference interferometer being analyzed has two signal terms of the same amplitude and two independent, identically distributed complex noise terms in the two received signals u_1 and u_2, each noise term with variance σ_n^2 per component. Because each complex noise is circularly symmetric, it is invariant under phase rotation. The two signals u_1 and u_2 each have an independent phase error with the same variance. The error in u_1 and u_2 have the same form. Therefore, it is enough to evaluate the phase error in the term $u_1 = A + n_{1R} + jn_{1I}$. The phase error in u_1 is

$$\theta_e = \tan^{-1}\frac{n_{1I}}{A + n_{1R}} \approx \frac{n_{1I}}{A},$$

where the approximation holds asymptotically well as the noise becomes small in comparison to the signal. The random phase errors in u_1 and u_2 are independent and have the same variance. That variance is $\sigma_\theta^2 = \sigma_n^2/A^2$.

Hence, the phase difference $\Delta\theta = \theta_1 - \theta_2$ has a differential phase error $\theta_{1e} - \theta_{2e}$ involving two error terms. The variance of the differential phase error is

$$\sigma_{\Delta\theta}^2 = 2\frac{\sigma_n^2}{A^2}.$$

In the presence of phase error, the phase difference $\theta_1 - \theta_2$ between the two signals received at the two antennas is related to the estimate of the angle of arrival ϕ by

$$\Delta\theta + \Delta\theta_e = 2\pi\frac{d}{\lambda}\sin(\phi + \phi_e)$$
$$\approx 2\pi\frac{d}{\lambda}(\cos\phi\sin\phi_e + \sin\phi),$$

where d is the separation in the phase centers of the two antennas and $\cos\phi_e \approx 1$. The error in the angle of arrival for small values of the error then satisfies

$$\theta_e \approx 2\pi \frac{d}{\lambda} \phi_e \cos\phi.$$

Within this approximation with $\sin\phi_e \approx \phi_e$, the variance in θ_e satisfies

$$\left(\frac{2\pi d}{\lambda}\cos\phi\right)^2 \sigma_\phi^2 = 2\frac{\sigma_n^2}{A^2}.$$

The statement of the theorem follows directly. This completes the proof of the theorem. ∎

The proposition states that the variance of the measured angle depends on the received signal and the receiver noise through the ratio of noise power to signal power. Near broadside, where $\phi \approx 0$, this approximation reduces to

$$\sigma_\phi^2 \approx 2\left(\frac{\lambda}{2\pi d}\right)^2 \frac{\sigma_n^2}{A^2}$$

by setting $\cos\phi$ equal to one.

When the noise is external, the signal power A^2 and the noise power σ_n^2 both depend directly on the sizes of the two apertures. Then the signal-to-noise ratio does not depend on the size of the apertures. When only the signal power depends on the size of its aperture but the noise arises in the receivers, the aperture size does matter to the signal-to-noise ratio.

Array of Subapertures

Suppose now that the total available area is divided into multiple subapertures with α_i for $i = 1, \ldots, I$ being the fraction of the total area used by the ith subaperture, where $\sum_i \alpha_i = 1$. This amounts to a finer spatial quantization of that aperture. Suppose that the signal received in the ith subaperture is contaminated by an additive noise term $n_i(t)$ that is independent of noise terms in other subapertures and independent of the size of the ith subaperture.

The total signal energy received depends on the size of the full aperture. The ith subaperture receives a share of the energy proportional to α_i. The total signal is

$$\sum_{i=1}^{I} \int_{-\infty}^{\infty} \alpha_i |s(t)|^2 dt = \int_{-\infty}^{\infty} |s(t)|^2 dt.$$

The total signal at the ith subaperture is

$$v_i(t) = \alpha_i s(t) e^{j\theta_i} + n_i(t),$$

where the additive receiver noise terms $n_i(t)$ are independent, identically distributed, white gaussian-noise processes, each with the power density spectrum $N_0/2$. The phases θ_i depend on the positions of the phase centers of the subapertures. At each subaperture, the receiver noise is independent, so the phase is measured with an independent phase error. At the phase center of the ith subaperture, the phase θ_i is measured with the variance

$$\sigma_{\theta_i}^2 = \frac{\sigma_n^2}{\alpha_i A^2}.$$

The noise variance σ_n^2 is the same for all receivers and is independent of the number of subapertures. When all subapertures are the same size with the same noise variance, this simplifies as

$$\sigma_{\theta_i}^2 = \frac{\sigma_n^2}{A^2/I}.$$

The phase depends linearly on the position x, so it has the form

$$\theta = ax + b.$$

A straight line can be fit to the I phase measurements to first estimate the slope a. The slope satisfies

$$a = \frac{d\theta}{dx} = \frac{2\pi}{\lambda}\sin\phi,$$

so an estimate of the coefficient a gives the desired estimate of ϕ.

The coefficients a and b of a straight line fit to a set of measured data points can be estimated using the standard method of least-squares. The least-squares straight-line fit, $y = ax + b$, to any set of points (x_i, y_i), $i = 1, \ldots, I$, is derived from the fact that for any parameters a and b, the squared error in the y direction at the ith point is $(y_i - ax_i - b)^2$. The total squared error is $\sum_i (y_i - ax_i - b)^2$. To find the least-squares fit, set the partial derivatives with respect to a and to b equal to zero. This gives the two equations for the least-squares straight line,

$$\sum_i x_i(y_i - ax_i - b) = 0,$$

$$\sum_i (y_i - ax_i - b) = 0.$$

For the problem at hand, the x_i are the deterministic phase centers of the apertures and the y_i are replaced by the phase errors θ_i. Without loss of generality, the origin of the x axis can be chosen so that $\sum_i x_i = 0$. Then the equations reduce to the single equation of interest,

$$a\sum_i x_i^2 = \sum_i x_i \theta_i.$$

The error terms da and $d\theta_i$ satisfy

$$\sum_i x_i^2 \, da = \sum_i x_i \, d\theta_i.$$

By assumption, the $d\theta_i$ have zero mean, are independent and identically distributed. The variance is found as follows:

$$E\left[\left(\sum_i x_i^2 \, da\right)^2\right] = E\left[\left(\sum_i x_i \, d\theta_i\right)\left(\sum_{i'} x_{i'} \, d\theta_{i'}\right)\right]$$

$$= \sum_i \sum_{i'} x_i x_{i'} E[d\theta_i d\theta_{i'}]$$

$$= \sum_i x_i^2 E[d\theta_i^2].$$

Let $\overline{x^2} = (\sum_i x_i^2)/I$. Then the expression reduces to

$$(\overline{Ix^2})^2 \sigma_a = (\overline{Ix^2})^2 \sigma_\theta^2.$$

This expression reduces to $\sigma_a^2 = \sigma_\theta^2/\overline{Ix^2}$. Adjusting notation for the application to phase interferometry,

$$\sigma_a^2 = \frac{\sigma_\theta^2}{\overline{Ix^2}} = \frac{\sigma_n^2}{A^2 \overline{x^2}}.$$

The slope of the phase is related to spatial angle ϕ by

$$a = \frac{2\pi}{\lambda} \sin \phi.$$

Therefore,

$$\sigma_a^2 = \left(\frac{2\pi}{\lambda}\right)^2 \cos^2 \phi \, \sigma_\phi^2.$$

At boresight, $\phi = 0$, and

$$\sigma_\phi^2 = \left(\frac{\lambda}{2\pi}\right)^2 \frac{\sigma_n^2}{\overline{x^2} A^2}.$$

This is similar to the formula with only two subapertures spaced by d except that $d^2/2$ is replaced by $\overline{x^2}$. When the full aperture has length $2d$ and the number of subapertures is large, then

$$\overline{x^2} \approx \tfrac{2}{3} d^2.$$

Thus, when the noise is external, partitioning the aperture into multiple subapertures does give a modest improvement over using two subapertures. When the noise entering each signal is internal thermal noise, the performance is worse when the noise enters through each receiver as is the usual case.

13.6 Lidar Surveillance

A *lidar* is a source of illumination for active surveillance that operates at optical or infrared frequencies. It is used for image formation by measuring the optical signal scattered by an object of interest. The common form of a lidar uses only the intensity of the received signal. Therefore, it may be regarded as a baseband system. Because the wavelength is so small – on the order of a micron – a lidar has a narrow beam compared to a radar. A lidar imaging system may be a scanning-beam, real-aperture system with cross-beam resolution determined primarily by the width of the beam. The beam is scanned in azimuth and elevation to produce an image of the reflectance density of a reflecting object. A lidar can also be used to image the three-dimensional density of a nearly semitransparent gas such as regions of the atmosphere, or may image the density distribution of a particular species of atom or molecule within the gas. Images of pollutants in the upper atmosphere can be obtained in this way. The wavelength λ of the lidar may be chosen to target a particular spectral line of an atomic species, such as sodium. An individual volume cell is isolated in width by the lidar beamwidth and, in range, by the time delay. The return signal in each individual volume cell has the simple baseband form $v = r + n$.

The scattering of light by atomic species is a strong function of wavelength because the scattering involves the resonances within the atomic structure. This means that the image, in general, is a function of four variables: the three spatial variables and the wavelength. At a fixed spatial point, (x,y,z), the image $\rho(x,y,z)$ also becomes a function of λ. This function may now be denoted $\rho(x,y,z,\lambda)$. Near a known resonance, λ_0, the function $\rho(x,y,z,\lambda)$ gives a great deal of information about the composition of the gas in a volume cell at (x,y,z). The actual location of the peak of $\rho(x,y,z,\lambda)$ near λ_0 measures the radial velocity of that cell because of the doppler effect. By measuring this peak shift as a function of (x,y,z), one obtains a three-dimensional image of the radial velocity of the scattering gas.

The function $\rho(x,y,z,\lambda)$ near a known resonance λ_0 also gives information about the temperature within the reflecting cell. At any fixed spatial point (x,y,z), the temperature is caused by the thermal agitation of the gas molecules. This thermal agitation is manifested by a spread in velocity which leads to a spread in doppler shift, causing the spectral line to be broadened. By measuring the width of the resonance at position (x,y,z), the temperature at (x,y,z) is measured.

Problems

13.1 Two complex matched-filter output samples are:

$$u_1 = Ae^{j\theta_1} + n_1,$$
$$u_2 = Ae^{j\theta_2} + n_2,$$

where $n_1 = n_{1R} + jn_{1I}$ and $n_2 = n_{2R} + jn_{2I}$ are independent, complex gaussian random variables of variance σ^2 per component.

To determine the phase difference $\theta_1 - \theta_2$, it is proposed to estimate the phase of the product $u_1 u_2^*$. Explain why this method might be suitable. Compare the accuracy of this method with the accuracy of subtracting the phases of the two terms estimated separately.

13.2 Formulate the equations for a *weighted least-squares* straight-line fit to a set of data as follows. Given that $y = ax + b$ and that the value y_i is measured with the error variance σ_i^2 at location x_i, determine the estimates of a and b that minimize the value of $\sum_i (y_i - ax_i - b)/\sigma_i^2$. Can this be justified as a maximum-likelihood procedure?

13.3 A plane wave at angle ϕ is incident on a one-dimensional aperture rect(x/L) of length L. Show that the phase of the received signal uniformly integrated across the aperture has the phase angle of the signal received at the center of the aperture.

13.4 Which of the following pairs of apertures is better for estimating the direction of arrival of a plane wave in the presence of independent gaussian noise introduced by the receivers for the two apertures?

(i) rect$\left(x - \frac{1}{2}, y\right)$ and rect$\left(x + \frac{1}{2}, y\right)$
(ii) rect$\left(2x - \frac{3}{2}, y\right)$ and rect$\left(2x + \frac{3}{2}, y\right)$

Does the conclusion change when the noise is external noise with the noise power linearly dependent on the aperture area?

13.5 An n by n array of antennas occupies a total aperture of size L by L.

a. What is the angular resolution of the antenna beam at wavelength λ?
b. Suppose that the antenna elements of a radio telescope are placed in a pattern of size L by L such that all correlation values in an n by n grid are obtained. What is the angular resolution of the image?
c. Is it true that correlation processing is equivalent to synthesizing an aperture from the individual elements?

This comparison is a variation of the van Cittert–Zernike theorem, which states a parallel between the correlation function on an aperture and a Fourier transform on that aperture.

13.6 Two stationary zero-mean gaussian random processes have the two by two covariance matrix Σ. Estimate Σ from a block of pairwise synchronized samples of duration T. What is the error variance in the estimate of the correlation coefficient?

13.7 (**Phase Closure**) Phase errors (or gain errors) can be introduced into a received radio astronomy signal by flaws in the antenna or the receiver circuitry. A cross-correlation sample $c_{12} = \int_0^T v_1(t) v_2^*(t) dt$, will have a phase error caused by phase errors in both $v_1(t)$ and $v_2(t)$. Because the image is known to be real and nonnegative, and perhaps nearly black, it may be possible to remove the phase errors, in part, by imposing consistency conditions on the cross-correlation samples. Suppose that the true cross-correlation coefficient c_{ij} has phase ϕ_{ij}, and the computed cross-correlation coefficient $\widehat{c}_{ij}$ has phase $\widehat{\phi}_{ij}$, which differs from ϕ_{ij} because of phase errors in the receivers. Show that

$$\widehat{\phi}_{ij} + \widehat{\phi}_{jk} + \widehat{\phi}_{ki} = \phi_{ij} + \phi_{jk} + \phi_{ki}.$$

That is, the sum of the correlation phases is correct even when the individual correlation phases are in error. Can this fact be used to improve a radio astronomy image? Is a similar statement true for amplitudes?

13.8 A hypothetical radio telescope consists of an array of circular antennas each of diameter D. It is decided to arrange the antennas so that all position differences are on a uniformly spaced grid. Given that two antennas cannot share the same space (or otherwise block each other), what can be said about grating lobes? How is the situation changed when the antennas are carelessly placed so that the grid is slightly perturbed from a uniform grid?

13.9 To account for missing radio-telescope data samples, let $H(f_x,f_y)$ be an indicator function for a virtual "aperture" in the f_x,f_y plane. (An indicator function only takes values 0 and 1.) Suppose that a radio telescope uniformly samples $H(f_x,f_y)C(f_x,f_y)$ on a square grid in the f_x,f_y plane. How does the missing data, as modeled by $H(f_x,f_y)$ affect the image when simple inverse Fourier transform processing is used? Describe the degradation when $H(f_x,f_y)$ has the form of a plus sign (a square with the corners missing). Are there any signal-processing techniques that will recover the missing data? Why?

13.10 a. Describe the support of the correlation function of $s(x,y)$ in terms of the support of $s(x,y)$ itself.
b. Bound the area of the support of the correlation function in terms of the area of the support of $s(x,y)$.
c. The *set sum* of sets $\mathcal{A} \subset \mathbb{R}^2$ and $\mathcal{B} \subset \mathbb{R}^2$ is the set $\mathcal{A} + \mathcal{B} = \{x + y | x \in \mathcal{A}; y \in \mathcal{B}\}$. Can the support of a correlation function be described as a set sum?

13.11 A hypothetical radio telescope at wavelength λ produces all correlation samples on a uniform n by n square grid with adjacent samples horizontally and vertically spaced by d.

a. Near boresight, where the two direction cosines α and β are approximately zero, what is the angular size of objects that can be observed without aliasing?
b. Near boresight, what is the resolution in angle of the telescope?
c. What might be the reason for amplitude weighting of the correlation samples? Describe this.

13.12 Suppose that a radio telescope has a resolution cell that is 0.1 arc second on a side. Approximately how many pixels (resolution cells) are there on the full celestial sphere? How long would it take to image the entire celestial sphere (in one radio frequency band) when a scene of 1,000 by 1,000 pixels is imaged in 1 hour? Repeat the calculation for a resolution cell that is 10^{-4} arc second on a side. Repeat that calculation for the situation where the process is to be repeated for each of N different frequency intervals.

13.13 A source emits white gaussian noise with power density spectrum $N_S(f) = N_{0S}/2$. At each of two surveillance receivers with identical transfer functions $H(f)$, the radiation is contaminated by white gaussian noise with power density spectrum $N_{01}/2$ and $N_{02}/2$. Find the variance of a noncoherent estimator of the relative delay based on finding the peak of a cross-ambiguity surface.

13.14 Under thermal agitation, the velocity distribution of the molecules of a gas in a box that consists of a single species of molecule is a three-dimensional gaussian distribution. What is the distribution of radial velocity? (This probability distribution is called the maxwellian probability density function.) Give an expression relating the temperature to the width of the three-dimensional gaussian distribution describing the velocity spread.

13.15 Prove a one-dimensional version of the van Cittert–Zernike theorem in the following form: The antenna pattern of a uniformly illuminated one-dimensional aperture has the same functional form as the correlation function of a spatially white random process in that aperture. Then extend this statement to two dimensions.

13.16 Three points (x_1, y_1), (x_2, y_2), (x_3, y_3) in the x, y plane are known. A fourth point (x_0, y_0) is unknown. The unknown distances from the known points to the unknown point are denoted R_{10}, R_{20}, and R_{30}. Differential distances $\Delta_{12} = R_1 - R_2$ and $\Delta_{13} = R_1 - R_3$ are known. Write out an explicit flow diagram for computing (x, y) from $(\Delta_{12}, \Delta_{13})$. The computation should not break down when the three known points are on a straight line.

13.17 A square aperture is subdivided into four subapertures by connecting each of the four vertices by a straight line to the point at the center of the square.

a. Find the phase center of each of the four subapertures fully illuminated.
b. Find expressions for the vertical and horizontal direction cosines.
c. Find expressions for the error in these direction cosines due to additive noise in each receiver.
d. Describe a method of resolving ambiguities.

Notes

Radio astronomy has been a spectacular success and is the principal source of information about the distant universe. Following Jansky's discovery in 1931 of cosmic radiation in the radio bands, Ryle (1952) first proposed the idea of a radio telescope based on the interferometric techniques originally introduced into optical astronomy by Michelson (1890, 1920). The first radio telescope to use the earth's rotation to synthesize an aperture was the 1964 Cambridge One-Mile Radio Telescope developed by Ryle. Earth's rotation had been used earlier to form images of the sun by Christiansen and Warburton (1955).

Several mathematical theorems are important to radio astronomy. The fundamental van Cittert–Zernike theorem, which describes the image in terms of a two-dimensional correlation function, was stated by van Cittert (1934) and refined by Zernike (1938) for study of coherence in optical systems. The van Cittert–Zernike theorem is often developed in connection with an application of that theorem and so burdened in those derivations with the approximations of that application. The approach herein, apparently uncommon, states and proves a formal, and so precise, mathematical theorem. All approximations are left with the applications of the theorem and are not within the theorem itself.

The van Vleck–Middleton theorem, which describes the effect of a hardlimiter, was described by van Vleck (1943) in an unpublished laboratory report. A fuller account was published later in the open literature by van Vleck and Middleton (1966). The Clean algorithm was introduced into radio astronomy by Högbom (1974) as a way to accommodate the sparsity of Fourier data by exploiting the sparseness and nonnegativity of the images of galaxies. The Clean algorithm has had an important and successful role in the history in radio astronomy. The method of phase closure was introduced into radio astronomy by Jennison (1958), and was further discussed by Readhead and Wilkinson (1978).

Magnetic anomaly detection is a baseband technique for detecting large metallic objects within an environment that is transparent to magnetic fields but opaque to other sensors. Such objects may be a submarine beneath the ocean, land mines buried in sand, or water pipes buried in earth. The challenging task of inverting the inversion of the equations of the magnetic dipole was presented early on by Wynn (1972) and Frahm (1972). The description given by Wynn et al. (1975) is followed herein.

14 Data Combination and Tracking

Some large data sets have a hierarchal substructure that leads to new kinds of algorithms for tasks such as data combination or tracking. The term *data combination* typically refers to a task in which several estimates of a scene or object are combined, with each estimate based on partial data. Perhaps several snapshots, or data sets, of the same scene or object are available and multiple measurements or estimates are to be combined or averaged in some way. The topics of this chapter refer to multiple sources of data that are developed separately, then combined in a postprocessing step. The postprocessing step would be used after those processing steps such as correlation or tomographic reconstruction have been completed.

Various sets of data may be combined either to merge several data sets either after detection or before detection. The combination of data from multiple sensors, usually in the form of parameter estimation from the multiple sensors, is called *data fusion*. This term often conveys the understanding that there is a diversity of types of data.

Another form of data combination is called *tracking*. Tracking follows the movement of one or more objects as a function of time as by using a sequence of detections such as by radar or sonar. Each detection may be the position of an object of interest. Sometimes only tentative detections of various targets are made before tracking, while a hard detection is deferred until after a potential track of an object is established. In some applications, this is called "track before detect." This chapter studies the interplay between the functions of detection, data fusion, and object tracking. These topics consider sensor processing on a much longer time scale than the topics in previous chapters.

A post-detection data record may contain a large number of detected data points. Such data points consist of observations of a number of stationary or moving objects made by a single sensor, such as a radar, or of observations of several objects from multiple sensors. The objects detected at each observation time are described as a number of points moving in space, possibly with each point labeled by one or more components of its velocity vector. The data may be contaminated by many kinds of impairments, such as measurement errors, false alarms, missing samples, or ambiguities. A common task is to sort the data into individual targets, suppressing both ambiguities and false alarms. After the task of association is complete, the data is partitioned and smoothed into estimates of the individual trajectories of each target.

The trajectory of an individual moving point-target is called a *track*. A radar or a sonar system may estimate the several tracks of several targets from the postdetection data. This can be an easy problem when a target is continually under surveillance, or when the density of the targets is low. There are situations, however, such as a radar with a scanning antenna, in which a target is only observed intermittently and tentatively with a density of targets that is high. Then a method is needed to sort the successive detections and string them together to form estimated tracks. This is the task of *multitarget tracking*. The task of multitarget tracking can be compared to the mathematical task of data fusion. For multitarget tracking, one type of sensor is used many times. For data fusion, data from many kinds of sensors may be used.

A typical multitarget environment consists of a set of M trajectories of M targets in three-dimensional space, denoted $(x_m(t), y_m(t), z_m(t))$ for $m = 1, \ldots, M$. A measurement of a trajectory parameter may provide partial knowledge of the trajectory positions and velocities at some instant of time but need not fully measure that position or velocity for all relevant time. For example, a radar with a scanning antenna will produce a range and a bearing angle to the Mth trajectory, $(x_m(t_{km}), y_m(t_{km}), z_m(t_{km}))$, at each time, t_{km}, that the antenna beam sweeps over the mth target. When the scan rate is slow, a target may move a considerable distance between samples, in which case the samples of the trajectory will be sparse. Identifying a trajectory from within the measured points of several intermingled trajectories may be difficult.

14.1 Noncoherent Integration

An elementary instance of combining multiple measurements is the topic of noncoherent integration. This is the task of finding a noncoherent pulse train in noise. Let I be the number of received waveforms and consider the formal hypothesis-testing problem given by

$$H_0: v_i(t) = n_i(t) \qquad i = 0, \ldots, I-1,$$
$$H_1: v_i(t) = s_i(t)e^{j\theta_i} + n_i(t) \quad i = 0, \ldots, I-1,$$

where the θ_i are independent random variables on the interval $[0, 2\pi]$ and the $n_i(t)$ are independent, complex gaussian random processes. When I equals one, the optimal detector is simply the noncoherent detector as derived in Section 12.5 in terms of the modified Bessel function $I_0(x)$. When I is larger than one and all $s_i(t)$ are copies of the same pulse $s(t)$, the binary hypothesis-testing problem is described as the detection of a noncoherent pulse train. In this problem, the noncoherent pulse train is said to be nonfluctuating in the sense that the probability distribution on the amplitude remains the same from pulse to pulse. The detection of a noncoherent pulse train is different than the noncoherent detection of a coherent pulse train.

14.1 Noncoherent Integration

Because the noise and the phase errors are independent, the probability density function for the composite measurement is the product of the probability density functions for the I component measurements. This means that the loglikelihood function or loglikelihood ratio for the composite measurement is the sum of the individual loglikelihood functions or loglikelihood ratios. The decision rule, analyzed in the next theorem, is to compare the loglikelihood ratio to a threshold Θ.

Theorem 14.1.1 *The Neyman–Pearson decision rule for the nonfluctuating, noncoherent pulse train with single-pulse matched-filter samples r_i is:*
Decide H_0 when:

$$-\sum_{i=0}^{I-1} \log I_0 \left(\frac{r_i A}{\sigma^2}\right) < \Theta.$$

Decide H_1 when:

$$-\sum_{i=0}^{I-1} \log I_0 \left(\frac{r_i A}{\sigma^2}\right) \geq \Theta,$$

where $(A/\sigma)^2 = 2E_p/N_0$.

Proof:
The likelihood function for the ith measurement, maximized over θ_i, is given in terms of the ricean and rayleigh probability density functions

$$p_0(r) = \frac{r}{\sigma^2} e^{-r^2/2\sigma^2} \qquad \text{for } r \geq 0,$$

$$p_1(r) = \frac{r}{\sigma^2} e^{(r^2+A^2)/2\sigma^2} I_0 \left(\frac{rA}{\sigma^2}\right) \qquad \text{for } r \geq 0.$$

The loglikelihood function for the ith measurement r_i is

$$\Lambda_0(r_i) = \log \frac{r_i}{\sigma^2} e^{-r_i^2/2\sigma^2},$$

$$\Lambda_1(r_i) = \log e^{-(r_i^2+A^2)/2\sigma^2} I_0 \left(\frac{r_i A}{\sigma^2}\right).$$

The loglikelihood function for the block of I independent measurements $r = (r_0, \ldots, r_{I-1})$ is

$$\Lambda_0(r) = \log \prod_{i=0}^{I-1} \frac{r_i}{\sigma^2} e^{-r_i^2/2\sigma^2},$$

$$\Lambda_1(r) = \log \prod_{i=0}^{I-1} \frac{r_i}{\sigma^2} e^{-(r_i^2+A^2)/2\sigma^2} I_0 \left(\frac{r_i A}{\sigma^2}\right).$$

14 Data Combination and Tracking

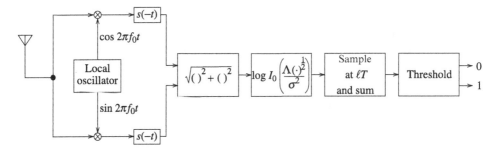

Figure 14.1 Detection of a nonfluctuating, noncoherent pulse train

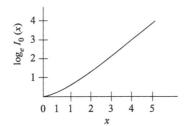

Figure 14.2 The function $\log I_0(x)$

The loglikelihood ratio is

$$\Lambda(r) = \log \frac{\prod_{i=0}^{I-1} \frac{r_i}{\sigma^2} e^{-r_i^2/2\sigma^2}}{\prod_{i=0}^{I-1} \frac{r_i}{\sigma^2} e^{-(r_i^2+A^2)/2\sigma^2} I_0\left(\frac{r_i A}{\sigma^2}\right)}$$

$$= \sum_{i=0}^{I-1} \left[\frac{A^2}{2\sigma^2} - \log I_0\left(\frac{r_i A}{\sigma^2}\right)\right].$$

The loglikelihood function function $\Lambda(r)$ is compared to a threshold to decide between H_0 and H_1. Because the term $A^2/2\sigma^2$ is a constant, it can be absorbed into the choice of threshold. Therefore, the loglikelihood ratio is redefined as

$$\Lambda(r) = -\sum_{i=0}^{I-1} \log I_0\left(\frac{r_i A}{\sigma^2}\right).$$

The Neyman–Pearson rule simply applies the modified loglikelihood ratio $\Lambda(r)$ to a threshold Θ. By the choice of Θ, the probability p_{FA} of a false alarm is traded against the probability p_{MD} of a missed detection. ∎

An optimal receiver for a nonfluctuating, noncoherent pulse train is shown in Figure 14.1. The structure of this receiver is an immediate consequence of Theorem 14.1.1. The function $\log I_0(x)$ that appears in the receiver is shown in Figure 14.2.

This decision rule includes the modified log-Bessel function $\log I_0(x)$ for each pulse. The decision rule for the optimal noncoherent detection of a single pulse or a coherent pulse train does not include the log-Bessel function. In that case, the Bessel function was needed only to compute the performance curves for the optimal detector, but it does not appear in the threshold test. In contrast, for noncoherent integration, every pulse of a received pulse train is passed through the log-Bessel function $\log I_0(x)$ at the output of the matched filter before the summation.

The series expansion

$$\log I_0(x) = \tfrac{1}{4}x^2 - \tfrac{1}{64}x^4 + \cdots$$

is well-approximated by the quadratic function for small values of x. Then, the function $\log I_0(x)$ can be approximated as

$$\log I_0(x) \approx \tfrac{1}{4}x^2,$$

when x is small. This approximation can also be used for large values of x, accepting that the detector will then be suboptimal. When the signal is strong, some loss of performance may be acceptable.

The quadratic approximation has an important consequence in the structure of the receiver. The detection rule now has the form

$$\left(\frac{A}{\sigma^2}\right) \sum_{i=0}^{I-1} r_i^2 \geq \Theta.$$

By redefining the threshold, this becomes

$$\sum_{i=0}^{I-1} r_i^2 \geq \Theta.$$

The *structure* of the detector itself now does not depend on A. The *performance* of the detector, however, does depend on A as compared to Θ.

14.2 Sequential Detection

The Neyman–Pearson detection rule processes a set of data to form the likelihood statistic, then decides between H_0 and H_1 representing target present and target absent, respectively, based on whether the likelihood statistic exceeds a fixed threshold. The probability of false alarm and the probability of missed detection both depend on the threshold. To reduce the probability of a false alarm, increase the threshold. To reduce the probability of a missed detection, decrease the threshold.

To simultaneously do both, use two thresholds. This results in a region between the two thresholds that corresponds to no decision. It is straightforward to calculate the probability of no decision as a function of the threshold under each of the two hypotheses, as well as the probability of missed detection and the probability of a false alarm.

A sequential detection rule describes a procedure to use this gap between the two thresholds. This procedure is to collect more data, changing the two thresholds accordingly. The sequential rule decides H_0 when the statistic is below the lower threshold, and decides H_1 when the statistic is above the upper threshold. When the statistic is between the two thresholds, the decision is left open, pending the collection of more data. As more data is gathered, the statistic is reapplied to two thresholds in order to detect target present or target absent. This process will continue until one of the two thresholds is crossed.

A sequential decision rule almost surely makes a decision eventually. However, the amount of time that it takes to make a decision is not predetermined. The time varies with the *actual* data realization rather than with the *typical* data realization. The advantage of a sequential procedure is better performance. The disadvantage of a sequential procedure is a variable wait time to a decision.

14.3 Multitarget Tracking

A detection sensor, such as a radar or a sonar, may observe the same scene multiple times in succession calling each observation of a common scene a *scan*. During each scan, the radar detects and locates those targets that lie within that scene, but each position measurement has some position error. It may be that a single target has a unique characteristic, such as size or shape, that allows that target to be recognized within each scan. In many cases, however, the detected targets are not distinguishable. Then one has the task of target association which is the task of matching the targets from one scan with targets in the next scan based only on position. This section is interested in associating targets using only consistency of position.

In the general case, the targets may be densely distributed and moving, and there may be false alarms and missed detections, whereby the problem of target association by consistency of position can appear formidable. The discussion here begins with a simple case with no target motion, no false alarms, and no missed detections.

Figure 14.3 shows a two-dimensional situation in which there are four targets and three scans. The targets are stationary and there are no false detections or missed detections. Each scan locates the same four targets without identification. Each target is marked with a symbol unique to that scan as shown in Figure 14.3. Because of measurement errors, each target appears displaced from its true position. The task of association is to sort the detections into bins, one detection from each scan to a bin so that all detections in the same bin correspond to the same target. Because the four targets in Figure 14.3 are widely separated compared to the error dispersion, this instance of the problem is easy to solve. The solution is visually apparent in Figure 14.3. After the four targets are partitioned into four bins, the cluster of three measurements in each bin can be averaged or smoothed in some way to give a better estimate of that target location. Our only purpose in this section, however, is only to study the assignment of targets to bins.

Figure 14.4 shows another instance of the problem, now with a larger dispersion in the measurements. The problem again is to determine the most likely partition into four sets with each set containing one symbol from each scan. Now the best partition is not apparent, nor is the criterion apparent for judging a particular solution to be the

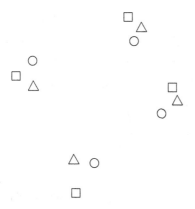

Figure 14.3 A simple association problem

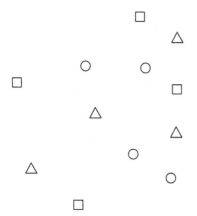

Figure 14.4 A more difficult association problem

best solution. As will be seen, when the errors are gaussian distributed, the optimal association for three scans minimizes the sum of pairwise euclidean distances between points in the same bin.

This kind of problem is called a *three-dimensional assignment problem*. The term "three-dimensional" here refers to the fact that there are three scans; it does not refer to the two physical dimensions of the spatial problem. In general, an n-dimensional assignment problem refers to n scans of a scene. The scene itself may be a two-dimensional or a three-dimensional scene.

The two-dimensional assignment problem with n objects in a two-dimensional space is treated as follows. The n objects are at true positions (x_i, y_i) for $i = 1, \ldots, n$. During each scan, the (x, y) positions of the n objects in the x, y plane are measured in a random order. The random order is independent from scan to scan. The measurements of the positions in each scan have no preferred order. The true positions in the x, y plane during the first scan are denoted $\left(x_i^{(1)}, y_i^{(1)}\right)$ for $i = 1, \ldots, n$. Thus $\left(x_i^{(1)}, y_i^{(1)}\right) =$

$(x_{\pi_1(i)}, y_{\pi_1(i)})$, where $\pi_1(i)$ denotes a permutation of the indices. The locations during the second scan are $\left(x_i^{(2)}, y_i^{(2)}\right)$ for $i = 1, \ldots, n$. Thus $\left(x_i^{(2)}, y_i^{(2)}\right) = (x_{\pi_2(i)}, y_{\pi_2(i)})$.
When there are only two scans, there are two hypotheses given by

$$H_0: \begin{cases} (x_1^{(2)}, y_1^{(2)}) = (x_1^{(1)}, y_1^{(1)}), \\ (x_2^{(2)}, y_2^{(2)}) = (x_2^{(1)}, y_2^{(1)}), \end{cases}$$

$$H_1: \begin{cases} (x_1^{(2)}, y_1^{(2)}) = (x_2^{(1)}, y_2^{(1)}), \\ (x_2^{(2)}, y_2^{(2)}) = (x_1^{(1)}, y_1^{(1)}). \end{cases}$$

Hypothesis H_0 says that the indices in scan 1 and scan 2 are the same, while hypothesis H_1 says that the indices in scan 1 and scan 2 have been interchanged. The measured data $(u_i^{(j)}, v_i^{(j)})$ is the actual position contaminated by measurement error. When the errors are gaussian random variables, the probability density functions on the data on scan one and scan two under the two hypotheses are

$$H_0: \begin{cases} p_0(u_1^{(1)}, v_1^{(1)}, u_2^{(1)}, v_2^{(1)}) = \left(\frac{1}{2\pi\sigma^2}\right)^2 e^{-\left[(u_1^{(1)}-x_1)^2+(v_1^{(1)}-y_1)^2\right]/2\sigma^2} e^{-\left[(u_2^{(1)}-x_2)^2+(v_2^{(1)}-y_2)^2\right]/2\sigma^2}, \\ p_0(u_1^{(2)}, v_1^{(2)}, u_2^{(2)}, v_2^{(2)}) = \left(\frac{1}{2\pi\sigma^2}\right)^2 e^{-\left[(u_1^{(2)}-x_1)^2+(v_1^{(2)}-y_1)^2\right]/2\sigma^2} e^{-\left[(u_2^{(2)}-x_2)^2+(v_2^{(2)}-y_2)^2\right]/2\sigma^2}, \end{cases}$$

$$H_1: \begin{cases} p_1(u_1^{(1)}, v_1^{(1)}, u_2^{(1)}, v_2^{(1)}) = \left(\frac{1}{2\pi\sigma^2}\right)^2 e^{-\left[(u_1^{(1)}-x_1)^2+(v_1^{(1)}-y_1)^2\right]/2\sigma^2} e^{-\left[(u_2^{(1)}-x_2)^2+(v_2^{(1)}-y_2)^2\right]/2\sigma^2}, \\ p_1(u_1^{(2)}, v_1^{(2)}, u_2^{(2)}, v_2^{(2)}) = \left(\frac{1}{2\pi\sigma^2}\right)^2 e^{-\left[(u_1^{(2)}-x_2)^2+(v_1^{(2)}-y_2)^2\right]/2\sigma^2} e^{-\left[(u_2^{(2)}-x_1)^2+(v_2^{(2)}-y_1)^2\right]/2\sigma^2}. \end{cases}$$

The negatives of the loglikelihood functions are sums of squared distances given by

$$-\Lambda(x_1, y_1, x_2, y_2 | H_0) = \left[(u_1^{(1)} - x_1)^2 + (v_1^{(1)} - y_1)^2\right] + \left[(u_2^{(1)} - x_2)^2 + (v_2^{(1)} - y_2)^2\right]$$
$$+ \left[(u_1^{(2)} - x_1)^2 + (v_1^{(2)} - y_1)^2\right] + \left[(u_2^{(2)} - x_2)^2 + (v_2^{(2)} - y_2)^2\right],$$

$$-\Lambda(x_1, y_1, x_2, y_2 | H_1) = \left[(u_1^{(1)} - x_1)^2 + (v_1^{(1)} - y_1)^2\right] + \left[(u_2^{(1)} - x_2)^2 + (v_2^{(1)} - y_2)^2\right]$$
$$+ \left[(u_1^{(2)} - x_2)^2 + (v_1^{(2)} - y_2)^2\right] + \left[(u_2^{(2)} - x_1)^2 + (v_2^{(2)} - y_1)^2\right].$$

The likelihood is now maximized by first minimizing the right side of each equation over (x_1, y_1, x_2, y_2) and then choosing the hypothesis for which the resulting term is a minimum. This minimization can be done by geometric reasoning. In the first equation, only the first and third terms depend on (x_1, y_1), and the sum of these two terms is the sum of two squared distances of the form

$$d(P^{(1)}, Q^{(1)})^2 + d(P^{(2)}, Q^{(1)})^2,$$

where the measured points $P^{(i)}$ are equal to $(u_1^{(i)}, v_1^{(i)})$, and the actual point $Q^{(1)}$ is equal to (x_1, y_1). Clearly, the minimum occurs when $Q^{(1)}$ is midway between $P^{(1)}$ and $P^{(2)}$ so each distance is equal to $d(P^{(1)}, P^{(2)})/2$. Therefore,

$$\min_{Q^{(1)}} \left[d(P^{(1)}, Q^{(1)})^2 + d(P^{(2)}, Q^{(1)})^2\right] = \tfrac{1}{2} d^2(P^{(1)}, P^{(2)}).$$

After this minimization is also applied to (x_2, y_2), the negative loglikelihoods are

$$-\Lambda_0 = \tfrac{1}{2}d(P_1^{(1)}, P_1^{(2)})^2 + \tfrac{1}{2}d(P_2^{(1)}, P_2^{(2)})^2,$$
$$-\Lambda_1 = \tfrac{1}{2}d(P_1^{(1)}, P_2^{(2)})^2 + \tfrac{1}{2}d(P_2^{(1)}, P_1^{(2)})^2.$$

The decision between the two hypotheses is now made based on which of the two quantities is smaller. This question can be expressed by defining the matrix

$$M = \begin{bmatrix} d_{11}^2 & d_{12}^2 \\ d_{21}^2 & d_{22}^2 \end{bmatrix},$$

where the elements d_{ij}^2 of M are the four squared distances $d(P_i^{(1)}, P_j^{(2)})^2$. The decision is then based on a comparison between the trace of M and the trace of the matrix with the two columns of M interchanged.

When there are n targets and two scans, there are $n!$ ways of assigning the targets in the second scan to targets in the first scan. One can again form the n by n matrix of pairwise squared distances, given by

$$M = \begin{bmatrix} d_{11}^2 & d_{12}^2 & \cdots & d_{1n}^2 \\ d_{21}^2 & d_{22}^2 & \cdots & d_{2n}^2 \\ \vdots & & & \\ d_{n1}^2 & d_{n2}^2 & \cdots & d_{nn}^2 \end{bmatrix},$$

where $d_{ij} = d(P_i^{(1)}, P_j^{(2)})$. The natural generalization of the decision rule to n targets is to minimize the trace of the matrix M by permutation of its columns. There are $n!$ permutations of the columns and the permutation that minimizes the trace corresponds to maximum-likelihood association of targets in the first and second scans.

To see that it is the maximum-likelihood decision rule, first observe that there are $n!$ hypotheses corresponding to the $n!$ permutations of the points in the second scan. For a hypothesis that assigns $P_i^{(1)}$ and $P_j^{(2)}$ to the same target, the maximum-likelihood rule requires that the position of that target be estimated at the midpoint of $P_i^{(1)}$ and $P_j^{(2)}$. This means that the distance $d(P_i^{(1)}, P_j^{(2)})$ is an appropriate statistic. The sum of such distances should be minimized.

In the case of multiple scans, an appropriate statistic is more complicated. For example, with three scans, the statistic is

$$d_{ijk} = \min_Q \left[d(P_i^{(1)}, Q)^2 + d(P_j^{(2)}, Q)^2 + d(P_k^{(3)}, Q)^2 \right].$$

This minimum occurs at the point $Q(x, y)$ where

$$x = \frac{1}{3}(x_1 + x_2 + x_3), \quad y = \frac{1}{3}(y_1 + y_2 + y_3).$$

The three-dimensional matrix M with elements equal to the distances d_{ijk} is then defined. The permutation of the j and k indices that minimizes the generalized trace[1] of M is the maximum-likelihood solution.

---

[1] The *generalized trace* of the three-dimensional n by n matrix is the sum of the elements on the generalized diagonal. This notion extends immediately to L-dimensional arrays.

This decision procedure can be extended to moving targets when the data set is rich enough and when the motion in the model is appropriately constrained. For example, with three scans, the velocity would best be modeled as constant. For a hypothesis that assigns $P_i^{(1)}, P_j^{(2)}$, and $P_k^{(3)}$ to the same target, a least-squares straight line is fit to the three points. Then d_{ijk}^2 is the cumulative squared error from the straight-line fit. Finally, minimize the trace of the three-dimensional matrix with elements d_{ijk}^2 by choice of permutation of any two of the three dimensions. The minimizing permutation defines the assignment.

14.4 The Assignment Problem

The assignment problem is a combinatorial optimization problem. It has application to many situations and is described here in the context of the tracking problem.

The simplest instance of the assignment problem deals with two scans, each scan detecting the same n targets, as illustrated in Figure 14.4. Index the targets in the first scan by i. Index the targets in the second scan by j. The task is to pair each target in the first scan with a target in the second scan so that the sum of the pairwise euclidean distances is minimized. The pairing that yields this minimum total distance is the assignment of interest. The pairing itself is the primary conclusion of the algotithm. The value of the minimum euclidean distance itself may not be of primary interest.

The problem can be stated as a standard optimization problem called the *assignment problem*. Let d_{ij} be the euclidean distance from the ith point of the first scan to the jth point of the second scan, and let M be the matrix whose ij entry is the squared euclidian distance d_{ij}^2 between the two points. The *trace* of a square matrix is the sum of its diagonal elements. Another matrix can be obtained by permuting the rows of M, and that matrix also has a trace. Thus, the rows of M can be permuted to minimize the trace. The solution of the assignment problem is given by the row permutation that minimizes the trace. An equivalent statement of this problem is to find the column permutation that minimizes the trace.

For example, for a set of three objects, let

$$M = \begin{bmatrix} 1 & 4 & 8 \\ 12 & 5 & 3 \\ 5 & 2 & 1 \end{bmatrix},$$

where matrix element d_{ij} is the observed distance between the ith object and the jth object.

The trace of the matrix M is 7. There are 3! ways to permute the columns (or the rows) of M. By interchanging the second column and third column of M, the trace becomes 6. This is the minimum value of the trace over all permutations of columns, as can be seen by examining each of the six possible column permutations. (Equivalently, the trace can be minimized by interchanging the second and third rows.) This permutation is described by the matrix

$$P = \begin{bmatrix} 1 & 0 & 0 \\ 0 & 0 & 1 \\ 0 & 1 & 0 \end{bmatrix}.$$

The general case is based on the following definition.

Definition 14.4.1 *An n by n permutation matrix is an n by n array of zeros and ones such that every row of the array contains a single one and every column of the array contains a single one.*

An L-dimensional n by n by ... by n permutation matrix is an L-dimensional n by n by ... by n array of zeros and ones such that every generalized "row" of the array contains a single one.

In a large two-dimensional assignment problem with an n by n matrix, there are $n!$ possible permutations of columns. For large n, such as $n = 100$, it is not practical to exhaustively try all possibilities. Instead, fast computational algorithms are needed to find the minimizing permutation by efficient methods.

More generally, for the L-ary assignment problem of dimension n, described by an L-ary n by n by ... by n array M with elements, $d_{i_1 i_2 \ldots i_L}$, each of any set of $n - 1$ indices should be permuted to minimize the sum of the diagonal elements. The minimizing permutation is the solution to the n-dimensional assignment problem. Thus there are $(n!)^{L-1}$ possible permutations to be minimized over. When n is large, it is not possible to try all possibilities. An efficient computational procedure is needed to find the minimum.

An equivalent statement of the problem is given as follows. Let $C = \{C_{i_1 i_2 \ldots i_n}\}$ be an n-dimensional matrix of nonnegative real numbers. Let $Z = \{Z_{i_1 i_2 \ldots i_n}\}$ be a permutation matrix, that is, every element of Z is either a zero or a one. The problem is to find the zero-one matrix Z that minimizes the expression

$$\sum_{i_1=0}^{I_1-1} \sum_{i_2=0}^{I_2-1} \cdots \sum_{i_n=0}^{I_n-1} C_{i_1 i_2 \ldots i_n} Z_{i_1 i_2 \ldots i_n},$$

and subject to

$$\sum_{i_2=0}^{I_2-1} \sum_{i_3=0}^{I_3-1} \cdots \sum_{i_n=0}^{I_n-1} Z_{i_1 i_2 \ldots i_n} = 1 i_1 = 0, \ldots, I_1 - 1,$$

$$\sum_{i_1=0}^{I_1-1} \sum_{i_3=0}^{I_3-1} \cdots \sum_{i_n=0}^{I_n-1} Z_{i_1 i_2 \ldots i_n} = 1 i_2 = 0, \ldots, I_2 - 1,$$

$$\vdots$$

$$\sum_{i_1=0}^{I_1-1} \sum_{i_2=0}^{I_2-1} \cdots \sum_{i_{n-1}=0}^{I_{n-1}-1} Z_{i_1 i_2 \ldots i_n} = 1 i_n = 0, \ldots, I_n - 1.$$

The constraints say that each "row" must contain exactly a single one.

The two-dimensional version of the assignment problem in this form is to find the permutation matrix $Z = \{Z_{ij}\}$ that minimizes the term $\sum_{i=0}^{I-1} \sum_{j=0}^{J-1} C_{ij} Z_{ij}$ subject to the constraints that all rows of Z sum to one and all columns of Z sum to one. Equivalently, every row of Z is to have a single one, and every column is to have a single one.

Munkres Algorithm

A useful algorithm for the two-dimensional assignment problem is the *Munkres algorithm*. The key to this algorithm is the fact that any constant can be added to all elements of any row (or column) of M without affecting which permutation minimizes the trace. Consequently, by subtracting the appropriate constant from all elements of any row (or any column), the minimum element of that row (or that column) can be changed to a zero with no essential change in the problem. The Munkres algorithm combines two elementary operations: adding and subtracting constant rows or constant columns, and permutations.

The Munkres algorithm starts with the matrix M and iteratively makes changes in various elements to eventually obtain the desired permutation matrix Z. In the description of the Munkres algorithm, each row or column of the array may be flagged for further reference. This row or column is referred to as a *covered row* or *covered column*. Each element in a covered row or covered column is referred to as a *covered element*. Individual elements of the array may also be marked by being starred ($*$) or by being primed ($'$). The locations of the stars designate the locations in which the matrix Z will contain a one. The elements marked by a prime are candidates for a star, but not yet starred. Eventually each prime is either replaced by a star or is removed. An element, once starred, remains starred.

The Munkres algorithm consists of a number of iterations. The result of each iteration consists of one or more elements of C being marked by stars, or one or more elements of C being marked by primes. The algorithm iterates until there is exactly one star in every row and one star in every column. The set of locations of these stars constitutes the solution to the problem.

Proposition 14.4.2 *Let c'_i and r'_j be the smallest elements of any column and row of the nonnegative square matrix A whose elements are denoted a_{ij}. Then the matrix B with elements $b_{ij} = a_{ij} - c'_i - r'_j$ has its trace minimized by the same row permutation as does A.*

Proof:
Let π be a permutation of indices. Then

$$\sum_{i=1}^{n} b_{i\pi(i)} = \sum_{i=1}^{n} a_{i\pi(i)} - \sum_{i=1}^{n} c'_i - \sum_{i=1}^{n} r'_{\pi(i)}.$$

Because the second two terms do not depend on π, the proposition is proved. ∎

14.4 The Assignment Problem

It is clear that if each row and each column of the nonnegative matrix $\hat{c}$ has only one zero, then the location of these zeros mark the location of the ones in the permutation matrix Z. The computations defined in Proposition 14.4.2 lead to the following example:

$$\begin{bmatrix} 1 & 4 & 8 \\ 12 & 5 & 3 \\ 5 & 2 & 1 \end{bmatrix} \to \begin{bmatrix} 0 & 3 & 7 \\ 9 & 2 & 0 \\ 4 & 1 & 0 \end{bmatrix} \to \begin{bmatrix} 0 & 2 & 7 \\ 9 & 1 & 0 \\ 4 & 0 & 0 \end{bmatrix}.$$

Because the right side does not have a single zero in each row and column, it does not yet reveal the permutation. A different example is

$$\begin{bmatrix} 1 & 4 & 8 \\ 12 & 5 & 3 \\ 5 & 1 & 2 \end{bmatrix} \to \begin{bmatrix} 0 & 3 & 7 \\ 9 & 2 & 0 \\ 4 & 0 & 1 \end{bmatrix} \to \begin{bmatrix} 0 & 3 & 7 \\ 9 & 2 & 0 \\ 4 & 0 & 1 \end{bmatrix}.$$

Because the right side has a single zero in each row and column, the permutation is revealed.

Two elements of a matrix will be called unaligned if they do not lie in the same row or column. The Munkres algorithm will produce a matrix of starred elements in which all the starred elements are unaligned. These stars mark the ones in the permutation matrix Z.

Theorem 14.4.3 *When the largest set of unaligned zero elements in a matrix A has size n, all zero elements of A lie in a set of m rows or columns.*

Proof:
By permutation of rows, the m rows containing these zeros can be made to be the first m rows. By column permutation these zeros can be made to lie in the first m columns. Clearly all zeros must lie in the first m rows or the first m columns because any zero that did not could be used to enlarge the set of m unaligned zeros, contrary to the assumption. ■

The Munkres algorithm is as follows:

Step 0
To prepare the matrix, subtract the smallest element of each row from all elements of that row, then subtract the smallest element of each column from all elements of that column.

Step 1
Star every zero of M that is unaligned with any other zero. If n zeros are starred, halt. Otherwise flag every column with a starred zero and go to Step 2.

Step 2
Choose a zero in an unflagged column and mark it. If there is none, go to Step 4. If there is no starred zero in the row of this marked zero, go to Step 3. Otherwise, flag this row and unflag the column of of this marked zero. Repeat Step 2.

Step 3
Find the smallest element of the matrix not in a flagged row or flagged column. Add this element to each flagged row, then subtract it from each flagged column. Return to Step 1.

Step 4
Starting with the uncovered, primed zero, find a starred zero (if any) in that column. Then, in the row of that starred zero, find a primed zero. Continue to alternate between primed zeros and starred zeros in this way by alternating moves along columns and rows until a primed zero is reached with no starred zero in its column. Unstar each starred zero of this sequence and star each primed zero of the sequence. Remove all primes and covers and return to Step 1.

For an example of the Munkres algorithm, consider the matrix

$$M = \begin{bmatrix} 7 & 5 & 13 \\ 5 & 4 & 1 \\ 9 & 3 & 2 \end{bmatrix}.$$

The algorithm first makes the following changes:

$$\begin{bmatrix} 7 & 5 & 13 \\ 5 & 4 & 1 \\ 9 & 3 & 2 \end{bmatrix} \to \begin{bmatrix} 2 & 0 & 8 \\ 4 & 3 & 0 \\ 7 & 1 & 0 \end{bmatrix} \to \begin{bmatrix} 0 & 0 & 8 \\ 2 & 3 & 0 \\ 5 & 1 & 0 \end{bmatrix}.$$

Upon reaching Step 4, the algorithm computes

$$\begin{bmatrix} \overline{0^*} & \overline{0'} & 6 \\ 2 & 3 & \overline{0^*} \\ 5 & 1 & 0 \end{bmatrix} \to \begin{bmatrix} \overline{0^*} & \overline{0'} & \overline{8} \\ 1 & 2 & \overline{0^*} \\ 4 & 0 & \overline{0} \end{bmatrix}.$$

Upon returning to Step 1, the algorithm finds

$$\begin{bmatrix} \overline{0^*} & \overline{0} & \overline{8} \\ \overline{1} & \overline{2} & \overline{0^*} \\ \overline{4} & \overline{0^*} & \overline{0} \end{bmatrix}.$$

Because all elements are covered, the algorithm terminates. The three starred zeros determine the locations of the ones in Z.

14.5 Lagrangian Relaxation

A three-dimensional assignment problem can be reduced to a two-dimensional assignment problem by using a lagrange multiplier λ to constrain one dimension. This results in a two-dimensional assignment problem that can be solved by the Munkries algorithm.

The corresponding alternative formulation of the assignment problem replaces the notion of a column permutation by the notion of a permutation matrix. The locations of the ones in the array define the permutation. These elements are the elements that would be permuted into the diagonal to obtain the permutation with the minimum trace. This formulation facilitates the introduction of optimization methods based on lagrange multipliers. Such methods reduce an L-dimensional assignment problem reduce to an $(L-1)$-dimensional assignment problem by appending a constraint with the aid of a lagrange multiplier. This is called *lagrangian relaxation*. The $(L-1)$-dimensional problem is solved for many values of the lagrange multiplier in order to find the particular value that gives a solution satisfying the constraint. In this way, the L-dimensional assignment problem is replaced by a search over a large number of $(L-1)$-dimensional assignment problems. In turn, by recursion, the $(L-1)$-dimensional problem is solved in the same way. Lagrangian relaxation reduces the L-dimensional assignment problem to a large stack of two-dimensional assignment problems. An algorithm for the two-dimensional assignment problem is then needed, but this algorithm is executed a great number of times.

14.6 Multilateration

Multilateration is a form of triangulation in which multiple lines or curves of position are intersected to determine the location of a point. These lines or curves of position may be of various kinds, such as lines of bearing, shown in Figure 14.5; circles of constant range, shown in Figure 14.6; or hyperbolas or ellipses. Other situations, such as curves of constant differential doppler, may use the intersections of curves of position that do not correspond to standard forms of an elementary curve. Figure 14.6 and Figure 14.5 show how a target is located at the intersection of several curves.

A multilateration system must cope with the occurance of ambiguities. Whenever there are multiple targets, as shown in Figure 14.6, there will be the need to associate a particular curve of position from one family with the correct line of position of the other family. In the absence of any side information to resolve an ambiguity, there will be one or more unresolved ambiguities, as indicated in the figures. There are two pairs of solutions shown in Figure 14.5 and Figure 14.6 coming from the data itself, and there is no way to distinguish the two true targets from the two false targets or ghosts. When there are multiple targets, the situation becomes even more complicated. When there are N targets, there may be as many as N^2 points of intersection. Of these N describe true target positions and the others describe false target positions.

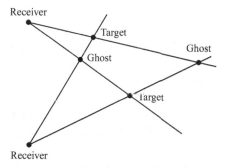

Figure 14.5 Bistatic angle ambiguities

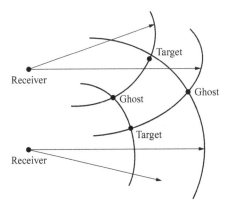

Figure 14.6 Bistatic range ambiguities

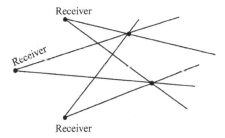

Figure 14.7 Resolution of ambiguities

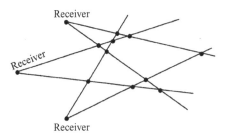

Figure 14.8 Perturbed lines of position

One way of resolving the ambiguities is to use another sensor to form a third line of position, as shown in Figure 14.7. The extra sensor completely resolves the problem; the true targets are at the triple intersections, while the ambiguities are at double intersections.

In a realistic problem, the measurements are not perfect. This means that the lines of position will be perturbed, and the targets will not be at a triple intersection. Instead, targets will be near a small cluster of three double intersections, as shown in Figure 14.8. This, then, leads to an instance of the assignment problem.

Problems

14.1 A radar detects and locates a set of four targets on each of two scans.

a. On each scan, each target is located with position error variance σ^2 on the x axis and with position error variance $2\sigma^2$ on the y axis. Find an optimum assignment rule.
b. On each scan, each target is located with position error covariance matrix Σ on the x, y plane. Find an optimum assignment rule.
c. On the first and second scan, respectively, each target is located with position error covariance matrix Σ_1 and Σ_2, respectively, on the x, y plane. Find an optimum assignment rule.

14.2 A given radar detects an object at the point $x = (x, y, z)$ in three-dimensional space. The error in the three-dimensional location is a vector gaussian random variable with the three-by-three covariance matrix Σ_1. The joint probability density function is

$$p(x|x_1) = \frac{1}{\sqrt{\det(2\pi \Sigma_1)}} e^{-(x-x_1)^\dagger \Sigma_1^{-1} (x-x_1)},$$

where x is the measured position and x_1 is the true position of the object. A second radar with independent errors also locates an object with gaussian-distributed errors with the covariance matrix Σ_2. A decision is to be made as to whether these two objects are the same object. Set up the appropriate hypothesis-testing problem and give expressions for the probabilities of error as a function of the threshold. Does the problem require an assumption about the vector difference $x_1 - x_2$? Can the problem be made parametric in this difference?

14.3 A two-dimensional gaussian random variable with the probability density function $q(x, y)$ has zero mean and covariance matrix Σ. The *circular error probability* defined as $R = \text{cep}(p)$ is the value of R satisfying

$$\int_{-\infty}^{\infty} \int_{-\infty}^{\infty} \text{circ}\left(\frac{x}{2R}, \frac{y}{2R}\right) q(x, y) dx dy = p.$$

The circular error probability is one way of compressing a two-parameter zero-mean gaussian density described by the two eigenvalues λ_1 and λ_2 into a single figure of merit.

a. How does the circular error probability relate to the rayleigh distribution?
b. Calculate the circular error probability as a function of λ_1 and λ_1/λ_2, where λ_1 and λ_2 are the largest and the smallest eigenvalues of Σ, respectively.
c. Generalize the rayleigh distribution to the radial marginal of an elliptical two-dimensional gaussian distribution.

14.4 Show that the maximum-likelihood solution to the target-sorting problem in the presence of independent gaussian position errors for each of n targets and m scans reduces to the m-ary n-dimensional assignment problem.

14.5 Given a three-target, two-scan, target-sorting problem in the presence of gaussian noise, suppose that one data point from the third scan is not detected and so is missing. The third scan has only two points. Find a maximum-likelihood solution to this sorting problem.

14.6 Given the three points $P^{(1)} = (x_1, y_1)$, $P^{(2)} = (x_2, y_2)$, and $P^{(3)} = (x_3, y_3)$ in the plane $\mathbb{R}^2$, let
$$d(P, Q) = \sqrt{(x_i - x)^2 + (y_i - y)^2},$$
where Q denotes the arbitrary point (x, y). Find the point $\widehat{Q} = (\widehat{x}, \widehat{y})$ defined by
$$(\widehat{x}, \widehat{y}) = \operatorname*{argmin}_{(x,y)} \left[\sum_{i=1}^{3} d(P^{(i)}, Q)^2 \right].$$
Does this point have a clear geometric description.

14.7 The three points $P^{(1)}, P^{(2)}, P^{(3)}$ are given, and $\widehat{Q}$ is defined as in Problem 14.6. Evaluate whether the ratio of $d(P^{(1)}, P^{(2)})^2 + d(P^{(2)}, P^{(3)})^2 + d(P^{(3)}, P^{(1)})^2$ to $\sum_{i=1}^{3} d(P^{(i)}, \widehat{Q})^2$ can be bounded.

Notes

The need to identify tracks in fragmentary numerical data is at least as old as radar. This need to identify tracks existed even in prehistoric human activities. As a formal mathematical topic, however, track assignment took shape only recently. The papers of Wax (1955) and Sittler (1964) can be considered to start the development of a formal theory. Nahi (1969) studied the problem of false measurements in a recursive estimator. Singer, Sea, and Housewright (1974) presented an optimal, maximum-likelihood tracking algorithm that requires data memory to grow with time as data is accumulated. The maximum-likelihood approach was further studied by Stein and Blackman (1975). Morefield (1977) posed the tracking problem as an integer programming problem that clarified the connection with the general assignment problem. The tracking problem has been extensively studied in a series of papers by Bar-Shalom (1978). Iterative methods of maintaining an established track are closely related to the filtering methods of Kalman (1960) and Kalman and Bucy (1961) and may sometimes employ the Kalman filter.

The assignment problem has its own literature as a topic in optimization theory. When treated by the methods of complexity theory, it is found to be nonpolynomial hard in the number of data points. Radar applications, however, do not normally require that the worst-case assignment problem be solved – only that typical or nearly typical problems be solved and, for these typical problems, occasional failed attempts at an assignment can be tolerated.

The best-known algorithm for the two-dimensional assignment problem is the Munkres (1957) algorithm, which was based on earlier work by the Hungarian mathematician Egervary, as reported by Kuhn (1955), and later augmented to nonsquare matrices by Bourgeois and La Salle (1971). An alternative to the Munkres algorithm has been developed by Bertsekas (1988).

15 Phase Noise and Phase Distortion

The theory of image formation, as developed in this book, emphasizes the roles of probability theory, the Fourier transform, and the theory of coherence. These topics are the foundation and the scaffolding for the unification that is sought herein. The notion of coherence is an outgrowth of probability theory. This final chapter compiles various facts relating to the topic of coherence, both temporal coherence and spatial coherence.

An image-formation system that is based on a coherent waveform is degraded by any loss of the coherence of the received waveform. This reduced coherence has the form of an anomalous phase angle in the received signal. This anomalous phase angle is referred to as a *phase error*. A random phase error may be a simple constant phase error or it may be a phase error varying in some way. A phase error can be varying in the time domain or it can be varying in the frequency domain. Random phase errors in the time domain arise when the phase varies randomly with the time. Random phase errors in the frequency domain arise when the phase varies randomly with the frequency. Random phase errors considered herein are primarily described in the time domain as random functions of time.

A random phase error in the time domain appears as the exponent of a complex time-varying exponential $e^{j\theta(t)}$ multiplying the received complex baseband signal. When the phase error is time-varying, it is called *phase noise*. Phase noise limits the maximum waveform duration that can be processed coherently.

The source of phase noise is any unintentional and unknown phase modulation that is introduced into the received signal, either by the transmitter, by the propagation medium, or by the receiver. There are many sources of phase noise. These include motion of the transmitter or motion of the receiver such as antenna vibration, phase errors arising in the local oscillators and the mixing circuits in the transmitter and receiver, phase errors arising in the carrier recovery circuitry, and phase errors due to inhomogeneities and aberrations in the propagation medium such as may be caused by atmospheric turbulence.

The effect of phase errors is studied in several ways. First, Section 15.1 models the phase error $\theta(t)$ as a slowly varying error that is described by several unknown parameters. The performance degradation is studied as a function of the unknown parameters. Then, in Section 15.2, the expected performance degradation is based on a probabilistic model of $\theta(t)$ as a random process.

15 Phase Noise and Phase Distortion

15.1 Quadratic-Phase Errors

Let the signal $s(t)$ be a pulse or waveform, possibly complex, of finite energy. Let $\theta(t)$ be an unknown real function of time. The signal $s(t)$ contaminated by the phase error $\theta(t)$ is given by

$$v(t) = s(t)e^{j\theta(t)}.$$

This section models $\theta(t)$ as a slowly varying deterministic–though unknown–function. The next section models $\theta(t)$ as a random noise process.

Approximate the deterministic signal over the relevant time interval by the first terms of a series expansion. The series expansion is

$$\theta(t) = \theta_0 + \dot\theta_0 t + \tfrac{1}{2}\ddot\theta_0 t^2 + \cdots.$$

The coefficients of this series are unknown deterministic parameters. Suppose that the phase $\theta(t)$ is adequately approximated by the first three terms of this series. The performance as a function of the values of the parameters θ_0, $\dot\theta_0$, and $\ddot\theta_0$ is studied by studying the effect of these parameters on the Fourier transform, on the matched filter, and on the sample cross-ambiguity function.

The first parameter θ_0 is an arbitrary constant in $[0, 2\pi]$. The term $e^{j\theta_0}$ can be brought outside the integral defining the Fourier transform, the matched filter, or the sample cross-ambiguity function. In each case, this phase term is suppressed by taking the magnitude of that function. In a detection or an estimation problem, the effect of the constant θ_0 is to require that the detector or estimator be noncoherent, such as by observing the magnitude of that quantity. Noncoherent detection and estimation are studied in some detail in Chapter 12. The parameter θ_0 is not studied further in this chapter.

The second parameter $\dot\theta_0$ is simple to analyze. This slope parameter $\dot\theta_0$ appears as a false frequency shift. The Fourier transform is offset by $\dot\theta_0$ on the frequency axis. The effect on the cross-ambiguity function is translated by $\dot\theta_0$ in the doppler direction.

The third parameter $\ddot\theta_0$ requires more effort to analyze. The term $\tfrac{1}{2}\ddot\theta_0 t^2$ is called the *quadratic-phase error*. It is also written as $\tfrac{1}{2}\alpha t^2$. The quadratic-phase error is the first term in the series expansion of the phase error that affects the shape of the cross-ambiguity function. The quadratic phase term is also the dominant term among the terms neglected in the Fresnel and Fraunhofer approximations. An analysis of the effect of quadratic-phase error is needed to justify those approximations.

The effect of quadratic-phase error on the Fourier transform is studied first. The rectanglular pulse

$$s(t) = \text{rect}(t),$$

typifies the effect of this phase error for many pulse shapes. A quadratic-phase error on a rectangular pulse is

$$v(t) = s(t)e^{j\pi\alpha t^2}$$
$$= \text{rect}(t)e^{j\pi\alpha t^2}.$$

15.1 Quadratic-Phase Errors

The Fourier transform $V(f)$ is

$$V(f) = \int_{-1/2}^{1/2} e^{j\pi\alpha t^2} e^{-j2\pi ft} dt,$$

noting that rect(t) is nonzero for $|t| \leq 1/2$. The relative reduction in amplitude at $f = 0$ as given by

$$L = \frac{V(0)}{S(0)},$$

is

$$L = \int_{-1/2}^{1/2} e^{j\pi\alpha t^2} dt.$$

Over an interval of duration T, this is

$$L = 2 \int_0^{T/2} \cos(\pi\alpha t^2) dt.$$

This integral cannot be integrated in closed form but can be expressed in terms of standard functions.

The Fresnel Integrals
The *Fresnel cosine integral* and the *Fresnel sine integral* are defined as

$$C(\eta) = \int_0^\eta \cos\frac{\pi t^2}{2} dt \qquad S(\eta) = \int_0^\eta \sin\frac{\pi t^2}{2} dt.$$

Numerically evaluated tables of $C(\eta)$ and $S(\eta)$ are widely available. Figure 15.1 shows an attractive and popular graph known as the *Cornu spiral*. This is a graph of $S(\eta)$ versus $C(\eta)$ parameterized by η. The limit points are given by

$$\int_0^\eta \cos\frac{\pi t^2}{2} dt = \int_0^\eta \sin\frac{\pi t^2}{2} dt = \sqrt{\pi/8} = 0.6267.$$

The two limit points suggested in the graph are at $\{\pm 0.6267, \pm 0.6267\}$.

The amplitude reduction L due to the quadratic-phase error is plotted in Figure 15.2. For a half-interval phase change less than 30°, the loss due to the quadratic-phase term is less than one decibel. This may be acceptable as a loss tolerance in many applications.

Degradation of the Ambiguity Function
In the presence of a quadratic-phase error, the cross-ambiguity function is

$$\chi_c(\tau, \nu) = \int_{-\infty}^\infty v(t) s^*(t - \tau) e^{-j2\pi\nu t} dt$$

$$= \int_{-\infty}^\infty s(t) s^*(t - \tau) e^{j\pi\alpha t^2} e^{-j2\pi\nu t} dt.$$

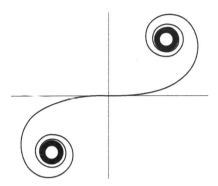

Figure 15.1 The Cornu spiral

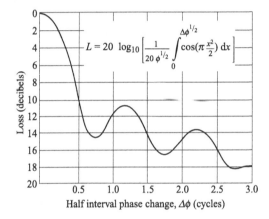

Figure 15.2 Loss due to quadratic-phase error versus phase change

The presence of the quadratic-phase term means that $\chi_c(\tau,\nu)$ is not a simple offset of $\chi(\tau,\nu)$ by a τ error or a ν error. For small α, $\chi_c(\tau,\nu)$ is approximated as a translation of $\chi(\tau,\nu)$ and has a peak similar to the peak of $\chi(\tau,\nu)$. The peak of $\chi_c(\tau,\nu)$ has a smaller amplitude and its curvature is less sharp. For larger values of α, the peak may even lose its appearance as a peak.

Define the proportional reduction in the peak amplitude as the ratio:

$$L = \frac{|\chi_c(0,0)|}{\chi(0,0)}$$

$$= \frac{1}{E_p}\left|\int_{-\infty}^{\infty} |s(t)|^2 e^{j\pi\alpha t^2}\,dt\right|.$$

For an analysis of the representative case, choose $|s(t)|^2$ equal to one for $t \in [-T/2, T/2]$ and otherwise choose $|s(t)|^2$ equal to zero. This representative case is not as restrictive as it might seem. As long as the fluctuations in a typical $|s(t)|^2$ are rapid compared to the rate at which the phase changes, the fluctuations may have little effect on the value of L. Even for a pulse train $s(t)$ of low duty cycle, the value of L may be adequately approximated by setting $|s(t)|^2$ equal to an average value.

In addition to attenuating the peak, phase noise may also change the location and the shape of the peak. The primary change in the curvature is in the ν direction. The change in the curvature in the τ direction is small. The location of the peak in the ν direction is given by

$$\bar{t} = \frac{1}{E_p} \int_{-T/2}^{T/2} t|s(t)|^2 e^{j\pi\alpha t^2} dt,$$

and the curvature is described by

$$\overline{t^2} - \bar{t}^2 = \frac{1}{E_p} \int_{-T/2}^{T/2} t^2 |s(t)|^2 e^{j\pi\alpha t^2} dt - \bar{t}^2.$$

Again, as a representative case, setting $s(t) = 1$ for $|t| \leq T/2$ and otherwise zero is adequate for purposes of the present analysis. Then $E_p = T$. By symmetry, $\bar{t} = 0$ any the curvature at the origin is described by

$$\overline{t^2} = \frac{1}{T} \int_{-T/2}^{T/2} t^2 |s(t)|^2 \cos(\pi\alpha t^2) dt.$$

For any nonzero value of α, the curvature is reduced.

15.2 Phase Noise and Coherence

This section studies the effect of a stationary time-varying phase-noise random process on the coherence of the signal. A pulse contaminated by both phase noise and additive noise is given by

$$v(t) = s(t)e^{j\theta(t)} + n(t).$$

The pulse contaminated only by phase noise is

$$v(t) = s(t)e^{j\theta(t)}.$$

The time-varying phase-noise process $\theta(t)$ is taken to be a real covariance-stationary random process with mean $\bar{\theta}$ and with the known correlation function

$$\phi(\tau) = E\big[\theta(t)\theta(t+\tau)\big]$$
$$= \sigma_\theta^2 \rho(\tau).$$

This second-order description is not a full description of the phase-noise process. A full description requires that higher-order moments are specified. A second-order description is a full description when the phase noise is gaussian noise because then the higher-order properties of a gaussian process are implied by the second-order properties. When the phase noise is not gaussian, the second-order description may still be an adequate description when a small-angle approximation for the phase noise can be made within the analysis. When the phase noise is nongaussian and large, higher-order moments may need to be considered.

Suppose that the phase noise $\theta(t)$ is a gaussian random process with probability density function

$$p(\theta) = \frac{1}{\sqrt{2\pi\sigma_\theta^2}} e^{-\theta^2/2\sigma_\theta^2}$$

15 Phase Noise and Phase Distortion

at each value of t. For a gaussian random variable of mean $\bar{\theta}$ and variance σ_θ^2, the characteristic function $E[e^{j\theta(t)}]$ has the form of a Fourier transform. The following theorem is central to the analysis of gaussian phase noise.

Theorem 15.2.1 *Let θ be a gaussian random variable with mean $\bar{\theta}$ and variance σ_θ^2. Then*

$$E[e^{j2\pi a\theta}] = e^{j2\pi a\bar{\theta}} e^{-2\pi^2 \sigma_\theta^2 a^2},$$

with θ in radians.

Proof:
The expectation of $e^{j2\pi a\theta}$ is defined as

$$E[e^{ja\theta}] = \int_{-\infty}^{\infty} p(\theta) e^{j2\pi a\theta}\, d\theta$$

$$= \frac{1}{\sqrt{2\pi}\sigma_\theta} \int_{-\infty}^{\infty} e^{-(\theta-\bar{\theta})^2/2\sigma_\theta^2} e^{j2\pi a\theta}\, d\theta = \frac{1}{\sqrt{2\pi}\sigma_\theta} e^{j2\pi a\bar{\theta}} \int_{-\infty}^{\infty} e^{-\theta^2/2\sigma_\theta^2} e^{ja\theta}\, d\theta.$$

The integral on the right has the form of the Fourier transform

$$e^{-ct^2} \leftrightarrow \sqrt{\frac{\pi}{c}} e^{-\pi^2 f^2 / c}.$$

This Fourier transform pair with the substitution $c = 1/2\sigma^2$ and using the variables θ and a in place of t and f then gives

$$\int_{-\infty}^{\infty} e^{-\theta^2/2\sigma_\theta^2} e^{ja\theta}\, d\theta = \sqrt{2\pi}\sigma_\theta e^{-2\pi^2 \sigma_\theta^2 a^2}.$$

Substituting this expression into the expression for $E[e^{ja\theta}]$ completes the proof of the theorem. ∎

Corollary 15.2.2 *Let $\theta(t)$ be a stationary, gaussian random process with zero mean. Then*

$$E[e^{j\theta(t)}] = e^{-\sigma_\theta^2/2},$$

for all values of t.

Proof:
Set $2\pi a$ equal to one in the theorem. Because the mean and the variance are independent of time, the statement of the theorem is independent of time. ∎

Definition 15.2.3 *The mutual coherence Γ of two complex random processes $v_1(t) = e^{j[\theta_1(t)]}$ and $v_2(t) = e^{j[\theta_2(t)]}$ is given by the cross-correlation*

$$\Gamma = E[v_1(t) v_2^*(t)] = E\left[e^{j(\theta_1(t) - \theta_2(t))}\right].$$

Thus coherence is another name for correlation when that correlation is applied to complex exponentials. When $\theta_2(t) = \theta_1(t)$, the coherence Γ is equal to 1. When $\theta_2(t)$ is equal to $\theta_1(t) + \pi$, the coherence Γ is equal to -1.

Corollary 15.2.4 *Let $\theta_1(t)$ and $\theta_2(t)$ be gaussian random processes of zero mean, equal variance σ^2, and correlation $\rho = \mathrm{E}[\theta_1(t)\theta_2(t)]/\sigma^2$. The mutual coherence is*

$$\Gamma = e^{-\sigma^2(1-\rho)/2}.$$

Proof:
The difference $\theta_1(t) - \theta_2(t)$ is itself a gaussian random process with the variance $\mathrm{E}[\theta_1(t) - \theta_2(t)]^2 = 2\sigma^2(1-\rho)$. Corollary 15.2.2 then gives

$$\Gamma = e^{-\sigma^2(1-\rho)/2}$$

for the coherence.

15.3 Phase Noise and the Fourier Transform

Let $s(t)$ be any pulse with energy E_p equal to one and with $\int ts(t)dt$ equal to zero. Let

$$v(t) = s(t)e^{j\theta(t)},$$

where $\theta(t)$ is a stationary gaussian random process. The Fourier transform of $v(t)$ is

$$V(f) = \int_{-\infty}^{\infty} v(t)e^{-j2\pi ft}dt$$

$$= \int_{-\infty}^{\infty} s(t)e^{j\theta(t)}e^{-j2\pi ft}dt.$$

The phase noise $\theta(t)$ causes distortion in $V(f)$. When $\theta(t) = 0$, there is no phase noise and $V(f)$ is equal to $S(f)$.

The effect of phase noise on the Fourier transform $S(f)$ of the pulse $s(t)$ can be characterized in various ways by several scalar parameters. These parameters are the loss of amplitude at the origin, the change in the location of the maximum or the mean of the transform, and the width of the transform. Because the phase noise $\theta(t)$ is random, $V(f)$ is also random. The expectation is

$$\mathrm{E}[V(f)] = \mathrm{E}\left[\int_{-\infty}^{\infty} s(t)e^{j\theta(t)}e^{-j2\pi ft}dt\right]$$

$$= e^{-\sigma_\theta^2/2}S(f)$$

by Corollary 15.2.2. Thus, in expectation, the phase noise attenuates the Fourier transform.

The energy in $v(t)$ is the same as the energy in $s(t)$, so the energy theorem says that the energy in $V(f)$ is the same as the energy in $S(f)$. Because $V(f)$ has less amplitude than $S(f)$ on average, it must be wider on average. This is quantified in the next proposition.

15 Phase Noise and Phase Distortion

Proposition 15.3.1 *Let $s(t)$ be a real-valued pulse with Gabor bandwidth B_G^2. The expected squared Gabor bandwidth of the pulse contaminated by stationary phase noise $\phi(t)$ of power density spectrum $\Phi(f)$ is*

$$E[B_G^2(v)] = \int_{-\infty}^{\infty} f^2 \Phi(f) df + B_G^2.$$

Proof:
Because the energy of pulse $v(t)$ is equal to one, its squared Gabor bandwidth is

$$B_G^2(v) = \int_{-\infty}^{\infty} |\dot{v}(t)|^2 dt$$

$$= \int_{-\infty}^{\infty} \left| j\dot{\theta}(t)s(t)e^{j\theta(t)} + \dot{s}(t)e^{j\theta(t)} \right|^2 dt$$

$$= \int_{-\infty}^{\infty} \left| j\dot{\theta}(t)s(t) + \dot{s}(t) \right|^2 dt.$$

Because $E[\dot{\theta}(t) = 0]$, the expectation is

$$E[B_G^2(v)] = \int_{-\infty}^{\infty} E\left[\left| j\dot{\theta}(t)s(t) + \dot{s}(t) \right|^2 \right] dt$$

$$= \int_{-\infty}^{\infty} |s(t)|^2 E\left[\dot{\theta}(t)\right]^2 dt + \int_{-\infty}^{\infty} |\dot{s}(t)|^2 dt.$$

The second term is the Gabor bandwidth B_G^2 of $s(t)$. The first term is evaluated by using

$$E[\dot{\theta}(t)]^2 = \phi''(0) = \int_{-\infty}^{\infty} f^2 \Phi(f) df.$$

This completes the proof of the proposition. ∎

15.4 Phase Noise and the Matched Filter

The expected degradation in the output of the filter $g(t)$ due to phase noise is easily calculated. Let $u_s(t) = g(t) * v(t)$ be the output of the filter $g(t)$ in the presence of phase noise. The reduction in amplitude is defined as

$$L = \frac{E[u_s(t)]}{u(t)},$$

where $u(t) = g(t) * s(t)$ is the output of the filter in the absence of phase noise.

Proposition 15.4.1 *Let $v(t) = s(t)e^{j\theta(t)}$ where $s(t)$ is a known pulse and $\theta(t)$ is a zero-mean, stationary, gaussian random noise process with variance σ_θ^2. Then the expected output of the filter is*

$$E[u_s(t)] = e^{-\sigma_\theta^2/2} u(t),$$

for which $L = e^{-\sigma_\theta^2/2}$, which is independent of t.

15.4 Phase Noise and the Matched Filter

Proof:

$$E[u_s(t)] = E\left[\int_{-\infty}^{\infty} g(t-\xi)s(\xi)e^{j\theta(\xi)}d\xi\right]$$

$$= \int_{-\infty}^{\infty} g(t-\xi)s(\xi)E[e^{j\theta(\xi)}]d\xi.$$

Corollary 15.2.2 then gives

$$E[e^{j\theta(\xi)}] = e^{-\sigma_\theta^2/2},$$

which is independent of ξ, so that term comes out of the integral. The proof of the proposition is thereby complete. ∎

The conclusion drawn from Proposition 15.4.1 is that the expectation of the output of the filter in the presence of phase noise is simply an attenuated version of the filter output in the absence of phase noise. The effect of the phase noise is a reduction in the expectation of the effective signal strength. This expected value of the effective signal strength degrades quickly with σ_θ^2. This means that situations with large phase noise are rarely of practical interest. Problems that are not otherwise amenable to solution may become amenable only under the condition that σ_θ^2 is small.

Whereas Proposition 15.4.1 describes the expected value of the matched-filter output in the presence of gaussian phase noise, the following proposition studies the variance of the matched-filter output in the presence of any stationary random phase noise with small variance.

Proposition 15.4.2 *For small values of σ_θ^2, the variance in the output of a matched filter at the nominal peak output due to the phase noise of power density spectrum $N_\theta(f)$ is approximated by*

$$\sigma^2 \approx e^{-\sigma_\theta^2/2} \int_{-\infty}^{\infty} N_\theta(f)|W(f)|^2 df,$$

where $s(t)$ is the filter input and $W(f) = S(f) * S^*(-f)$ is the Fourier transform of $|s(t)|^2$.

Proof:
In the presence of phase noise $\theta(t)$, the output of the matched filter at the nominal peak is given by

$$u_s(0) = \int_{-\infty}^{\infty} |s(\xi)|^2 e^{j\theta(\xi)} d\xi.$$

Noting that $\phi(\tau) = \sigma_\theta^2 \rho(\tau)$ where $\phi(\tau)$ is the correlation function of $\theta(t)$, refer to Corollary 15.2.2 to give

$$E[u_s(0)] = e^{-\sigma_\theta^2/2} \int_{-\infty}^{\infty} |s(\xi)|^2 d\xi = e^{-\phi(0)/2} \int_{-\infty}^{\infty} |s(\xi)|^2 d\xi.$$

Again referring to Corollary 15.2.2, but now with $\theta(t)$ replaced by $\theta(\xi) - \theta(\xi')$, the expectation of the magnitude-squared is

$$E[|u(0)|^2] = E\left[\int_{-\infty}^{\infty}\int_{-\infty}^{\infty} |s(\xi)|^2 |s(\xi')|^2 e^{j(\theta(\xi)-\theta(\xi'))} d\xi d\xi'\right]$$

$$= \int_{-\infty}^{\infty}\int_{-\infty}^{\infty} |s(\xi)|^2 |s(\xi')|^2 e^{-(2\sigma_\theta^2 - 2\sigma_\theta^2 \rho(\xi-\xi')/2)} d\xi d\xi'$$

$$= \int_{-\infty}^{\infty}\int_{-\infty}^{\infty} |s(\xi)|^2 |s(\xi')|^2 e^{-(\phi(0)-\phi(\xi-\xi'))} d\xi d\xi'.$$

The variance is

$$\sigma^2 = E[|u(0)|^2] - |E|u(0)||^2$$

$$\sigma^2 = \int_{-\infty}^{\infty}\int_{-\infty}^{\infty} |s(\xi)|^2 |s(\xi')|^2 e^{-\phi(0)}(e^{\phi(\xi-\xi')} - 1) d\xi d\xi'.$$

Let $\eta = \xi - \xi'$. Then

$$\sigma^2 = e^{-\phi(0)}\int_{-\infty}^{\infty}(e^{\phi(\eta)}-1)\int_{-\infty}^{\infty}|s(\xi)|^2|s(\xi-\eta)|^2 d\xi d\eta.$$

Under the condition that σ_θ^2 is small, $\phi(n)$ is small as well. Then $e^{\phi(\eta)} - 1$ can be approximated by $\phi(\eta)$. Finally, Parseval's theorem completes the proof. ∎

The exponential in the equation for σ^2 in Proposition 15.4.2 can be approximated by one. This gives

$$\sigma^2 \approx \int_{-\infty}^{\infty} N_\theta(f)|W(f)|^2 df.$$

This expression is the same as the expression for additive thermal noise passed through the filter $W(f)$. The same expression can be obtained in another way by starting with the approximation

$$v(t) = s(t)e^{j\theta(t)}$$
$$\approx s(t) + j\theta(t)s(t).$$

With this approximation, the output of the matched filter $s(t)$ is

$$u(t) = \int_{-\infty}^{\infty}|s(t)|^2 dt + j\int_{-\infty}^{\infty}\theta(t)|s(t)|^2 dt.$$

The variance of the matched-filter output is

$$\sigma^2 = E\left[\int_{-\infty}^{\infty}\int_{-\infty}^{\infty}\theta(\xi)|s(\xi)|^2\theta(\xi')|s(\xi')|^2 d\xi d\xi'\right]$$

$$= \int_{-\infty}^{\infty}\phi(\eta)\int_{-\infty}^{\infty}|s(\xi)|^2|s(\xi-\eta)|^2 d\xi d\eta$$

under the change of variables $\eta = \theta - \theta''$ followed by Parsival's formula. This again leads to the approximation

15.4 Phase Noise and the Matched Filter

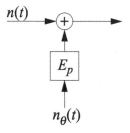

Figure 15.3 Modeling phase noise as an additive noise

$$\sigma^2 \approx \int_{-\infty}^{\infty} N_\theta(f)|W(f)|^2 df,$$

which is the same expression given earlier.

Example

For example, when $|s(t)|^2$ is a rectangular pulse of duration T, then

$$\sigma^2 = \int_{-\infty}^{\infty} N_\theta(f) T^2 \text{sinc}^2(fT) df.$$

Moreover, when the phase noise is white with $N_\theta(f) = N_\theta/2$, then

$$\sigma^2 = \frac{N_\theta}{2} T = \frac{N_\theta}{2} E_p.$$

The signal at the output of the filter is equal to E_p, so the signal-to-phase-noise power ratio is

$$\frac{S}{N} = \left(\frac{N_\theta}{2}\right)^{-1} E_p.$$

Although phase noise is not the same as additive noise, its effect on the peak output of the matched filter can be compared to the effect of additive white noise with power density spectrum $E_p N_\theta$. When a system has both additive thermal noise and phase noise, both of their effects may be combined by introducing an additive noise whose effect on signal-to-noise ratio is equivalent to the effect of the phase noise, as shown in Figure 15.3:

$$\frac{S}{N} = \frac{2E_p}{N_0 + E_p N_\theta}.$$

This summary formula combines the effects of two kinds of noise that are not themselves interchangeable.

The formula for the combined effect on the signal-to-noise ratio makes it obvious that increasing E_p will not increase performance indefinitely. Eventually, the phase noise will dominate the additive noise. For large values of E_p, this effective signal-to-noise ratio is always smaller than $2/N_0$ and approaches this limiting value for large values of E_p.

15.5 Phase Noise and the Ambiguity Function

The effect of phase noise on the sample cross-ambiguity function $\chi_c(\tau,\nu)$ is given by

$$\chi_c(\tau,\nu) = \int_{-\infty}^{\infty} s(t)v^*(t+\tau)e^{-j2\pi\nu t}\,dt$$

$$= \int_{-\infty}^{\infty} s(t)s^*(t+\tau)e^{-j\theta(t+\tau)}e^{-j2\pi\nu t}\,dt,$$

or on the sample cross-ambiguity surface $|\chi_c(\tau,\nu)|$. Because $\theta(t)$ is a random process, the sample cross-ambiguity function $\chi_c(\tau,\nu)$ becomes a random process through its dependence on $\theta(t)$. As illustrated in one dimension in Figure 15.4, the main lobe can become degraded by phase noise in several ways. As the variance of the phase noise is increased, the maximum of the peak of the main lobe may be displaced, its amplitude reduced, and its width broadened. The reduction in amplitude is referred to as a *coherence loss*. The increase in width is referred to as a *resolution loss*. The change in shape may be quantified in terms of an equivalent noise power. The phase noise may also produce additional random sidelobes near the main lobe.

A radar image formed from the received pulse $s(t)$ as a function of increasing noise variance σ_θ^2 shows increasing degradation. As σ_θ^2 increases from zero, the image intensity begins to slowly fade because the main lobe is attenuated. The position of the maximum may move from the correct coordinates $(0,0)$ to the error coordinates (τ_e,ν_e). The peak does not initially lose its sharpness, but eventually the resolution loss becomes stronger because the main lobe widens. The image not only continues to fade as phase noise is further increased, but the sharpness is dissipated. While the desired image is fading, undesired background clutter is increasing because of spurious sidelobes caused by the phase noise.

The effect of phase noise can be measured in many ways. Depending on the application, various expectations such as $E[\chi_c(0,0)]$, $E[|\chi_c(0,0)|]$, $E[|\chi_c(0,0)|^2]$, $E[\tau_e]$, $\text{var}[\tau_e]$, $E[\nu_e]$, or $\text{var}[\nu_e]$ may be considered. Another significant effect is the effective loss in signal-to-noise ratio, defined as

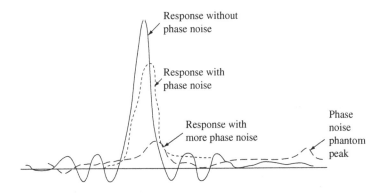

Figure 15.4 Illustrating the effects of phase noise

$$L = \frac{\mathrm{E}[\chi_c(0,0)]}{\chi_c(0,0)}.$$

Because $\chi_c(\tau, \nu)$ is related nonlinearly to $\theta(t)$, the first and second moments of $\chi_c(\tau, \nu)$ depend on all moments of $\theta(t)$. A complete statistical analysis is not possible when only the first and second moments of $\theta(t)$ are known. An analysis is practical when the noise is gaussian or when the phase noise is small. When θ is small, a small-angle approximation to the complex exponential can be used to linearize the relationship. When θ is gaussian, the first two moments determine all other moments..

15.6 Effect of Phase Distortion

A random phase error in the frequency domain appears as a complex exponential multiplying the Fourier transform of the received complex baseband signal. It is called *phase distortion*. Phase distortion may set a limit on the maximum bandwidth that a single signal can occupy.

Let $s(t)$ be a pulse, possibly complex with Fourier transform $S(f)$. The pulse $s(t)$ contaminated by phase distortion in the frequency domain is given by

$$V(f) = S(f) e^{j\theta(f)}.$$

The phase distortion $\theta(f)$ is a frequency-dependent phase shift that may be introduced by phase errors in the transmission medium or in the transfer function of a filter that the pulse has passed through. Phase distortion can be studied in a way that is parallel to the study of phase noise.

Only time-domain phase noise is studied in this chapter. Some of the lessons learned from studying phase noise can be used to study phase distortion.

15.7 Array Errors

A uniform pulse train consisting of N equispaced, identical pulses centered at the origin has the Fourier transform

$$\begin{aligned} P(f) &= S(f) \frac{\sin N\pi f T}{\sin \pi f T} \\ &= S(f) \mathrm{dirc}_N(fT). \end{aligned}$$

The dirichlet function $\mathrm{dirc}_N t$ was studied in Section 2.3. This section studies the effect of timing errors in the pulse spacing on the dirichlet function. Let

$$p(t) = \sum_{\ell=0}^{N-1} s(t - \ell T + T_{\ell e} + \tfrac{1}{2}(N-1)T),$$

where $T_{\ell e}$ is a small error in the time of the ℓth pulse. Then

$$P(f) = S(f)e^{j\pi f(N-1)T} \sum_{\ell=0}^{N-1} e^{-j2\pi f\ell T} e^{-j2\pi f T_{\ell e}}.$$

Now take the error $T_{\ell e}$ to be zero-mean gaussian random variables with variance σ_T^2. The expectation of $P(f)$ is

$$E[P(f)] = S(f)e^{j2\pi f(N-1)T} \sum_{\ell=0}^{N-1} e^{-j2\pi f\ell T} E[e^{-j2\pi f T_{\ell e}}].$$

To evaluate the expectation, we use Theorem 15.2.1 to write

$$E[P(f)] = S(f)e^{j\pi f(N-1)T} e^{-\sigma_\theta^2/2} \sum_{\ell=0}^{N-1} e^{-j2\pi f\ell T}$$

$$= [S(f)\mathrm{dirc}_N(fT)]e^{-2\pi^2 f^2 \sigma_T}.$$

The grating lobes are attenuated in expectation by $e^{-2\pi^2 f^2 \sigma_T}$.

Problems

15.1 The variance in the value of the peak of the ambiguity function $\chi(\tau,\nu)$ due to phase noise is approximated by

$$\sigma^2 \approx \int_{-\infty}^{\infty} N_\theta(f)|W(f)|^2 df,$$

where $W(f)$ is the Fourier transform of $|s(t)|^2$.

a. For the pulse train $s(t)$ given by

$$s(t) = \left(\frac{T_r}{T}\right)^{\frac{1}{2}} \sum_{n=0}^{N-1} p(t - nT_r),$$

where $p(t)$ is a rectangular pulse of width T much smaller than T_r. Find an approximate expression for σ^2.

b. Suppose that the spectrum of the phase noise is confined to low frequencies, that is, $N_\theta(f)$ is approximately zero for large frequencies. Show that σ^2 could be calculated just as easily by treating $s(t)$ as a single rectangular pulse of width approximately NT_r and height 1. In terms of T_r and T, what does it mean to say that $N_\theta(f)$ is confined to low frequencies?

15.2 A binary test is to decide between the two possible signs of $\widetilde{s}(t)$ given by

$$\widetilde{s}(t) = \pm\mathrm{rect}(t)\cos 2\pi f_0 t,$$

based on the observation

$$\widetilde{v}(t) = \widetilde{s}(t)e^{j2\pi\theta(t)},$$

where $\theta(t)$ is a gaussian process with the power density spectrum $N(f)$.

a. Give a test for determining the sign of $\widetilde{s}(t)$ from the received signal $\widetilde{v}(t)$.
b. Give an approximate formula for the probability of error as a function $N(f)$.

Notes

The study of the effect of random phase errors has a long history, with early work on this topic coming primarily from the field of optics, usually described under the term *coherence* and often studied and quantified in the frequency domain. Many classical experiments of physics such as the Michelson interferometer and the Michelson–Morley experiment involve the notion of coherence.

The coherence of two waves describes the ability of those two waves to interfere, either constructively or destructively. The Cornu spiral is named after French physicist Marie Alfred Cornu (1841–1902). Cornu introduced this construction within his studies of the effect of the phase approximation on the analysis of Fresnel diffraction.

Bibliography

Abbe, E. (1873) "Beitrage zur Theorie des Mikroskops und der Mikroskopischen Wahrnehmung," *Archiv für Mikroskopische Anatomie.*, vol. 9, pp. 413–468.

Ables, J. C. (1968) "Fourier Transform Photography: A New Method for X-ray Astronomy," *Proceedings of the Astronomical Society of Australia*, vol. 4, pp. 172–173.

Ali, S. M. and Silvey, S. D. (1966) "A General Class of Coefficients of Divergence of One Distribution From Another," *Journal of the Royal Statistical Society*, vol. 28, pp. 131–142.

Arimoto, S. (1972) "An Algorithm for Computing the Capacity of an Arbitrary Discrete Memoryless Channel," *IEEE Transactions on Information Theory*, vol. IT-18, pp. 14–20.

Armitage, J. D. and Lohmann, A. W. (1965) "Character Recognition by Incoherent Spatial Filtering," *Applied Optics*, vol. 4, pp. 461–467, 1965.

Armstrong, E. H. (1936) "A Method of Reducing Disturbances in Radio Signaling by a System of Frequency Modulation," *Proceedings of the IRE*, vol. 24, pp. 689–740.

Arridge, S. R. and Schotland, J. C. (2009) "Optical Tomography: Forward and Inverse Problems, *Inverse Problems*, vol. 28, pp. 1–29

Auslander, L. and Tolimieri, R. (1985) "Radar Ambiguity Functions and Group Theory," *Siam Journal of Mathematical Analysis*, vol. 16, pp. 577–601.

Ayers, G. R. and Dainty, J. C. (1982) "Iterative Blind Deconvolution Method and its Applications," *Applied Optics*, vol. 21, pp. 2758–2769.

Bader, T. R. (1960) "Wideband Signal Processing for Emitter Location," *Proceedings of SPIE Symposium, Advances in Optical Information Processing*, vol. 1296.

Baggeroer, A. B., Kuperman, W. A., and Mikhalevsky, P. N. (1993) "An Overview of Matched Field Methods in Ocean Acoustics," *IEEE Journal of Oceanic Engineering*, vol. OE-18, pp. 401–424.

Banerjee, P. P. (1985) "A Simple Derivation of the Fresnel Diffraction Formula," *Proceedings of the IEEE*, vol. 73, pp. 1859–1860.

Banzhaf, S. and Waldschmidt, C. (2021) "Phase-Code-Based Modulation for Coherent Lidar," *IEEE Transactions on Vehicular Technology*, vol. VT-70, pp. 9886–9897.

Barabell, A. J. et al. (1984) "Performance Comparison of Superresolution Array Processing Algorithms," MIT Lincoln Laboratory Project Report, TST-72.

Barbarosa, S. and Farina, A. (1990) "A Novel Procedure for Detecting and Focusing Moving Objects with SAR Based on the Wigner-Ville Distribution," *Proceedings of the 1990 IEEE International Radar Conference*, pp. 44–50.

Bar-Shalom, Y. (1978) "Tracking Methods in a Multitarget Environment," *IEEE Transactions on Automatic Control*, vol. AC-23, pp. 618–626.

Bates, R. H. T. (1984) "Uniqueness of Solutions to Two-dimensional Fourier Phase Problems of Localized and Positive Images," *Computer Vision, Graphics, and Image Processing*, vol. 24, pp. 205–207.

Bates, R. H. T. and McDonnell, M. J. (1986) *Image Restoration and Reconstruction*, Oxford University Press, Oxford.

Bello, P. (1960) "Joint Estimation of Delay, Doppler and Doppler Rate," *IRE Transactions on Information Theory*, vol. IT-6, pp. 330–341.

Bernfeld, M. (1984) "Chirp Doppler Radar," *Proceedings of the IEEE*, vol. 72, pp. 540–541.

Bertsekas, D. P. (1988) "The Auction Algorithm; A Distributed Relaxation Method for the Assignment Problem," *Annuals of Operations Research*, vol. 14, pp. 105–123.

Bhargava, R. (2012) "Infrared Spectroscopic Imaging: The Next Generation," *Applied Spectroscopy*, vol. 66. pp. 1091–1120.

Blackman, S. S. and Poopoli, R. (1999) *Design and Analysis of Modern Tracking Systems*, Artech House, Dedham, MA.

Blahut, R. E. (1972) "Computation of Channel Capacity and Rate Distortion Functions," *IEEE Transactions on Information Theory*, vol. IT-18, pp. 460–473.

Bloch, F. (1946) "Nuclear Induction," *Physics Review*, vol. 70, pp. 460–474.

Boerner, W. M., Ho, C.-M., and Foo, B.-Y. (1981) "Use of Radon's Projection Theory in Electromagnetic Inverse Scattering," *IEEE Transactions on Antennas and Propagation*, vol. AP-29, pp. 336–341.

Bojarski, N. M. (1982) "A Survey of Physical Optics Inverse Scattering Identity," *IEEE Transactions on Antennas and Propagation*, vol. AP-30, pp. 980–989.

Born, M. and Wolf, E. (1997) *Principles of Optics*, Cambridge University Press, Cambridge.

Bouman, C. A. (2022) *Foundations of Computational Imaging*, SIAM, Philadelphia, PA.

Bouman, C. A. (2023) "Digital Image Processing Class Notes," Purdue University.

Bourgeois, F. and La Salle, J. C. (1971) "An Extension of the Munkres Algorithm for the Assignment Problem to Rectangular Matrices," *Communications of the ACM*, vol. 14, pp. 802–806.

Bracewell, R. N. (1956) "Strip Integration in Radio Astronomy," *Australian Journal of Physics*, vol. 9, pp. 198–217.

Bracewell, R. N. (1958a) "Restoration in the Presence of Errors," *Proceedings of the IRE*, vol. 46, pp. 106–111.

Bracewell, R. N. (1958b) "Radio Interferometry of Discrete Sources," *Proceedings of the Institute of Radio Engineers*, vol. 46, pp. 97–105.

Bracewell, R. N. (1995) *Two-Dimensional Imaging*, Prentice-Hall, Upper Saddle River, NJ.

Bracewell, R. N. and Riddle, A. C. (1967) "Inversion of Fan-Beam Scans in Radio Astronomy," *The Astrophysics Journal*, vol. 150, pp. 427–434.

Bragg, W. L. (1929) "The Determination of Parameters in Crystal Structures by Means of Fourier Analysis," *Proceedings of the Royal Society, A*, vol. 123, pp. 537–559.

Bragg, W. L. (1942) "The X-Ray Microscope," *Nature*, vol. 149, pp. 470–471.

Bresler, Y. and Macouski, A. (1987) "Three-Dimensional Reconstruction from Projections with Incomplete and Noisy Data by Object Estimation," *IEEE Transactions on Acoustics, Speech, and Signal Processing*, vol. ASSP-35, pp. 1139–1152.

Bricogne, G. (1984) "Maximum Entropy and the Foundations of Direct Methods," *Acta Cystallographia*, vol. A40, pp. 410–445.

Brown, W. M. and Fredericks, R. (1969) "Range-Doppler Imaging with Motion Through Resolution Cells," *IEEE Transactions on Aerospace and Electronic Systems*, vol. AES-5, pp. 98–102.

Brown, W. M. and Palermo, C. J. (1963) "Effects of Phase Errors on the Ambiguity Function," *IEEE International Convention Record*, Pt. 4, pp. 118–123.

Brown, W. M. and Porcello, L. J. (1969) "An Introduction to Synthetic Aperture Radar," *IEEE Spectrum*, p. 52.

Bruck, Y. M. and Sodin, L. G. (1979) "On the Ambiguity of the Image Reconstruction Problem," *Optical Communications*, vol. 30, pp. 304–308.

Bruyant, P. P. (2002) "Analytic and Iterative Reconstruction Algorithms in SPECT," *The Journal of Nuclear Medicine*, vol. 43, pp. 1343–1358

Budinger, T. F. (1980) "Physical Attributes of Single-Photon Tomography," *Journal of Nuclear Medicine*, vol. 21, pp. 579–592.

Burckhardt, C. B. (1978) "Speckle in Ultrasound B-Mode Scans," *IEEE Transactions on Sonics and Ultrasonics*, vol. SU-25, pp. 1–6.

Burg, J., Luenberger, D., and Weaver, D. (1982) "Estimation of Structured Covariance Matrices," *Proceedings of the IEEE*, vol. 70, pp. 963–974.

Butler, J. and Lowe, R. (1961) "Beam-forming Matrix Simplifies Design of Electronically Scanned Antennas," *Electronic Design*, vol. 9, pp. 170–173.

Byrne, C. L. (1993) "Iterative Image Reconstruction Algorithms Based on Cross-Entropy Minimization," *IEEE Transactions on Image Processing*, vol. IP-2, pp. 96–103.

Campbell, D. B. (1971) *Radar Interferometric Observations of Venus*, Ph.D. dissertation, Cornell University.

Capon, J. (1964) "Optimum Weighting Functions for the Detection of Sampled Signals in Noise," *IRE Transactions on Information Theory*, vol. IT-10, pp. 152–159.

Capon, J. (1969) "High-Resolution Frequency-Wavenumber Spectrum Analysis," *Proceedings of the IEEE*, vol. 57, pp. 1408–1418.

Carney, P. S. and Schotland, J. C. (2003) "Near-Field Tomography," *Inside Out: Inverse Problems*, MSRI Publications, vol. 47, pp. 133–168.

Chen, C. C. and Andrews, H. C. (1980) "Multi-Frequency Imaging of Radar Turntable Data," *IEEE Transactions on Aerospace and Electronic Systems*, vol. AES-16, pp. 15–22.

Chen, V. C. (1994) "Radar Ambiguity Function, Time-Varying Matched Filter, and Optimum Wavelet Correlator," *Optical Engineering*, vol. 33, pp. 2212–2217.

Chen, V. C. and Qian, S. (1998) "Joint Time-Frequency Transform for Radar Range-Doppler Imaging," *IEEE Transactions on Aerospace and Electronic Systems*, vol. AES-34, pp. 486–499.

Chen, Z. Zhao, Y. Srinivas, S. M., Nelson, J. S., Prakash, N., and Frostig, R. D. (1999) "Optical Doppler Tomography," *IEEE Journal of Selected Topics in Quantum Electronics*, vol. QE-5, pp. 1134–1141.

Chestnut, P. C. (1982) "Emitter Location Accuracy Using TDOA and Differential Doppler," *IEEE Transactions on Aerospace and Electronic Systems*, vol. AES-18, pp. 214–218.

Cho, Z. H. (1974) "General View on 3-4 Image Reconstruction and Computerized Transverse Axial Tomography," *IEEE Transactions on Nuclear Science*, vol. NS-21, pp. 44–71.

Choi, K., Lanterman, A. D., and Raich, R. (2006) "Convergence of the Schulz-Snyder Phase Retrieval Algorithm to Local Minima," *Journal of the Optical Society of America*, vol. 23, pp. 1835–1845.

Choi, H. and Munson, Jr., D. C. (1998) "Direct Fourier Reconstruction in Tomography and Synthetic Aperture Radar," *International Journal on Imaging Systems Technology*, vol. 9, pp. 1–13.

Christiansen, W. N. and Warburton, J. A. (1955) "The Distribution of Brightness over the Solar Disk at a Wavelength of 21 cm III. The Quiet Sun–Two Dimensional Observations," *Australian Journal of Physics*, vol. 8, pp. 474–486.

Classen, T. A. C. M. and Mecklenbrauker, W. F. (1980) "The Wigner distribution: A Tool for Time-Frequency Signal Analysis," *Phillips Journal of Research*, vol. 35, Parts 1, 2, and 3, pp. 217–250, 276–300, and 372–389.

Clem, T. R. (1995) "Superconducting Magnetic Sensors Operating from a Moving Platform," *IEEE Transactions on Applied Supercomputing*, vol. AS-5, pp. 2124–2128.

Coble, M. R. (1992) *High Resolution Radar Imaging of a Rotating Sphere*, M.S. Dissertation, University of Illinois, Urbana, IL.

Cochran, W., Crick, F. H. C., and Vand, V. (1952) "Structure of Synthetic Polypeptides.I. The Transform of Atoms on a Helix," *Acta Cystallographia*, vol. 5, pp. 581–586.

Cook, C. and Bernfeld, M. (1967) *Radar Signals: An Introduction to Theory and Applications*, Academic Press, New York.

Cormack, A. M. (1963) "Representation of a Function by its Line Integrals, with some Radiological Applications," *Journal of Applied Physics*, vol. 34, pp. 2722–2727.

Costas, J. P. (1975) "Medium Constraints on Sonar Design and Performance," *EASCON Convention Record*, pp. 68A–68L.

Costas, J. P. (1984) "A Study of a Class of Detection Waveforms Having Nearly Ideal Range-Doppler Ambiguity Properties," *Proceedings of the IEEE*, vol. 72, pp. 996–1009.

Crowther, R. A., DeRosier, D. J., and Klug, A. (1970) "The Reconstruction of a Three Dimensional Structure from Projections and Its Application to Electron Microscopy," *Proceedings of the Royal Society of London*, vol. A317, pp. 319–340.

Csiszár, I. (1991) "Why Least Squares and Maximum Entropy–An Axiomatic Approach to Inverse Problems," *Annals of Statistics*, vol. 19, pp. 2033–2066.

Csiszár, I. and Tusnady, G. (1984) "Information Geometry and Alternating Decisions," *Statistical Decisions*, Supplementary Issue #1, pp. 205–207.

Cutrona, L. J. (1975) "Comparison of Sonar System Performance Achievable Using Synthetic-Aperture Techniques with the Performance Achievable with More Conventional Means," *Journal of the Acoustical Society of America*, vol. 58, pp. 336–348.

Cutrona, L. J., Leith, E. N., Palermo, C. J. and Porcello, L. J. (1960) "Optical Data Processing and Filtering Systems," *IRE Transactions on Information Theory*, vol. IT-6, pp. 386–400.

Cutrona, L. J., Vivian, W. E., Leith, E. N. and Hall, G. O. (1961) "A High-Resolution Radar Combat-Surveillance System," *IRE Transactions Military Electronics*, vol. MIL-5, pp. 127–131.

Dainty, J. C. and Fienup, J. R. (1987) "Phase Retrieval and Image Reconstruction for Astronomy," *Image Recovery: Theory and Applications*, pp. 231–275.

Davila, J. M. (2011) "High Resolution Solar Imaging with a Photon Sieve," *SPIE Optical Engineering and Applications*, pp. 81480O-81480O.

DeBuda, R. (1970) "Signals That Can Be Calculated from Their Ambiguity Function," *IEEE Transactions on Information Theory*, vol. IT-16, pp. 195–202.

DeGraaf, S. R. (1998) "SAR Imaging via Modern 2D Spectral Estimation Methods," *IEEE Transactions on Image Processing*, vol. IP-7, pp. 729–761.

Delong, D. F. (1983) "Multiple Signal Direction Finding with Thinned Linear Arrays," MIT Lincoln Laboratory Project Report, TST-68.

Dempster, A. D., Laird, N. M., and Rubin, D. B. (1977) "Maximum Likelihood from Incomplete Data via the EM Algorithm," *Journal of the Royal Statistical Society*, vol. B39, pp. 1–38.

DeRosier, D. J. and Klug, A. (1968) "Reconstruction of Three-Dimensional Structures from Electron Micrographs," *Nature*, vol. 217, pp. 130–134.

Dicke, R. H. (1968) "Scatter-Hole Cameras for X-Rays and Gamma Rays," *Journal of Astrophysics*, vol. 153, pp. L101–L106.

Dines, K. A. and Lytle, R. J. (1979) "Computerized Geophysical Tomography," *Proceedings of the IEEE*, vol. 67, pp. 1065–1073.

Dolph, C. L. (1946) "A Current Distribution for Broadband Arrays Which Optimizes the Relationship Between Beam Width and Sidelobe Level," *Proceedings of the IRE*, vol. 34, pp. 335–348.

Donoho, D. L., Johnstone, I. M., Hoch, J. C., and Stern, A. S. (1992) "Maximum Entropy and the Nearly Black Object," *Journal of the Royal Statistical Society B*, pp. 41–81.

Dowski, E. and Cathey, W. (1995) "Extended Depth of Field Through Wavefront Coding," *Applied Optics*, vol. 34, pp. 1859–1866.

Dragone, C. (1987) "Use of Imaging with Spatial Filtering in Reflector Antennas," *IEEE Transactions on Antennas and Propagation*, vol. AP-35, pp. 258–267.

Duane, W. (1925) "The Calculation of the X-Ray Diffracting Power at Points in a Crystal," *Proceedings of the National Academy of Science*, vol. 11, pp. 489–493.

Duffieux, P. M. (1946) "L'Integrale de Fourier et ses Applications a L'optique," in *Faculte des Sciences*, Besancon, Paris.

Dugundji, J. (1958) "Envelopes and Pre-envelopes of Real Waveforms," *IRE Transactions on Information Theory*, vol. IT-4, pp. 53–57.

Durnin, J. (1987) "Exact Solutions for Nondiffracting Beams I. The Scalar Theory," *Journal of the Optical Society of America*, vol. 4, pp. 651–654.

Dziewonski A. and Woodhouse, J. (1987) "Global Images of the Earth's Interior," *Science*, vol. 236, pp. 37–48.

Elias, P. (1953) "Optics and Communication Theory," *Journal of the Optics Society of America*, vol. 43, pp. 229–232.

Elias, P., Grey, D. S., and Robinson, D. Z. (1952) "Fourier Treatment of Optical Processes," *Journal of the Optical Society of America*, vol. 42, pp. 127–134.

Emerson, R. C. (1954) "Some Pulsed Doppler, MTI, and AMTI Techniques," Rand Corp. Rep. R-274.

Emslie, A. G. (1946) "Moving Target Indication on MEW," MIT Radiation Laboratory Report 1080.

Ermert, H. and Karg, R. (1979) "Multi-Frequency Acoustical Holography," *IEEE Transactions on Sonics and Ultrasonics*, vol. SU-26, pp. 279–286.

Ewald, P. P. (1921) "Die Berechnung Optischer und Electrostatischer Gitterpotentiale," *annalen der Physik*, vol. 369, pp. 253–287.

Fan, H. and Sanz, J. L. C. (1985) "Comments on 'Direct Fourier Reconstruction in Computer Tomography,'" *IEEE Transactions on Acoustics, Speech, and Signal Processing*, vol. ASSP-33, pp. 446–449.

Feig, E. and Grünbaum, F. A. (1986) "Tomographic Methods in Range-Doppler Radar," *Inverse Problems*, vol. 2, pp. 185–195.

Fenimore, E. E. and Cannon, T. M. (1978) "Coded Aperture Imaging with Uniformly Redundant Arrays," *Applied Optics*, vol. 17, pp. 337–347.

Fienup, J. R. (1978) "Reconstruction of an Object from the Modulus of its Fourier Transform," *Optics Letters*, vol. 3, pp. 27–29.

Fienup, J. R. (1981) "Reconstruction and Synthesis Applications of an Iterative Algorithm," *SPICE Transformations in Optical Signal Processing*, vol. 373, pp. 147–160.

Fienup, J. R. (1982) "Phase Retrieval Algorithms: A Comparison," *Journal of Applied Optics*, vol. 21, pp. 2758–2769.

Fisher, M. L. (1981) "The Lagrangian Relaxation Method for Solving Integer Programming Problems," *Management Science*, vol. 27, pp. 1–18.

Fisher, R. A. (1925) "Theory of Statistical Estimation," *Proceedings of the Cambridge Philosophical Society*, vol. 22, pp. 700–725.

Fomalont, E. B. (1973) "Earth-Rotation Aperture Synthesis," *Proceedings of the IEEE*, vol. 61, pp. 1211–1218.

Foucault, L. (1858) *Ann. de l'Observ. Imp de Paris*, vol. 5, p. 203.

Frahm, C. P. (1972) "Inversion of the Magnetic Field Gradient Equations for a Magnetic Dipole Field," Naval Coastal Systems Laboratory Informal Report 135-72.

Franklin, R. E. and Klug, A. (1955) "The Splitting of Layer Lines in X-Ray Fibre Diagrams of Helical Structures: Application to Tobacco Mosaic Virus," *Acta Crystallographia*, vol. 8, pp. 777–780.

Frieze, A. M. and Yadegar, J. (1981) "An Algorithm for Solving 3-Dimensional Assignment Problems with Application to Scheduling a Teaching Practice," *Journal of the Operational Research Society*, vol. 32, pp. 989–995.

Friis, H. T. (1946) "A Note on a Simple Transmission Formula," *Proceedings of the IRE*, vol. 34, pp. 254–256.

Gaarder, N. T. (1968) "Scattering Function Estimation," *IEEE Transactions on Information Theory*, vol. IT-14, pp. 684–693.

Gabor, D. (1946) "Theory of Communication," *Journal of the Institute of Electrical Engineers*, pt. III, vol. 93, pp. 429–441.

Gabor, D. (1948) "A New Microscope Principle," *Nature*, vol. 161, pp. 777–778.

Gabor, D. (1949) "Microscopy by Reconstructed Wavefronts, Part I," *Proceedings of the Royal Society*, vol. A197, pp. 454–487.

Gabor, D. (1951) "Microscopy by Reconstructed Wavefronts, Part II," *Proceedings of Physical Society*, vol. B64, pp. 449–469.

Geoffrion, A. M. (1974) "Lagrangian Relaxation for Integer Programming," in M. L. Balinski, ed., *Mathematical Programming Study 2: Approaches to Integer Programming*, North Holland Publishing Company, Amsterdam.

Gerchberg, R. W. and Saxton, W. O. (1972) "A Practical Algorithm for the Determination of Phase from Image and Diffraction Plane Pictures," *Optics*, vol. 35, pp. 237–246.

Ghiglia, D. C., Romero, L. A., and Mastin, G. A. (1996) "Systematic Approach to Two-Dimensional Blind Deconvolution by Zero-Sheet Separation," *Journal of the Optical Society of America*, vol. 10, pp. 1024–1036.

Gilmour, G. A. (1978) "Synthetic Aperture Side-Looking Sonar System," US. Patent 4 088 978.

Goldstein, G. B. (1973) "False Alarm Regulation in Log-Normal and Weibull Clutter," *IEEE Transactions on Aerospace and Electronic Systems*, vol. AES-9, pp. 84–92.

Golay, M. J. E. (1961) "Complementary Series," *IRE Transactions on Information Theory*, vol. IT-7, pp. 82–87.

Golomb, S. W. and Taylor, H. (1982) "Two-Dimensional Synchronization Patterns for Minimum Ambiguity," *IEEE Transactions on Information Theory*, vol. IT-28, pp. 600–604.

Golomb, S. W. and Taylor, H. (1984) "Constructions and Properties of Costas Arrays," *Proceedings of the IEEE*, vol. 72, pp. 1143–1163.

Goodman, J. W. (1968) *Introduction to Fourier Optics*, McGraw-Hill, New York.

Goodman, J. W. (1976) "Some Fundamental Properties of Speckle," *Journal of the Optical Society of America*, vol. 66, pp. 1145–1150.

Goodman, J. W. (1985) *Statistical Optics*, John Wiley and Sons, New York.

Goodman, N. R. (1963) "Statistical Analysis Based on a Certain Multivariate Complex Gaussian Distribution," *Annals of Mathematical Statistics*, vol. 34, pp. 152–177.

Gorden, R. (1974) "A Tutorial on ART (Algebraic Reconstruction Techniques)," *IEEE Transactions on Nuclear Science*, vol. NS-21, pp. 78–93.

Graham, L. C. (1974) "Synthetic Interferometer Radar for Topographic Mapping," *Proceedings of the IEEE*, vol. 62, pp. 763–768.

Green, Jr., P. E. (1962) "Radar Astronomy Measurement Techniques," Technical Report No. 282, Lincoln Laboratory, MIT.

Grenander, U. (1981) *Abstract Inference*, Wiley, New York.

Green, T. J. and Shaprio, J. H. (1994) "Maximum-Likelihood Laser Radar Range Profiling with the Expectation-Maximization Algorithm," *Optical Engineering*, vol. 33, pp. 865–873.

Grünbaum, F. A. (1984) "A Remark on Radar Ambiguity Functions," *IEEE Transactions on Information Theory*, vol. IT-30, pp. 126–127.

Gull, S. F. and Daniell, G. J. (1979) "The Maximum Entropy Method," in C. van Schooneveld, ed., *Image Formation from Coherence Functions in Astronomy*, pp. 219–225, D. Reidel, Dordrecht, Holland.

Hahn, E. L. (1960) "Detection of Sea-Water Motion by Nuclear Precession," *Journal of Geophysical Research*, vol. 65, pp. 776–777.

Hansen, V. G. (1970) "Detection Performance of Some Non-Parametric Rank Tests and an Application to Radar," *IEEE Transactions on Information Theory*, vol. IT-16, pp. 609–618.

Harker, D. and Kasper, J. S. (1948) "Phases of Fourier Coefficients Directly from the Crystal Diffraction Data," *Acta Cystallographia*, vol. 1, pp. 70–75.

Harris, F. J. (1978) "On the Use of Windows for Harmonic Analysis with the Discrete Fourier Transform," *Proceedings of the IEEE*, vol. 66, pp. 51–83.

Hauptman, H. (1986) "The Direct Methods of X-Ray Crystallography," *Science*, vol. 233, pp. 178–183.

Hauptman, H. (1991) "The Phase Problem of X-Ray Crystallography," *Reports on Progress in Physics*, vol. 54, pp. 1427–1454.

Hauptman, H. and Karle, J. (1953) *Solution of the Phase Problem I: The Centrosymmetric Crystal*, American Crystallographic Association, Monograph 3, Polycrystal Book Service, Western Springs, IL.

Hayes, M. H. (1982) "The Reconstruction of a Multidimensional Sequence from the Phase or Magnitude of its Fourier Transform," *IEEE Transactions on Acoustics, Speech, and Signal Processing*, vol. ASSP-30, pp. 140–154.

Hayes, M. H. and McClellan, J. H. (1982) "Reducible Polynomials in More Than One Variable," *Proceedings of the IEEE*, vol. 70, no. 2, pp. 197–198.

Hayes, M. P. and Gough, P. T. (2009) "Synthetic Aperture Sonar: A Review of Current Status," *IEEE Journal on Oceanic Engineering*, vol. 34, pp. 207–224.

Helleseth, T. and Kumar, P. V. (1998) "Sequences With Low Correlation," *Handbook of Coding Theory*, North Holland, pp. 1765–1855.

Helstrom, C. W. (1960) *Statistical Theory of Signal Detection*, Pergamon Press, Oxford.

Hero, A. O. and Fessler, J. A. (1995) "Convergence in Norm for Alternating Expectation-Maximization (EM) Type Algorithms," *Statistica Sinica*, vol. 5, pp. 41–54.

Hirosawa, H. and Kobayashi, N. (1986) "Terrain Height Measurement by Synthetic Aperture Radar with an Interferometer," *International Journal on Remote Sensing*, vol. 7, pp. 339–348.

Högbom, J. A. (1974) "Aperture Synthesis with a Nonregular Distribution of Interferometer Baselines,"*Astronomy and Astrophysics Supplement Series*, vol. 15, pp. 417–426.

Hounsfield, C. N. (1972) "A Method of and Apparatus for Examination of a Body by Radiation such as X-Ray or Gamma Radiation," British Patent Number 1283915, London.

Hounsfield, C. N. (1973a) "Method and Apparatus for Measuring X or-Radiation Absorption or Transmission at Plural Angles and Analyzing the Data," US Patent 3 778 614.

Hounsfield, C. N. (1973b) "Computerized Transverse Axial Scanning (Tomography): Part 1. Description of System," *British Journal of Radiology*, vol. 46, pp. 1016–1022.

Ishii, M., Leigh, J. and Schotland, J. (1995) "Optical Diffusion Imaging Using a Direct Inversion Method," *Physics Review E*, vol. 52, p. 4361.

Izraelevitz, D. and Lim, J. S. (1987) "A New Direct Algorithm for Image Reconstruction from Fourier Transform Magnitude," *IEEE Transactions on Acoustics, Speech, and Signal Processing*, vol. ASSP-35, pp. 511–519.

Jackson, P. L. (1965) "Diffractive Processing of Geophysical Data," *Applied Optics*, vol. 4, pp. 419–427.

Jain, A. K. and Ranganath, S. (1981) "Applications of Two Dimensional Spectral Estimation in Image Restoration," *Proceedings of the IEEE International Conference on Acoustics, Speech, and Signal Processing*, pp. 1113–1116.

Jakowitz, Jr., C. V. and Thompson, P. A. (1992) "A Three-Dimensional Tomographic Formulation for Spotlight Mode–Synthetic Aperture Radar," Sandia National Laboratories, Albuquerque, NM.

Jakowatz, C. V. and Thompson, P. A. (1995) "A New Look at Spotlight Mode Synthetic Aperture Radar as Tomography," *IEEE Transactions on Image Processing*, vol. IP-4, pp. 699–703.

Jaynes, E. T. (1957) "Information Theory and Statistical Mechanics," *Physics Review*, vol. 106, pp. 620–630.

Jennison, R. C. (1958) "A Phase Sensitive Interferometer Technique for the Measurement of the Fourier Transforms of Spatial Brightness Distributions of Small Angular Extent," *Monthly Notices of the Royal Astronomical Society*, vol. 118, pp. 276–284.

Jensen, H., Graham, L. C., Porcello, L. J., and Leith, E. N. (1977) "Side-Looking Airborne Radar," *Scientific American*, pp. 84–95.

Johnson, D. H. (1982) "The Application of Spectral Estimation Methods to Bearing Estimation Problems," *Proceedings of the IEEE*, vol. 70, pp. 1018–1028.

Kak, A. C. (1979) "Computerized Tomography with X-Ray, Emission, and Ultra-Sound Sources," *Proceedings of the IEEE*, vol. 67, pp. 1245–1271.

Kak, A. C. and Slaney, M. (1988) *Principles of Computerized Tomographic Imaging*, IEEE Press, New York.

Kalman, R. E. (1960) "A New Approach to Linear Filtering and Prediction Problems," *Transactions of the American Society of Mechanical Engineers*, Series D, Journal of Basic Engineering, vol. 82D, pp. 35–46.

Kalman, R. E. and Bucy, R. S. (1961) "New Results in Linear Filtering and Prediction Theory," *Transactions of the American Society of Mechanical Engineers*, Series D, Journal of Basic Engineering, vol. 83, pp. 95–108.

Karle, J. and Hauptman, H. (1950) "The Phases and Magnitudes of the Structure Factors," *Acta Crystallographia*, vol. 3, pp. 181–187.

Kashyap, R. L. and Rao, A. R. (1976) *Dynamic Stochastic Models from Empirical Data*, Academic, New York.

Kay, I. and Silverman, R. A. (1957) "On the Uncertainty Relation for Real Signals," *Information and Control*, vol. 1, pp. 64–75.

Kell, R. R. (1965) "On the Derivation of Bistatic RCS from Monostatic Measurements," *Proceedings of the IEEE*, vol. 53, pp. 983–988.

Khintchine, A. (1934) "Korrelationstheorie der Stationaren Stochastischen Prozesse, *Mathematische Annalen*, vol. 109, pp. 604-615.

Klauder, J. R. (1960) "The Design of Radar Signals Having Both High Range Resolution and High Velocity Resolution," *Bell System Technical Journal*, vol. 39, pp. 809–820.

Klauder, J. R., Price, A. C., Darlington, S., and Albersheim, W. J. (1960) "The Theory and Design of Chirp Radars," *Bell System Technical Journal*, vol. 39, pp. 745–808.

Klug, A., Crick, F. H. C., and Wyckoff, H. W. (1958) "Diffraction by Helical Structures," *Acta Crystallographia*, vol. 11, pp. 199–213.

Klug, A. (1978) "Image Analysis and Reconstruction in the Electron Microscopy of Biological Macromolecules," *Chemica Scripta*, vol. 14, pp. 245–256.

Klug, A. and Berger, J. E. (1964) "An Optical Method for the Analysis of Periodicities in Electron Micrographs, and Some Observations on the Mechanism of Negative Staining," *Journal of Molecular Biology*, vol. 10, pp. 565–569.

Knox, K. T. and Thompson, B. J. (1974) "Recovery of Images from Atmospherically Degraded Short-Exposure Photographs," *Astrophysics Journal Letters*, vol. 193, pp. L45–L48.

Kotel'nikov, V. A. (1933) "O Propusknoj Sposobnosti 'efira'i provoloki v elektrosvjazi," ("On the transmission capacity of 'ether' and wire in electro-communications"), *First All-Union Conference on Questions of Communications*.

Kozma, A. and Kelly, D. L. (1965) "Spatial Filtering for Detection of Signals Submerged in Noise," *Applied Optics*, vol. 4, p. 387.

Kozma, A. (1966) "Photographic Recording of Spatially Modulated Coherent Light," *Journal of the Optical Society of America*, vol. 56, pp. 428–432.

Kuhl, D. E. and Edwards, R. Q. (1963) "Image Separation Radioisotope Scanning," *Radiology*, vol 80, pp. 653–661.

Kuhn, H. W. (1955) "The Hungarian Method for the Assignment Problem, *Naval Research Logistics Quarterly*, vol. 2, pp. 83–97.

Kulkarni, M. D., Thomas, C. W., and Izatt, J. A. (1997) "Image Enhancement in Optical Coherence Tomography Using Deconvolution," *Electronics Letters*, vol. 33, pp. 1365–1467.

Kullback, S. (1968) *Information Theory and Statistics*, Wiley, New York.

Kumar, A., Welti, D. and Ernst, R. (1975) "NMR Fourier Zeugmatography," *Journal of Magnetic Resonance*, vol. 18, pp. 69–83.

Labeyrie, A. (1970) "Stellar Interferometry Methods," Annual Review of Astronomy and Astrophysics, vol. 16, pp. 77–102.

Lane, R. G. and Bates, R. H. T. (1987) "Automatic Multidimensional Deconvolution," *Journal of the Optical Society of America*, vol. 4, pp 180–188.

Lane, R., Fright, W., and Bates, R. (1978) "Direct Phase Retrieval," *IEEE Transactions on Acoustics, Speech, and Signal Processing*, vol. ASSP-35, pp 520–526.

Lange, K. and Carson, R. (1984) "EM Reconstruction Algorithms for Emission and Transmission Tomography," *Journal of Computer Assisted Tomography*, vol. 8, pp. 306–316.

Lanterman, A. D. (2000) "Statistical Radar Imaging of Diffuse and Specular Targets Using an Expectation-Maximization Algorithm," *Algorithms for Synthetic Aperture Radar Imagery VII*, SPIE Proceedings 4053, Orlando.

Lauterbur, P. C. (1973) "Image Formation by Induced Local Interactions: Examples Employing Nuclear Magnetic Resonance," *Nature*, vol. 242, pp. 190–191.

Leith, E. N. and Upatnieks, J. (1962) "Reconstructed Wavefronts and Communication Theory," *Journal of the Optical Society of America*, vol. 52, pp. 1123–1130.

Leith, E. N. and Upatnieks, J. (1964) "Wavefront Reconstruction with Diffused Illumination and Three-dimensional Objects," *Journal of the Optical Society of America*, vol. 54, pp. 1295–1301.

Lerner, R. M. (1958) "Signals with Uniform Ambiguity Functions," *IRE National Convention Record*, Part 4, pp. 27–33.

Lewis, R. M. (1969) "Physical Optics Inverse Diffraction," *IEEE Transactions on Antennas and Propagation*, vol. AP-17, pp. 308–314.

Liang, Z.-P. and Lauterbur, P. C. (2000) *Principles of Magnetic Resonance Imaging – A Signal Processing Perspective*, IEEE Press.

Lighthill, M. J. (1958) *Introduction to Fourier Analysis and Generalized Functions*, Cambridge University Press, Cambridge.

Lockwood, G., Talman, J., and Brunke, S. (1998) "Real-Time 3-D Ultrasound Imaging Using Sparse Synthetic Aperture Beamforming," *IEEE Transactions on Ultrasonics, Ferroelectronics, and Frequency Control*, vol. UFFC-45, pp. 1077–1087.

Lohmann, A. (1977) "Incoherent Optical Processing of Complex Data," *Applied Optics*, vol. 16, pp. 261–263.

Lorentz, H. A. (1896) "The Theorem of Poynting Concerning the Energy in the Electromagnetic Field and two General Propositions Concerning the Propagation of Light," *Amsterdammer Akademie der Wetenschappen*, vol. 4, p. 176.

Lu, J.-Y. and Greenleaf, J. F. (1990) "Ultrasonic Nondiffracting Transducer for Medical Imaging," *IEEE Transactions on Ultrasonics, Ferroelectronics, and Frequency Control*, vol. UFFC-37, pp. 438–447.

Lucy, L. (1974) "An Iterative Technique for the Rectification of Observed Distributions," *The Astronomical Journal*, vol. 79, pp. 745–754.

Makris, N. C., Ingenito, F., and Kuperman, W. A. (1994) "Detection of a Submerged Object Insonified by Surface Noise in an Ocean Waveguide," *Journal of the Acoustical Society of America*, vol. 96, pp. 1703–1724.

Mallart, R. and Fink, M. (1991) "The van Cittert-Zernike Theorem in Pulse Echo Measurements," *The Journal of the Acoustical Society of America*, vol. 90, pp. 2718–2727.

Manasse, R. (1959) "The Use of Radar Interferometric Measurements to Study Planets," Group Report No. 312–324, Lincoln Laboratory, MIT, Cambridge, MA.

Manasse, R. (1961) "The Use of Pulse Coding to Discriminate Against Clutter," Technical Report 312-12, Lincoln Laboratory MIT, Cambridge, MA.

Marcum, J. R. (1960) "A Statistical Theory of Target Detection of Pulsed Radar," *IRE Transactions on Information Theory*, vol. IT-6, pp. 145–267.

McEwan, N. J. and Goldsmith, P. F. (1989) "Gaussian Beam Techniques for Illuminating Reflector Antennas," *IEEE Transactions on Antennas and Propagation*, vol. AP-37, pp. 297–303.

McPherson, A. (1989) "Macromolecular Crystals," *Scientific American*, March, pp. 62–69.

Mersereau, R. M. (1979) "The Processing of Hexagonally-Sampled Two-Dimensional Signals," *Proceedings of the IEEE*, vol. 67, pp. 930–949.

Mersereau, R. M. and Oppenheim, A. V. (1974) "Digital Reconstruction of Multidimensional Signals from Their Projections," *Proceedings of the IEEE*, vol. 62, pp. 1319–1338.

Mersereau, R. M. (1973) "Recovering Multi-Dimensional Signals from their Projections," *Computer Graphics and Image Processing*, vol. 1, pp. 179–195.

Michelson, A. A. (1890) "On the Application of Interference Methods to Astronomical Measurements," *Philosophical Magazine*, vol. 30, pp. 1–21.

Michelson, A. A. (1920) "On the Application of Interference Methods to Astronomical Measurements," *Astrophysics Journal*, vol. 53, pp. 249–259.

Middleton, D. and Van Meter, D. (1955) "On Optimum Multiple-Alternative Detection of Signals in Noise," *IRE Transactions on Information Theory*, vol. IT-1, pp. 1–9.

Millane, R. P. (1990) "Phase Retrieval in Crystallography and Optics," *Journal of the Optical Society of America*, vol. 7, no. 3, pp. 394–411.

Miller, M. I. and Snyder, D. L. (1987) "The Role of Likelihood and Entropy in Incomplete-Data Problems: Applications to Estimating Point-Process Intensities and Toeplitz Constrained Covariances," *Proceedings of the IEEE*, vol. 75, pp. 892–907.

Miller, M. I., Snyder, D. L., and Miller, T. R. (1985) "Maximum Likelihood Reconstruction for Single Photon Emission Computed Tomography," *IEEE Transactions on Nuclear Science*, vol. NS-32, pp. 769–778.

Moran, P. R. (1982) "A Flow Velocity Zeugmatographic Interlace for NMR Imaging in Humans," *Magnetic Resonance Imaging*, vol. 2, pp. 555–566.

Morefield, C. L. (1977) "Application of 0-1 Integer Programming to Multitarget Tracking Problems," *IEEE Transactions on Automatic Control*, vol. AC-22, pp. 302–312.

Moulin, P. (1990) *A Method of Sieves for Radar Imaging and Spectral Estimation*, D.Sc. Dissertation, Washington University, St. Louis.

Moulin, P., O'Sullivan, J. A., and Snyder, D. L. (1992) "A Method of Sieves for Multiresolution Spectrum Estimation and Radar Imaging," *IEEE Transactions on Information Theory*, vol. IT-38, pp. 801–813.

Mouyan, Z. and Unbehauen, R. (1997) "Methods for Reconstruction of 2-D Sequences from Fourier Transform Magnitude," *IEEE Transactions on Image Processing*, vol. IP-6, pp. 222–233.

Mueller, R. K., Kaveh, M., and Wade, G. (1979) "Reconstructive Tomography and Applications to Ultrasonics," *Proceedings of the IEEE*, vol. 67, pp. 567–587.

Munkres, J. (1957) "Algorithm for the Assignment and Transportation Problems," *Siam Journal*, vol. 5, pp. 32–38.

Munson, Jr., D. C. (1998) "Computational Imaging," in A. Vardy, ed., *Codes, Curves, and Signals: Common Threads in Communications*, Kluwer Academic.

Munson, Jr., D. C., O'Brien, J. D., and Jenkins, W. K. (1983) "A Tomographic Formulation of Spotlight-Mode Synthetic Aperture Radar," *Proceedings of the IEEE*, vol. 71, pp. 917–925.

Munson, Jr., D. C. and Sanz, J. L. C. (1984) "Image Reconstruction from Frequency-Offset Fourier Data," *Proceedings of the IEEE*, vol. 72, pp. 661–699.

Munson, D. C. and Visentin, R. L. (1989) "A Signal Processing View of Strip-Mapping Synthetic Aperture Radar," *IEEE Transactions on Acoustics, Speech, and Signal Processing*, vol. ASSP-37, pp. 2131–2147.

Murino, V. and Trucco, A. (2000) "Three-Dimensional Image Generation and Processing in Underwater Acoustic Vision," *Proceedings of the IEEE*, vol. 88, pp. 1903–1948.

Nahi, N. E. (1969) "Optimal Recursive Estimation with Uncertain Observations," *IEEE Transactions on Information Theory*, vol. IT-15, pp. 457–462.

Natterer, F. (1986) *The Mathematics of Computerized Tomography*, John Wiley and Sons, New York.

Nayar, S. K. and Nakagawa, Y. (1994) "Shape from Focus," *IEEE Transactions on Pattern Analysis and Machine Intelligence*, vol. PAMI-16, pp. 824–831.

Neumaier, A. (1998) "Solving Ill-Conditioned and Singular Linear Systems: A Tutorial on Renormalization," *Siam Review*, vol. 40, pp. 636–666.

Newell, A. C. (1988) "Error Analysis Technique for Planar Near Field Measurements," *IEEE Transactions on Antennas and Propagation*, vol. AP-36, pp. 754–768.

Neyman, J. and Pearson, E. (1933) "On the Problem of the Most Efficient Tests of Statistical Hypotheses," *Philosophical Transactions of the Royal Society*, series A, vol. 231, p. 289.

Nilsson, N. J. (1961) "On the Optimum Range Resolution of Radar Signals in Noise," *IRE Transactions on Information Theory*, vol. IT-7, pp. 245–253.

North, D. O. (1943) "An Analysis of the Factors which Determine Signal/Noise Discrimination in Pulsed-Carrier Systems," RCA Technical Report PTR-6C.

Nyquist, H. (1928) "Certain Topics in Telegraph Transmission Theory," *Transactions of the AIEE*, vol. 47, pp. 617–644.

Oktem, F. S. (2014) *Computational Imaging and Inverse Techniques for High Resolution and Instantaneous Spectral Imaging*, PhD Dissertation, University of Illinois.

Oktem, F. S. and Blahut, R. E. (2011) "Schulz-Snyder Phase Retrieval Algorithm as an Alternating Minimization Algorithm," *Computational Optical Sensing and Imaging*, page CMC3, Optical Society of America.

Oktem, F. S., Kamalabadi, F., and Davila, J. M. (2013) "High-Resolution Computational Spectral Imaging with Photon Sieves," *Proceedings of the IEEE International Conference on Image Processing*, pp. 2373–2377.

Oktem, F. S., Kamalabadi, F., and Davila, J. M. (2014) "High-Resolution Computational Spectral Imaging with Photon Sieves," *Proceedings of the IEEE International Conference on Image Processing*, pp. 5122–5126.

Oktem, F. S., Kamalabadi, F., and Davila, J. M. (2018) "Analytical Fresnal Imaging Models for Photon Sieves," *Optics Express*, vol. 26, pp. 32259–32279.

Ollinger, J. M. and Fessler, J. A. (1997) "Positron-Emission Tomography," *IEEE Signal Processing Magazine*, vol. 14, pp. 43–55.

O'Neill, E. L. (Ed.) (1962) *Communication and Information Theory Aspects of Modern Optics*, General Electric Co., Electronics Laboratory, Syracuse, NY.

O'Neill, E. L. and Walther, A. (1963) "The Question of Phase in Image Formation," *Optica Acta*, vol. 10, pp. 33–40.

O'Neill, E. L. (1956) "Spatial Filtering in Optics," *IRE Transactions on Information Theory*, vol. IT-2, pp. 56–65.

O'Sullivan, J. A. (1998) "Alternating Minimization Algorithms: From Blahut-Arimoto to Expectation-Maximization," in A. Vardy, ed., *Codes, Curves, and Signals*, Kluwer.

O'Sullivan, J. A. (2002) "Iterative Algorithms for Maximum Likelihood Sequence Detection," in R. E. Blahut and R. Koetter, eds., *Codes, Graphs, and Systems*, Kluwer.

O'Sullivan, J. A. and Benac, J. (2007) "Alternating Minimization Algorithms for Transmission Tomography," *IRE Transactions on Medical Imaging*, vol. MI-26, pp. 283–297.

O'Sullivan, J. A., Blahut, R. E., and Synder, D. L. (1998) "Information-Theoretic Image Formation," *IEEE Transactions on Information Theory*, vol. IT-44, pp. 2094–2123.

Papoulis, A. (1974) "Ambiguity Functions in Fourier Optics," *Journal of the Optical Society of America*, vol. 64, pp. 779–788.

Pasedach, K. and Haase, E. (1981) "Random and Guided Generation of Coherent Two-Dimensional Codes," *Optics Communications*, vol. 36, pp. 423–428.

Patterson, A. L. (1934) "A Fourier Series Method for the Determination of the Components of Interatomic Distance in Crystals," *Physical Review*, vol. 46, pp. 372–376.

Patterson, A. L. (1935) "A Direct Method for the Components of the Interatomic Distances in Crystals," *Zeitschrift Furkristallographie, Kristallgeometrie, Kristalphysik, Kristallchemie*, vol. A90, pp. 517–542.

Pauling, Ł. (1925) *The Determination with X-Rays of the Structures of Crystals*, Ph.D. Dissertation, California Institute of Technology.

Peterson, W. W. and Birdsall, T. G. (1953) "The Theory of Signal Detectability, Parts I and II," Technical Report 13, University of Michigan. (Also with W. Fox, *IRE Transactions on Information Theory*, vol. IT-4, pp. 171–212, 1954.)

Petersen, D. P. and Middleton, D. (1962) "Sampling and Reconstruction of Wave-Number-Limited Functions in N-dimensional Euclidean Space," *Information and Control*, vol. 5, pp. 279–323.

Politte, A. G. and Snyder, D. L. (1991) "Corrections for Accidental Coincidences in Maximum-Likelihood Image Reconstruction for Positron-Emission Tomography," *IEEE Transaction on Medical Imaging*, vol. MI-10, pp. 82–89.

Poor, H. V. (1994) *An Introduction to Signal Detection and Estimation*, 2nd ed., Springer-Verlag, New York.

Poore, A. B. (1994) "Multidimensional Assignment Formulation of Data Association Problems Arising from Multitarget and Multisensor Tracking," *Computational Optimization and Applications*, vol. 3, pp. 27–57.

Poore, A. B. and Robertson III, A. J. (1997) "A New Lagrangian Relaxation Based Algorithm for a Class of Multidimensional Assignment Problems," *Computational Optimization and Applications*, vol. 8, pp. 127–150.

Poore, A. B. and Rijavec, N. (1993) "A Lagrangian Relaxation Algorithm for Multidimensional Assignment Problems Arising from Multitarget Tracking," *SIAM Journal on Optimization*, vol. 3, pp. 554–563.

Porcello, L. J. (1970) "Turbulence-Induced Phase Errors in Synthetic-Aperture Radars," *IEEE Transactions on Aerospace and Electronic Systems*, vol. AE6-6, pp. 636–644.

Porter, A. B. (1906) "On the Diffraction Theory of Microscopic Vision," *London, Edinburgh, Dublin Philosphical Magazine*, vol. 11, pp. 154–166.

Price, R. and Hofstetter, E. M. (1965) "Bounds on the Volume and Height Distribution of the Ambiguity Function," *IEEE Transactions on Information Theory*, vol. IT-11, pp. 207–214.

Radon, J. (1917) "Uber die Bestimmung von Funktionen durch ihre Integralwerte langs gewisser Mannigfaltigkeiten," Berichte Sachsische Akademie der Wissenschaften, Leipzig, Math–Phys Kl, vol. 69, pp. 262–267. (English translation: "On the Determination of Functions from their Integrals Along Certain Manifolds," Appendix A of *The Radon Transform and Some of its Applications*, S. R. Deans, Wiley, New York.)

Rattey, R. A. and Lindgren, A. G. (1981) "Sampling the 2-D Radon Transform," *IEEE Transactions on Acoustics, Speech, and Signal Processing*, vol. ASSP-29, pp. 994–1002.

Lord Rayleigh (1879) *Philosophical Magazine*, vol. 8, p. 261.

Readhead, A. C. S. and Wilkinson, P. N. (1978) "The Mapping of Compact Radio Sources from VLBI Data," *Astrophysics Journal*, vol. 223, pp. 25–36.

Reimers, P. and Goebbels, J. (1983) "New Possibilities of Nondestructive Evaluation by X-Ray Computed Tomography," *Materials Evaluation*, vol. 41, pp. 732–737.

Rhodes, D. R. (1974) *Synthesis of Planar Antenna Sources*, Clarendon Press, Oxford.

Richardson, W. H. (1972) "Bayesian-Based Iterative Method of Image Restoration," *Journal of the Optical Society of America*, vol. 62, pp. 55–59.

Rihaczek, A. W. (1965) "Radar Signal Design for Target Resolution," *Proceedings of the IEEE*, vol. 53, pp. 116–128.

Rihaczek, A. W. (1967) "Radar Resolution of Moving Targets," *IEEE Transactions on Information Theory*, vol. IT-13, pp. 51–56.

Rockmore, A. J. and Macovski, A. (1976) "A Maximum Likelihood Approach to Emission Image Reconstruction from Projections," IEEE Transactions on Nuclear Science, vol. NS-23, pp. 1428–1432.

Röentgen, W. K. (1896) "On a New Kind of Rays," (English Translation) *Nature*, vol. 53, pp. 274–276.

Ron, M. Y. and Unbehauen, R. (1995) "New Algorithms of Two-Dimensional Blind Convolution," *Optical Engineering*, vol. 34, pp. 2945–2956.

Root, W. L. (1987) "Ill-posedness and Precision in Object-Field Reconstruction Problems," *Journal of the Optical Society of America*, vol. 4, pp. 171–179.

Ryle, M. (1952) "A New Radio Interferometer and Its Application for the Observation of Weak Radio Stars," *Proceedings of the Royal Society of London*, vol. A211, pp. 351–375.

Ryle, M. and Hewish, A. (1960) "The Synthesis of Large Radio Telescopes," *Monthly Notices of the Royal Astronomical Society*, vol. 120, pp. 220–230.

Sayre, D. (1952a) "Some Implications of a Theorem Due to Shannon," *Acta Crystallographia*, vol. 5, p. 843.

Sayre, D. (1952b) "The Squaring Method: A New Method for Phase Determination," *Acta Crystallographia*, vol. 5, pp. 60–65.

Schotland, J. (1997) "Continuous-Wave Diffusion Imaging," *Journal of the Optical Society of America*, vol. 14, p. 275.

Schmitt, J. M. (1999) "Optical Coherence Tomography (OCT): A Review," *IEEE Journal of Selected Topics in Quantum Electronics*, vol. QE-5, pp. 1205–1215.

Schulz, T. J. (1993) "Multiframe Blind Convolution of Astronomical Images," *Journal of the Optical Society of America A*, vol. 10, pp. 1064–1073.

Schulz, T. J. and Snyder, D. L. (1991) "Imaging a Randomly Moving Object from Quantum-Limited Data: Applications to Image Recovery from Second- and Third-Order Autocorrelations," *Journal of the Optical Society of America A*, vol. 8, pp. 801–807.

Schulz, T. J. and Snyder, D. L (1992) "Image Recovery from Correlations," *Journal of the Optical Society of America A*, vol. 9, pp. 1266–1272.

Schwartz, L. (1952) "Theorie des Distributions," *Bulletin of the American Mathematical Society*, vol. 58, pp. 75–85.

Schwarz, U. J. (1978) "Mathematical-Statistical Description of the Iterative Beam Removing Technique (Method Clean)," *Astronomy and Astrophysics*, vol. 65, pp. 345–356.

Schwarz, U. J. (1979) "The Method 'CLEAN'-Use, Misuse, and Variations," in C. van Schoonveld, ed., *Image Formation from Coherence Functions in Astronomy*, Reidel, Dordrecht, pp. 261–275.

Schweppe, F. C. (1968) "Sensor-Array Data Processing for Multiple Signal Sources," *IEEE Transactions on Information Theory*, vol. IT-14, pp. 294–305.

Scudder, H. J. (1978) "Introduction to Computer Aided Tomography," *Proceedings of the IEEE*, vol. 66, no. 6, pp. 628–637.

Sekine, M., Ohtani, S., and Muska, T. (1981) "Weibull-Distributed Ground Clutter," *IEEE Transactions on Aerospace and Electronic Systems*, vol. AES-17, pp. 596–598.

Selin, I. (1965) "Detection of Coherent Radar Returns of Unknown Doppler Shift," *IEEE Transactions on Information Theory*, vol. IT-11, pp. 396–400.

Shannon, C. E. (1948) "A Mathematical Theory of Communication," *Bell System Technical Journal*, vol. 27, pp. 379–423, pp. 623–656.

Shapiro, J. (1982) "Target Reflectivity Theory for Coherent Laser Radars," *Applied Optics*, vol. 21, pp. 3398–3407.

Shapiro, J., Capron, B. A., and Harney, R. C. (1981) "Imaging and Target Detection with a Heterodyne-Reception Optical Radar," *Applied Optics*, vol. 20, pp. 3292–3313.

Shepp, L. A. (1980) "Computerized Tomography and Nuclear Magnetic Resonance," *Journal of Computer Assisted Tomography*, vol. 4, no. 1, pp. 94–107.

Shepp, L. A. and Logan, B. F. (1974) "The Fourier Reconstruction of a Head Section," *IEEE Transactions on Nuclear Science*, vol. NS-21, pp. 21–43.

Shepp, L. A. and Vardi, Y. (1982) "Maximum Likelihood Reconstruction for Emission Tomography," *IEEE Transactions on Medical Imaging*, vol. MI-1, pp. 113–122.

Sherwin, C. W., Ruina, J. P., and Rawcliffe, R. D. (1962) "Some Early Developments in Synthetic-Aperture Radar Systems," *IRE Transactions on Military Electronics*, vol. MIL-6, pp. 111–115.

Shore, J. E. and Johnson, R. W. (1980) "Axiomatic Derivation of the Principle of Maximum Entropy and the Principle of Minimum Cross Entropy," *IEEE Transactions on Information Theory*, vol. IT-26, pp. 26–37.

Siebert, W. M. (1956) "A Radar Detection Philosophy," *IRE Transactions on Information Theory*, vol. IT-2, pp. 204–221.

Siebert, W. M. (1958) "Studies of Woodward's Uncertainty Function," *Massachusetts Institute of Technology, Research Laboratory Electronics Quarterly Progress Report*, April.

Silva, M. T. and Robinson, E. A. (1979) *Deconvolution of Geophysical Time Series in the Exploration of Oil and Natural Gas*, Elsevier, Amsterdam.

Simpson, R. G., Barrett, H. H., Suback, J. A., and Fisher, H. D. (1975) "Digital Processing of Annual Coded-Aperture Imagery," *Optical Engineering*, vol. 14, pp. 490–494.

Singer, R. A., Sea, R. G., and Housewright, K. B. (1974) "Derivation and Evaluation of Improved Tracking Filters for Use in Dense Multitarget Environments," *IEEE Transactions on Information Theory*, vol. IT-20, pp. 423–432.

Singer, Jr., G. T. (1971) "NMR Spin-Echo Flow Measurements," *Journal of Applied Physiology*, vol. 42, p. 938.

Singer, J. R. Grünbaum, F. A., Kohn, P., and Zubelli, J. P. (1990) "Image Reconstruction of the Interior of Bodies that Diffuse Radiation," *Science*, vol. 244, pp. 990–993.

Sittler, R. W. (1964) "An Optimal Data Association Problem in Surveillance Theory," *IEEE Transactions on Military Electronics*, vol. MIL-8, pp. 125–139.

Skinner, G. K. (1988) "X-ray Imaging with Coded Masks," *Scientific American*, pp. 66–71.

Slepian, D. (1967) "Restoration of Photographs Blurred by Image Motion," *Bell System Technical Journal*, vol. 46, pp. 2353–3362.

Slepian, D. and Pollak, H. O. (1961) "Prolate Spheroidal Wave Functions, Fourier Analysis and Uncertainty, Part I," *Bell System Technical Journal*, vol. 40, pp. 43–64.

Slichter, C. P. (1990) *Principles of Magnetic Resonance*, 3rd edn, Springer-Verlag, New York.

Smith, B. D. (1985) "Image Reconstruction from Cone-Beam Projections: Necessary and Sufficient Conditions and Reconstruction Methods," *IEEE Transactions on Medical Imaging*, vol. MI-4, pp. 14–25.

Snyder, D. L. and Cox, Jr., J. R. (1977) "An Overview of Reconstructive Tomography and Limitations Imposed by a Finite Number of Projections," in M. Ter-Pogossian et al., eds., *Reconstruction Tomography in Diagnostic Radiology and Nuclear Medicine*, University Park Press, Maryland.

Snyder, D. L. and Fishman, P. M. (1975) "How to Track a Swarm of Fireflies by Observing Their Flashes," *IEEE Transactions on Information Theory*, vol. IT-22, pp. 692–695.

Snyder, D. L. and Miller, M. I. (1985) "The Use of Sieves to Stabilize Images Produced with the EM Algorithm for Emission Tomography," *IEEE Transactions on Nuclear Science*, vol. NS-32, pp. 3864–3871.

Snyder, D. L. and Miller, M. I. (1991) *Random Point Processes in Time and Space*, Springer-Verlag, New York.

Snyder, D. L, Miller, M. I., Thomas, Jr., J. J., and Politte, D. G. (1975) "Noise and Edge Artifacts in Maximum-Likelihood Reconstructions for Emission Tomography," *IEEE Transactions on Medical Imaging*, vol. MI-6, pp. 228–238.

Snyder, D. L, O'Sullivan, J. A., and Miller, M. I. (1989) "The Use of Maximum Likelihood Estimation for Forming Images of Diffuse Radar Targets from Delay-Doppler Data," *IEEE Transactions on Information Theory*, vol. IT-35, pp. 536–548.

Snyder, D. L. and Politte, D. G. (1983) "Image Reconstruction from List-Mode Data in an Emission Tomography System Having Time-of-Flight Measurements," *IEEE Transactions on Nuclear Science*, vol. NS-30, pp. 1843–1849.

Snyder, D. L. and Schulz, T. J. (1990) "High-Resolution Imaging at Low-Light Levels Through Weak Turbulence," *Journal of the Optical Society of America A*, vol. 7, pp. 1251–1265.

Snyder, D. L, Schulz, T. J., and O'Sullivan, J. A. (1992) "Deblurring Subject to Nonnegativity Constraints," *IEEE Transactions on Signal Processing*, vol. SP-40, pp. 1143–1150.

Solomon, D. C. (1976) "The X-ray Transform," *Journal of Mathematical Analysis and Applications*, vol. 56, pp. 61–83.

Sorenson, H. W. (1980) *Parameter Estimation: Principles and Problems*, Marcel-Dekker, New York.

Soumekh, M. (1997) "Moving Target Detection in Foliage Using Along Track Monopulse Synthetic Aperture Radar Imaging," *IEEE Transactions on Image Processing*, vol. IP-6, pp. 1148–1163.

Southwell, W. H. (1981) "Validity of the Fresnel Approximation in the Near Field," *Journal of the Optical Society of America*, vol. 71, pp. 7–14.

Spiess, F. and Anderson, V. C. (1983) "Wise Swath Precision Echo Sounder," US Patent 4 400 803.

Stark, H. (1979) "Sampling Theorems in Polar Coordinates," *Journal of the Optical Society of America,*" vol. 69, pp. 1519–1525.

Stark, L. (1974) "Microwave Theory of Phased-Array Antennas–A Review," *Proceedings of the IEEE*, vol. 62, pp. 1661–1701.

Staudaher, F. M. (1970) "Airborne MTI," Chapter 18 in M. I. Skolnik, ed., *Radar Handbook*, McGraw-Hill, New York.

Stein, S. (1981) "Algorithms for Ambiguity Function Processing," *IEEE Transactions on Acoustics, Speech, and Signal Processing*, vol. ASSP-29, pp. 588–599.

Stein, J. J. and Blackman, S. S. (1975) "Generalized Correlation of Multi-Target Track Data," *IEEE Transactions on Aerospace and Electronic Systems*, vol. AES-11, pp. 1207–1217.

Stout, G. H. and Jensen, L. H. (1989) *X-Ray Structure Determination*, Wiley, New York.

Stutt, C. A. (1964) "Some Results on Real-Part/Imaginary-Part and Magnitude-Phase Relations in Ambiguity Functions," *IEEE Transactions on Information Theory*, vol. IT-10, pp. 321–327.

Sullivan, III, W. T., ed., *The Early Years of Radio Astronomy*, Cambridge University Press, Cambridge, MA.

Sussman, S. M. (1962) "Least-Square Synthesis of Radar Ambiguity Functions," *IRE Transactions on Information Theory*, vol. IT-8, pp. 246–254.

Swenson, Jr., G. W. and Mathur, N. C. (1968) "The Interferometer in Radio Astronomy," *Proceedings of the IEEE*, vol. 56, pp. 2114–2130.

Swerling, P. (1957) "Detection of Fluctuating Pulsed Signals in the Presence of Noise," *IRE Transactions on Information Theory*, vol. IT-3, pp. 175–178.

Swerling, P. (1960) "Probability of Detection for Fluctuating Models," *IRE Transactions on Information Theory*, vol. IT-6, pp. 269–308.

Swerling, P. (1964) "Parameter Estimation Accuracy Formulas," *IRE Transactions on Information Theory*, vol. IT-10, pp. 302–314.

Synge, E. (1928) "A Suggested Method for Extending Microscopic Resolution into the Ultramicroscopic Region," Philosophical Magazine, vol. 6, pp. 356–362.

Synge, E. (1932) "An Application of Piezoelectricity to Microscopy," *Philosophical Magazine*, vol. 13, pp. 297–300.

Tagfors, T. and Campbell, D. (1973) "Mapping of Planetary Surfaces by Radar," *Proceedings of the IEEE*, vol. 61, pp. 1219–1225.

Taxt, T. (1995) "Restoration of Medical Ultrasound Images Using Two-Dimensional Homomorphic Deconvolution," *IEEE Transactions on Ultrasonics, Ferroelectronics, and Frequency Control*, vol. UFFC-42, pp. 543–554.

Ter-Pogossian, M. M., Raichle, M. E., and Soble, B. E. (1980) "Positron-Emission Tomography," *Scientific American*, vol. 243, p. 140.

Thompson, A. R., Moran, J. M., and Swenson, Jr., G. W. (2001) *Interferometry and Synthesis in Radio Astronomy*, John Wiley and Sons, New York (2nd edn., 2001).

Thomas, J. B. and Wolf, J. K. (1962) "On the Statistical Detection Problem for Multiple Signals," *IRE Transactions on Information Theory*, vol. IT-8, pp. 274–280.

Tikhonov, A. N. (1943) "On the Stability of Inverse Problems," *Doklady Akademii Nauk SSSR*, vol. 39, pp. 195–198.

Tikhonov, A. N. (1963) "Solution of Incorrectly Formulated Problems and the Regularization Method," *Soviet Mathematics: Doklady*, vol. 4, pp. 1035–1038.

Tikhonov, A. N. and Arsenin, V. (1977) *Solutions of Ill-Posed Problems*, Winston, Washington.

Titchmarsh, E. C. (1937) *Introduction to the Theory of Fourier Integrals*, Clarendon Press, Oxford.

Tolimieri, R. and Winograd, S. (1985) "Computing the Ambiguity Surface," *IEEE Transactions on Acoustics, Speech, and Signal Processing*, vol. ASSP-33, pp. 1239–1245.

Tur, M., Chin, K. C., and Goodman, J. W. (1982) "When Is Speckle Noise Multiplicative?" *Applied Optics*, vol. 21, pp. 1157–1159.

H. K. (1983) "An Inversion Formula for Cone-Beam Reconstruction," *SIAM Journal of Applied Mathematics*, vol. 43, pp. 546–551.

Urkowitz, H. (1953) "Filters for Detection of Small Targets in Clutter," *Journal of Applied Physics*, vol. 24, pp. 1024–1031.

van Cittert, P. H. (1934) "Die wahrscheinliche Schwingungsverteilung in einer von einer ... oder mittels einer Linse beleuchteten Ebene," *Physica*, vol. 1, pp. 201–210.

Vander Lugt, A. B. (1964) "Signal Detection by Complex Spatial Filtering," *IEEE Transactions on Information Theory*, vol. IT-10, pp. 139–145.

Van Trees, H. L. (1971) *Detection, Estimation and Modulation Theory, Part III: Radar-Sonar Signal Processing and Gaussian Signals in Noise*, John Wiley and Sons, New York.

Van Vleck, J. H. (1943) "The Spectrum of Clipped Noise," Radio Research Laboratory, Harvard University, Report RRL-51.

Van Vleck, J. H. and Middleton, D. (1966) "The Spectrum of Clipped Noise," *Proceedings of the Institute of Electrical Engineers*, vol. 54, pp. 2–19.

Vardi, Y. and Lee, D. (1993) "From Image Deblurring to Optimal Investments: Maximum Likelihood Solutions for Positive Linear Inverse Problems," *Journal of the Royal Statistics Society, B*, vol. 55, pp. 569–612.

Vardi, Y., Shepp, L. A., and Kaufman, L. (1985) "A Statistical Model for Positron Emission Tomography," *Journal of the American Statistical Association*, vol. 80, pp. 8–35.

Ville, J. (1948) "Theory and Application of the Notion of the Complex Signal," *Cables et Transmission*, vol. 2, pp. 67–74.

von Neumann, J. (1932) "Mathematische Grundlagen der Quantenmechanik," Springer. English translation: "Mathematical Foundations of Quantum Mechanics," R. T. Beyer, Princeton University Press, Princeton, NJ, 1955 and 1996.

Walker, J. L. (1980) "Range-Doppler Imaging of Rotating Objects," *IEEE Transactions on Aerospace and Electronic Systems*, vol. AES-16, pp. 23–52.

Watson, J. D. and Crick, F. H. C. (1953) "A Structure for Deoxyribose Nucleic Acid," *Nature*, No. 4356, pp. 737–738.

Wax, N. (1955) "Signal-to-Noise Improvement and the Statistics of Tracking Populations," *Journal of Applied Physics*, vol. 26, pp. 586–595.

Webb, J. L. H. and Munson, Jr., D. C. (1995) "Radar Imaging of Three-dimensional Surfaces Using Limited Data," *Proceedings of the IEEE International Conference on Image Processing*, Washington, D.C., pp. 136–139.

Webb, J. L. H., Munson, Jr., D. C., and Stacy, N. J. (1998) "High-resolution Planetary Imaging Via Spotlight-mode Synthetic Aperture Radar," *IEEE Transactions on Image Processing*, vol. IP-7, pp. 1571–1582.

Whittaker, E. T. (1915) "On the Functions which are Represented by the Expansions of the Interpolation Theory," *Proceedings of the Royal Society*, Edinburgh, vol. 35, pp. 181–184.

Wiener, N. (1930) "Generalized Harmonic Analysis," *Acta Mathematica*, vol. 55, pp. 117–258.

Wiener, N. (1949) *Extrapolation, Interpolation, and Smoothing of Stationary Time Series with Engineering Applications*, Wiley, New York.

Wilcox, C. H. (1960) "The Synthesis Problem for Radar Ambiguity Functions," University of Wisconsin, Mathematical Research Center, Technical Summary Report 157.

Wiley, C. A. (1965) "Pulsed Doppler Radar Methods and Apparatus," US Patent 3 196 436.

Wolf, E. (1969) "Three-Dimensional Structure Determination of Semi-Transparent Objects from Holographic Data," *Optics Communications*, vol. 1, pp. 153–156.

Wolf, E. (1966) "Principles and Development of Diffraction Tomography," in A. Consortini, ed., *Trends in Optics*, pp. 83–110, Academic Press, San Diego, CA.

Wolfe, P. (1975) "A Method of Conjugate Subgradients for Minimizing Nondifferentiable Functions," *Mathematical Programming*, Study 3, pp. 147–173.

Woodward, P. M. (1953) *Probability and Information Theory, with Applications to Radar*, McGraw-Hill, New York and Pergamon Press, Oxford.

Woolfson, M. and Fan, H. (1995) *Physical and Nonphysical Methods of Solving Crystal Structures*, Cambridge University Press, Cambridge.

Wu, C. F. T. (1983) "On the Convergence Properties of the EM Algorithm," *Annals of Statistics*, vol. 11, pp. 95–103.

Wynn, W. M. (1972) "Dipole Tracking with a Gradiometer," NSRDL/PC Informal Report 3493.

Wynn, W. M., Frahm, C. P., Caroll, P. J., Clark, R. H., Wellhoner, J., and Wynn, M. J. (1975) "Advanced Superconducting Gradiometer/Magnetometer Arrays and a Novel Signal Processing Technique," *IEEE Transactions on Magnetics*, vol. MAG-11, pp. 701–707.

Zadeh, L. A. and Ragazzini, J. R. (1952) "Optimum Filters for the Detection of Signals in Noise," *Proceedings of the IRE*, vol. 40, pp. 1123–1131.

Zebker, H. A. and Goldstein, R. M. (1986) "Topographic Mapping from Interferometric Synthetic Aperture Radar Observations," *Journal of Geophysical Research*, vol. 91, pp. 4993–4999.

Zernike, F. (1935) "Das Phasekontrastverfahren bei der Mikroskopischen Beobachtung," *Z. Tech. Physics*, vol. 16, pp. 454–457.

Zernike, F. (1938) "The Concept of Degree of Coherence and its Application to Optical Problems," *Physica*, vol. 5, pp. 785–795.

Zernike, F. (1942) "Phase Contrast, a New Method for the Microscopic Observation of Transparent Objects, Par II," *Physica*, vol. 9, pp. 686–698, 974–986.

Index

Page numbers in **bold** indicate primary references.

f-number, 141, 161

Abbe diffraction limit, 110, 150, 166, 292
Abbe resolution criterion, 150
Absorption tomography, 208
Acoustic wave, 11, 26
Active surveillance, 2
Actual data, 311
Actual data space, 333
Airy disk, 87, 117, 150
Airy function, 149
Algebraic reconstruction, 217
Algorithm
    alternating maximization, 309
    alternating minimization, 349
    clean, 262, 528
    expectation-maximization, 333
    Fienup, 274
    Gerchberg–Saxton, 273, 299
    Munkres, 568
    phase-retrieval, 273
    Richardson–Lucy, 264, 321
    Schulz–Snyder, 323
Ali–Silvey distance, 348
Aliasing, **53**, 116, 278, 359
    color, 295
    in one dimension, 53
    radar image, 448
    two-dimensional, 99
Alternating expectation-maximization, 333, 345
Alternating maximization, **309**
Alternating minimization, 310
    Schulz–Snyder algorithm, 323
    transmission tomography, 253
    variance estimation, 310
Alternating projection, 219, 250
Alternative hypothesis, 483
Ambiguity function, 3, 28, **399**, 400
    chirp pulse train, 417
    Costas pulse, 419
    cross-ambiguity, 423

    in Fourier optics, 435
    in optics, 166
    pulse train, 414
    rectangular pulse, 403
    spatial, 435
    volume property, 406
Ambiguity surface, 400
Angle
    eulerian, 13
Angle of arrival, 185
Angular spectrum, 15, **123**, 123, 198
    vector wave, 198
Antenna, 167
    Cassegrain, 204
    phase center, 187
Antenna array, 175, 370
    hexagonal, 178
Antenna element, 175
Antenna gain, 174, 482
Antenna pattern, 167, 173
    one-dimensional, **170**
    two-dimensional, 171
Antenna quality factor, 175, 204
Antenna radiation pattern, 167
Aperture, 113, 121, 135, 168, 269
    circular, 149
    real, 457
    subwavelength, 300
    virtual, 149
Aperture synthesis, 9
Apodization, 67, 109, 434
    two-dimensional, 110
Apodizing window, 434
Approximation
    Born, 362
    delay-doppler, 444
    Fraunhofer, 130
    Fresnel, 128
    geometrical optics, 132
    narrowband, 170
    paraxial, 130, 145

Approximation (Cont.)
  radar imaging, **424**, 446, 447
  small-angle, 173
  thin lens, 144
Array
  antenna, 175
  phased, 180
  sequentially formed, 179
  steered, 180
Array radiation pattern, 177
Assignment problem, 566
Asymptotically unbiased, 345
Atomicity assumption, 378
Autocorrelation function, 35, 68
  three-dimensional, 367
  two-dimensional, 435
Axial image, 207

Babinet's principle, 163
Back projection, 24, 212
Back-projection theorem, 211
Backscatter, 445
Bandwidth
  effective, 70
  Gabor, **50**, 409
  noise, 67
  Woodward, 67
Barker-coded pulse, 431
Baseband representation, 10
  complex, 439
Baseband signal
  complex, 9
  real, 9
Bayes formula, 20, 264
Bayesian estimator, 314
Beam, 164, 167
  antenna, 171, 185
  gaussian, 164
  nondiffracting, 184
Beamforming, 176
  near-field region, 183
Beamsteering, 180
Beamwidth, 203
Bessel function, 84, 115, 185, 203, 366
  nth-order, 84, 357, 366
  first kind, 84, 357
  first-order, 85
  modified, 493, 515
  zero-order, 84, 184
Bistatic, 441
Bistatic radar, 441, 442
  imaging, 455
Bistatic reflection, 441
Bivariate gaussian density function, 29
Bivariate polynomial, 58
Blind deblurring, **254**

Blind deconvolution, 254, 277, 279, 328
Blurring, 256
Boltzmann constant, 225
Boltzmann probability distribution, 225
Boresight, 180
Born approximation, 238, 244, 289, 290, **362**, 365, 397
Bound
  Cramer–Rao, 500
Bounded support, 31
Bragg–Laue equations, 178, 373
  and sampling theorem, 375
  nonrectangular lattice, 394
Bravais lattice cell
  in three dimensions, 369
  in two dimensions, 369
Brownian motion, 242
Butterworth filter
  two-dimensional, 108

Calculus of variations, 260, 318, 320, 331, 339
Camcorder, 116
Carrier frequency, 10
Cassegrain antenna, 204
Causal filter, 58, 107, 256, 259, 483, 496
Cell, 104
Central dark ground method, 151
Central limit theorem, 151, 523
Centrosymmetric property, 117
Charge-coupled device, 116
Chirp filter, 72
Chirp pulse, 38, 404, 405, 410, 417
Chirped pulse, 463
Circle function, 85, 90
Circular aperture, 149
Circular error probability, 573
Circular hat function, 112, 149, 394
  three-dimensional, 394
  two-dimensional, 394
Circular symmetry, 83
Circularly symmetric complex gaussian, 21, 499
Circularly symmetric, gaussian noise, 548
Clean algorithm, 262, 300, 528, 556
Clean point-spread function, 263
Clipped signal, 531
Clutter, 509, 512
Coded aperture, 270, 299
Coded-aperture imaging, 268
Coherence, 4, 579, 580
Coherence function
  spatial, **529**
Coherence loss, 586
Coherence tomography, 25, 244
Coherent, 17
  spatial, 17, 145, 146
  temporal, 17, 146

Coherent detection, 482
Coherent estimation
  pulse arrival time, 496
Coherent imaging, 140
Coherent processing, 17, 27
Coherent surveillance, 4
Color samples, 298
Color spectrum, 295
Comb function
  one-dimensional, 43
  three-dimensional, 354
  two-dimensional, 81
Complete data, 311
Complete data space, 333
Complex baseband representation, 10, 14
  wavefront, 12
Complex baseband signal, 10
Complex modulation property, 33
Complex passband representation, 10
Complex random variable, 21
Complimentary code, 431
Computational image formation, 117
Conditional, 20, 264
Cone-beam tomography, 253
Conjection, 251
Conjugation, 161
Continuous real random variable, 20
Convex combination, 414
Convex decomposition lemma, 318, 325, **345**
Convolution
  cyclic, 68
Convolution theorem
  one-dimensional, 34
  three-dimensional, 352
  two-dimensional, 77
Coordinate rotation
  three-dimensional, 393
  two-dimensional, 78
Coordinate transformation
  three-dimensional, 352, 393
  two-dimensional, 78
Cornu spiral, 577
Correlation, 35
Correlation coefficient, 49, 63
Correlation function, 35, 58, **63**
  spatial, 529
Correlation theorem, 35
Cost, 567
Costas array, 419
Costas pulse, 418, 419
Covariance estimation, 334
Covariance function, **63**, 63
Covariance matrix, 21
Cramer–Rao bound, 501
Cramer–Rao inequality, 500
Cross-ambiguity function, 422, 423, 432

Cross-correlation function, 64
Cross-section density, 441
Crystal, 350, 371
Crystallography, 7, 370, 377
Csiszár distance, 316, 324, 341, 344
  and triangle inequality, 346
Cyclic convolution, 68
Cyclic difference set, 270, 299
Cylinder function, 354

Data
  complete, 311
  incomplete, 311
Data fusion, 557
de Broglie wave, 294
de Broglie wavelength, 294
Deblurring, 254, 256, 257
  blind, **254**
Dechirping, 73
Decimation algorithm, 182
Decision rule, 302, 483
  maximum-likelihood, 303
Deconvolution, 159, **255, 256**, 257, 448
  blind, 254, 279, 328
  nonnegative images, 263
Delay ambiguity, 417
Delay property, 33
Delay resolution, 450
Delta function, 37
  two-dimensional, 81
Density function
  multivariate gaussian, 29
Detection
  coherent, 482
  noncoherent, 482, **494**
  sequential, 561
Deterministic reflection, 444
Differentiation property, 35
Diffraction, 120, 127
  scalar, 120
  vector, **194**
Diffraction grating, 161
Diffraction imaging, 350, 366, 377
Diffraction limit, 166, 293
  Abbe, 150
Diffraction tomography, 5, 25, 236
Diffraction-limited imaging, 150
Diffractive lens, 163, 266, 298
Diffuse reflection, 134, 444, 467
Diffusion coefficient, 242
Diffusion equation, 242
Diffusion tomography, 25, 241
Digital processor, 463
Digitization, 53
Dilation, 77
Dirac delta function, 37

Index

Direct method, 380
Direction cosine, 12, 172, 537
Direction of arrival, 185, 547
Directivity, 173
Dirichlet function, 39, 177, 359, 373, 417
   one-dimensional, 41
   three-dimensional, 354
   two-dimensional, 82
Dirty point-spread function, 262
Discrete Fourier transform, 67, 181
Discrete random variable, 20
Discrete real random variable, 20
Distance
   Ali–Silvey, 348
   Csiszár, 316, 341
   euclidian, 316
   Kullback, 315
Doppler, 8, **27**
Doppler ambiguity, 417
Doppler grating lobe, 417
Doppler radar, 460
Doppler resolution, 450
Doppler shift, **443**
Down-conversion, 438
Dual aperture, 465
Dual aperture imaging, 465
Dual space, 351

Effective area, 174
Effective bandwidth, 70
Effective gain, 174
Effective timewidth, 69
Efficient estimator, 502
Electron illuminated microscopy
   reflection, 294
   transmission, 294
Electron-beam microscopy, 294
Emission tomography, 5, 25
   from decay events, 233
   from magnetic-resonance, 221
   photon, 233
   positron, 233
Emitter location
   passive, 539
Energy, 30
Energy relation, 33, 78
   one-dimensional, 33
   three-dimensional, 352
   two-dimensional, 78
Entropy, 301
Equation
   Friis, 481
   radar imaging, 445
   radar range, 480, 481
   wave, 123
Error function, 485

Estimate
   direction of arrival, 547
   pulse arrival time, 496
Estimation
   coherent, 504
   differential, 545
   noncoherent, 506
Estimator
   bayesian, 314
   efficient, 502
   linear, 502
   maximum-posterior, 314
   robust, 499
Euclidian distance, 316
Eulerian angles, 13, 173
Evanescent wave, 15, 124, 246, 293
Ewald data ball, 368
Ewald sphere, **367**, 374, 377
   in three dimensions, 367
   in two dimensions, 241
Excitation field, 227
Expectation-maximization
   radar imaging, 469
Expectation-maximization algorithm, 333
Expectation-maximization method, 342

False alarm, 488
False negative, 488
False positive, 488
Fan-beam tomography, 210, 251
Far-field diffraction, 350
   Fraunhofer, 128, 168
Fast Fourier transform, 182
Fast lens, 161
Fiber array, 389, 390
Fienup algorithm, **274**, 289, 376
Filter
   Butterworth, 108
   causal, 60, 256
   Hilbert, 153
   knife-edge, 153
   linear, 256
   matched, 263
   North, 74
   passband, 45
   space-invariant, 256
   time-invariant, 256
   time-variant, 256
   two-dimensional, 107, 256
   whitening, 306
   Zernicke, 152
Filtered back-projection, 214, 251
Finite comb function
   one-dimensional, 43
   three-dimensional, 354
   two-dimensional, 82

Fluoroscope, 5
Focal length, 135, 145
Focal plane, 141
Focused antenna, 183
Focusing, 162
　antenna, 183
　optical, 162
　radar, 452
Formula
　Bayes, 20
　Friis, 481
　Karle–Hauptman, 381
　Nyquist–Shannon, 54
　　two-dimensional, 100, 358
　Parseval, 34, 73
　Sayre, 381
Forward problem, 4, 254
　three-dimensional, 370
Fourier antenna theory, 117, 168
Fourier optics, 117, 121
Fourier transform, 30, 31
　discrete, 67, 181
　generalized, 39
　inverse, 31
　multidimensional, 114
　one-dimensional, 30
　short-time, 423
　three-dimensional, 350
　two-dimensional, 75
Fraunhofer approximation, 128, 130, 175
　diffraction, 362
Fraunhofer diffraction, 128, 168
Frequency compression, 463
Frequency domain, 30
　one-dimensional, 31
　three-dimensional, 351
　two-dimensional, 76
Frequency shift, 28
Fresnel approximation, 128, 129, 137
Fresnel diffraction, 128, 435
Fresnel integrals, 577
Fresnel phase plate, 163, 267
Fresnel point-spread function, 129, 162
Fresnel zone, 268
Fresnel zone plate, 163, 266, 268
Friedel's property, 352, 371
Friis equation, 481
Fringe pattern, 524
Fubini's theorem, 76
Function
　ambiguity, 399
　circle, 85, 90
　circular hat, 112
　comb, 44
　cylinder, 354
　helix, 356

　impulse, 354
　jinc, 86, 355
　lazy pyramid, 111
　rectangular, 36
　ring, 83
　sinc, 355
　sphere, 355
　tinc, 355
　trapezoid, 112
　triangle, 111
Fundamental cell, 104

Gabor bandwidth, **51**, 69, 392, 409, 410, 506, 507, 582
Gabor hologram, 156
Gabor parameter, **51**, 409
　bandwidth, 50
　skew, 51, 71
　timewidth, 50
Gabor parameters, 51
Gabor resolution, 67
Gabor resolution criterion, 109
Gabor timewidth, 50, **51**, 409, 410, 507
Gain, 173
　antenna, 174
　effective, 174
　peak-of-beam, 173
Gain pattern, 173
Gamma-ray astronomy, 299
Gaussian beam, 164, 205
Gaussian density function, 21, 311, 343
　multivariate, 29
　three-dimensional, 516
　two-dimensional, 492
Gaussian pulse, 38, 90, 404
　one-dimensional, 38
　three-dimensional, 354, 516
　two-dimensional, 81, 90
Gaussian random variable, 21, 485
Generalized function, 37, 44, 212, 249
　two-dimensional, 81, 83
Generalized trace, 565
Generator matrix, 104
Geometrical image, 137
Geometrical optics, 11, 132, 162, 289
Geometrical optics approximation, 132, 208
Geophysical tomography, 25
Gerchberg–Saxton algorithm, 273, 299
Gibbs inequality, 316, 322
Gibbs phenomena, 31
Golay-coded pulse pair, 431
Gradient field, 229
Gradiometer, 536
Grating lobe, **42**, 97, 388
　three-dimensional, 350
　two-dimensional, 97, 178

Green's function, 120
Grenander Sieve, 344
Grocer's lattice, 395
Guard interval, 55
Gyromagnetic ratio, 223

Halftone imagery, 159, 165
Hankel transform
 of order zero, 86
Harker–Kasper inequality, 381
Heisenberg uncertainty relationship, 73
Helical scan, 252
Helix, 356, 365
 discontinuous, 396
 double, 398
 finite, 396
 Fourier transform, 357
 scattering function, 365
Helix function, 356
 finite, 396
Heterodyne, 156
 spatial, 156
Hexagonal array, 105, 178
Hexagonal lattice, 104
Hexagonal sampling, 103, 106, 178
Hilbert filter, 153
Hologram, 7, 155
 Gabor, 156
 Leith–Upatnieks, 156
Holography, 7, 155
Huygens point-spread function, 164
 simplified, 128
Huygens wavelet, 127, 168
Huygens–Fresnel point-spread function, **125**, 160, 161, 198
 diffusion, 242
Huygens–Fresnel principle, 120, 122, **125**, 293
Hydraphone, 167
Hydrophone, 26, 167, 547
Hypothesis testing, 302, 494

Ideal lens, 135, 266
Ill-conditioned, 255
Ill-conditioned problem, 257
Illumination, 291
Illumination function, 121, 167
Image
 ultrasound, 291
Image deconvolution
 blind, 279
Image doubling property, 352
Image formation, 1, 235
 coded-aperture, 268
 diffraction, 377
 diffusion, 241
 lidar, 551
 magnetic, 534
 magnetic-resonance, 5, 221
 photon emission, 235
 positron emission, 235
 radar, 437
Image restoration, 254
Imaging
 diffraction-limited, 150
 phase-contrast, 151
Imaging equation
 radar, 445
Imaging radar, 437, 454
 bistatic, 455
 dual aperture, 465
 interferometric, 465
 monostatic, 455
 sidelooking, 475
 spotlight-mode, 455
 swath-mode, 455
Impulse
 ring, 83, 88
Impulse function
 one-dimensional, 37
 three-dimensional, 354
 two-dimensional, 81
Impulse response, 256
In-phase component, 9
In-phase modulation, 45
Incomplete data, 311
Incomplete data space, 333
Inequality
 Cramer–Rao, 500
 Gibbs, 322
 Harker–Kasper, 381
 Pinsker, 346
 Schwarz, 48, 429
 triangle, 49
Inference principle, 314
Informatics, xiii
Infrared, 5
Inner product, 47
Insonification, 291
Intensity
 of an image, 137
Interference, 120
Interferometer, 202, 465, 522, 547
 microwave, 185
Interferometry, **185**
Interpolating function, 54
Interpolation, 54, 100, 358
 hexagonal, 106
 jinc, 102
 sinc, 55, 102
 tinc, 395
Interpolation spectrum, 100

Intrrpolation
  soft, 55
Inverse filter, 257
Inverse Fourier transform
  one-dimensional, 30
  three-dimensional, 351
  two-dimensional, 76
Inverse problem, 3, 255
Inverse Radon transform, 211
Invisible
  antenna pattern, 174
  grating lobe, 377
Isserlis theorem, 503
Iterative projection, 218

Jaynes maximum-entropy principle, 315
Jinc function, 86, 172, 355
  two-dimensional, 87
Jinc interpolation, 102
Jinc pulse, 355

Karhunen–Loeve expansion, 306
Karle–Hauptman formula, 381, 385
Karle–Hauptman matrix, 386
Knife-edge filter, 153
Kuhn–Tucker condition, 320
Kullback distance, **315, 316**, 316, 328, 344
Kullback minimum-relative-entropy principle, 316

Lagrangian relaxation, 571
Larmor frequency, 223, 228
Lattice, **104**, 368
  three-dimensional, 104, 370
    grocer's, 395
    rectangular, 370
  two-dimensional, 104
    hexagonal, 104
Lattice cell, 104
Laue scattering pattern, 373
Layer line, 358
Lazy pyramid function, 111
Least-squares
  weighted, 552
Least-squares fit, 550
Leith–Upatnieks hologram, 7, 156
Lens, 135
  diffractive, 163, 266, 298
  ideal, 135
  reflection, 268
  refractive, 163, 266
Lens law, 72, **137**, 141, 435
Lens maker's equation, 145
Lidar, 478, **551**
Likelihood function, 302
  waveform, 305
Likelihood ratio, 305, 489

Likelihood statistic, 304
Line impulse, 83
Linear estimator, 496, 502
Linear filter, 256
Linear phase function, 203
Lithography, 246
Lobe
  grating, 42
  main, 41
Log-Bessel function, 561
Log-likelihood function, 302
Log-likelihood ratio, **305**, 494
Lower bound
  Cramer–Rao, **500**

Macroscopic magnetization vector, 226
Magnetic anomaly detection, 534
Magnetic dipole, 537
Magnetic imaging, 534
Magnetic moment, 223, 224
Magnetic monopole, 535
Magnetic resonance, **227**
Magnetic resonance imaging (MRI), 5, 25, 221
Magnetic surveillance, 534
Magnetometer, 535
Magnification, 77, 137
Main beam, **174**
  antenna, 171
Main lobe, 37, 41
Marginal, 20, 264
Matched filter, **60**
  complex baseband, 62
  real passband, 62
  transform domain, 62
  whitened, 62
  with phase noise, 583
Matched-field processing, 205
Matrix
  Toeplitz, 386
Maximum-entropy principle, 301
Maximum-likelihood decision, 303
Maximum-likelihood estimate, 503
Maximum-likelihood principle, **301**, 338
Maximum-posterior estimator, 314
Maxwell's equations, 439, 536
Maxwellian probability density function, 516, 555
Mean, 20
Measurement, 487
Medical imaging, 299
Method of sieves, 349
Microscopy, 7
  electron-beam, 294
  near-field, 246, 293, 300
  scanning, 292
Mills cross, 202

Index

Minimum-relative-entropy principle
   Kullback, 316
Missed detection, 488
Model-based imaging, 221
Modified Bessel function, 493, 515, 560
Modulation component, 10
   in-phase, 45
   quadrature, 45
Modulation property, 33
   one-dimensional, 33
   three-dimensional, 352
   two-dimensional, 77
Monic polynomial, 283
Monochromatic wave, 11, 13
Monodirectional wave, 11, 13
Monostatic imaging radar, 455
Monostatic radar, 440
Monostatic reflection, 440
Mostly black image, 262, 528
Motion compensation, 452
Moving-target detection, 512
MRI (see magnetic resonance), 5, 25, 221
Multilateration, 571
Multispectral image, 296
Multitarget tracking, 558
Multivariate gaussian density function, 29
Multivariate random variable, 21
Munkres algorithm, 568

Narrowband approximation, 170
Nats, 315
Near-field diffraction, 128
Near-field microscopy, 246, 293, 300
Negative lens, 135
Neyman–Pearson decision rule, 559
Neyman–Pearson theorem, 487
Noise
   additive, 438
   complex, 58
   white, 65
Noise bandwidth, 67
Noise power, **58**
Noncoherent, 17, 145
   spatially, 134, 145
   temporally, 145
Noncoherent detection, 482, 494
Noncoherent estimation, 506
   pulse arrival frequency, 506
   pulse arrival time, 506
Noncoherent imaging, 145
Noncoherent integration, 558
Noncoherent light, 134
Noncoherent processing, 17
Noncoherent pulse, 494
Noncoherent pulse train, 18, 558
Nondiffracting beam, 184, 204

Nonnegativity constraint, 319
Norm, 47
North filter, 74
Nuclear magnetic resonance, 5
Null hypothesis, 483
Numerical aperture, 141
Nyquist rate, 54
Nyquist sample, **54**
   three-dimensional, 361
   two-dimensional, 100
Nyquist–Shannon interpolation, 314
   one-dimensional, 54
   three-dimensional, 394
      for slices, 251
   two-dimensional, 100, 101, 358
      circular support, 102
Nyquist–Shannon sampling theorem
   one-dimensional, 54
   three-dimensional, 358, 375
   two-dimensional, 100

Observable data ball, 368
Omnidirectional antenna, 168
One-dimensional Fourier transform, 30
Optical coherence tomography, 290
Optical filtering, 158
Optical point-spread function, 140
   coherent, 140
Optical processor, 4, 7, 159, 463
Optical transfer function, 149
   coherent, 140
   noncoherent, 149
Optical-coherence tomography, **244**
Orthogonal, 47

Parallelepiped, 368
Parameter
   Gabor, 51, 409
Parametric model, 378
Paraxial approximation, 130, 145, 164
Paraxial ray, 145
Parseval's formula
   one-dimensional, 34
   three-dimensional, 352
   two-dimensional, 78
Passband filter, 45
Passband pulse, 10
Passband representation, 10, 13
   complex, 439
   wavefront, 13
Passband sampling, 56
Passband signal, 9, 44
Passband waveform, 10
Passive emitter location, 539
Passive surveillance, 2, 521
Patterson map, 366

Peak-of-beam gain, 173
Periodic sinc, 42
Permutation matrix, 567
Phase center, 201, 549
    of an antenna, 187
Phase centers, 547
Phase closure, 556
Phase distortion, 587
Phase error, 575
Phase extension formula, 388
Phase noise, 19, 575
Phase retrieval, 254, **272**, 323, 332
    Fienup algorithm, 274
    Gerchberg–Saxton algorithm, 273
    Schulz–Snyder algorithm, 323
    three-dimensional, 367
Phase shift, 443
Phase-contrast imaging, 6, 151, 166
Phase-contrast microscope, 166
Phased array, 180
Photography, 4, 7
Photon, 285
Photon differencing, 236, 288
Photon position difference, 329
Photon sieve, 267
Pinhole camera, 163, 268
Pinhole pattern, 270
Pinsker inequality, 346
Pixel, 109, 116
Pixelization, 109, 342
Planck constant, 22
Plane wave, 11, 12
    vector-valued, 195
Point-spread function, **107**, **257**, 262
    clean, 263
    coherent optical, 140
    Fresnel, 129, 162
    Huygens, 164
    Huygens–Fresnel, 120, **125**, 160, 161
    noncoherent optical, 148
    of free space, 120
Poisson process, 286, 296, 330, 345
    three-dimensional, 390
Polarization, 26, 195, 438, **442**
Polynomial
    monic, 283
Position differencing, 288, 329
Positive lens, 135
Positron-emission tomography, 25, 285
Power density spectrum, 64, 496
Primitive cell, 368
Primitive element, 421
Principle
    Babinet, 163
    Huygens–Fresnel, 119, 125
    Jaynes maximum-entropy, 315
    maximum-likelihood, 301, 338
    reciprocity, 191
    stationary phase, 132
Principle of inference, 314
Principle of stationary phase, 132
Probabilistic reflection, 444
Probability density function, 20, 302
    conditional, 20
    gaussian, 20, 485
    maxwellian, 516, 555
    Poisson, 330
    rayleigh, 492, 515
    ricean, 493, 515
    two-dimensional gaussian, 491
Probability distribution, 302
Probability mass function, 302
Probability of false alarm, 484, 485, 560
Probability of missed detection, 484, 485, 560
Probability vector, 20
Projection, 24, **88**, 88, 250, 252
Projection tomography, 208
    cone-beam, 253
    fan-beam, 210
    magnetic excitation, 221
    parallel-beam, 209
Projection-slice theorem, 3, 23, 24, **88**, 89, 116, 210, 250
    diffracting, 239
    three-dimensional, 115, 250
Property
    complex modulation, 33
    modulation, 33
    quadratic-phase, 402
Pseudoscopic, 165
Pulse
    chirp, 38
    Costas, 418
    gaussian, 38, 90
    Golay-coded, 431
    lazy pyramid, 111
    quadratic phase, 38
    rectangular, 36
    sinc, 36
    tinc, 355
    triangle, 115
Pulse arrival time
    noncoherent estimator, 517
Pulse detection, 483
    noncoherent, 494
Pulse radar, 460
Pulse repetition interval, 18, 414, 416
Pulse train, 39, 414

Quadratic-phase error, 576
Quadratic-phase property, 402, 410, 432, 434

Quadratic-phase pulse, 38, 404, 463
  two-dimensional, 81
Quadratic-phase term, 129
Quadrature component, 9
Quadrature modulation, 45
Quality factor
  antenna, 175
Quantization, 53
Quantizer
  scalar, 53
  vector, 53
Quantum mechanics, 435
Quarter-square multiplier, 71

Radar, 25
  bistatic, 441
  doppler, 460
  imaging, 437, 454
  monostatic, 26
  moving-target detection, 512
  pulse, 460
  search, 8, 480
  synthetic-aperture, 7, **454**
Radar astronomy, 458
Radar cross section, 440, 481
  bistatic, 442
Radar cross-section density, 441
Radar imaging
  diffuse reflector, 466
  dual aperture, 465
  planetary objects, 8
Radar imaging approximation, **424**, 446, 447, 473
Radar imaging equation, 428, 445, 446, 509, 511
Radar range equation, 480, 481
Radar waveform, 17
Radiation pattern, 167
  wideband, 188
Radio astronomy, 9, 521
Radio telescope, 9, 522
Radiography, 5
Radon transform, 23, 90, 209
  inverse, 211
Raleigh resolution, 162
Random process, 19, 21, 63, 297
  stationary, 63
  two-dimensional, 259
Random variable, 19, 20
  complex, 21
  discrete, 20
  gaussian, 21
  multivariate, 21
  Rayleigh, 492
  real, 20
  ricean, 493
  vector, 21
Ray of light, 132

Ray tracing, 11, 132, 161, 163
Rayleigh probability density function, **492**, 515, 559
Rayleigh random variable, 151, **492**
Rayleigh resolution, 67, 110
Rayleigh resolution criterion, 66, 109, 150, 162
Reactive term
  Huygens–Fresnel principle, 125
Real aperture, 457
Real image, 157
Real random variable, 20
  continuous, 20
  discrete, 20
Reciprocal lattice, 104, 114, **371**
Reciprocal matrix, 395
Reciprocal space, 351, 367
Reciprocity principle, 167, 191, **193**
Reciprocity theorem, **176**
Rectangle function, 353
  one-dimensional, 36
  three-dimensional, 353
  two-dimensional, 80
Rectangular pulse, 36
Reflected signal, 441
Reflection
  diffuse, 134, 444
  specular, 134, 444
Reflectivity, 438, **440**, 440
Reflectivity density, 438, 441, 480
Refractive index, 144
Refractive lens, 163, 266
Regularization, 255, 341, 342
  Tikhonov, 343
Relative aperture, 141
Relative entropy, **315**
Remote image formation, xiii
Remote surveillance, xiii
Representation
  complex baseband, 10
  complex passband, 10
  passband, 10
Resolution, 65, 109
  antenna, 174
  cross-range, 450, 451, 453
  range, 451
Resolution criterion, 67
  Abbe, 150
  half-power, 67
  Rayleigh, 67
  Sparrow, 67
  Woodward, 67
Resolution loss, 586
Ricean probability density function, 493, 515, 559
Ricean random variable, 493
Richardson–Lucy algorithm, 264, 321, 328, 528

Ring function, 83
Ring impulse, 83, 88

Sample cross-ambiguity function, 426
Sampling, **53**, 99
    hexagonal, 103
    in one dimension, 53, 359
    in three dimensions, 358
    passband, 56
    polar, 118
Sampling image, 53
Sampling theorem
    one-dimensional, 54
    three-dimensional, 360, 394
    two-dimensional, 99, 342
Sayre formula, 381, 397
Scalar diffraction, 120
Scalar wave, 11
Scaling property
    one-dimensional, 32
    three-dimensional, 351, 354
    two-dimensional, 77
Scan, 562
Scanning aperture, 188
Scanning microscopy, 292
Scattering function, 364, 365
Scattering matrix, 442
Scene, 2
Schlieren method, 6, 153, 166
Schulz–Snyder algorithm, 273, 323, 349, 376
Schwarz inequality, 48, 382, 429, 498
    expectation, 48
Search radar, 437, 480, 481
Seismic processing, 4, 25
Seismic wave, 292
Self-noise, 524
Sequential detection, 561
Sequentially formed array, 179
Set sum, 376, 554
Shadow transform, 90
Shannon entropy, 315
Shift property, 33
Short-time Fourier transform, 423
Sidelobe, 37, 171, 407
Sidelooking radar, 450
Sieve, 344, 349
Sifting property, 37, 63
sifting property
    two-dimensional, 81
Signal
    one-dimensional, 30
    three-dimensional, 350
    two-dimensional, 75
Signal processing
    discrete, 56
Signal space, 46

one-dimensional, 400
two-dimensional, 400
Signal-to-noise ratio, 59, 61
Sinc function
    one-dimensional, 36
    three-dimensional, 353
    two-dimensional, 80
Sinc interpolation, 55
Sinc pulse, 36, 355
Skew parameter
    Gabor, 51, 409
Slice, **88**, 207, 208, 252
Soft interpolation, 102
Sonar, 25
Sonogram, 291
Space-invariant filter, 256
Sparrow resolution criterion, 66, 109
Spatial ambiguity function, 435
Spatial coherence function, 529
Spatial equalization, 159
Spatially coherent, 17
Spatially noncoherent, 134, 146
Speckle, 147, 150, 300
Speckled pattern, 467
Specular reflection, 134, 444
Sphere function, 115, **355**
Spherical wave, 126
Spin, 222
Spin polarization, 224
Spin quantum number, 222
Spotlight-mode imaging, 455, 458
Stagnation, 276
Stagnation point, 276
Staring aperture, 188
Stationary phase
    Principle of, 132
Stationary random process, 63
Statistic, 487, 500
    likelihood, 304
    sufficient, 304
Steered array, 180
Stochastic process, 21
Structure factor, 372, 377, 388
Structure invariant, 397
Subclutter visibility, 509
Successive projections, 219
Sufficient statistic, 304
Superresolution, 262, 300
Support, **30**, 554
    bounded, **31**, 45
Surveillance, 1
Swath-mode imaging, 455, 458
Synthetic aperture, 451
Synthetic-aperture length, 451, 457
Synthetic-aperture radar, 7, 27, **454**
    sidelooking, 476

Talbot distance, 164
Talbot effect, 164
Tangent formula, 388
Temporally coherent, 17
Temporally noncoherent, 145
Tessellation, 368
Texture function, 467
Theorem
  back-projection, 211
  convolution, 34
  correlation, 35
  Cramer–Rao, 500
  energy relation, 33
  Isserlis, 503
  Neyman–Pearson, 487
  Nyquist–Shannon, 54
    two-dimensional, 100, 358
  projection-slice, 88, 89
  reciprocity, 167, 176, 193
  van Cittert–Zernike, 529, 555
  van Vleck–Middleton, 532
  Wiener–Khintchine, 64, 529
Thin lens, 144, 163, 266
Thin-lens approximation, 144
Thomson scattering, 362
Three-dimensional
  comb function, 354
  cylinder function, 354
  dirichlet function, 354
  Fourier transform, 350
  gaussian pulse, 354
  helix function, 356
  impulse function, 354
  rectangle function, 353
  signal, 350
  sinc function, 353
  tinc function, 355
Three-dimensional function, 350
Three-dimensional sphere function, 355
Threshold, 483
Tikhonov regularization, 343
Time delay, 28
Time difference of arrival, 545
Time domain, 30
Time-invariant filter, 256
Time-variant filter, 256
Timewidth
  effective, 69
  Gabor, 50
  root-mean-squared, 50
  Woodward, 67
Tinc function, 355, 368
Tinc pulse, 355
Toeplitz matrix, 386

Tomography, 23, 207, 254
  absorption, 208
  coherence, 25, 244
  cone-beam, 253
  diffraction, 25, 236
  diffusion, 25, 241
  emission, 25, 221
  fan-beam, 210
  geophysical, 25
  magnetic-resonance, 221
  optical coherence, 290
  photon-emission, 233
  positron-emission, 233, 285
  projection, 208
  radar, 472
Trace
  generalized, 565
Track, 558
Tracking, 557
Transfer function
  coherent optical, 140
  noncoherent optical, 149
  of free space, **125**, 127
Transform
  Fourier, 30
  Hankel, 86
  Radon, 23, 90, 209
  shadow, 90
Translation property
  one-dimensional, 33
  three-dimensional, 352
  two-dimensional, 77
Transmission tomography, 208
Transmittance, 121, 144, 164
Transparency, 121
Transverse vector-valued wave, 194
Transverse wave, 195
Transverse-vector wave, 16
Trapezoid function, 112
Triangle function, 111
Triangle inequality, 49
Triangle pulse, 115
Two-dimensional
  comb function, 81
  Fourier transform, 75
  gaussian function, 92
  gaussian pulse, 81
  impulse function, 81
  jinc function, 87
  pulse array, 94
  rectangle function, 80
  signal, 75
  sinc function, 80
Two-dimensional filter, 256
Type I error, 488
Type II error, 488

Ultrasound, 26, 291
Ultrasound transducer, 179, 205
Uncertainty ellipse, 410, 433, 507
Uncertainty principle, 51, 71, 433
 for antennas, 205
 two-dimensional, 115
Unfocused beam, 183
Unit sphere function, 115
Up-conversion, 438
Urkowitz filter, 511

van Cittert–Zernike theorem, 146, 529, 553, 555
van Vleck–Middleton theorem, 532, 556
Variance, 20
Vector diffraction, 120, **194**
Vector random variable, 21
Vector space, 47
Vector wave number, 13
Vector-valued wave, 11, 194
VHF band, 23, 444
Virtual image, 156, 157
Visibility function, 366
Visible
 antenna pattern, 174
 grating lobe, 377
Volume property
 ambiguity function, 406, 429

Waist of a beam, 164
Watson–Crick model, 357
Wave
 evanescent, 15, 124
 narrowband, 14
 plane, 11
 scalar-valued, 11
 transverse, 16
 vector-valued, 11, 194
Wave equation, 12, 13, **123**, 166, 184, 194, 206, 242
 cylindrical coordinates, 184

Wave number, 11, 13
 vector, 13
Waveform, 30
Wavefront
 monodirectional, 13
Wavefront reconstruction, 153
Wavelength, 11
 de Broglie, 294
Wavelet
 Huygens, 127
Webb space telescope, 273
Weighted least-squares, 552
Welch–Costas array, 422
Well-posed problem, 255
White noise, 62, 65
Whitened matched filter, **62**, 258, 298, 484, 510
Whitening filter, 306
Wiener filter, **259**
 multichannel, 296
 two-dimensional
  basic form, 261
  whitened, 259
Wiener–Hopf equation, 259
Wiener–Khintchine theorem, 64, 529
Wigner distribution, 435
Wigner function, 403, 411, 434
Wind tunnel, 152
Woodbury matrix identity, 472
Woodward function, 28, 399, 403, 407, 411
Woodward resolution criterion, 67, 109, 410
Woodward timewidth, 67

X-ray astronomy, 299
X-ray band, 268
X-ray diffraction, 6, 350
X-ray tomography, 5

Zee transform, 58
Zernicke filter, 152, 166
Zero sheets, 299